Springer Tracts in Advanced Robotics

Volume 7

Editors: Bruno Siciliano · Oussama Khatib · Frans Groen

Springer

Berlin
Heidelberg
New York
Hong Kong
London
Milan
Paris
Tokyo

J.-D. Boissonnat · J. Burdick · K. Goldberg · S. Hutchinson (Eds.)

Algorithmic Foundations of Robotics V

With 226 Figures

Springer

Professor Bruno Siciliano, Dipartimento di Informatica e Sistemistica, Università degli Studi di Napoli Federico II, Via Claudio 21, 80125 Napoli, Italy, email: siciliano@unina.it

Professor Oussama Khatib, Robotics Laboratory, Department of Computer Science, Stanford University, Stanford, CA 94305-9010, USA, email: khatib@cs.stanford.edu

Professor Frans Groen, Department of Computer Science, Universiteit van Amsterdam, Kruislaan 403, 1098 SJ Amsterdam, The Netherlands, email: groen@science.uva.nl

STAR (Springer Tracts in Advanced Robotics) has been promoted under the auspices of EURON (European Robotics Research Network)

Editors

Prof. Jean-Daniel Boissonnat
INRIA
2004 Route des Lucioles
06902 Sophia-Antipolis
France

Prof. Joel Burdick
California Inst. of Technology
Dept. of Mechanical Engineering
East California Boulevard 12000
Pasadena CA 91125
USA

Prof. Ken Goldberg
University of California at Berkeley
Dept. of Mechanical Engineering
4189 Etcheverry Hall
Berkeley CA 94720
USA

Prof. Seth Hutchinson
University of Illinois
Beckman Inst.
North Matthews Avenue 405
Urbana IL 61801
USA

ISSN 1610-7438

ISBN 3-540-40476-7 Springer-Verlag Berlin Heidelberg New York

Library of Congress Cataloging-in-Publication Data
International Workshop on the Algorithmic Foundations of Robotics (5th : 2002 : Nice, France)
Algorithmic foundations of robotics V / J.-D. Boissonnat ... [et al.](eds.)
p. cm. -- (Springer tracts in advanced robotics ; 7)
Includes bibliographical references and index.
ISBN 3-540-40476-7 (alk. paper)
1. Robotics--Congresses. 2. Algorithms--Congresses. I. Boissonnat, J.-D. (Jean-Daniel), 1953- II. Title. III. Series.

TJ210.3.I635 2002
629.8'92--dc21 2003054426

Springer-Verlag Berlin Heidelberg New York
a member of BertelsmannSpringer Science + Business Media GmbH

http://www.springer.de

Printed in Germany

Typesetting: Digital data supplied by author.
Data-conversion by PTP-Berlin Protago-TeX-Production GmbH, Berlin
Cover-Design: design & production GmbH, Heidelberg

Printed on acid-free paper 62/3020Yu - 5 4 3 2 1 0

Foreword

At the dawn of the new millennium, robotics is undergoing a major transformation in scope and dimension. From a largely dominant industrial focus, robotics is rapidly expanding into the challenges of unstructured environments. Interacting with, assisting, serving, and exploring with humans, the emerging robots will increasingly touch people and their lives.

The goal of the new series of *Springer Tracts in Advanced Robotics (STAR)* is to bring, in a timely fashion, the latest advances and developments in robotics on the basis of their significance and quality. It is our hope that the wider dissemination of research developments will stimulate more exchanges and collaborations among the research community and contribute to further advancement of this rapidly growing field.

Since its inception about a decade ago, WAFR, the *Workshop on Algorithmic Foundations of Robotics* has published archival volumes of high reference value. With the launching of STAR, a more suitable home is found for this and other thematic symposia devoted to excellence in robotics research.

The Fifth edition of *Algorithmic Foundations of Robotics* edited by Jean-Daniel Boissonnat, Joel Burdick, Ken Goldberg and Seth Hutchinson offers in its nine-chapter volume a collection of a broad range of topics in advanced robotics. The contents of these contributions represent a cross-section of the current state of robotics research from one particular aspect: algorithms, and how they reflect on the theoretical basis of subsequent developments. Validation of algorithms, design concepts, or techniques is the common thread running through this focused collection.

Rich by topics and authoritative contributors, WAFR culminates with this unique reference on the current developments and new directions in the field of algorithmic foundations. A fine addition to the series!

Naples, Italy
May 2003

Bruno Siciliano
STAR Editor

Preface

Robot algorithms are abstractions of computational processes that control or reason about motion and perception in the physical world. Because actions in the physical world are subject to physical laws and geometric constraints, the design and analysis of robot algorithms raises fundamental questions in computer science, computational geometry, mechanical modelling, operations research, control theory, and associated fields. The biannual Workshop on the Algorithmic Foundations of Robotics was established in 1994 as a single-track meeting to focus on algorithmic issues related to robotics and automation.

The Fifth international Workshop on the Algorithmic Foundations of Robotics was held from 15–17 December 2002 at the Westminster Hotel in Nice, France. Approximately sixty international researchers participated. Two plenary talks were presented: Algorithms for Motion and Navigation in Virtual Environments and Games, by Mark Overmars of Utrecht University, and Surgery Simulation: Integrating Geometry, Physics and Physiology into the Virtual Patient, by Nicholas Ayache of INRIA.

The highly selective program highlighted significant new results in probabilistic and sampling-based path planning, modelling of knots, reconfigurable robots, coordination of multiple robots, distributed manipulation, graphics and visualization, molecular modeling, and mobile robotics. Thirty-three papers presented at WAFR 2002 are collected in the present volume.

In addition to the editors of this volume, the Program Committee for WAFR 2002 included: Pankaj Agarwal, Nancy Amato, Nicholas Ayache, Antonio Bicchi, Robert-Paul Berretty, Karl Bohringer, Howie Choset, Leonidas J. Guibas, Kamal Gupta, Dan Halperin, Hirohisa Hirukawa, Makoto Kaneko, Lydia Kavraki, Jean-Paul Laumond, Matt Mason, Joe Mitchell, Dinesh Pai, Jean Ponce, and Frank van der Stappen.

As the co-chairs of WAFR 2002, we are grateful to Jean-Claude Latombe and the past WAFR organizers for their leadership and guidance. We also wish to thank Bruno Siciliano for encouraging us to publish this volume with Springer-Verlag, and Thomas Ditzinger of Springer. We also thank Terry Peterson for her efforts in collecting and formatting the text, Dany Sergeant and Agnes Clement-Bessiere for their help with local arrangements, Kevin Clarke for poster design, and INRIA for additional financial support.

Finally, our thanks to all of the participants and authors for their enthusiasm in making WAFR 2002 such a lively research meeting.

Nice, France
December 2002

Jean-Daniel Boissonnat
Joel Burdick
Ken Goldberg
Seth Hutchinson

Contents

Session III

Session IV

Session V

Session VIII

Algorithms for Motion and Navigation in Virtual Environments and Games

Mark H. Overmars

Institute of Information and Computing Sciences, Utrecht University,
P.O. Box 80.089, 3508 TB Utrecht, The Netherlands
Email: markov@cs.uu.nl.

1 Introduction

Motion planning was traditionally studied in the area of robotics. In recent years the techniques are increasingly used in virtual environments and games. In many such applications entities move around and their motion must be planned. In particular, we can distinguish the following types of motion:

navigation Entities must find their route to a particular goal while avoiding collisions with obstacles and other entities.

animation The internal (often articulated) movements of the entities must be computed. Such movements must match the navigation through the environment.

manipulation Entities manipulate objects in the environment. We need to compute the motion of the manipulated objects, in relation to the manipulation itself.

camera motion Also the camera through which the user inspects the environment moves and its motion must be controlled (when it is not under the direct control of the user).

Virtual environments and games offer a challenging problem setting because of the following aspects:

complexity The scenes are very complex with easily up to millions of obstacles.

dynamic The scenes tend to be dynamic, that is, obstacles can appear or disappear (for example when closing or opening a door).

real-time Motions must be computed real-time. One cannot temporarily stop the entities.

many degrees of freedom Many motion planning problems must deal with a large number of degrees of freedom. A reasonable accurate structure of the human body has over a hundred degrees of freedom. And often multiple entities move in the same environment.

natural motion To give the user the feeling of immersion, the resulting motions must be natural, that is, visually convincing.

Though a lot of research has been done on motion planning, current techniques cannot adequately solve these problems. This short paper lists some of the challenges and indicates possible directions for their solution.

J.-D. Boissonnat et al. (Eds.): Algorithmic Foundations of Robotics V, STAR 7, pp. 1–5, 2004.

2 Navigation

The navigation problem is most closely related to the traditional motion planning problem. An entity (character, car, etc.) must move from its current position to a given goal position. As indicated above though there are a number of aspects that complicate the problem. Because the scenes are huge and motions must be computed real-time we need an approach that constructs a data structure which can then be used to perform motion planning queries. A roadmap seems an adequate structure for this. Because most paths are easy while the scenes are complex an adaptive approach is required. The probabilistic roadmap (PRM) approach seems to form a good basis for this (see [2] and many papers since). The approach though has a number of drawbacks that need to be addressed. In particular, the resulting paths are ugly. Smoothing techniques have been proposed but these are too slow to perform real-time. Hence, it is necessary to incorporate such smoothing in the preprocessing phase. Also there is no guarantee for success. This might be a major concern in real-time virtual environment applications. Some form of guaranteed performance will be required.

Dynamic changes in the environment form a different challenge. Most of the scene will be static and preprocessing is possible here but some changes will occur in the scene. The roadmap must be able to cope with that. Some ideas that are currently being investigated are the inclusion of alternative routes in the roadmap, taking known dynamic objects (like doors) into account when creating the roadmap, and adapting the roadmap locally when changes occur.

3 Animation

Besides navigational motion through the environment, many characters also have an internal articulated motion. A typical human character has over a hundred degrees of freedom which must all be controlled to obtain a good walking animation. Currently such animations are largely scripted (although packages for this use concepts from robotics and assist the animator in obtaining good motions) or are obtained through motion capture techniques. Some experiments have been done in automatically computing such internal body motions. A lot more research is required though to make such motions realistic. It is expected that these techniques will first be used in the animation design packages rather than in the actual application.

Because only a limited number of animations can be designed or captured, the actual motions in the applications must be composed from such building blocks. This puts limits on the possible navigational motions. To obtain natural motion it is important that the two types of motion are combined in an adequate way. Also it is interesting to study how generic motion building blocks can be created that can be adapted to the local situation.

Fig. 1. A scene from the game Age of Empires II by Microsoft in which many units must pass through a narrow gate.

4 Multiple Entities

In many virtual environment applications there are multiple entities that move around and we need to coordinate their motions. There are two basic approaches to this problem [3]. When applying ***centralized planning*** the entities together are considered as one robotic system with many degrees of freedom. When applying ***decoupled planning*** we first compute the individual motions of the entities and then try to coordinate these over time. This is faster but can lead to deadlock. In a recent paper Sánchez and Latombe[6] show that with a number of improvements the PRM approach can be successfully applied to solve complicated motion planning with up to 6 robot arms. When the number of entities is much larger, like in Fig. 1, the problem remains largely unsolved. An additional complication is that the coordinated motion should be visually convincing. This might be achieved by incorporating flocking techniques[1,5].

5 Manipulation

Virtual characters will manipulate virtual objects. The motion of such virtual objects must be computed, often using some form of physical simulation. To be able to automatically plan such manipulation (instead of scripting it), we need to take the physical behavior of the objects into account. The result must be convincing (although in many applications it is not important that it is completely accurate). Research in this domain will require a combination of motion planning, (robotic) manipulation, and (real-time) physical simulation. This is a very challenging domain in which only first steps have been made.

6 Camera Motion

A virtual world is inspected through a virtual camera. The virtual camera is a special kind of moving entity. Currently, in many applications the camera

Fig. 2. Moving a camera through part of Rotterdam.

is directly under control of the user (for example in first-person games or architectural walk-throughs), using the arrow keys, mouse or joystick. In other applications the user indirectly controls the camera (for example in third-person games, where the camera follows a person under control of the user). This type of control though is difficult and results in ugly motions, which easily lead to motion sickness (in particular under observers). It would be easier if the user could specify goals and subgoals and the camera automatically moves to such positions.

Automatic camera motion is a special case of motion planning. The camera must avoid obstacles and obey certain rules from camera theory. When the camera follows an object there is the additional problem that the object must remain visible from the camera. We recently implemented a system for automatic camera motion which is being used for architectural walk-through (see Fig. 2). I will briefly describe the approach here (for details see [4]).

A camera is described by its position (p_x, p_y, p_z) in space, a view direction, indicated by a view point (v_x, v_y, v_z), and the twist of the camera, indicated by an up-vector. For most practical cases the up-vector should remain the vertical axis. The motion planning problem we want to solve is as follows: given a starting point and view direction and a goal point and view direction for the camera, compute a natural motion for the robot from start to goal, avoiding obstacles. Such a motion must have the following properties:

- The path for the camera is collision free and has some minimal clearance.
- The path is relatively short.
- The path is smooth.
- The speed along the path changes continuously, depending on the curvature, with a bounded maximal acceleration and deceleration.
- The viewing direction changes smoothly.

Also, it turns out that the choice of viewing point is crucial. It should give the viewer an indication of what is going to happen such that he can anticipate on changes in the motion. Experiments show that a good choice is to keep the viewing point on the camera path itself, at the place where the camera will be some time in the future (e.g. in 1 second time). To achieve these properties we use a three step approach:

First we compute a path for the camera using PRM. We use a ball to guarantee a minimal clearance. To make sure that the path is short we allow additional edges in the roadmap, creating cycles.

In the second step we turn this into a C^1 continuous path using circular blends between pairs of edges.

We use the curvature of the path to compute a maximal speed function along the path. Introducing a maximal acceleration and deceleration, this diagram can be converted into an actual speed diagram. Adapting the speed of the moving object according to this diagram, results in smooth motions.

It can be shown that with this approach the viewing direction is second-order continuous. Experiments indicate that the resulting camera motion is natural and comfortable to watch. We are currently extending this result. In particular we want to add restrictions to the motion, for example keeping the camera at eye-height. Also we are applying the technique to situations in which you want to follow an object.

7 Conclusions

We can conclude that motion and navigation in virtual environments and games poses many interesting challenges. The MOVIE (MOtion planning in VIrtual Environments) project is an EU-funded long term research project that aims at tackling these challenges. The project starts in january 2003. It is a collaboration between Utrecht University, CNRS-LAAS, Tel-Aviv University, and the company Kineo. More information can be found on the project website `http://www.cs.uu.nl/research/projects/movie/`.

References

1. O.B. Bayazit, J-M. Lien, N.M. Amato, Better flocking behaviors using rule-based roadmaps, this proceedings.
2. L. Kavraki, P. Švestka, J-C. Latombe, M.H. Overmars, Probabilistic roadmaps for path planning in high-dimensional configuration spaces, *IEEE Trans. on Robotics and Automation* **12** (1996), pp. 566–580.
3. J-C. Latombe, *Robot motion planning*, Kluwer Academic Publishers, Boston, 1991.
4. D. Nieuwenhuisen, M.H. Overmars, Motion planning for camera movements in virtual environments, 2002, in preparation.
5. C. W. Reynolds, Flocks, herds, and schools: A distributed behavioral model, *Computer Graphics* (1987), pp. 25–34.
6. G. Sánchez, J.-C. Latombe, A single-query bi-directional probabilistic roadmap planner with lazy collision checking, *Int. Journal of Robotics Research*, 2002, to appear.

Motion Planning for Knot Untangling

Andrew M. Ladd and Lydia E. Kavraki

Department of Computer Science, Rice University, Houston, TX 77005, USA

Abstract. This paper investigates the application of motion planning techniques to the untangling of mathematical knots. Knot untangling can be viewed as a high-dimensional planning problem in reparameterizable configuration spaces. In the past, simulated annealing and other energy minimization methods have been used to find knot untangling paths. We develop a probabilistic planner that is capable of untangling knots described by over four hundred variables and known difficult benchmarks in this area more quickly than has been achieved with minimization in the literature. The use of motion planning techniques was critical for the untangling. Our planner defines local goals and makes combined use of energy minimization and randomized tree-based planning. We also show how to produce candidates with minimal number of segments for a given knot. Finally, we discuss some possible applications of our untangling planner in computational topology, in the study of DNA rings and protein folding and for planning with flexible robots.

1 Introduction

Some of the current challenges at the frontiers of modern motion planning applications stem from high dimensional or even non-parametric configuration spaces, complex but 'soft' constraints on allowable motions imposed by the energetics of an underlying physical system and the possibility of innovative uses of planners for discovery. Recent problems of interest are daunting but may have a great deal of structure that can be exploited to sidestep the inherent complexity of high dimensional planning in constrained spaces. Motivating examples can be found in recent work on planning schemes for modular and reconfigurable robots [8,38,39], planning for deformable robots [28,5], abstractions of random planning analysis to non-parametric, non-manifold spaces [27], approximate dimensionality reduction of configuration space [37], determining protein folding pathways [36], planning in the energy landscapes of molecules [4], and computing stationary distributions of Markov chains defined over roadmaps for analysis of Monte Carlo simulations of protein folding phenomena [3].

This paper deals with the knot untangling problem. A knot is a simple, closed piecewise linear curve and is called an unknot if there exists an ambient isotopy[1] transforming it to a triangle. Otherwise it is a non-trivial knot [29]. The induced equivalence relation seperate the knots into different classes which can be called knot types. Our approach applies techniques from

[1] Piecewise linear homeomorphism of the space around the knot.

J.-D. Boissonnat et al. (Eds.): Algorithmic Foundations of Robotics V, STAR 7, pp. 7–23, 2004.

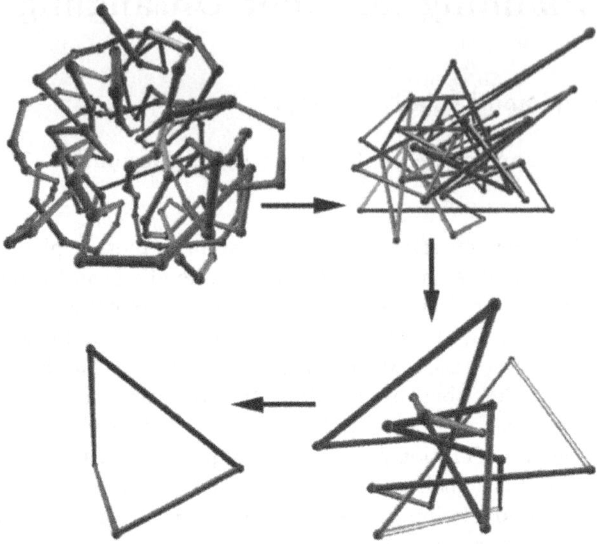

Fig. 1. Knot untangling of a tangled 133 segment unknot.

motion planning to find a sequence of motions and transformations that reduces the number of segments in a knot to the minimal number possible for that knot type. During these motions and transformations the topological type of the knot is preserved. When cast as a robotics problem, untangling is about manipulation of a high-dimensional, reparametrizable, deformable object possibly requiring many parameters to describe important configurations. Figure 1 illustrates some intermediate steps in the untangling of a tricky 133 segment knot.

We describe the design and implementation of a novel motion planner for knot untangling. The planner formulates a sequence of local goals that, when taken together, achieve a global untangling. Planning towards the local goals is achieved by combining energy minimization and randomized tree-based planning in the spirit of those employed by the RRT framework [23,24,26,25] and expansive spaces framework [19,17,18,33]. Our approach is minimalistic and we show that a simple planner suffices for solving this problem. We demonstrate the effectiveness of this planner by applying it to some standard benchmarks used in other implementations, some random knots and by finding minimal segment candidate for a sixteen crossing knot. Our experimental results are significantly better than comparable implementations in the literature. We also discuss how the techniques employed might be applied to other high dimensional motion planning problems such as those found in robotics approaches to molecular biology or in problems with deformable robots.

2 Background

Classification of knots is often attempted by the computation of knot invariants [29,2,16]. Invariants are typically computable constants or polynomials that are invariant over all embeddings of a knot type. If two knots differ over some particular invariant then they are different knots, however the converse does not necessarily hold.

We first describe two classic problems in computational knot theory [6].

Problem 1 (UNKNOT). Given a knot K, determine if K is the same knot type as the unknot.

Problem 2 (KNOT EQUIVALENCE). Given two knots K_1 and K_2, determine if K_1 is the same type as K_2.

One possible proof that a given knot is the unknot is to exhibit an explicit isotopy transforming it to the triangle, in other words, to find a path transforming the given knot into a triangle. In the remainder of the background section, we will discuss in more detail some of the previous work on the untangling problem.

There are many interesting invariant quantities defined for knots. The crossing number is the minimum number of crossings over all regular drawings of the knot on the plane. The unknotting number of a knot is the minimum number of the crossing flips required to transform the knot to the unknot. The stick number is the minimum number of segments required to draw a given knot in 3-space. The genus is the minimum genus over all surfaces that the knot can be drawn on. For a knot K, we write $c(K)$ for crossing number, $u(K)$ for unknotting number, $s(K)$ for stick number and $g(K)$ for the genus. An unknot K has $c(K) = u(K) = g(K) = 0$ and $s(K) = 3$. The simplest non-trivial knot, the trefoil, has $c(K) = 3$, $u(K) = g(K) = 1$ and $s(K) = 6$.

The UNKNOT problem is known to be in **NP** and KNOT EQUIVALENCE is in **PSPACE** [15]. More recently the MAXIMUM GENUS problem, $g(K) \leq k$, has been shown to be **NP**-complete [1]. Another recent unknot recognition algorithm works by enumeration of unknot diagrams and is polynomial in some cases [7]. Hass notes in [1] that if MINIMUM GENUS, $g(K) \geq k$, is in **NP** then UNKNOT is in both **NP** and **co − NP**.

There are several polynomial invariants that are computed on knots. Each has various distinguishing powers. The Alexander-Conway invariant can be computed in **PTIME** but has limited distinguishing power [29]. Computing the Jones, Kauffmann and HOMFLY polynomials are #**P**-hard [20]. These polynomials can be computed in practice for small knots although is not known if there are multiple knots with the same polynomial as the unknot. Jenkins gives an efficient implementation of a HOMFLY calculator, although the size of knots it can handle is limited by bit arithmetic in the code. If c is the number of crossings in a link, then the HOMFLY polynomial can be computed in time $O(n!2^n c^3)$ and in space $O(n!c^2)$, where n is bounded above by $\sqrt{c}+1$ [21]. By using combinations of invariants [2], attempts to tabulate all distinct knots have been made. A recent list consists of all prime knots with at most 16 crossings and consists of 1701936 distinct knot types [16].

A key tool used in polynomial invariant calculation is a regular plane projection of a knot or knot diagram. These graph-like objects can be manipulated by three simple moves, called the Reidemeister moves, to obtain all knot diagrams for that knot type [29]. Hass and Lagarias derive an upper bound on the number of Reidemeister moves to transform a tangled unknot into a triangle. If k is the number of crossings in a given unknot diagram, it takes at most $O(2^{ck})$ where $c = 10^{11}$ [14] to transform this diagram to an unknot. Some intuition for the difficulty here comes from the fact that for some diagrams a non-minimizing Reidemeister move must be made, that is to say a crossing must be added before others can be removed.

2.1 Energy Functions

Knot energy was originally defined as a way of arriving at "ideal" knot configurations. Existing untanglers are essentially energy minimizers so this issue is of critical importance. A good survey of this area can be found in Scharein's thesis [34].

There are several different varieties of energy functions in the literature. Many of the early equations were for the case where the function was a smooth curve and hence the energy was infinite for piecewise linear knots. For further details on energy for smooth curves refer to [9]. Due to the computational nature of our work, we restrict ourselves to energy defined over piecewise linear knots and give only two important examples from the literature. Other examples of energy functions can be found in [30,11,13,34] and include geometric energy, Möbius energy, spring energy, electrostatic energy and others. In our work, we used a modified version of MD energy.

Minimum Distance Energy Minimum distance (MD) energy is an approximation of the distance integral over all pairs of points. The approximation is for the case of a piecewise linear knot and uses the minimum distance between pairs of segments (d_{MD}). Given a knot K with segments $s_1, ..., s_n$, we define MD energy of the knot as follows

$$E_{MD}(K) = \sum_{s_i, s_j \text{ disjoint}} \frac{|s_i||s_j|}{d_{MD}(s_i, s_j)^2}.$$

This energy function can be computed in $O(n^2)$ time. The energy is bounded from below by $2\pi \cdot c(K)$ [35]. To our knowledge, no convergence guarantees have been proved for a gradient descent energy minimization and experimental evidence suggests the presence of local minima, large plateaus and nearly non-unimodal behaviour [40,13].

A key problem with MD energy is that there is a tendency for tangled portions of the knot to become small. This is a result of the length normalizer term in the equation. In particular, the global minimum for non-trivial knots do not conform well to intuitive notions of "ideal" configurations. It has been suggested that this causes convergence problems during energy minimization [30].

Symmetric Energy Another approach is to view the knot as a radiating tube of small radius. Symmetric energy measures the amount of self-illumination of the knot. This energy can be calculated for a knot K parametrized by $x(t)$, $y(t)$,

$$E_S(K) = \int\int \frac{|dx \times r||dy \times r|}{|x - y|^2},$$

where $dx = \dot{x}(t)dt$ and $r = (x - y)/|x - y|$. This energy function does not shrink tangled portions of the knot and has been used to make beautiful drawings of knots with the software KnotPlot [34].

2.2 Previous Work on Untangling Knots

Given an energy function defined over the space of knots, there are various techniques to find its global minimum that have been explored. The objective is to optimize the rate of convergence to the global minimum or at least to a good local minimum. In knotting, the methods employed have been random perturbation, simulated annealing, gradient descent, and the method of stochastic energy functions.

Random perturbation has been used by Simon's untangler [35], KED and KnotPlot [34]. Random perturbation methods are slowed by convergence problems and many self-intersection checks.

The annealing schedule for knots proposed by Ligocki and Sethian generates new configurations with a standard deviation proportional to an inverse logarithmic function of time [30]. The high-dimensionality of the problem seems to necessitate slow cooling and this method suffers from convergence problems and many self-intersection checks.

Another standard energy minimization approach is to use gradient descent which will quickly find a local minimum. Few self-intersection checks are needed assuming an energy function that goes to infinity as the knot approaches a singular embedding. To escape a local minimum, heuristic perturbation methods can be used after convergence occurs. This technique has been successfully implemented in the MING knot evolver [40].

The stochastic energy untangler [12,13] uses spring energy, electrostatic energy, randomly positioned point charges and reparametrization of the knot to create a random family of energy functions. The magnitude of the perturbation is determined by an annealing schedule. These energy functions are then minimized by gradient descent. The authors report their method to be roughly twice as fast as MING on a difficult benchmark.

We would like to point out the relation between efforts in knot energy minimization and potential field efforts in motion planning. Many of the same difficulties arise. In motion planning, potential field methods have largely been replaced by global probabilistic planners [22,19,18,17,23,24,26,25,28,5]. It is noteworthy to observe that Scharein, in his thesis, remarks that by manually dragging vertices around, minimizations of some very tangled knots could be achieved considerably more quickly [34].

3 Geometry of Knots

A given knot K with n pieces is defined by a sequence of vertices $p_1, ..., p_n$. For convenience, we sometimes say p_{n+1} to mean p_1 and p_0 to mean p_n. Formally the knot consists of the subset of $\mathbf{R}^3$ defined by the union of the segments $s_i = \overline{p_i p_{i+1}}$, for $1 \leq i \leq n$. We say two segments belonging to a knot are disjoint when they do not share an endpoint.

The clearance of a knot K with vertices $p_1, ..., p_n$ is the function

$$clr(K) = \min_{s_i,\ s_j \text{ disjoint}} \{d_{MD}(s_i, s_j)\}.$$

d_{MD} is the minimum distance between a pair of line segments in $\mathbf{R}^3$. This expression measures the distance between disjoint segments. When $clr(K) = 0$, the knot K is called singular. When $clr(K) > \epsilon$ for some $\epsilon > 0$, K is ϵ-non-singular. The self-intersection equations we use are not numerically stable for small values of $clr(K)$ when evaluated with floating point precision. Also, it is assumed that all knots we consider are non-singular.

Given two knots of n pieces, K_a with vertices $p_1, ..., p_n$ and K_b with vertices $q_1, ..., q_n$, we can define a linear interpolation between K_a and K_b parametrized by time t, $K(t)$, which is the knot defined by vertices $r_i(t) = p_i + t \cdot (q_i - p_i)$. This interpolation preserves the topology of the knots in question when the knots $K(t)$ are non-singular for all $t \in [0, 1]$. In particular, existence of such an interpolation between K_a and K_b is a sufficient condition for concluding topological equivalence of K_a and K_b as it implies an explicit ambient isotopy. Detecting the existence of zeroes for $clr(K(t))$ for $t \in [0, 1]$ is the self-intersection detection problem for knot manipulation.

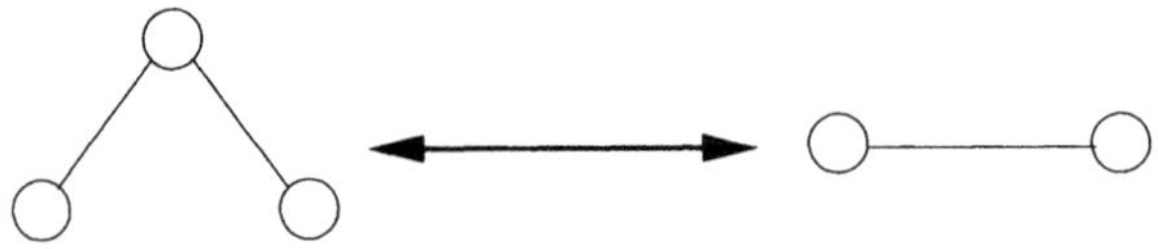

Fig. 2. An elementary move.

Adding and removing vertices can be accomplished via a so-called elementary or triangle move, illustrated in Figure 2. A vertex p_i can be removed from knot K with vertices $p_1, ..., p_n$ to obtain knot K' with vertices

$p_1, ..., p_{i-1}, p_{i+1}, ..., p_n$ if the triangle $\Delta p_{i-1}p_ip_{i+1}$ is unpunctured by all segments disjoint from s_{i-1} and s_i. Alternately, when three vertices are collinear, the middle vertex can be removed. This move and its inverse are the elementary moves and any two equivalent knots can be related by a sequence of elementary moves. The best known upper bound on the number of moves required to transform a given unknot by elementary moves to a triangle is $O(2^{ct})$, where c is 10^7 and t is the number of tetrahedrons in a space triangulation where the knot can be drawn on the 1-skeleton [14]. The authors know of no non-trivial lower bounds in the literature.

4 Untangling Planner

The untangling planner works to reduce the number of vertices in a given knot. In this section, we will describe the design and implementation of our untangling planner. Essentially it is a tree-based planner similar to an RRT [23,24,26,25] or expansive space planner [19,17,18,33]. Our planner is simpler and does not use nearest neighbour queries. Although on longer runs our planner converges to the uniform distribution more slowly, our approach calls the planner often and generates a small number of nearby nodes. Since the dimension of the space is quite high on the interesting benchmarks, we expected the asymptotic advantages of more systematic schemes like RRT or expansive spaces would not have had sufficiently significant impact to make it worth building the necessary geometric data structures. Our experiments in this work and with some toy problems have partially validated this assumption.

Rather than planning a path towards a particular configuration such as the unknot, the planner is used to find a configuration wherein a single vertex can be removed via an elementary move. When no progress is being made, the energy of the knot is optimized instead. If no optimization was found, a random move is made instead. Other important design decisions that were made involved the sampling, energy and weighting functions used in the planner and the choice to avoid configurations where the clearance is very low and that are prone to numerical instability in the self-intersection detection.

Configuration Space Let $\mathcal{K}_n$ be the set of all non-singular knots K with n vertices and let $\mathcal{K} = \bigcup_{n \geq 3} \mathcal{K}_n$. This is the configuration space for the planner. In contrast to many planning problems, configurations will change over the course of the plan as vertices are removed.

Let K be a knot with n vertices $p_1, ..., p_n$ and let K' be a knot with vertices $q_1, ..., q_n$. We say that $K \sim K'$ if the linear interpolation between them has no intermediate configuration that is singular. In particular, this proves that K and K' are the same topological knot.

Given a knot K with vertices $p_1, ..., p_n$ and $\epsilon > 0$ such that $clr(K) \geq \epsilon$, we define two local planners.

Linear Interpolation Planner The first planner moves a single, given vertex as close as possible towards a target point. Given $1 \leq i \leq n$ and a point $q \in \mathbf{R}^3$, we define a parametric knot $K(t)$ with vertices $p_1, ..., p_{i-1}, p_i + t \cdot (q - p_i), p_{i+1}, ..., p_n$. We find all the zeroes of the intersection equation applied to all pairs of disjoint segments in $K(t)$ where one segment contains the ith vertex. Suppose the zeroes occur at $0 < t_1 < \cdots < t_m < 1$. Beginning at the first zero, they are processed in order. The jth zero, t_j, will occur between some segment s with fixed endpoints and a segment $s'(t)$ which has p_i as endpoint. We calculate $d_{MD}(s, s'(t_j))$ and verify that it is larger than ϵ. If not, we find the smallest integer $r > 0$ such that $clr(K(\frac{t_j}{2^r})) > \epsilon$ and set $t^* = \frac{t_j}{2^r}$. If after processing all the zeroes, t^* is undefined then we calculate the smallest integer $r \geq 0$ such that $clr(K(2^{-r})) > \epsilon$ and set $t^* = 2^{-r}$. This allows us to find a knot $K' = K(t^*)$ where $K \sim K'$ and $clr(K') > \epsilon$. We summarize this construction as $K' = lip_i(K, q)$, lip is a mnemonic for linear interpolation planner. Since we are only moving a single vertex, this calculation can occur in linear time.

Elementary Move Planner The second planner removes a vertex by an elementary move. Given $1 \leq i \leq n$, we take the midpoint q of p_{i-1} and p_{i+1} and compute t^* as with the previous section. If $t^* = 1$, then the move is successful and we can safely remove the ith vertex with an elementary move to obtain K' with vertices $p_1, ..., p_{i-1}, p_{i+1}, ..., p_n$. The binary function $elm_i(K)$ takes the value 1 if $t^* = 1$ and the value 0 otherwise.

Vertex Freeing Algorithm 1 below is called with some knot K_0, the input knot, and, N, which is the number of iterations to run the main loop before aborting. N should be chosen to balance the amount of energy optimization and the aggressiveness of the planner in trying to free a vertex. Implicitly there is a minimum tolerated knot clearance as a parameter as well.

The algorithm is split into three sections. Lines 1-5 are the first section, lines 6-15 the second and lines 16-19 are the third section. Let n be the number of vertices in K_0. The first section exhaustively searches for a vertex that can be removed by an elementary move. The vertices are considered in a random order to eliminate bias. The first section runs in time $O(n^2)$: each call to elm costs $O(n)$ geometric comparisons and n calls are made.

The second section consists of probabilistic tree expanding planner which searches for a move that frees a vertex. This section is more complex and requires some simple data structures for a correct implementation. The tree is stored as a directed graph. The distribution of weights and associated probabilities for the tree nodes can be stored in binary heap-like structure which supports $O(\log N)$ weight insertion, weight modification and choose operation. Therefore lines 8, 13 and 14 cost $O(\log N)$. Lines 10 and 11 each require $O(n)$ operations as they require a single traversal of the knot. Line 14 can be implemented in $O(n)$ by maintaining a count of how many segments conflict with each triangle in the knot. The remaining lines in section two

Algorithm 1 Untangle knot K_0

1: Generate a permutation of $1, \ldots, n$ called $j_1, \ldots, j_n$.
2: **for** i ranges from 1 to n **do**
3: Check to see if $elm_{j_i}(K_0) = 1$.
4: If so return K_0 with the j_ith vertex removed.
5: **end for**
6: Create a weighted tree T with root K_0 having weight 1.
7: **while** some number of iterations N **do**
8: Choose a knot K at random from T using the weight distribution.
9: Choose a vertex p_i at random from K.
10: Choose a point q distributed uniformly in the spanning sphere of K.
11: Compute $K' = lip_i(K, q)$ and the associated t^*.
12: Divide the weight of K in T by 2.
13: Add K' to T with parent K and weight t^*.
14: Check if for any j, $elm_j(K') = 1$, if so return the path from K_0 to K' and K'' which is K' with the jth vertex removed.
15: **end while**
16: Computes the energies of all nodes of K.
17: If there is a node with energy less than K_0 return it and the path to it from K_0 in T.
18: Choose a knot K at random from T using the weight distribution.
19: Return the path from K_0 to K in T and K.

can be implemented in constant time. Thus the section can be implemented in time $O(Nn + N \log N)$. The space usage for this section is dominated by the storage for the tree nodes and can be seen as $O(nN)$. By keeping only incremental updates and memoizing periodically the space can be reduced to closer to $O(N)$ without sacrificing much of the runtime.

Finally, the third section requires energy calculations over all nodes in the tree. We are assuming that the time to calculate the energy function is dominated by an operation on pairs of the disjoint edges. The energy of K_0 can be calculated in $O(n^2)$. Since each other node is constructed by an incremental change to one vertex, the energy can be calculated from the previous node in time $O(n)$ with appropriate book-keeping in the tree. Thus line 17 can be done in time $O(n^2 + nN)$ and this cost dominates section three.

Global Scheme The untangling algorithm is called repeatedly some number of times until no more vertices can be removed from the knot. Depending on the application in mind, it may be enough to simplify a certain amount or it might be better to search aggressively for a minimal embedding.

Choosing an Energy Function The energy function that we used was based on MD energy but we dropped the length normalizer to obtain

$$E(K) = \sum_{s_i, s_j \text{ disjoint}} \frac{1}{d_{MD}(s_i, s_j)^2}.$$

The energy function we used is non-standard and was chosen for two reasons: it is easy to compute and it does not shrink tangled portions of the knot like MD energy does for certain tangled sections of the knot [30,13]. This tendency is very undesirable and is one of the sources of convergence problems for MD energy. The problem of the knot becoming very large is dealt with by sampling in a sphere. We ran 50 trials on a large unknot with MD energy instead of unnormalized MD energy and very little progress was made. We found it surprising that the difference between the two functions was so striking because they superficially seem similar. When visualizing intermediate configurations during the untangling we observed that MD energy tended to shrink the tangled portions of the knot. The unnormalized MD energy tended to separate some large triangles or "ears" from the knot. The planner would have ample space to pass other parts of the knot through these ears. That particular move seemed to be critical to making progress in the untangling.

5 Experiments and Results

This section will describe experiments using our untangler. We will show that our planner is significantly faster than the earlier approaches in the literature. The software packages we are comparing against were the simulated annealer of Ligocki-Sethian [30], KnotPlot [34], MING knot evolver [40] and stochastic energy optimizer [12,13]. Each of these packages either report some results in a related paper or are publicly distributed.

Ligocki-Sethian Unknot The second Ligocki-Sethian Unknot shown in Figure 3 is a nearly flat weave with 160 vertices that was used as a test case for a simulated annealer. It was reported that it took more than 16000000 iterations to obtain an embedding that was recognizably a circle[30]. Each iteration involved $O(n^2)$ collision checks for each configuration generated. Our untangler solves this problem nearly instantaneously.

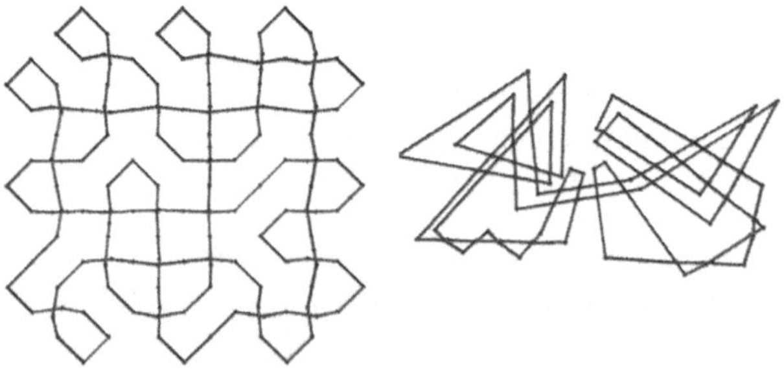

Fig. 3. Ligocki-Sethian Example 2 and Twisted Freedman Unknot.

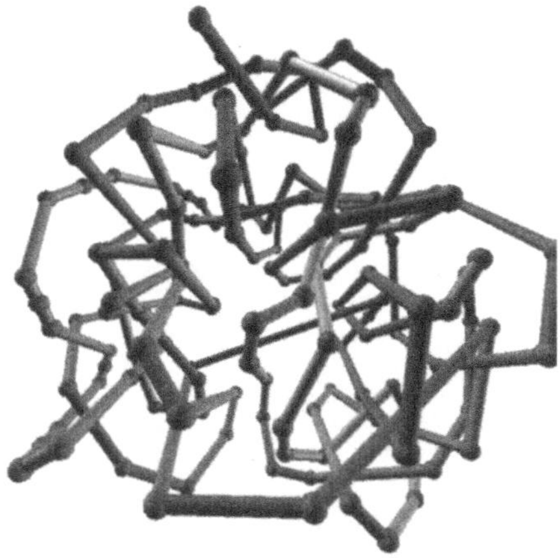

Fig. 4. The Ochiai unknot.

Twisted Freedman Unknot The Twisted Freedman unknot is another nearly flat weave with 122 vertices that was used to show that MING evolver was faster and more precise than perturbation methods [40]. It is shown in Figure 3. We ran MING for more than 12 hours on a 300MhZ SGI O2 without the knot untangling to a circular embedding. Our untangler ran on a single cpu of a dual AMD 1900MP and spent an average of approximately 8 minutes for the runs that completed in less than 25 minutes. About a third of the runs ran for more than 25 minutes and were terminated.

Ochiai unknot The Ochiai unknot is Figure 4 is an unknot with 139 vertices that has foiled many energy minimizers. It was originally given as an example of a general construction in a paper by M. Ochiai written in 1990. Since then it has often been used to test unknotting programs. It is reported in [13] that MING running on an SGI O2 untangled it in 108 hours and stochastic minimization took 48 hours. KnotPlot was not able to reduce it [34]. Our untangler averaged about 10 minutes on a single cpu of a dual AMD 1900MP and had roughly one sixth of its runs terminated after 35 minutes.

Random unknots We generated several random unknots in hopes of obtaining more interesting examples to test on. We generated unknots by beginning with a circle of vertices and then making a large number of random moves. Even by running the tangler for many hours, we were not able to generate any configurations that were not solved instantaneously by the untangler.

Summary of unknot experiments We summarize the results of our untangling experiments in Figure 5. Once we factor in the difference in processor speeds, it is clear that our method is orders of magnitude faster than the energy minimizations in the literature. A second advantage is the probabilistic nature

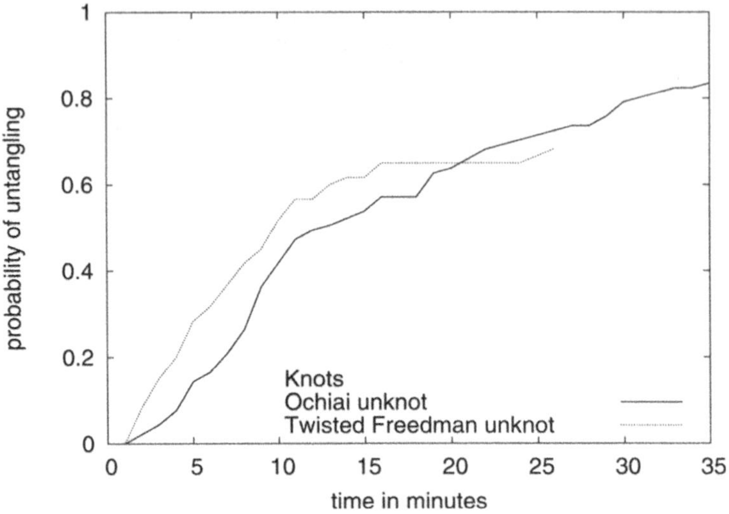

Fig. 5. Probability of finding solution versus time in minutes

of the algorithm allows it to run many instances in parallel to obtain better times. In the case of the Twisted Freedman and Ochiai this would mean solution times closer to two or three minutes.

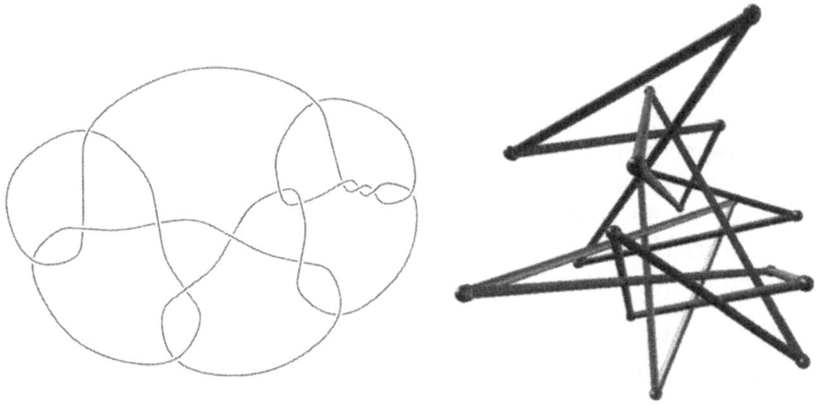

Fig. 6. The knot diagram for 16_{148} rendered by Knotscape and 3D drawing with 17 sticks.

Stick number of non-trivial knots The results we have reported up to now have been for unknots. Another application of the untangler is to minimize the number of segments used to draw a given knot. As an example we took the non-alternating knot 16_{148} shown in Figure 6 from the database of knots stored in the software Knotscape by Hoste and Thistlethwaite and found a 3d presentation of it with 17 sticks after a few minutes of computation. The initial embedding had 50 segments.

6 Discussion

The algorithm we gave engendered several design choices. In this section, we explain the motivations for these choices and discuss implications for other path planning problems.

Use of trees Knot space has many properties that make it hostile to PRM planners that build a roadmap [22]. It is very difficult to sample uniformly from a single connected component. Determining whether two knots are in the same component is an instance of UNKNOT or KNOT EQUIVALENCE and is challenging for examples with many pieces. Even finding two random knots K, K' such that $K \sim K'$ for 30 segment knots requires the generation of several million random knots. It is also difficult to find fast algorithmic solutions to knot matching that take into account the symmetries of knots. For these reasons, we use a tree expanding planner in the spirit of other such planners [19,18,17,23,24,26,25,33] to avoid the difficulties with global sampling.

Tree expansion In tree-based probabilistic planning, it has been observed that it is undesirable for the tree to be concentrated in a small area of the space. The planner will be much faster if it covers the space more quickly and other tree-based planners have employed a number of heuristics for this reason [19,18,17,23,24,26,25,33]. Our planner's expansion scheme is simplistic but seems to produce similar distributions initially until the more sophisticated heuristics of other methods begin to kick in. Since our planner was a hybrid with an energy minimizer, we found it was more efficient to prune the tree fairly aggressively by energy than rely on long term space covering heuristics to explore a very high dimensional space.

When moving all vertices simulataneously along some linear path, the knot is constrained by the most constrained vertex. When moving a single vertex at random n times, we observed that the final position of the knot was typically much further from the start position. Since these two schemes are roughly the same amount of work computationally, moving one vertex at a time has a clear advantage. A further benefit is that the intersection equations have lower degree.

Freeing a vertex Finding a path that takes a given tangled unknot to a triangle requires finding a very long path. Worse, when trying to simplify a knot of unknown type, the final configuration may not be known. One possibility for finding a path to an untangled configuration is to construct a very large search tree. This is impractical as it is very space intensive. In addition, it is also very slow if backtracking after performing an elementary move is possible, even for completely trivial input such as n vertices arranged in a planar polygon. This problem occurs because there can be at most $n!$ ways to reduce the knot to a triangle. We disallow backtrack, essentially throwing away the tree but for a path ending in an elementary move, we speed the planner greatly at the cost of having some probability of putting the knot in a bad configuration. We believe that such bad configurations are responsible for the occasional long run times observed in the experiments section. The general approach here was to decompose the problem of finding an untangling into a sequence of local goals consisting of elementary moves. This has two advantages: the local goal is much easier to plan for and the final configuration does not need to be known. This idea appears in robotics literature in task planning and in hierarchical planning. In our case, formulating the planning problem in terms of sequence of local goals was essential to obtain a usable planner. When it is possible to find such formulation, we suggest that it will be often a major improvement to the speed of the planner.

Energy hints versus random There are several planning applications with natural energy functions such as protein folding, molecular docking and realistic deformable robots. Also, there has been some investigation in using energy when planning for such systems [36,4]. After running the loop in Algorithm 1 N times the next configuration to consider is chosen. In our implementation we used energy to make this choice. By eliminating lines 17 and 18 from the final section of Algorithm 1, thus always executing the random guess on line 19, we obtained very poor results. We ran 50 trials on the Ochiai unknot for nearly an hour with almost no progress. We observed that the knot tended to stay in high energy configurations when moving at random. This is consistent with our observation that the planner seems to hampered by high energy knot configurations. We believe that the reason for this is that high energy configurations are more constrained and it is more difficult for the tree to escape from a high energy region of a high dimensional configuration space. Energy functions with similar properties, either natural or artificially designed, might exist for other planning instances. A more general observation is that using an energy function to bias sampling can be extremely effective. Further investigation and awareness of hybrid energy and planning algorithms may lead to improvements in a variety of planning instances.

7 Future Work and Applications

Probabilistic completeness Although we have not shown probabilistic completeness for our planner, we give a conjecture which is both necessary and sufficient for our planner to be probabilistically complete.

Conjecture 1. Let K be a piecewise linear knot drawn with n line segments. Suppose $s(K) < n$. Then there is a sequence of knots with n pieces $K = K_0, K_1, ..., K_m$ with $K_i \sim K_{i+1}$ and K_m has three collinear vertices.

The proof of probabilistic completeness for the planner follows from this conjecture and proceeds according to the scheme described in [27].

Applications to Computational Topology Our planner can be used for stick number calculation of a knot where unknot detection is a special case. Another application is to calculate the unknotting number of a knot: given some knot, the task is to find a path containing as few crossings as possible that transforms that knot to the unknot.

Applications to DNA knotting Knots in DNA rings play an important role in the life-cycle of a cell [32,10]. Determining low energy paths for DNA knotting and unknotting might be used to learn how certain processes occur in cell biology. However, the energetics of DNA rings are considerably more complex than MD energy.

Flexible Robots There is a strong similarity with untangling and the general problem of planning with flexible robots: high dimensional configuration spaces that can change parameters over the course of the plan. In this work, we introduced the use of local goals and hybrid planning with potential fields, either of which may be used in a more general setting. An interesting direct extension of this work would be study the problem of tying knots in physically realistic rope such as that described in [31].

Acknowledgements. Work on this paper by A. Ladd and L. Kavraki has been supported in part by NSF 9702288, NSF 0114796, a Whitaker Biomedical Engineering Grant and a Sloan Fellowship to L. Kavraki. A. Ladd is also supported by FCAR. We would like to thank N. Dean for discussions and insights. We also thank R. Scharein for providing us with the Ochiai and Ligocki-Sethian unkots. We thank the anonymous WAFR reviews for their suggestions on improving the paper. Finally, we thank J. Phillips for our fruitful collaboration in our earlier work and for discussions about this work.

References

1. I. Agol, J. Hass, and W. Thurston. The computational complexity of knot genus and spanning area. (preprint.).
2. C. N. Anerizis. *The Mystery of Knots - Computer Programming for Knot Tabulation.* Series on Knots and Everything. World Scientifical Publishing Co. Pte. Ltd., 1999.
3. M. Apaydin, D. Brutlag, C. Guestrin, D. Hsu, and J. Latombe. Stochastic roadmap simulation: An efficient representation and algorithm for analyzing molecular motion. In *International Conference on Computational Molecular Biology (RECOMB)*, April 2002.

4. M. Apaydin, A. Singh, D. Brutlag, and J. Latombe. Capturing molecular energy landscapes with probabilistic conformal roadmaps. In *IEEE International Conference on Robotics and Automation (ICRA)*, May 2001.
5. O. Bayazit, J.-M. Lien, and N. Amato. Probabilistic roadmap motion planning for deformable objects. In *IEEE International Conference on Robotics and Automation*, 2002.
6. M. Bern, D. Eppstein, and al. Emerging challenges in computational topology, 1999.
7. J. S. Birman, P. Boldi, M. Rampichini, and S. Vigna. Towards an implementation of the b-h algorithm for recognizing the unknot. In *KNOTS-2000*, 2000.
8. Z. Butler, K. Kotay, D. Rus, and K. Tomita. Cellular automata for decentralized control of self-reconfigurable robots. In *IEEE International Conference on Robotics and Automation*, 2001.
9. X. Dai and Y. Diao. The minimum of knot energy functions. *Journal of Knot Theory and its Ramifications*, 9(6):713–724, 2000.
10. R. Deibler, S. Rahmati, and E. Zechiedrich. Topoisomerase iv, alone, unknots dna in escherichia coli. *Genes and Development*, 15:748–761, 2001.
11. Y. Diao, C. Ernst, and J. Rensburg. In search of a good polygonal knot energy. *Journal of Knot Theory and its Ramifications*, 6(5):633–657, 1997.
12. R. Grzeszczuk, M. Huang, and L. Kauffman. Untangling knots by stochastic energy optimization. *IEEE Visualization*, pages 279–286, 1996.
13. R. Grzeszczuk, M. Huang, and L. Kauffman. Physically-based stochastic simplification of mathematical knots. *IEEE Transactions on Visualization and Computer Graphics*, 3(3):262–278, 1997.
14. J. Hass and J. Lagarias. The number of reidemeister moves needed for unknotting. (preprint.).
15. J. Hass, J. C. Lagarias, and N. Pippenger. The computational complexity of knot and link problems. In *IEEE Symposium on Foundations of Computer Science*, pages 172–181, 1997.
16. J. Hoste and M. Thistlethwaite. The first 1,701,936 knots. *Math. Intelligencer*, 20(4):33–48, 1998.
17. D. Hsu. *Randomized Single-Query Motion Planning In Expansive Spaces.* PhD thesis, Department of Computer Science, Stanford University, 2000.
18. D. Hsu, R. Kindel, J. Latombe, and S. Rock. Control-based randomized motion planning for dynamic environments. In *Algorithmic and Computational Robotics: New Directions: The Fourth International Workshop on the Algorithmic Foundations of Robotics*, pages 247–264, 2001.
19. D. Hsu, J. Latombe, and R. Motwani. Path planning in expansive spaces. In *Proc. IEEE Int'l Conf. on Robotics and Automation*, pages 2719–2726, 1997.
20. F. Jaeger, D. L. Vertigan, and D. Welsh. On the computational complexity of the jones and tutte polynomials. In *Math. Proc. Camb. Phil. Soc.*, 108, pages 35–53, 1990.
21. R. Jenkins. A dynamic approach to calculating the homfly polynomial for directed knots and links. Master's thesis, Carnegie Mellon University, 1989.
22. L. E. Kavraki, P. Svestka, J.-C. Latombe, and M. H. Overmars. Probabilistic roadmaps for path planning in high-dimensional configuration spaces. *Transaction on Robotics and Automation*, 12(4):566–580, June 1996.
23. J. J. Kuffner and S. M. LaValle. Randomized kinodynamic planning. In *Proc. IEEE Int'l Conf. on Robotics and Automation*, pages 473–479, 1999.

24. J. J. Kuffner and S. M. LaValle. RRT-connect: An efficient approach to single-query path planning. In *Proc. IEEE Int'l Conf. on Robotics and Automation*, 2000.
25. J. J. Kuffner and S. M. LaValle. Randomized kinodynamic planning. *International Journal of Robotics Research*, 20(5):378–400, May 2001.
26. J. J. Kuffner and S. M. LaValle. Rapidly exploring random trees: Progress and prospects. In *Algorithmic and Computational Robotics: New Directions: The Fourth International Workshop on the Algorithmic Foundations of Robotics*, pages 293–308, 2001.
27. A. Ladd and L. Kavraki. A measure theoretic analysis of prm. In *IEEE International Conference on Robotics and Automation*, May 2002.
28. F. Lamiraux and L. Kavraki. Planning paths for elastic objects. *International Journal of Robotics Research*, 20(3), 2001.
29. W. Lickorish. *An Introduction to Knot Theory*. Springer, 1997.
30. T. Ligocki and J. A. Sethian. Recognizing knots using simulated annealing. *Journal of Knot Theory and Its Ramifications*, 3(4):477–495, 1994.
31. J. Phillips, A. Ladd, and L. Kavraki. Simulated knot tying. In *IEEE International Conference on Robotics and Automation*, May 2002.
32. L. Postow, B. Peter, and N. Cozzarelli. Knot what we thought before: The twisted story of replication. *BioEssays*, 21:805–808, 1999.
33. G. Sanchez and J.-C. Latombe. A single-query bi-directional probabilistic roadmap planner with lazy collision checking. In *ISRR*, 2001.
34. R. Scharein. *Interactive Topological Drawing*. PhD thesis, University of British Columbia, 1988.
35. J. Simon. Energy functions for polygonal knots. *J. Knot Theory and its Ramif.*, 3:299–320, 1994.
36. G. Song and N. Amato. Using motion planning to study protein folding pathways. In *International Conference on Computational Molecular Biology (RECOMB)*, pages 287–296, April 2001.
37. M. Teodoro, G. Phillips, and L. Kavraki. A dimensionality reduction approach to modeling protein flexibility. In *International Conference on Computational Molecular Biology (RECOMB)*, April 2002.
38. S. Vassilvitskii, J. Suh, and M. Yim. A complete, local and parallel reconfiguration algorithm for cube style modular robots. In *IEEE International Conference on Robotics and Automation (ICRA)*, May 2002.
39. J. Walter, B. Tsai, and N. Amato. Choosing good paths for fast distributed reconfiguration of hexagonal metamorphic robots. In *IEEE International Conference on Robotics and Automation*, 2002.
40. Y.-Q. Wu. Ming user manual. www.math.uiowa.edu/ ~wu/ming/ming.pdf, 1996.

Exact Collision Checking of Robot Paths

Fabian Schwarzer, Mitul Saha, and Jean-Claude Latombe

Dept. of Computer Science, Stanford University, Stanford CA 94305, USA

Abstract. This paper describes a new exact and efficient collision checker for testing single straight-line segments in c-space or sequences of such segments. This checker is particularly suited for probabilistic roadmap (PRM) planners applied to manipulator arms and multi-robot systems. Unlike the commonly used fixed-resolution method, the new checker does not require manual tuning of the resolution parameter and cannot fail to find existing collisions. To achieve this, it automatically and dynamically adjusts the required local c-space resolution by relating distances and path lengths in workspace. Several additional techniques are presented in this paper to further increase the checker's efficiency in PRM planners and scenarios with high geometric complexity. Extensive experiments, first on segments randomly generated in c-space, next as part of an existing PRM planner, and finally as part of a path smoother/optimizer have shown that the new checker is faster than the common resolution-based approach (with suitable resolution).

1 Introduction

Collision checking is a fundamental operation in robot motion planning, graphic animation and physical simulation [5,13,18]. While *static* collision checking amounts to testing a single configuration for spatial overlaps, *dynamic* collision checking requires showing that all configurations on a continuous path in c-space are collision-free. Four major families of methods can be used for dynamic collision checking: feature-tracking, bounding-volume, swept-volume and trajectory parameterization methods.

- *Feature-tracking* methods rely on the following coherence assumption: the pair of closest features between two objects in relative motion changes only at discrete points of time and, when it changes, computing the new pair from the old one can be done efficiently [3,8,16,17,19]. However, to be practical, these methods require each object to be made of few convex components and to have relatively small geometric complexity. Furthermore, they test a path by small increments, from one end to the other, which is usually not the fastest way to detect if the path collides. The coherence assumption is particularly problematic for kinematic chains and may require tiny steps, especially for the links closer to the end of the chain.
- *Bounding-volume* (BV) methods precompute, for each object, a hierarchy of BVs (e.g., spheres, boxes) that approximate the geometry of the object at varying resolutions [11,15,21,24]. BVs then speed up collision checking

J.-D. Boissonnat et al. (Eds.): Algorithmic Foundations of Robotics V, STAR 7, pp. 25–41, 2004.

by making it possible to quickly discard large portions of objects that cannot possibly collide. These methods have been successfully applied to objects with surfaces described by several 100,000 triangles, and more. But they are fundamentally static methods. To test a path for collision, the common approach is to test intermediate configurations along the path until a collision is found or any two successive configurations are less than some prespecified ε apart, according to some metric in c-space. In the second case, the path is declared collision-free, despite the fact that this answer has a slight chance of being incorrect.

- *Swept-volume and space-time volume intersection* methods consist of computing the volumes swept by the objects in the workspace, possibly with a time dimension added, and testing these volumes for overlap [6,10]. However, exact computation of such volumes is time consuming, especially, when objects undergo rotations and are geometrically complex. Moreover, the overlap test can no longer be speeded up by exploiting precomputed data structures (such as BV hierarchies). Another problem is that, unless either the time dimension is added or relative motions are considered explicitly for all pairs, swept volumes for pairs of moving objects may overlap although the objects themselves do not collide.
- *Trajectory parameterization* methods, e.g., [7,23], describe the motions as algebraic polynomials in a single variable t. These polynomials are then used to derive collision constraints on t. In principle, this allows for computing the exact intervals of collision along a given motion. However, in general the degrees of the involved polynomials may be arbitrarily high and thus solving the constraints for t can be quite expensive and furthermore pose numerical problems.

Hence, dynamic collision checking remains a major bottleneck in many applications. In particular, probabilistic roadmap (PRM) planners heavily rely on the availability of efficient dynamic checkers to test "local paths" between randomly sampled configurations for collision [1,2,4,12,14,22]. Most such planners use a static BV method to test intermediate configurations along the local paths at some resolution ε (these configurations are usually obtained by recursively bisecting the path [9,22]). Choosing ε is difficult, especially in scenarios with articulated arms and/or multiple robots. It typically involves several trial-and-error experiments, which must be repeated for each new type of robot and environment. Large values of ε are acceptable when both robot links and obstacles are fat. But when objects are thin, collisions are easy to miss. An example is shown in Fig. 1a, which contains two skinny linkages, with 20 revolute joints each, and thin obstacles, all with significant geometric complexity. In this example, even small changes in the joint angles can make a linkage jump over an obstacle or over the other linkage. The value of ε needed to reliably detect such collisions must be very small, resulting in an inefficient PRM planner. Further examples illustrating this problem are shown in Figs. 1b and 5.

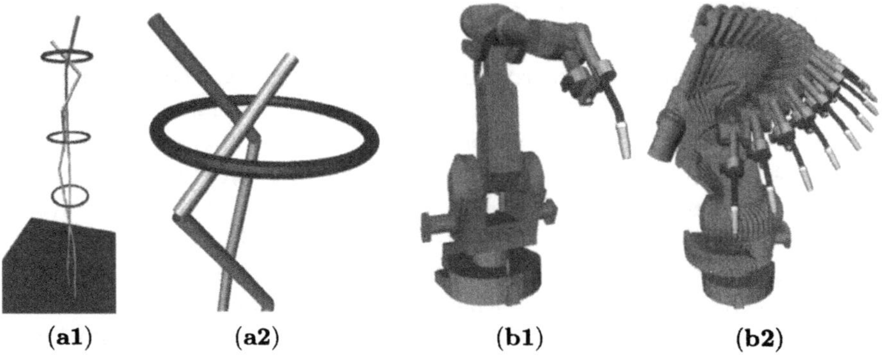

Fig. 1. (a1) Two skinny 20-DOF arms (320 triangles each) which have to retract from three fixed thin tori (6300 triangles each); (a2) Detail showing arms in top ring; (b1) IRB 2400 robot with thin arc welding gun (3594 triangles); (b2) Snapshots along motion on a line segment in c-space

Some planners use expanded models for collision checking. If the expanded models do not collide at the chosen discrete steps, then the original models are guaranteed not to collide on their motions between the steps. However, this further complicates choosing an appropriate resolution ε: a large value for ε will require to "grow" the objects significantly and thus increase the rejection rate and make planning harder. A small value will again lead to an inefficient checker.

The new checker described in this paper addresses these problems by automatically and dynamically adjusting the local c-space resolution at which the segment is tested. To achieve this, it relates distances between objects and lengths of paths traced by points on these objects in workspace. This approach frees the user from manually tuning a constant ε and at the same time guarantees not to miss any collisions. The new checker is particularly suited for manipulator arms and multi-robot scenarios and can check for collisions between robot links and obstacles as well as self-collisions and collisions between links of different robots. While the basic idea of relating distances and path lengths has been proposed before (e.g., in [2,5]), several new techniques are introduced in this paper to make the method more efficient in PRM planners and applicable to scenarios with high geometric complexity. These techniques include a greedy distance computation algorithm that is as efficient as pure collision checking, a method for bounding lengths of paths traced out in workspace, and a heuristic for ordering collision tests.

The rest of the paper is organized as follows. Section 2 describes our approach and presents in detail the new adaptive segment checking algorithm. Section 3 reports on our experiments with random segments, a lazy PRM planner and a randomized path smoothing algorithm. Section 4 concludes the paper and points to possible future work.

2 Exact Adaptive Segment Checking

We consider the robot (or the robots) and all obstacles as a collection of moving and static rigid bodies $\mathcal{A}_1, \ldots, \mathcal{A}_n$ whose poses are uniquely determined by a vector $\boldsymbol{q} = (q_1, \ldots, q_d)$, the configuration of the system.

Let $\mathcal{A}_i(\boldsymbol{q})$ denote body $\mathcal{A}_i$ at configuration $\boldsymbol{q}$ and $\eta_{ij}(\boldsymbol{q})$ be the workspace distance (or a non-trivial lower bound of it) between $\mathcal{A}_i(\boldsymbol{q})$ and $\mathcal{A}_j(\boldsymbol{q})$. If $\mathcal{A}_i(\boldsymbol{q})$ and $\mathcal{A}_j(\boldsymbol{q})$ intersect, we set $\eta_{ij}(\boldsymbol{q}) = 0$. For each body $\mathcal{A}_i$, let $\lambda_i(\boldsymbol{q}_a, \boldsymbol{q}_b)$ denote an upper bound on the length of the curve segments traced (in workspace) by all points on $\mathcal{A}_i$ when the configuration of the system is linearly interpolated from some configuration $\boldsymbol{q}_a$ to another configuration $\boldsymbol{q}_b$. Clearly, $\lambda_i(\boldsymbol{q}_a, \boldsymbol{q}_b) = 0$ if $\mathcal{A}_i$ is a (static) obstacle. Note that although the connection of $\boldsymbol{q}_a$ and $\boldsymbol{q}_b$ is a straight line segment in c-space, the curves traced by points on the robot in workspace are usually non-linear due to rotations. For example, if $\mathcal{A}_i$ rotates around a fixed axis, then points on $\mathcal{A}_i$ trace out circular arcs.

We can now state a sufficient condition to guarantee that two bodies $\mathcal{A}_i$ and $\mathcal{A}_j$ cannot collide during a motion along a line segment in c-space:

Lemma 1. *Two objects $\mathcal{A}_i$ and $\mathcal{A}_j$ do not collide at any configuration $\boldsymbol{q}$ on the straight path segment joining $\boldsymbol{q}_a$ and $\boldsymbol{q}_b$ in c-space, if:*

$$\lambda_i(\boldsymbol{q}_a, \boldsymbol{q}_b) + \lambda_j(\boldsymbol{q}_a, \boldsymbol{q}_b) < \eta_{ij}(\boldsymbol{q}_a) + \eta_{ij}(\boldsymbol{q}_b). \tag{1}$$

Proof. Assume that Inequality (1) is verified. If $\mathcal{A}_i$ and $\mathcal{A}_j$ collide, then a point $\boldsymbol{p}_i$ of $\mathcal{A}_i$ must coincide with a point $\boldsymbol{p}_j$ of $\mathcal{A}_j$ at some intermediate configuration $\boldsymbol{q}_c$ along the straight path segment between $\boldsymbol{q}_a$ and $\boldsymbol{q}_b$. Let $\ell_i(\boldsymbol{q}, \boldsymbol{q}')$ be the length traced by $\boldsymbol{p}_i$ between any two configurations $\boldsymbol{q}$ and $\boldsymbol{q}'$ in this segment. Define $\ell_j(\boldsymbol{q}, \boldsymbol{q}')$ in the same way for point $\boldsymbol{p}_j$. For $\boldsymbol{p}_i$ and $\boldsymbol{p}_j$ to coincide at $\boldsymbol{q}_c$, we must have $\ell_i(\boldsymbol{q}_a, \boldsymbol{q}_c) + \ell_j(\boldsymbol{q}_a, \boldsymbol{q}_c) \geq \eta_{ij}(\boldsymbol{q}_a)$ and $\ell_i(\boldsymbol{q}_c, \boldsymbol{q}_b) + \ell_j(\boldsymbol{q}_c, \boldsymbol{q}_b) \geq \eta_{ij}(\boldsymbol{q}_b)$. Since $\ell_i(\boldsymbol{q}_a, \boldsymbol{q}_c) + \ell_i(\boldsymbol{q}_c, \boldsymbol{q}_b) = \ell_i(\boldsymbol{q}_a, \boldsymbol{q}_b)$ and $\ell_j(\boldsymbol{q}_a, \boldsymbol{q}_c) + \ell_j(\boldsymbol{q}_c, \boldsymbol{q}_b) = \ell_j(\boldsymbol{q}_a, \boldsymbol{q}_b)$, summing the previous two inequalities yields $\ell_i(\boldsymbol{q}_a, \boldsymbol{q}_b) + \ell_j(\boldsymbol{q}_a, \boldsymbol{q}_b) \geq \eta_{ij}(\boldsymbol{q}_a) + \eta_{ij}(\boldsymbol{q}_b)$. Using $\lambda_i(\boldsymbol{q}_a, \boldsymbol{q}_b) \geq \ell_i(\boldsymbol{q}_a, \boldsymbol{q}_b)$ and $\lambda_j(\boldsymbol{q}_a, \boldsymbol{q}_b) \geq \ell_j(\boldsymbol{q}_a, \boldsymbol{q}_b)$, we get $\lambda_i(\boldsymbol{q}_a, \boldsymbol{q}_b) + \lambda_j(\boldsymbol{q}_a, \boldsymbol{q}_b) \geq \eta_{ij}(\boldsymbol{q}_a) + \eta_{ij}(\boldsymbol{q}_b)$ which contradicts our initial hypothesis that Inequality (1) is verified. So, $\mathcal{A}_i$ and $\mathcal{A}_j$ do not collide. •

Note that the reverse of Lemma 1 is not true: $\lambda_i(\boldsymbol{q}_a, \boldsymbol{q}_b) + \lambda_j(\boldsymbol{q}_a, \boldsymbol{q}_b)$ can exceed $\eta_{ij}(\boldsymbol{q}_a) + \eta_{ij}(\boldsymbol{q}_b)$ without introducing a collision.

2.1 Core Algorithm

The new segment checking algorithm uses Lemma 1 to decide whether a segment is collision-free or whether it requires further examination.

It is important to note that (1) applies to an individual pair of bodies. We can therefore check each pair of objects independently of the other pairs. For example, links closer to the base of a manipulator usually move less than links closer to the tip and thus often require fewer tests along the segment (e.g., see Fig. 1b). Furthermore, the pairs can be processed in an order that depends on their relative distances and motions and thus decreases the expected time to find possible collisions. For example, pairs whose bodies are close and move a lot will be examined before pairs of slowly moving bodies that are far apart. To realize this, our algorithm maintains a priority queue of ***pair-specific (sub-)segments.***

An entry $[\mathbf{q}_a, \mathbf{q}_b]_{ij}$ in the queue indicates that the pair $(\mathcal{A}_i, \mathcal{A}_j)$ needs to be tested for collisions along $[\mathbf{q}_a, \mathbf{q}_b]$. (Section 2.4 describes how we assign priorities to these entries.) Initially, the queue is filled with entries $[\mathbf{q}, \mathbf{q}']_{ij}$ spanning the entire given segment for each pair of rigid objects which we would like to check. The algorithm then processes the queue by removing the first element, say $[\mathbf{q}_a, \mathbf{q}_b]_{ij}$, and tests if it satisfies (1). If successful, which means that $\mathcal{A}_i$ and $\mathcal{A}_j$ cannot collide along their motions defined by $[\mathbf{q}_a, \mathbf{q}_b]$, it continues with the next element in the queue. Otherwise, it checks if the indexed pair collides $(\mathcal{A}_i, \mathcal{A}_j)$ at $\mathbf{q}_{mid} = (\mathbf{q}_a + \mathbf{q}_b)/2$. If this reveals a collision, the whole procedure can stop and report the colliding pair of objects. Otherwise, two new entries $[\mathbf{q}_a, \mathbf{q}_{mid}]_{ij}$ and $[\mathbf{q}_{mid}, \mathbf{q}_b]_{ij}$ are inserted into the queue and the algorithm continues with the first element in the queue. When the queue is empty, the input segment $[\mathbf{q}, \mathbf{q}']$ can be reported to be collision-free.

Algorithm CHECK-SEGMENT$(\mathbf{q}, \mathbf{q}')$

Initialize priority queue Q with $[\mathbf{q}, \mathbf{q}']_{ij}$ for all pairs
  of bodies $(\mathcal{A}_i, \mathcal{A}_j)$ that need to be tested.
while Q not empty do
  $[\mathbf{q}_a, \mathbf{q}_b]_{ij} \leftarrow$ remove-first(Q)
  if $\lambda_i(\mathbf{q}_a, \mathbf{q}_b) + \lambda_j(\mathbf{q}_a, \mathbf{q}_b) \geq \eta_{ij}(\mathbf{q}_a) + \eta_{ij}(\mathbf{q}_b)$ then
    $\mathbf{q}_{mid} \leftarrow (\mathbf{q}_a + \mathbf{q}_b)/2$
    if *distance*$(\mathcal{A}_i(\mathbf{q}_{mid}), \mathcal{A}_j(\mathbf{q}_{mid})) < \delta$ then return *collision*
    else insert $[\mathbf{q}_a, \mathbf{q}_{mid}]_{ij}$ and $[\mathbf{q}_{mid}, \mathbf{q}_b]_{ij}$ into Q
return *no collision*

Fig. 2. Segment checking core algorithm

Figure 2 summarizes this basic algorithm, including a small modification: we replace the collision check at $\mathbf{q}_{mid}$ by a test if the distance between $\mathcal{A}_i(\mathbf{q}_{mid})$ and $\mathcal{A}_j(\mathbf{q}_{mid})$ is smaller than some small constant $\delta > 0$. This modification is required to bound the running time, for the following reason. The maximum bisection depth for an entry $[\mathbf{q}, \mathbf{q}']_{ij}$ in the queue depends on the minimum workspace distance between $\mathcal{A}_i$ and $\mathcal{A}_j$ on their motions given by the segment $[\mathbf{q}, \mathbf{q}']$. Without a lower bound on this distance, the recursion could go very deep if the two bodies intersect only for an extremely short interval on $[\boldsymbol{q}, \boldsymbol{q}']$ or happen to come extremely close during the motion but do not collide at all. Although this modification may seem like a serious limitation, we found that in practice it is not. In all of our experiments, δ was set to the same very small constant value (see Sect. 3).

The major advantage of this algorithm over the simple fixed-resolution approach is that it dynamically adapts the required resolution at which individual (sub-)segments are checked. Pairs of objects that are well-separated and move only little will require less bisections than pairs that move more and are closer. Lemma 1 guarantees that no collisions are missed.

Inequality (1) and the above algorithm can be illustrated schematically by drawing a segment of length proportional to $\lambda_i(\boldsymbol{q}_a, \boldsymbol{q}_b) + \lambda_j(\boldsymbol{q}_a, \boldsymbol{q}_b)$ and centering discs with radii proportional to $\eta_{ij}(\boldsymbol{q}_a)$ and $\eta_{ij}(\boldsymbol{q}_b)$ at the endpoints of this segment. If the drawn segment is covered by these discs then the motions of the two bodies are collision-free (see Fig. 3a). Otherwise, if there are no collisions along the segment, additional discs (placed at intermediate configurations $\boldsymbol{q}_{mid}$) will finally allow covering it (see Fig. 3b).

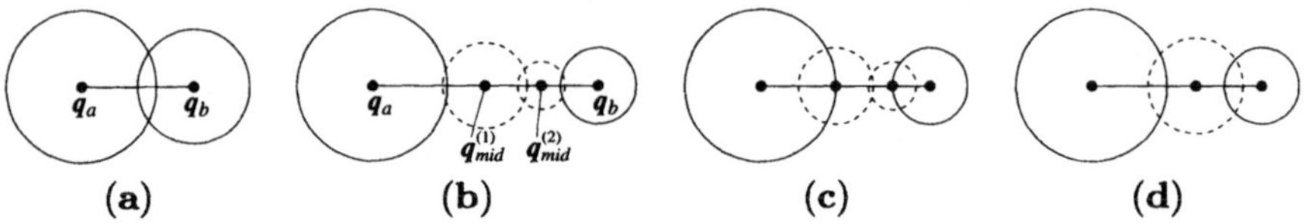

Fig. 3. The segment checking algorithm tries to "cover" a segment with discs with radii proportional to workspace distances. (a) Discs at endpoints already cover segment; (b) Disc at first bisection point $\boldsymbol{q}_{mid}^{(1)}$ covers left subsegment while right subsegment requires one more bisection at $\boldsymbol{q}_{mid}^{(2)}$; (c-d) Two alternative covering strategies (see text)

Instead of bisecting and testing configuration $\boldsymbol{q}_{mid}$ exactly at the midpoint of the (sub-)segment (as shown in Fig. 3b), one could come up with different strategies to refine a yet uncovered (sub-)segment. For example, we could add two new configurations and place them at the intersection points of the two existing circles with the segment, as shown by the two dashed circles in Fig. 3c. This approach would require re-inserting only one new sub-segment into the priority queue instead of two. However, half of the new distance information would be wasted for redundantly covering parts of the segment. Another strategy would be to bisect in the middle of the uncovered section of the segment (see Fig. 3d) instead of symmetrically between the endpoints. In practice, however, we could not notice a significant difference between (d) and (b). Furthermore, (d) has the drawback of bisecting at different points for different pairs of bodies. In fact, we favor (b) because it allows for an additional caching and indexing mechanism for rigid body transforms at the inserted configurations (see Sect. 3).

2.2 Greedy Distance Computation

Our approach requires an efficient algorithm to compute non-trivial lower distance bounds (or exact distances) between pairs of complex polyhedral bodies and to test for $\geq \delta$ separation. Tighter bounds allow the segment to be covered with less bisections but they are also more expensive to compute.

Bounding volume (BV) hierarchies are commonly used for checking collision and/or computing distances between bodies (see Sect. 1). However, computing exact distances has proven significantly more expensive than pure collision checking. The efficiency gained by our adaptive bisection can easily be dwarfed by the higher cost of computing exact distances. Some BV algorithms are able to compute approximate distances much faster than exact distances (e.g., [15,21]), but still slower than they can check for collision. Furthermore, their efficiency depends on the allowed relative or absolute error, a parameter that may be difficult to set.

We therefore propose a new algorithm that greedily computes lower distance bounds between general non-convex polyhedra while checking collision or minimal separation δ (as indicated in Sect. 2.1). This algorithm, called GREEDY-DIST, is also based on BV hierarchies. In practice it returns good lower bounds most of the time. It is independent of the choice of BV, as long as the distance between pairs of BVs can be computed efficiently. In our implementation, we use rectangle swept spheres (RSS) as BVs because distance computation between them has been found to be faster than for OBBs [15].

We assume that each BV hierarchy is an approximately balanced binary tree of BVs, with each leaf being a triangle of the surface of the polyhedral body represented by the hierarchy. To compute a lower bound on the distance between two bodies, we call GREEDY-DIST(B_i, B_j), where B_i and B_j denote the BVs at the roots of the hierarchies representing the two bodies. The algorithm is shown in Fig. 4, where $\delta \geq 0$ is the required minimum separation between the two bodies and B_{j1} and B_{j2} are the two children of B_j in the BV hierarchy.

If the original call GREEDY-DIST(B_i, B_j) returns 0, then the distance between the two bodies is less than or equal to δ, otherwise it is greater than δ and the returned value is a lower bound.

This algorithm is much more efficient than exact distance computation because it terminates as soon as a separation greater than δ has been shown. In fact, when δ is set to 0, the algorithm visits exactly the same pairs of BVs and triangles as a standard recursive collision checker using the same BV type would do. But it additionally returns a non-zero lower bound on the distance if the two bodies do not intersect. The experiments reported below show that in practice this bound is quite good.

Table 1 compares the performance of GREEDY-DIST with $\delta = 0$ to algorithms as described in [15] for pure collision checking, exact distance computation and approximate distance computation (with 0.5 relative error in this case). Since the original approximate distance algorithm in [15] reports

Algorithm GREEDY-DIST(B_i, B_j)

$d \leftarrow distance(B_i, B_j)$
if B_i **and** B_j **are both triangles then return** d
if $d > \delta$ **then return** d
if B_i **is bigger than** B_j **then switch** B_i **and** B_j
$\alpha \leftarrow$ GREEDY-DIST(B_i, B_{j1})
if $\alpha > \delta$ **then**
  $\beta \leftarrow$ GREEDY-DIST(B_i, B_{j2})
  if $\beta > \delta$ **then return** $\min\{\alpha, \beta\}$
return 0

Fig. 4. Greedy distance computation algorithm

Table 1. Comparison of numbers of BV/triangle pairs examined by GREEDY-DIST, pure collision checking, and exact and approximate distance algorithms. Columns μ and μ' show the fraction of the exact distance returned by GREEDY-DIST and 0.5-Dist, respectively

Fig.	Coll (OBB)	Greedy (RSS)	Dist (RSS)	0.5-Dist (RSS)	μ	μ'
1a	1.7/0.1	2.1/0.1	926/156	3.7/0.1	0.99	0.53
5a	47/0.6	54/0.8	1017/92	105/2.1	0.82	0.64
5b	26/1.0	25/0.8	903/81	58/1.9	0.81	0.61
6a	14/0.4	14/0.3	2303/450	70/2.4	0.65	0.60
6b	41/0.9	44/0.9	1398/227	129/4.8	0.51	0.58
6c	25/1.3	25/1.4	2093/261	111/3.2	0.57	0.62

only an upper approximation, we slightly modified it along the lines of [21] to compute a lower approximation instead. The number returned by the algorithm is then guaranteed to be between $0.5d$ and d, where d is the exact distance. Columns 2 through 5 give average numbers of pairs of BVs and triangles tested by each of the four implemented algorithms. These numbers were generated as follows, for the six examples shown in Figs. 1a, 5 and 6. In each example, we generated 1,000 random configurations of the robot(s) and at each configuration all robot links were tested against all fixed obstacles. The numbers reported in Table 1 are averages over all configurations and all tested pairs. They show that the performance of GREEDY-DIST is similar to the pure collision checker (note that, in this case, computing the distance between two triangles or two BVs costs only a small constant factor more than checking whether they intersect). But the performance of GREEDY-DIST is much better, by large factors, than the algorithm performing exact distance computation. It is also significantly better than the algorithm computing an 0.5-approximate distance.

The last two columns of Table 1 compare the lower distance bounds returned by GREEDY-DIST with the distances returned by the other two algorithms. The column labelled by μ describes the lower bounds from GREEDY-DIST by their average fractions of the exact distances. Here, the averages were computed only for the collision-free configurations. Indeed, with $\delta = 0$, when there is a collision, GREEDY-DIST always returns the exact distance. All the factors in column μ are larger than 0.5; half of them are significantly larger. The column labelled μ' evaluates the distances returned by the approximate-distance algorithm by their average fractions of the exact distances. In most cases, the μ' factor is smaller than the μ factor, indicating that GREEDY-DIST usually provides tighter bounds than the approximate-distance algorithm with 0.5 relative error, at smaller computational cost.

As is often the case with such comparison, there are subtleties that we must be aware of. Several types of BVs (e.g., spheres, AABBs, OBBs, RSSs, k-DOPs) have been proposed and discussed in the literature [18]. The BVs used by GREEDY-DIST, RSSs, provide reasonably tight approximations for a wide range of bodies found in robotic scenarios and allow fast distance computation. However, a distance query on two RSSs is slightly more expensive than an intersection query on two OBBs (the close relatives of RSSs). The implementation that was used for the pure collision checker uses OBBs. Since OBBs and RSSs are geometrically related, in most cases OBB and RSS hierarchies are very similar. This explains why the results in columns 2 and 3 of Table 1 are almost the same (hence, comparable). On the other hand, in our experiments, we found that performing an intersection query on two OBBs is, on average, only less than twice faster than computing the distance between two RSSs.

2.3 Bounding Manipulator Link Motions

As the configuration of the system changes by linear interpolation along a (sub-)segment $[\boldsymbol{q}_a, \boldsymbol{q}_b]$, every point of each moving body $\mathcal{A}_i$ traces its own specific curve segment in 3-space. Our algorithm needs to efficiently compute a good upper bound $\lambda_i(\boldsymbol{q}_a, \boldsymbol{q}_b)$ on the lengths of all curve segments traced by all points on $\mathcal{A}_i$.

It can be shown that one of the vertices of a convex bounding polyhedron for $\mathcal{A}_i$ traces a curve segment that is not shorter than the curve segment traced by any point on $\mathcal{A}_i$. Our problem can thus be reduced to computing bounds for the motion of a small number of points. Assume that the motion of one such point is given by $\boldsymbol{p}(t)$ for $t \in [0, 1]$. The exact length of the curve segment traced by $\boldsymbol{p}(t)$ is given by $L = \int_0^1 \|\dot{\boldsymbol{p}}(t)\| dt$. For general linkages, this integral cannot be solved analytically, and numerical evaluation with the required precision can be costly. Obviously, the straightforward approach of sampling the curve traced by $\boldsymbol{p}(t)$ and approximating it with a sequence of straight line segments yields an underestimate of the length, while we need an upper bound to guarantee that no collisions will be missed.

Most practical manipulators contain either revolute or prismatic joints and a few of them contain both types. For robots with only prismatic joints, traced curves are straight segments whose exact lengths are easy to compute.

To illustrate how we deal with revolute joints we consider the following simplified example. Let the robot be a planar linkage, consisting of m straight line segments connected by m revolute joints with angles $\varphi_1, \dots, \varphi_m$. Link 1 is directly connected to the environment and controlled by angle φ_1, link 2 is connected to the endpoint of link 1 and affected by angles φ_1 and φ_2, etc. Let each link have unit length and assume a simple correspondence between joint angles and c-space parameters, i.e. $\boldsymbol{q} = (q_1, \dots, q_m)$ and $\varphi_i = q_i$.

A simple upper bound of the length of the curve traced by any point on link i is

$$\lambda_i(\boldsymbol{q}_a, \boldsymbol{q}_b) = \sum_{k=1}^{i} (i - k + 1) \left| q_k^{(a)} - q_k^{(b)} \right| \tag{2}$$

In general, we use (2) to compute both $\lambda_i(\boldsymbol{q}_a, \boldsymbol{q}_b)$ and $\lambda_j(\boldsymbol{q}_a, \boldsymbol{q}_b)$ in (1). However, for two links i and j $(j < i)$ on the same kinematic chain, we can derive tighter bounds by considering motions in the local frame of link j. Thus, $\lambda_j(\boldsymbol{q}_a, \boldsymbol{q}_b) = 0$ and for $\lambda_i(\boldsymbol{q}_a, \boldsymbol{q}_b)$, we simply sum over $k = j+1, \dots, i$ instead of $k = 1, \dots, i$ in (2).

Although (2) does not yield the tightest possible bounds, they will be cut in half by each bisection step, thus requiring only a few additional calls to our fast greedy distance algorithm. It is doubtful whether exact methods for computing the lengths of traced curves would result in an overall faster checker. Note that in probabilistic planners, a large fraction of the segments are colliding and these segments will not have to be covered completely. In these cases, the computational effort for deriving tighter bounds would clearly be wasted. Furthermore observe that the error is smaller for links closer to the base. In our experiments in Sect. 3, $\lambda_i(\boldsymbol{q}_a, \boldsymbol{q}_b) + \lambda_j(\boldsymbol{q}_a, \boldsymbol{q}_b)$ appeared to be frequently dominated by large workspace distances $\eta_{ij}(\boldsymbol{q}_a)$ and $\eta_{ij}(\boldsymbol{q}_b)$.

The above approach can be extended to robots with both prismatic and revolute joints, and also to more realistic robots with polyhedral links but we omit the details here because of space limitations.

2.4 Ordering of the Priority Queue

We now discuss the ordering of the entries $[\boldsymbol{q}_a, \boldsymbol{q}_b]_{ij}$ in the priority queue used by the core algorithm presented in Sect. 2.1. For a collision-free segment, the ordering has no impact on the efficiency of the checker, since all entries have to be processed anyway. However, for a colliding segment, an appropriate ordering of the entries can be very effective to discover a collision quicker. This is important because in PRM planning, a significant amount of time is "wasted" for showing that candidate segments between sampled milestones are actually colliding.

In [9,22], it was shown that in practice, the prior probability of a segment to be colliding increases with its length in c-space. The planners in [20,22] exploit this result to test multi-segment paths, by maintaining a priority queue of (sub-)segments sorted by decreasing lengths.

Our new checker can take additional advantage of the fact that it also knows bounds on workspace distances and on lengths of traced curves for each (sub-)segment. Intuitively, two bodies are more likely to collide if they are closer at one or both segment endpoints and/or the points in these bodies trace longer curves. This intuition is directly related to inequality (1). This leads us to proceed as follows. For each entry $[\boldsymbol{q}_a, \boldsymbol{q}_b]_{ij}$ in the priority queue, we compute the length of the subsegment that is not covered (in the sense defined by the schema in Fig. 3). This is done by subtracting the right-hand side of (1) from the left-hand side. We call the result the ***non-covered length*** of the segment between $\boldsymbol{q}_a$ and $\boldsymbol{q}_b$ for the bodies $\mathcal{A}_i$ and $\mathcal{A}_j$. If this length is negative or null, then the two bodies are guaranteed not to collide along the segment, hence the entry is not inserted into the queue. All other entries are sorted by decreasing non-covered length.

This heuristic ordering can be extended from single segments and their sub-segments to multi-segment paths [22]. Instead of checking all segments on such a path, one after the other in some fixed sequence, the checker can enter all of them in the same priority queue and thus switch dynamically among them, as dictated by the uncovered lengths. In most cases, if a segment collides, this collision will be detected before having spent much time toward the validation of collision-free segments, hence making it possible to reject the entire path quickly. Actually, a better implementation is to maintain two levels of priority queues, one priority queue for the path (with one entry per segment that has not yet been shown free of collision), and one priority queue for each segment. The priority queue for a segment is the same as above. The priority of a segment in the path priority queue is the uncovered length of the first entry of this segment's priority queue. With these two levels of queues, if the path is eventually found to collide, we can then cache the priority queues of the segments that have neither been shown to collide nor to be collision-free. So, if these segments are later parts of other paths, the prior collision-checking work does not have to be repeated.

3 Experiments

All experiments were performed on a 1GHz Pentium III PC with 1GB RAM, using a single processor. Our implementation caches rigid-body transforms to avoid redundantly computing them: since we bisect the subsegments in the same uniform way for all pairs of bodies, partial results of kinematic computations can be shared across pairs. Especially in scenarios with long kinematic chains or several robots this has been found to considerably reduce the running times. Pairs of rigid bodies that cannot possibly collide (e.g., links

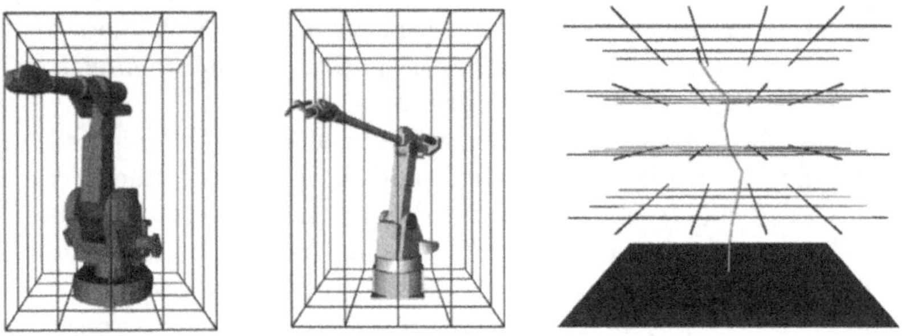

(a) (b) (c)

Fig. 5. Examples with thin obstacles (cages with 432 triangles each). (a) IRB 2400 robot (2,991 triangles), (b) Fanuc 200 robot (2,502 triangles) with arc welding gun, (c) a skinny 20-DOF linkage (320 triangles) amongst thin lattices (384 triangles)

(a) (b) (c)

Fig. 6. (a) Fanuc 200 with arc welding gun and a machine tool (robot: 2,502 triangles, obstacles: 34,171 triangles), (b) IRB 2400 robot in workshop (robot: 5,450 triangles, obstacles: 74,681 triangles), (c) Six IRB 2400 robots working on a car body (per robot: 3,594 triangles, car: 19,668 triangles)

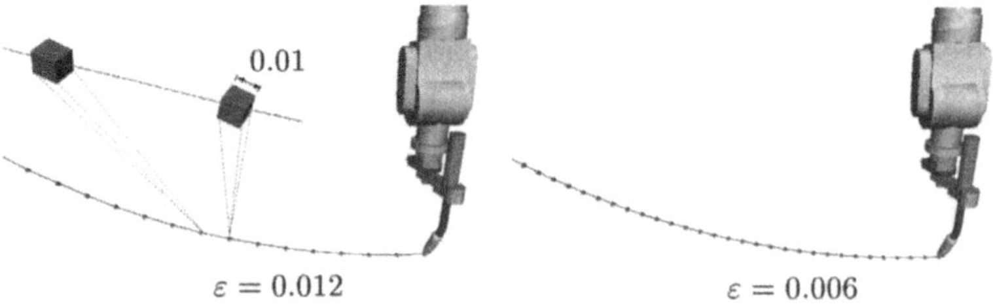

Fig. 7. Example segment: points visited by tip of end-effector on sample segment for robot in Figs. 5b/6a for different values of the c-space resolution ε. For comparison, the points themselves are represented by tiny cubes with side length 0.01 (twice the actual value of $\delta = 0.005$ that was used for the experiments with the new checker)

Table 2. Average times per segment (in ms) of simple and new collision checker for 1,000 random colliding/free segments (colliding segments have free endpoints)

Colliding segments		
Fig.	Simple $\varepsilon = 0$	New $\delta = 0$ (0.005)
5a	0.44	0.33 (0.29)
5b	0.46	0.43 (0.38)
6a	0.67	0.40 (0.37)
6b	1.21	0.57 (0.40)
6c	3.11	2.44 (2.23)

Free segments	Simple		New	
Fig.	ε	t (ms)	δ	t (ms)
5a	0.012	15.8	0.005	10.8
5b	0.006	14.6	0.005	11.3
6a	0.006	5.1	0.005	3.9
6b	0.012	28.2	0.005	11.2
6c	0.012	89.6	0.005	24.8

close to the bases of far-apart robots in Fig. 6c) were identified and pruned in a pre-processing step and not tested in any of the experiments below.

3.1 Random Segments

We first consider random segments to compare the performance of our new segment checker to a simple fixed-resolution approach that is used in most PRM planners. The simple approach bisects the segments and orders the (sub-)segment tests until their (Euclidean) lengths in c-space become smaller than a fixed ε as described in [22]. OBBs were used for checking individual configurations in the simple segment checker because they were previously reported to be more efficient for pure collision checks [15]. RSSs were used for the greedy distance computation in our new adaptive checker.

Table 2a shows the results for 1,000 random colliding segments with collision-free endpoints. Since the segments were known to be colliding, we ran the simple checker with $\varepsilon = 0$ (to assure that the checker would find all collisions) and our new checker with $\delta = 0$ (to prevent early rejection as soon as objects come closer than $\delta > 0$; for comparison, the speedup obtained by this effect is shown by the numbers in parentheses which were obtained with $\delta = 0.005$). Our new segment checker performed faster than the simple checker, although distance computation between RSSs is slightly slower than collision checking for OBBs. This shows that the new heuristic ordering of the priority queue is an improved way of directing the checker towards the collisions.

Table 2b compares the average times per segment for 1,000 random collision-free segments which were generated using the new checker. The values of ε for the simple checker were chosen empirically to achieve a reasonable tradeoff between performance and accuracy (see also Sect. 3.2). For illustration, Fig. 7 shows the points visited by the tip of the end-effector of the robot in Figs. 5b/6a on a particular segment in c-space for discretization with two different values of ε. To prevent deep recursion of our new checker, as described in Sect. 2.1, we set the minimum workspace clearance to $\delta = 0.005$ (for all examples). For comparison, this value equals half the side lengths of the tiny cubes that are used to represent the points on the paths in Fig. 7.

Again, the new checker performed faster than the simple ε-checker. Note that for the example in Fig. 6c, the L_∞-norm was used (instead of the Euclidean norm) in the simple checker to test the subsegment lengths against ε (clearly, with the Euclidean norm, the checker would be even slower).

3.2 Application in a PRM Path Planner

The PRM planner described in [22], called SBL, bisects each segment up to a given resolution ε. Its lazy collision checking strategy significantly reduces the amount of collision checking work spent on free segments that are not on the final solution path. SBL assumes that the path is collision-free if all intermediate points are collision-free. This approach does not prove that a path is actually collision-free, but the error probability can be made arbitrarily small by decreasing ε. However, reducing ε also means that more bisection points have to be checked. It thus requires some trial-and-error experimentation to determine a reasonable value of ε and this has to be done for every environment anew.

We have replaced this discrete segment checking algorithm in the original SBL implementation by our new adaptive segment checker. We call the resulting planner "adaptive SBL" (A-SBL). This planner is always guaranteed to return a collision-free path. As in Sect. 3.1, we set $\delta = 0.005$ for A-SBL in all examples.

Table 3. Original SBL planner vs A-SBL variant with new segment checker

	SBL		A-SBL	
Fig.	ε	t (s)	δ	t (s)
5a	0.006	4.4	0.005	4.4
5b	0.003	12.2	0.005	11.7
6a	0.003	14.2	0.005	11.4
6b	0.006	5.8	0.005	6.7
6c	0.006	3.6	0.005	1.5

In Table 3 we compare the running times of SBL and A-SBL on examples in the five environments shown in Fig. 5 and 6. For each environment, a value of ε was determined for SBL such that the planner returned a collision-free path for 10 successive runs on the respective problem. The times reported in the table are averages over these 10 runs of SBL and over 10 independent runs of A-SBL, respectively. The values required for ε in this experiment turn out to be even smaller than the ones determined empirically in Sect. 3.1 by looking at the resulting discretization in workspace for some segments. This further emphasizes the need for an exact and self-calibrating collision checker.

The advantage of A-SBL is that it does not require such experimental tuning of the resolution parameter and at the same time never misses a collision.

3.3 Application to Path Smoothing

We implemented a simple but effective randomized path smoother that iteratively tries to guess "shortcuts" and replace subsections of the path by single collision-free line segments. The new collision checker was used to test the potential shortcut segments for collisions. Since the smoother is only an application example, we omit further details of this algorithm here.

With this path smoother, we were able to find quite good paths for a set of difficult examples (Figs. 1a/5c and similar examples), even with a remarkably simple planner. The "planner" that we used here performs a random walk in c-space by changing a single DOF at each step. The starting configurations are given by the fully extended linkage(s) which are attached with one end to the base plate. The goal is to retract the linkage(s) downward out of the obstacles but no goal positions are given explicitly.

All possible collisions of links with obstacles and other links of the same linkage or the other linkage are checked. These examples are not only difficult because they describe narrow channels in high-dimensional c-spaces but also because it is easy to miss collisions. A segment checker based on discrete static collision checks would require an extremely small step size.

Table 4 shows the obtained results for "planning" and smoothing, respectively. Note that each smoothing step can consist of several segment checks since the smoother tries to maximize the length of the bypassed path section. The column σ shows the fraction of the path length after smoothing over the initial length, respectively. The example (*) is similar to Fig. 5c but contains nine identical linkages, each threaded through a different set of grid holes.

Table 4. Results for random walks and smoothing.

	Random walk		**Smoothing**		
Fig.	# segs.	t (s)	# steps	t (s)	σ
1a	40,439	82	1,000	152	0.06
5c	4,208	5.9	500	16	0.11
(*)	47,744	823	6,000	5,952	0.25

4 Conclusion and Future Work

We have presented a new exact collision checker for straight line segments in c-space and sequences of such segments that is particularly suited for PRM planners applied to manipulator arms and multi-robot systems. Unlike the commonly used approach of checking discrete configurations up to a fixed, user-specified c-space resolution, it has the advantage of never missing a collision. It also obviates the need for experimentally determining a reasonable constant c-space resolution parameter which can be quite unintuitive.

Exactness and efficiency of our new checker are obtained by dynamically adjusting the local resolution at which configurations along a path are tested by relating the distances between objects in the workspace to the maximum lengths of the paths traced out by points on these objects. While this basic idea has been proposed before, we introduced several new techniques that make the method applicable to PRM planners and scenarios with high geometric complexity. These techniques include a greedy distance computation algorithm that is nearly as efficient as pure collision checking, a simple and efficient method for bounding lengths of paths traced out in workspace, and a heuristics for ordering collision tests to reveal collisions as quickly as possible. In our experiments, the new checker was faster than the fixed-resolution method with an appropriately set resolution.

To bound the running time of our checker in practice, we require a minimum workspace clearance $\delta > 0$ between objects. However, unlike a c-space resolution parameter, such a minimum clearance in workspace is intuitive, and we found that this value can be set to a very small constant that works well for a wide variety of scenarios of comparable sizes.

Future work includes further optimization of the tradeoff between bisection and refinement of bounds on distances and curve lengths.

Acknowledgements. This work was partially funded by a gift from General Motors and a grant from ABB.

References

1. N. M. Amato, O. B. Bayazit, L. K. Dale, C. Jones, and D. Vallejo. OBPRM: An obstacle-based PRM for 3d workspaces. In *Proc. of the Workshop on Algorithmic Foundations of Robotics (WAFR'98)*, pages 155–168, March 1998.
2. J. Barraquand, L. Kavraki, J. C. Latombe, T.Y. Li, R. Motwani, and P. Raghavan. A random sampling scheme for path planning. *Inernational Journal of Robotics Research*, 16(6):759–774, 1996.
3. J. Basch, L. Guibas, and J. Hershberger. Data structures for mobile data. In *SODA: ACM-SIAM Symposium on Discrete Algorithms (A Conference on Theoretical and Experimental Analysis of Discrete Algorithms)*, 1997.
4. R. Bohlin and L. Kavraki. Path planning using lazy PRM. In *Proc. of the Int. Conf. on Robot. & Autom. (ICRA)*, pages 521–528, 2000.
5. S. Cameron. A study of the clash detection problem in robotics. In *Proc. IEEE Int. Conf. on Robotics & Automation*, volume 1, pages 488–493, 1985.

6. S. A. Cameron. Collision detection by four-dimensional intersection testing. *IEEE Trans. Robotics Automat.*, 6:291–302, June 1990.
7. J. F. Canny. Collision detection for moving polyhedra. *IEEE Trans. Patt. Anal. Mach. Intell.*, 8(2):200–209, Mar 1986.
8. J. D. Cohen, M. C. Lin, D. Manocha, and M. Ponamgi. I-COLLIDE: An interactive and exact collision detection system for large-scale environments. In *Sym. on Interactive 3D Graphics*, pages 189–196, 218, 1995.
9. L. Dale, G. Song, and N. Amato. Faster, more effective connection for probabilistic roadmaps. Technical Report TR00-005, Department of Computer Science, Texas A&M University, 20, 2000.
10. A. Foisy and V. Hayward. A safe swept volume method for collision detection. In *The Sixth International Symposium of Robotics Research*, pages 61–68, Pittsburgh (PE), Oct. 1993.
11. S. Gottschalk, M. C. Lin, and D. Manocha. OBBTree: A hierarchical structure for rapid interference detection. *Comp. Graphics*, 30(Annual Conf. Series):171–180, 1996.
12. D. Hsu, J. C. Latombe, and R. Motwani. Path planning in expansive configuration spaces. *International Journal of Computational Geometry and Applications*, 9(4&5):495–512, 1999.
13. P. Jiménez, F. Thomas, and C. Torras. 3D collision detection: A survey. *Computers and Graphics*, 25(2):269–285, 2001.
14. L. E. Kavraki, P. Svestka, J. C. Latombe, and M. Overmars. Probabilistic roadmaps for path planning in high-dimensional configuration spaces. *IEEE Transactions on Robotics and Automation*, 12(4):566–580, 1996.
15. E. Larsen, S. Gottschalk, M. C. Lin, and D. Manocha. Fast distance queries with rectangular swept sphere volumes. In *IEEE Conf. on Rob. and Auto.*, 2000.
16. M. Lin, D. Manocha, J. Cohen, and S. Gottschalk. Collision detection: Algorithms and applications. In J. P. Laumond and M. Overmars, editors, *Algorithms for Robotic Motion and Manipulation: WAFR 1996*, pages 129–142. A. K. Peters, 1996.
17. M. C. Lin and J. F. Canny. A fast algorithm for incremental distance calculation. In *IEEE Int. Conf. on Rob. and Auto.*, pages 1008–1014, 1991.
18. M. C. Lin and S. Gottschalk. Collision detection between geometric models: a survey. In *IMA Conference on Mathematics of Surfaces*, volume 1, pages 602–608, San Diego (CA), 1998.
19. B. Mirtich. V-clip: Fast and robust polyhedral collision detection. *ACM Transactions on Graphics*, 17(3):177–208, July 1998.
20. Ch. Nielsen and L. E. Kavraki. A two-level fuzzy PRM for manipulation planning. In *Proc. of the IEEE/RSJ International Conference on Intelligent Robots and Systems (IROS)*, Japan, 2000.
21. S. Quinlan. Efficient distance computation between non-convex objects. In *IEEE Intern. Conf. on Rob. and Auto.*, pages 3324–3329, 1994.
22. G. Sanchez and J. C. Latombe. A single-query bi-directional probabilistic roadmap planner with lazy collision checking. In *Int. Symposium on Robotics Research (ISRR'01)*, Lorne, Victoria, Australia, 2001.
23. A. Schweikard. Polynomial time collision detection for manipulator paths specified by joint motions. *IEEE Trans. Rob. Autom.*, 7(6):865–870, 1991.
24. G. Van der Bergen. Efficient collision detection of complex deformable models using AABB trees. *Journal of Graphic Tools*, 2(4):1–13, 1997.

A Comparative Study of Probabilistic Roadmap Planners

Roland Geraerts and Mark H. Overmars

Institute of Information and Computing Sciences, Utrecht University,
P.O. Box 80.089, 3508 TB Utrecht, The Netherlands
Email: [roland,markov]@cs.uu.nl.

Abstract. The probabilistic roadmap approach is one of the leading motion planning techniques. Over the past eight years the technique has been studied by many different researchers. This has led to a large number of variants of the approach, each with its own merits. It is difficult to compare the different techniques because they were tested on different types of scenes, using different underlying libraries, implemented by different people on different machines. In this paper we provide a comparative study of a number of these techniques, all implemented in a single system and run on the same test scenes and on the same computer. In particular we compare collision checking techniques, basic sampling techniques, and node adding techniques. The results should help future users of the probabilistic roadmap planning approach to choose the correct techniques.

1 Introduction

The basic motion planning problem asks for computing a collision-free, feasible motion for an object (or kinematic device) from a given start to a given goal placement in a workspace with obstacles. Originally the problem was studied within the robotics community, but in recent years many new applications arise in such areas as animation, virtual environments, computer games, computer aided design and maintenance, and computational chemistry.

Over the years many different approaches to solving the motion planning problem have been suggested. See the book of Latombe[21] for an extensive overview of the situation up to 1991 and e.g. the proceedings of the yearly IEEE International Conference on Robotics and Automation (ICRA) or the Workshop on Foundations of Robotics (WAFR) for many more recent results.

The *probabilistic roadmap planner (PRM)* is a relatively new approach to motion planning, developed independently at different sites [3,4,17,18,23,28]. It turns out to be very efficient, easy to implement, and applicable for many different types of motion planning problems (see e.g. [10,13,15,19,20,26,24,27–30]).

Globally speaking, the PRM approach samples the configuration space (that is, the space of all possible placements for the moving object) for collision-free placements. These are added as nodes to a roadmap graph.

J.-D. Boissonnat et al. (Eds.): Algorithmic Foundations of Robotics V, STAR 7, pp. 43–57, 2004.

Pairs of promising nodes are chosen in the graph and a simple local motion planner is used to try to connect such placements with a path. This process continues until the graph covers the connectedness of the space.

The basic PRM approach leaves many details to be filled in, like how to sample the space, what local planner to use and how to select promising pairs. Over the past eight years researchers have investigated these aspects and developed many improvements over the basic scheme (see e.g. [1,6,7,16,19,22,24,29,31]). Unfortunately, the different improvements suggested are difficult to compare. Each author used his or her own implementation of PRM and used different test scenes, both in terms of environment and the type of moving device used. Also the effectiveness of one technique sometimes depends on choices made for other parts of the method. So it is still rather unclear what is the best technique under which circumstances. (See [12] for a first study of this issue.)

In this paper we make a first step toward a comparison between the different techniques developed. We implemented a large number of the techniques in a single motion planning system and added software to compare the approaches. In particular we concentrated on the sampling technique and the choice of promising pairs of nodes. This comparison gives insight in the relative merits of the techniques and the applicability in particular types of motion planning problems. Also we hope that in the longer run our results will lead to improved (combinations of) techniques and adaptive approaches that choose techniques based on observed scene properties.

The paper is organized as follows. In Section 2 we review the basic PRM approach. In Section 3 we describe our experimental setup and the scenes we use. In Section 4 we compare different ways of performing collision checks of the paths produced by the local planner. We conclude that a binary approach performs best. In Section 5 we consider different sampling strategies. We conclude that, except for very special cases, one best uses a deterministic approach based on Halton points, with a small amount of added randomness. In Section 6 we study different strategies for choosing promising pairs of nodes to connect. We conclude that one best picks a few nodes in each connected component of the roadmap. These results partially contradict earlier claims.

2 The PRM Method

The motion planning problem is normally formulated in terms of the *configuration space* $\mathcal{C}$, the space of all possible placements of the moving object. Each degree of freedom of the object corresponds to a dimension of the configuration space. Each obstacle in the workspace, in which the object moves, transforms into an obstacle in the configuration space. Together they form the forbidden part $\mathcal{C}_{\text{forb}}$ of the configuration space. A path for the moving object corresponds to a curve in the configuration space connecting the start and the goal configuration. A path is collision-free if the corresponding curve

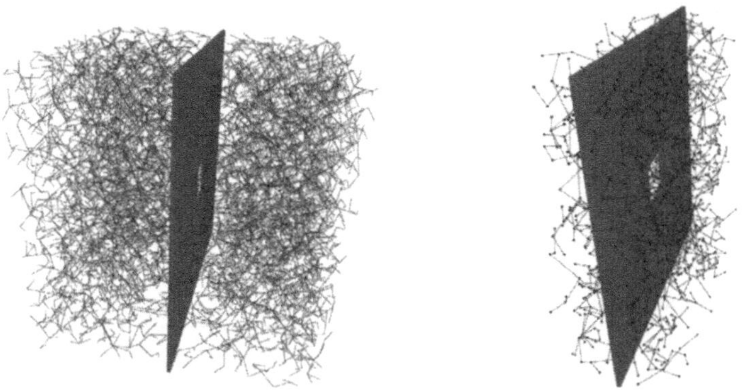

Fig. 1. The roadmap graph we get for the difficult hole test scene used in this paper. The left image shows the graph using halton sampling and the right image uses gaussian sampling.

does not intersect $\mathcal{C}_{\text{forb}}$, that is, it lies completely in the free part of the configuration space, denoted with $\mathcal{C}_{\text{free}}$.

The probabilistic roadmap planner samples the configuration space for free configurations and tries to connect these configurations into a roadmap of feasible motions. There are a number of versions of PRM, but they all use the same underlying concepts.

The global idea of PRM is to pick a collection of (useful) configurations in the free space $\mathcal{C}_{\text{free}}$. These free configurations form the nodes of a graph $G = (V, E)$. A number of (useful) pairs of nodes are chosen and a simple local motion planner is used to try to connect these configurations by a path. When the local planner succeeds an edge is added to the graph. The local planner must be very fast, but is allowed to fail on difficult instances. A typical choice is to use a simple interpolation between the two configurations, and then check whether the path is collision-free. So the path is a straight line in configuration space.

Once the graph reflects the connectivity of $\mathcal{C}_{\text{free}}$ it can be used to answer motion planning queries. (See Figure 1 for an example of the roadmap graphs computed.) To find a motion between a start configuration and a goal configuration, both are added to the graph using the local planner. (Some authors use more complicated techniques to connect the start and goal to the graph, e.g. using bouncing motion.) Then a path in the graph is found which corresponds to a motion for the object. The pseudo code for the algorithm for constructing the graph is shown in Algorithm ConstructRoadmap.

Note that in this version of PRM we only add an edge between nodes if they are not in the same connected component of the roadmap graph. This saves time because such a new edge will not help solving motion planning queries. On the other hand, to get short paths such extra edges are useful (see e.g. [14,25]). For this comparative study we will not add these additional edges.

Algorithm 1 ConstructRoadmap

Let: $V \leftarrow \emptyset$; $E \leftarrow \emptyset$;
1: **loop**
2: $c \leftarrow$ a (useful) configuration in $\mathcal{C}_{\text{free}}$
3: $V \leftarrow V \cup \{c\}$
4: $N_c \leftarrow$ a set of (useful) nodes chosen from V
5: **for all** $c' \in N_c$, in order of increasing distance from c **do**
6: **if** c' and c are not connected in G **then**
7: **if** the local planner finds a path between c' and c **then**
8: add the edge $c'c$ to E

In this study we concentrate on the various choices for picking useful samples, in line 2 of the algorithm, and for picking useful pairs of nodes for adding edges, that is, on the choice of N_c in line 4. These are the most crucial steps and they strongly influence the running time and the structure of the roadmap graph.

3 Experimental Setup

The probabilistic roadmap approach has been applied to many different types of motion planning problems involving different kinds of robotic devices. In this comparative study we restrict ourselves to free-flying objects in a three-dimensional workspace. Such objects have six degrees of freedom, three translational degrees and three rotation degrees. In a later study we will consider other devices.

In all experiments we use the most simple local method that consists of a straight-line motion in configuration space, interpolating between the start and goal configuration. In a later study we will consider the effect of using more sophisticated local planners.

An important distinction is between single shot methods and preprocessing methods. A single shot method asks to solve a particular motion planning query for a given start and a given goal configuration. A preprocessing method builds a data structure that can then be used to solve different motion planning queries efficiently. We feel that the PRM approach is particularly useful as a preprocessing method and, hence, we choose that method in this paper. This means that we won't consider techniques like bidirectional methods and lazy PRMs [6,24] which are particularly suited for single shot situations and situations were just a few queries occur. We aim at computing a roadmap that is checked for collisions and covers the free space adequately. To this end, in each scene (see below) we defined a relevant query that can only be solved when there is an adequate covering of the free space. We continue building the roadmap until the query configurations lie in the same connected component.

All techniques were integrated in a single motion planning system, called SAMPLE (System for Advanced Motion PLanning Experiments), implemented in Visual C++ under Windows XP. All experiments were run on a 2.4 GHz Pentium 4 processor with 1 GB of internal memory. In all experiments we report the running time in seconds. Because the experiments were conducted under the same circumstances, the running time is a good indication of the efficiency of the technique. For those techniques where there are random choices involved we report the average time over 20 runs. When there is a large difference between runs we will indicate this.

For the experiments we used the following five scenes (see Figure 2):

corridor This is the most simple scene. An L-shaped object must translate and rotate through a winding corridor to the other end. Some obstacles force rotation of the object. We expect that all methods can solve this problem very quickly.

rooms In this scene there are three rooms with lots of space and with narrow doors between them. So the density of obstacles is rather non-uniform. The table must move through the two narrow doors to the other room. The table is rather complex because it has four cylindrical legs, leading to more time-consuming collision checking.

clutter This scene consists of 500 uniformly distributed tetrahedra. An L-shaped object must move among them from one corner to the other. The configuration space will consist of many narrow corridors. There are multiple solutions to the query.

hole A rather famous test scene. The moving object consists of four legs and must rotate in a complicated way to get through the hole. The hole is placed off-center to avoid that certain grid-based method have an advantage. The configuration space will have two large open areas with (most likely) two narrow winding passages between them.

house The house is a complicated scene consisting of 1600 polygons. It has many small rooms. As a result a large roadmap will be required to cover the free space. Also, because walls are thin, the collision checker must make rather small steps along the paths, resulting in much higher collision checking times.

4 Collision Checking

The most time-consuming steps in the probabilistic roadmap planner are the collision checks that are required to decide whether a sample lies in $\mathcal{C}_{\text{free}}$ and whether the motion produced by the local planner is collision free. In particular the second type of checks is time consuming. In this section we investigate some techniques for collision testing of the paths.

As basic collision checking package we use Solid[5]. The advantage of this package is that it considers objects like blocks, tetrahedra, spheres, and

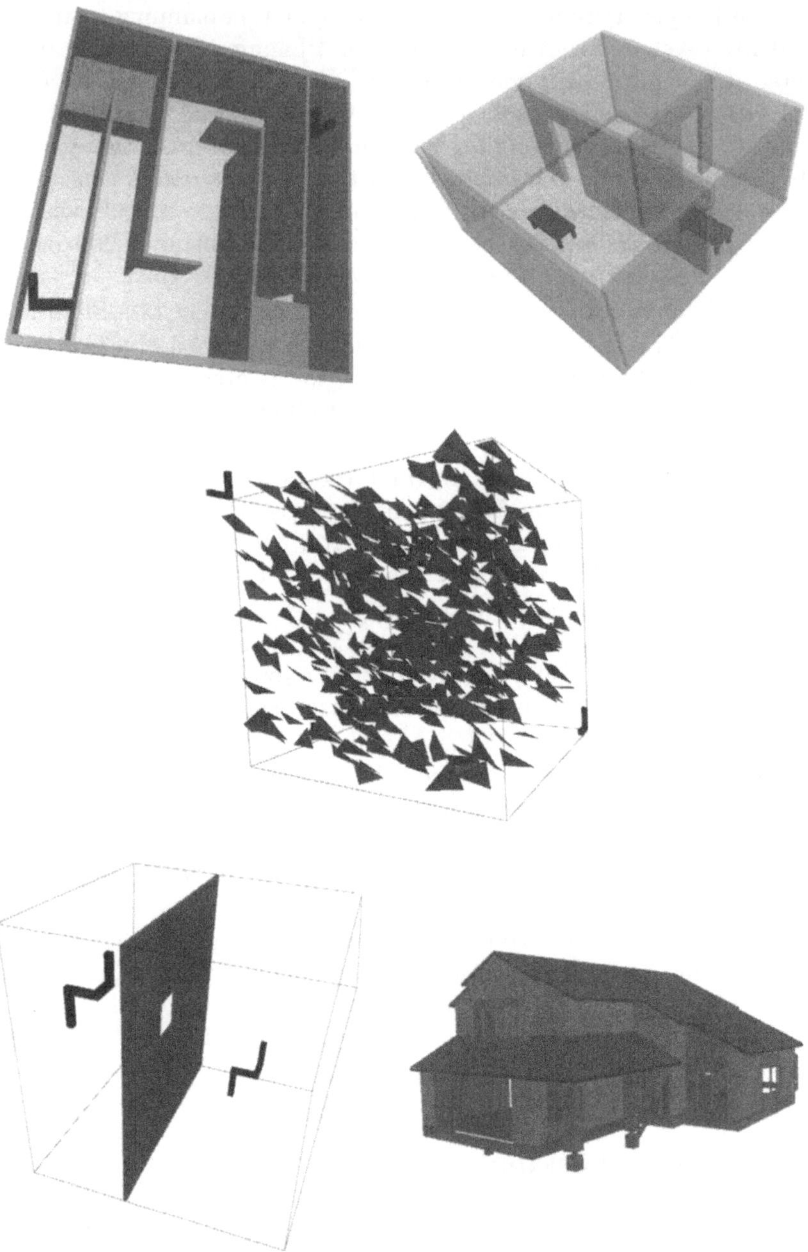

Fig. 2. The scenes used for testing, together with the start and goal configuration for the moving object.

cylinders as solids rather than boundary representations. This avoids the generations of samples inside obstacles. Also it reduces the number of obstacles required to describe complicated scenes. Solid builds a data structure based on bounding boxes for fast query answering.

When testing a path for collisions we can use the following techniques:

incremental In the incremental method we take small steps along the path from start to goal. For each placement we check for collision with the scene.

binary In the binary method we start by checking the middle position along the path. If it is collision free we recurse on both halves of the path checking the middle positions there. In this way we continue until either a collision is found or the checked placements lie close enough together (again determined by a given step size).

In early papers on PRM the incremental method was used. Later papers suggest that the binary method works better[24]. The reason is that the middle position is the one that has the highest chance of not being collision free. This means that, when the path is not collision free, this collision will most likely be found earlier, that is, after fewer collision checks.

We tested both approaches. To make the results comparable we used a deterministic version of PRM using Halton points for sampling and a simple nearest-n node adding strategy (see Sections 5 and 6). As a result both version create exactly the same roadmap. The following table shows the results for the five test scenes:

collision checking		
	incremental	binary
corridor	0.5	**0.2**
rooms	0.8	**0.3**
clutter	8.7	**4.0**
hole	44.2	**43.3**
house	380.0	**207.4**

As can be seen from the table, in all scenes the binary approach is faster although the improvement varies over the type of scene. This can be explained from the fact that in the hole scene path checks are more often successful than in the other scenes. The binary approach only helps when the check fails.

It has been suggested that one should try to compute sweep volumes and use these for collision tests. As a result, a path check would require just one collision test. The problem is that it is very difficult to compute sweep volumes for three-dimensional moving objects with six degrees of freedom. A much simpler technique is to first check with the sweep volume introduced by the origin of the object, that is, with a line segment between start and goal position (see [11]).

line check In this method we first perform a collision test with the line segment between the start and goal position in the workspace. Only if it is collision-free we perform either the incremental or the binary method. (This assumes that the origin of the object lies inside it.)

We would expect that this test will quickly discard many paths that have a collision, leading to an improvement in running time. The following table summarizes the results:

binary collision checking		
	no line	line check
corridor	**0.2**	**0.2**
rooms	**0.3**	**0.3**
clutter	**4.0**	4.3
hole	**43.3**	44.5
house	**207.4**	208.1

As can be seen, against our expectation, the line check did not lead to any improvement. (It does help for the incremental method but the bounds are still worse than for the binary method.) Sometimes it is even slightly worse. We believe that this is due to the way Solid performs the collision checking with the line segment. The line segment will normally have a rather large bounding box. As a result the bounding box test will not be enough to discard many obstacles. This requires more precise tests, which are more expensive. Also, Solid will always report all intersections, not just the first one. As the line might intersect multiple obstacles this will lead to additional (redundant) work. Other collision checking packages might treat such line intersections differently. We will investigate this in the future. Also, in all test scenes, except for the rooms scene, our moving objects are rather simple. When the moving object gets more complex, collision tests with it will become more expensive. In such a case the line check might lead to an improvement. In [11] it was suggested to only apply the line check when the distance between the endpoints is large. We tried this but did not see any significant improvement in performance. We conclude that, with Solid, we best use binary collision checking without line checks. This is the technique we will use in the remainder of this paper.

5 Basic Sampling

The first papers on PRM used uniform random sampling of the configuration space to select the nodes to add to the graph. In recent years other approaches have been suggested, either to create more samples in difficult regions or to remedy certain disadvantages of the random behavior. We will first study and compare some of the second type of approaches. In particular we will study the following techniques:

random In the random approach a sample is created by choosing random values for all its degrees of freedom. The sample is added when it is collision-free. This was the traditional way of creating samples.

grid It has often been suggested to simply choose samples on a grid. Because we don't know the required grid resolution in advance we start with a coarse grid and refine this grid in the process, halving the cell size. Grid points on the same level of the hierarchy are added in random order.

halton In [8] it has been suggested to use so-called Halton point sets as samples. Halton point sets have been used in discrepancy theory to obtain a good coverage of a region with points that is better than using a grid (see e.g. [9]). Even though the criteria in PRM are rather different than in discrepancy theory it has been suggested in [8] that the method is well suited for PRM. The method is deterministic!

cell-based In this approach we take random configuration within cells of decreasing size in the workspace. The first sample is generated randomly in the whole space. Next we split the workspace in 2^3 equally sized cells. In a random order we generate a configuration in each cell. Next we split each cell into 2^3 sub-cells and repeat this for each sub-cell, etc. The rationale behind this is that we get a much better spread of the positions over the configuration space. (Also choosing cells for the rotational degrees of freedom would lead to too many cells.)

Note that we do not include visibility sampling[22] in this test. We consider visibility sampling as a node adding technique rather than a sampling technique and, hence, it is described in the next section.

The following table summarizes the results.

basic sampling strategy				
	random	grid	halton	cell-based
corridor	0.4	1.4	**0.2**	0.4
rooms	0.8	0.6	**0.3**	0.7
clutter	**2.7**	7.3	4.0	3.1
hole	33.9	**26.2**	43.3	42.3
house	210.0	262.5	**207.4**	225.9

It can be seen that the results are rather varying and that the deterministic halton approach is not always the fastest. In the hole scene it is even considerably slower. The numbers contradict the results reported in [8] were considerable improvements are reported. This might be due to the much smaller connection distance chosen in [8] which seems to favor the halton sampling. Such a smaller connection distance though increases the total time bound. A big advantage of the halton approach though is that the results are always the same while for the random approach and to a lesser extend for the cell-based approach there is a large variation in running time. A few bad samples added at the beginning can complicate the connection process considerably. It has been suggested that this can be remedied by from time to time restarting the process. We did not test this further. In general we were surprised by the small differences. It seems the sampling approach is not as critical as people have suggested before.

So we can conclude that the major advantage of using halton points does not lie in the faster running time but in the deterministic nature that guarantees that the running time is always the same. But at the same time this is the weakness of the method. In the hole scene this can be seen. Here halton seems to be unlucky in the configurations it picks. For any deterministic sampling technique one can construct scenes for which it will take a huge amount of running time. We suggest to remedy this by adding a small amount of randomness to the method:

random halton In this approach we use halton points. But when adding the nth sample point we choose an area around this point (in configuration space) and choose a random configuration in this area. As size of the area we choose kA/n where A is the area of (the relevant part of) the configuration space and k is a small constant (in the order of 0.002 for the experiments we conducted). So the area becomes smaller when more and more points are added.

The following table shows the difference between halton and random halton.

sampling with halton points		
	deterministic	random
corridor	**0.2**	0.3
rooms	0.3	**0.2**
clutter	4.0	**2.2**
hole	43.3	**15.8**
house	207.4	**152.9**

We conclude that adding a little bit of randomness does improve the running time in almost all cases. The approach is better than any of the other sampling techniques considered. So we recommend this method to be used. (In the next section we do use halton sampling without randomness. The reason is that this is easier for testing because we do not have to average over different runs.)

As indicated above other sampling techniques have been suggested that try to add more samples in difficult regions of the environment. It has for example been suggested to add more samples near to obstacle boundaries[1,7], to allow samples inside obstacles and push the to the free space[1,16] and to sample near to the Voronoi diagram of the obstacles[31]. Here we compare three techniques:

gaussian Gaussian sampling is meant to add more samples near obstacles. The idea is to take two random samples, where the distance between the samples is chosen according to a Gaussian distribution. Only if one of the samples lies in the $\mathcal{C}_{\mathrm{free}}$ and the other lies in $\mathcal{C}_{\mathrm{forb}}$ do we add the free sample. It has been shown that this leads to a favorable sample distribution[7].

obstacle based This technique, based on [1], has a similar goal. We pick a random sample. If if lies in $\mathcal{C}_{\text{free}}$ we add it to the graph. Otherwise, we pick a random direction and move the sample in that direction with increasing steps until it becomes free (or lies outside the scene). We add the free sample. This will normally lie close to an obstacle boundary.

obstacle based * This is a variation of the previous technique where we throw away a sample if it initially lies in $\mathcal{C}_{\text{free}}$. This will avoid many samples in large open regions.

We expect these techniques to be useful only in scenes where there are large open areas (in configuration space) and some narrow passages. The results are as follows, comparing them to the random halton approach (see also Figure 1).

sampling around obstacles				
	gaussian	obstacle	obstacle*	halton
corridor	0.4	0.5	0.5	**0.3**
rooms	0.4	0.7	0.5	**0.2**
clutter	5.6	3.7	7.9	**2.2**
hole	3.1	8.0	**2.1**	15.8
house	268.0	197.1	209.4	**152.9**

As expected the techniques only perform better than halton for the hole scene. As the techniques can be quite a bit worse for other scenes they cannot be used as a general approach. Because all three techniques use a lot of randomness they suffer from the same problem as random sampling that some runs are very bad. We are currently investigating combinations of these techniques with halton sampling.

6 Node Adding

In this section we will study the second important choice to be made in PRM: how to select the set of neighbors to which we try to make connections. As each test for a connection is expensive we should try to avoid these as much as possible. On the other hand, if we try too few connections we will fail to connect the graph. Connected components in the graph play a crucial role here. We like to connect such components into larger components whenever we can. Also, we need to create new components in unexplored parts of the configuration space.

There are two global observations that can be made. First of all, as indicated in Section 2, it is not useful (from a complexity point of view) to make connections to nodes that are already in the same connected component. Secondly, it is never useful to connect to nodes that lie too far away. The chance of success for such a connection is minimal while the collision checks required for testing the path are expensive. An important question here is what we use as a distance measure (see also [2]). In particular, it is unclear how to take the rotational distance into account. In this paper we use $d_t + d_r$ where d_t denotes the translational distance of the origin of the object and d_r denotes the distance traveled by the point of the object furthest from its origin, while performing the rotation. This measure is easy to compute and gives a reasonably estimate.

We consider the following techniques:

nearest-*n* Here we try to connect the new configuration c to the nearest n nodes in the graph. The rationale is that nearby nodes lead to short connections that can be checked efficiently. If many nearby nodes lie in the same connected component there is no other component in the neighborhood so it is acceptable that we only connect to a single component.

component Here we try to connect the new configuration to the nearest node in each connected component that lies close enough. The rationale is that we prefer to connect to multiple connected components.

component-*n* Here we try to connect the new configuration to at most n nodes in each connected component. Still we keep the total number of connections tried small (the same number as for nearest-n). The rationale is that when the number of components is small we prefer to spend some extra time on trying to make connections. Otherwise the time required for adding the node will become the dominant factor.

visibility This method is based on the visibility sampling technique described in [22]. We try to connect the new configuration to useful nodes. Usefulness is determined as follows: When a new node can be connected to no other nodes it forms a new connected component and is labeled useful. If it connects two or more components it is also labeled useful. If it can be connected to just one component it is not labeled useful. (In [22] useless nodes are deleted but there is actually no reason for this. They won't take any time and they can be useful in query answering.) It has been observed in [22] that the number of useful nodes remains small, making it possible to try connections to all of them.

The following table summarizes our test results. In all tests we use deterministic halton sampling. As maximal connection distance we use a box that is about half the size in each dimension of the configuration space. For the nearest-n method we picked a value for n of about 15 (depending a bit on the scene). For component-n we used a value of 3 in the experiments.

node adding strategy				
	nearest-n	comp	comp-n	visibility
corridor	**0.2**	0.8	0.5	1.5
rooms	**0.3**	0.5	0.7	2.3
clutter	4.0	4.4	**3.4**	7.6
hole	43.3	> 120	31.9	**20.9**
house	207.4	279.6	**189.1**	> 600

We were disappointed with the results for visibility adding. Only for the hole scene did it perform better (but here the obstacle based sampling techniques performed even better). From previous papers we had expected the method to perform better. We believe that this is due to the fact that other papers compare it with a method that is worse than nearest-n or component-n. As can be seen, the results are a bit varying. But the component-n method is in most cases the fastest and in the other cases almost the fastest. We conjecture that the reason is that it combines the following important aspects

- It aims at connecting different components.
- It never tries too many connections.
- It never tries too few connections.

The other methods lack in at least one of these aspects.

We plan to investigate node adding strategies further over the coming months. We believe that a better definition of the notion of useful nodes is required to improve the performance. Such a notion should probably change during the construction process and might also depend on local scene properties.

7 Conclusions

In this paper we presented the results of a comparative study of various PRM techniques. The results confirm previously made claims that the binary approach for collision checking works well. Previous claims on the quality of halton points could not be confirmed but a variation of halton, adding some randomness, gave a better performance in most scenes. For node adding it turned out that visibility sampling did not perform as well as expected. A technique based on choosing a number of nodes per component seemed to perform best. We plan to further investigate the issue of selecting promising nodes for connections. Also we plan to experiment with a number of additional techniques, in particular techniques aimed at narrow passages.

One thing that is clear from the study is that a careful choice of techniques is important. Also, it is not necessarily true that a combination of good techniques is good. For example, for the hole scene one might expect that a combination of obstacle based and visibility works best. But experiments show that this combination is actually about twice as bad (4.5 seconds).

We hope that our studies shed some more light on the questions what technique to use in which situation. A major challenge is to create planners that automatically choose the correct combination of techniques based on scene properties or that learn the correct settings while running.

References

1. N. Amato, O. Bayazit, L. Dale, C. Jones, D. Vallejo, OBPRM: An obstacle-based PRM for 3D workspaces, in: P.K. Agarwal, L.E. Kavraki, M.T. Mason (eds.), *Robotics: The algorithmic perspective*, A.K. Peters, Natick, 1998, pp. 155–168.
2. N. Amato, O. Bayazit, L. Dale, C. Jones, D. Vallejo, Choosing good distance metrics and local planners for probabilistic roadmap methods, *Proc. IEEE Int. Conf. on Robotics and Automation,* 1998, pp. 630–637.
3. N. Amato, Y. Wu, A randomized roadmap method for path and manipulation planning, *Proc. IEEE Int. Conf. on Robotics and Automation,* 1996, pp. 113–120.
4. J. Barraquand, L. Kavraki, J.-C. Latombe, T.-Y. Li, R. Motwani, P. Raghavan, A random sampling scheme for path planning, *Int. Journal of Robotics Research* **16** (1997), pp. 759–774.
5. G. van den Bergen, *Collision detection in interactive 3D computer animation,* PhD thesis, Eindhoven University, 1999.
6. R. Bohlin, L.E. Kavraki, Path planning using lazy PRM, *Proc. IEEE Int. Conf. on Robotics and Automation,* 2000, pp. 521–528.
7. V. Boor, M.H. Overmars, A.F. van der Stappen, The Gaussian sampling strategy for probabilistic roadmap planners, *Proc. IEEE Int. Conf. on Robotics and Automation,* 1999, pp. 1018–1023.
8. M. Branicky, S. LaValle, K. Olson, L. Yang, Quasi-randomized path planning, *Proc. IEEE Int. Conf. on Robotics and Automation,* 2001, pp. 1481–1487.
9. B. Chazelle, *The discrepancy method,* Cambridge University Press, Cambridge, 2000.
10. J. Cortes, T. Simeon, J.P. Laumond, A random loop generator for planning the motions of closed kinematic chains using PRM methods, *Proc. IEEE Int. Conf. on Robotics and Automation,* 2002, pp. 2141–2146.
11. L. Dale, *Optimization techniques for probabilistic roadmaps,* PhD thesis, Texas A&M University, 2000.
12. L. Dale, N. Amato, Probabilistic roadmaps – Putting it all together, *Proc. IEEE Int. Conf. on Robotics and Automation,* 2001, pp. 1940–1947.
13. L. Han, N. Amato, A kinematics-based probabilistic roadmap method for closed chain systems, *Proc. Workshop on Algorithmic Foundations of Robotics (WAFR'00),* 2000, pp. 233–246.
14. O. Hofstra, D. Nieuwenhuisen, M.H. Overmars, Improving the path quality for probabilistic roadmap planners, to appear.
15. C. Holleman, L. Kavraki, J. Warren, Planning paths for a flexible surface patch, *Proc. IEEE Int. Conf. on Robotics and Automation,* 1998, pp. 21–26.
16. D. Hsu, L. Kavraki, J.C. Latombe, R. Motwani, S. Sorkin, On finding narrow passages with probabilistic roadmap planners, in: P.K. Agarwal, L.E. Kavraki, M.T. Mason (eds.), *Robotics: The algorithmic perspective*, A.K. Peters, Natick, 1998, pp. 141–154.

17. L. Kavraki, *Random networks in configuration space for fast path planning,* PhD thesis, Stanford University, 1995.
18. L. Kavraki, J.C. Latombe, Randomized preprocessing of configuration space for fast path planning, *Proc. IEEE Int. Conf. on Robotics and Automation,* 1994, pp. 2138–2145.
19. L. Kavraki, P. Švestka, J-C. Latombe, M.H. Overmars, Probabilistic roadmaps for path planning in high-dimensional configuration spaces, *IEEE Trans. on Robotics and Automation* **12** (1996), pp. 566–580.
20. F. Lamiraux, L.E. Kavraki, Planning paths for elastic objects under manipulation constraints, *Int. Journal of Robotics Research* **20** (2001), pp. 188–208.
21. J-C. Latombe, *Robot motion planning,* Kluwer Academic Publishers, Boston, 1991.
22. C. Nissoux, T. Siméon, J.-P. Laumond, Visibility based probabilistic roadmaps, *Proc. IEEE Int. Conf. on Intelligent Robots and Systems,* 1999, pp. 1316–1321.
23. M.H. Overmars, *A random approach to motion planning,* Technical Report RUU-CS-92-32, Dept. Comput. Sci., Utrecht Univ., Utrecht, the Netherlands, October 1992.
24. G. Sánchez, J.-C. Latombe, A single-query bi-directional probabilistic roadmap planner with lazy collision checking, *Int. Journal of Robotics Research,* 2002, to appear.
25. E. Schmitzberger, Probabilistic approach to list all non homotopic solutions of a motion planning problem, unpublished notes, 2002.
26. S. Sekhavat, P. Švestka, J.-P. Laumond, M.H. Overmars, Multilevel path planning for nonholonomic robots using semiholonomic subsystems, *Int. Journal of Robotics Research* **17** (1998), pp. 840–857.
27. T. Simeon, J. Cortes, A. Sahbani, J.P. Laumond, A manipulation planner for pick and place operations under continuous grasps and placements, *Proc. IEEE Int. Conf. on Robotics and Automation,* 2002, pp. 2022–2027.
28. P. Švestka, *Robot motion planning using probabilistic roadmaps,* PhD thesis, Utrecht Univ. 1997.
29. P. Švestka, M.H. Overmars, Motion planning for car-like robots, a probabilistic learning approach, *Int. Journal of Robotics Research* **16** (1997), pp. 119–143.
30. P. Švestka, M.H. Overmars, Coordinated path planning for multiple robots, *Robotics and Autonomous Systems* **23** (1998), pp. 125–152.
31. S.A. Wilmarth, N.M. Amato, P.F. Stiller, MAPRM: A probabilistic roadmap planner with sampling on the medial axis of the free space, *Proc. IEEE Int. Conf. on Robotics and Automation,* 1999, pp. 1024–1031.

On the Relationship between Classical Grid Search and Probabilistic Roadmaps

Steven M. LaValle[1] and Michael S. Branicky[2]

[1] Dept. of Computer Science, University of Illinois, Urbana, IL 61801 USA
[2] Dept. of Electrical Engineering and Computer Science, Case Western Reserve University, Cleveland, OH 44106 USA

Abstract. *We present, implement, and analyze a spectrum of closely-related planners, designed to gain insight into the relationship between classical grid search and probabilistic roadmaps (PRMs). Building on quasi-Monte Carlo sampling literature, we have developed deterministic variants of the PRM that use low-discrepancy and low-dispersion samples, including lattices. Classical grid search is extended using subsampling for collision detection and also the optimal-dispersion Sukharev grid, which can be considered as a kind of lattice-based roadmap to complete the spectrum. Our experimental results show that the deterministic variants of the PRM offer performance advantages in comparison to the original PRM and the recent Lazy PRM. This even includes searching using a grid with subsampled collision checking. Our theoretical analysis shows that all of our deterministic PRM variants are resolution complete and achieve the best possible asymptotic convergence rate, which is shown superior to that obtained by random sampling. Thus, in surprising contrast to recent trends, there is both experimental and theoretical evidence that some forms of grid search are superior to the original PRM.*

1 Introduction

The main motivation of this paper is to provide insight into fundamental questions that arise in the development of sampling-based motion planning algorithms. What factors lead to good computational performance in practice? Is randomization really important to breaking the curse of dimensionality? While the number of samples required for a grid is known to increase exponentially in dimension, is it true that the probabilistic roadmap (PRM), which was "primarily developed for robots with many dofs" [19], overcomes this difficulty through random sampling? By building on existing sampling literature, developing theoretical analysis, and performing experimental studies, we conclude that the *original* PRM does not offer clear advantages over some forms of grid-based search. Furthermore, we even provide theoretical and experimental evidence that suggests the superiority of grids and other deterministic sampling schemes. One theoretical result, which is expressed in terms of a measure from the sampling literature (dispersion), implies that neither the PRM, grids, nor other sampling schemes can avoid the need for an exponential number of samples in dimension.

J.-D. Boissonnat et al. (Eds.): Algorithmic Foundations of Robotics V, STAR 7, pp. 59–75, 2004.

This paper builds on our previous work [9], in which well-known deterministic sampling techniques, such as Halton/Hammersley points and low-discrepancy lattices, were shown to offer advantages as replacements to pseudo-random sampling in the PRM context. Improved performance using Halton points was also observed recently in the context of the Visibility PRM [33]. An interesting deterministic alternative to PRMs was also proposed in [5].

Given the tremendous variety of possible path planning examples, and the fact that algorithm performance often depends greatly on parameter settings, it is a daunting task to compare path planners. Our approach to this problem is to define a spectrum of planners that has classical grid search and the original PRM at its ends. Neighboring planners in this spectrum vary only slightly, which helps us understand and assess how one small variation affects each algorithm. By chaining these variations together and drawing conclusions along the way, we provide a clearer understanding of how the planners at opposite ends are related.

There are important qualifications to the assertions made in this paper:

- We believe that importance sampling is important. Many recent probabilistic roadmap methods [1,2,8,14,23,30,32,36] have demonstrated improved performance by concentrating samples in a nonuniform way, such as along C-space boundaries [2,8], or the medial axis [14,30,36]. Our work can be viewed as complementary to importance sampling. In many of these approaches it may be possible to obtain performance improvements by replacing random sampling with deterministic schemes; however, given the variety of methods and heuristics often involved, this task is beyond the scope of the present paper.
- We are not recommending any of the planners presented here as the fastest available. Our intent is to gain insight into planning issues, as opposed to delivering the best planner.
- We believe that randomization is useful in many contexts. Its value, however, depends greatly on the paradigm within which it is used.

2 Grids and PRMs

This section introduces a few influential planners that will be discussed in detail in this paper, rather than providing a complete survey of techniques.

The path planning problem Let $\mathcal{C}$ denote the configuration space (or *C-space*) of a d-DOF robot in a 2D or 3D world that contains static obstacles. For convenience, assume that the configuration space is parameterized so that $\mathcal{C} = [0,1]^d \subset \mathbb{R}^d$, and some boundary points are identified to respect topology. Let $\mathcal{C}_{free}$ denote the set of all collision-free configurations. A path planning *query* is a pair of configurations, $\langle q_{init}, q_{goal} \rangle$. The path planning problem is to find a continuous path, $\tau : [0,1] \rightarrow \mathcal{C}_{free}$ such that $\tau(0) = q_{init}$ and $\tau(1) = q_{goal}$. It is assumed to be computationally prohibitive to construct an explicit representation of $\mathcal{C}_{free}$; however, a *collision detection* algorithm is available, which can quickly indicate whether a given configuration lies in $\mathcal{C}_{free}$ (distance information could also be computed).

Classical grid-based search Grid-based search is considered by many to be the most straightforward form of path planning. Each component of $\mathcal{C}$ is quantized, and a d-dimensional bitmap representation can be precomputed by iterating a collision detector over all quantized configuration values. A neighborhood structure must be defined, such as the set of $2d$ neighbors for each interior element of the bitmap ("up", "down", "left", "right" in the case of $d = 2$). A given query is quantized, and the bitmap can then be searched using a classical AI or graph search algorithm, such as dynamic programming, A^*, best-first, or bidirectional search, to connect q_{init} to q_{goal}. In fact, the bitmap could also be searched using recent path planning methods that are based on incremental search, such as randomized potential fields [4], Ariadne's clew [27,26], RRTs [22], and the planner in [16,31]. It is well-known that only *resolution completeness* can be obtained, and that for a fixed resolution, the number of samples (bitmap size) increases exponentially in d.

The original PRM The Probabilistic Roadmap (PRM) was introduced in [19] as a way to overcome the well-known curse of dimensionality that exists in grid search. The primary philosophy behind the PRM was to perform substantial preprocessing so that *multiple queries* for the same environment could be handled efficiently. This is analogous to the bitmap precomputation in classical grid based search. First, a *roadmap* encoded as an undirected graph, G, is constructed in a *preprocessing phase*. In a *query phase*, G is used to solve a particular path planning question for a given q_{init} and q_{goal}. Each vertex in G represents an element of $\mathcal{C}_{free}$, and each edge represents a collision-free path between two configurations.

The algorithm outlined in Figure 1 constructs a PRM with N vertices. In Step 3, a random configuration in $\mathcal{C}_{free}$ is found by repeatedly picking a random configuration until one is determined by a collision detection algorithm to be in $\mathcal{C}_{free}$.[1] The NBHD function in Line 5 is a range query in which all vertices within a specified distance of q are returned, sorted by distance from q. Other variations are possible, of course. The variant considered here is the original PRM algorithm described in [18], page 25. The CONNECT function in Line 7 uses a fast local planner to attempt a connection between q and v. Usually, a "straight line" path in $\mathcal{C}_{free}$ is evaluated between q and v by stepping along incrementally with a collision detection algorithm. A heuristic, node-enhancement phase, described in [19], is not considered here.

Once the PRM has been constructed, the query phase attempts to solve planning problems. Essentially, q_{init} and q_{goal} are treated as new nodes in the PRM, and connections are attempted. Then, standard graph search is performed to connect q_{init} to q_{goal}. If the method fails, then either more vertices are needed in the PRM, or there is no solution. This is analogous to the problem of insufficient resolution in classical grid search.

[1] In practice, of course, pseudo-random samples are generated.

```
BUILD_PRM
1 G.init();
2 for i = 1 to N
3     q ←RAND_FREE_CONF(q);
4     G.add_vertex(q);
5     for each v ∈ NBHD(q,G)
6         if CONNECT(q, v) then
7             G.add_edge(q, v);
```

Fig. 1. The preprocessing phase: build a PRM.

The Lazy PRM A recent PRM variant called the Lazy PRM has been proposed for the problem of answering single planning queries efficiently, as opposed to building an extensive roadmap prior to consideration of a planning query [6]. The resulting planner is sometimes very efficient in comparison to the original PRM. This represents a shift from the *multiple query* philosophy of the original PRM [19], and returns to the *single query* philosophy which was used in some earlier planners [4,13,27].

The key idea in the Lazy PRM is to build the roadmap initially without the use of a collision detector. The difference with respect to the algorithm in Figure 1 is that the condition in Line 6 is dropped, and Line 7 is executed every time. This allows the PRM to be constructed quickly; however, more burden is placed on searching in the query phase. Once an initial-goal query is given, the planner performs A^* search on the roadmap to find a solution. If any of the solution edges are in collision, they are removed from the roadmap, and the A^* search is repeated. Eventually, all edges may have to be checked for collision, but often the problem is solved well before this happens. Alternatively, it might be preferable to run the search only once on the initial graph, while validating edges during the search (instead of waiting for a solution and then validating it) [9]. If no solution is found, then more nodes may need to be added to the roadmap. The advantage of the Lazy PRM is that the collision checking is only performed as needed. Thus, all edges do not have to be collision checked as in the case of the original PRM. In classical grid search, this philosophy implies that the bitmap is not precomputed; collision checking is performed only as needed during the search.

3 A Spectrum of Sampling Techniques

A first step toward constructing a spectrum of planners from PRMs to classical grid search is to characterize a spectrum of sampling techniques to cover $\mathcal{C}$, from pseudo-random sampling to grids. A brief introduction to sampling is given here. See [9,25,28] for more details. The key philosophy is that sampling is viewed as an optimization problem in which a finite point set or infinite sequence is designed that minimizes some performance criterion relevant to a particular application.

Motivated by applications in numerical integration and optimization, the most common criteria are discrepancy and dispersion. Let $X = [0,1]^d \subset \mathbb{R}^d$ define a space over which to generate samples. Consider evaluating the uniformity of a set, P, of N d-dimensional sample points $\{p_0, \ldots, p_{N-1}\}$. The *discrepancy* is defined as

$$D(P, \mathcal{R}) = \sup_{R \in \mathcal{R}} |\mu(R) - |P \cap R|/N| \tag{1}$$

in which $|\cdot|$ of a set denotes its cardinality, μ denotes the Lebesgue measure, and $\mathcal{R}$ is a *range space*, which will be taken by default in this paper to be the set of all axis-aligned rectangular subsets of X. The *dispersion* is defined as

$$\delta(P, \rho) = \sup_{x \in X} \min_{p \in P} \rho(x, p), \tag{2}$$

in which ρ denotes any metric; unless otherwise stated, the default metric in this paper will be ℓ^∞. Dispersion can also be considered as the "radius" of the largest empty ρ-ball, among all balls whose centers lie in X. Both discrepancy and dispersion seem very relevant in the PRM context because range queries are repeatedly performed, and these criteria ensure that either the appropriate number of samples (discrepancy) or at least one sample (dispersion) will fall within a range.

A *low-discrepancy* point set or sequence is one that yields the best-possible asymptotic discrepancy, which is $O(N^{-1} \log^d N)$ for infinite sequences and $O(N^{-1} \log^{d-1} N)$ for finite point sets. The simplest low-discrepancy point sets and sequences are Hammersley and Halton points, respectively, which were applied to motion planning in [9]. Other low-discrepancy techniques exist that produce smaller constants in the asymptotic convergence rate. The best family of methods are the (t,m,s)-nets and (t,s)-sequences [28], and the current best within this family are the Niederreiter-Xing sequences [29].

Regarding ℓ^∞ dispersion, the *Sukharev sampling criterion* [34] states that for any point set P, $\delta(P) \geq \frac{1}{2}\lfloor N^{\frac{1}{d}} \rfloor$. Thus, to keep dispersion fixed, it is impossible to avoid exponentially-many samples in dimension. A *low-dispersion* point set or sequence is one that produces the best possible asymptotic dispersion, which is $O(N^{-1/d})$. For a fixed N, if $N^{\frac{1}{d}}$ is an integer, k, the *Sukharev grid* yields the best possible dispersion, which is precisely $\frac{1}{2}N^{-1/d}$. In this case, the grid is constructed by partitioning $[0,1]^d$ into N cubes of width $1/k$ so that a tiling of $k \times k \times \cdots \times k$ is obtained, and a sample is placed at the center of each cube. Nongrid, low-dispersion infinite sequences exist that have $\frac{1}{\ln 4}$ as the constant in the asymptotic convergence rate [28].

Grids fall under a more-general class of sample sets known as *lattices*, which have a regular neighborhood structure (they are essentially grids that may have nonorthogonal alignments). Most low-discrepancy sequences, such as Halton points, are not lattices. Surprisingly, lattices exist that have very low discrepancy; one was used in the PRM context in [9]. These offer the additional advantage of avoiding nearest-neighbor computations in the PRM.

4 A Spectrum of Planners

We take a bidirectional approach to relating grid search to PRMs. Section 4.1 extends and adapts grid search, and Section 4.2 presents variations of the PRM. Section 4.3 discusses the overlap between the two families of planners, which includes a planner based on the Sukharev grid.

4.1 Extensions of Grid Search

Let CGS refer to *classical grid search*, as described in Section 2; for simplicity, assume that A^* search is used. As a simple first extension, consider making a Lazy CGS planner. The grid implicitly encodes a roadmap because the location of all samples and the neighborhood structure is implied in its definition. A Lazy CGS planner is simply obtained by not precomputing the collision bitmap, and deferring all collision checks until each is required in the search (either the multi-iteration searching approach in [6], or the single-iteration approach in [9] can be used). This straightforward idea (undoubtedly considered previously) avoids performing unnecessary collision checks. Just as in the case of a PRM, the original CGS is best suited for multiple queries, and the Lazy CGS is suitable for single-query problems.

The next extension may be applied to either the CGS or Lazy CGS. Consider allowing two different sampling resolutions: one is the grid resolution, and the other is the sampling rate required for collision checking. We assume that the latter sampling rate is much higher than the grid resolution. If we consider a motion from one grid point, q, to an adjacent grid point, q', then collision checking must be performed along a sequence of points obtained by linear interpolation (respecting topology) between q and q'. The grid can now be interpreted as a kind of trellis of paths. A similar idea was applied long ago for grid search in [10]. The advantage of this approach is that we might be able to solve a query using a low resolution for the grid, yet still be able to check for collisions at the required level of resolution. Good performance can be obtained if there are no narrow corridors in the configuration space. If q_{init} and/or q_{goal} do not lie on the grid points, then attempts can be made to connect them (again by interpolation) to a set of nearby grid points. We will refer to this extension as *subsampled grid search*. If a bitmap is precomputed, then the approach will be called SGS; otherwise, it is called a Lazy SGS.

One final point of concern is: where should the grid be placed? It seems reasonable to align the grid axes with the coordinate axes, but the translation remains to be chosen. Typically, the "origin" of a grid is at the coordinate origin (i.e, a sample appears at $(0, 0, \ldots, 0)$). Let k denote the number of points per axis, and let d denote the dimension. Suppose $k = 3$ and $d = 2$. A classical grid would place one point in the center and the rest along the boundary (assuming no topological identifications). The maximum ℓ^∞ distance possible from a point $x \in [0, 1]^d$ to a grid point is $1/4$ (in a unit cube). Now, consider the Sukharev grid, which provides optimal dispersion. In this case, the maximum distance is only $1/6$. Using only two points per axis ($k = 2$), the Sukharev grid yields a maximum distance of $1/4$, which is equivalent to that of a classical grid for which $k = 3$. These differences might appear small; however, for high-dimensional problems for which only low-resolution sampling is possible, the difference is dramatic. For example, suppose $d = 10$. In this case, a Sukharev grid with $2^{10} = 1024$ points provides the same quality coverage (in terms of dispersion) as a classical grid with $3^{10} = 59049$ points. Of course, as k becomes large, the improvement diminishes; however, for large d, it is impractical to make k large. Thus, the difference remains significant. Note also that the improvement diminishes in other topological spaces. In fact, on a toroidal manifold, the two grids are equivalent. That said, the Sukharev grid is never worse than the classical grid, and it is usually much better. Careful grid placement, as opposed to naive grids, is a theme also evident in kinodynamic planning [11].

4.2 Variations of the PRM

This section briefly reviews some of the PRM variants and experiments from [9], and also presents new experiments that use the Sukharev grid. For any sampling method, we can consider two variations: 1) a version that precomputes the roadmap, G, and is suitable for multiple queries, and 2) a lazy version that performs collision checking during the search, and is suitable for single-query problems.

First, consider constructing a *quasi-random roadmap*, QRM, by using the d-dimensional low-discrepancy sequence (such as Halton) instead of points generated from a pseudo-random number generator. The operation of the QRM should appear identical to that of the PRM; there are no new requirements. If the samples happen to form a lattice, costly neighborhood range queries can be avoided because the neighborhood structure is implicitly defined by the lattice rules. For many problems this can result in substantial performance improvement, particularly for lazy planners (because the time required to precompute the initial graph in the Lazy PRM represents a sizable fraction of the total running time for some problems [6,9]). Let LRM refer to a roadmap in which lattice points are used.

The experiments presented here can be considered as a continuation of those in [9]. Section 3 of [9][2] showed improvement factors between 1.02 and 4.08 for a series of "bent corridor" examples (see Figure 2 from dimension 2 to 10, using Hammersley points in the place of pseudo-random samples (obtained from a nonlinear congruential generator). We recently performed hundreds of new experiments based on the Sukharev grid that consider many different radii, corridor thicknesses, dimensions, and random rotations of samples. We generally observed that the Sukharev grid performs better than

[2] Regrettably, the original figure of bent corridor experiments contains a typographical error: the width values given are actually $1/2$ of the actual width used.

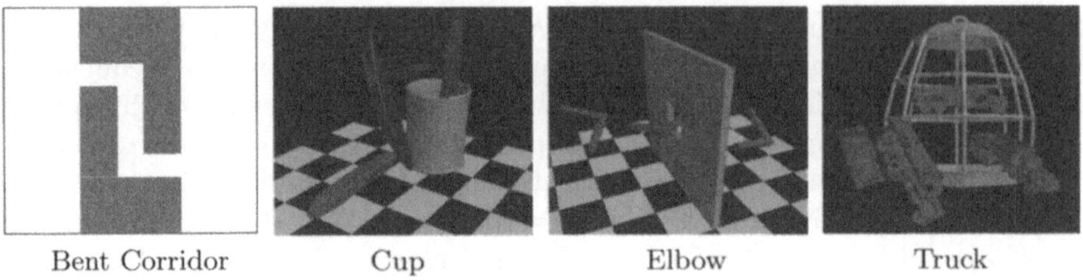

Fig. 2. The first frame shows a 2D rendering of the bent corridor series of experiments. The cup, elbow, and truck are 6-DOF problems that were solved using a Sukharev grid with 3, 4, and 5 points per axis, respectively, which is an improvement over PRM performance, and comparable to the low-discrepancy lattices used in [9].

pseudo-random sampling, especially in the case of a narrow corridor and small radius. As the radius increases, the pseudo-random samples are able to overcome much of their high-dispersion flaws; however, Suhkarev grids still remain advantageous. The following table

Dim.	Width	Rad	Min	Max	Mean	Sukh-1st	Last-fail
2	0.02	0.5	65	652	461.97	25	400
3	0.05	0.6	131	1861	757.52	125	1000
6	0.15	0.6	195	1444	1237.85	729	64

shows a summary of the experiments for large radius values; with smaller values the results were even more in favor of Sukharev grids. The columns marked Min, Max, and Avg, give the number of nodes used in the PRM. The "Sukh-1st" column gives the smallest Sukharev grid that solved the problem. In some cases, the next-largest Sukharev grid does not solve the problem; thus, the last column gives the largest grid that failed to solve the problem. To avoid problems of axis alignment, we performed numerous experiments in which the Sukharev grid was rotated at random. This had no significant effect on the outcomes. For example, for 5 random rotations of the Sukharev grid, the same number of points, 729 (3 points per axis), was sufficient to solve the 6D problem listed in the table. Several other 6-DOF examples were solved using Sukharev grids, as indicated in the caption of Figure 2.

4.3 Sukharev Grid: Closing the Gap

Good performance was obtained using the Sukharev grid, but what kind of method is this? It can be considered as an extension of subsampled grid search that uses subsampling and Sukharev samples. The Sukharev grid is also a kind of lattice; therefore, the method can be considered as a special kind of LRM, and the spectrum of planners appears as:

$$CGS - SGS - LRM - QRM - PRM$$

Alternatively, the "Lazy" term could be added to each of these, to make single-query versions. The entire spectrum can be considered as families of *sampling-based motion planning* algorithms.

We now describe some basic relationships between the planners in the spectrum; additional conclusions and insights are derived from the theory in Section 5 and the discussion in Section 6. Two problems with classical grid search (CGS) were: high sampling density were required, and too many points were concentrated along the boundary. The first problem was fixed by subsampling (SGS), and the second was fixed by using a Sukharev grid. Once these changes are made, grid search appears to be comparable to other lattice-based roadmaps (LRM). Based on our experiments presented here and in [9], the LRMs (including grids) yield performance that is comparable to that of QRMs; however, LRMs have the additional advantage of known neighborhood structure. Finally, the deterministic planners appear superior to the original PRM in performance because they avoid the clumpiness and sparseness that must occur for a set of pseudo-random samples. This will be captured theoretically by Proposition 6 in Section 5.

5 Theoretical Considerations

We provide some theoretical analysis for all of the deterministic planners presented in Section 4, which includes the QRM, LRM, CGS, and SGS, and their Lazy counterparts. Let DRM refer to this collection of planners (for *deterministic roadmaps*), under the condition that they use asymptotically dispersion-optimal samples. This includes Halton, Hammersley, and many other low-discrepancy samples, and also the Sukharev grid.

Deterministic sampling enables the DRM planners to be *resolution complete*, in the sense that if it is possible to solve the query at a given sampling resolution, they will solve it. The resolution can be increased arbitrarily to ensure that any problem can be solved, if a solution exists. This is in contrast to the original PRM and other randomized variants, which are only probabilistically complete [21] (the probability tends to one that a solution will be found as the number of samples grows to infinity).

We exploit dispersion bounds to characterize the set of configuration spaces over which all queries can be correctly answered. This characterization is in terms of a parameter that measures the narrowest corridor width, in a manner similar to that of [3,7,15]. We define a cylindrical tube, and the "width" of $\mathcal{C}_{free}$ is expressed in terms of the largest possible cross section of the tube, over all possible queries. Measuring this parameter may be as difficult as the planning problem; however, the expression of planner performance in terms of parameters that are difficult to measure is common in randomized planning analysis [3,6,15,16,20]. If a solution does not exist, our deterministic planners are able to declare that either the solution path must travel through a narrow passage that has a width smaller than a specified value, or there is no solution. Such a result might be useful in applications because once the corridor is known to be narrower than a reasonable precision level, the solution might be impractical anyway.

We assume that for any PRM-like method, the radius parameter used to select neighbors in the NBHD function from Line 5 of Figure 1 is always sufficiently large. In theory, the radius can be made large enough so that an attempt is made to connect every vertex to every other vertex. In practice, this often becomes inefficient; therefore, a smaller value is typically used.

Let $\gamma = \langle q_{init}, q_{goal} \rangle$ denote a *query*. The set, $\Gamma(\mathcal{C}_{free})$, of all queries in which $q_{init} \in \mathcal{C}_{free}$, $q_{goal} \in \mathcal{C}_{free}$, and $q_{init} \neq q_{goal}$, for a given $\mathcal{C}_{free}$ is called the *query space* of $\mathcal{C}_{free}$. Let $\Gamma_s(\mathcal{C}_{free}) \subseteq \Gamma(\mathcal{C}_{free})$ denote the set of all queries for which a solution exists.

Let $\mathcal{C}$ represent any d-dimensional configuration space, parameterized to yield $\mathcal{C} = [0,1]^d \backslash \sim$, in which $\backslash \sim$ denotes appropriate topological identifications along the boundary of the unit cube. Let Ψ represent the subset of the power set of $\mathcal{C}$ corresponding to all open subsets that can be constructed with algebraic constraints, as formulated in [21].

Let a *tube*, $\mathcal{B}$, represent an uncountable collection of balls of equal radius whose centers are generated by a continuous path, $\tau : [0,1] \rightarrow \mathcal{C}_{free}$. For each $s \in [0,1]$ there exists an open ball $B \in \mathcal{B}$ that is centered at $\tau(s)$ and has radius r, which is fixed for all $B \in \mathcal{B}$; let $B(s)$ denote the ball centered at $\tau(s)$. We call $2r$ the *width*, $w(\mathcal{B})$, *of the tube.*

Let $V(q)$ denote the set of all points visible from a set $q \in \mathcal{C}_{free}$ (i.e., for each $q' \in V(q)$, $\lambda q + (1-\lambda)q' \in \mathcal{C}_{free}$ for all $\lambda \in [0,1]$, assuming that topology is respected in the interpolation).

Suppose that a query $\gamma \in \Gamma_s(\mathcal{C}_{free})$ is given. Among all possible tubes, let $\mathcal{B}(\gamma)$ denote the tube with the largest width such that $B(0) \subset V(q_{init})$ and $B(1) \subset V(q_{goal})$. In other words, the entire first ball is visible from q_{init}, and the entire last ball is visible from q_{goal}. Denote this largest-width tube as the $\mathcal{B}(\gamma)$, and call its width the *width*, $w(\gamma)$, *of the query.* For any query $\gamma \in \Gamma(\mathcal{C}_{free}) \backslash \Gamma_s(\mathcal{C}_{free})$, we say that its width is zero because no tube exists.

Define the *width of* $\mathcal{C}_{free}$ as

$$w(\mathcal{C}_{free}) = \inf_{\gamma \in \Gamma_s(\mathcal{C}_{free})} w(\gamma). \tag{3}$$

Let $\Psi(x)$ for $x \in (0,\infty)$ denote the set of all $\mathcal{C}_{free} \in \Psi$ such that $w(\mathcal{C}_{free}) \geq x$. Intuitively, this can be considered as the set of problems for which the width of the narrowest corridor is at least x.

Suppose that the roadmap, G, is constructed for a particular $\mathcal{C}_{free}$. Then, the algorithm is said to be *complete for* $\mathcal{C}_{free}$ if all queries in $\Gamma(\mathcal{C}_{free})$ are answered correctly in the query phase. A solution path must be reported if one exists; otherwise, failure is reported.

Our first two results establish the resolution completeness and complexity of all of the DRM planners. Therein, we only assume that sampling is accomplished using a set P of *low-dispersion points*, for which $\delta(P,\rho) < b(d)N^{-1/d}$, in which $b(d)$ is a constant that may depend on the dimension d. For Halton/Hammersley points, b is a function of the largest prime used in the construction. For many low-dispersion sequences, including a multiresolution Sukharev grid, $b(d) \equiv 1$, which is a small constant that is independent of dimension. If N is restricted to values such that $N^{1/d}$ is an integer, then $b(d) \equiv 1/2$. A low-dispersion sequence for which $b(d) \equiv \frac{1}{\ln 4}$ and no restriction is placed on N is given in [28] (surprisingly, it is not uniform).

Propositions 1 through 3 below hold for any metric, provided that both tube width and dispersion are measured using the same metric, ρ. Propositions 4 through 6 hold for norms (which are all related by constants in subsets of $\mathbb{R}^d$) under the same provision.

Proposition 1 *After N iterations, the DRM planners are complete for all $\mathcal{C}_{free} \in \Psi(4b(d)N^{-1/d})$, in which N is the number of points, d is the dimension of $\mathcal{C}$, and $b(d)$ is a factor that depends on the sampling method.*

Proof: Suppose first that $\mathcal{C} = [0,1]^d$ (ignoring any identifications). Assume that $\mathcal{C}_{free} \in \Psi(2b(d)N^{-1/d})$. To show completeness, we establish that for any solvable query, a solution path will be found; let $\gamma \in \Gamma_s(\mathcal{C}_{free})$ be such a query. Because $\mathcal{C}_{free} \in \Psi(2b(d)N^{-1/d})$, there exists a tube, $\mathcal{B}$, of width at least $2b(d)N^{-1/d}$, such that $B(0) \subset V(q_{init})$ and $B(1) \subset V(q_{goal})$.

Let P denote the set of sample points, which is also the set of vertices in the roadmap, G. Each ball $B \in \mathcal{B}$ must contain at least one $q \in P$. This follows from the fact that N samples were generated, and the asymptotic dispersion bound is $\delta(P,\rho) < b(d)N^{-1/d}$. If any ball of radius $b(d)N^{-1/d}$ is empty, then the dispersion would violate this upper bound.

First, consider connecting q_{init} and q_{goal} to the roadmap. Since $B(0) \subset V(q_{init})$, all configurations found by linear interpolation between q_{init} and any point in $B(0)$ are collision free. Therefore, q_{init} will be connected to a configuration in P (either one contained in $B(0)$, or at least one in the same connected component of G as a configuration of P that lies in $B(0)$). Using a similar argument for $B(1)$, q_{goal} will also be connected to a roadmap vertex.

It finally remains to show that there exists a path in G between the two configurations in P to which q_{init} and q_{goal} are connected. Consider the balls of $\mathcal{B}$ as parameterized using $B(s)$ for $s \in [0,1]$. We construct a sequence, $q_0, \ldots, q_{k-1}$ of k configurations as follows. Let q_0 be any element of $P \cap B(0)$. Let $s_1 \in [0,1]$ denote the last point at which the ball $B(s)$ contains q_0, by starting with $B(0)$ and increasing s continuously. Let q_1 be any element of $P \cap B(s_1) \setminus \{q_0\}$. Note that the $B(s_1)$ must contain at least two points in P because q_0 lies on its boundary. Inductively, let q_i be any element of $P \cap B(s_i) \setminus \{q_{i-1}\}$, where s_i is the first point at which $B(s)$ does not contain q_{i-1}. Note that the induction is finite, and let q_{k-1} denote the final configuration in the sequence.

We argue that there must exist a path in G between each pair, q_i, q_{i+1}, of configurations for $i \in \{0, \ldots, k-2\}$. The point q_i must lie on the boundary of $B(s_{i+1})$; therefore, $B(s_{i+1})$ contains two points of P. Furthermore, all points between q_i and q_{i+1} via linear interpolation must be collision free. The algorithm in Figure 1 would either have produced an edge between them, or failed to because both were already part of the same connected component of G. Either way, there exists a path in G between q_i and q_{i+1}. By applying this for each configuration in the sequence, there exists a path in G between q_0 and q_{k-1}. Furthermore, q_0 is connected to q_{init}, and q_{k-1} is connected to q_{goal}. Therefore, the query is correctly answered by returning a solution path.

We now turn to the case in which $\mathcal{C} = [0,1]^d \backslash \sim$, in which $\sim$ denotes boundary identifications needed to appropriately reflect the topology of transformation groups that arise in motion planning: S^1, P^3, etc. For the dispersion measurements in $[0,1]^d$, balls near the boundary have to be contained entirely inside the unit cube. Once identifications are considered, some balls are allowed to overflow as long as their center lies in $[0,1]^d$. Since $\delta(P, rho) < b(d)N^{-1/d}$ for $[0,1]^d$, these overflowing empty balls cannot have radius larger than $2b(d)N^{-1/d}$. Thus, the dispersion in the part of the proof for $[0,1]^d$ is simply scaled by two for the case of $[0,1]^d \backslash \sim$ by assuming $\mathcal{C}_{free} \in \Psi(4b(d)N^{-1/d})$ in the first step, which establishes the proposition. ∎

Proposition 1 can be reworked to bound the width of the query:

Proposition 2 *After N iterations, the DRM planners, for a query γ, either report a solution path or correctly declare that one of the following is true: there is no solution path, or $w(\gamma) < 4b(d)N^{-1/d}$.*

Proof: This follows directly from Proposition 1. Since the DRM is complete for $\Psi(4b(d)N^{-1/d})$, if no solution is found after N iterations, then $w(\mathcal{C}_{free}) < 4b(d)N^{-1/d}$ and $w(\gamma) < 4b(d)N^{-1/d}$. ∎

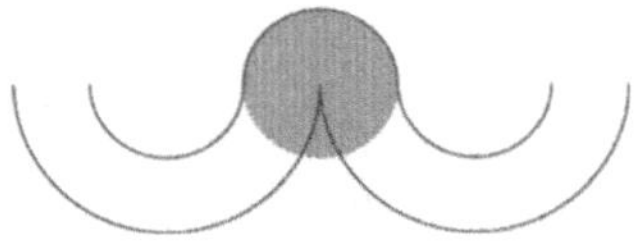

Fig. 3. A narrow corridor in $\mathcal{C}_{free}$, used in the proof of Proposition 3.

The next proposition indicates that if the dispersion is at least δ, then a DRM or PRM planner might miss solutions in corridors of width δ.

Proposition 3 *For any sample set, P, that has dispersion at least δ, no roadmap constructed using the algorithm in Figure 1 can be complete for $\Psi \setminus \Psi(\delta)$.*

Proof: We argue that completeness is lost by producing a $\mathcal{C}_{free}$ and query $\gamma \in \Gamma_s(\mathcal{C}_{free})$ that will be answered incorrectly. If the dispersion is δ, then there exists a ball, $B \subset [0,1]^d$ with radius δ such that $P \cap B = \emptyset$. Consider the corridor shown in Figure 3. Assume that no tube greater than width δ can be placed in the corridor. If $\mathcal{C}_{free}$ is chosen so that B is located as shown in the shaded area, then there will be no path in G that traverses the corridor. For any point on one side of the corridor outside of B, the straight-line path to any point on the other side of the corridor outside of B will intersect $\mathcal{C} \setminus \mathcal{C}_{free}$. Thus, a solution path will not be found. ∎

From this the next proposition follows, which establishes that any PRM approach will require an exponential number of samples. It is assumed that the sampling scheme generates samples independently of the obstacle region.

Proposition 4 *Under any sampling scheme (including pseudo-random), a roadmap requires a number of samples exponential in dimension, d, to be complete for* $\Psi(\delta)$.

Proof: This follows immediately from Proposition 3 and the Sukharev sampling criterion [34]. ∎

We now consider asymptotic bounds for DRM planners. The next proposition indicates that the DRM planners do the best possible, asymptotically.

Proposition 5 *The number of samples required by each DRM planner to be complete for* $\Psi(\delta)$ *is asymptotically optimal.*

Proof: By Proposition 3, to be complete for $\Psi(\delta)$, the dispersion must be less than δ. Thus, the goal of a DRM algorithm should be to reduce δ using as few samples as possible. The low-dispersion sequences achieve the best possible asymptotic dispersion. Therefore, the number of samples used in the DRM planners is asymptotically optimal. ∎

The following proposition gives some indication that random sampling does not yield the best possible asymptotic convergence in the PRM; in fact, it is significantly worse than using deterministic sampling.

Proposition 6 *For a fixed dimension, d, the PRM with random sampling requires* $O((\log N)^{\frac{1}{d}})$ *times as many samples (with probability one) as the DRM planners to achieve the same* ℓ^∞ *dispersion.*

Proof: It was shown by Deheuvels [28] that ℓ^∞ dispersion for random samples is $O((\log N)^{\frac{1}{d}} N^{-\frac{1}{d}})$ with probability one. The asymptotic dispersion in a DRM achieves $O(N^{-\frac{1}{d}})$. The factor difference between the two is $O((\log N)^{\frac{1}{d}})$. ∎

Relating the PRM to grid search We now discuss the implications of the previous propositions. Consider Propositions and 1 and 2. If we use ℓ^∞ dispersion and select N such that $N^{1/d}$ is an integer, then the Sukharev grid yields $b(d) \equiv 1/2$, which is the best possible performance that can be obtained. An exponential number of samples in d is required, but according to Proposition 4, this is unavoidable. Note that the PRM cannot even provide these deterministic guarantees. The PRM at least has asymptotic analysis that establishes probabilistic completeness and convergence [3,19]. However, Propositions 5 and 6 indicate that the asymptotic rate of convergence obtained by the best deterministic sequences (including the Sukharev grid) is asymptotically optimal, and the use of random samples is a significant factor worse (with probability one). Proposition 6 quantifies the problems with random sampling. To satisfy statistical tests for uniformity, a pseudo-random number generator must clump too many points in some places, and not enough in others. If the distribution of points is "too uniform," then it will fail statistical tests. The deterministic samplers, on the other hand, are able to carefully *disperse* the points without this wasteful concern. Thus, in addition to the experimental advantages discussed in Section 4, there are strong theoretical indications that deterministic sampling methods, including grids, offer advantages over the original PRM.

6 Discussion

We have provided a spectrum of planners that ranges from classical grid search to the probabilistic roadmap. Surprisingly, the original PRM does not appear to be advantageous over deterministic approaches, including grid search, according to our experiments and theoretical analysis.[3] Based on our work, we believe one of the main factors for the success of the original PRM was the excellent use of subsampling, which for grid sampling, enables challenging problems to be solved with only a few points per axis. In the original PRM, this idea also contributed greatly to the performance. Although there appears at first to be no exponential dependency on dimension because N is chosen directly, the exponential dependency reveals itself once again if we try to hold dispersion fixed. The notion of fixing the dispersion is general enough to mean keeping the resolution the same for a grid, while also applying to any sampling scheme. Using the Sukharev sampling criterion, $N^{\frac{1}{d}}$ can be considered as the best possible "points per axis" for any sampling scheme, whether or not it is a grid, and also whether it is random or deterministic.

Very high dimensions We note that our experiments focused mostly on six dimensions, with some examples in [9] up to ten dimensions. This includes many problems of interest in robotics, but examples exist in robotics and computational biology in which there are dozens or hundreds of dimensions. Obviously, even a Sukharev grid with two points per axis would be impossible to manage for some problems, while pseudo-random samples would appear to have no trouble. One straightforward way to use deterministic sampling would be to use a non-lattice sequence in the place of pseudo-random samples. In very high dimensional problems we expect, however, that the performance differences between using pseudo-random sampling and deterministic sampling would be negligible for most problems. This is due primarily to the fact that spaces of this dimension will be severely undersampled, regardless of the sampling scheme used (i.e., the "points per axis" for any sampling scheme would be approximately one).

[3] Since the "probabilistic" theme of PRMs does not seem advantageous, it is perhaps better to refer to these, grid search, and other related planners as instances of *sampling-based motion planning*.

Search issues Searching is one of the oldest and most fundamental issues in motion planning. Although most PRM-based planners use A^* search, it is important to revisit some of the basic search issues in this context. Given the close connection drawn between classical grid search and the PRM in this paper, it should be clear that classical search issues which apply to grids should also apply to PRMs, QRMs, Sukharev grids, etc. that have a large number of samples. If a PRM has thousands of nodes in a high-dimensional space, then the common local minima problems most likely exist. Since optimality is not important in this context, many motion planning ideas that are based directly on incremental search could be used to search the roadmaps obtained in the family of planners described in this paper. Potential field approaches have improved search by using heuristics to escape local minima [4,13,17]. For the randomized potential field planner in [4], the search is even performed over an implicit grid. Thus, it can be considered as a Lazy CGS, with A^* search replaced by the potential field method. The planner in [10] can also be considered as a Lazy CGS planner. Of course many other planning approaches can be considered. Incremental searching methods, such as Ariadne's clew [27], bidirectional RRTs [22], or the planners in [16,31], can be considered as alternatives to A^* search in any roadmap approach. As the implicit grid resolution in a Lazy CGS planner becomes higher, it is clear that the sampling itself is not important. The key is how to search in a bounded metric space.

Multiresolution approaches One disadvantage of the Sukharev grid is that for each increment of the points per axis, the total number of points increases dramatically for higher-dimensional problems. In general, for sample sets that require N to be specified, what happens when N needs to be increased? Many planning algorithms involve a repeated alternation between path searching and improving the sampling resolution. Of course, numerous multiresolution approaches have been considered in planning literature. For example, in [12], the sampling can be improved locally. For the Lazy LRM, the number of lattice points could be iteratively doubled, while still remaining low discrepancy [9]. In [5], a trellis is iteratively maintained (similar to SGS), and a hyperplane of samples is added in each iteration before search is performed again. In the original PRM, multiresolution issues appear to be equivalent to the problem of determining how many samples are necessary before a solution can be found. Any number of points can be easily added before search is performed again. A similar behavior can be obtained for grid-based sequences by adding points one at a time and maintaining low discrepancy along the way [24].

Bringing back randomization A recent trend in quasi-Monte Carlo literature is to consider randomized versions of Halton sequences, lattices, and (t,s)-sequences. There exist simple techniques that preserve the low-discrepancy/low-dispersion properties of deterministic sequences while at the same time each sample is uniform randomly distributed.[4] An elegant example of this work appears in [35], in which a randomized Halton sequence is obtained by cleverly constructing the first element at random, and computing the appropriate continuations of classical van der Corput sequences for each coordinate. Such sequences can be used in motion planning to obtain *both* probabilistic analysis and the deterministic guarantees from Section 5!

Acknowledgments. We are grateful for the funding provided in part by NASA grant NAG2-1333 (Branicky), NSF awards 9875304, 0208891 (LaValle), NSF awards 0208891 and 0208919 (Branicky, Frazzoli, and LaValle). Steve Lindemann performed the bent corridor series of experiments. We thank Robert Bohlin, Francesco Bullo, Bruce Donald, Mike Erdmann, Leo Guibas, Blake Hannaford, Lydia Kavraki, Steve Lindemann, Kari Olson, Shai Sachs, Nicola Simeon, and Libo Yang for helpful discussions, comments, and suggestions. We also thank Eric Larsen, Stefan Gottschalk, Ming Lin, and Dinesh Manocha for making the PQP collision detection package available.

References

1. N. M. Amato, O. B. Bayazit, L. K. Dale, C. Jones, and D. Vallejo. OBPRM: An obstacle-based PRM for 3D workspaces. In *Proceedings of the Workshop on Algorithmic Foundations of Robotics*, pages 155–168, 1998.
2. N. M. Amato and Y. Wu. A randomized roadmap method for path and manipulation planning. In *IEEE Int. Conf. Robot. & Autom.*, pages 113–120, 1996.
3. J. Barraquand, L. Kavraki, J.-C. Latombe, T.-Y. Li, R. Motwani, and P. Raghavan. A random sampling scheme for robot path planning. In G. Giralt and G. Hirzinger, editors, *Proc. of the 7th International Symposium on Robotics Research*, pages 249–264. Springer, New York, NY, 1996.
4. J. Barraquand and J.-C. Latombe. A Monte-Carlo algorithm for path planning with many degrees of freedom. In *IEEE Int. Conf. Robot. & Autom.*, pages 1712–1717, 1990.
5. R. Bohlin. Path planning in practice; lazy evaluation on a multi-resolution grid. In *IEEE/RSJ Int. Conf. on Intelligent Robots & Systems*, 2001.
6. R. Bohlin and L. Kavraki. Path planning using Lazy PRM. In *IEEE Int. Conf. Robot. & Autom.*, 2000.
7. R. Bohlin and L. Kavraki. A randomized algorithm for robot path planning based on lazy evalaution. In S. Rajasekaran, P. Pardalos, J. Reif, and J. Rolim, editors, *Handbook on Randomized Computation*. Kluwer Academic, 2001.
8. V. Boor, N. H. Overmars, and A. F. van der Stappen. The gaussian sampling strategy for probabilistic roadmap planners. In *IEEE Int. Conf. Robot. & Autom.*, pages 1018–1023, 1999.
9. M. Branicky, S. M. LaValle, K. Olsen, and L. Yang. Quasi-randomized path planning. In *Proc. IEEE Int'l Conf. on Robotics and Automation*, pages 1481–1487, 2001.
10. B. R. Donald. A search algorithm for motion planning with six degrees of freedom. *Artif. Intell.*, 31:295–353, 1987.
11. B. R. Donald, P. G. Xavier, J. Canny, and J. Reif. Kinodynamic planning. *Journal of the ACM*, 40:1048–66, November 1993.

[4] However, the samples in the sequence are not uncorrelated, as in the case of a uniform, random sequence.

12. B. Faverjon. Obstacle avoidance using an octree in the configuration space of a manipulator. In *IEEE Int. Conf. Robot. & Autom.*, pages 504–512, 1984.
13. B. Faverjon and P. Tournassoud. A local based method for path planning of manipulators with a high number of degrees of freedom. In *IEEE Int. Conf. Robot. & Autom.*, pages 1152–1159, 1987.
14. C. Holleman and L. E. Kavraki. A framework for using the workspace medial axis in PRM planners. In *IEEE Int. Conf. Robot. & Autom.*, pages 1408–1413, 2000.
15. D. Hsu, L. E. Kavraki, J.-C. Latombe, R. Motwani, and S. Sorkin. On finding narrow passages with probabilistic roadmap planners. In et al. P. Agarwal, editor, *Robotics: The Algorithmic Perspective*, pages 141–154. A.K. Peters, Wellesley, MA, 1998.
16. D. Hsu, J.-C. Latombe, and R. Motwani. Path planning in expansive configuration spaces. *Int. J. Comput. Geom. & Appl.*, 4:495–512, 1999.
17. Y. K. Hwang and N. Ahuja. A potential field approach to path planning. *IEEE Trans. Robot. & Autom.*, 8(1):23–32, February 1992.
18. L. E. Kavraki. Computation of configuration-space obstacles using the Fast Fourier Transform. *IEEE Trans. Robot. & Autom.*, 11(3):408–413, 1995.
19. L. E. Kavraki, P. Svestka, J.-C. Latombe, and M. H. Overmars. Probabilistic roadmaps for path planning in high-dimensional configuration spaces. *IEEE Trans. Robot. & Autom.*, 12(4):566–580, June 1996.
20. F. Lamiraux and J.-P. Laumond. On the expected complexity of random path planning. In *IEEE Int. Conf. Robot. & Autom.*, pages 3306–3311, 1996.
21. J.-C. Latombe. *Robot Motion Planning*. Kluwer Academic Publishers, Boston, MA, 1991.
22. S. M. LaValle and J. J. Kuffner. Rapidly-exploring random trees: Progress and prospects. In B. R. Donald, K. M. Lynch, and D. Rus, editors, *Algorithmic and Computational Robotics: New Directions*, pages 293–308. A K Peters, Wellesley, MA, 2001.
23. P. Leven and S. Hutchinson. Real-time motion planning in changing environments. In *Proc. International Symposium on Robotics Research*, 2000.
24. S. R. Lindemann and S. M. LaValle. Incremental low-discrepancy lattice methods for motion planning. 2003. Submitted to IEEE International Conference on Robotics and Automation.
25. J. Matousek. *Geometric Discrepancy*. Springer-Verlag, Berlin, 1999.
26. E. Mazer, J. M. Ahuactzin, and P. Bessière. The Ariadne's clew algorithm. *J. Artificial Intell. Res.*, 9:295–316, November 1998.
27. E. Mazer, G. Talbi, J. M. Ahuactzin, and P. Bessière. The Ariadne's clew algorithm. In *Proc. Int. Conf. of Society of Adaptive Behavior*, Honolulu, 1992.
28. H. Niederreiter. *Random Number Generation and Quasi-Monte-Carlo Methods*. Society for Industrial and Applied Mathematics, Philadelphia, USA, 1992.
29. H. Niederreiter and C. P. Xing. Nets, (t,s)-sequences, and algebraic geometry. In P. Hellekalek and G. Larcher, editors, *Random and Quasi-Random Point Sets, Lecture Notes in Statistics, Vol. 138*, pages 267–302. Springer-Verlag, Berlin, 1998.
30. C. Pisula, K. Hoff, M. Lin, and D. Manoch. Randomized path planning for a rigid body based on hardware accelerated Voronoi sampling. In *Proc. Workshop on Algorithmic Foundation of Robotics*, 2000.
31. G. Sánchez and J.-C. Latombe. A single-query bi-directional probabilistic roadmap planner with lazy collision checking. In *Int. Symp. Robotics Research*, 2001.
32. T. Simeon, J.-P. Laumond., and C. Nissoux. Visibility based probabilistic roadmaps for motion planning. *Advanced Robotics Journal*, 14(6), 2000.
33. Thierry Simeon, 2002. Personal communication.
34. A. G. Sukharev. Optimal strategies of the search for an extremum. *U.S.S.R. Computational Mathematics and Mathematical Physics*, 11(4), 1971. Translated from Russian, *Zh. Vychisl. Mat. i Mat. Fiz.*, 11, 4, 910-924, 1971.
35. X. Wang and F. J. Hickernell. Randomized halton sequences. *Math. Comp. Modelling*, 32:887–899, 2000.
36. S. A. Wilmarth, N. M. Amato, and P. F. Stiller. MAPRM: A probabilistic roadmap planner with sampling on the medial axis of the free space. In *IEEE Int. Conf. Robot. & Autom.*, pages 1024–1031, 1999.

Algorithms for Rapidly Dispersing Robot Swarms in Unknown Environments

Tien-Ruey Hsiang[1], Esther M. Arkin[1], Michael A. Bender[1], Sándor P. Fekete[2], and Joseph S. B. Mitchell[1]

[1] Stony Brook University, Stony Brook, NY 11794, USA
[2] TU Braunschweig, 38106 Braunschweig, Germany

Abstract. We develop and analyze algorithms for dispersing a swarm of primitive robots in an unknown environment, R. The primary objective is to minimize the *makespan*, that is, the time to *fill* the entire region. An environment is composed of *pixels* that form a connected subset of the integer grid. There is at most one robot per pixel and robots move horizontally or vertically at unit speed. Robots enter R by means of $k \geq 1$ *door pixels*. Robots are primitive finite automata, only having local communication, local sensors, and a constant-sized memory.

We first give algorithms for the single-door case (i.e., $k = 1$), analyzing the algorithms both theoretically and experimentally. We prove that our algorithms have optimal makespan $2A - 1$, where A is the area of R.

We next give an algorithm for the multi-door case ($k \geq 1$), based on a wall-following version of the leader-follower strategy. We prove that our strategy is $O(\log(k + 1))$-competitive, and that this bound is tight for our strategy and other related strategies.

1 Introduction

The objective of swarm robotics is to program a huge number of simple, tiny robots to perform complex tasks collectively. A typical application scenario for a robot swarm may involve two phases: first, the robot swarm fills an environment as quickly as possible while keeping the swarm "connected" (a term made more precise later). Once the robots are distributed in the environment, the robots perform *world-embedded* computational tasks, such as computing distances and short paths between points in the environment, finding chokepoints, mapping the environment, and searching for intruders.

This paper develops efficient algorithms for filling an environment with a swarm of robots, thus focusing on the first phase described above. Because target applications require thousands to millions of robots and current demonstration scenarios have at most hundreds of robots, it is crucial to develop algorithms that have explicitly stated performance guarantees, thus ensuring scalability.

There has already been work on dispersion algorithms, but most dispersion algorithms rely on greedy strategies such as *go-for-free space*, where robots move to fill unoccupied space, or *artificial physics* strategies, where neighboring robots exert "forces" on each other: repulsion forces if the robots

J.-D. Boissonnat et al. (Eds.): Algorithmic Foundations of Robotics V, STAR 7, pp. 77–93, 2004.

are closer than the target separation, and attraction forces if the neighboring robots are too far away. Although such robot behaviors often lead to reasonable dispersion, there are cases in which the swarm enters infinite loops, never filling the domain. Even in artificial physics strategies, which seem to disperse robots most efficiently, there are no guarantees on filling time. Furthermore, the swarm behavior under artificial-physics rules results in dispersions that are more reminiscent of molecular dynamics simulations (with large amounts of "bouncing" and "jitter") than of the ideal coordinated behavior of cooperating teams of robots.

The dispersion algorithms in this paper represent a departure from previously studied dispersion strategies. First of all, we test our algorithms in arbitrary polygonal environments, rather than restricting ourselves to rectangles or the infinite plane. It is important to study the dispersion algorithms in complicated polygonal domains because the target domains may be indoor environments with many bottlenecks or systems of pipes or ductwork.

Our algorithms are based on follow-the-leader local rules, where the robots form chains of leaders and followers emanating from central sources of robots (the "doors"). The emergent behavior of our main algorithm is a left-hand-on-the-wall strategy, where the entire robot swarm hugs the left wall of the domain as the filling proceeds. We provide matching lower and upper bounds on the amount of time to fill as a function of the total size of the source. The difficulty in dispersing efficiently is in sustaining a large flow of robots coming in through the doors; the "chains" of robots coming in through the doors have a tendency to "cut each other off"; see Figure 1(left). Thus, we devise a set of local rules for splitting and splicing the robot chains. Our experience with our robot simulator shows that unless the local rules are precisely implemented, the flow of robots out of the source diminishes and perhaps even grinds to a halt prematurely.

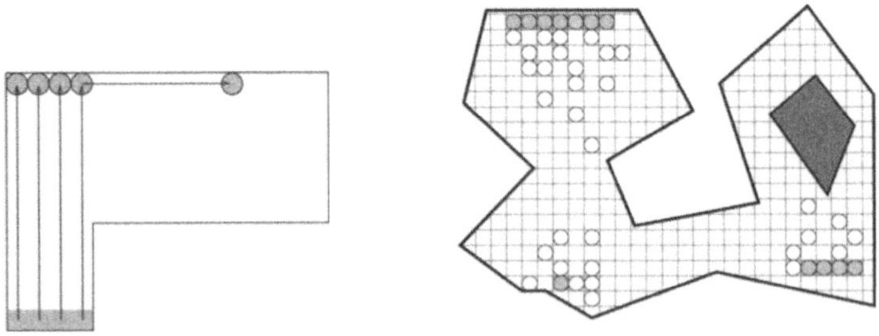

Fig. 1. Left: One turn can cut off most of the flow lines leading out of the door pixels (shaded grey): The left three columns of robots are trapped. Right: A polygonal domain encloses a discrete set of possible robot locations (*pixels*); special door pixels are shown in grey.

1.1 Our Results

We develop and analyze algorithms for dispersing robot swarms in an unknown environment R. The primary objective is to minimize the *makespan*, that is, the time to *fill* the entire region. An environment is composed of *squares* or *pixels* that form a connected subset of the integer grid; such an environment is also called a *maze*. There is at most one robot per pixel and robots move horizontally or vertically at unit speed. There is a source of robots entering the environment through $k \geq 1$ *doors*. Robots are primitive finite automata, only having local communication, local sensors, and a constant-sized memory. The challenge is in devising purely local strategies distributed over a swarm of agents that result in a predictable and provable behavior for the collective.

This paper presents the following results:

- We give filling algorithms for the single-door case (i.e., $k = 1$), analyzing the algorithms both theoretically and experimentally in terms of natural performance metrics, such as makespan and total traveled distance by all robots. Our algorithms are based on leader-follower strategies that are adapted from depth-first and breadth-first search to apply to robot swarms. We prove that our algorithms have optimal makespan $2A - 1$, where A is the area (number of pixels) of R.
- We give a filling algorithm for the multi-door case ($k \geq 1$), also based on a leader-follower strategy. The emergent behavior of the swarm is a filling strategy, called *laminar flow*, that generalizes the "left-hand-on-wall" strategy to multiple streams of robots coming in through multiple doors. An important contribution of the algorithm is the formulation of splitting and splicing strategies that maintain a large flow of robots out of the doors. We prove that the laminar flow algorithm is $O(\log(k+1))$-competitive, that is, the ratio of the makespan of our algorithm to the optimal makespan is $O(\log(k+1))$.
- We give a matching lower bound of $\Omega(\log(k+1))$ on the competitiveness of the natural class of "simple-minded" strategies that use only strictly local information.
- We report on experiments with a Java simulator that we built for swarms in grid environments. All of the algorithms in this paper are implemented in this simulator.

While our results here are given purely in terms of robots moving synchronously on discrete grids, the results apply also to dispersing swarms of robots within an arbitrary connected planar domain, in the ideal setting in which the robots have perfect motion control and synchrony. More importantly, our results may form the foundation for theoretical work on more realistic models of swarm robotics in the continuum, with asynchronous motion, sensor errors, slippage, and hardware failures.

1.2 Prior and Related Work

There has been considerable study recently of distributed control and coordination of a set of autonomous robots. Gage [14,15] has proposed the development of command and control tools for arbitrarily large swarms of microrobots and has proposed coverage paradigms in the context of robot dispersal in an environment. Payton et al. [20,21] propose the notion of "pheromone robotics" for world-embedded computation. Wagner et al. [9,27–29] developed multi-robot algorithms, directly inspired by ant behaviors, for searching and covering. Principe et al. [13,22] and Suzuki et al. [12,25,26] have studied algorithmic aspects of pattern formation in distributed autonomous robotics under various models of robots with minimal capabilities. The related flocking problem, which requires that a set of robots follow a leader while maintaining a formation, has been studied in several recent papers; see, e.g., [3,6,16]. Balch [2] has developed "Teambots", a Java-based general-purpose multi-robot simulator.

There is a vast literature on algorithms for one or several robots to explore unknown environments. The environments can be modeled as graphs (directed or undirected), mazes, or geometric domains, and the robots can have a range of computing powers; see, e.g., [1,7,8,10,11]. Mitchell [19] includes a survey of many results on exploring and navigating in geometric environments.

Dispersion algorithms have been recently studied both in the context of multi-robot coverage and sensor network deployment; see [4,5,17,30]. Methods of this type include potential fields [17,23], "artifical physics" [24], attraction-repulsion [18], and algorithms based on molecular dynamics [5]. While these studies have examined empirically the merits of heuristics, they have not addressed formally the algorithmic efficiency of the dispersion problem.

2 Preliminaries

For an arbitrary connected region $\mathcal{R}$ in the plane, we let R denote the set of *pixels* that lie entirely within $\mathcal{R}$. A *pixel* is an axis-aligned unit square, $\{(x, y) : i \leq x \leq i+1, j \leq y \leq j+1\}$, whose lower left corner is given by the integer grid point (i, j). Two pixels are *neighbors* if they share a common edge; thus, pixel (i, j) has four neighbors, $(i-1, j)$, $(i+1, j)$, $(i, j-1)$, and $(i, j+1)$. We refer to R as the *environment*, and we assume that it is connected. A subset, $D \subset R$, of the pixels are *doors*, which serve as sources of incoming robots. See Figure 1(right).

A *robot* r is said to *occupy* a pixel $p = (i, j)$ when it is located in that pixel. We assume that at most one robot can occupy any one pixel at any given time t. We let $p(r, t)$ denote the pixel occupied by robot r at time t. We let $prev(r, t) = p(r, t-1)$ denote the pixel *previously* occupied by robot r.

Time is discretized into steps $t = 0, 1, 2, 3, \ldots$. The positions of the robots at time t are given by the set $S(t)$, where $D \subseteq S(t) \subseteq R$. At time 0, there is one robot in each door: $S(0) = D$.

Each pixel is in one of three states at time t: (1) the pixel is an *obstacle* if it does not lie in R; (2) the pixel is *occupied* if it is in R and a robot occupies it; or (3) the pixel is *unoccupied* if it is in R and no robot occupies it at time t. An unoccupied pixel at time t is classified as either *previously occupied*, if it was occupied by some robot at some time prior to t, or *frontier*, if it has never been occupied.

Robots have *sensors* that detect information about the local environment. In particular, if a robot occupies pixel (i, j) at time t, then we assume that it can detect the state of each pixel within a distance r_S of (i, j), where r_S is the (fixed) *sensor radius*. Robots have no global sensor (e.g., a GPS) and no knowledge of the coordinates of the pixels they occupy.

We assume that each robot has a small finite memory that allows it to remember the sensor readings of the last $T \geq 1$ time steps; i.e., a robot knows the sensor readings it has taken at time $t - T, t - T + 1, \ldots, t - 1, t$. In particular, each robot knows which nearby pixels have been occupied recently by other robots. The algorithms in this paper have an $O(1)$-bit memory; thus, the robots have the computational power of finite automata.

We assume that robots have a limited ability to communicate with nearby robots. In particular, at any given time step a robot is able to exchange a constant-size message with any robot that lies within a prescribed *communication radius*, r_C, of the robot. The *communication graph*, $\mathcal{G}(t)$, of the swarm $S(t)$ at time t is defined to be the undirected graph whose nodes are the robots $S(t)$ and whose edges link pairs of robots that communicate. The swarm is said to be *connected* at time t if each connected component of $\mathcal{G}(t)$ contains at least one door pixel.

Robots move discretely on the grid of pixels. If at time t a robot occupies pixel (i, j) of R, and a neighboring pixel (say, $(i + 1, j)$) of R is unoccupied at time t, then the robot may take a step to the right, so that at time $t + 1$ it occupies pixel $(i + 1, j)$. Naturally, no two robots can move into the same pixel, since no pixel can be occupied by more than one robot. Thus, if two robots are occupying pixels that neighbor an empty pixel p of R, then the robots must *negotiate* which one of them will have the right of way to enter p. A simple way to handle this negotiation is to establish a priority among the four directions, e.g., north, south, east, west. The robot occupying the pixel with the highest priority with respect to p has first rights for moving to p.

The above rules on occupying and vacating pixels force a constant *delay* in the motion. Suppose that robot r occupies pixel (i, j) at time t and vacates (i, j) at time $t + 1$. No other robot r' can occupy pixel (i, j) during time $t + 1$; the earliest possible time that the pixel can be occupied is $t + 2$. In principle, we could vary these local rules. For example, we could allow robot r' to enter pixel (i, j) at timestep $t + 1$, if it enters in one direction while the current robot r leaves in the opposite direction. However, these rules would require a much higher degree of coordination among the robots, and the movement decisions of robots would have to be less local. Specifically, whether a given robot can enter one pixel may depend on whether a different robot far away leaves another pixel. Alternatively, we could build a longer delay into the movement rules, but the resulting filling algorithms would be essentially the same. Note that the delay has an immediate impact on the necessary sensor range: If we require a delay of two time steps, robots need at least a sensor range of two, in order to be able to keep track of predecessors and successors.

A similar issue arises in the way new robots enter through the door. For simplicity, we assume that a door pixel is always occupied by a robot; as soon as the robot occupying a door pixel moves to an adjacent pixel, a new robot materializes at the door. One can consider there to be an infinite supply of robots on the "other side" of a door pixel, so that a new robot steps into the door whenever the robot that was there previously moves to another pixel of R. When a robot first appears at the door, we assume that the robot points north.

We say that the robots *have filled* the region R at time t if $S(t) = R$; in this case, the robots are *well dispersed*. A *strategy* is a set of rules by which robots move, basing their movement decisions solely on the constant amount of information they sense and remember. The *makespan* of a strategy is the minimum time, t^*, until the robots have filled the region R.

3 Dispersion Strategies for a Single Door

We begin with the case of a single door pixel s, which we call the *source*. We describe strategies based on the notion of "leaders" and "followers".

A robot r at time t at position $p(r, t)$ is classified as *moving* (meaning that it is "active") or *stopped* (meaning that it no longer moves). If r is moving, it is classified as either a *leader* or a *follower*.

In our leader-follower strategies, each robot r at time t has a successor robot, $succ(r, t)$, that is following r and a predecessor robot, $pred(r, t)$, that is leading r. If robot r is at the door at time t ($p(r, t) = s$), then we define $succ(r, t)$ to be NIL; if r is a leader, then we define $pred(r, t)$ to be NIL.

At time $t = 0$ there is a robot at the door s, which is designated as moving and as a leader.

3.1 Depth-First Leader-Follower

The depth-first leader-follower strategy (DFLF) is inspired by depth-first search in a graph. Specifically, at any given time $t \in \{0, 1, 2, 3, \ldots\}$, there is exactly one leader, r, which is on pixel $p(r, t)$ and r is looking for a frontier pixel (one that has never been occupied by a robot). (Thus, necessarily the pixel from which the robot came, $prev(r,t)$, is not frontier at time t.) If there are two or more frontier pixels neighboring $p(r,t)$, the leader selects any one of them as its next destination.

If the leader has no frontier pixel next to it, it stops (its state changes to "stopped"), and it tells its successor, $succ(r)$, to assume the leadership ($succ(r)$'s state changes to "leader") at the conclusion of this time step. (The successor completes its move before taking over the leadership.) If the leader has no successor (i.e., it is a robot at the door), then the algorithm halts. (In this section, successors and (non-Nil) predecessors are invariant over time, and thus t is removed from the argument.)

Any robot r that is a follower (i.e., not the leader) simply follows its predecessor: at time t it steps next to $prev(pred(r),t)$, the pixel previously occupied by the predecessor of r. Note that once a robot stops, it never moves again. Furthermore, at any point in time there is exactly one leader.

When dealing with finite automata, we have to use particular care when implementing the leader-follower strategy. As each robot has only a finite number of states, it cannot keep track of the identities of other robots, nor carry its own "ID label". Instead, we use spatial information to pursue the immediate predecessor. Each robot r has a "heading", indicating the direction in which it is moving. The predecessor $pred(r)$ is the nearest robot in that direction. Similarly, each robot keeps track of its "tailing", the direction from which it came. Thus, the successor is the nearest visible robot in that direction. Whenever a robot is about to change its heading, it signals this turn to its immediate successor. Heading and tailing are updated according to delay and sensor-range. This is reflected in the following lemma and corollary, which are proved by induction on time.

Lemma 1. *If r is not the leader at time t, then $prev(pred(r),t)$ is an unoccupied pixel neighboring $p(r,t)$. Furthermore, the maximum (L_1) distance between r and $pred(r)$ is 2.*

Corollary 1. *If a pixel is not occupied in two consecutive time steps t and $t+1$, then it was never occupied before time t. (i.e., there are no "large gaps" of time in the occupation of a pixel.)*

A further consequence of the above lemma and corollary is that the DFLF strategy can be implemented with a sensor radius of $r_S = 2$ and a memory of $T = 1$ (i.e., each robot recalls its previous sensor reading). Now it is not hard to derive the following result:

Theorem 1. *The DFLF algorithm halts when all pixels are occupied by robots. The makespan of the DFLF algorithm is $2A-1$, where A is the number of pixels of the environment R.*

Proof. At each time step there is exactly one leader and that leader will do one of two things: (1) move to a frontier pixel, or (2) change its status to stopped and transfer leadership to its successor. (If the leader has no successor, that is, it is at the door, the algorithm halts.) Since each non-door pixel can be changed from frontier to non-frontier at most once, and each pixel can have a robot "park" itself on top of it at most once, the number of steps before the algorithm halts is at most $2A - 1$. The algorithm does not stop earlier since there is always a leader and the leader must do one of the two actions ((1) or (2)) above. □

Note that the factor 2 is a result of using a delay of 2. It is easy to see that this is best possible.

The total distance traveled by all robots in a filling using DFLF is at most A^2, since each robot steps at most A. We note that there are examples in which *any* dispersion strategy will require $\Omega(A^2)$ total travel. One trivial example is a $1 \times A$ rectangle, with the door pixel at one end. A trivial lower bound on the total travel for any given instance is the sum of the distances from each pixel of R to the door.

DFLF may use substantially more total travel than an optimal (minimum makespan) strategy that minimizes total travel; consider, for example, a square with side lengths $A^{1/2}$, for which DFLF requires total travel of A^2 while an optimal strategy achieves total travel $O(A^{3/2})$.

3.2 Breadth-First Leader-Follower

We now describe an alternative dispersion algorithm , the breadth-first leader-follower (BFLF) strategy. BFLF often has advantages over the DFLF strategy in terms of other metrics of performance (total travel of the robots, maximum travel of any one robot, total relative distance, etc), while still being optimal in terms of makespan ($2A - 1$).

The BFLF strategy is substantially more complex than the DFLF, as it is not trivial to implement breadth-first search with local decision rules on (finite-automata) robots; indeed, our strategy does not perform breadth-first search, but it "approximates" breadth-first search sufficiently well to have some similar properties.

As before, a robot can be in a "moving" state or a "stopped" state; however, we now introduce a third state, the *waiting* state, which is an interim state when a robot pauses temporarily and waits to be able to move. Once a robot is in a stopped state, it never moves again. In the BFLF strategy we can have multiple leaders, while in DFLF we have only one. As described for DFLF, we use a limited amount of spatial information to keep track of successors and predecessors. Other local rules are more complex and are described in the following.

Initially, there is one leader robot — the robot at the door, s. When a robot r materializes at the door, it chooses to follow the previous robot that left the door. A leader always attempts to go to a neighboring frontier pixel, but makes sure it does not stray too far from its successor (which follows it).

If there are no neighboring frontier pixels, the leader waits for the immediate successor to catch up. As soon as the immediate successor catches up, the leader becomes stopped and passes on the leadership role to the immediate successor. If the leader r at pixel u has a choice among neighboring frontier pixels, it picks any one of them to be its heading. If other frontier pixels adjacent to u remain unclaimed by other robots following different branches of dispersion, then the follower r' of r will choose one of these pixels as its heading when it arrives at u; if there remains a frontier pixel adjacent to u, then r''s follower r'' chooses this pixel as its heading when it arrives at u. Here, r' and r'' are relabeled "leaders" and pixel u is marked as a branching point.

A robot r is only allowed to move at time t, if it satisfies the *Movement Criterion* (MC): A robot currently occupies $prev(r,t)$ (its "parent" position) or a robot is currently *heading for* $prev(r,t)$. If the MC is satisfied, the robot r moves to its heading; otherwise, it remains at its current position, $p(r,t)$ but is still *heading for* its target pixel. As we will see below, a robot r is never blocked by its immediate predecessor *except* for the time step at which r first appears at the door or a time when its immediate predecessor is a leader with nowhere to go, meaning that the leadership will be passed to r.

The BFLF strategy tries to create as many paths as possible at all times. The visited pixels form a *tree* that guides the directions that robots move; thus, pixels have *parents* and *grandparents*. At branches in the tree (pixels with ≥ 2 children), robots alternate the direction that they travel. Specifically, when a robot r leaves a pixel at time t, it tells its immediate follower $r' = succ(r,t)$ what r''s immediate destination should be. Branching enables robots arriving at the same pixel at different times to go in different directions, thereby balancing the flow.

Lemma 2. *If not all pixels are occupied, then some robot can move.*

The following structural lemma shows that the BFLF algorithm is non-blocking, i.e., a robot r is never "delayed" by its predecessors, only by its successors; i.e., only the MC holds a robot back, not temporary blockages downstream. Correctness follows by induction on the height of the tree.

Lemma 3. *If a robot is at pixel v and a robot is at pixel u, the non-root parent of v, then the robot at v is stopped.*

Corollary 2. *A robot moves from the root every other time step, until all pixels are occupied. I.e., if a robot did not pop up at the root at time t, then one must pop up at time $t+1$, unless all pixels (including the root) are occupied.*

Lemma 4. *The robots maintain a communication distance of 3, meaning that the algorithm works for $r_S \geq 3$. The maximum L_1-distance from a robot to a follower is 3. Furthermore, at least one of the following pixels is occupied when a robot is at pixel u: (1) the parent of u, (2) the grandparent of u, or (3) the uncle of u.*

Putting all the claims together and using similar reasoning as for Theorem 1, we get the same kind of bound on the makespan:

Theorem 2. *The BFLF algorithm halts when all pixels are occupied by robots. The makespan of the BFLF algorithm is $2A - 1$, where A is the number of pixels of the environment R.*

3.3 Experimental Comparison

While the DFLF and BFLF strategies both have optimal makespan, they differ with respect to other performance metrics. We have implemented in Java a simulator for testing our dispersion strategies, while measuring various performance measures, including: (1) the *depth* of the tree of all paths from the source (i.e., the length of the longest path of a robot); (2) the average distance traveled by a robot during the dispersion; and (3) the average of the ratios, $\rho_i = l_i/d_i$, of the (L_1) distance l_i from the door to robot i's final location to the total distance d_i traveled by robot i during the dispersion.

We have compared the DFLF and BFLF strategies for the case of a single door, in a variety of different environments. Our results show that the average depth of the tree grows substantially more rapidly, as a function of the number of pixels, for DFLF than it does for BFLF; e.g., for an n-by-n square grid, BFLF tree depth grows linearly in n while DFLF depth grows close to quadratically in n. Similarly, the average distance traveled by a robot using DFLF grows more rapidly than using BFLF. The average of the ratios $\rho_i = l_i/d_i$ decreases with increasing problem size for DFLF, while it increases for BFLF. Details of the experimental results are deferred to the full paper.

4 Dispersion with Multiple Doors

We now consider the case in which there are $k \geq 1$ door pixels. In order to achieve rapid filling, the objective is to maintain a flow of robots out of as many doors as possible. The difficulty is that one robot chain out of one door pixel may unnecessarily block the flow of robots out of other door pixels. (See the introduction and Figure 1(left) for examples.)

4.1 Laminar Flow Leader-Follower (LFLF) Algorithm

This section describes the *Laminar Flow Leader-Follower (LFLF)* algorithm. The LFLF algorithm maintains rapid flows by careful use of cutting and splicing. The emergent behavior of the swarm is a left-hand-on-wall strategy. The behavior of the algorithm is complex because bottlenecks in the environment may force the flow to divide into smaller flows that fill different regions and may merge and split.

In the LFLF algorithm there are k door pixels, $s_1, s_2, \cdots, s_k$. Note that LFLF matches DFLF algorithm if $k = 1$. As in the $k = 1$ case, our dispersion strategies are based on leaders and followers. A robot r at time t at position $p(r,t)$ is classified as *active* or *inactive* (stopped). If r is active, it is either a *leader* or a *follower*. However, unlike for $k=1$, robots *revert* between leaders and followers. (In contrast, for $k=1$ a robot changes roles from follower to leader but only relinquishes the leader role by stopping and becoming inactive.) Furthermore, in DFLF once a robot stops, it becomes inactive. In contrast, in LFLF a robot may temporarily stop while still remaining active.

In LFLF (as in DFLF), robots have successors and predecessors, but in LFLF a robot's immediate successor or predecessor may change over time. Thus, each robot r at time t has a predecessor robot denoted *robot-pred*(r,t) and a successor robot denoted *robot-succ*(r,t). If robot r is at a door pixel at time t ($p(r,t)=s_i$, for $i \in [1,k]$), then *robot-succ*(r) = NIL; if r is a leader at time t, then *robot-pred*(r) = NIL.

Here we further refine the roles of the pixels. A pixel can be: (1) an *obstacle*, (2) a *frontier* pixel (a pixel that has never been occupied), (3) an *inactive* pixel, that is, an (occupied) pixel that hosts an inactive robot, (4) an *active* pixel, (which may be occupied or not). If a pixel is unoccupied but previously occupied, then it is always active; if a pixel is occupied, then it is active if and only if the robot it hosts is active.

Note that door pixels may be active or inactive, as described later. Active pixels play the role of transporting robots through the domain whereas inactive pixels are essentially obstacles. Analogous to the definition of predecessor and successor robots, each active pixel has predecessor and successor pixels. In particular, pixel u at time step t has successor pixel *pixel-succ*(u,t) and predecessor pixel *pixel-pred*(u,t). For door pixels s_i at any time $t \geq 0$ *pixel-succ*(u,t) = NIL.

A *leader pixel* u at time t has *pixel-pred*(u,t) = NIL. As we will see, if a leader pixel contains a robot, then the robot is a leader, however the leader robot may not be on a leader pixel and the leader pixel may not contain robots. We call a chain of predecessor pixels starting from the door pixel s_i and ending at a leader pixel, a *flow line*.

We now define formally the *left side* and *right side* of the flow. To do so, we first present some intermediate definitions. Consider a pixel v at time t, and let u = *pixel-succ*(v,t) and w = *pixel-pred*(v,t). Let the *incoming direction* for v at time t be the vector $\overrightarrow{vu}$ and let the *outgoing direction* for v at time t be the vector $\overrightarrow{vw}$.

Consider the (vertical or horizontal) neighboring pixels, when we start from the incoming direction $\overrightarrow{vu}$ and rotate *clockwise* until reaching the outgoing direction $\overrightarrow{vw}$. Those neighboring pixels at intermediate (vertical or horizontal) orientations are defined to be on the *left-hand side* of v time t. Thus, the left-hand side may contain 0, 1, or 2 pixels. Any neighboring pixel that is *not* on the left-hand side, is on the *right-hand side*; this includes the pixels at the incoming and outgoing directions.

Note that leader pixels do not have outgoing edges, invalidating this definition. For a leader pixel, we define all neighboring pixels (except the successor pixel) to be on the left-hand side.

The *left side of a flow line* is the union of the left sides of the non-leader pixels in the flow line.

The essential idea of the LFLF algorithm is to treat the left side of the flow line completely differently from the right side of the flow line. Specifically, we splice into the left-hand side but we cannot splice into the right.

4.2 Splicing and the LFLF Algorithm

Suppose that at time t a leader robot r_a reaches a pixel $u = p(r_a, t)$. If u has an (unoccupied) predecessor pixel *pixel-pred*(u, t), then robot r_a moves to this pixel. (Since r_a is a leader robot, it will never have an occupied predecessor pixel.)

Otherwise, pixel $u = p(r_a, t)$ is a leader pixel. If $u = p(r_a, t)$ has no neighboring frontier pixels, then robot r_a tries to pass on the leadership. Robot r_a first looks for a neighboring active pixel v. If no such v exists, then r_a becomes inactive and passes the leadership to its successor robot *robot-succ*(r_a, t). When r_a becomes inactive, pixel u also becomes inactive and its successor pixel becomes the new leader pixel.

If robot r_a looks for a neighboring active pixel v, and such a v exists, then r_a checks whether $u = p(r_a, t)$ is a left-hand neighbor of v. If this is not the case for any possible v, then again r_a becomes inactive and passes the leadership to its successor robot *robot-succ*(r_a, t). As before, when r_a becomes inactive, pixel u also becomes inactive and its successor pixel becomes the new leader pixel.

If there exists such a v, robot r_a chooses the first such v sweeping *clockwise*, starting from the incoming direction. We redefine the predecessor pixel *pixel-pred*$(u, t+1) := v$ and thus successor pixel *pixel-succ*$(v, t+1) := u$. Thus, the previous successor pixel of v becomes a new leader pixel. (Note that this new leader pixel may or may not have a robot on it.) Now r_a identifies its new leader; it follows predecessor pixels starting at v until it finds a robot r_b and sets *robot-pred*$(r_a, t+1) = r_b$ and *robot-succ*$(r_b, t+1) = r_a$. Robot r_a then passes the leadership to $r_c =$ *robot-succ*(r_b, t).

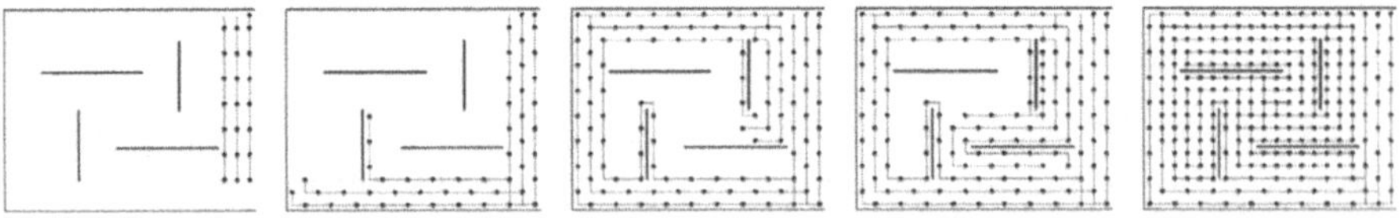

Fig. 2. The laminar flow leader-follower algorithm in action. There are three door pixels in the upper right corner and four line segment obstacles arranged in the middle of the room.

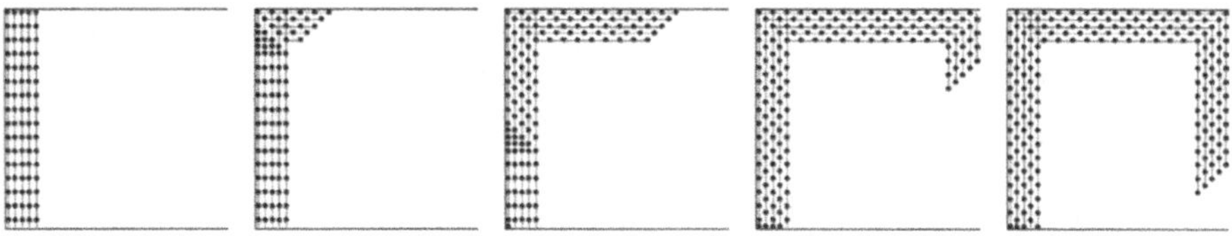

Fig. 3. The LFLF algorithm: There are six door pixels in the lower left corner. Splicing enables the flow to turn a corner.

4.3 Correctness of LFLF

We now show that LFLF always fills an environment. During LFLF, we claim that the following invariant is maintained: *The left side of a flow line consists of visited pixels and obstacles.* This follows from the fact that, whenever a robot enters a pixel, it turns as far left as it can (left-hand-on-the-wall). The invariant implies the following lemma, which we use to show that the visited regions propagate out from obstacle pixels:

Lemma 5. *The union of boundary points separates the visited region from obstacles and the right side of flows. In other words, the unvisited region only touches a flow from its right.*

Another useful property of flow lines is the following:

Lemma 6. *Flows do not self-intersect.*

Lemma 7. *Leadership passing does not change the total number of leader robots. Thus, the number of leader robots at any time step t is equal to the number of active doors.*

Theorem 3. *The robots always fill environment R.*

Proof. When a robot becomes inactive, it is a leader and its surrounding pixels must all not be frontier, for otherwise r would advance to such a pixel and still be active. Therefore, if an active pixel u is next to a frontier pixel v, any robot advancing into u will not become inactive and some flow lines can advance into the empty region so it will be filled. □

In the next sections we analyze the makespan for this filling strategy.

4.4 Competitive Analysis for Online Strategies

For any strategy $\mathcal{S}$, let $n_{\mathcal{S}}(t)$ be the number of robots that enter through the doors during timestep t; we define $n_{\mathcal{S}}(0) = k$.

Note that the inherent delay of 2 in robot movement means that $n_{\mathcal{S}}(t)$ is a highly erratic function, so that in one timestep k robots might enter the doors, implying that in the next timestep no robots enter the doors.

In LFLF additional delays are caused by splicings, and therefore we cannot even say that a robot enters an active door pixel every other timestep. To understand why, consider Figure 3. When one flow line splices into a different flow line, a robot may get delayed by one additional time step in order to obtain the required separation between robots. If many splicing are happening next to each other, this can cause an accumulated delay of up to k time units, which causes a "wave" of packed robots that propagates back towards the door pixels. Thus, a door pixel might temporarily hold off ejecting robots for many units without the door pixel becoming inactive. Whenever an active door pixel does not eject a robot we say there is a *door-pixel delay*.

Fortunately, the following lemma bounds the number of splicings (and also the total number of door-pixel delays).

Lemma 8. *A pixel can only be used to initiate $O(1)$ splicings.*

Theorem 4. *The LFLF strategy is $O(\log(k+1))$-competitive.*

Proof. Consider the times $\{t_i\}_{i=0}^{\lceil\log(k+1)\rceil}$, where t_i is the latest timestep during which at least $k/2^i$ door pixels are active. We call times $t_0, t_1, t_2, \ldots, t_{\lceil\log(k+1)\rceil}$ *significant events*. Consider the time interval from t_i to t_{i+1}, when at least $k/2^{i+1}$ door pixels are active. Thus, the total number of robots leaving the door plus door-pixel delays during this interval is at least $(k/2^{i+1})(t_{i+1}-t_i)$.

Now consider an optimal strategy $\mathcal{S}^*$, and consider the cut χ associated with the significant event t_i of LFLF. Cut χ is the boundary between the occupied and frontier pixels and has size at most $k/2^i$. A lower bound on the makespan of any strategy is the area "behind" χ (i.e., the side of χ not containing the doors) divided by the length of χ plus the amount of time required to send the flow from the door to reach χ. Thus, the makespan, OPT, of $\mathcal{S}^*$ satisfies

$$\text{OPT} \geq \frac{k/2^{i+1}(t_{i+1}-t_i)}{k/2^i} = (t_{i+1}-t_i)/2.$$

Summing over all significant events we obtain that the total filling time of LFLF is $O(\text{OPT} \cdot \log(k+1))$. □

If we disregard the issue of delays caused by pixels having to be fully vacated before being re-entered, we get a whole class called *sensible strategies*, defined by the following two conditions:

(1) $n_{\mathcal{S}}(t)$ is non-increasing in t; and
(2) If $n_{\mathcal{S}}(t) > n_{\mathcal{S}}(t+1)$, then at time t the number of occupied pixels bordering on a frontier pixel is $n_{\mathcal{S}}(t+1)$.

The above proof also implies the following; quite clearly, this result still holds if we extend the notion of sensible strategies to allow for some delays, as long as these only add a constant factor to the filling times.

Theorem 5. *Any sensible strategy is $O(\log(\frac{k}{b}+1))$-competitive.*

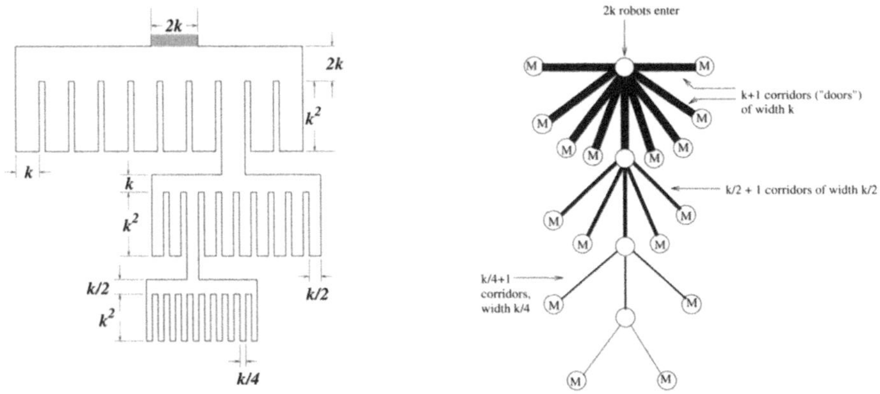

Fig. 4. Left: An environment that is hard for both depth-first and breadth-first filling strategies. This example shows that LFLF and related strategies are $\Omega(\log(k+1))$ competitive. Right: A scenario in which any simple-minded strategy is $\Omega(\log(k+1))$ competitive.

4.5 Competitive Analysis of Simple-Minded Strategies

Under certain natural assumptions that limit the power of strategies, it is possible to give matching lower bounds on the competitiveness. To give some intuition, consider first the class of examples shown in Figure 4(left), which shows that the laminar flow algorithm may take as much as $\Omega(\log(k+1))$ times OPT.

Theorem 6. *The LFLF algorithm is* $\Omega(\log(k+1))$*-competitive.*

This lower bound can be generalized to a much larger class of strategies. We say that strategy $\mathcal{S}$ is *simple-minded* if the following condition holds: *There is a constant* C*, such that we can only tell the number* A' *of pixels in a region* $R' \subseteq R$ *when a set* $R'' \subseteq R$ *has been visited that has all pixels in* R' *within* L_1 *distance* C*.*

Considering simple-minded strategies is quite natural in the context of robot swarms: If we assume we only "know" a region R' after each pixel has been seen by one of the robots, then the constant C corresponds precisely to the sensor range. The example of Figure 4(left) can be generalized as shown schematically in Figure 4(right), which forms the basis for the proof of the following theorem:

Theorem 7. *Any simple-minded strategy is* $\Omega(\log(k+1))$*-competitive.*

A particular subclass of simple-minded strategies is one that arises quite naturally in exploration problems: We say that a strategy is *conservative* if any pixel that has been "discovered", i.e., come within sensor range of a robot, must stay within sensor range of some robot. For a constant-size sensor range, it is easy to prove the following statement, which has been claimed for a long time in a different context:

Proposition 1. *All conservative strategies are simple-minded.*

As the LFLF strategy is conservative by design, we conclude that the lower bound of Theorem 7 applies to LFLF.

Acknowledgements. We thank N. Jovanovic and M. Sztainberg for contributions to this research. This research was partially supported by HRL Labs (a DARPA subcontract), DFG grant FE 407/6-1, Honda Fundamental Research Labs, NASA (NAG2-1325), NSF (CCR-0098172, EIA-0112849, CCR-0208670), U.S.-Israel BSF, and Sandia National Labs.

References

1. B. Awerbuch, M. Betke, R. L. Rivest, and M. Singh. Piecemeal graph exploration by a mobile robot. In *Proc. 8th Annual ACM Conference on Computational Learning Theory*, pages 321–328, 1995.
2. T. Balch. http://www.teambots.org.
3. T. Balch and R. C. Arkin. Behavior-based formation control for multi-robot teams. *IEEE Transactions on Robotics and Automation*, 14(6):926–939, 1998.
4. M. A. Batalin and G. S. Sukhatme. Multi-robot dynamic coverage of a planar bounded environment. Technical report, Robotic Embedded Systems Laboratory, University of Southern California, 2002.
5. M. A. Batalin and G. S. Sukhatme. Spreading out: A local approach to multi-robot coverage. In H. Asama, T. Arai, T. Fukuda, and T. Hasegawa, editors, *Proc. 6th International Symposium on Distributed Autonomous Robotic Systems*, Fukuoka, Japan, 2002. Springer-Verlag.
6. O. B. Bayazit, J.-M. Lien, and N. M. Amato. Better flocking behaviors using rule-based roadmaps. In *Proc. 5th International Workshop on Algorithmic Foundations of Robotics*, 2002.
7. M. A. Bender, A. Fernández, D. Ron, A. Sahai, and S. Vadhan. The power of a pebble: Exploring and mapping directed graphs. *Information and Computation*, 176:1–21, 2002.
8. M. A. Bender and D. Slonim. The power of team exploration: Two robots can learn unlabeled directed graphs. In *Proc. 35th Annual Symposium on Foundations of Computer Science*, pages 75–85, 1994.
9. A. M. Bruckstein, C. L. Mallows, and I. A. Wagner. Probabilistic pursuits on the grid. *American Mathematical Monthly*, 104(4):323–343, 1997.
10. X. Deng and C. H. Papadimitriou. Exploring an unknown graph. In *Proc. 31st Annual Symposium on Foundations of Computer Science*, pages 356–361, 1990.
11. G. Dudek, M. Jenkin, E. Milios, and D. Wilkes. Map validation and robot self-location in a graph-like world. *Robotics and Autonomous Systems*, 22(2):159–178, 1997.
12. A. Dumitrescu, I. Suzuki, and M. Yamashita. High speed formations of reconfigurable modular robotic systems. In *Proc. IEEE International Conference on Robotics and Automation (ICRA '02)*, pages 123–128, 2002.

13. P. Flocchini, G. Prencipe, N. Santoro, and P. Widmayer. Distributed coordination of a set of autonomous mobile robots. In *Proc. IEEE Intelligent Vehicle Symposium (IV 2000)*, pages 480–485, 2000.
14. D. Gage. Many-robot systems. `http://www.spawar.navy.mil/robots/research/manyrobo/manyrobo.html`.
15. D. Gage. Sensor abstractions to support many-robot systems. In *Proc. of SPIE Mobile Robots VII, Boston MA, 1992, Vol. 1831*, pages 235–246, 1992.
16. V. Gervasi and G. Prencipe. Flocking by a set of autonomous mobile robots. Technical Report TR-01-24, University of Pisa, Oct. 2001.
17. A. Howard, M. J. Mataric, and G. S. Sukhatme. Mobile sensor network deployment using potential fields: A distributed scalable solution to the area coverage problem. In H. Asama, T. Arai, T. Fukuda, and T. Hasegawa, editors, *Proc. 6th International Symposium on Distributed Autonomous Robotic Systems*, Fukuoka, Japan, 2002. Springer-Verlag.
18. N. Jovanovic, T-R. Hsiang, and M. Sztainberg. T-R. Hsiang, N. Jovanovic, and M. Sztainberg. Experimental Comparison of Dispersion Protocols for Robot Swarms. Technical Report, Stony Brook University, 2002.
19. J. S. B. Mitchell, Geometric shortest paths and network optimization, In *Handbook of Computational Geometry* (J.-R. Sack and J. Urrutia, editors), pages 633–701. Elsevier Science Publishers B.V. North-Holland, Amsterdam, 2000.
20. D. Payton, R. Estkowski, M. Howard. Progress in Pheromone Robotics. In *Proc. 7th International Conference on Intelligent Autonomous Systems*, 2002.
21. D. Payton, M. Daily, R. Estkowski, M. Howard, C. Lee. Pheromone Robotics. *Autonomous Robots*, Vol. 11, No. 3, pages 319-324, 2001.
22. G. Prencipe. Distributed Coordination of a Set of Autonomous Mobile Robots. Ph.D. thesis Dipartimento di Informatica, Università degli Studi di Pisa, 2002. `http://www.di.unipi.it/phd/tesi/tesi_2002/PhDthesis_Prencipe.ps.gz`.
23. J. H. Reif and H. Wang. Social potential fields: A distributed behavioral control for autonomous robots. *Robotics and Autonomous Systems*, 27(3):171–194, 1999.
24. W. Spears and D. Gordon. Using artificial physics to control agents. In *Proc. IEEE Internat. Conf. on Information, Intelligence, and Systems*, 1999.
25. K. Sugihara and I. Suzuki. Distributed algorithms for formation of geometric patterns with many mobile robots. *Journal of Robotic Systems*, 13(3):127–139, 1996.
26. I. Suzuki and M. Yamashita. Distributed anonymous mobile robots: Formation of geometric patterns. *SIAM Journal on Computing*, 28(4):1347–1363, 1999.
27. I. Wagner and A. Bruckstein. Cooperative cleaners: a study in ant robotics. Technical Report CIS9512, Technion, 1995.
28. I. Wagner, M. Lindenbaum, and A. Bruckstein. Distributed covering by ant-robots using evaporating traces. *IEEE Transactions on Robotics and Automation*, 15(5):918–933, 1999.
29. I. A. Wagner, M. Lindenbaum, and A. M. Bruckstein. Efficiently searching a graph by a smell-oriented vertex process. *Annals of Mathematics and Artificial Intelligence*, 24(1-4):211–223, 1998.
30. A. Winfield. Distributed sensing and data collection via broken ad hoc wireless connected networks of mobile robots. In L.E. Parker, G.W. Bekey, and J. Barhen, editors, *Distributed Autonomous Robotic Systems*, Vol. 4, pp. 273–282, 2000. Springer-Verlag.

Better Group Behaviors Using Rule-Based Roadmaps

O. Burchan Bayazit, Jyh-Ming Lien, and Nancy M. Amato

Texas A&M University, College Station TX 77843, USA

Abstract. While techniques exist for simulating group behaviors, these methods usually only provide simplistic navigation and planning capabilities. In this work, we explore the benefits of integrating roadmap-based path planning methods with flocking techniques. We show how group behaviors such as exploring can be facilitated by using dynamic roadmaps (e.g., modifying edge weights) as an implicit means of communication between flock members. Extending ideas from cognitive modeling, we embed behavior rules in individual flock members and in the roadmap. These behavior rules enable the flock members to modify their actions based on their current location and state. We propose new techniques for three distinct group behaviors: homing, exploring (covering and goal searching) and passing through narrow areas. Animations of these behaviors can be viewed at `http://parasol.tamu.edu/dsmft`.

1 Introduction

Coordinating the movement of a group of robots plays an important role in robotics. Although techniques to achieve coordinated movements have attracted the attention of many researchers, most research has focused on techniques for modeling individual behavior of flock members inspired by Reynolds' *boids* [13]. Boids exhibit so-called *emergent behavior* in which characters only react to immediate events. Although initially developed for computer animation, this strategy has been used to control flocks of robots. Usually, the flock shows homogeneous behavior. Yu and Terzopoulos extended Reynolds' flocking system to create more realistic, self-animating characters. They use biomechanical modeling with a behavioral finite state machine, the intention generator [18], and propose the idea of *cognitive modeling* [5], which controls how characters gather knowledge and how they act. They develop the *cognitive modeling language, CML*, which eases an animator's task of specifying character behaviors and also enables heterogeneous behaviors to be included in flock members. Nevertheless, the ability to generate animations in complex environments is poor even if more elaborate behaviors are encoded in the CML. For example, in Figure 1(a), an alien (center) is trying to escape from a predator hiding in one of the rooms. Without memory, she cannot remember which room contains the predator she just encountered.

J.-D. Boissonnat et al. (Eds.): Algorithmic Foundations of Robotics V, STAR 7, pp. 95–111, 2004.

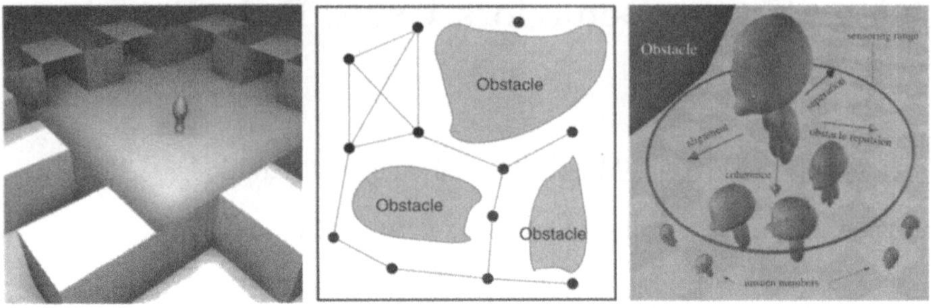

(a) (b) (c)

Fig. 1. (a) A character with emergent behaviors has difficulty navigating in symmetric environments. (b) A roadmap. (c) Basic flocking dynamics with obstacle avoidance.

Although, they can be coupled with simple methods for guiding global flock movement [14], existing methods have difficulty if complex navigation is required, such as in cities, through crowded rooms, or over rough terrain. In contrast, path planning algorithms developed in the robotics community are capable of navigation in complex environments [7]. In particular we note the *roadmap*-based methods which construct, usually during preprocessing, a network of representative feasible paths in the environment. While roadmap methods can efficiently support complex navigation, they have generally not been customized to support coordinated group behavior.

In this work, we explore the benefits of integrating roadmap-based path planning techniques with flocking techniques. We extend ideas from cognitive modeling [5], and embed behavior rules in individual flock members and in the nodes and edges of the roadmap. We find that the global information provided by our *rule-based roadmaps* improves the behavior of autonomous characters and enables more sophisticated group behaviors than are possible using traditional flocking methods. Key features of our approach include:

- The roadmap provides a convenient abstract representation of global environmental information.
- Dynamic roadmaps (e.g., modifiable node and edge weights) enable communication between agents.
- Associating rules with roadmap components enables local customization of behaviors.

We illustrate our approach by proposing new techniques for four behaviors: homing, covering, goal searching, and traversing narrow passages. Our techniques can be applied to an entire flock, to individual flock members, or to an external agent that may influence the flock (e.g., a sheep dog) [2].

To our knowledge, this is the first time global maps have been used to support group behavior. However, Parker's work supports our use of global information to enable sophisticated group behaviors [12]. In particular, she concluded that global knowledge should be used to provide general guidance for the longer-term actions of an agent, whereas local knowledge influences the more short-term, reactive actions. She also suggested that local information should be used to ground global knowledge in the current situation. This allows agents to remain focused on the overall goals of their group while reacting to the dynamics of their current situations

2 Related Work

Reynolds' influential flocking simulation [13] showed that flocking is a dramatic example of emergent behavior in which global behavior arises from the interaction of simple local rules. Each individual member of the flock (boid), has a simple rule set stating that it should move with its neighbors. This concept has been used successfully by researchers both in computer graphics and robotics. Tu and Terzopoulos [18] used flocking behaviors with intention generators to simulate a school of fish. They also demonstrated shepherding behavior in which a T-Rex herds raptors out of its territory.

A number of related methods for achieving group behaviors have been proposed [1,3,4,10,11,17,16,20]. An interesting approach by Vaughan et al. [19] used a robotic external agent to steer a flock of real geese.

Although there is little research on path planning for flocks, many methods have been proposed for planning for multiple robots. These methods can be characterized as centralized or decoupled. Centralized methods consider all robots as one entity, while decoupled methods first find a path for each robot independently and then resolve conflicts. In work from Li et al. [8], each group of crowds is guided by a leader and the paths of the leaders are generated using a decoupled approach.

The observation of the behavior of ant colonies has inspired the ant colony optimization (ACO) meta-heuristic for discrete optimization. Dorigo et al. [9] exploit this ant-like behavior to optimize solutions for several NP-Complete problems. In our work, the flock's ability to explore comes from using an ACO-like approach to adaptively adjust roadmap edge weights

3 Roadmap Methods and Flocking Systems

A roadmap is a connectivity graph encoding representative feasible paths in the environment. Each node of the graph is a configuration of the robot that satisfies certain feasibility requirements (e.g., collision-free). The connections between nodes of the roadmap graph represent feasible (sub)paths. Several algorithms have been proposed to automatically generate roadmaps. Figure 1(b) shows an example of an automatically generated roadmap. We call a roadmap *adaptive* if its edge weights are updated according to new information gathered by flock members during simulation.

Basic flocking systems [13] model simple group behavior by providing individual members with simple rules that implement separation (to avoid collision with nearby neighbors), alignment (to move in the same direction as its neighbors) and coherence (to stay close to neighbors) maneuvers based on the positions and velocities of the flockmates inside the sensing range. Constraints are satisfied by generating forces for each rule and applying an integrated force to change the state of the flock member. In the presence of obstacles, force is also generated to push the flock member away. This basic system can be seen in Figure 1(c). More complicated behavior is usually simulated by adding other forces. As we will see in Section 6, flocking members sometimes are trapped in local minima of the potential field of these forces (see Figure 2(a)).

Flocking systems can be classified into three categories based on agent knowledge level (Figure 2(b)). In the basic system, an agent can only access information within its sensing range. On the other extreme, agents have complete information about the current state of the environment and can plan their motion. However, it is usually expensive to maintain complete information, and, more importantly, realistic simulation should not assume creatures always have such information and abilities. Like the basic system, agents in our roadmap-based system can only access local information. The global information (roadmap) is only locally available. This means each flock member can only access the environment and update the portion of the roadmap inside its sensing range. The agent's reasoning capabilities enable the agent to decide its next step based only on the local information. In addition, the agent is able to remember the places it has visited. With these simple mechanisms, we are able to simulate complex behaviors that are hard (expensive) to accomplish using the basic (complete information) system.

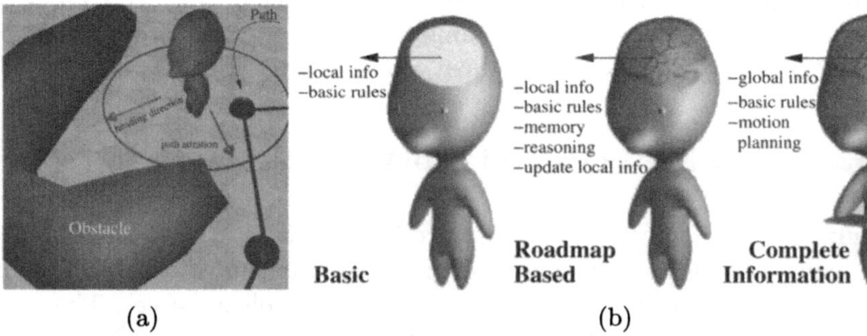

(a) (b)

Fig. 2. (a) This agent will be trapped in a local minimum of the path's attractive force. (b)Comparison between agents of different information and behaviors.

4 Rule-Based Roadmaps

The roadmaps encode global navigation information, but they usually do not include 'local' information for influencing behavior differently in different regions. We propose providing such information in the form of *behavior rules* which can be stored in the roadmap. When in the region of influence of a roadmap node (or edge), the agent's behavior is governed by the rule, if any, associated with that node. If no rule is specified, a default behavior will be followed. The embedded rules in the roadmap are analogous to the traffic signs on the streets which control cars.

Rules may be as simple as "GO TO NEXT NODE IN YOUR PATH" or as complex as "WAIT FOR THE OTHERS, SELECT A LEADER, FOLLOW THE LEADER." Rules are assigned to nodes automatically in our implementation. For example, narrow passage behaviors (rules) are assigned to nodes on path segments that have small clearance. This process can be done manually as well, enabling customization.

4.1 System Overview

The general components of our system are one or more AGENTs and an ENVIRONMENT. Environment-aware AGENT behavior is specified in RULEs.

System Components Each AGENT is a flock member, and it has several attributes associated with it, such as `position`, `goal`, `role`, and `path`. These are stored in variables called `AgentVars`. When an AGENT reaches a node, it starts executing that node's rule script. These rules might be executed only once, or at each time step that the AGENT is in the region of influence of that node. AGENTs may also have dynamic (temporary) variables which are instantiated when rules are executed.

The ENVIRONMENT stores global information about the environment in the roadmap which is shared by all AGENTs. The roadmap nodes store NODE attributes, such as node `position` and `edges` and functions to assist navigation (e.g., returning the minimum weight edge leaving the node) in `NodeVars`. The `NodeVars` and functions are shared by all AGENTs running the NODE's script (when they are within the NODE's influence area). A NODE may also have dynamic `NodeVars` instantiated by the NODE's script.

RULEs describe AGENT behavior. They are associated with regions of the ENVIRONMENT and are stored in roadmap NODEs. They can contain variables, assignments, control statements and loops. TRIGGERs are special rules which are invoked to initialize and finalize variables governing behaviors. For example, in the narrow passage behavior, the first AGENT to reach the passage is selected as the leader by the TRIGGER for that narrow passage. TRIGGERs may be run only once, the first time an AGENT reaches the associated NODE, or they may run continuously until all AGENTs have left the region.

System Control Structure Our system is a single program. There is a main loop in the program which, at each time step, assigns AGENTs to NODEs, runs any necessary TRIGGERs, and then calls individual agent rule scripts. Since all the executions are sequential, the system is thread-safe, i.e., an AGENT's rule script won't be executed while another AGENT's rule script is running. For parallel or distributed implementation, synchronization issues have to be considered.

Rule scripts are stored with roadmap nodes, and are executed by AGENTs that are assigned to those nodes by the control loop (i.e., that are in the sphere of influence of those nodes). The structure of a rule script is as follows:

```
Definition of NodeVars
Definition of AgentVars
Definition of TRIGGERS
RULES....
```

In our scripting language, the predefined variable `agent` refers to the AGENT running the script. All AGENTs have `agent.position` and `agent.goal` attributes providing their current position and goal position, and an `agent.role` attribute defining their current role, which is *boid* by default. The AGENT attribute `agent.path` has functions such as `currentPosition,nextPosition, index, push, pop`, etc. The predefined variable `node` in the script represents the NODE with which that script is associated. Each edge belongs to an `edge` structure made up of `start, end` and `weight` values. AGENTs can access the roadmap (or ENVIRONMENT) information through the predefined `node` variable or by calling functions which return information available in the `node` or `edge` structures, e.g., `node.edge(i)` which returns the ith edge.

5 Roadmap-Based Group Behavior

In this section, we show how roadmap-based techniques can be used to achieve different behaviors. We consider several behaviors: *homing*, *exploring (covering and goal searching)* and *traversing narrow passages*. The first two behaviors influence *where* the flock goes – reaching a pre-defined goal (homing), attempting to cover (visit) all reachable areas of the environment (covering) or search for a goal whose location is not known. The narrow passage behavior influences *how* the flock members position themselves relative to each other when they move through the passage.

5.1 Homing Behavior

Homing behavior is usually simulated by adding an attractive force toward the goal [6]. However, this method may easily be trapped in a local minimum even in a simple environment. A method commonly used in computer games requiring motion of a group of objects is a grid-based A^* search [15]. In this approach, the environment is discretized to small grid cells and the search for the flock's path is based on expanding toward the most promising neighbor of already visited positions. Although A^* search finds shortest paths and it is usually fairly fast, it does have drawbacks. Of particular note here is the necessity of finding a completely new path for each new goal which reduces efficiency and increases the computation time for complex environments.

In contrast, roadmap-based path planning methods work on a global scale and once the roadmap is generated, finding new paths is fast and efficient. In our approach, we use MAPRM [21] to find a path for the flock. One of the advantages of MAPRM is that the paths we find tend to have large clearances from obstacles. Once a path is found, individual flock members follow the path. The path is partitioned into a set of subpaths (identified by subgoals) based on the individual flock member's sensor range. Each member keeps track of subgoals and as soon as a subgoal comes within sensory range the next subgoal becomes the steering direction for the global goal.

With other interacting forces from neighboring flock members and obstacles, steering toward the subgoal has the lowest priority, so individual members still move together while moving toward goal. This results in a flocking toward the goal and avoids getting trapped in local minima. The homing behavior is shown in Algorithm 5.1. The script to implement this algorithm in our system is Script 5.1.

Algorithm 5.1 Homing

```
if (goal is in view range) then
    set goal as target.
else if (target is in view range) then
    set next subgoal as the target.
end if
steer toward the target.
```

(a)

Script 5.1 Homing

```
agent.goal = agent.path.next();

//don't run again at
//this node

stop;
```

(b)

Fig. 3. Homing Behavior: (a) algorithm, (b) script to implement algorithm.

5.2 Exploring Behavior

There are several variants of exploring. For example, flocks may have good *a priori* data about the environment and they may be looking for objects whose locations are unknown, or the environment may be unknown and the objective of the exploring behavior is to gather information about the environment. For our work, we assume the environment is known. We consider two

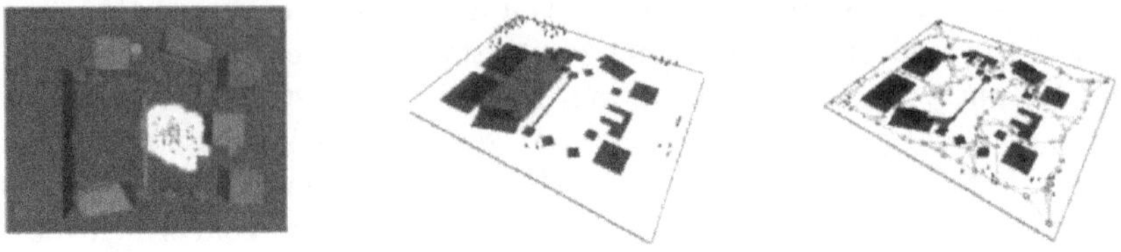

(a) (b) (c)

Fig. 4. Covering behavior. Unexplored regions are shaded. (a) Initial configuration of flock. (b) Normal flocking system. (c) Roadmap-based method.

different exploring behaviors: (i) covering the environment, and (ii) searching for a goal and moving toward the goal once it is found. To achieve these behaviors, we use a roadmap with adaptive edge weights. Each agent moves independently in the roadmap. At each roadmap node, they choose an edge to follow based on the weights of the incident edges. The weights represent how relevant the edge is to the current task, either covering the environment or searching for and reaching the goal. This is similar to *ant colony optimization (ACO)* where the ants deposit pheromones to indicate routes.

Covering Behavior The objective in this behavior is for some member of the flock to have visited every location in the environment. We assume that we start with a roadmap covering all relevant portions of the environment, and our goal is to have some flock member visit every edge and vertex of the roadmap. We implement this behavior using a roadmap graph with adaptive edge weights. Initially, all edges have the same weight. As they traverse a roadmap edge they increase its weight – this is similar to ant pheromones which increase as more ants follow the same path. Since our goal is to cover the environment, individual flock members are biased toward relatively unexplored areas of the roadmap, i.e., to roadmap edges that have smaller weights. That is, they select low weight edges to follow with some higher probability. The covering behavior is shown in Algorithm 5.2 and the corresponding script is Script 5.2. The first two lines of Script 5.2 define local variables. The third line calls a function of the predefined **node** variable which returns, with some probability, the minimum weight outgoing edge. Then, the next goal is set to that edge's second endpoint and the edge weight is increased.

Goal Searching Behavior Although the individual flock members have access to environmental information through the roadmap (at least locally), they don't know the location of the goal. If an individual reaches a location where the goal is within sensor range, its location is communicated to the other members, perhaps indirectly, and they then attempt to reach the goal as well. We implemented this behavior using adaptive roadmap edge weights.

Algorithm 5.2 Covering the Env.

```
if (reach target t) then
    Push t onto agent's stack
    if (no outgoing edge) then
        t = topmost stack node with un-
        visited edge
    else
        select edge e and increase wt(e)
        t = endpoint(e)
    end if
end if
steer toward the target
```

(a)

Script 5.2 Covering the Env.

```
AgentVar NODE nextNode;
AgentVar EDGE edge;

edge = node.probMinEdge();
nextNode = edge.end;
agent.goal = nextNode.position;
edge.weight = edge.weight+1;
```

(b)

Fig. 5. Covering Behavior: (a) algorithm, (b) script to implement algorithm.

(a) (b) (c) (d) (e)

Fig. 6. 10 flock members are searching an unknown goal. (a) flock is facing a split point, (b) since two branches have same edge weight, flock splits into two groups, (c) members on the lower (and upper right) part encountered dead end, and weight of edges leading to dead end will decrease, (d) some members find goal, weight of edges leading to the goal will increase, (e) rest of members reach the goal.

As an agent traverses a path in the roadmap, it remembers the route it has taken. Then, when it reaches a goal, it increases the weight of the edges on the route it took. If the individual reaches a roadmap node without any outgoing connections (i.e., with only one edge) or a node already contained in the current path (i.e., a cycle), the weight of the edges it followed will be decreased. The algorithm for this approach is summarized in Algorithm 5.3.

This algorithm can be implemented using two different rule sets. The first rule set is for all roadmap nodes except the (hidden) goal, it adds the visited nodes to the AGENT's path. The second rule set will only be executed if the agent reaches the goal, and the task of this set is to increase the weight of edges this agent used to get there. Script 5.3.I. implements the first set of rules. The first two lines define local variables. The third line calls a function of the predefined `node` variable which returns, with some probability, the maximum weight outgoing edge. Then, the next goal is set to that edge's second endpoint and this information is added to path so that the promising

Algorithm 5.3 Goal Searching

```
if (goal found) then
    increase weight of edges to goal.
end if
if (reach target t) then
    Push t onto agent's stack
    if (no outgoing edge) then
        t = topmost stack node with
        unvisited edge. Decrease edge
        weights leading to t
    else
        select edge e and increase wt(e)
        t = endpoint(e).
    end if
end if
steer toward target t.
```

(a)

Script 5.3.I Goal Searching

```
AGENTVAR Node nextNode;
AGENTVAR Edge edge;
TRIGGER checkDeadEnd;
edge = node.probMaxEdge();
nextNode = edge.end;
agent.goal = nextNode.position;
agent.addPath(nextNode.position);
```

(b)

Script 5.3.II Hidden Goal

```
// Update the passed edges
agent.path.increaseEdgeWeights();
stop;
```

(c)

Fig. 7. Goal Searching: (a) algorithm, scripts for (b) searching, and (c) goal nodes.

edges can be updated if the AGENT reaches the goal. The TRIGGER function checkDeadEnd will run once for every node and would check the path so far to see if there is a dead end, and if so decrease the weight of edges and remove them from the path. Script 5.3.II, which only resides in the node closest to the goal, will increase the weight of the edges on the path AGENT took to the goal.

5.3 Narrow Passage Behavior

Sometimes the flock's behavior should depend on the surrounding environment. For example, different group formations may be used in relatively open areas than are used when passing through narrow regions.

A naive way to achieve narrow passage traversal by the flock is to use the homing behavior and to select two nodes as goals, first a node in front of the entrance to the passage and then a node outside the exit from the passage. One drawback of this approach is that flock members may bunch up and conflict with each other as they try to move through the passage.

A *follow-the-leader* strategy may avoid the congestion problems of the naive strategy. In this strategy, we first assemble the flock in front of the narrow passage, and then select the closest agent to the narrow passage entrance as the leader. Then, the remaining flock members are arranged into a queue that follows the leader. Their position in the queue depends on their distance to the entrance of the narrow corridor. They can be kept from crowding each other by selecting appropriate values for the repulsive force from other flock members.

Note that different behaviors can be achieved by using a different criterion to select the next flock member in line 6 of Algorithm 5.4. For example, instead of selecting the next closest flock member to the narrow passage, one might select the farthest, which would create a 'milling around' effect at the entrance to the passage.

Algorithm 5.4 Narrow Passage

```
1: while (not all flock members in gathering area) do
2:    set individual members' goal to gathering area
3: end while
4: set leader to NIL
5: while (there are flock members outside passage) do
6:    select the closest unselected member as Current
7:    if (Leader is NIL) then
8:       set Leader to Current and set Leader's goal to next step in the path
9:    else
10:      set Current's goal to Previous
11:   end if
12:   set Previous to Current
13:   increase neighbor avoidance threshold
14: end while
```

Fig. 8. Narrow Passage Behavior algorithm.

A script for the Narrow Passage behavior can be found in [2]. Initially, all agents are in the same class. After they all have gathered outside the passage, one of them is selected as the *leader* and the others become *followers*; these roles are maintained while traversing the passage. There is a Trigger function that continuously checks if all Agents have reached the gathering area. When they are there, the Trigger updates a `NodeVar` variable which keeps track of the state so that the follow behavior starts. A second Trigger function continuously finds the closest Agent to the passage and updates its role (the first time a closest Agent is found, it is the *leader*, the successive ones are *followers*). Then, the Agents running the script compare themselves with the closest, if they are closest then they either follow their pre-assigned path (the leader) or follow the previous Agent.

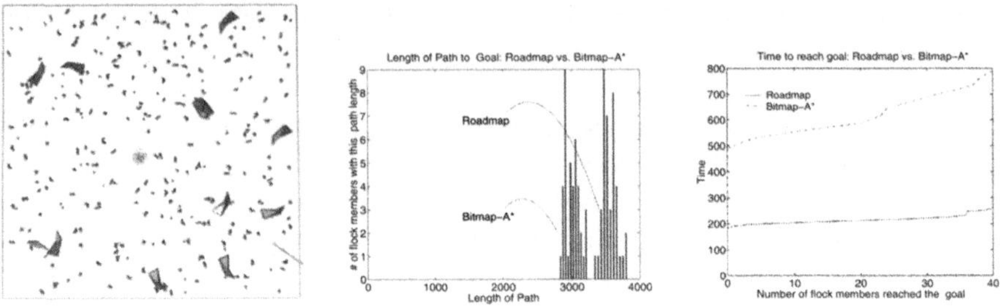

(a) (b) (c)

Fig. 9. Homing behavior: the number of flock members reaching (a) goals with respect to the length of the paths they took, and (b) homes with the evolution of time. Although the grid-based A^* behavior finds shorter paths, the flock spends less time to reach the goals with the roadmap-based behavior.

Table 1. The number of the 40 flock members reaching their home within 30 seconds, the initialization time, the average time to find a path, and the total time spent by all flockmates escaping local minima using basic flocking, grid-based A^*, and rule-based roadmap behaviors.

Homing behavior: Comparison of different methods

BEHAVIOR METHOD	#flockmate reaching the goal	init time	find path time	local minima #	local minima escape(s)
Basic	10	N/A	N/A	N/A	N/A
grid-based A*	40	6.02	5.757	2005	1035.43
roadmap-based	40	0.88	0.652	255	22.99

6 Experimental Results

In this section we evaluate our rule-based roadmap techniques for the homing, covering, and narrow passage behaviors described previously. Movies can be found on our webpage (http://parasol.tamu.edu/dsmft/).

Our experiments are designed to compare our roadmap-based techniques with more traditional approaches for simulating flocking behavior and to study the improvements possible by incorporating global information about the environment as encoded in a roadmap. To study the efficiency of our covering technique, we also compare our roadmap-based method with an 'ideal' variant in which the location of all unvisited nodes and edges in the (known) roadmap is known at all times.

All of our experiments were run on a Linux system with Athlon 1.33 processor and 256MB memory.

6.1 Homing Behavior

For the homing behavior, our rule-based roadmap technique, described in Section 5.1, is compared with a basic flocking behavior using a potential field and a grid-based A^* search behavior.

The 420×420 environment is populated with 301 randomly placed obstacles (of six types) (see Figure 9(a)). At any given time there is one goal, and when all flock members reach it, a new goal is randomly generated; this process continues until eight goals have been generated and reached. There are 40 flock members which are initially placed according to a Gaussian distribution around the center of the environment. The simulation is updated every 100 ms.

For the grid-based A^* behavior, a bitmap of the environment of 914×914 cells is constructed; the length of a side of each square cell is equal to the diameter of a flock member. Cells are classified as free cells and collision cells. Paths are found in this bitmap using A^* search. For the rule-based roadmap behavior, the roadmap is built using the MAPRM method to generate 400 roadmap nodes and we attempt to connect each node to its 4 nearest neighbors.

In our experiments we first tried to evaluate the effect of the global information. In order to do that, we first compared different methods to see how many flock members can reach a given goal with predetermined time (See Table 1, second column). It is clear from Table 1, without global information, that only 25% of the flock members reach the goal and most of them are trapped in local minima. On the other hand, when global navigation information is utilized, either with the grid-based A^* method or our rule-based roadmap method, all flock members reach the goal. Based on these results, we further evaluated the grid-based A^* and the roadmap-based method. In Table 1, columns 3–6, we show the time spent searching for paths, the number of local minima encountered along all paths, and the total time spent escaping from local minima. This offers some insight into the methods studied, as can be seen more clearly in Figure 9. Although the flock takes a shorter path with the grid-based A^* search than with the roadmap-based method (Figure 9(b)), the flock reaches the final goal faster with the roadmap-based method (Figure 9(c)). As A^* search is known to be fast and always finds shortest paths, this example illustrates that our roadmap-based method indeed is a competitor for grid-based A^* methods – while the paths found are a bit longer, they are found faster.

6.2 Covering Behavior

We next consider the covering behavior. This behavior is tested on the environment shown in Figure 4, which requires flock members to pass through narrow passages to access undiscovered areas. In this experiment, we compare basic flocking behavior, rule-based roadmap behavior, and an ideal variant of the roadmap-based behavior which has dynamic knowledge of the undiscovered regions. The former comparison enables us to establish the value of the global information stored in the roadmap, while the latter comparison enables us to determine how far our methods are from achieving optimal performance.

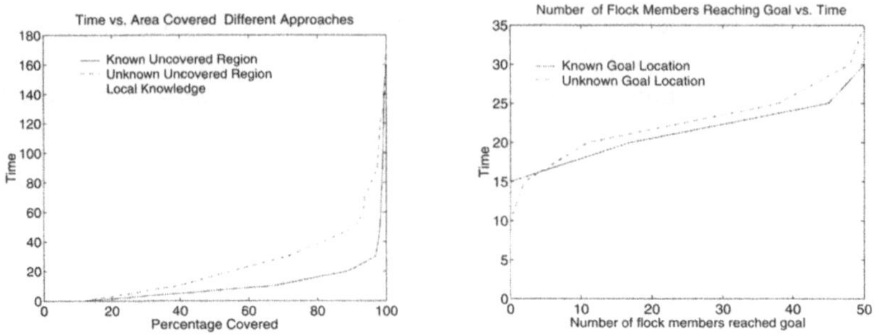

(a) (b)

Fig. 10. Exploring behavior: (a) Covering: the percentage of the environmental covered in terms of time, (b) Goal Searching: the number of flock members reaching the goal area in terms of time.

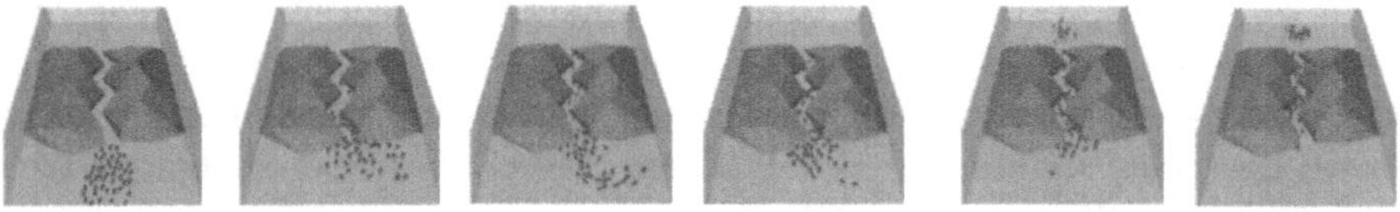

Fig. 11. Passing through a narrow passage using the FOLLOW THE LEADER behavior (Algorithm 5.4).

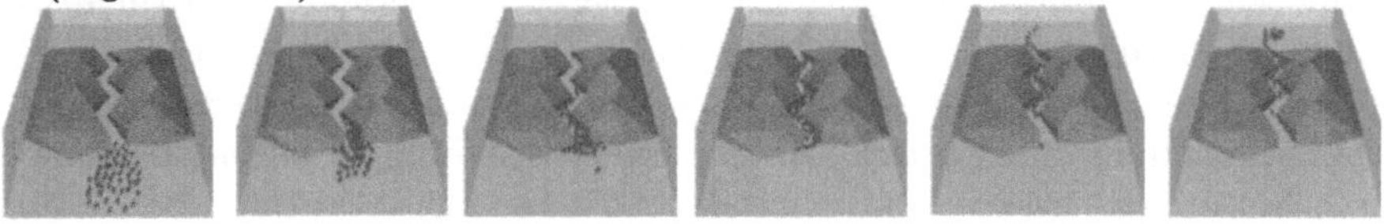

Fig. 12. Passing through a narrow passage using the naive strategy – homing to the exit node of the passage.

The environment (80×100) is populated with 16 obstacles (6 types of obstacles) and in total 24% of the environment is occupied by obstacles. 50 flock members are simulated and states are updated every 100ms. A bitmap is built to record discovered/undiscovered information. A bitmap cell is discovered when it is inside the sensory range of any flock member. We set the radius of the sensory circle at 5m. For the roadmap, 120 nodes are sampled and connections are attempted to each node's 4 nearest neighbors.

The roadmap-based covering behavior is as described previously. The basic behavior uses only local information, and is essentially a random walk through the environment. It shows that the lack of global knowledge results in some areas never being discovered, especially those nearly surrounded by obstacles. The ideal behavior with complete knowledge of the undiscovered locations uses the roadmap to find paths from a flockmate's current position to the closest unexplored spot. Although such knowledge is generally not available, this variant indicates how fast an environment can be covered in the best case.

As seen in Figure 10(a), the ideal behavior rapidly covered all of the free environment in the first 30 seconds. The rule-based roadmap behavior, using indirect communication (adaptive edge weights) takes about three times as long (90 seconds), to reach a similar coverage. Nevertheless, like the ideal behavior, the rule-based roadmap behavior found most reachable areas, while the basic flocking behavior had difficulty covering more than 89% of the free environment. However, it is interesting to note that the basic flocking behavior found more undiscovered areas than the rule-based roadmap approach in the first 40 seconds; this is due to the flockmates' with basic behavior which tend to bounce around and discover 'easy' areas very quickly.

6.3 Goal Searching Behavior

In this experiment, the rule-based roadmap behavior is compared with simple flocking behavior that has only local information about the environment and no knowledge of the goal position and with an ideal variant of the roadmap-based behavior that has *a priori* knowledge of the position of the goal. The environment is the same as in the covering experiment (See Figure 4).

We are interested in the number of flock members that reach the goal and how fast they get there. As mentioned before, the behavior with complete knowledge is used to establish a best case (lower bound) for the simulation efficiency, and the basic behavior using only local information is used to illustrate the importance of global knowledge.

The results of some experiments are shown in Figure 10(b). Flocks using the basic behavior did not discover any goals within 35 seconds, and hence this behavior is not shown in the plot. Overall, the roadmap-based behavior is competitive with the ideal roadmap-based behavior – only taking 5 seconds longer than the method in which the position of the goal is known *a priori*. In addition, it is surprising to note that two of the flock members in the roadmap-based method reach the goal earlier than any of their flockmates in the ideal roadmap-based behavior. While we expect the roadmap-based method to continue to perform well in more complex environments, we expect its efficiency relative to the ideal method to decline somewhat.

6.4 Narrow Passage Behavior

The narrow passage environment in Figures 11 and 12 contains 50 flock members and two mountain-like obstacles. Agents are asked to reach the goal on the other side of environment. The only way to reach their destination is to pass through the narrow passage between these two obstacles. As before, the roadmap is generated using MAPRM [21] which attempts to generate high clearance nodes on the medial axis of the free space. To demonstrate our rules in Section 5.3, we consider two experiments.

In the first experiment, a node is automatically assigned the narrow passage rule described in Section 5.3 if its clearance is below some threshold. All other nodes are assigned homing rules as described in Section 5.1 with the goal set as a node outside the exit of the narrow passage. The homing rule script also sets `agent.role = boid`, so that all AGENTs are boids when they gather outside the entrance to the narrow passage. Snapshots of the agent movement are shown in Figure 11. Note how the agents maintain clearance from each other as they move through the passage. The amount of clearance is adjusted by modifying the repulsive force between agents, which can be set in the rule scripts stored in narrow passage nodes.

In the second experiment, homing rules are assigned to all nodes, with the goal of the homing rules set to the node at the exit of the narrow passage. This is the naive narrow passage behavior mentioned in Section 5.3, which is essentially a homing behavior. Snapshots of the agent movement are shown in Figure 12. Note how the agents clump together and interfere with each other as they move through the passage.

7 Conclusion

In this paper, we have shown that complex group behaviors can be generated if some global environmental information is available in a roadmap. The information the roadmap contains, such as topological information and adaptive edge weights, enables the flock to achieve behaviors that cannot be modeled with local information alone. Moreover, since in many cases global knowledge involves high communication costs between individuals, indirect communication though dynamic updates of the roadmap's edge weights provides a less expensive means of obtaining global information. The behavior rules embedded in our roadmaps and agents enable the agents to modify their actions based on their current location and state. For example, the flock can move as an unordered group in open regions and in a follow-the-leader fashion through narrow passages. Our simulation results for the three types of behaviors studied show that the performance of the rule-based roadmap behaviors is very close to ideal behaviors that have complete knowledge.

References

1. T. Balch and M. Hybinette. Social potentials for scalable multirobot formations. In *Proc. IEEE Int. Conf. Robot. Autom. (ICRA)*, pages 73–80, 2000.
2. O. B. Bayazit, J.-M. Lien, and Nancy M. Amato. Better group behaviors in complex environments using global roadmaps. In *Artif. Life*, Dec 2002. To appear.
3. D. C. Brogan and J. K. Hodgins. Group behaviors for systems with significant dynamics. In *Autonomous Robots*, pages 137–153, 1997.
4. T. Fukuda, H. Mizoguchi, K. Sekiyama, and F. Arai. Group behavior control for MARS (micro autonomous robotic system). In *Proc. IEEE Int. Conf. Robot. Autom. (ICRA)*, pages 1550–1555, 1999.

5. J. Funge, X. Tu, and D. Terzopoulos. Cognitive modeling: Knowledge, reasoning and planning for interlligent characters. In *Computer Graphics*, pages 29–38, 1999.
6. O. Khatib. Real–time obstacle avoidance for manipulators and mobile robots. *Int. J. Robot. Res.*, 5(1):90–98, 1986.
7. J. C. Latombe. *Robot Motion Planning.* Kluwer Academic Publishers, Boston, MA, 1991.
8. T. Y. Li, Y. J. Jeng, and S. I. Chang. Simulating virtual human crowds with a leader-follower model. In *Proceedings of 2001 Computer Animation Conference*, 2001.
9. Dorigo M., G. Di Caro, and L. M. Gambardella. Ant algorithms for discrete optimization. In *Artificial Life*, pages 137–172, 1999.
10. M. J. Mataric. *Interaction and Intelligent Behavior.* PhD thesis, MIT EECS, 1994.
11. S.I. Nishimura and T. Ikegami. Emergence of collective strategies in prey-predator game model. *Artif. Life*, 3:243–260, 1997.
12. L. E. Parker. Designing control laws for cooperative agent teams. In *IEEE International Conference on Robotics and Automation*, pages 582–587, 1993.
13. C. W. Reynolds. Flocks, herds, and schools: A distributed behaviroal model. In *Computer Graphics*, pages 25–34, 1987.
14. C. W. Reynolds. Steering behaviors for autonomous characters. In *Game Developers Conference*, 1999.
15. Stuart Russell and Peter Norvig. *Artificial Intelligence: A Modern Approach.* Prentice Hall, 1th edition, 1994.
16. N. Saiwaki, T. Komatsu, T. Yoshida, and S. Nishida. Automatic generation of moving crowd using chaos model. In *IEEE Int. Conference on System, Man and Cybernetics*, pages 3715–3721, 1997.
17. S.-J. Sun and D.-W Lee K.-B. Sim. Artificial immune-based swarm behaviors of distributed autonomous robotic systems. In *Proc. IEEE Int. Conf. Robot. Autom. (ICRA)*, pages 3993–3998, 2001.
18. X. Tu and D. Terzopoulos. Artificial fishes: Physics, locomotion, perception, behavior. In *Computer Graphics*, pages 24–29, 1994.
19. R. T. Vaughan, N. Sumpter, J. Henderson, A. Frost, and S. Cameron. Experiments in automatic flock control. *J. Robot. and Autonom. Sys.*, 31:109–117, 2000.
20. C.R. Ward, F. Gobet, and G. Kendall. Evolving collective behavior in an artificial ecology. *Artif. Life*, 7:191–209, 2001.
21. S. A. Wilmarth, N. M. Amato, and P. F. Stiller. MAPRM: A probabilistic roadmap planner with sampling on the medial axis of the free space. In *Proc. IEEE Int. Conf. Robot. Autom. (ICRA)*, pages 1024–1031, 1999.

A Delaunay Approach to Interactive Cutting in Triangulated Surfaces

Han-Wen Nienhuys and A. Frank van der Stappen

Institute of Information and Computing Sciences, Utrecht University
PO Box 80089, 3508 TB Utrecht, The Netherlands
{hanwen,frankst}@cs.uu.nl

Abstract. We present a method for producing cuts in triangulated surfaces. This method keeps the mesh size low and element quality high. We show the method for triangle meshes in two dimensions, and then generalize it to three dimensional curved surfaces, where bifurcations and annihilations of incisions may occur. This method could be applied to simulating surgery of membrane-like structures, such as veins or intestine.

1 Introduction

This paper is about simulating cuts in triangulated surfaces. The motivation of this work is formed by interactive surgery simulation [2, 8, 11, 15, 16]: computer simulations of surgical procedures can bring down the costs and risks associated with learning to perform difficult operations.

Such simulations combine interactive deformation of virtual tissue with simulations of surgical procedures. The Finite Element Method (FEM) seems suited for computing deformation since it is physically accurate. It is a technique for solving partial differential equations (PDEs) of various types. Meshes form the basis of this technique: the domain of the problem is subdivided in geometric primitives ('elements') such as tetrahedra or triangles. With this subdivision the PDE can be transformed in a system of numerical equations. For the rest of the discussion, we will assume that the PDE is linear and derives from an elasticity problem. The equations can then be discretized into a linear system, where the matrix is called *stiffness matrix.*

Finding a solution is a two step process: first, the stiffness matrix is constructed, then the unknowns (a set of deformations) are computed from the stiffness matrix and given loads. The characteristics of a mesh influence both steps. The accuracy of the discretization process is determined by element shapes. For triangle meshes, large elements and large angles decrease the accuracy of the FEM discretisation (analogously, large dihedral angles do so for tetrahedra). The accuracy of the solution process in finite precision arithmetic is determined by the condition number of the stiffness matrix, where a lower condition number is better. The convergence speed of iterative solution methods also depends on the condition number. This holds for both

J.-D. Boissonnat et al. (Eds.): Algorithmic Foundations of Robotics V, STAR 7, pp. 113–129, 2004.

the conjugate gradient algorithm (a static approach), and for damped dynamic relaxations using explicit time integration. The condition number of a complete stiffness matrix can be bounded by the condition numbers for separate elements. In contrast to the FEM discretisation error, high condition numbers are caused by elements with small sizes and small angles. So, the requirements for optimal accuracy in the discretisation and for optimal speed and accuracy in the numerical process are contradictory. However, it is safe to say that a mesh with 'round' elements, i.e. no angles close to π and 0, and bounded element sizes is a good general purpose mesh.

Iterative algorithms seem well suited as numerical solution strategies for surgery simulations, since they allow on-line mesh changes [8, 16]. If such an iterative approach is used, then mesh quality becomes an issue of vital importance: the maximum size of simulations is largely determined by mesh characterics: better meshes yield faster convergence, enabling larger and more accurate simulations. It follows that simulated surgical manipulations (such as needle insertions, cuts, and cauterizations) should be designed to keep mesh quality high. Secondly, the manipulations should also keep the complexity of the mesh low, since the cost of a single iteration step is proportional to the size of the mesh.

Simulation of cuts in surgery simulation is related to simulation of other destructive surgical procedures. The first operation to have been simulated on volumetric meshes is cauterization. This was done by removing elements in contact with a virtual cauterization tool [8]. A disadvantage of element removal is that it produces a jagged surface on the virtual tissue.

For cutting, subdivision methods are the norm [2, 11, 15]: elements that are in contact with the scalpel are marked active, and subdivided to produce a cut conforming to the scalpel position. The subdivision moves along with the scalpel during its stay in an active tetrahedron. When the scalpel leaves an active element, the subdivision is entered in the mesh permanently. A 2D example of subdivision is in Figure 10.

Subdivision methods always increase the size of the mesh. Moreover, these methods tend to produce degeneracies. This is caused by the use of an *active region*. Mesh modification is done only within a fixed region of the mesh, and if the scalpel moves close to the boundary of that region, poorly-shaped elements are inevitable.

Ganovelli and O'Sullivan [12] try to counter the the degeneracies caused by subdivision cutting. They deal with these degeneracies by collapsing short edges of the mesh. This approach does improve the quality of the mesh, but this solution does not repair all inconsistencies: not all edges may be contracted, and flat triangles and tetrahedrons, which do not contain short edges but are still degenerate, are not dealt with.

We have previously tried an approach where scalpel nodes are snapped to the trajectory swept by the scalpel [16]. The advantage of this method is that the mesh size remains small, and few short edges are created. However, there are a number of disadvantages: since no new nodes are created, the resolution of the cut is bounded by the mesh resolution. The incision does not reach up to the position of the scalpel, but lags behind it. A more serious problem is that snapping can result in degenerate elements in the mesh. Such degeneracies are dealt with by subdividing flat elements, and collapsing the resulting short edges, effectively removing the flat element. Unfortunately, not all edges can be collapsed. Using existing mesh features as a basis for mesh modification is problematic: when the scalpel is not especially close to a mesh feature, it may not be possible to match the mesh topology to the scalpel path without introducing degeneracies.

In summary, producing high quality cuts in tetrahedral meshes is a difficult problem. For cutting in surfaces, there is an analogous problem, which has the same difficulties as volumetric cutting approaches. Serby et al. [19] propose a method which also relies on a form of node snapping: the scalpel is modeled as a line segment, and nodes from the mesh are projected onto that segment. In a post-processing step, the edge lengths and element volumes are optimized using a particle system. The result is a good looking cut, without a decrease in mesh quality and size. However, in reality, the path of a scalpel is not a large line segment, but a concatenation of several small ones, and their approach does not seem to address this issue. Bruyns and Senger [5] use surface cutting with subdivision in large-scale simulations of various procedures.

In light of the complications of volumetric cutting, we decided to take a step back, and analyze the cutting problem for surface meshes first. In this paper we will present a method that produces cuts in a triangulated surface which does not decrease the mesh quality and keeps the mesh size low. We have analyzed this problem primarily to gain insight into the cutting problem for 3D tetrahedral meshes. Nevertheless, this technique could be applied in surgery simulation for membrane-like structures, such as skin or intestine.

2 Cutting in 2D

We state the general mesh cutting problem as follows: given a starting mesh, and positions of a user-controlled scalpel, modify the mesh at every moment to show an incision that represents the past trajectory of the scalpel, and ends exactly at the scalpel. The challenge in this problem is to produce a well-shaped mesh with few elements. The simplest case is the two-dimensional form. Here the mesh is a triangulation in the plane, and the virtual scalpel is a point. The challenge is to produce elements that have no large angles and no short edges.

2.1 Delaunay Triangulation

Since we are putting emphasis on the quality of the mesh, we briefly review the Delaunay Triangulation (DT), a popular technique for generating a well-shaped triangulation of a given set of points. The DT of a set of points is defined as a triangulation of that set where the circumcircle of every triangle does not contain any other points from the set. This property is also referred to as the empty-circle property. The empty-circle property can also be defined for edges of the mesh: an edge is called *legal* or *Delaunay* when the circumcircles of its incident triangles do not contain the opposite node of the other triangle. An example of an illegal edge is in Figure 1. Delaunay triangulations and Delaunay edges are intimately related: a Delaunay triangulation only has Delaunay edges.

Non-Delaunay or illegal edges of a triangulation can always be flipped: the two triangles incident with the edge always form a convex quadrilateral, and the diagonal may be switched. The flipped diagonal is always Delaunay, and the minimum angle of the pair of triangles is always increased, thus improving mesh quality. Flipping illegal edges only affects a single part of the mesh, so it can be seen as local improvement strategy. The Delaunay triangulation can be constructed by starting with an arbitrary triangulation and flipping illegal edges until none are left. The final result maximizes the minimum element angle. This is called the maxmin-angle property.

There are various ways to measure element quality: for example, minimum angle, circumradius to shortest edge ratio, circumradius to inscribed radius ratio, etc. In 2D all these measures are equivalent up to constant factors [14], so the maxmin-angle property of the DT means that it is a reasonable meshing technique for virtually all element quality measures. When coupled with algorithms for point insertion it is a basis form many refinement meshing techniques [1, 7, 18].

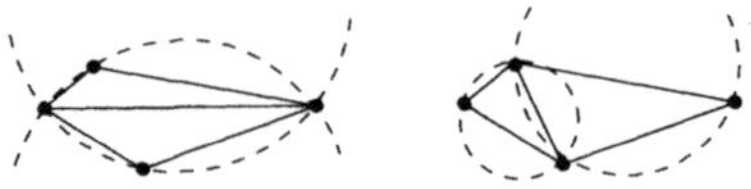

Fig. 1. An illegal edge has a triangle whose circumcircle contains the opposite vertex of a neighboring triangle (left). By reconnecting ('flipping') the edge, the circumcircles become disjoint (right).

2.2 Cutting and Delaunay Flips

Our problem, cutting in meshes, is different from the standard meshing problem. A starting mesh is given, and the shape of our domain is variable: as the cut progresses, the shape of the boundary is changed. The Delaunay Triangulation assumes a given set of points, and it is always convex. Cuts result in non-convex boundaries, so the standard DT is not directly usable. Nevertheless, we can retain the idea of using edge flips to locally improve the mesh.

Our approach works as follows: when cutting, the scalpel is attached to a node, the *active node*, so moving the scalpel moves the active node. The active node is always part of the boundary, so it is incident to two boundary edges, the *cut edges*. During a cut, these edges almost coincide geometrically, and form an *incision*. We call triangles incident with the active node *active triangles*. The active node is moved, and after each movement, local remeshing is applied to the active triangles. The technique is demonstrated in Figure 2; a more elaborate example is in Figure 4. The remeshing process consists of the following actions.

- Edges are flipped to improve the element angles.
- If another node is found close to the active node, it is removed.
- When the incision becomes too large, it is split thus introducing new nodes to approximate the scalpel path.

During a cut, nodes are removed in front of the scalpel, and inserted behind the scalpel.

The rationale for flipping and removing nodes is that the Delaunay criterion tends to flip away triangles with angles approaching 180°, since these have large circumcircles. Nodes that are close together get connected. Such triangles have a small angle opposite a short edge. These elements are dealt with by removing nodes that are close to the active node. This can only be done for nodes that are not part of the boundary: removing boundary nodes would change the shape of the object represented.

During a cut, the cut edges are in almost in the same location. From a geometric point of view, we can identify both cut edges into a single edge, and check if this edge satisfies the empty-circle criterion. If it does not, new nodes are inserted, where the line connecting nodes opposite the cut edges intersect the scalpel path. We call this procedure a *incision split*, and it is demonstrated in Figure 3. When the scalpel passed that point previously, close nodes were removed, so the split will not lead to short edges. The newly inserted nodes are dilated slightly, to prevent numerical problems when a path self-intersects. The new nodes always lie on a line that connects two existing mesh nodes, hence the accuracy of the represented trajectory is determined by the resolution of the starting mesh.

When the scalpel enters or exits the mesh, it is close to the boundary of the mesh, so a realistic incision would contain very short edges. To prevent these short edges, special precautions are taken: incisions are postponed and exits are done in advance.

When the scalpel enters the mesh, an active node is created. If the entry happens close to an existing node, we move it and label it active. Otherwise, a new node is inserted at at the entry point. The active node moves along with the scalpel, creating a temporary dent. When the active node is sufficiently far from the entry point, it is "fixed up": nodes are added at the entry point, creating an incision. If the scalpel is *retracting* (moving away) before this

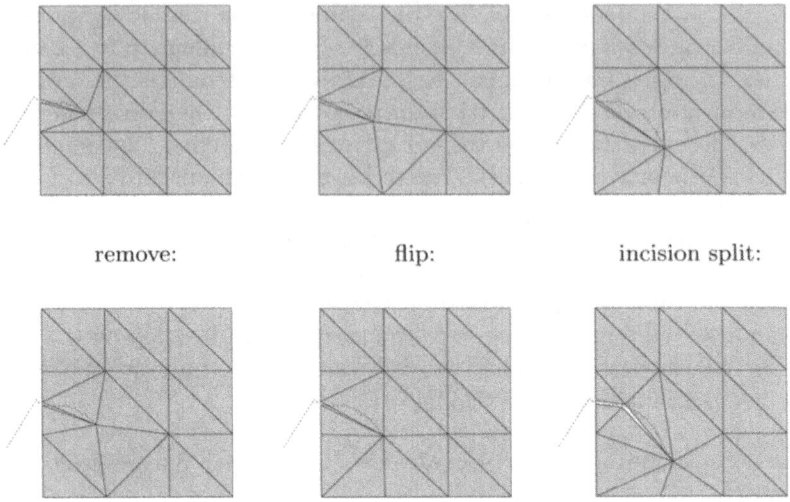

Fig. 2. The evolution of a simple 2D cut is depicted from left to right; the dotted line indicates the scalpel trajectory. The key steps are to flip edges (center), to remove close nodes (left), and to insert nodes behind the scalpel (right).

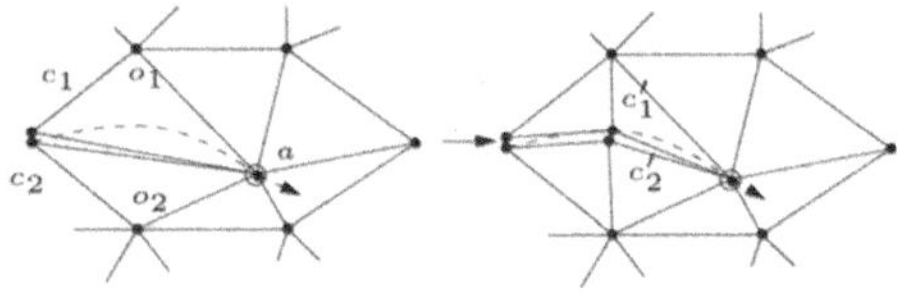

Fig. 3. During an incision split, nodes are inserted in the cut edges. This happens when ac, with $c = (c_1 + c_2)/2$, fails the empty-circle criterion as diagonal of aco_1o_2.

fix-up happens, the dent is left in the mesh permanently. Before a fix-up, no incision splits take place.

When exiting, the the cut is finished before the scalpel comes too close to the boundary. At every step the movement of the scalpel is extrapolated. When this extrapolation hits the boundary, and is close to the active node, the cut is finished: a node is inserted at the extrapolation, and the cut is dissected. This exit procedure still leaves relatively short edges permanently in the mesh. To rectify this, these edges are contracted. The boundary edges that are created with this exit operation are added to a list of forbidden edges. These edges are not tested for collisions during ensuing movements. This prevents artifacts when the actual scalpel movement differs from the predicted movement. If the scalpel is sufficiently far from the exit point, they are eligible for collision checking again.

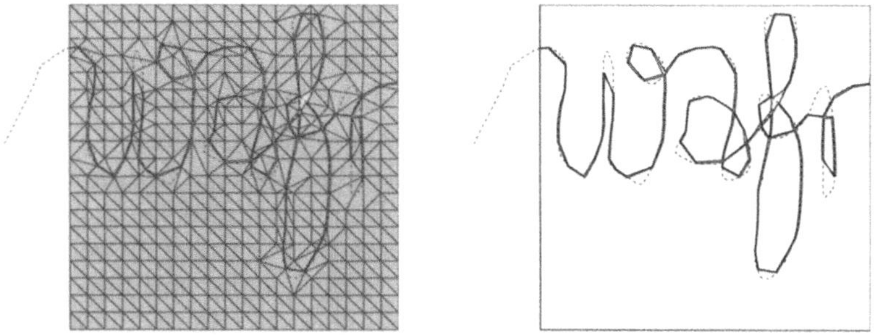

Fig. 4. Delaunay cutting on a triangle mesh. Both the mesh itself and the boundary are shown. The scalpel trajectory is indicated with a dotted curve. Notice how strongly curved path segments are cut short in the realized cut. The starting mesh was regular and consisted of 722 triangles. The cut increased mesh size by 41 triangles.

The cutting routines must maintain references to active parts of the mesh, while changes are applied to the mesh. This implies that certain mesh features must be persistent across such change operations. This persistence was enable by using a data structure that resembles standard data structures for subdivision storage, like the half-edge data structure [13]. The difference is that our data structure is only used for manifold simplicial complexes. It also works for tetrahedral subdivisions, and can be manipulated with only two very generic operations, being replace-triangles and change-triangles. Oriented simplexes are associated explicitly with the Edge and Triangle objects representing them. This allows mesh changes to be specified formally in terms of simplexes, and it makes it easy catch programming errors that inadvertently create non-manifold meshes. Pointers linking Triangle and Edge objects allow for efficient traversal of the subdivision. In this framework, edge flips are specified in terms of replace-triangles, can be performed in constant time, and Edge objects that are also present in the new flipped mesh are retained, making pointers to these objects persistent. Similarly, dissecting internal faces of the mesh can be specified concisely in terms of change-triangles, and pointers to the faces involved are persistent across this change.

3 Surface Cuts in 3D

Triangulated surfaces in 3D are a common concept in computer graphics. They have been used for surgery simulations [4, 6]. In these cases elastic behavior was simulated with damped mass-spring systems instead of the FEM. Nevertheless, the concerns for mesh quality continue to hold: flat elements are inverted more easily, and small elements correspond to short springs with small masses. Their high vibration frequencies translate into expensive integration schemes.

In the 2D scheme, the scalpel is a point, and it is attached to a single active node. Triangles are remeshed in the vicinity of the active node. There are two generalizations of the 2D scheme: first, the remeshing process around an active node can be done for curved instead of flat surfaces, assuming a line-shaped scalpel. Second, such a line-shaped scalpel can intersect a curved 3D surface in multiple points, so a consistent model of cutting allows multiple incisions, each with an active node. These active nodes can interact: incisions may meet, leading to annihilations, or incisions may hit folds, leading branches.

We assume that the surface is given as a triangle mesh with a boundary, and that no further information on the surface shape is known. The virtual scalpel is a line segment, and its movement is given by sampled positions of its endpoints. The endpoints are assumed to move with constant velocity between the samples. During a cut, active nodes are part of the boundary of the surface. The two boundary edges incident with an active node again are called cut edges, and define an incision.

3.1 Single Incision

A scalpel movement is handled as follows: the line segment representing the new scalpel position is intersected with all active triangles. If a single intersection is found, then the active node is moved to that point. The active triangles are subjected to flipping. The 2D criterion is used to determine whether an edge is flipped: an edge is considered illegal when the two triangles incident to that edge would be illegal in a 2D triangulation. Conceptually, we could say that the incident triangles are unfolded to be coplanar, and then the two dimensional criterion is used. It is not clear whether this flipping criterion leads to a terminating algorithm when applied to the edges of an arbitary 3D surface mesh. In this sense, this technique is now truly a heuristic.

Flipping on 3D surfaces is a delicate operation: some flips are topologically impossible (as demonstrated in Figure 5). This means that all flips must be checked in advance for failure cases.

Incisions are split analogously to the 2D case: during a cut, the last path of the scalpel is stored. When the cut edges violate the 3D empty-circle criterion, new nodes are inserted where the path is closest to the line connecting the nodes opposite the cut edges.

Node removal is done in a heuristic manner. Suppose that we want to remove a node v that is incident with n triangles. We say that two adjacent triangles incident with v form an *ear*, if the sum of the angles opposite v is less than π. When removing v, all ears are flipped until v is incident with only 3 triangles. Then v is removed, and the involved edges are flipped back if they violate the empty-circle criterion.

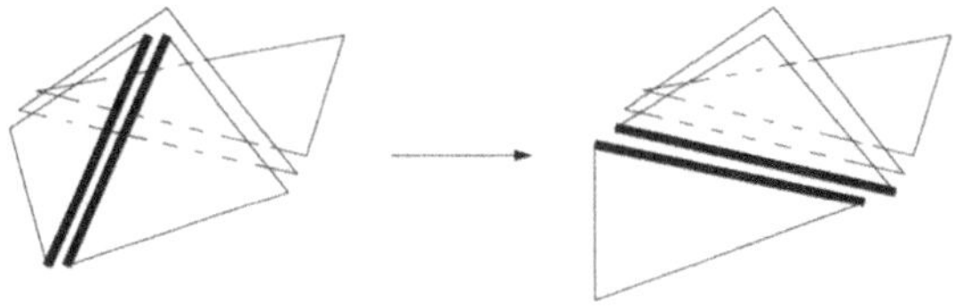

Fig. 5. Not all flips in 3D are topologically valid: when flipping the bold edge on the left, the new edge (bold) occurs twice in the resulting complex. The triangles are displayed in an exploded view for clarity.

When the scalpel is almost parallel to the surface, small movements of the scalpel can result in large movements of an active node. For this reason it is necessary to control large movements. Large movements are subdivided using a maximum distance. This distance is computed as the minimum for all distances between the line spanned by the scalpel and lines spanned by the edge opposite the active node, as indicated in Figure 6. A similar mechanism for large movement control is also used in our prototype of the 2D mechanism, where it compensates for the slowness of the implementation.

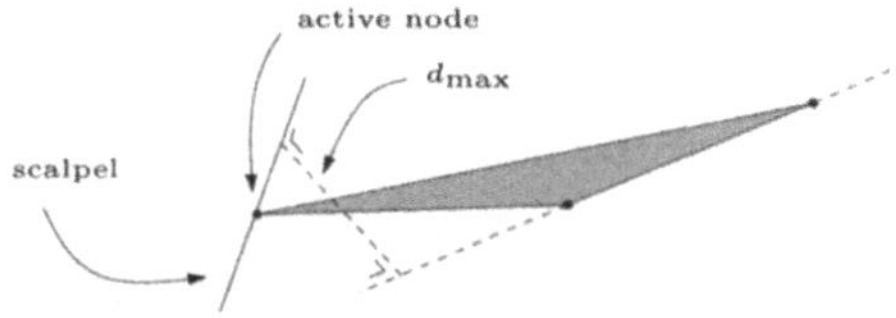

Fig. 6. The maximum movement d_{max} for a scalpel and a single triangle. The total maximum distance is given by the minimum over all triangles incident with the active node.

3.2 Multiple Incisions

In 3D, the scalpel can interact with the entire surface, and the scalpel may enter the mesh in any place. We can distinguish three cases, as demonstrated in Figure 7: the scalpel hits a boundary edge of the mesh, the scalpel tip enters through a triangle, or the scalpel hits an internal edge of the mesh. Only the first case also occurs in 2D. It is handled completely analogous to the 2D case.

In the second case, a new active node is inserted in the incised triangle. When the active node is far enough from the entry point, a node is inserted at the original entry point. The result is an edge that connects the entry node with the active node. During the next incision split a real incision is formed. Until that time, the edge is constrained, so it cannot be flipped away. The third case is when the scalpel hits an exposed edge of the mesh. Then a single

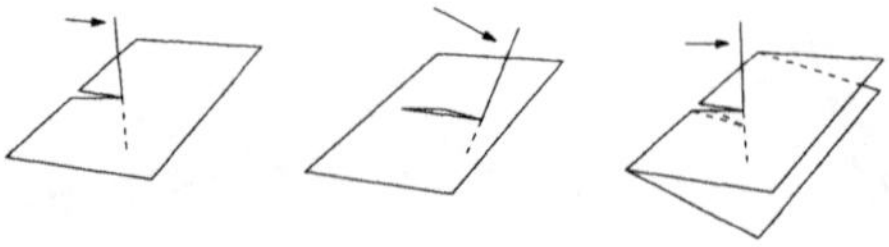

Fig. 7. There are three ways for the scalpel to enter: by hitting the boundary, entering with the tip or hitting an exposed internal edge.

active node is inserted in the edge. During the next movement, the cut will branch into two incisions. We now discuss branching and annihilations.

The scalpel is represented by a line segment, and line segments can interact with curved surfaces in many places: the scalpel may incise the surface in multiple locations, and during a cut a single incision may branch into multiple incisions, as is shown in Figure 9. We could apply the 3D remeshing from the previous section, if we could rule out any interactions between different incisions. Fortunately, this seems possible: we can forbid interactions by ensuring that every triangles is incident with at most one active node. This is achieved by the following restrictions.

- Incisions in edges or triangles that already active are rejected.
- Nodes that separate active nodes cannot be removed.
- Edges that separate two active nodes cannot be flipped.

When an active node moves towards a restricted node, an annihilation is performed: all edges from the restricted node to an active node are dissected. This is demonstrated in Figure 8. When an active node comes close to an edge separating active nodes, then its movement is extrapolated, and a new node is inserted at the extrapolated point. The annihilation now proceeds with the newly inserted node. A normal boundary exit is a special case of an annihilation.

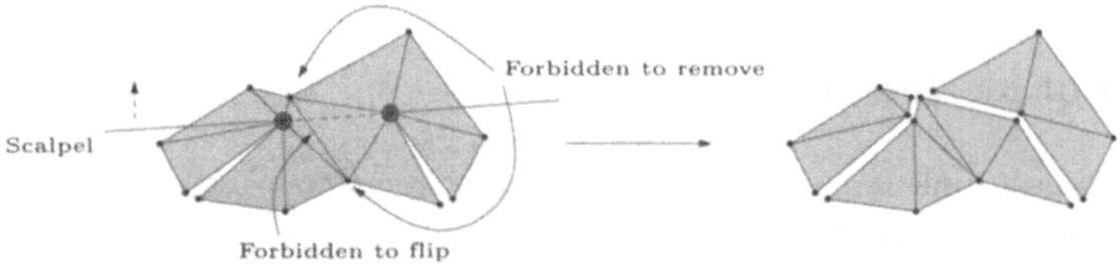

Fig. 8. Edges that separate incisions may not be flipped, nodes separating incisions may not be removed. Instead, when an active node comes too close to such a forbidden node or edge, an annihilation is performed (right).

When a scalpel movement is processed, the next position of the active node is determined by intersecting all triangles incident with the new scalpel position. If multiple active triangles are intersected, as demonstrated in Figure 9, then the cut will branch. New incisions are inserted, and the old incision is marked as no longer active. The edges connecting old and new active nodes

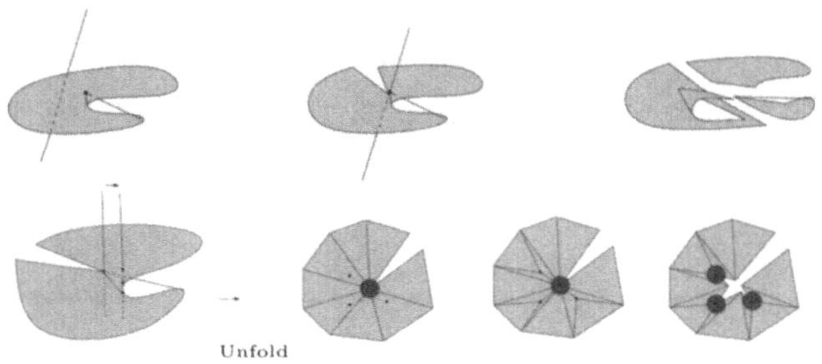

Fig. 9. Folded surfaces may lead to branching cuts (top). In such cases, a movement will cause the scalpel to intersect multiple triangles (bottom; intersection points are indicated by dots). When this happens, multiple incisions replace the old active node.

are dissected. During a branch, the new nodes are close to the original incision. This leads to short edges and degenerate triangles. When the scalpel progresses further, these short edges disappear.

4 Results

The 2D version of this algorithm has been implemented in a small application written in Python. The virtual scalpel is controlled by the mouse. A sample is shown in Figures 4 and 10. The mesh was uniform on a square grid.

Fix-up, node removal and exits depend on points being close to or far from the active node. These notions are expressed in global constant thresholds for the distance. These thresholds are denoted by $\varepsilon_{\text{edge}}$ for node removal, $\varepsilon_{\text{entry}}$ for entry fixup, and $\varepsilon_{\text{exit}}$ for predicting exits. They were set to fraction of the average global edge length $\bar{h}$. For Figure 4 we used $\varepsilon_{\text{edge}} = \bar{h}/3$, and for Figure 10 we used $\bar{h}/3$ and $\bar{h}/8$. The entry and exit tolerances were set to $\bar{h}/6$. This choice is somewhat arbitrary, but if $\varepsilon_{\text{edge}}$ is set larger than $\bar{h}$, mesh complexity will be *reduced* during a cut. If it is larger than $\bar{h}/2$, then most nodes in the vicinity of the scalpel will be removed. Lower values than $\bar{h}/2$ set a threshold for edge length. This does not directly control the accuracy of the cut trajectory represented in the mesh: the example in Figure 10 shows that lowering the threshold can decrease accuracy.

In Figure 10 the new cutting scheme for 2D is compared with a subdivision approach, where each sliced triangle is replaced by three triangles. Elements shapes and edge lengths are much worse in the subdivision mesh, as evidenced by the histograms of minimum edge length and minimum triangle angle in Figures 11. The price for this better and smaller mesh is a decrease in resolution: the subdivision cut follows the scalpel path more closely.

The 3D version of this algorithm has been implemented in a small application written in C++. It also uses the mouse to obtain scalpel movements. Samples are shown in Figures 12 and 13. The 3D version assumes that all

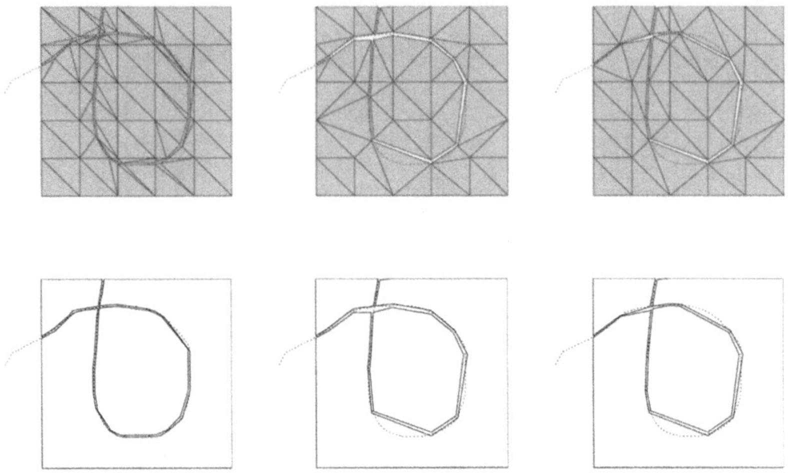

Fig. 10. The same cut performed with both subdivision (left) and Delaunay cutting (center, $\varepsilon_{\text{edge}} = \bar{h}/3$ and right, $\varepsilon_{\text{edge}} = \bar{h}/8$). The starting mesh contains 50 triangles. The subdivision cut increases size by 62 triangles, the Delaunay cut in the center by 10, and on the right by 12 triangles. The scalpel trajectory is indicated with a dotted line. Notice that the mesh in the center picture matches the trajectory better than the mesh on the right, although $\varepsilon_{\text{edge}}$ was smaller.

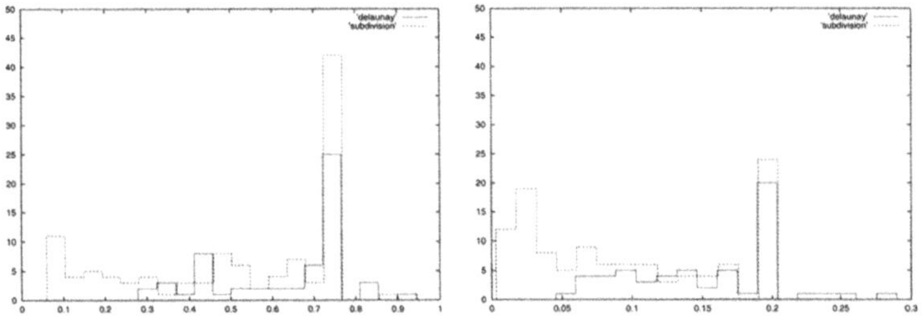

Fig. 11. Histogram of minimum element angle (left) and minimum element edge for the subdivision and $\bar{h}/3$ Delaunay cut from Figure 10. The peak corresponds to the unaltered triangles (minimum angle $\pi/4$, minimum length 0.2). The Delaunay cut (solid line) yields larger minimum angles, and longer edges.

events are strictly ordered in time, and does not take special precautions for degenerate cases. When the scalpel enters or exit in a movement parallel to the surface in, multiple events happen simultaneously, which leads to various failures.

Flips can be implemented in constant time. Node removals can be accomplished in $\mathcal{O}(d \log d)$, where d is the degree of the node [9]. The expected value of d in surface meshes is 6, so the total remeshing process for a sin-

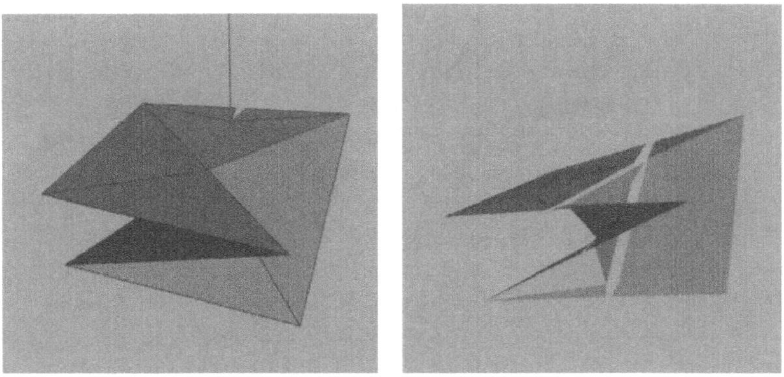

Fig. 12. In principle, our approach also works for a cut branching into three cuts. The mesh has very sharp angles and a limited resolution, so edge flips produce an odd end result.

gle movement can be implemented in constant time, which guarantees that the remeshing process itself is scalable to larger meshes. However, in the 3D version the scalpel can interact with all parts of the mesh at every step. In practice, efficient collision detection will be needed for a scalable implementation.

5 Discussion

We have presented an approach to cutting in triangulations that produces measurably smaller and better-shaped meshes than previous methods. The method is based on a model of a point-shaped scalpel that moves an active node through a static mesh. During movements, the area around the active node is remeshed. The approach generalizes to 3D surfaces. The 3D surface model supposes a line-shaped scalpel, and it is possible to generalize the single-incision technique to multiple incisions consistently.

The technique bears some resemblance to interactive mesh dragging, a technique proposed by Suzuki et al. [21] to allow dragging operations on triangle surfaces in interactive geometric modelers.

The technique that we have described is heuristic: it employs more or less arbitrary constant tolerances $\varepsilon_{\text{edge}}$, $\varepsilon_{\text{exit}}$ and $\varepsilon_{\text{entry}}$. This implies that the technique is only usable for meshes with a constant density, and that suitable values must be computed beforehand. In retrospect, it might have been wiser to consider an edge too short when the opposite angle in incident triangles is shorter than some threshold. This criterion is also local but scale independent.

We have used a static model for mesh cutting. In the context of surgery simulation and deformable models, this is not realistic: surgical instruments always interact with the deformed mesh, and exert force on the tissue that is

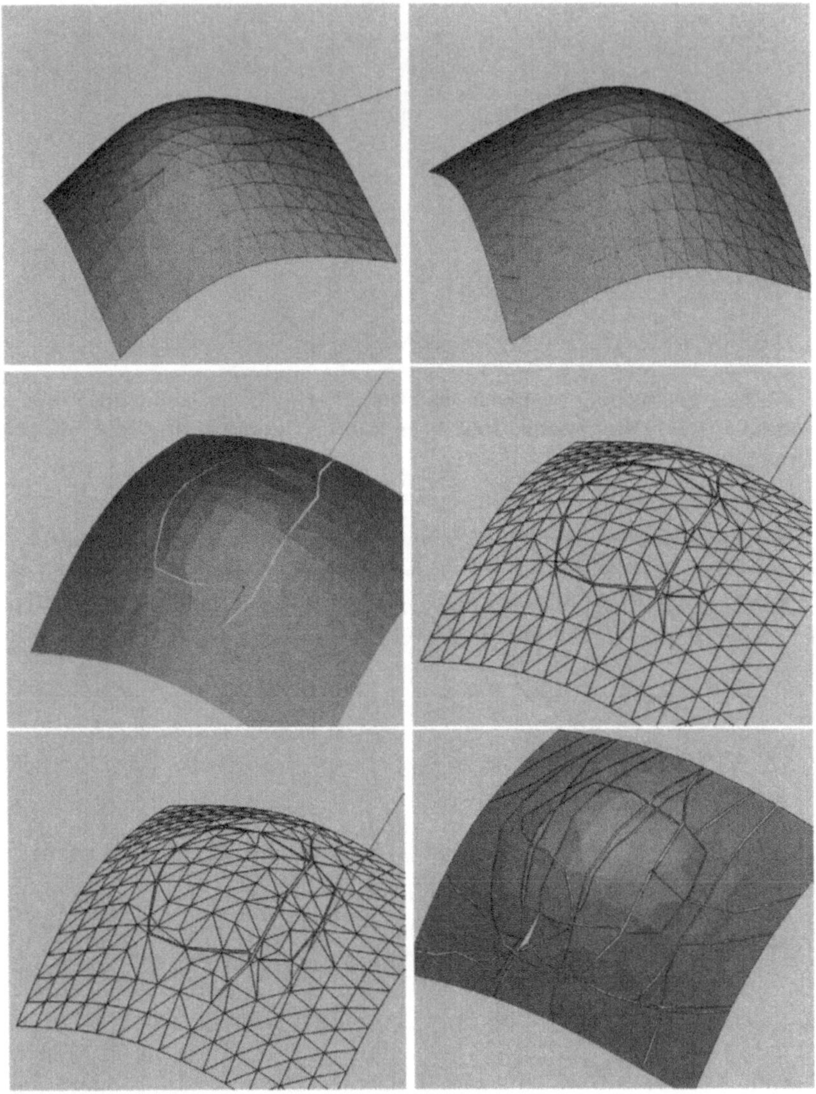

Fig. 13. Cuts in a 392 triangle 3D surface mesh of a a Gaussian bell-curve. The first 5 images show the evolution of two cut movements. The final image demonstrates more self-intersecting paths.

cut. A full-fledged simulation would be able to respond to movements with reaction forces, which could be relayed to a force-feedback device. However, the remeshing processs is not affected by the presence of deformation. Both the 2D and 3D algorithm include a step where the new position of the scalpel is intersected with parts of the mesh. When deformation is present, these intersections are done with the deformed mesh. The deformation of the triangles involved can be used to translate the point back to the reference situation. Our technique can proceed as described using reference locations.

Entry and exit are changed by the presence of deformation: surgical instruments must overcome a threshold in force to puncture membranes covering organs [3, 10]. Cutting requires less force than puncture, so after the instrument enters the tissue, it will make an incision immediately, ensuring that entry does not lead to arbitrary small incision depths. This is a physical variant of our entry-fixup, and it would make our own entry-control superfluous. A similar observation might also hold for a scalpel finishing a cut. Removing entry control is attractive, since our approach depends on a heuristically chosen quantitity $\varepsilon_{\mathrm{entry}}$, and it is possible to create a deadlock situation: if a scalpel entry has to be fixed up, the active node is part of the boundary. It is not always possible to preclude element inversion when moving boundary nodes, so the step-size control for large movements may halt further movements in some cases.

During a cut, nodes are added and removed; the nodes are added on line segments connecting existing nodes, which implies that the overall resolution of the mesh does not increase. Some interactive simulations of deformable objects refine meshes on demand to provide more accurate results in the region of interest [10, 17, 22]. Delaunay refinement algorithms [7, 18] seem to fit our framework of using Delaunay Triangulations, however more research is needed before this can be used in practice. Refinement algorithms need input geometries without small angles. Moreover, in 3D we are operating on a discretization of a curved surface, so it is not clear where to insert new points. Existing surface meshing algorithms, such as Chew's guaranteed quality surface Delaunay refinement [7], assume that a smooth description of the surface is available.

We have demonstrated an extension of our single incision 3D approach to multiple incisions. The extension is consistent with our model of the scalpel as a moving line segment. It remains to be seen how branching cuts should be combined with deformations.

Unfortunately, both the rationale for Delaunay and our heuristics do not readily generalize to tetrahedral meshes. Delaunay tetrahedralizations of well-spaced points admit *slivers*, flat tetrahedrons that have four nearly planar and nearly cocircular vertices (see Figure 14). Moreover, the higher dimensional equivalent of edge flipping (face flipping), does not always work: there exist configurations of non-Delaunay faces which can not be flipped. This implies that flipping as a local improvement strategy does not always work. If the tetrahedralization is already Delaunay, then flipping non-Delaunay faces during insertion or removal of points in a mesh is always possible. Since cuts are non-convex, this suggests that a tetrahedral generalization must build a constrained or conforming Delaunay tetrahedralization with a moving boundary. A first step towards enabling such cuts would be to extend the current 3D surface cutting approach to cuts of closed surfaces: when cutting a closed surface, surface triangles are added to the inside of the incision, so the surface remains closed during cuts. This surface could then be used as a basis for making a tetrahedralization. An additional complication is that a constrained Delaunay tetrahedralization only exists if edges have no non-incident nodes close by: the surface should be *ridge-protected* [20].

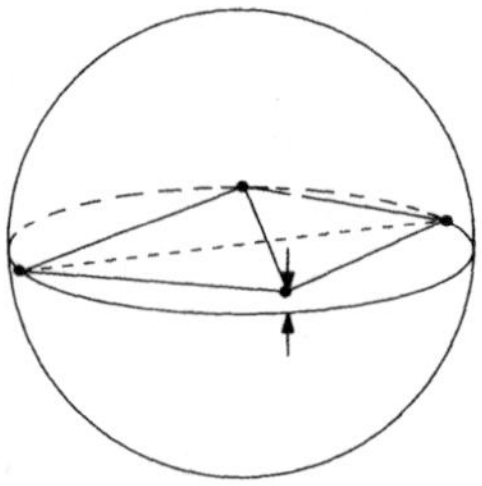

Fig. 14. Slivers have an excellent circumradius to shortest-edge ratio, so they can be present in Delaunay tetrahedralizations of well-spaced point sets. They are degenerate nevertheless.

Acknowledgements.
We would like to thank the anonymous referees for suggesting improvements in the presentation.

References

1. M. Bern and D. Eppstein. *Computing in Euclidian Geometry*, chapter Mesh generation and optimal triangulation, pages 47–123. World Scientific, 1995.
2. D. Bielser and M. H. Gross. Interactive simulation of surgical cuts. In *Proc. Pacific Graphics 2000*, pages 116–125. IEEE Computer Society Press, October 2000.
3. P. Brett, T. Parker, A. Harrison, T. Thomas, and A. Carr. Simulation of resistance forces acting on surgical needles. *J. of Engineering in Medicine*, 211(13):335–347, September 1997.
4. M. Bro-Nielsen, D. Helfrick, B. Glass, X. Zeng, and H. Connacher. VR simulation of abdominal trauma surgery. In *Medicine Meets Virtual Reality 6*, pages 117–123. IOS Press, 1998.
5. C. Bruyns and S. Senger. Interactive cutting of 3-D surface meshes. *Computer & Graphics*, 25(4):635–642, 2001.

6. C. D. Bruyns, S. Senger, A. Menon, K. Montgomery, S. Wildermuth, and R. Boyle. A survey of interactive mesh-cutting techniques and a new method for implementing generalized mesh cutting using virtual tools. *The journal of visualization and computer animation*, 13:21–42, 2002.
7. L. P. Chew. Guaranteed quality mesh generation for curved surfaces. In *Annual ACM Symposium on Computational Geometry*, pages 274–280, 1993.
8. S. Cotin, H. Delingette, and N. Ayache. A hybrid elastic model allowing real-time cutting, deformations and force-feedback for surgery training and simulation. *The Visual Computer*, 16(8):437–452, 2000.
9. O. Devillers. On deletion in delaunay triangulations. In *Annual ACM Symposium on Computational Geometry*, June 1999.
10. S. DiMaio and S. Salcudean. Needle insertion modelling and simulation. In *IEEE International Conference Robotics and Automation (ICRA)*, 2002.
11. F. Ganovelli, P. Cignoni, C. Montani, and R. Scopigno. A multiresolution model for soft objects supporting interactive cuts and lacerations. *Computer Graphics Forum*, 19(3):271–282, 2000.
12. F. Ganovelli and C. O'Sullivan. Animating cuts with on-the-fly re-meshing. In J. C. Roberts, editor, *EuroGraphics Short Presentations*, 2001.
13. M. Mäntylä. *An introduction to solid modeling.* Computer Science Press, College Park, MD, 1988.
14. G. Miller, D. Talmor, S. Teng, N. Walkington, and H. Wang. Control volume meshes using sphere packing: Generation, refinement and coarsening. In *Fifth International Meshing Roundtable*, pages 47–61, October 1996.
15. A. B. Mor and T. Kanade. Modifying soft tissue models: Progressive cutting with minimal new element creation. In *Medical Image Computing and Computer Assisted Intervention (MICCAI)*, number 1935 in LNCS, pages 598–607. Springer-Verlag, 2000.
16. H.-W. Nienhuys and A. F. van der Stappen. A surgery simulation supporting cuts and finite element deformation. In *Medical Image Computing and Computer Assisted Intervention (MICCAI)*, number 2208 in LNCS, pages 153–160. Springer-Verlag, October 2001.
17. C. Paloc, F. Bello, R. I. Kitney, and A. Darzi. Online multiresolution volumetric mass spring model for real time soft tissue deformation. In T. Dohi and R. Kikinis, editors, *Medical Image Computing and Computer Assisted Intervention (MICCAI)*, number 2489 in LNCS, pages 291–226. Springer-Verlag, 2002.
18. J. Ruppert. A Delaunay refinement algorithm for quality 2-dimensional mesh generation. *J. Algorithms*, 18(3):548–585, 1995.
19. D. Serby, M. Harders, and G. Székely. A new approach to cutting in finite element models. In *Medical Image Computing and Computer Assisted Intervention (MICCAI)*, number 2208 in LNCS, pages 425–433. Springer-Verlag, October 2001.
20. J. R. Shewchuk. A condition guaranteeing the existence of higher-dimensional constrained delaunay triangulations. In *Annual ACM Symposium on Computational Geometry*, 1998.
21. H. Suzuki, T. Kanai, Y. Sakurai, and F. Kimura. Interactive mesh dragging with adaptive remeshing technique. In *Pacific Graphics*, 1998.
22. X. Wu, M. S. Downes, T. Goktekin, and F. Tendick. Adaptive nonlinear finite elements for deformable body simulation using dynamic progressive meshes. *Computer Graphics Forum*, 20(3), 2001.

Stochastic Conformational Roadmaps for Computing Ensemble Properties of Molecular Motion

Mehmet Serkan Apaydın[1], Douglas L. Brutlag[1], Carlos Guestrin[1], David Hsu[2*], and Jean-Claude Latombe[1]

[1] Stanford University, Stanford, CA, USA
[2] National University of Singapore, Singapore

Abstract. A key intuition behind probabilistic roadmap planners for motion planning is that many collision-free paths potentially exist between two given robot configurations. Hence the connectivity of a robot's free space can be captured effectively by a network of randomly sampled configurations. In this paper, a similar intuition is exploited to preprocess molecular motion pathways and efficiently compute their ensemble properties, i.e., properties characterizing the average behavior of many pathways. We construct a directed graph, called *stochastic conformational roadmap*, whose nodes are randomly sampled molecule conformations. A roadmap compactly encodes many molecular motion pathways. Ensemble properties are computed by viewing the roadmap as a Markov chain. A salient feature of this new approach is that it examines all the paths in the roadmap simultaneously, rather than one at a time as classic methods such as Monte Carlo (MC) simulation would do. It also avoids the local-minima problem encountered by the classic methods. Tests of the approach on two important biological problems show that it produces more accurate results and achieves several orders of magnitude reduction in computation time, compared with MC simulation.

1 Introduction

Probabilistic roadmap (PRM) [1,8,9,13,16,17,25] planners have been successfully used in recent years to compute collision-free paths for robots with many degrees of freedom. A classic PRM planner [16] samples at random a robot's configuration space to construct a network that approximates the connectivity of the free space, and then searches the roadmap to process path planning queries. A key intuition behind PRM planners is that many collision-free paths potentially exist between two given robot configurations. Hence the connectivity of a robot's free space can be captured effectively by a network of randomly sampled configurations connected by collision-free curves. In this paper, a similar intuition is exploited to develop an efficient approach for analyzing the motion pathways of molecules during vital biological processes, such as protein folding and ligand-protein binding.

Molecules can be modeled approximately as articulated structures in 3-D space. They move under the influence of an energy field that includes van der Waals, electrostatic, and other potentials. For instance, to carry out biological functions, protein molecules remarkably assemble themselves, or *fold*, into unique 3-D structures called *native folds*. Protein folding plays a central role in biological processes essential to life. Failing to fold into the correct structures has serious consequences, including well-known diseases such as the Creutzfeldt-Jacobs (mad cow)

* Work completed at the University of North Carolina at Chapel Hill.

J.-D. Boissonnat et al. (Eds.): Algorithmic Foundations of Robotics V, STAR 7, pp. 131–147, 2004.

or Alzheimer's disease. Despite its importance, the protein folding process remains a mystery. While it is traditionally studied through tedious and costly laboratory experiments, computer simulation plays an increasingly important role.

Classic techniques for simulating molecular motion, including Monte Carlo [15] and molecular dynamics [12] methods, have two major drawbacks:

1. They compute individual pathways, one at a time; however, many interesting properties of molecular motion, in particular, the *ensemble properties*, are best characterized statistically over many pathways. For instance, the "new view" of protein folding asserts that proteins fold in a multi-dimensional energy funnel by following a myriad of pathways, all leading to the same native structure.
2. A typical molecular energy function contains many local minima, and the classic simulation techniques waste considerable computation time trying to escape from these local minima. This is similar to the behavior of potential field planners (see, e.g., [6]) in robot motion planning.

The high computational cost of these existing techniques prevents them from being used to analyze many pathways.

The *Stochastic Roadmap Simulation* (SRS) framework described in this paper overcomes both drawbacks [3]. In SRS, we build a network, called *stochastic conformational roadmap*, or just *roadmap* for short. A roadmap compactly encodes many pathways and captures the stochastic nature of molecular motion. More precisely, a roadmap is a directed graph, whose nodes are randomly sampled molecule conformations. The *conformation* of a molecule specifies its 3-D structure; the concept is similar to that of the *configuration* of a robot. An edge between two nodes v_i and v_j in the roadmap carries a weight P_{ij}, which estimates the probability for the molecule to transition from v_i to v_j. Every path in the roadmap corresponds to a potential motion pathway of the molecule. A roadmap thus combines a huge number of pathways, weighted by the probabilities that the molecule may follow these pathways. The construction of the roadmap also circumvents the local-minima problem encountered by the classic simulation techniques.

The probabilities attached to the edges of a roadmap directly express the stochastic nature of molecular motion. We view the motion of molecules on the roadmap as a random walk similar to a Monte Carlo (MC) simulation run. At each step of the random walk, a molecule either stays at the current node or moves to a neighboring node according to the assigned transition probabilities. However, to compute efficiently the ensemble properties of molecular motion, we avoid explicit simulation. Instead, we treat the roadmap as a Markov chain and apply the first-step analysis [31] from the Markov chain theory to process all pathways in the roadmap simultaneously, rather than one at a time as classic methods such as MC simulation would do. Conceptually, this is equivalent to performing infinitely many simulation runs simultaneously and extracting statistics from them, but it results in tremendous gain in computational efficiency.

SRS is by necessity more coarse-grained in sampling than MC simulation. While a MC simulation run focuses on one individual pathway, SRS must spread the samples over the entire conformation space. On the other hand, SRS examines many motion pathways at once and obtains information not easily accessible by classic methods. Tests of SRS on a number of protein folding and ligand-protein binding examples indicate empirically that SRS computes ensemble properties satisfactorily even with rather coarse roadmaps. In addition, it can be shown formally that, with appropriately defined edge probabilities, SRS and MC simulation converge to the same sampling distribution—the Boltzmann distribution—in the limit.

SRS is inspired by the PRM methods for robot motion planning. The stochastic conformational roadmap is a generalization of the probabilistic roadmap in PRM planners. In motion planning, the configuration space of a robot is the domain of a binary function specifying whether a configuration is collision-free. The conformation space of a molecule or a collection of molecules is the domain of a continuous energy function governing the motion of the molecules. As a result, the edges of a stochastic conformational roadmap are weighted with transition probabilities, while the edges of a probabilistic roadmap are unweighted.

Singh et al. introduced the PRM methods to the study of molecular motion in their work on ligand-protein binding [27]. Their approach has since been applied to protein folding as well [2,5,29]. Earlier work treats the roadmap as a deterministic graph, with heuristic edge weights based on the energy difference between molecule conformations. The heuristic edge weights measure the energetic difficulty of transitioning along the edges. Graph search techniques are used to extract "low-energy" paths from the roadmap. Our stochastic conformational roadmap is fundamentally different: it defines a Markov chain that captures the stochastic nature of molecular motion and enables us to analyze globally all the pathways contained in a roadmap by applying tools from Markov Chain theory. It also allows us to establish a formal relationship between SRS and the well-established MC method.

This paper combines and extends results presented in [3,4]. It provides additional experimental results and give new ideas on how efficient sampling strategies can be eventually designed to extend SRS to very high dimensional conformation spaces. In the following, we first cover some preliminaries (Sect. 2). We then describe how to construct a roadmap (Sect. 3) and query it to compute ensemble properties (Sect. 4). We used SRS to address two biological problems: computing the probability of folding in protein folding and estimating the escape time in ligand-protein binding. These experimental results are reported in Sects. 5 and 6.

2 Preliminaries

2.1 Conformation Space

The conformation of a molecule determines its 3-D structure. Conformations can be specified in various ways. For a protein molecule, we can specify the positions of the constituent atoms or the backbone torsional angles ϕ and ψ. SRS applies to many different representations, provided that the conformation of a molecule is specified by a finite number of parameters that uniquely determine the 3-D position of every atom in the molecule. Formally, a conformation q of d parameters is specified by a vector $(q_1, q_2, \ldots, q_d)$. The set of all conformations form the *conformation space* $\mathcal{C}$.

By determining the molecule's 3-D structure, the conformational parameters also determine the interactions between the atoms of the molecule and between the molecule and the medium, e.g., van der Waals and electrostatic interactions. These interactions give rise to the attractive and repulsive forces that govern molecular motion. SRS assumes that these interactions are described by an energy function $E(q)$ that depends only on the conformation q of the molecule; it does not require E to have any particular properties or functional forms.

2.2 Monte Carlo Simulation

MC simulation—more precisely, the Metropolis algorithm [20]—is one of the most common methods for simulating molecular motion. It samples the conformation space $\mathcal{C}$ of a system of molecules in order to study how they relax to or fluctuate around the equilibrium state. A key property of MC simulation is that, in the limit, the conformation space is sampled according to the Boltzmann distribution [18].

MC simulation starts at some initial conformation and performs a random walk in $\mathcal{C}$. Let q be the conformation at the current step of this random walk. To obtain the next conformation, a conformation q' is chosen from a small neighborhood of q, with a uniform or Gaussian distribution centered at q. The move to q' is accepted with a probability A that depends on the energy difference $\Delta E = E(q') - E(q)$. Define the *Boltzmann factors* $\varepsilon = \exp(-E(q)/k_B T)$ and $\varepsilon' = \exp(-E(q')/k_B T)$, where k_B is the Boltzmann constant, and T is the temperature of the system. The Metropolis criterion prescribes the acceptance probability as

$$A = \begin{cases} \exp(-\Delta E/k_B T) & \text{if } \varepsilon'/\varepsilon < 1 \\ 1 & \text{otherwise.} \end{cases} \tag{1}$$

Since $\varepsilon'/\varepsilon = \exp(-\Delta E/k_B T)$, the condition $\varepsilon'/\varepsilon < 1$ holds if and only if $\Delta E > 0$. So, if a move decreases the energy, it is always accepted; otherwise, it is accepted with probability $\exp(-\Delta E/k_B T)$. If the move from q to q' is accepted, the simulation transitions to q'; otherwise, it stays at q. The procedure then repeats to generate more sampled conformations, until termination conditions are met (e.g., the maximal number of steps is reached, or the quantity being computed stabilizes).

This procedure guarantees that if the number of simulation steps is sufficiently large, the sampled conformations are distributed according to the Boltzmann distribution with the density function

$$\beta(q) = \frac{1}{Z_\beta} \exp(-E(q)/k_B T),$$

where $Z_\beta = \int_{\mathcal{C}} \exp(-E(q)/k_B T)\, dq$ is a normalization constant. So any subset $S \subset \mathcal{C}$ is sampled with probability $\beta(S) = \int_S \beta(q)\, dq$.

MC simulation is an important tool for studying molecular motion. However, it is computationally intensive. Each simulation run yields a single pathway, and the simulation must be run many times over extended durations in order to produce accurate statistical results. Moreover, the energy function E typically contains many local minima. A MC simulation run spends most of its time overcoming energy barriers to escape from these local minima. Many similar conformations are sampled near the same local minimum, but they contain little new information.

2.3 Stationary Distribution of a Markov Chain

A Markov chain is a stochastic process that takes values from a finite or countable set of states $v_1, v_2, \ldots$ [31]. The probability of going from state v_i to v_j is P_{ij}, which depends only on v_i and v_j. Under suitable conditions, a Markov chain has a limit distribution $\pi = (\pi_1, \pi_2, \ldots)$ that can be obtained as follows. Starting at an initial state, perform a random walk over the set of all possible states. At each step of the random walk, make a move to the next state with the transition probability P_{ij}. If the random walk continues infinitely, then under the condition that the Markov chain is *ergodic*, each node v_i is visited with a fixed probability π_i in the limit, regardless of the starting state [31]. So π describes the limit behavior of *all* possible random walks. The probability π_i gives the fraction of the time that v_i is visited in the limit.

The limit distribution π satisfies the following equations [31]:

$$\pi_i = \sum_j \pi_j P_{ji} \quad \text{for all } i. \tag{2}$$

With the additional constraints $\pi_i \geq 0$ for all i and $\sum_i \pi_i = 1$, the solution to (2) is guaranteed to be a well-defined probability distribution. Equation (2) says that, in the limit, the distribution π no longer changes from one step of the random walk to the next. For this reason, π is called the *stationary distribution.*

If a conformation space is discretized into a finite set of states, MC simulation over this discretized space can be described by a Markov chain with appropriately defined transition probabilities. The stationary distribution of the Markov chain then gives the limit behavior of the MC simulation.

3 Stochastic Conformational Roadmaps

In Stochastic Roadmap Simulation, we preprocess molecular pathways by precomputing a roadmap, which provides a discrete representation of molecular motion. A roadmap compactly encodes a large number of MC simulation paths simultaneously and enables us to perform key computation efficiently.

3.1 Roadmap Construction

A roadmap G is a directed graph. Each node of G is a randomly sampled conformation in $\mathcal{C}$. Each (directed) edge from node v_i to node v_j carries a weight P_{ij}, which represents the transition probability of the molecule to move from conformation v_i to v_j. The probability P_{ij} is 0 if there is no edge from v_i to v_j. Otherwise, it depends on the energy difference $\Delta E_{ij} = E(v_j) - E(v_i)$.

To construct a roadmap, our algorithm first samples conformations independently at random from $\mathcal{C}$. Our current implementation samples uniformly by picking a value for each conformational parameter $q_1, q_2, \ldots$ uniformly at random from its allowable range (see Section 7 for a discussion of more efficient sampling strategies). Next, for each node v_i, the algorithm finds its nearest neighbors according to a suitable metric such as the root mean squared distance [18]. It then creates an edge between v_i and every neighboring node v_j and attaches to it the transition probability

$$P_{ij} = \begin{cases} \frac{1}{d_j} \exp(-\Delta E_{ij}/k_B T) & \text{if } \frac{\varepsilon_j/d_j}{\varepsilon_i/d_i} < 1 \\ \frac{1}{d_i} & \text{otherwise,} \end{cases} \tag{3}$$

where ε_i and ε_j are the Boltzmann factors at v_i and v_j, and d_i and d_j are the numbers of neighbors of v_i and v_j. If there is no edge between v_i and v_j, then they are considered too far apart for their energy difference to be a good basis for estimating the transition probability, and we set $P_{ij} = 0$. The molecule can still move from v_i to v_j, but the move must traverse at least one other node of the roadmap. Finally, a self-transition probability $P_{ii} = 1 - \sum_{j \neq i} P_{ij}$ is attached to each node v_i, ensuring that the transition probabilities from any node sum up to 1.

In contrast to the heuristic edge weights used in [5,27,29], the transition probabilities used in SRS allow us to establish a formal relationship between SRS and MC simulation [3]. We now describe this important relationship.

3.2 Relationship with Monte Carlo Simulation

In MC simulation, we perform a random walk in the conformation space $\mathcal{C}$. We can perform a similar random walk on the roadmap G: at node v_i of G, we choose a node v_j uniformly at random from the set of neighbors of v_i and propose a move to v_j. The move is accepted with probability

$$A_{ij} = \begin{cases} \frac{d_i}{d_j} \exp(-\Delta E_{ij}/k_B T) & \text{if } \frac{\varepsilon_j/d_j}{\varepsilon_i/d_i} < 1 \\ 1 & \text{otherwise.} \end{cases} \tag{4}$$

Expressions (1) and (4) are similar, except for the extra factor d_i/d_j, which is needed because the neighborhood sizes of all conformations are the same in MC simulation, but the number of neighbors varies from one node to another for the random walk on the roadmap. Since node v_i has d_i neighbors and each one is chosen with probability $1/d_i$, the transition probability from v_i to v_j is $(1/d_i)A_{ij}$, which, after simplification, is equal to P_{ij} given in (3). Hence, with our choice of transition probabilities, every path in the roadmap corresponds to a MC simulation run.

We have also stated in Sect. 2.2 that MC simulation samples conformations with a distribution that converges to the Boltzmann distribution β. So, in the limit, the probability of sampling any subset $S \subset \mathcal{C}$ is

$$\beta(S) = \frac{1}{Z_\beta} \int_S \exp(-E(q)/k_B T)\, dq.$$

Now we ask the question: what is the limit behavior of SRS? In other words, if we perform an arbitrary long random walk on the roadmap as described above, what is the probability of sampling a subset $S \subset \mathcal{C}$? Since a roadmap defines a Markov chain with transition probabilities P_{ij}, the limit behavior of SRS is governed by the the stationary distribution of the Markov chain:

Lemma 1. *A stochastic conformational roadmap defines a Markov chain with stationary distribution*

$$\pi_i = \frac{1}{Z_\pi} \exp(-E(v_i)/k_B T) \quad \textit{for all } i, \tag{5}$$

where $Z_\pi = \sum_i \exp(-E(v_i)/k_B T)$ *is a normalization constant.*

See Appendix A for the proof. To estimate the probability of sampling a set S, we simply sum the stationary distribution π over all the nodes v_i that lie in the set S:

$$\pi(S) = \sum_{v_i \in S} \pi_i = \frac{1}{Z_\pi} \sum_{v_i \in S} \exp(-E(v_i)/k_{\mathrm{B}}T).$$

If SRS represents the stochastic motion of a molecule with the same limit behavior as MC simulation, we expect the limit distributions of these two methods to converge. In other words, given a suitably dense roadmap, $\pi(S)$ should approximate $\beta(S)$ to any arbitrary precision, a result proven in our earlier paper [3].

Although SRS is closely related to MC simulation, it is far more efficient. A roadmap constructed by SRS combines many MC simulation paths, which can be processed together using tools from Markov chain theory, as we will see next.

4 Roadmap query

A roadmap G encodes considerable information on molecular motion. Given two nodes v_i and v_j in G, we can compute the most likely pathway from v_i to v_j by searching for a minimum-weight path from v_i to v_j in a graph similar to G, but with $-\ln P_{ij}$ as edge weights. This leads to results similar to those in earlier work [5,27,29]. However, since a roadmap explicitly represents the stochastic nature of molecular motion, it allows us to take advantage of powerful tools from the Markov chain theory. We now focus on one such tool, the *first-step analysis*.

To illustrate our description with a concrete example, consider a roadmap G built in the conformation space of a protein whose native fold is known. Let $\mathcal{F}$ be a set of nodes of G in the folded state. In other words, they are structurally close to the native fold. We are interested in finding, for every node v_i in G, the expected number of transitions t_i needed to go from v_i to the folded state, i.e., any node in $\mathcal{F}$. A naive way to compute t_i would be to perform many MC simulation runs, starting from v_i, and average the number of transitions taken by each run. Due to the potentially high variance among independent runs, it takes a large number of simulation runs for each node v_i in order to get accurate results.

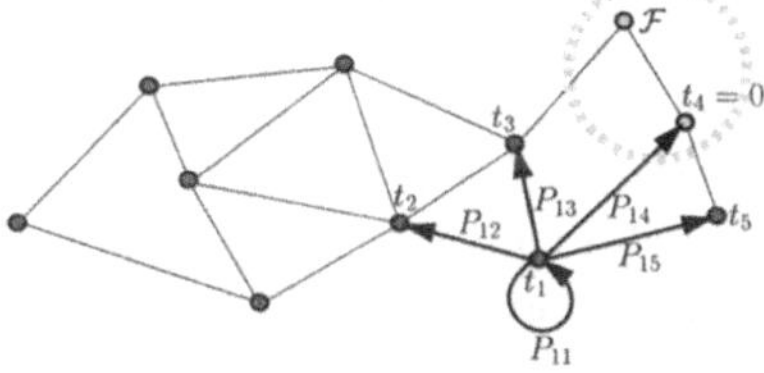

$$t_1 = 1 + P_{11} \cdot t_1 + P_{12} \cdot t_2 + P_{13} \cdot t_3 + P_{15} \cdot t_5$$

Fig. 1. First-step analysis.

In contrast, the first-step analysis proceeds by conditioning on the first transition. Suppose that we start at some node $v_i \notin \mathcal{F}$. After one step of transition, t_i is increased by one, and we either enter the folded state or reach another node $v_j \notin \mathcal{F}$. In the former case, we simply stop. In the latter case, the expected number of steps from then on is t_j. So we have the following system of linear equations:

$$t_i = 1 + \sum_{v_j \in \mathcal{F}} P_{ij} \cdot 0 + \sum_{v_j \notin \mathcal{F}} P_{ij} \cdot t_j \quad \text{for every } v_i \notin \mathcal{F}. \tag{6}$$

In the second term of (6), P_{ij} is multiplied by zero, because we stop as soon as we enter the folded state. See Fig. 1 for an illustration. The linear system (6) contains one equation and one unknown for each node $v_i \notin \mathcal{F}$. By solving this system, we obtain t_i for all the nodes simultaneously, without any explicit simulation.

To solve the linear system, let us rewrite (6) in the matrix form:

$$\mathbf{t} = \mathbf{Q} \cdot \mathbf{t} + \mathbf{b}, \tag{7}$$

where $\mathbf{Q}$ is a matrix with the transition probabilities P_{ij} as the entries, $\mathbf{t}$ is the vector of unknowns t_i, and $\mathbf{b}$ is a vector collecting the remaining constant terms in (6). Since a roadmap usually contains many nodes, the size of the coefficient matrix $\mathbf{I} - \mathbf{Q}$ is large ($\mathbf{I}$ is the identity matrix). Direct methods for solving (7), based on, e.g., Gaussian elimination, are impractical. However, it can be shown that a unique solution to (7) exists. So iterative methods can be used instead. In fact, the naive iteration $\mathbf{t}^{(k+1)} = \mathbf{Q} \cdot \mathbf{t}^{(k)} + \mathbf{b}$ is guaranteed to converge to the unique solution. This simple iterative method amounts to performing many simulation runs simultaneously using matrix multiplication. More efficient iterative methods, such as the conjugate gradient method [23], can also be used. Finally every roadmap node is, by construction, connected to only a small number of neighboring nodes, resulting in a sparse matrix $\mathbf{Q}$. The sparsity can be exploited to accelerate the iterative solver.

5 Computing the Probability of Folding

In this and next section, we use SRS to compute two ensemble properties: the *probability of folding* in protein folding and the *escape time* in ligand-protein binding.

Protein folding is one of the most marvelous processes in nature. Under suitable conditions, proteins go through a series of geometric transformations and arrive at the native folds where they perform intricate biological functions. There are large on-going efforts to understand the folding process (e.g., [14,22]): What geometric transformations does a protein go through during folding? Which conformations are "closer" to the native fold along the folding pathways?

To address this type of questions, the probability of folding (P_{fold})—also known as the transmission coefficient—has been introduced to measure how far away a protein conformation is from the native fold kinetically [11]. For a folding process dominated by two stable states, a folded state $\mathcal{F}$ and an unfolded state $\mathcal{U}$, the P_{fold} value τ for a conformation q is the probability of reaching $\mathcal{F}$ before $\mathcal{U}$, starting from q. If $\tau > 0.5$, then the protein is more likely to fold first than to unfold first, and therefore q is kinetically closer to the folded state [11]. Trivially, if q is in $\mathcal{F}$, then $\tau = 1$, and if q is in $\mathcal{U}$, then $\tau = 0$. The P_{fold} value at q is not associated with any particular folding pathway, but depends on all possible pathways from q. It is thus an ensemble property that describes the average behavior of the folding process.

5.1 Algorithmic Details

Using SRS, we can compute P_{fold} as follows. Let $v_i, i = 1, 2, \ldots$ be the nodes in a roadmap, and τ_i be the P_{fold} value for v_i. After constructing the roadmap, first-step analysis yields the following equation for every node v_i not in $\mathcal{F}$ or $\mathcal{U}$:

$$\tau_i = \sum_{v_j \in \mathcal{F}} P_{ij} \cdot 1 + \sum_{v_j \in \mathcal{U}} P_{ij} \cdot 0 + \sum_{v_j \notin (\mathcal{F} \cup \mathcal{U})} P_{ij} \cdot \tau_j, \tag{8}$$

which is obtained by conditioning on the first transition. After one step of transition, we have three cases. In the first case, we reach a node in $\mathcal{F}$. Then, we have reached $\mathcal{F}$ before $\mathcal{U}$ with probability 1. In the second case, we reach a node in $\mathcal{U}$. Then, we have reached $\mathcal{U}$ before $\mathcal{F}$, and the probability of reaching $\mathcal{F}$ before $\mathcal{U}$ is 0. In the third case, we reach a node v_j in neither $\mathcal{F}$ nor $\mathcal{U}$. The probability τ_i then depends on the value of τ_j. Linear system (8) has the same matrix form as the example in Sect. 4. A unique solution exists and can be obtained by an iterative solver.

We can improve the accuracy and potentially the speed of the iterative solver by setting all the self-transition probabilities in the roadmap to 0 and renormalizing the other probabilities. Set

$$\begin{aligned} P'_{ii} &= 0 && \text{for all } i, \\ P'_{ij} &= P_{ij} / \textstyle\sum_{k \neq i} P_{ik} && \text{for all } i \neq j \end{aligned} \tag{9}$$

and solve the linear system

$$\tau_i = \sum_{v_j \in \mathcal{F}} P'_{ij} \cdot 1 + \sum_{v_j \in \mathcal{U}} P'_{ij} \cdot 0 + \sum_{v_j \notin (\mathcal{F} \cup \mathcal{U})} P'_{ij} \cdot \tau_j . \tag{10}$$

If we think in terms of performing a random walk on the roadmap as described in Sect. 3.2, then setting the self-transition probabilities to 0 means never staying at the current node. It is easy to verify that linear systems (8) and (10) have the same solution by substituting (9) into (10). However, if we write (10) in the matrix form, the coefficient matrix $\mathbf{I} - \mathbf{Q}$ contains 1 in all its diagonal entries, which are greater than or equal to the corresponding entries in the matrix for (8). So (10) tends to be better conditioned, resulting in more stable, faster solution by iterative methods.

5.2 Experimental Results

We now present results on three examples. The first one uses a relatively simple synthetic energy function in a 2-D conformation space, and the other two use real protein data. We compare the results from SRS to those from MC simulation, and demonstrate that SRS reduces the running time by several orders of magnitude and is more accurate. The main reason for using synthetic data in the first example is that MC simulation takes extremely long computation time on real proteins. The simpler synthetic energy function allows us to perform more extensive comparisons.

Synthetic data The synthetic energy function in a 2-D conformation space is constructed from a linear combination of radially symmetric Gaussians, with a paraboloid centered at the origin. The centers, the decay rates, and the heights of the Gaussians are picked at random. The function has two global minima, one of which represents the native fold. When constructing a roadmap in this hypothetical conformation space, we use the Euclidean distance in the 2-D space for finding neighboring nodes.

We used SRS to compute P_{fold} for 101 sampled conformations with a roadmap of 10102 nodes, and

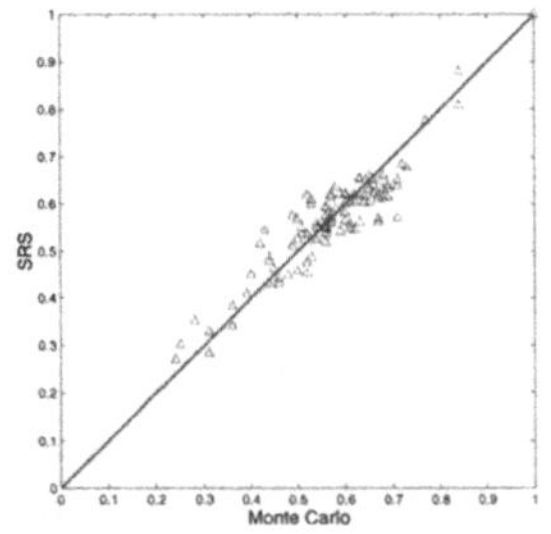

Fig. 2. The scatter plot of P_{fold} values from MC simulation and from SRS on a synthetic energy function.

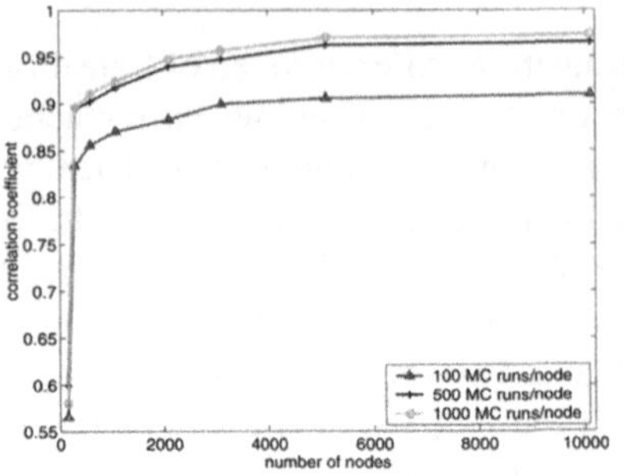

Fig. 3. The correlation coefficient κ as a function of the number of nodes in the roadmap. The three curves correspond to MC simulation with $r = 100, 500, 1000$ independent runs per node. As r increases, the correlation between the results from the two methods improves.

then used MC simulation to compute the results for the same conformations. In the MC simulation, we performed 500 independent runs for each conformation. Each run stops as soon as it enters a small neighborhood of a conformation in the folded or the unfolded state. The results computed with the two methods are plotted along the two axes in Fig. 2. All the points in the plot lie close to the diagonal line, indicating that the results from the two methods are in good correspondence.

We conducted further tests by varying the number of nodes sampled by SRS and the number of independent MC simulation runs per node. In each test, we summarize the correspondence between the results from the two methods by their (normalized) correlation coefficient, which is defined as

$$\kappa(x,y) = \frac{\langle xy\rangle - \langle x\rangle\langle y\rangle}{\sqrt{(\langle x^2\rangle - \langle x\rangle^2)(\langle y^2\rangle - \langle y\rangle^2)}}$$

for two vectors x and y, where $\langle\cdot\rangle$ denotes the operation of taking the average. Note that κ is always between -1 and 1, with 1 indicating perfect correlation, -1 indicating perfect inverse correlation, and 0 indicating no correlation. Fig. 3 shows the results of these additional tests. The horizontal axis of the graph is the number of nodes in the roadmap, and the vertical axis is the correlation coefficient κ. The graph contains three curves, corresponding to different numbers of independent MC simulation runs per node. These curves show a generally similar trend. Initially κ improves quickly as the number of nodes in the roadmap increases. The curves then flatten out after a certain point. It is not immediately clear whether they will reach 1, which indicates perfect correlation. Since κ measures only the correspondence between the two methods and we do not know the ground truth, these general trends do not tell us whether the discrepancy is due to the inaccuracy in SRS or the variance inherent in MC simulation. However, we can get a hint by comparing the three curves. For a roadmap of a given size, κ generally improves as we increase the number of independent MC simulation runs per node. This seems to indicate that SRS is more accurate: when the number of independent MC simulation runs per node increases, the variance of MC simulation decreases, and simultaneously, the results get closer to those obtained from SRS.

We also compared the running time of the two methods. The running time of SRS consists of the time to construct the roadmap and the time to solve a linear system of equations. On real proteins, the first part is dominated by the time to evaluate the energy of sampled nodes and the time to find neighboring nodes. The

Table 1. Running times of 100 MC simulation runs per conformation on the synthetic energy function. The first row gives the number of conformations for which P_{fold} is computed. The second row gives the corresponding running times.

No. Conf.	10	20	30	40	50	60	70	80	90	100
Time (sec.)	866	1588	2356	3191	4026	4913	5621	6404	7203	8077

second part depends on the size of the linear system, hence on the number of nodes in the roadmap. The running time of MC simulation is dominated by the time to compute the energy of sampled conformations.

In our current implementation, the roadmap construction part of SRS is coded in C++, and the linear system solver, in Matlab. MC simulation is implemented entirely in C++. The timing results reported here were obtained on a 1GHz Pentium-III PC with 1GB of memory. In a typical run on the synthetic energy function, SRS took about 8 seconds to construct a roadmap of 10,000 nodes, and 3 seconds to solve the linear system and obtain P_{fold} values for all the nodes. The running times of MC simulation is tabulated in Table 1. As expected, the running time of MC simulation is linear in the number of conformations processed. Although we did not perform MC simulation on all 10,000 conformations, we can easily infer that that the running time would be around 800,000 seconds.

1ROP and 1HDD We also tested SRS on two real proteins (Fig. 4), repressor of primer and engrailed homeodomain, which are identified as 1ROP and 1HDD, respectively, in the Protein Data Bank [7]. 1ROP is a dimer that consists of two identical parts called monomers. As in [30], we study one monomer in isolation. The monomer contains 56 residues forming two α helices connected by a loop. 1HDD contains 57 residues forming three α helices packed against each other.

Our implementation represents a protein as a sequence of vectors, each representing a secondary structure element [5,26]. In this vector-based representation, 1ROP has 6 degrees of freedom (dofs), and 1HDD has 12 dofs. Our energy function uses an H-P model [30] consisting of two terms, measuring the hydrophobic interaction and the excluded volume, respectively. In both SRS and MC simulation, we discard conformations that cause steric hindrance, i.e., self-collision of atoms in the protein. We define the folded state to contain all conformations within a small root mean squared distance (RMSD) of the native fold (3 Å for 1ROP and 5 Å for 1HDD), and the unfolded state to contain all the conformations within 10 Å of the fully-extended conformation. The roadmap construction software uses the RMSD as a metric to find neighboring nodes, as it better measures the similarity between two protein conformations than the Euclidean distance.

For each protein, we computed the P_{fold} values at about 45 randomly selected conformations using both SRS and MC simulation. With SRS, we computed the estimates with roadmaps having increasing numbers of nodes. In MC simulation, we performed up to 300 independent runs for each of the selected conformations. The results, shown in Fig. 5, suggest conclusions similar to those obtained from the synthetic energy function. First, SRS estimates generally improve very fast as

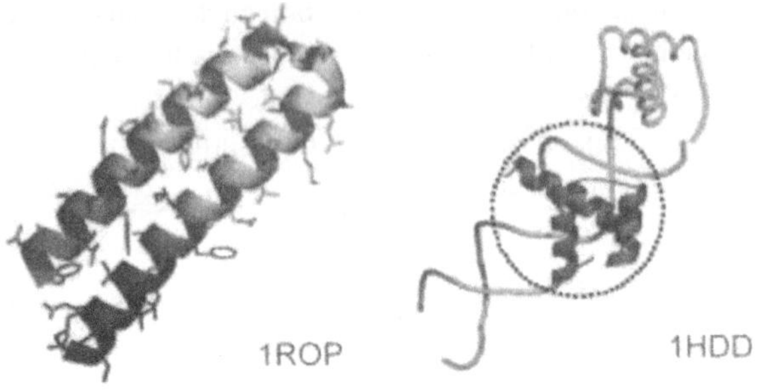

Fig. 4. Two proteins used in the experiments: 1ROP and 1HDD (circled) in complex with DNA.

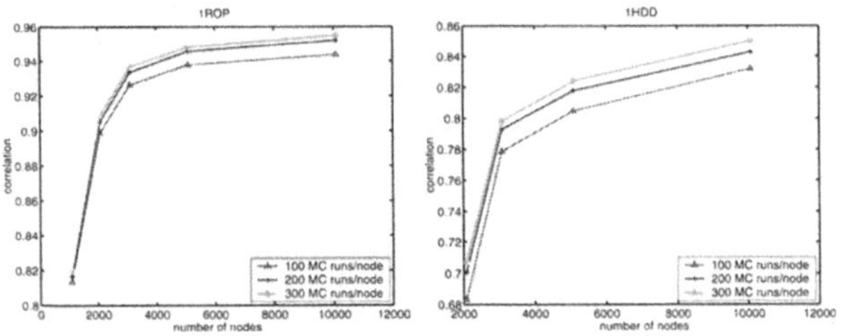

Fig. 5. The correlation of P_{fold} values for 1ROP and 1HDD, computed by SRS and MC simulation.

the roadmap size increases. Second, the correlation tends to increase as we perform more MC runs per node. We also compared the P_{fold} values obtained from the two methods using their average absolute differences, instead of their correlation coefficients. The conclusion is similar.

The total time to generate a roadmap of 5000 nodes and compute the P_{fold} values for all these nodes was about 1.5 hours. In comparison, it took an average of five to six hours on the same machine to execute 300 MC simulation runs required to estimate P_{fold} at just *one* conformation. To compute P_{fold} at the 45 selected conformations, MC simulation took about 250 hours, while SRS took only *1.5 hours* for all 5000 conformations. So SRS reduced the running time by at least four orders of magnitude in these examples.

6 Estimating the Escape Time from a Ligand-Protein Binding Site

Ligand-protein binding is another important biological process, in which a small molecule, called a *ligand*, is bound to a specific site, usually a cavity on the surface of a receptor protein in order to inhibit or enhance activities of the protein. Studying ligand-protein binding helps in discovering new pharmaceutical drugs. For example, drug molecules have been designed to bind to the active site of the enzymatic protein HIV-1 protease. They block access to the active site of the amino acid chains forming part of the HIV virus and thus disable the activation of the mature virus.

A receptor protein may have several potential binding sites. Therefore, it is important to be able to predict which is the active site, the site that enables specific biological functions, e.g., inhibition or catalysis. Let us consider the conformation space $\mathcal{C}$ of a ligand-protein complex with a suitably defined energy function. A bound conformation $q \in \mathcal{C}$ generally corresponds to a local energy minimum and has a "funnel of attraction" around q to stabilize the ligand. At the active site, the ligand is usually bound with very high affinity, and so it takes much longer to escape from the corresponding funnel, compared with other potential binding sites. We hypothesize that the longer escape time results from higher energy barriers around the active site and it may serve as a basis for prediction. We thus examine the ligand-protein binding process using a dynamic model. In contrast, most existing methods for analyzing ligand-protein binding use static models and consider only the energy of the final bound conformation of the ligand-protein complex (e.g., see [21,33]).

Following [10], we define the funnel of a bound conformation q as the set of conformations within 10 Å of q in RMSD. We then use SRS to compute the expected number of transitions for the ligand to reach a conformation outside of the funnel and use it as an approximation to the escape time.

6.1 Algorithmic Details

Given a suitable energy function for the ligand-protein complex, we first construct a roadmap to capture the motion of the ligand and then apply first-step analysis to obtain a system of equations similar to (6). Let $\mathcal{A}$ be the set of roadmap nodes in the funnel of the bound conformation q_{b}. Let t_i be the expected number of transitions to reach a conformation outside of $\mathcal{A}$, starting from a node $v_i \in \mathcal{A}$. We have

$$t_i = 1 + \sum_{v_j \notin \mathcal{A}} P_{ij} \cdot 0 + \sum_{v_j \in \mathcal{A}} P_{ij} \cdot t_j \quad \text{for every } v_i \in \mathcal{A}. \tag{11}$$

The solution to this system of equations gives an approximation of the escape time for every node in the funnel, including the bound conformation q_{b}.

6.2 Experimental Results

We applied our method to seven different ligand-protein complexes whose active sites are known. They are listed in Table 2. For each complex, we assume that the protein is rigid, and that the ligand is flexible with varying number of torsional dofs listed in column 3 of the table. Our energy function models the ligand-protein interactions by incorporating terms for van der Waals and electrostatic interactions [28].

To find potential binding sites, we picked 10,000 conformations at random and performed descent from them until we reached local energy minima. For each complex, four low-energy conformations that are less than 5 Å to the protein surface and are separated by greater than 10 Å in RMSD were selected along with the known active site as the potential bound conformations.

We computed 20 roadmaps of 10,000 nodes each for every potential binding site. The nodes were uniformly sampled in a region within 15 Å in RMSD of the bound conformation. We then solved for the escape times using (11). The results were averaged and listed in columns 5–8 of Table 2. Every row of the table shows the estimates of escape times for the various binding sites of a ligand-protein complex. In four of the seven cases, the escape time for the active site is larger than those for the other binding sites by at least two orders of magnitude, clearly distinguishing the active site. In two other cases (1LDM and 1CJW), the escape time for

Table 2. Escape times from potential binding sites for seven ligand-protein complexes.

Protein	Ligand	dofs	Binding Sites				
			Active	1	2	3	4
1LDM	oxamate	7	5.8e+06	1.6e+07	1.1e+06	3.7e+06	4.5e+05
1AO5	IPM	10	4.1e+10	1.2e+07	7.9e+06	1.2e+05	2.9e+04
3TPI	PTI	13	1.0e+10	1.1e+06	1.8e+05	1.0e+05	6.6e+05
4TS1	hydroxylamine	9	2.4e+10	5.4e+06	4.2e+07	7.2e+05	2.2e+06
1CJW	COT	21	6.3e+06	8.2e+06	5.6e+05	1.5e+05	1.9e+05
1AID	THK	14	1.4e+06	2.8e+07	5.0e+05	1.2e+05	2.1e+06
1STP	streptavidin	11	7.0e+08	6.4e+06	2.2e+06	8.5e+05	2.0e+06

the active site is close to the largest. In one case (1AID), the escape time fails to give a clear indication on the active site. The failure may have several causes. The size of the roadmaps may be too small to estimate the escape times accurately. The energy function that we use may not be detailed enough to capture all significant interactions between the ligand and the protein. Finally, it is possible that the active site often, but not always, has the highest escape time in nature.

For each binding site, our software took about 7 minutes in total to construct the roadmap and solve the linear system to obtain an estimate of the escape time.

Using a similar approach, we also studied the effect of mutating individual amino acids in the catalytic site of enzymatic proteins. These additional experimental results and others are available in [4,32].

7 Conclusion and Future Work

Stochastic Roadmap Simulation is a new framework for computing ensemble properties of molecular motion. It is closely related to MC simulation. A path in a stochastic conformational roadmap can be interpreted as a MC simulation run. Furthermore SRS converges to the same sampling distribution as MC simulation. A salient feature of SRS is that it compactly encodes many motion pathways and processes them together by solving linear equations, rather than considering them one at a time as the classic Monte Carlo and molecular dynamic methods would do. It also avoids the local-minima problem that plagues the existing methods. As a result, SRS gains tremendous computational efficiency, as demonstrated by our experiments.

We applied SRS to two biological problems. In the first problem, we computed the probability of folding, which measures the "kinetic distance" between a protein conformation and the native fold. Our experiments show that SRS reduced the running times by several orders of magnitude and obtained arguably more accurate results, compared with MC simulation. In the second problem, we computed estimates of the expected time for a ligand to escape from a binding site and used it to compare the active site of a protein to other potential binding sites.

SRS can be extended in several ways. An important algorithmic question is to develop sampling strategies that allow us to study larger molecules with more complex energy models. Currently we sample the conformation space $\mathcal{C}$ or a subset of it uniformly at random. As the dimension of $\mathcal{C}$ increases, it becomes more difficult to obtain biologically interesting conformations with uniform sampling, and the quality of results obtained from uniformly sampled roadmaps is likely to degrade. This is similar to the development of PRM planners for motion planning. To solve complex motion planning problems in high-dimensional configuration spaces, efficient sampling strategies are needed (see [24] for a review).

To address the problem, one approach is to construct a sampling distribution that favors low-energy conformations over high-energy ones, because molecules are more likely to stay in low-energy states. We achieve this by first sampling conformations uniformly at random and then retaining them with probability that decreases as the the energy of the conformations increases. We may also resample near known low-energy conformations to boost the density of roadmap nodes in these potentially interesting regions. Equally important is to identify energy barriers between neighboring nodes in a roadmap, when computing transition probabilities. To this end, we may sample the straight-line path between two neighboring nodes and compute the energy of intermediate conformations along the path. Finally, biologically interesting conformations are known to be often located in regions where the energy function undergoes significant variations, e.g., protein conformations in the transition state. To increase the sampling density in these regions, techniques such as the Gaussian sampling [9] can be used to sample a pair of conformations and retain a sample with higher probability when the pair exhibits very different energies.

Assume that a good sampling distribution σ can be constructed. We must then adjust the transition probabilities to account for the non-uniformity of the roadmap nodes so that SRS still converges to the Boltzmann distribution in the limit. One possibility is to define the new transition probability

$$P_{ij} = \begin{cases} \frac{\sigma_i}{d_j \sigma_j} \exp(-\Delta E_{ij}/k_B T) & \text{if } \frac{\varepsilon_j/d_j\sigma_j}{\varepsilon_i/d_i\sigma_i} < 1 \\ \frac{1}{d_i} & \text{otherwise,} \end{cases}$$

where σ_i and σ_j are the probabilities of sampling nodes v_i and v_j. Work is underway to investigate the validity of this transition probability assignment.

Another interesting extension is to add nodes incrementally to the roadmap and refine the solution of SRS, thus avoiding the need to choose the number of roadmap nodes in advance.

We are also interested in applying SRS to other important problems related to molecular motion. For instance, our transition probabilities are closely related to the master equation [19], which has been used to study the rate of protein folding. Earlier work on this problem exhaustively enumerates all conformations and is thus limited to very small proteins on a lattice (in the plane). Our experience indicates that SRS will likely remove the need for such costly enumeration.

Acknowledgements. This work has been funded in part by NSF ITR grants ACI-0082554 and CCR-0086013 and a grant from Stanford Bio-X program. Apaydin was supported by the D.L. Cheriton Stanford Graduate Fellowship. Brutlag was supported by NHGRI grant HGF02235. Guestrin was supported by a Siebel Scholarship and by the Sloan Foundation. We thank D. Koller, V. Pande, A. Singh, and J. Snoeyink for helpful discussions. We also thank C. Varma for preparing the energy files for some of the ligand-protein complexes.

References

1. N.M. Amato, O.B. Bayazit, L.K. Dale, C. Jones, and D. Vallejo. OBPRM: An obstacle-based PRM for 3D workspaces. In P.K. Agarwal et al., editors, *Robotics: The Algorithmic Perspective: 1998 Workshop on the Algorithmic Foundations of Robotics*, pages 155–168. A. K. Peters, Wellesley, MA, 1998.
2. N.M. Amato, K.A. Dill, and G. Song. Using motion planning to map protein folding landscapes and analyze folding kinetics of known native structures. In *Proc. ACM Int. Conf. on Computational Biology (RECOMB)*, pages 2–11, 2002.
3. M.S. Apaydin, D.L. Brutlag, C. Guestrin, D. Hsu, and J.C. Latombe. Stochastic roadmap simulation: An efficient representation and algorithm for analyzing molecular motion. In *Proc. ACM Int. Conf. on Computational Biology (RECOMB)*, pages 12–21, 2002.
4. M.S. Apaydin, C.E. Guestrin, Chris Varma, D.L. Brutlag, and J.C. Latombe. Stochastic roadmap simulation for the study of ligand-protein interactions. *Bioinformatics*, 18(supplement 2):18S–26S, 2002.
5. M.S. Apaydin, A.P. Singh, D.L. Brutlag, and J.C. Latombe. Capturing molecular energy landscapes with probabilistic conformational roadmaps. In *Proc. IEEE Int. Conf. on Robotics & Automation*, pages 932–939, 2001.
6. J. Barraquand and J.C. Latombe. Robot motion planning: A distributed representation approach. *Int. J. Robotics Research*, 10(6):628–649, 1991.
7. F.C. Bernstein et al. The protein data bank: A computer-based archival file for macromolecular structure. *J. Mol. Biol.*, 112(3):535–542, 1977.
8. R. Bohlin and L.E. Kavraki. Path planning using lazy PRM. In *Proc. IEEE Int. Conf. on Robotics & Automation*, pages 521–528, 2000.
9. V. Boor, M.H. Overmars, and F. van der Stappen. The Gaussian sampling strategy for probabilistic roadmap planners. In *Proc. IEEE Int. Conf. on Robotics & Automation*, pages 1018–1023, 1999.
10. C.J. Camacho and S. Vajda. Protein docking along smooth association pathways. *Proc. Nat. Acad. Sci. USA*, 98(19):10636–10641, 2001.
11. R. Du, V. Pande, A.Y. Grosberg, T. Tanaka, and E. Shakhnovich. On the transition coordinate for protein folding. *J. Chem. Phys.*, 108(1):334–350, 1998.
12. J.M. Haile. *Molecular Dynamics Simulation: Elementary Methods.* John Wiley & Sons, New York, 1992.
13. D. Hsu, J.C. Latombe, and R. Motwani. Path planning in expansive configuration spaces. *Int. J. Computational Geometry & Applications*, 9(4 & 5):495–512, 1999.
14. IBM Blue Gene Team. Blue gene: A vision for protein science using a petaflop supercomputer. *IBM Systems Journal*, 40(2):310–327, 2001.
15. M.H. Kalos and P.A. Whitlock. *Monte Carlo Methods*, volume 1. John Wiley & Son, New York, 1986.
16. L.E. Kavraki, P. Švestka, J.C. Latombe, and M.H. Overmars. Probabilistic roadmaps for path planning in high-dimensional configuration space. *IEEE Trans. on Robotics & Automation*, 12(4):566–580, 1996.
17. S.M. LaValle and J.J. Kuffner. Randomized kinodynamic planning. *Int. J. Robotics Research*, 20(5):278–400, 2001.
18. A.R. Leach. *Molecular Modelling: Principles and Applications*. Longman, Essex, England, 1996.
19. M.Cieplak, M.Henkel, J. Karbowski, and J.R.Banavar. Master equation approach to protein folding and kinetic traps. *Phys. Rev. Let.*, 80:3654, 1998.
20. N. Metropolis, A.W. Rosenbluth, M.N. Rosenbluth, A.H. Teller, and E. Teller. Equations of state calculations by fast computing machines. *J. Chem. Phys.*, 21:1087–1092, 1953.

21. G.M. Morris, D.S. Goodsell, R.S. Halliday, R. Huey, W.E. Hart, R.K. Belew, and A.J. Olson. Automated docking using a Lamarckian genetic algorithm and an empirical binding free energy function. *J. Comput. Chem.*, 19(14):1639–1662, 1998.
22. V.S. Pande et al. Atomistic protein folding simulations on the hundreds of microsecond timescale using worldwide distributed computing. *Biopolymers*, to appear.
23. Y. Saad. *Iterative Methods for Sparse Linear Systems*. PWS, New York, 1996.
24. G. Sanchez and J.C. Latombe. On delaying collision checking in PRM planning—application to multi-robot coordination. *Int. J. Robotics Research*, 21(1):5–26, 2002.
25. T. Siméon, J.P. Laumond, and F. Lamiraux. Move3D: A generic platform for motion planning. In *Proc. IEEE Int. Symp. on Assembly & Task Planning*, 2001.
26. A.P. Singh and D.L. Brutlag. Hierarchical protein structure superposition using both secondary structure and atomic representations. In *Proc. Int. Conf. on Intelligent Systems for Molecular Biology*, pages 284–293, 1997.
27. A.P. Singh, J.C. Latombe, and D.L. Brutlag. A motion planning approach to flexible ligand binding. In *Proc. Int. Conf. on Intelligent Systems for Molecular Biology*, pages 252–261, 1999.
28. K. Smith and B. Honig. Evaluation of the conformational free energies of loops in proteins. *Proteins: Structure, Function, and Genetics*, 18:119–132, 1994.
29. G. Song and N.M. Amato. Using motion planning to study protein folding pathways. In *Proc. ACM Int. Conf. on Computational Biology (RECOMB)*, pages 287–296, 2001.
30. S. Sun, P.D. Thomas, and K.A. Dill. A simple protein folding algorithm using a binary code and secondary structure constraints. *Protein Engineering*, 8:769–778, 1995.
31. H. Taylor and S. Karlin. *An Introduction to Stochastic Modeling*. Academic Press, New York, 1994.
32. C. Varma. Computing protein-ligand interaction kinetics using Markov methods. Master's thesis, Dept. of Computer Science, Stanford University, Stanford, CA, 2002.
33. J. Wang, P.A. Kollman, and I.D. Kuntz. Flexible ligand docking: A multiple strategy approach. *Proteins: Structure, Function, and Genetics*, 36(1):1–19, 1999.

A Proof of Lemma 1

Proof. We would like to show that the distribution π given in (5) is the stationary distribution for the Markov chain induced by the roadmap G. First, note that it suffices to show that π satisfies the detailed balance [31]:

$$\pi_i P_{ij} = \pi_j P_{ji}, \tag{12}$$

because if (12) holds, then $\sum_j \pi_j P_{ji} = \sum_j \pi_i P_{ij} = \pi_i \sum_i P_{ij} = \pi_i$, as required by the condition for a stationary distribution, given in (2). Now consider two nodes v_i and v_j from the roadmap. Without loss of generality, assume $\frac{\varepsilon_j/d_j}{\varepsilon_i/d_i} < 1$. We have

$$P_{ij} = \frac{1}{d_j}\exp(-\Delta E_{ij}/k_{\mathrm{B}}T) \quad \text{and} \quad P_{ji} = \frac{1}{d_j}.$$

Substituting these expressions into (12), we can easily verify that (12) is satisfied, after simplification. □

View Planning via Maximal C-space Entropy Reduction

Pengpeng Wang and Kamal Gupta

School of Engineering Science, Simon Fraser University, Burnaby, BC, CANADA, V5A 1S6 {pwangf, kamal}@cs.sfu.ca

Abstract. We introduced the concept of C-space entropy recently in [1–3] as a measure of knowledge of C-space for sensor-based path planning and exploration for general robot-sensor systems. The robot plans the next sensing action to maximally reduce the expected C-space entropy, also called the maximal expected entropy reduction, or MER criterion. The expected C-space entropy computation, however, made two idealized assumptions. The first was that the sensor field of view (FOV) is a point; and the second was that no *occlusion (or visibility) constraints* are taken into account, i.e., as if the obstacles are transparent. We extend the expected C-space entropy formulation where these two assumptions are relaxed, and consider a generic range sensor with non-zero volume FOV and occlusion constraints, thereby modelling a real range sensor. Planar simulations show that (i) MER criterion results in significantly more efficient exploration than the naive physical space based criterion (such as maximize the unknown physical space volume), (ii) the new formulation with non-zero volume FOV results in further improvement over the point FOV based MER formulation.

1 Introduction

While most research in sensor-based path planning and exploration has concerned itself with mobile robots, our recent work has concentrated on general robot-sensor systems, where the sensor is mounted on a robot with non-trivial geometry and kinematics [1–3,5]. See also [6–10]. This class of robots is broad and includes robots ranging from a simple polygonal robot to articulated arms, mobile-manipulator systems, and humanoid robots [11]. The sensor is assumed to be an "eye" type sensor that is capable of providing distances from a given vantage point (actual implementation may be a laser range scanner, passive stereo vision, etc.). Figure 1 shows a simple example of such a robot-sensor system — an eye-in-hand system — an articulated arm with a wrist mounted range sensor. The robot must simultaneously plan paths and sense its environment for obstacles. Unlike for a simple mobile robot (often modelled as a point, see [13] for next best view planning for mobile robots), where the robot can move (path planning) and what it should sense (view planning), has a much more complex relationship here [5]. "Where to move" is best posed and answered in configuration space, the natural space for path planning. In [1,2], we showed that the view planning problem is also appropriately posed in the configuration space of the robot — the next view should

J.-D. Boissonnat et al. (Eds.): Algorithmic Foundations of Robotics V, STAR 7, pp. 149–165, 2004.

be planned to give maximum knowledge or information (whether a configuration is free or in collision with an obstacle) of the C-space of the robot. What this implies is that the sensing action, which obviously senses physical space, must be (implicitly or explicitly) transformed to the configuration space. Treating the unknown environment stochastically, we introduced the notion of C-space entropy, measure of the robot's knowledge of C-space. The next best view is then the one that maximizes the expected entropy reduction (MER criterion) or, equivalently expected information gain. In contrast, earlier approaches had simply used a naive criterion, such as maximize unknown physical space volume (MPV) in the sensor FOV, to choose the next best view [6].

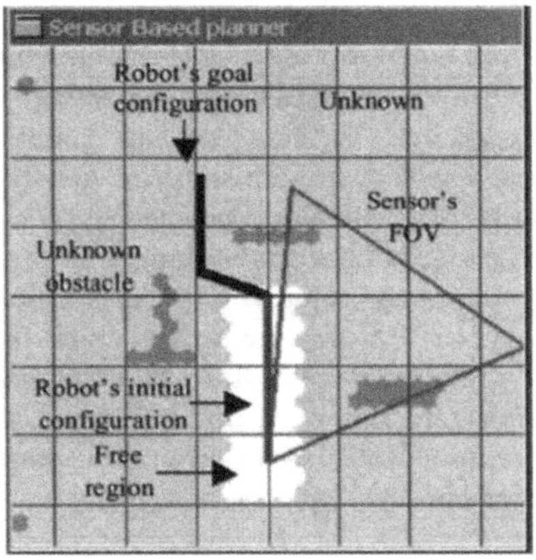

Fig. 1. An eye-in-hand system — a two-link robot with a wrist mounted range sensor (with triangle FOV) moving in an unknown environment. A key question (the view planning problem) is: where should the robot sense next?

We derived closed form expressions for expected C-space entropy reduction, or information gain under a Poisson point process model of the environment [12]. However, two idealized assumptions were made regarding the sensor in that paper: (i) the sensor has a point field of view (FOV), i.e., it senses a single point and (ii) no occlusion constraints were taken into account, i.e., as if the sensor would "see" (get range measurement) through the obstacles. The next best view is planned using this formulation, i.e., the algorithm computes the point (say, x_{max}) which, if sensed, would yield maximum expected information gain and places the sensor so that the center of the actual FOV (a cone) coincides with x_{max}.

In this paper, we relax the above two assumptions and present the MER criterion computation for a generic range sensor with a non-zero volume FOV while respecting occlusion constraints, thereby modelling real range sensors. This computation is valid for a Poisson point process model of the environment, admittedly a simplification, but the resulting closed form expressions give us insights and are useful at least as approximations. We present simulations that show clear improvement in the efficiency of exploration with the new formulation. Our initial simulations are planar for ease of visualization. We emphasize that our formulations and results are valid for 3D environments and are currently being implemented on a real six-dof SFU eye-in-hand system consisting of a PUMA 560 with a wrist mounted area-scan laser range sensor that has been developed in our lab and was reported in [5]. We expect to report these experimental results in the near future.

2 Notation

Let $\mathcal{A}$ denote the robot and q denote a point in its configuration space, $\mathcal{C}$. $\mathcal{A}(q)$ then denotes the region in physical space, $\mathcal{P}$, occupied by the robot. Let $\mathcal{S}$ denote a sensor attached to the robot. We attach a coordinate frame to the sensor's origin. Let s denote the vector of parameters that completely determine the sensor frame, i.e., sensor's configuration. For instance, assuming the sensor is attached to the end-effector of the robot, for planar case, $s = (x, y, \theta)$; for 3D case, $s = (x, y, z, \alpha, \beta, \gamma)$. Let $\mathcal{V}(s) \in \mathcal{P}$ denote the region to be sensed (sensor FOV) by the sensor at configuration s. We discuss three different sensor FOV's: (i) an idealized point FOV sensor, i.e., $\mathcal{V}(s) = x$, a point $\in \mathcal{P}$, (ii) a beam FOV sensor, i.e., $\mathcal{V}(s) = L$, a line segment $\in \mathcal{P}$, and (iii) a non-zero volume FOV sensor, such as an area scan range sensor, i.e., $\mathcal{V}(s) = \mathcal{B}$, an open set $\in \mathcal{P}$. Subscripts $free$, *obs*, [1] and u denote the known free, known obstacle and unknown regions, respectively in physical and configuration space. So, for example, $\mathcal{P}_{obs}$ denotes the known obstacles in physical space, $\mathcal{A}_u(q)$ denotes the part of robot lying in unknown physical space at configuration q, and $\mathcal{C}_{free}$ denotes the known free configuration space.

3 C-space Entropy and MER Criterion

In the following, first we overview the notion of C-space entropy and the closed form expression for point FOV sensor derived in [1,2], followed by the closed form expression for the beam FOV sensor. We then present the closed form expression for the generic FOV sensor.

We assume that the obstacles' distribution in the physical environment is modelled with an underlying stochastic process (e.g., the Poisson model used later). The kinematics and geometry of the robot, embodied by function $\mathcal{A}(q)$ map the probability distribution in physical space to a probability distribution over the C-space. Shannon's Entropy then provides a measure of the robot's ignorance of the status of C-space [2]. One can then compute the expected entropy reduction (or, equivalently, expected information gain) if a region $\mathcal{V}(s) \in \mathcal{P}$ was sensed (obstacle/free). The next best view (the region

[1] Ideally, these two subscripts should be *known-free* and *known-obs*. But we omit *known* for brevity.

to be sensed) is then the one that maximizes the expected entropy reduction (MER criterion), or equivalently, expected information gain (IG).

The information gain (IG) function is defined as

$$IG_{\mathcal{C}}(s) = -\underset{s}{E}\{\triangle H(\mathcal{C})\}$$

where $H(\mathcal{C})$ denotes the current C-space entropy, $\underset{s}{E}\{\triangle H(\mathcal{C})\} = E\{H(\mathcal{C} \mid \mathcal{V}(s)\} - H(\mathcal{C})$ denotes the expected entropy change after $\mathcal{V}(s)$ is sensed.

3.1 Environment Model

We use a simple probabilistic model of physical space — the Poisson point process, essentially characterized by uniformly distributed points in space [12]. From a motion planning point of view, these points, denoted by *pt*, are obstacles in the physical space of the robot. Given the density parameter of this model, λ, the void probability of an arbitrary set $\mathcal{B} \in \mathcal{P}$ — the probability that there is no *pt* (obstacle) in $\mathcal{B}$ — denoted by $p(\mathcal{B})$, is given by

$$p(\mathcal{B}) = \Pr[\mathcal{B} \subseteq \mathcal{P}_{free}] = e^{-\lambda \cdot vol(\mathcal{B})} \tag{1}$$

3.2 Point FOV Sensor

For a point FOV sensor model, which only senses a point (or an infinitesimal ball) in physical space, rather than compute information gain, it is more appropriate to compute the corresponding density function, i.e., expected entropy reduction (or, equivalently, expected information gain) per unit volume if a point $x \in \mathcal{P}$ was sensed (obstacle/free). The reason is as follows [1,2]. The information gain occurs either because the status of an unknown point becomes obstacle or free. While the absolute information gain due to a single point becoming free is zero almost everywhere, the corresponding density function is non zero and finite! Thus, it gives a more complete picture of the information gain.

The information gain density (IGD) function for a point FOV sensor is defined as

$$IGD_{\mathcal{C}}(x) = \lim_{vol(\mathcal{B}(x)) \to 0} \frac{-\underset{\mathcal{B}(x)}{E}\{\triangle H(\mathcal{C})\}}{vol(\mathcal{B}(x))}$$

where, consistent with earlier notation, $H(\mathcal{C})$ denotes the current C-space entropy, $\underset{\mathcal{B}(x)}{E}\{\triangle H(\mathcal{C})\} = E\{H(\mathcal{C}|\mathcal{B}(x))\} - H(\mathcal{C})$ denotes the expected entropy change after $\mathcal{B}(x)$, a ball centered at point x, is sensed.

In order to get efficiency in computing, we neglect the mutual entropy terms[2], essentially treating each (discretized) configuration as an independent

[2] Please see section 6 for more discussion on this.

random variable, i.e., $\widetilde{H}(\mathcal{C}) = \sum_{q \in C} H(Q)$. In this equation, Q denotes the binary random variable (r.v.) corresponding to configuration q being free (=0) or in collision (=1); $H(Q)$ denotes the entropy of r.v. Q, i.e.,

$$H(Q) = p(q) \cdot \log(p(q)) + (1 - p(q)) \cdot \log(1 - p(q)) \tag{2}$$

where $p(q) = \Pr[q = free] = e^{-\lambda \cdot vol(A_u(q))}$ is the marginal probability that configuration q is collision-free, also called the void probability of q. With this simplification one can show that:

$$\widetilde{IGD}_{\mathcal{C}}(x) = \lim_{vol(\mathcal{B}(x)) \to 0} \frac{-E\{\underset{\mathcal{B}(x)}{\triangle} \widetilde{H}(\mathcal{C})\}}{vol(\mathcal{B}(x))} = \sum_{q \in \mathcal{C}} igd_q(x)$$

where $igd_q(x)$ is given by:

$$igd_q(x) = \lim_{vol(\mathcal{B}(x)) \to 0} \frac{-E\{\underset{\mathcal{B}(x)}{\triangle} H(Q)\}}{vol(\mathcal{B}(x))} \tag{3}$$

When $\mathcal{B}(x)$ is sensed, the sensed information affects the C-space entropy via each configuration q. $igd_q(x)$ is then the marginal contribution to information gain density via configuration q, if a point $x \in \mathcal{P}_u$ were to be sensed. Furthermore, $igd_q(x)$ equals 0 when $\mathcal{A}_u(q)$ does not contain x. So we need only compute the above summation over those q's such that $x \in \mathcal{A}_u(q)$, also called the unknown C-zone of x [2,3], and denoted by $\chi_u(x)$. Intuitively, C-zone of x is the set of configurations such that the robot when placed in such a configuration contains point x. One could think of this as a generalization of inverse kinematics that applies to the entire robot body.[3] Therefore one can write:

$$\widetilde{IGD}_{\mathcal{C}}(x) = \sum_{q \in \chi_u(x)} igd_q(x) \tag{4}$$

In [1,2], we showed that

$$igd_q(x) = \begin{cases} \lambda \cdot H(Q) - \lambda \cdot p(q) \cdot \frac{dH(Q)}{dp} & x \in \mathcal{A}_u(q) \\ 0 & \text{otherwise} \end{cases}$$

The above expression consists of two terms that contribute to expected entropy reduction per unit volume. The first term, $\lambda \cdot H(Q)$ is the expected contribution when $\mathcal{B}(x)$ contains an obstacle. Note that in this case, for each such event, the entropy would reduce to zero, since robot in configuration q would now be in collision. The change in entropy is therefore $H(Q)$, the

[3] [14] used a similar concept but in a different context. Their aim was to come up with a C-space representation that is easy to modify.

entropy before scanning. The scaling factor λ simply reflects the probability (per unit volume) of the event that $\mathcal{B}(x)$ contains an obstacle under Poisson model. The second term, $\lambda \cdot p(q) \cdot \frac{dH(Q)}{dp}$ is the contribution when $\mathcal{B}(x)$ changes to free. In this event, the absolute amount of entropy reduction is zero almost everywhere, however, the derivative w.r.t. $vol(\mathcal{B}(x))$ is non-zero and is equal to $\lambda \cdot p(q) \cdot \frac{dH}{dp}$. The probability of this event under Poisson model is very high and approaches unity in the limit.

Substituting for $H(Q)$ and its derivative (differentiating expression 2 w.r.t. $p(q)$), the final expression for igd_q is

$$igd_q(x) = \begin{cases} -\lambda \log(1 - p(q)) & x \in \mathcal{A}_u(q) \\ 0 & \text{otherwise} \end{cases} \tag{5}$$

Combining Eq. (4) and (5), we get

$$\widetilde{IGD}(x) = \sum_{q \in \mathcal{X}_u(x)} -\lambda \cdot \log(1 - p(q)) \tag{6}$$

This expression, combined with the expression for void probability $p(q)$ in Eq. (1) completely determines the IGDF for a point FOV sensor assuming a Poisson point process for obstacle distribution. The point to be scanned next is the one that maximizes $\widetilde{IGD}_c(x)$.

The closed form expression for expected entropy reduction for a point FOV sensor assumes that there are no occlusion constraints, i.e., a point to be sensed will always result in a free/obstacle status. We now discuss the beam FOV sensor that models the occlusion constraints in the expected entropy formulation.

3.3 Beam Sensor Model: Occlusion Constraint

The beam sensor, as the name implies, senses along a beam (ray) of finite length, L, emanating from the sensor origin (See Figure 2). It returns the distance of the first hit point (obstacle) along the beam. Points along the beam that are in front of the hit point (i.e., closer to the origin than the hit point) are in free space. Points along the beam behind the hit point (i.e., farther from origin than the hit point) are deemed to be un-sensible (by occlusion constraints) in this particular sensing action and their status (obstacle/free/unknown) remains the same as it was before the sensing action. Therefore, for a particular sensing action, a point along the beam may acquire one of three possible states: 0 (free), 1 (obstacle) or u (unsensible).

For this more general case, when a region $\mathcal{V}(s)$ is being sensed, the IGD function is now defined over the space of all sensor configurations. Note that since the volume of the sensor FOV is still zero, it is the density function that is still relevant. For each sensor configuration, s, the IGD function assigns a real value that corresponds to the expected information gain per unit volume

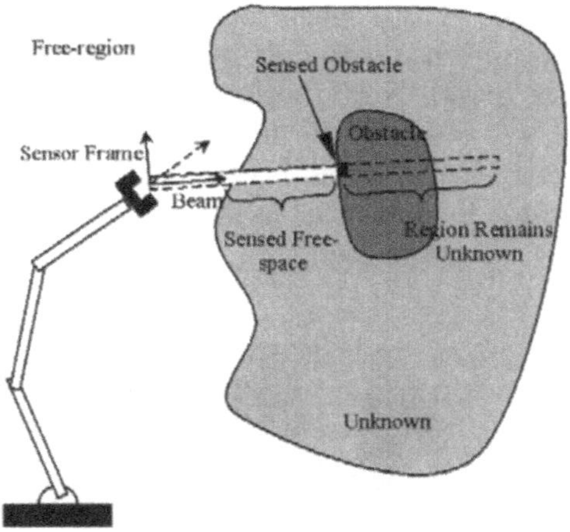

Fig. 2. The beam sensor model.

of sensor FOV with the sensor placed at configuration s. Using the same simplifications as for a point sensor (i.e., ignoring mutual entropy terms),

$$\widetilde{IGD}_{\mathcal{C}}(s) = \lim_{vol(\mathcal{V}(s))\to 0} \frac{E\{\underset{s}{\triangle \widetilde{H}(\mathcal{C})}\}}{vol(\mathcal{V}(s))} = \sum_{q\in\chi_u(s)} igd_q(s)$$

and

$$igd_q(s) = \lim_{vol(\mathcal{V}(s))\to 0} \frac{-E\{\underset{s}{\triangle H(Q)}\}}{vol(\mathcal{V}(s))} \tag{7}$$

As before, $igd_q(s)$ is the marginal contribution to IGD via configuration q when the sensor senses at configuration s. Similar to the point FOV sensor, the summation is restricted to the unknown C-zone of $\mathcal{V}_u(s)$ (defined in the next paragraph), denoted by $\chi_u(s)$, and defined as the set of q's such that $\mathcal{A}_u(q) \cap \mathcal{V}_u(s) \neq \phi$. [4]

For mathematical formulation, we will assume that the sensor FOV, $\mathcal{V}(s)$, is a thin cylinder of infinitesimal radius (or equivalently, infinitesimal cross sectional area denoted by $\triangle a$) and length L. We discretize this cylinder into n "disks", each of length $\triangle l$, by planes orthogonal to its axis, as shown in Figure 3. As $\triangle a$ and $\triangle l$ approach zero, the cylinder becomes an ideal beam.

Let $\mathcal{V}_u(s)$ [5] denote the portion of sensor FOV that lies inside $\mathcal{P}_{unk}$ and is *in front of the first known obstacle* along the sensing direction, i.e., $\mathcal{V}_u(s)$ denotes the largest possible sensing region the beam sensor can sense at

[4] This extends the notion of C-zone to that of a set in physical space rather than a single point x.

[5] There is a slight abuse of notation for simplicity. Subscript u in $\mathcal{V}_u(s)$ denotes not only unknown part of $\mathcal{V}(s)$ but also further excludes those portions of unknown environment in $\mathcal{V}(s)$ that are occluded by known obstacles.

s. $\mathcal{V}_u(s) \approx x_1 \cup x_2 \cup \ldots \cup x_m$ where $x_i, i = 1, \ldots, m$ are the disks lying inside $\mathcal{V}_u(s)$ (we are concerned with only the unknown portion of sensor FOV and label only those disks). Note that $\mathcal{V}_u(s)$ may be a multiply connected set. If a disk (say, x_i) contains the hit point, its status would become 1 (obstacle). All disks x_j, $j < i$, would become 0 (free), and all disks x_k, $k > i$ would keep their status u (unknown) as shown in Figure 3. So we will get a "$0, 0, \ldots, 0, 1, u, u, \ldots, u$" sequence.

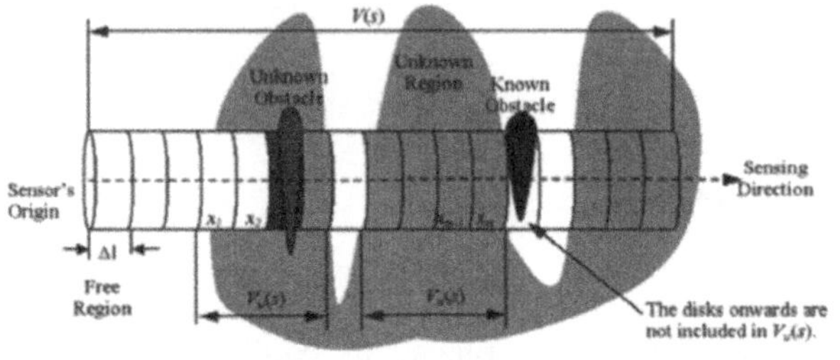

Fig. 3. The beam sensor model with infinitesimal width and discretized into "disks". The hit point lies in disk x_i with $i = 3$. Disks x_1 and x_2 become free and disks x_4 onwards remain unknown.

The location of the hit point is a random variable. The event that the ith disk contains the hit point, denoted by $x_i = h$, corresponds to first $(i-1)$ disks, $x_1, x_2, \ldots, x_{i-1}$, being all in $\mathcal{P}_{free}$ and the ith disk, x_i, containing an obstacle point, i.e., $\{x_j = 0, j = 1, \ldots, i-1 \wedge x_i = 1\}$ where $i \in \{1, \ldots, m\}$. The corresponding probability, denoted by $p(x_i = h)$ is then given by using Eq. (1):

$$p(x_i = h) = e^{-\lambda \cdot (i-1) \cdot \Delta a \cdot \Delta l} \cdot (1 - e^{-\lambda \cdot \Delta a \cdot \Delta l}) \tag{8}$$

with $1 \le i \le m$, and $\Delta a \cdot \Delta l$ being the volume of each disk.

3.3.1 Compute $igd_q(s)$

The numerator (expectation) in $igd_q(s)$ in Eq. (7) is:

$$\begin{aligned} E\{\underset{s}{\Delta H(Q)}\} &= E\{\underset{x_i \in \mathcal{A}(q)}{\Delta} H\} + E\{\underset{x_i \notin \mathcal{A}(q)}{\Delta} H\} + E\{\underset{\exists \text{ no hit pt}}{\Delta} H\} \\ \Leftrightarrow igd_q(s) &= (igd_q)_1 + (igd_q)_2 + (igd_q)_3 \end{aligned} \tag{9}$$

The r.h.s. in the above expression consists of three terms: the first one corresponds to an x_i being a hit point and belonging to $\mathcal{A}(q)$ for the given q (event 1); the second term corresponds to an x_i being a hit point but not belonging to $\mathcal{A}(q)$ for the given q (event 2); the last term corresponds to there being no hit point in the sensed beam, i.e., all x_i's are sensed free (event 3). The three cases are shown schematically in Figure 4. The robot (solid line) is holding the sensor at a given sensor configuration s. For a given q, the region occupied by the robot, $\mathcal{A}(q)$ is shown in dotted. The hit point may lie inside $\mathcal{A}(q)$, outside $\mathcal{A}(q)$, or there may not be any hit point, i.e., the entire beam is free.

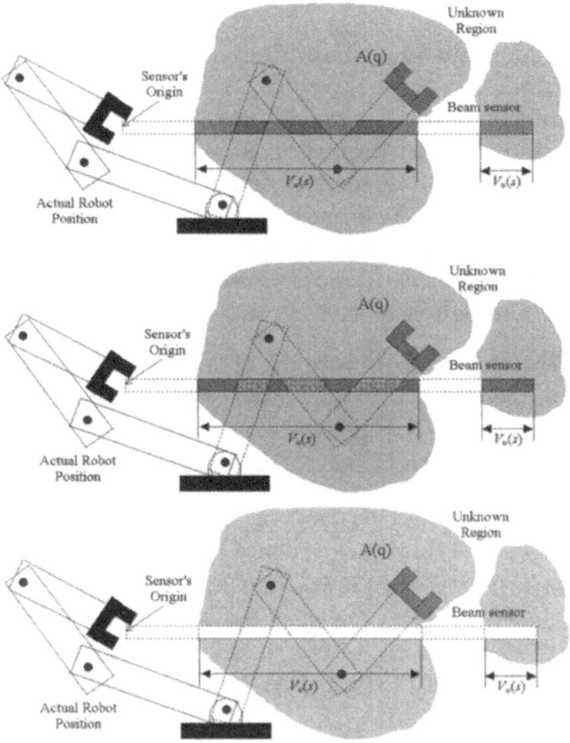

Fig. 4. Computation of $igd_q(s)$. The three terms on the r.h.s. in equation 9 correspond to the above three events. (top) When hit point is inside $\mathcal{A}(q)$. The hit point may lie anywhere inside the dark strip. (middle) When hit point is outside $\mathcal{A}(q)$. The hit point may lie anywhere inside the dark strip. (bottom) There is no hit point, i.e., the entire beam is sensed free.

Computing the 1st Component ($E\{\triangle H_1\}$) Derivation omitted.
When the hit point lies inside $\mathcal{A}(q)$, we have

$$(igd_q)_1 = \frac{\lambda}{L} \cdot len(\mathcal{A}(q) \cap \mathcal{V}_u(s)) \cdot H(Q) \tag{10}$$

The expression makes intuitive sense. Since it is the expected contribution from those cases where a sensed hit point lies inside $\mathcal{A}(q)$, we know that in each such outcome, the entropy would reduce to zero since the status of Q would become known (in collision) and hence the entropy reduction will be $H(Q)$, the entropy before sensing. The multiplying factor $\frac{\lambda}{L} \cdot len(\mathcal{A}(q) \cap$

$\mathcal{V}_u(s)$) simply represents the expectation (per unit length) of such an event happening under Poisson model.

Computing the 2nd Component ($E\{\triangle H_2\}$) Derivation Omitted.

When the hit point lies outside $\mathcal{A}(q)$, we have

$$(igd_q)_2 = 0 \tag{11}$$

This implies for outcomes comprising event 2 (the hit point is sensed but does not lie within $\mathcal{A}(q)$ for a given q), the marginal information gain density due to configuration q is zero. We can think of $(igd_q)_2$ as essentially a summation (over outcomes comprising event 2) of "derivatives" of entropy w.r.t. volume weighted by the corresponding probability of such an outcome happening. The derivative is finite but the probability of the outcome happening (a hit point lying outside the robot but within the sensing beam) is very low and approaches zero under Poisson model as Δa approaches zero. $(igd_q)_2$ being a summation of these product terms, is also zero.

Computing the 3rd Component ($E\{\triangle H_3\}$) Derivation omitted.

When the whole beam is sensed free of hit points, we have

$$(igd_q)_3 = -\frac{\lambda}{L} \cdot len(\mathcal{A}(q) \cap \mathcal{V}_u(s)) \cdot p(q) \cdot \frac{dH(Q)}{dp} \tag{12}$$

$(igd_q)_3$ is essentially the "derivative" of entropy w.r.t. volume weighted by the probability of the event 3 (when the sensor does not sense any hit point). The derivative is again finite, however, the probability event 3 is very high and approaches unity as Δa approaches zero under Poisson model. The $(igd_q)_3$ being the product of the two, is therefore finite.

We can easily get $igd_q(s)$ from Eq. (10), (11) and (12),

$$igd_q(s) = \begin{cases} \frac{\lambda}{L} \cdot len(\mathcal{A}(q) \cap \mathcal{V}_u(s)) \cdot (H(Q) - p(q) \cdot \frac{dH(Q)}{dp}) & \\ = -\frac{\lambda}{L} \cdot len(\mathcal{A}(q) \cap \mathcal{V}_u(s)) \cdot \log(1 - p(q)) & x \in \mathcal{A}_u(q) \\ 0 & \text{otherwise} \end{cases}$$

And finally,

$$\widetilde{IGD}_c(s) = -\frac{\lambda}{L} \cdot \sum_{q \in \mathcal{X}_u(s)} len(\mathcal{A}(q) \cap \mathcal{V}_u(s)) \cdot \log(1 - p(q))$$

Note that the beam sensor result is as if occlusion does not matter within entropy computation and the net result is effectively a "summation" of point sensors along the beam. Moreover, we can easily see that as L, the length of the sensing beam, goes to zero, this result agrees with the one obtained for the point FOV sensor, i.e., we have $\lim_{L \to 0} \frac{\mathcal{A}(q) \cap \mathcal{V}_u(s)}{L} = 1$, and therefore, $igd_q(s) = \lambda \cdot (H(Q) - p(q) \cdot \frac{dH(Q)}{dp})$, precisely the result we obtained in [1–3][6].

[6] Please note that there is an algebraic error in earlier papers for the point FOV case. This is the corrected result.

3.4 Generic Range Sensor with Non-zero Volume FOV

We now consider the general case, that of a range sensor whose FOV has non-zero volume, i.e., $\mathcal{V}(s)$ is an open set in R^3, and the the actual volume sensed is governed by occlusion constraints. Most commercially available range sensors that provide range images (such as the area scan laser range sensor used in SFU eye-in-hand system [5]) fall into this category. Fig. 5 shows a schematic diagram as the sensor senses an unknown region within its FOV. As before, let $\mathcal{V}_u(s)$ denote the portion of the FOV that intersects $\mathcal{P}_u$ and is not occluded by known obstacles. Note that $\mathcal{V}_u(s)$ might be a multiply-connected set. In the figure, $\mathcal{V}_u(s)$ consists of regions A, B, C and D (region E is excluded from $\mathcal{V}_u(s)$ since it is occluded by a known obstacle). After sensing, regions A, B and C become free; region D remains unknown because it is occluded by the sensed obstacles (shown in dark). Of course, the sensor also provides the distances from the sensor's origin to the sensed obstacles.

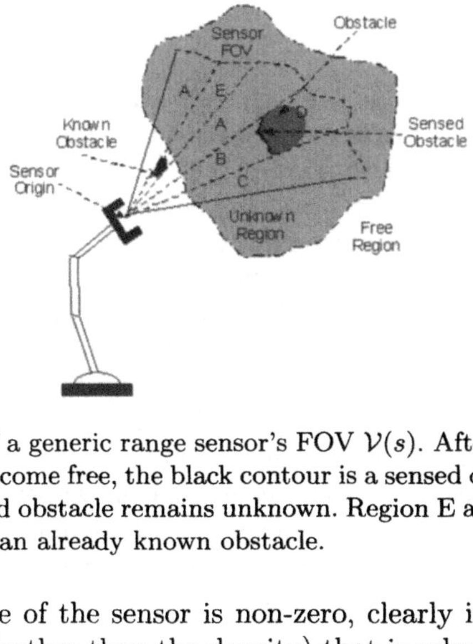

Fig. 5. Illustration of a generic range sensor's FOV $\mathcal{V}(s)$. After this sensing action, regions A, B and C become free, the black contour is a sensed obstacle and region D, occluded by the sensed obstacle remains unknown. Region E also remains unknown, but it is occluded by an already known obstacle.

Since the volume of the sensor is non-zero, clearly it is the information gain (IG) function (rather than the density) that is relevant. Making similar approximations as earlier,

$$\widetilde{IG}_{\mathcal{C}}(s) = -\underset{s}{E}\{\triangle \widetilde{H}(\mathcal{C})\} = \sum_{q \in \chi_u(s)} ig_q(s)$$

where $ig_q(s)$ is given by:

$$ig_q(s) = -\underset{s}{E}\{\triangle H(Q)\} \tag{13}$$

3.4.1 $IG_c(s)$ Computation

With occlusion constraint, the sensor can only detect the very first obstacle *pt*, called the hit point, along each sensing ray. Again, for a given q, $ig_q(s)$ is composed of two components, i.e., $ig_q = (ig_q)_1 + (ig_q)_2$.

The first component, denoted by $(ig_q)_1$, corresponds to those outcomes where the sensor would sense at least one hit point inside $\mathcal{A}(q) \cap \mathcal{V}_u(s)$, i.e., $hitpt \in \mathcal{A}(q) \cap \mathcal{V}_u(s)$, for a given configuration q. Let this set of outcomes be denoted event 1. After sensing, the robot, were it to be placed at configuration q, $\mathcal{A}(q)$, would be in collision with an obstacle (the sensed hit point). So $H(Q \mid \text{event } 1) = 0$ and $\underset{event\ 1}{\triangle} H(Q) = H(Q \mid \text{event } 1) - H(Q) = -H(Q)$.

It turns out (not unexpectedly in the light of beam sensor result) that the probability of event 1, $\Pr[hitpt \in \mathcal{A}(q) \cap \mathcal{V}_u(s)]$, is the same as $\Pr[pt \in \mathcal{A}(q) \cap \mathcal{V}_u(s)]$, as if occlusion does not matter! Theorem 1 states this result formally.

Theorem 1. $\Pr[hitpt \in \mathcal{B}] = \Pr[pt \in \mathcal{B}] = 1 - e^{-\lambda \cdot vol(\mathcal{B})}$ *where* $\mathcal{B} \subseteq \mathcal{V}_u(s)$ *is any open set and pt are point obstacles whose distribution is governed by a Poisson point process.*

Proof. Omitted. We outline the basic approach. We first discretize $\mathcal{V}(s)$ into N number of nearly-identical infinitesimal cones, $\mathcal{V}_1, \mathcal{V}_2, \ldots, \mathcal{V}_N$, as shown in Fig. 6. Consider a cone (with apex at sensor origin) which contains $\mathcal{V}(s)$. Lay a discrete grid of size ϵ and connect the boundary of these cells to the sensor's origin. Thus, we get N infinitesimal cones. As $\epsilon \to 0$, $N \to \infty$, discretized sensor approaches original range sensor.

Now consider an open set $\mathcal{B} \subseteq V_u(s)$. Indeed $\mathcal{B}$ will intersect $M \leq N$ of these infinitesimal cones. Wlog, we label them $\mathcal{VB}_1, \mathcal{VB}_2, \ldots, \mathcal{VB}_M$. Also we denote the intersection of these $\mathcal{VB}_i$'s with $\mathcal{B}$ by $\mathcal{B}_1, \mathcal{B}_2, \ldots, \mathcal{B}_M$ respectively and use $front(\mathcal{B}_i)$ to denote the part of $\mathcal{VB}_i$ that is in front of $\mathcal{B}_i$ along the sensing direction. We use these infinitesimal cones to compute $\Pr[hitpt \in \mathcal{B}]$. Note that the occlusion for each *pt* in $\mathcal{B}_i$ will happen in $front(\mathcal{B}_i)$. Again, since the obstacles are point obstacles, the probability of a *pt* being occluded approaches zero, although the actual proof is somewhat detailed and tedious and is omitted here for lack of space.

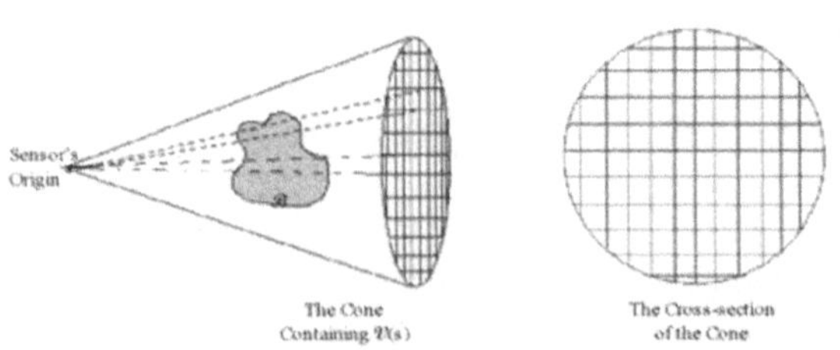

Fig. 6. The generic range sensor FOV discretized into N infinitesimal cones.

So we have

$$\begin{aligned}(ig_q)_1 &= \Pr[hitpt \in \mathcal{A}(q) \cap \mathcal{V}_u(s)] \cdot H(Q) \\ &= \Pr[pt \in \mathcal{A}(q) \cap \mathcal{V}_u(s)] \cdot H(Q) \\ &= (1 - e^{-\lambda \cdot vol(\mathcal{A}(q) \cap \mathcal{V}_u(s))}) \cdot H(Q)\end{aligned} \tag{14}$$

The second component, denoted by $(ig_q)_2$, corresponds to a set of outcomes in which there does not exist any hit point inside $\mathcal{A}(q) \cap \mathcal{V}_u(s)$. In this case, the status of $\mathcal{A}(q)$ would either remain unknown, albeit the unknown portion (volume) may have decreased, or it may become completely free; but it will not be known to be in collision. Let us denote this set of outcomes by event 2. Using the discretized FOV as in Figure 6, let us denote a possible state of $A(q) \cap \mathcal{V}_u(s)$ after sensing by $\underline{J}$. Event 2 then corresponds to the set of outcomes $\{\underline{J}:$ no $hitpt \in \mathcal{A}(q) \cap \mathcal{V}_u(s)\}$. By definition, then we have

$$(ig_q)_2 = \sum_{\underline{J} \in \text{event 2}} \Pr[\underline{J}] \cdot (H(Q) - H(Q \mid \underline{J})) \tag{15}$$

where $\Pr[\underline{J}]$ is the probability of $\mathcal{A}(q) \cap \mathcal{V}_u(s)$ being in state $\underline{J}$ after sensing.

We show that the above expectation turns out to be that of the event (let us call it event 3) that there does not exist any *pt* in $\mathcal{A}(q) \cap \mathcal{V}_u(s)$, or equivalently that the region $\mathcal{A}(q) \cap \mathcal{V}_u(s)$ is free! This implies that occlusion does not matter in the expectation computation! It is not entirely unexpected in the light of beam sensor result. Theorem 2 states this result formally.

Theorem 2.

$$\begin{aligned}&\sum_{\underline{J} \in \textit{event 2}} \Pr[\underline{J}] \cdot H(Q \mid \underline{J}) = \Pr[\textit{event 3}] \cdot H(Q \mid \textit{event 3}) \\ &= e^{-\lambda \cdot vol(\mathcal{A}(q) \cap \mathcal{V}_u(s))} \cdot H(Q \mid \textit{event 3})\end{aligned} \tag{16}$$

Proof. Omitted.

Thus, expanding the summation in Eq. (15), we have,

$$(ig_q)_2 = H(Q) \cdot \sum_{\underline{J} \in \text{event 2}} \Pr[\underline{J}] - \sum_{\underline{J} \in \text{event 2}} \Pr[\underline{J}] \cdot H(Q \mid \underline{J}) \tag{17}$$

The first term (the first summation) above is $1 - \Pr[hitpt \in \mathcal{A}(q) \cap \mathcal{V}_u(s)]$. The expression for it is given by Theorem 1 if we substitute $\mathcal{A}(q) \cap \mathcal{V}_u(s)$ for $\mathcal{B}$. Furthermore, substituting from Theorem 2 for the second term, we get

$$\begin{aligned}(ig_q)_2 &= H(Q) \cdot \Pr[\text{event 3}] - \Pr[\text{event 3}] \cdot H(Q \mid \text{event 3}) \\ &= e^{-\lambda \cdot vol(\mathcal{A}(q) \cap \mathcal{V}_u(s))} \cdot (H(Q) - H(Q \mid \text{event 3}))\end{aligned} \tag{18}$$

So summing the two components together, we will have

$$ig_q = (ig_q)_1 + (ig_q)_2 = H(Q) - e^{-\lambda \cdot vol(\mathcal{A}(q) \cap \mathcal{V}_u(s))} \cdot H(Q \mid \text{event 3}) \tag{19}$$

Both $H(Q \mid \text{event 3})$ and $H(Q)$ in the above equation are determined using Eq. (2) and that $p(q \mid \text{event 3}) = e^{-\lambda \cdot vol(\mathcal{A}_u(q) \setminus \mathcal{V}_u(s))}$, where $\setminus$ denotes set difference.

4 Algorithm for View Planning

Now that we have computed an expression for IG over sensor's configuration space, we can use the MER criterion to decide the next scan, i.e., choose the sensor configuration s_{max} such that $s_{max} = \max_{s}\{ \sum_{q \in \mathcal{X}_u(s)} ig_q(s)\}$. The algorithm then is as follows:

for every s /* *according to a certain resolution* */
  determine $\mathcal{V}_u(s)$
  $\widetilde{IG}(s) = 0$ /* *initialize* */
  for every q
    if $(\mathcal{A}_u(q)$ *overlaps with* $\mathcal{V}_u(s))$
      compute $ig_q(s)$
    $\widetilde{IG}(s) = \widetilde{IG}(s) + ig_q(s)$
$s_{max} = \max_{s}(\widetilde{IG}(s))$

Determining quantities such as $\mathcal{A}_u(q)$, $\mathcal{V}_u(s)$ involves straight forward geometrical computations[7]. For instance, determining $\mathcal{V}_u(s)$ corresponds to determining the intersection of the sensor FOV with $\mathcal{P}_u$ while excluding portions of $\mathcal{P}_u$ occluded by already known obstacles (before sensing action), a relatively simple geometric computation. $ig_q(s)$ for given s and q is therefore easily computed. The iteration over q, i.e., summation over C-space of the robot to determine $IG(s)$, may be prohibitive for robots with many degree of freedoms. In this case, the summation can be carried out over a large enough set of random samples [2]. The iteration over s, i.e., maximization over the sensor configuration space to determine s_{max} will be directly proportional to the number of discretized sensing configurations.

5 Simulation Results

In order to test the effectiveness of our formulae, we conducted a series of experiments on the simulated two-link eye-in-hand preliminary system shown in Figure 1. The task for the robot is to explore its environment, starting from its initial configuration. The overall planner used is SBIC-PRM (sensor-based incremental construction of probabilistic road map) reported in [3,5]. Briefly, SBIC-PRM consists of an incrementalized model-based PRM [15], that operates in the currently known environment; and a view planner that decides a reachable configuration within the currently known environment from which to take the next view. The two sub-planners operate in an interleaved manner. The simulation program, written in C++, runs on a Pentium III 800, and it takes about 53 seconds for each view planning iteration, corresponding to the algorithm in Section 4.

[7] The representation of the robot and the physical space in this algorithm is similar to [5,8].

We now compare the results of four different view planning criteria for efficiency of exploration of the physical and configuration space. The first strategy, denoted by RV (random views), is to randomly choose a viewpoint as the next scan. The second strategy, denoted by MPV (maximum unknown physical volume) is to choose the next viewpoint so as to maximize the unknown physical volume inside the scan [6]. The third strategy is to use point FOV based MER criterion for view-planning [1,2], and place the centre of the actual FOV (the cone) at x_{max}, the single point that results in maximum entropy reduction. The fourth is to use the generic non-zero volume FOV based MER criterion derived in this paper. In all these cases, the robot started off as in Figure 1.

As shown in Figure 7, the first two strategies expand the known C-space much less than the last two MER criterion based strategy. Using RV gives us about 8% expansion of known C-space in 5 scans, and the robot reached its goal in 36 scans. MPV results in C-space expansion by about 54% in 5 scans. The point FOV based MER criterion gives us much better results, resulting in about 73% expansion in 5 scans. The general FOV based MER criterion was the best, better than point FOV based MER. It made the C-space expand by about 82% in 5 scans. For reader's information, although not relevant here, black dots in all these figures are the nodes of the probabilistic roadmap, bulit by SBIC-PRM planner and used for planning paths for the robot.

Figure 8 plots known C-space vs. number of iterations for the above four view-planning criteria. We can easily see that the generic range sensor based MER is the most efficient one, which expanded known C-space to about 90% in 7 scans; point FOV based MER needed 11 scans; RV needed 33 scans; and MPV needed 19 scans respectively.

6 Conclusions

We presented closed form solutions for computing the expected C-space entropy reduction for a general non-zero volume FOV range sensor while taking into account the occlusion constraints inherent in range sensors. This extends our previous results that applied to a point FOV sensor and did not account for occlusion constraints. Planar simulations show that our new results lead to more efficient exploration of the robot configuration space. As mentioned in the introduction, our next step is to implement these results for six-dof SFU eye-in-hand system, a PUMA 560 with a wrist mounted area scan laser range finder, reported in [5].

There are two simplifications in the current formulation. The first is that it assumes a Poisson point process for obstacle distribution that treats obstacles as points. Extending our formulation for a Boolean stochastic model [12] where geometric shape of obstacles is taken into account would be the next step. The second simplification is that we ignored mutual entropy terms for computational efficiency. Computation of these terms will involve computing the intersection of robot's volume for pairs (and triplets, and so on) of unknown configurations. Even for second order mutual entropy terms, the computational complexity would be $O(N^2)$, where N is the number of robot configurations with the C-space discretized at a certain resolution. It will indeed be computationally expensive to calculate these terms [2]. The effect of these mutual entropy terms on the efficiency of exploration needs to be further investigated.

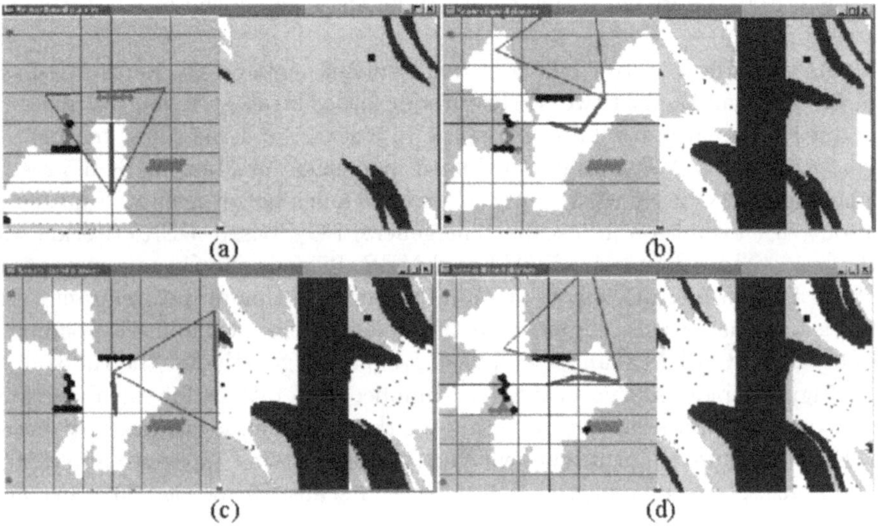

Fig. 7. Known Physical and C-space after 5 scans: (a) RV criterion, (b) MPV criterion, (c) Point FOV based MER criterion, (d) Generic non-zero volume FOV based MER criterion.

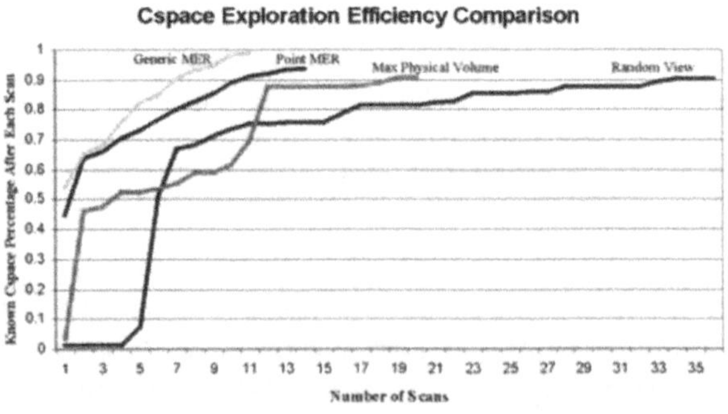

Fig. 8. The comparison of C-space exploration efficiency for the four view-planning algorithms: RV, MPV, Point FOV Based MER and Generic FOV Based MER

References

1. Y. Yu and K. Gupta. An Information Theoretic Approach to View Planning with Kinematic and Geometric Constraints. *Proc. IEEE Int. Conf. on Robotics and Automation*, Seoul, Korea, May 21-26, 2001, pp. 1948-1953.
2. Y. Yu. An Information Theoretical Incremental Approach to Sensor-based Motion Planning for Eye-in-Hand Systems. *Ph.D. Thesis*, School of Engineering Science, Simon Fraser University, 2000.
3. Y. Yu and K. Gupta. View Planning via C-Space Entropy for Efficient Exploration with Eye-in-hand Systems. *Proc. VII Int. Symp. on Experimental Robotics*, 2000. Available as lecture notes in *Control and Information Sciences*, LNCIS271, Springer. pp. 373-384.
4. Y. Yu and K. Gupta. Sensor-Based Motion Planning for Manipulator Arms: An Eye-in-Hand System. *Video Proc. IEEE Int. Conf. Robotics and Automation* , 2000.
5. Y. Yu and K. Gupta. Sensor-based Probabilistic Roadmaps: Experiments with an Eye-in-hand System. *Advance Robotics*,2000, Vol.14, No.6, pp.515-537.
6. E. Kruse, R. Gutsche and F. Wahl. Effective Iterative Sensor Based 3-D map Building using Rating Functions in Configuration Space. *Proc. IEEE Int. Conf. on Robotics and Automation.* 1996, pp.1067-1072.
7. P. Renton, M. Greenspan, H. Elmaraghy, and H. Zghal. Plan-n-Scan: A Robotic System for Collision Free Autonomous Exploration and Workspace Mapping. *Journal of Intell. And Robotic Systems*, 1999. 24:207-234.
8. J. Ahuactzin and A. Portilla. A Basic Algorithm and Data Structure for Sensor-based Path Planning in Unknown Environment. *Proc. Of IROS 2000*, Vol. 2, pp. 903-908.
9. E. Cheung and V. J. Lumelsky. Motion Planning for Robot Arm manipulators with Proximity Sensors. *Proc. IEEE Int. Conf. on Robotics and Automation*, 1988, Vol. 2, pp. 740-745.
10. H. Choset and J. W. Burdick. Sensor based Planning for a Planar Rod Robot. *Proc. of the IEEE Int. Conf. on Robotics and Automation*, 1996, Vol. 4, pp. 3584-3591.
11. K. Hirai, M. Hirose, M. Haikawa and T. Takenaka. The Development of HONDA Humanoid Robot. *Proc. Of ICRA 1998*, pp. 1321-1326.
12. D. Stoyan and W.S. Kendall. *Stochastic Geometry and Its Applications.* J. Wiley, 1995.
13. H. G. Banos and J. C. Latombe. Robot Navigation for Automatic Construction using Safe regions. *Preprints Proc. ISER 2000.* pp. 395-404.
14. P. Leven abd S. Hutchinson. Toward Real time Motion Planning in Changing Environmets. *New Directions in Algorithmic and Computational Robotics (WAFR 2001).* A. K. Peters, pp. 363-376.
15. L. Kavraki, P. Svestka, J. Latombe and M. Overmars, Probabilistic Roadmaps for Path Planning in High-dimensional Configuration Spaces, *IEEE Transactions on Robotics and Automation*, Aug. 1996, 12(4):556-580.

References

Exact and Distributed Algorithms for Collaborative Camera Control *

Dezhen Song[1], A. Frank van der Stappen[2], and Ken Goldberg[3]

[1] IEOR Department, UC Berkeley, USA
[2] ICS Department, Utrecht University, The Netherlands
[3] IEOR and EECS Department, UC Berkeley, USA

Abstract. We propose the ShareCam Problem: controlling a single robotic pan, tilt, zoom camera based on simultaneous frame requests from n online users. To solve it, we propose a new piecewise linear metric, Intersection Over Maximum (IOM), for the degree of satisfaction for each users. To maximize overall satisfaction, we present several algorithms. For a discrete set of m distinct zoom levels, we give an exact algorithm that runs in $O(n^2m)$ time. The algorithm can be distributed to run in $O(nm)$ time at each client and in $O(n \log n + mn)$ time at the server.

1 Introduction

Consider a robotic camera at a compelling location such as the Sydney boat harbor, United Nations, Academy Awards, or inside the International Space Station. There are $n >> 1$ viewers, and one camera. The camera frame is determined by pan, tilt, and zoom parameters that can be changed to observe details of the scene. The existing control method is based on queueing, where users have to wait patiently for their turn to operate the camera [20]. Can we eliminate the queue and allow many users to share control of the camera simultaneously?

We are developing network-based applications for education, journalism and entertainment where many users share control of a single physical resource. "Sharecam", is an example of Collaborative Telerobotics, where the camera is a telerobot with 3 degrees of freedom. In the taxonomy proposed by Chong et al. [9], this is a Multiple Operator Single Robot (MOSR) system. Our research is motivated by applications where groups of users desire simultaneous access to a single robotic resource. Inputs from each user are combined to generate a single control stream for the robot. There can be benefits to such collaboration: teamwork is a key element in education at all levels [10,36] and an ensemble of users may be more reliable than a single (possibly malicious) user.

As illustrated in Figure 1, the user views two windows, one for user input with a fixed image and the other with a variable image based on collective

* This work was supported in part by the National Science Foundation under IIS-0113147 and by Intel Corporation. For more information please contact dzsong@ieor.berkeley.edu or goldberg@ieor.berkeley.edu.

J.-D. Boissonnat et al. (Eds.): Algorithmic Foundations of Robotics V, STAR 7, pp. 167–183, 2004.

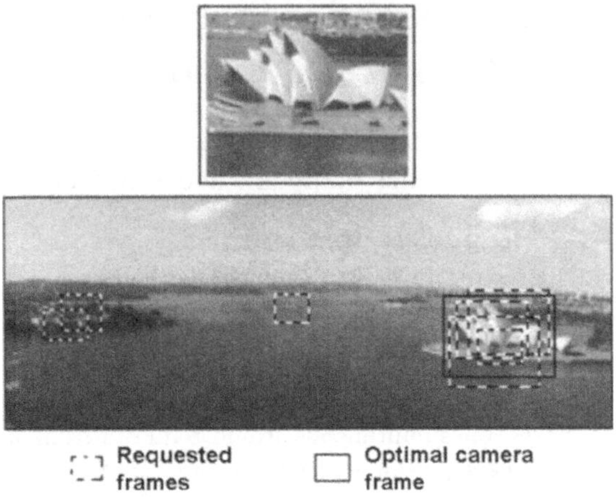

Fig. 1. Sharecam interface. Each Internet-based user views two image windows. The lower window is a fixed image of the camera's reachable range of view. Each user requests a camera frame by positioning a dashed rectangle in the lower window. Based on these requests, the algorithm computes an optimal camera frame (shown with solid rectangle), moves the camera accordingly, and displays the resulting live image in the upper window.

output. The input is a set of requested camera frames specified as desired fixed aspect-ratio iso-oriented rectangles from n users. The output is a single camera frame based on all inputs. We propose a new metric for user "satisfaction" based on how a user's requested frame compares with a candidate camera frame.

The metric for user satisfaction is proportional to the common area of the candidate frame and the requested frame. In addition, it is inversely proportional to the ratio of the sizes of the candidate and the request, to discourage excessively large candidates, such as an enclosing rectangle of all requests.

Finding the camera frame that maximizes total satisfaction is a non-linear optimization problem. We define a notion of "virtual corners" and prove that a global maximum must coincide with one of the virtual corners. We then present algorithms and complexity analysis. For n users and m zoom levels, the virtual corner search algorithm runs in time $O(n^2m)$. We then present distributed versions of the virtual corner search algorithm which run in time $O(n \log n + mn)$ on the server and $O(mn)$ on each client without loss of accuracy.

2 Related Work

The Internet provides a low-cost and widely-available interface that can make physical resources accessible to a broad range of participants. Online robots, controllable over the Internet, are an active research area. In addition to the challenges associated with time delay, supervisory control, and stability, online robots must be designed to be operated by non-specialists through intuitive user interfaces and to be accessible 24 hours a day; see [21,37,23,14] for examples of recent projects.

Chong et al. [9] proposed the following taxonomy for teleoperation systems: Single Operator Single Robot (SOSR), Single Operator Multiple Robot (SOMR), Multiple Operator Multiple Robot (MOMR). and Multiple Operator Single Robot (MOSR). Most online robots are SOSR, where control is limited to one operator at a time. One precedent of an online MOSR system is described by McDonald, Cannon, and colleagues [7,29]. In their work, several users assist in waste cleanup using Point-and-Direct (PAD) commands. Users point to cleanup locations in a shared image and a robot excavates each location in turn.

In [13,12], Goldberg and Chen described an Internet-based MOSR system that averaged multiple human inputs to simultaneously control a single industrial robot arm. In [15] Goldberg, Song, et al. propose the "Spatial Dynamic Voting" (SDV) interface. The SDV collects, displays, and analyzes a sequence of spatial votes from multiple online operators at their Internet browsers. The votes drive the motion of a single mobile robot or human "Tele-Actor".

We formulate the collaborative control problem as a nonlinear optimization problem with a non-differentiable objective function. The structure of the problem is closely related to the planar $p-$center problem, which has been proved to be NP-complete by Megiddo and Supowit [30]. Using a geometric approach, Eppstein [11] found an algorithm for the the planar 2-Center problem in $O(n \log^2 n)$. Halperin et al. [18] gave an algorithm for the 2-center problem with m obstacles that runs in randomized expected time $O(m \log^2(mn) + mn \log^2 n \log(mn))$.

In almost all nonlinear mathematical programming approaches, a constrained optimization problem is converted to a series of unconstrained problems using barrier or penalty methods. Line search is then used to solve the unconstrained optimization problems. Although there are many different ways of guiding search direction and step size, most of these methods are based on derivatives [34].

Rectangle-related problems in computational geometry include range searching and rectangle intersection. Agarwal and Erickson [1] provide a review of geometric range searching and its related topics. Grossi and Italiano [16,17] proposed the cross-tree data structure, a generalized version of balanced tree, to speed up range searching in high-dimensional space.

Kapelios et al. [25] developed an algorithm that reports the set of iso-oriented rectangles that intersect a query rectangle but do not enclose it and do not have one of their vertices inside it. They give solutions for unrestricted and restricted universe (grid) for R^d. Mount et al. [32] study, for a given pair of polygons P and Q, how to get the area of overlap of P and $Q + t$ as a function of the translation vector t.

It is also possible to view camera frame selection as a clustering problem [22,24]. The user satisfaction metric function we defined later in the paper is based on the resemblance and containment relationship between users' requested camera frames and real camera frame. The Symmetric Difference (SD) and "Intersection Over Union"(IOU) are well-known similarity metrics [3,5,39]. The comparison of these metrics and our metric will be discussed later in the paper.

There is also a connection with distributed manipulation. One branch of distributed manipulation uses potential fields defined as "potential-per-unit-area" acting on an object [4,31]. It is possible to interpret the satisfaction function as a special "lifted" potential field with some modifications.

Distributed computation has been used for sensor processing [33], multi-actuator control, and multi-robot systems. Sagawa et al. [38] developed a parallel algorithm to merge a set of range images into a volumetric implicit surface image representation, which is converted to a surface mesh. Safaric et al. [27] designed a distributed control system for an active surface device. The active surface device is a massive parallel micro-actuator array that can generate a pressure field on a planar surface. Applications of distributed algorithms include motion planing [6,35], localization [19,33], and task allocation [2,8].

In independent work, Kimber, Liu, Foote et al describe a multi-user robot camera in [26,28]. As we do in Sharecam, they formulate the frame selection for multiple simultaneous requests as an optimization problem based on position and area of overlap. To solve it, they propose an approximation based on comparing the bounding box of all combinations of user frames. This algorithm requires exponential time and does not provide formal bounds on approximation error.

3 Problem Definition

In this section we formulate the optimization problem: finding the camera frame that maximizes total user satisfaction.

3.1 Input and Assumptions

Let ϕ be a vector of system parameters. In the Sharecam system, $\phi = [x, y, z]^T$, where x, y specify the center point of the input rectangle corresponding to pan and tilt, and z specifies size of the rectangle, which can be used to control zoom. Since the camera has a fixed aspect ratio of 4:3, the length of the rectangle is $4z$ and width of the rectangle is $3z$. At each time increment, user i requests a desired frame, ϕ_i. Given requests from n users, the system computes a single global frame ϕ^* that will best satisfy the set of requests.

Let Θ be the set of all admissible $[x, y]$ pairs. Let Z be a small set of attainable discrete zoom levels. Hence the solution space is:

$$\Phi = \Theta \times Z = \{[x, y, z] | [x, y] \in \Theta, z \in Z\}.$$

3.2 Definition of a Metric for User Satisfaction

Recall that ϕ_i is the frame requested by user i, and let $\phi = [x, y, z]^T$ be a candidate camera frame. We define a scalar $s_i \in [0, 1]$ as the level of "satisfaction" that user i receives. User i should get no satisfaction if the candidate frame does not intersect ϕ_i, so $s_i = 0$ when $\phi \cap \phi_i = \emptyset$. User i should be perfectly satisfied when the candidate frame is identical to ϕ_i, so $s_i = 1$ when $\phi = \phi_i$. We define the satisfaction of user i provided by the candidate frame ϕ to be Intersection Over Maximum (IOM),

$$s_i(\phi_i, \phi) = p_i / \max(a, a_i)$$

where a is the area of the candidate ϕ ($a = 12z^2$), frame a_i is the area of frame ϕ_i ($a_i = 12z_i^2$), and p_i is the area of overlap between ϕ_i and ϕ. Furthermore,

$$s_i(\phi_i, \phi) = p_i / \max(a, a_i) = (p_i/a_i) \min((z_i/z)^2, 1)$$

is a special case of Generalized Intersection Over Maximum metric (GIOM) for $b = 2$,

$$s_i(\phi_i, \phi) = (p_i/a_i) \min((z_i/z)^b, 1)$$

where b is a small positive number.

If z is bigger, the candidate frame will be bigger. A sufficiently large z can define a candidate frame that covers all requested frames: $p_i/a_i = 1$ for $i = 1, ..., n$. However, user satisfaction is not necessarily high because a user wants to see the requested frame at a desired zoom level. The term $\min((z_i/z)^b, 1)$ characterizes this wish: it reaches its maximum of 1 if the candidate frame is the same or smaller than the requested frame. The $s_i \in [0, 1]$ is normalized.

During experiments, we found that if b is large, we have to pay more for increasing the size of the candidate frame in the objective function value. Therefore, the optimal frame tends to be small. We want to avoid cases where the optimal frame is too small such that it can only cover the intersected area among the requested frames. We thus set $b = 1$ in our settings to encourage the optimal frame to be a trade-off between the small intersected area of the requested frames and the big union area of the requested frames. Figure 2 shows five examples of requested frames and the corresponding satisfaction, for $b = 1$.

The total satisfaction for n users is

$$s(\phi) = \sum_{i=1}^{n} s_i(\phi_i, \phi). \tag{1}$$

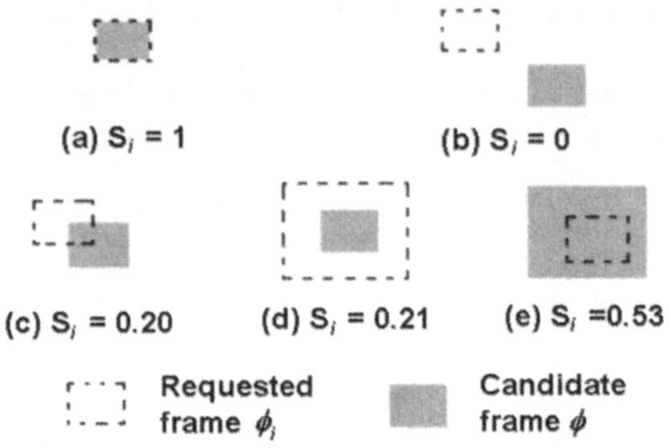

Fig. 2. Examples of the satisfaction metric for user i and candidate frame ϕ.

We want to find $\phi^* = \arg\max_\phi s(\phi)$, the frame that maximizes total satisfaction. To describe the objective function with respect to x, y, and z,

$$s(x, y, z) = s(\phi).$$

Equation 1 can be extended to a weighted sum format at time t:

$$s(\phi, t) = \sum_{i=1}^{n} \alpha_i(t) s_i(\phi_i(t), \phi(t)) \tag{2}$$

where the weight $\alpha_i(t)$ for user i is a function of the user's previous unsatisfaction levels: $u_i(t) = 1 - s_i(\phi_i(t), \phi(t))$. An example of $\alpha_i(t)$ is,

$$\alpha_i(t) = \sum_{k=0}^{t-1} \frac{u_i(k)}{2^{t-1-k}}$$

and its recursive formulation is very simple,

$$\alpha_i(t) = u_i(t-1) + \alpha_i(t-1)/2$$

If user i is not satisfied in previous runs, his/her $\alpha_i(t)$ will increase in the objective function. This formulation can avoid the situation that some user may not be satisfied all the time. From algorithm point of view, the difference between equation 1 and equation 2 is little. To simplify the discussion, we will use the objective function described by equation 1 in the rest of the paper.

3.3 Properties of the Satisfaction Metric

We show that the objective function s is nonsmooth and piecewise linear in both x and y. In addition we compare it to the intersection-over-union metric.

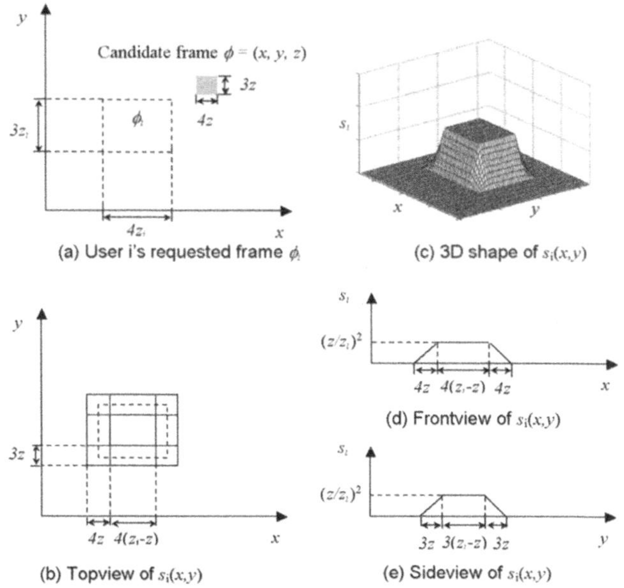

Fig. 3. Satisfaction function, $s_i(x, y)$, for a given candidate frame and $b = 1$. Since $z \leq z_i$ is given, we can move the candidate frame (gray rectangle) around the user i's requested frame to observe how $s_i(x, y)$ changes. The function is a plateau-like shape with a maximum height of $p_i/a_i = 12z^2/12z_i^2 = (z/z_i)^2$. The function consists of 5 planes and 4 quadratic surfaces at the corners.

Nonsmoothness. Recall that the objective function is

$$s(\phi) = \sum_{i=1}^{n} (p_i/a_i) \min((z_i/z)^b, 1).$$

Since we solve this problem for each attainable zoom level z, we treat z as a constant. The objective function becomes a function of the center point of the candidate frame,

$$s(x, y) = \sum_{i=1}^{n} \omega_i p_i(x, y) \tag{3}$$

where $\omega_i = (1/a_i) \min((z_i/z)^b, 1)$ is a constant for each user. We know that $p_i(x, y)$ is the area of the intersection of the requested frame of user i and the candidate frame (x, y, z). Therefore, the maximum value of $p_i(x, y)$ is $\min\{a_i, a\}$. This property determines that the shape of user i's satisfaction function is plateau-like. Figure 3 shows the shape of $s_i(x, y)$ given $z \leq z_i$, i.e. the candidate frame is smaller than the requested frame of user i. Note that s_i is non-differentiable with respect to x and y so we cannot use derivative-based approaches to solve this problem.

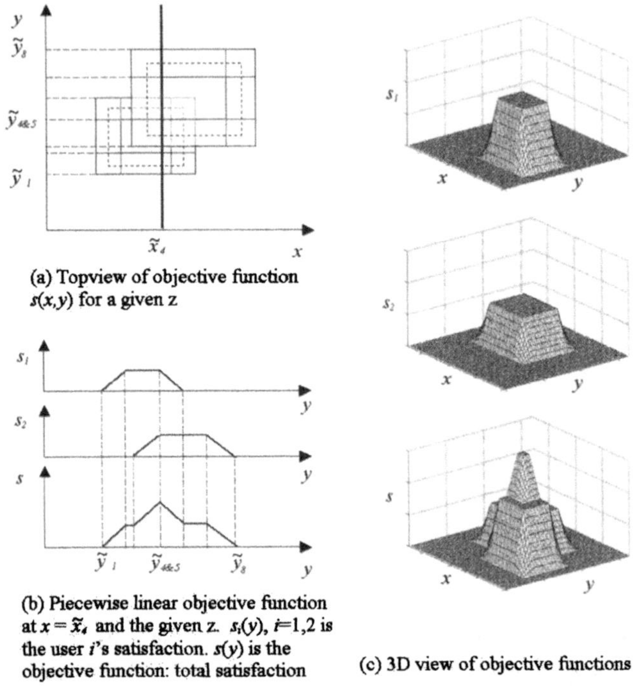

Fig. 4. Satisfaction function $s_i(y)$ for two-users. Ordered sets $\{\tilde{y}_k\}$ and $\{\tilde{x}_k\}$, $k = 1, ..., 8$ are corresponding to horizontal and vertical edges of plateaus. Note that $\tilde{y}_4$ and $\tilde{y}_5$ overlap in this case.

Piecewise linearity in x and y. Since all requested frames and the candidate frame are iso-oriented rectangles, the shape of any intersection between them is also a rectangle with its edges parallel to either x axis or y axis. Thus the term $p_i(x, y)$ in equation 3 is either 0 or the area of the rectangle formed by intersection between ϕ_i and $\phi = (x, y, z)$. This yields a nice property: the $p_i(x, y)$ is piecewise linear with respect to x if we fix y, and piecewise linear with respect to y if we fix x. Since the total satisfaction metric $s(x, y)$ is a linear combination of $p_i(x, y)$, $i = 1, ..., n$, it has the same property. Figure 4 shows an example for a case with two requested frames.

Comparison to other metrics. In pattern recognition and computational geometry standard similarity metrics are Symmetric Difference (SD) and Intersection Over Union (IOU)[3,5,39]. For a requested frame ϕ_i and a candidate frame ϕ, the SD metric is

$$SD = \frac{Area(\phi_i \cup \phi) - Area(\phi_i \cap \phi)}{Area(\phi_i \cup \phi)}.$$

The intersection-over-union metric is

$$IOU = \frac{Area(\phi_i \cap \phi)}{Area(\phi_i \cup \phi)} = 1 - SD.$$

Compared with IOU, our satisfaction metric has similar properties: (i) both attain their minimum value of 0 if and only if $\phi \cap \phi_i = \emptyset$, (ii) both attain their maximum value of 1 if and only if $\phi = \phi_i$, (iii) both are proportional to the area of $\phi \cap \phi_i$, and (iv) both depend—albeit differently—on the sizes of ϕ and ϕ_i. However, the intersection-over-union metric is not piecewise linear in x or y.

4 Algorithms

In this section we present algorithms for two versions of the problem of finding the frame that maximizes total satisfaction. We start with a version in which the pan (x) as well as the tilt (y) are restricted to a discrete set of equally-spaced values. Subsection 4.1 describes a brute-force algorithm for this discrete version of the problem. In Subsection 4.2 we allow the pan and tilt to vary continuously. This more general continuous version allows for an efficient exact algorithm. The algorithm exploits a geometric characteristic of the optimal solution (captured in the notion of a "virtual corner"). The exact algorithm can also be distributed across the client machines and the server. The distributed algorithm is given in Subsection 4.3.

4.1 Algorithms for Discrete Pan and Tilt

Theorem 1 *Let w be the width (in pixels) of the camera's total pan range, and h be the height (in pixels) of the camera's total tilt range. We can solve the discrete pan and tilt problem in $O(whmn)$.*

Proof. A brute force search for finding $\phi^* = \arg\max_\phi s(\phi)$ evaluates all whm candidate points. According to Equation 1, it takes $O(n)$ computing time to determine the satisfaction for a single candidate frame ϕ. The total amount of computation of the algorithm is $O(whmn)$.

Although this is linear in n, the constants are large (typically $w = 600, h = 200$).

4.2 An Algorithm for Continuous Pan and Tilt

We now focus on the more general problem where the camera's pan and tilt may vary continuously.

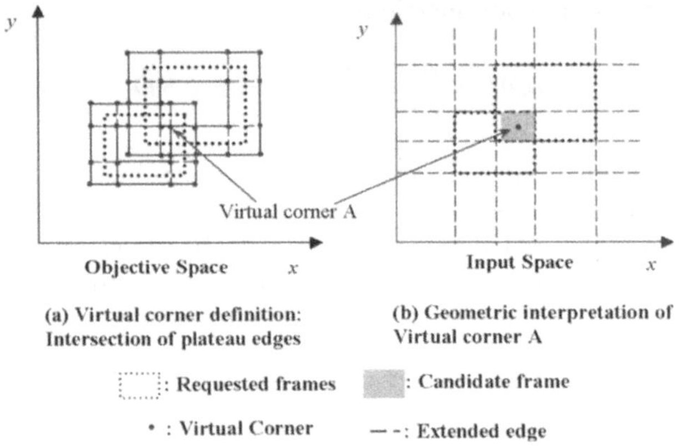

Fig. 5. Virtual corners and a geometric interpretation for two requested frames. A virtual corner corresponds to a candidate frame that has one corner at the intersection of one extended vertical edge of a requested frame and one horizontal extended edge of a requested frame.

Virtual corners. As shown in Figure 3, the $s_i(x, y)$ is a plateau-like function. For n users, there are n plateaus. Each plateau consists of 9 facets: 1 top plane, 4 side planes, and 4 quadratic surface at corners. There are two vertical boundaries and two horizontal boundaries at the bottom (bounding the entire plateau), the same numbers of similar edges at the top (bounding the plateau's flat top), and eight boundaries separating side planes and corner quadratic surfaces (see Figure 5a).

A *virtual corner* is an intersection between any two boundaries, which includes both intersections of facet boundaries induced by a single plateau or by two distinct plateaus. Since all plateaus are iso-oriented, one of the extensions has to be horizonal and the other has to be vertical. For n requested frames, there are $O(n^2)$ virtual corners. Figure 5b shows the geometric interpretation of virtual corner in the input space. If we map the virtual corner in the objective space back to input space, we see that the virtual corner corresponds to a candidate frame that has one corner overlapping with the intersection of two extensions of edges of requested frames. What makes a virtual corner important is the following theorem.

Lemma 1 *At least one optimal frame is centered at a virtual corner.*

Proof. Let $\phi^* = [x^*, y^*, z^*]$ be an optimal solution. As discussed earlier, for a fixed z and x, the objective function $s(y)$ is piecewise linear. So the optimum must be at a vertex $y = \tilde{y}$ such that $s(x^*, \tilde{y}, z^*) = s(x^*, y^*, z^*)$. We also know that line $y = \tilde{y}$ in (x, y) plane is one of the horizontal facet boundaries of the plateaus. Similarly, we can find another optimal frame $[\tilde{x}, \tilde{y}, z^*]$, where line $x = \tilde{x}$ is one of the vertical facet boundaries of the plateaus. Therefore, the optimal frame $[\tilde{x}, \tilde{y}, z^*]$ is centered at a virtual corner $(\tilde{x}, \tilde{y})$.

Brute force approach. Based on the theorem 1, we can solve the optimization problem by simply checking all combinations of zoom levels and corresponding virtual corners. We need to evaluate the objective function for each of the $O(n^2)$ virtual corners and repeat this for each of the m zoom levels. It takes $O(n)$ time to evaluate a candidate frame ϕ. Therefore, the brute force algorithm runs in $O(n^3m)$. However, we can do better if (for a fixed zoom level) we handle all virtual corners with the same x-coordinate consecutively in order of increasing y-coordinate, and take advantage of the fact that the objective function only changes slightly between two consecutive virtual corners.

Efficient traversal of virtual corners. For n requested frames, we have $4n$ horizontal plateau facet boundaries $\{\tilde{y}_1, \tilde{y}_2, ..., \tilde{y}_{4n}\}$ and $4n$ vertical plateau facet boundaries $\{\tilde{x}_1, \tilde{x}_2, ..., \tilde{x}_{4n}\}$ for plateaus. We can reduce the computation complexity from $O(n^3m)$ to $O(n^2m)$. Recall that $\phi_i = [x_i, y_i, z_i]$, $i = 1, ..., n$ are the requested frames.

$s^* = 0,$ $O(1)$
Sort $\{y_i + 1.5z_i\}$, $i = 1, ...n$ $O(n \log n)$
Sort $\{y_i - 1.5z_i\}$, $i = 1, ...n$ $O(n \log n)$
For each zoom level z (m in total)
i) Compute vertical extended plateau edges $\{\tilde{x}_1, \tilde{x}_2, ..., \tilde{x}_{4n}\}$ $O(n)$
For each user i, $i = 1, ..., n$,
$\tilde{x}_{4i-3} = x_i - 2(z_i + z),$
$\tilde{x}_{4i-2} = x_i - 2(z_i - z),$
$\tilde{x}_{4i-1} = x_i + 2(z_i - z),$
$\tilde{x}_{4i} = x_i + 2(z_i + z),$
ii) Compute the sorted sequence $\{\tilde{y}_1, \tilde{y}_2, ..., \tilde{y}_{4n}\}$, $O(n)$
For each i, $i = 1, ..., n$
$\tilde{y}_{4i-3} = (y_i - 1.5z_i) + 1.5z,$
$\tilde{y}_{4i-2} = (y_i - 1.5z_i) - 1.5z,$
$\tilde{y}_{4i-1} = (y_i + 1.5z_i) - 1.5z,$
$\tilde{y}_{4i} = (y_i + 1.5z_i) + 1.5z,$
Merge the 4 ordered sequences:
$\{\tilde{y}_{4i-3}\}$, $\{\tilde{y}_{4i-2}\}$, $\{\tilde{y}_{4i-1}\}$, and $\{\tilde{y}_{4i}\}$, $i = 1, ..., n$
to get the ordered sequence $\{\tilde{y}_1, \tilde{y}_2 ..., \tilde{y}_{4n}\}$,
where $\tilde{y}_1$ is the smallest.
iii) For $x = \tilde{x}_i$, $i = 1, ..., 4n$,
$s = \max_y\ s(\tilde{x}_i, y, z)$,
if $s > s^*$ then $s^* = s, x^* = \tilde{x}_i, y^* = y, z^* = z$.
Output s^* as optimal objective function value and (x^*, y^*, z^*) as optimal frame.

In step iii, we traverse the vertical facet boundaries of the plateaus one by one. For each vertical edge, we find the maximum by forcing the candidate frame to center at it. Using Theorem 1, we know that this procedure will find an optimal solution. It remains to show how much time is required to solve the resulting problem of finding

$$\max_{y} s(x, y, z)$$

for given x and z. This special optimization problem can be solved in $O(n)$ with a sorted sequence $\{\tilde{y}_1, \tilde{y}_2 ..., \tilde{y}_{4n}\}$. The objective function is a "summation" of n plateaus, which is shown in Figure 4. For fixed x and z, this piecewise linear function only changes slope at $\{\tilde{y}_i\}$, $i = 1, ..., 4n$. For each vertex $\tilde{y}_i$, we know how much the slope will change after crossing the vertex. We can find the maximum objective value by walking over all ordered vertices $\{\tilde{y}_i\}$ from the one side to the other side on the line $x = \tilde{x}_i$. This process only takes $O(n)$. Therefore, step iii) of the algorithm will take $O(n^2)$ and the following theorem is true.

Theorem 2 *We can solve the Sharecam problem in time $O(n^2 m)$ for n users and m zoom levels.*

4.3 A Distributed Algorithm

In the system, n is the number of users online, which is also the number of computers connecting to our server. The larger the value of n, the more computation power we have in our system. This suggests that a distributed computing strategy can further improve computational speed. The algorithm described in the previous section can be divided into client and server components.

Server Algorithm. The server should do the following.

i). Send all requested frames ϕ_i, $i = 1, ..., n$ to all clients,
ii). Sort sequence $\{y_i + 1.5z_i\}$ and sequence $\{y_i - 1.5z_i\}$, $i = 1, ...n$ and send them to all clients, $O(n \log n)$
iii). Wait until all clients send their solutions $\{s_1^*, ..., s_n^*\}$ back. Pick the largest. $O(n)$

Client Algorithm. Let us assume that $\phi_i = (x_i, y_i, z_i)$ is client i's requested frame. After client i receives the data from the server, it executes the following algorithm.

$s_i^* = 0$
For each zoom level z
  i) Compute the sorted sequence $\{\tilde{y}_1, \tilde{y}_2, ..., \tilde{y}_{4n}\}$, $O(n)$
  (Same as the centralized version.)

ii)a. $s_1 = \max_y \ s(x_i - 2(z_i + z), y, z)$, $O(n)$
if $s_1 > s_i^*$ then $s_i^* = s_1, x_i^* = x_i - 2(z_i + z), y_i^* = y, z_i^* = z$.
ii)b. $s_2 = \max_y \ s(x_i - 2(z_i - z), y, z)$, $O(n)$
if $s_2 > s_i^*$ then $s_i^* = s_2, x_i^* = x_i - 2(z_i - z), y_i^* = y, z_i^* = z$.
ii)c. $s_3 = \max_y \ s(x_i + 2(z_i - z), y, z)$, $O(n)$
if $s_3 > s_i^*$ then $s_i^* = s_3, x_i^* = x_i + 2(z_i - z), y_i^* = y, z_i^* = z$.
ii)d. $s_4 = \max_y \ s(x_i + 2(z_i + z), y, z)$, $O(n)$
if $s_4 > s_i^*$ then $s_i^* = s_4, x_i^* = x_i + 2(z_i + z), y_i^* = y, z_i^* = z$.
iii) Send s_i^* and (x_i^*, y_i^*, z_i^*) to server.

As we can see from the algorithm, the server runs at $O(n \log n + mn)$ and each client runs at $O(nm)$. The following theorem holds,

Theorem 3 *We can distribute the Sharecam algorithm among the server and clients resulting in a running time of $O(n \log n + mn)$.*

One can also see from the algorithm that the speed of computation is limited by the slowest client. One idea is to set a timeout for clients and have the server run the computation for clients that do not respond in time.

5 Results

We have implemented the algorithms on a PC with 950Mhz AMD Athlon CPU and 1GB memory. The machine runs under Redhat Linux 7.1. The algorithm is programmed in both matlab and Java.

Figure 6 shows the results for four different scenarios. As we can see from Figure 6(a) and (b), the optimal frame does not necessarily have one corner overlapping with a corner of a requested frame. However, one corner of the optimal frame does overlap with one of the virtual corners. Figure 6(b) has three requested frames exactly the same as those in (a) and one more big requested frame. It is interesting to see how the optimal frame changes after the big requested frame joined in the system. Figure 6(c) shows that if all input rectangles fall far way from each other, the algorithm functions as a chooser, which selects one input rectangle as the output. Since the algorithm searches optimum bottom-up, it picks the lowest requested frame as the solution. Figure 6(d) shows that a large requested frame does not necessarily yield a large optimal frame.

Figure 7 shows the relationship between the optimal frame size and the choice of the b value in GIOM satisfaction metric. It confirms that our analysis in Subsection 3.2: the bigger the b is, the smaller the optimal camera frame is. If $b \to 0^+$, then the optimal camera frame becomes the smallest frame that contains all request frame: $p_i = a_i \ \forall i$. If $b \to \infty$, then the optimal frame will converge to the rectangle area that most of requested frame insect each other. The parameter b allows us to find the best tradeoff between union and intersection.

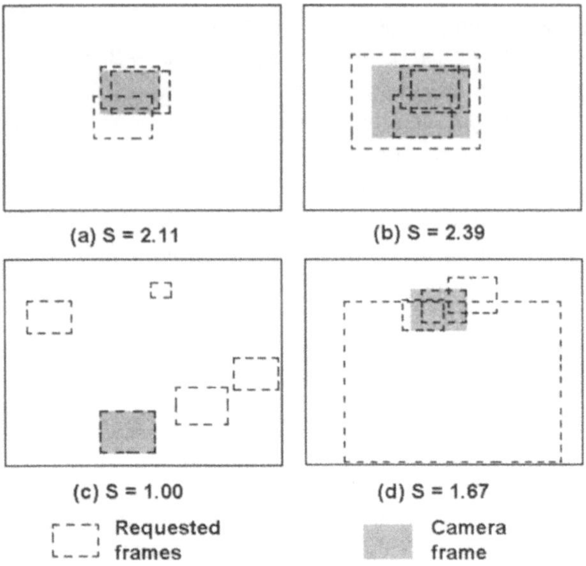

Fig. 6. Sample outputs of algorithm.

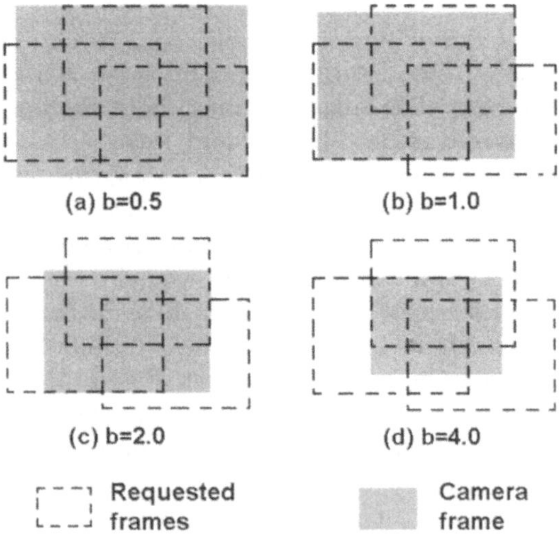

Fig. 7. Relationship between the optimal frame size and the choice of the b value in GIOM satisfaction metric in Subsection 3.2.

6 Conclusions and Future Work

This paper considers a new problem allowing multiple users to share control of a single remote robotic camera. Each user requests a desired camera frame by drawing a rectangle over a static global image. The problem is to find a camera frame that maximizes the overall user satisfaction. We define a new metric for user satisfaction and study algorithms for solving the nonlinear optimization problem.

We reviewed related work and studied properties of the objective function and inherent constraints on the location of extremal points. We defined "virtual corners" and proved that a global maximum must coincide with one of the virtual corners. We then presented algorithms and complexity analysis. For n users and m zoom levels, we described an efficient algorithm based on grouping and sorting of virtual corners running in time $O(n^2m)$. Finally we presented a distributed version of this algorithm that runs in time $O(n \log n + mn)$ on the server and $O(mn)$ on each client.

Unlike computing with multiple processors in a single supercomputer, distributed computing over the Internet requires input from a variety of heterogenous processors, each with different and varying communication delays. We are interested in distributed algorithms that optimize performance under such uncertainties.

In future work, we will also consider versions of the problem with continuous zoom levels ($m = \infty$). Preliminary results suggest that we can solve this in $O(n^3 \log n)$ time. We are also working on an non-uniform grid-based approximation algorithm. It employs a Branch and Bound-like approach and can solve the problems with continuous pan, tilt, and zoom in linear time. We will also consider extensions to p frames, with i) p sequential views from one camera or ii) from p different cameras. Scenario i) is also a scheduling problem. Scenario ii) is related to the p-center problem.

A preliminary version of the Sharecam system went online in Sep. 2002 at http://tele-actor.net/sharecam.

Acknowledgments.
Thanks to A. Pashkevich, V. Koltun, D. Halperin, A. Lim, S. Rao, J. Luntz, and S. Har-Peled, for insightful discussions and feedback. Thanks to M. Faldu, F. Hsu, and W. Guan for their contributions to the ShareCam system, and K. "Gopal" Gopalakrishnan, Ron Alterovitz, A. Levandowski, M. Mckelvin, A. Ho, and J. Vidales for helpful suggestions.

References

1. P. Agarwal and J. Erickson. Geometric range searching and its relatives. In B. Chazelle, J. E. Goodman, and R. Pollack, editors, *Advances in Discrete and Computational Geometry, volume 23 of Contemporary Mathematics.*

2. W. Agassounon, A. Martinoli, and R. Goodman. A scaleable distributed algorithm for allocating workers in embedded systems in embedded systems. In *IEEE Systems, Man and Cybernetics Conference*, 2001.
3. A.Z.Broder. On the resemblance and containment of documents. In *SEQS: Sequences '91*, 1998.
4. K. F. Bohringer and H. Choset, editors. *Distributed Manipulation.* Kluwer Academic Publishers, 2000.
5. A. Z. Broder, M. Charikar, A.M. Frieze, and M.Mitzenmacher. Min-wise independent permutations. *Journal of Computer and System Sciences*, 60(3):630–659, 2000.
6. Z. Bulter, S. Byrnes, and D. Rus. Distributed motion planning for modular robots with unit-compressible modules. In *IEEE/RSJ International Conference on Intelligent Robots and System (IROS)*, 2001.
7. D. J. Cannon. *Point-And-Direct Telerobotics: Object Level Strategic Supervisory Control in Unstructured Interactive Human-Machine System Environments.* PhD thesis, Stanford Mechanical Engineering, June 1992.
8. R. Chen, Z. Wu, Y. Wang, and D. Tan C. Dong. Cooperative assembly algorithm of multiple distributed agent based robotic system. In *30th International Symposium on Robotics*, 1999.
9. N. Chong, T. Kotoku, K. Ohba, K. Komoriya, N. Matsuhira, and K. Tanie. Remote coordinated controls in multiple telerobot cooperation. In *IEEE International Conference on Robotics and Automation*, April 2000.
10. C. H. Crouch and E. Mazur. Peer instruction: Ten years of experience and results. *Americian Journal of Physices*, 69(9), 2001.
11. D. Eppstein. Fast construciton of planar two-centers. In *Proc. 8th ACM-SIAM Sympos. Discrete Algorithms*, 1997.
12. K. Goldberg and B. Chen. Collaborative control of robot motion: Robustness to error. In *International Conference on Intelligent Robots and Systems (IROS)*, 2001.
13. K. Goldberg, B. Chen, R. Solomon, S. Bui, B. Farzin, J. Heitler, D. Poon, and G. Smith. Collaborative teleoperation via the internet. In *IEEE International Conference on Robotics and Automation (ICRA)*, April 2000.
14. K. Goldberg and R. Siegwart, editors. *Beyond Webcams: An Introduction to Online Robots.* MIT Press, 2002.
15. K. Goldberg, D. Song, Y. Khor, D. Pescovitz, A. Levandowski, J. Himmelstein, J. Shih, A. Ho, E. Paulos, and J. Donath. Collaborative online teleoperation with spatial dynamic voting and a human "tele-actor". In *IEEE International Conference on Robotics and Automation (ICRA)*, May 2002.
16. R. Grossi and G. F. Italiano. Efficient cross-trees for external memory. In James Abello and Jeffrey Scott Vitter, editors, *External Memory Algorithms and Visualization*, pages 87–106. American Mathematical Society Press, Providence, RI, 1999.
17. R. Grossi and G. F. Italiano. Revised version of "efficient cross-trees for external memory". Technical Report TR-00-16, 10 2000.
18. D. Halperin, M. Shairir, and K. Goldberg. The 2-center problem with obstacles. *Journal of Algorithms*, 32, 2002.
19. A.T. Hayes, A. Martinoli, and R.M. Goodman. Swarm robotic odor localization. In *IEEE/RSJ International Conference on Intelligent Robots and System (IROS)*, 2001.

20. **http://www.x-zone.canon.co.jp/WebView-E/all/list.htm.**
21. H. Hu, L. Yu, P. Tsui, and Q. Zhou. Internet-based robotic systems for teleoperation. *Assembly Automation Journal*, 21(2), 2001.
22. A.K Jain, M.N. Murty, and P.J. Flyn. Data clustering: A review. *ACM Computing Surveys*, 31(3), 1999.
23. S. Jia and K. Takase. A corba-based internet robotic system. *Advanced Robotics*, 15(6), 2001.
24. D.E. Johnson. *Applied Multivariate Methods for Data Analysis.* Duxbury Press, 1998.
25. V. Kapelios, G. Panagopoulou, G. Papamichail, S. Sirmakessis, and A. Tsakalidis. The cross rectangle intersection problem. *Computer Journal*, 38(3), 1995.
26. D. Kimber, Q. Liu, J. Foote, and L. Wilcox. Capturing and presenting shared multi-resolution video. In *SPIE ITCOM 2002. Proceeding of SPIE, Boston*, volume 4862, pages 261–271, Jul. 2002.
27. P. Ku, K.T. Winther, H.F. Stephanou, and R. Safaric. Distributed control system for an active surface device. In *IEEE International Conference on Robotics and Automation (ICRA)*, 2001.
28. Q. Liu, D. Kimber, J. Foote, L. Wilcox, and J. Boreczky. Flyspec: A multi-user video camera system with hybrid human and automatic control. In *ACM Multimedia 2002, France*, Dec. 2002.
29. M. McDonald, D. Small, C. Graves, and D. Cannon. Virtual collaborative control to improve intelligent robotic system efficiency and quality. In *IEEE International Conference on Robotics and Automation*, April 1997.
30. N. Megiddo and K.J. Supowit. on the complexity of some common geometric location problems. *SIAM J. Comput.*, 13, 1984.
31. H. Moon and J. Luntz. Computing equilibria on superpositions of logarithmic-radial potential fields. In *The Workshop on Algorithmic Foundations of Robotics, December*, Dec. 2002.
32. D.M. Mount, R. Silverman, and A.Y. Wu. On the area of overlap of translated polygons. *Computer Vision and Image Understanding*, 64:353–61, 1996.
33. E. Mumolo, M. Nolich, and G. Vercelli. Algorithms and architectures for acoustic localization based on microarray in service robotics. In *IEEE International Conference on Robotics and Automation (ICRA)*, 2000.
34. S.G. Nash and A. Sofer, editors. *Linear and Nonlinear Programming.* The McGraw-Hill Companies, Inc, 1996.
35. L.E. Parker. Distributed algorithms for multi-robot observation of multiple moving targets. *Autonomous Robots*, 12(3), 2002.
36. B. Rogoff, E. Matusov, and C. White. *Models of teaching and learning: Participation in a community of learners.* Oxford, England: Blackwell, 1996.
37. R. Safaric, M. Debevc, R. Parkin, and S. Uran. Telerobotics experiments via internet. *IEEE Transactions on Industrial Electronics*, 48(2):424–31, April 2001.
38. R. Sagawa, K. Nishino, M.D. Wheeler, and K. Ikeuchi. Parallel processing of range data merging. In *IEEE/RSJ International Conference on Intelligent Robots and System (IROS)*, 2001.
39. R. C. Veltkamp and M. Hagedoorn. Shape similarity measures, properties, and constructions. In *Advances in Visual Information Systems, 4th International Conference, VISUAL 2000, Lyon, France, November 2-4, 2000, Proceedings VISUAL*, volume 1929, pages 467–476. Springer, 2000.

Shape Complexes for Metamorhpic Robots*

Robert Ghrist

University of Illinois, Urbana, IL 61801, USA

Abstract. A **metamorphic** robotic system is an aggregate of identical robot units which can individually detach and reattach in such a way as to change the global shape of the system. We introduce a mathematical framework for defining and analyzing general metamorphic systems. This formal structure combined with ideas from geometric group theory leads to a natural extension of a configuration space for metamorphic systems — the **shape complex** — which is especially adapted to parallelization. We present an algorithm for optimizing reconfiguration sequences with respect to elapsed time. A universal geometric property of shape complexes — **non-positive curvature** — is the key to proving convergence to the globally time-optimal solution.

1 Introduction

In recent years, several groups in the robotics community have been modeling and building **reconfigurable** or, more specifically, **metamorphic systems** (*e.g.*, [6,11,13,19,20]). Such a system consists of multiple identical robotic cells in an underlying lattice structure which can disconnect/reconnect with adjacent neighbors, and slide, pivot, or otherwise locomote to neighboring lattice points following prescribed rules: see Fig. 1. There are as many models for such systems as there are researchers in the sub-field: 2-d and 3-d lattices; hexagonal, square, and dodecahedral cells; pivoting or sliding motion [6,11–15,19,20]. The common features of these systems are an aggregate of lattice-based cells having prescribed transitions from one shape to another.

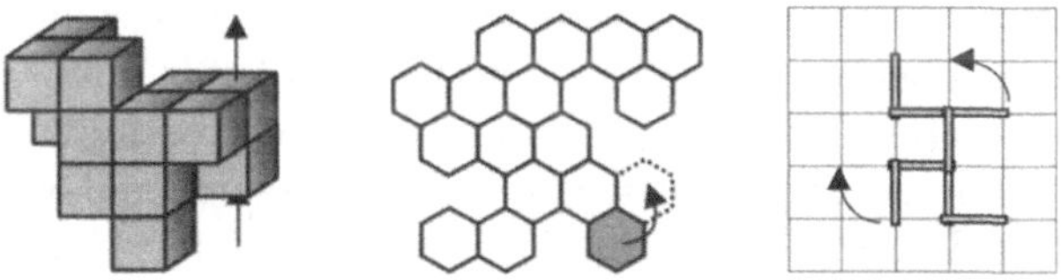

Fig. 1. Metamorphic systems may be built on a variety of lattice structures with sliding or pivoting motion.

The primary problem to be solved in such systems is the shape-planning problem: how to move from one shape to another via legal moves. One centralized approach [7,17] is to build a **transition graph** whose vertices are the

* Supported in part by National Science Foundation Grant DMS-0134408.

J.-D. Boissonnat et al. (Eds.): Algorithmic Foundations of Robotics V, STAR 7, pp. 185–201, 2004.

various shapes and whose edges are elementary legal moves from one shape to the next. It is easily demonstrated that the size of this graph is exponential in the number of cells.

We propose to extend the notion of a configuration space to metamorphic systems. The idea: consider the transition graph described above as the 1-skeleton of a **shape complex**. Assume that from a given shape there are two possible moves which are physically independent (or, more suggestively, "commutative"): *i.e.*, these moves can be executed simultaneously. In the transition graph, this corresponds to the four edges of a square. For any pair of commutative moves, fill in the four edges of the graph with an abstract square 2-cell. Continue inductively adding k-cubes corresponding to k commutative motions. The result is a cubical complex which has several advantages over the transition graph:

1. **Size.** The shape complex is often simpler than the transition graph: *e.g.*, the 1-d graph of an n-dimensional cube has $n2^{n-1}$ edges. This figure belies the simplicity of the single cube.
2. **Time.** Geodesics on this complex cut across the diagonals of cubes whenever possible: one performs all possible commutative motions simultaneously, yielding a speed-up by a factor of the dimension of the complex.
3. **Shape.** The global geometry/topology of the shape complex carries information about the metamorphic system. For certain examples, the topology of the shape complex "converges" upon refining the lattice. In addition, only special geometries can be realized as the shape complex of a local metamorphic system: commutativity in reconfiguration leads to an abhorrence of positive curvature in the shape complex.

Sections 2 through 4 give definitions and examples of [abstract] metamorphic systems and their shape complexes. The next two sections (sections 5-6) detail topological and geometric features of the shape complex.

For large systems, the problem of computing the shape complex and designing geodesics in order to perform shape planning is computationally infeasible for two reasons: (1) the size of the complex is often exponential in the number of robot cells; and (2) any control scheme induced by geodesic construction is necessarily centralized. Several researchers have begun building decentralized control algorithms for shape planning [4,5,18,21,22]. Such algorithms have the advantage of speed; however, the reconfiguration paths are rarely if ever optimal.

As an application of our techniques, we present in Sections 7-8 an algorithm for trajectory optimization which takes as its argument an arbitrary edge path in the transition graph. Algorithm 5 then performs a type of **curve shortening** within the shape complex. A deep theorem about the curvature of all shape complexes (Theorem 1) is then used to prove that this algorithm returns a shape trajectory which is the global minimum obtainable from this path with respect to elapsed time.

Our definitions and theorems are phrased for systems involving "discrete" reconfiguration. More general types of systems which employ continuous reconfiguration for locomotive gaits are not covered by our definitions. We note, however, that certain locomotive reconfigurable systems can be thought of as lattice-based tiles by amalgamating subsystems [12]. In addition, our definitions can easily be extended to general non-lattice reconfigurable systems [2].

2 A Mathematical Definition

While it is easy to generate examples of what is meant by a metamorphic system, it is more challenging to write a clean mathematical definition. We propose a set of definitions which is broad enough to include some non-obvious examples.

A local metamorphic system is a collection of states on a lattice, where each state is thought of as an indicator function for the aggregate. Any state can be modified by local rearrangements, these local changes being coordinated by a catalogue of models realized under the actions of isometries into the domain. The adjective "local" refers to legality criteria: anywhere in the domain at which a local change from the catalogue can be applied, it is legal to do so. To incorporate obstacles and basepoints into our systems, we distinguish between the amount of information needed to determine the legality of an elementary move (the "support" of the move) and the precise place in which modules are actually in motion (the "trace" of the move).

Definition 1. Let $\mathcal{L}$ denote a lattice in $\mathbb{R}^k$ and $\mathcal{D} \subset \mathcal{L}$ be some domain. The **catalogue** C for a local metamorphic system on $\mathcal{D}$ is a collection of **generators**. Each generator $\phi \in \mathsf{C}$ consists of (1) the **support**, $\sup(\phi) \subset \mathcal{L}$; (2) the **trace** of the move, $\mathrm{tr}(\phi) \subset \sup(\phi)$; and (3) an unordered pair of local states $\hat{U}_{0,1} : \sup(\phi) \to \{0,1\}$ satisfying[1]

$$\hat{U}_0\Big|_{\sup(\phi)-\mathrm{tr}(\phi)} = \hat{U}_1\Big|_{\sup(\phi)-\mathrm{tr}(\phi)}. \tag{1}$$

Otherwise said, the local states are equal on $\sup(\phi) - \mathrm{tr}(\phi)$.

Definition 2. An **action** of a generator $\phi \in \mathsf{C}$ is a rigid translation $\Phi : \sup(\phi) \hookrightarrow \mathcal{D}$. Given a state $U : \mathcal{D} \to \{0,1\}$, an action Φ is said to be **admissible** at U if $\hat{U}_0 = U \circ \Phi$. In this case, we write

$$\phi[U] := \begin{cases} U & : \text{ on } \mathcal{D} - \Phi(\sup(\phi)) \\ \hat{U}_1 \circ \Phi^{-1} & : \text{ on } \Phi(\sup(\phi)) \end{cases}.$$

Definition 3. A **local metamorphic system** on $\mathcal{D}$ is a collection of states $\{U_\alpha : \mathcal{D} \to \{0,1\}\}$ closed under all possible admissible actions of generators in the catalogue C.

[1] All generators are assumed to be **nondegenerate** in the sense that $\hat{U}_0 \neq \hat{U}_1$.

To repeat, the catalogue and the domain are the "seeds" for a local metamorphic system. From this pair, all possible translations of the supports into $\mathcal{D}$ yield the actions. Then, a collection of states on the domain is a local metamorphic system if, whenever an action Φ of a generator ϕ on a state U is admissible, then the corresponding state $\phi[U]$ is also included.

A metamorphic system with **obstacles** $\mathcal{O} \subset \mathcal{D}$ satisfies in addition the property that $U|_{\mathcal{O}} \equiv 1$ for all states U, and that for each admissible action of each $\phi \in \mathsf{C}$,

$$\Phi(\mathrm{tr}(\phi)) \cap \mathcal{O} = \emptyset. \tag{2}$$

Of course, the set $\mathcal{O}$ represents those portions of the domain $\mathcal{D}$ on which agents are forbidden to occupy or to pass through. They do count as "occupied" positions for determining the admissibility of a move (the support of an action may intersect $\mathcal{O}$), but no motion of metamorphic agents may incorporate the obstacle sites (the trace of an action must not intersect $\mathcal{O}$).

3 Examples

Example 1. [2-d hex with pivots] This reconfigurable system has a simple catalogue of six generators (or one, up to rigid rotations), represented in Fig. 2. The catalogue is chosen so that the aggregate does not change its topology but only its shape (a slightly smaller support allows for topology changes). To model a fixed "base" cell (which is, say, affixed to a power source as in [15]), one establishes this cell as an obstacle $\mathcal{O}$.

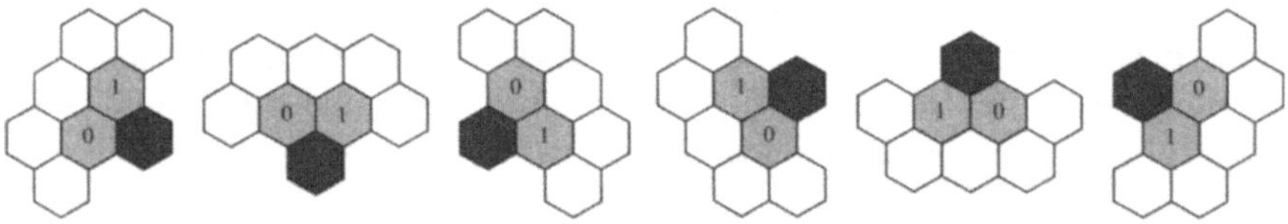

Fig. 2. The catalogue for a 2-d hexagonal lattice system with pivots. Black cells are occupied, white are unoccupied, and grey cells denote the trace. Cells turned on in local state $\hat{U}_0$ and $\hat{U}_1$ are indicated numerically.

Example 2. [2-d square lattice] In Fig. 3, we display a generator for a planar system in which rows and columns [not pictured] of an aggregate of square cells can slide. There are in fact several generators represented in "shorthand." The four white squares marked with dots can be turned on or off — but if on, their immediate neighbor (indicated by an arrow) must also be turned on. This condition guarantees that the aggregate does not disconnect (even locally) under slides. The trace of this set of generators is the entire middle row minus the two endpoints.

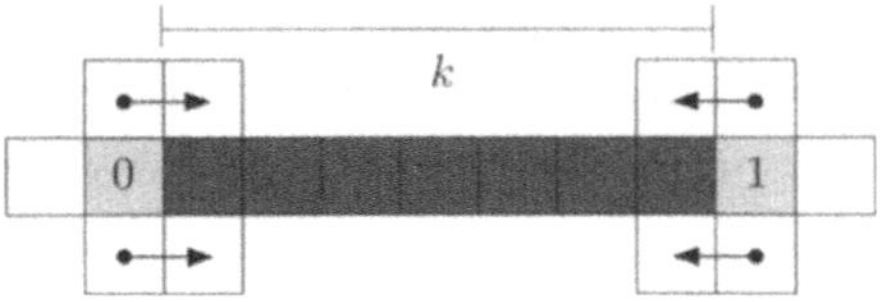

Fig. 3. One set of generators for a sliding-squares system.

Example 3. [2-d articulated planar arm] Consider as a domain $\mathcal{D}$ the set of edges in the planar integer lattice. The catalogue consists of two generators, pictured in Fig. 4. Beginning with a state having N vertical edges end-to-end (with the bottom one fixed as an obstacle), the metamorphic system thus generated models the position of an articulated robotic arm which can (1) rotate at the end and (2) flip corners as per the diagram.

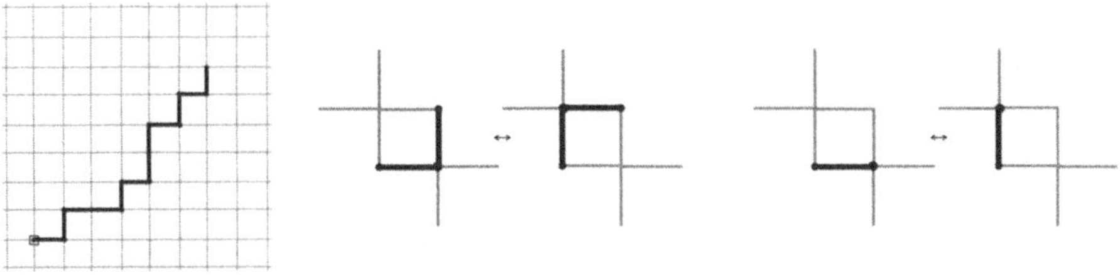

Fig. 4. A positive articulated robot arm example [left] with fixed endpoint. One generator [center] flips corners and has as its trace the central four edges. The other generator [right] rotates the end of the arm, and has trace equal to the two activated edges.

If rotations of these generator are not included, the robot arm is **positive** in the sense that the arm may extend up and to the right only. More general versions with multiple interacting arms can be achieved by adding an "attach-detach" generator to the catalogue.

One of the benefits of writing down a rigorous definition of a metamorphic system is the discovery of systems which have little resemblance to the systems of, say, Fig. 1: the following example is particularly interesting.

Example 4. Consider a graph Γ in the Euclidean lattice (all edges are edges in the Euclidean lattice in $\mathbb{R}^k$). The catalogue consists of a single generator whose support and trace is precisely the closure of an edge in this lattice. The local states of this generator consist of the pair $\hat{U}_0$ and $\hat{U}_1$ which evaluate to 1 on one of the endpoints and 0 on the other. The metamorphic system generated from a state U on Γ with N vertices evaluating to 1 mimics an ensemble of N non-colliding points on Γ.

More abstract examples of metamorphic examples include the space of triangulations of convex polygons with edge-flipping as the generator, examples arising from word representations in group theory, and certain multi-step assembly processes [2].

4 The Shape Complex

In the robotics literature, one often models a configuration space for a metamorphic system with a transition graph which coordinates actions of elementary moves on states. That is, the vertex set is the collection of all states $\{U_\alpha\}$ and the edges are unoriented pairs of states which differ by the action of one generator. Transition graphs are discussed for shape-planning in several particular cases in the literature (planar hex case: [7,17]). Our departure is to make the transition graph the 1-skeleton of a cubical complex (the analogue of a simplicial complex made out of abstract cubes) which coordinates "commutative" motions.

Definition 4. In a local metamorphic system, a collection of actions of (not necessarily distinct) generators $\{(\phi_{\alpha_i}, \Phi_{\alpha_i})\}$ is said to **commute** if

$$\Phi_{\alpha_i}(\mathrm{tr}(\phi_{\alpha_i})) \cap \Phi_{\alpha_j}(\sup(\phi_{\alpha_j})) = \emptyset \quad \forall i \neq j. \tag{3}$$

Example 5. Two simple examples suffice to illustrate the difference between commuting and noncommuting actions. First, consider the pair of commuting moves for a planar hexagonal pivoting system, as represented in Fig. 5 [left]. Compare this with a planar sliding block example as illustrated in Fig. 5

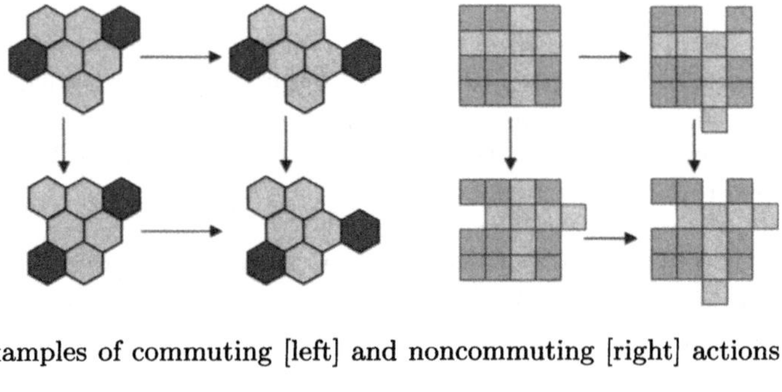

Fig. 5. Examples of commuting [left] and noncommuting [right] actions in planar systems.

[right]. Although the pair of moves illustrated forms a square[2] in the transition graph, this particular pair of actions does not commute. Physically, it is

[2] The individual square cells are not labeled: only the shape of the aggregate is recorded.

obvious why these moves are not independent: sliding the column part-way obstructs sliding a transverse row. Mathematically, this is captured by the traces of the actions intersecting.

The shape complex has an abstract k-cube for each collection of k admissible commuting actions:

Definition 5. The **shape complex** $\mathcal{S}$ of a local metamorphic system is the following abstract cubical complex. Each abstract k-cube $e^{(k)}$ of $\mathcal{S}$ is an equivalence class $[U; (\Phi_{\alpha_i})_{i=1}^k]$ where

1. $(\Phi_{\alpha_i})_{i=1}^k$ is a k-tuple of commuting actions of generators ϕ_{α_i};
2. U is some state for which all the actions $(\Phi_{\alpha_i})_{i=1}^k$ are admissible; and
3. $[U_0; (\Phi_{\alpha_i})_{i=1}^k] = [U_1; (\Phi_{\beta_i})_{i=1}^k]$ if and only if the list (β_i) is a permutation of (α_i) and $U_0 = U_1$ on the set $\mathcal{D} - \bigcup_i \Phi_{\alpha_i}(\sup(\phi_{\alpha_i}))$.

The boundary of each abstract k-cube is the collection of $2k$ faces obtained by deleting the ith action from the list and using U and $\phi_{\alpha_i}[U]$ as the ambient states. Specifically,

$$\partial[U; (\Phi_{\alpha_i})_{i=1}^k] = \bigcup_{j=1}^{k} \left([U; (\Phi_{\alpha_j})_{j\neq i}] \cup [\phi_{\alpha_i}[U]; (\Phi_{\alpha_j})_{j\neq i}] \right) \tag{4}$$

It follows easily that the k-cells are well-defined with respect to admissibility of actions. The proof of the following obvious lemma is given in detail to flesh out the previous definition.

Lemma 1. *(a) The 0-skeleton of $\mathcal{S}$, $\mathcal{S}^{(0)}$, is the set of states in the reconfigurable system. (b) The 1-skeleton of $\mathcal{S}$, $\mathcal{S}^{(1)}$, is precisely the transition graph.*

PROOF: (a) Vertices of $\mathcal{S}$ consist of equivalence classes consisting of zero (i.e., no) actions of generators up to permutation, together with a state defined on the complement of the supports of the actions. As there are no actions, each 0-cell is precisely a single state of the reconfigurable system.

(b) A 1-cell of $\mathcal{S}$ is an equivalence class of the form $[U; (\Phi)]$. The only other representative of the equivalence class is $[\phi[U]; (\Phi)]$; hence, the 1-cells are precisely the edges in the transition graph. Clearly, the boundary of $[U; (\Phi)]$ is the pair of 0-cells $[U; (\cdot)]$ and $[\phi[U]; (\cdot)]$. □

For small numbers of cells, it is easy to illustrate the shape complex.

Example 6. Consider the 2-d square lattice row/column sliding system whose catalogue is illustrated in Fig. 3. If we consider a system with obstacles in the form of a p-by-q rectangle generated from the state of Fig. 6 [left], one obtains a planar transition graph with $4(pq + 1) + 2(p + q)$ vertices and $8(pq + 1) - 2(p + q)$ edges. In contrast, the shape complex is that of Fig. 6 [right]: this is topologically a circle, corresponding to the fact that the pair of free squares can circulate about the obstacle set through a sequence of slides. The large 2-d regions correspond to states in which the two free squares are on separate (but adjacent) sides of the obstacle set.

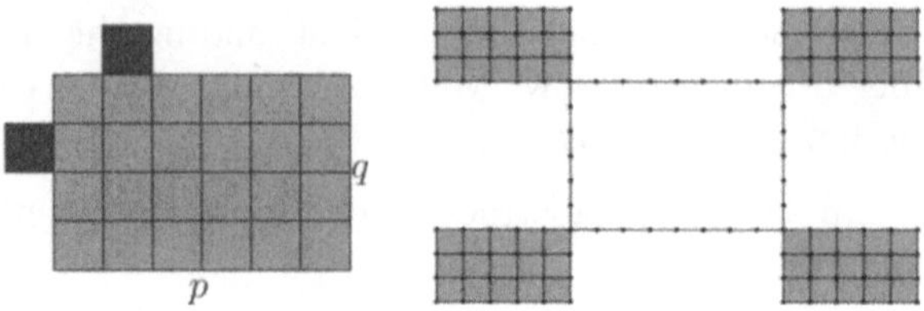

Fig. 6. The planar sliding square example [left] with two movable blocks [in black] and a p-by-q obstacle set [in grey] yields a shape complex $\mathcal{S}$ that is topologically a circle [right].

Example 7. The shape complex associated to the positive articulated robot arm of Example 3 in the case $N = 5$ is given in Fig. 7. Note that there can be at most three independent motions (when the arm is in a "staircase" configuration); hence the shape complex has top dimension three. Notice also that although the transition graph for this system is complicated, the shape complex itself is topologically trivial (contractible).

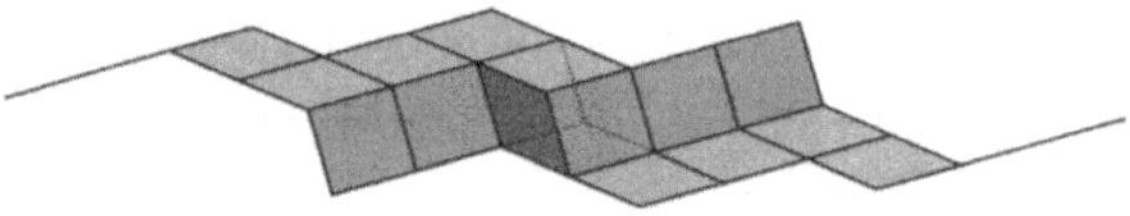

Fig. 7. The shape complex of a 5-link positive arm has one cell of dimension three, along with several cells of lower dimension.

Example 8. In the system of Example 4 with the graph being a K_5 (the complete graph on five vertices) and $N = 2$, the shape complex is a two-dimensional closed orientable surface. A simple combinatorial argument reveals that the Euler characteristic is -10, implying that the shape complex has genus six.

5 The Topology of $\mathcal{S}$

If one looks at a transition graph without knowing the particulars of the metamorphic system, very little information can be extracted. This section argues that completing the transition graph to the shape complex is "natural" — the shape complex simplifies the transition graph and endows it with geometric content.

Our first example of naturality is motivated by the desire to build metamorphic systems with large numbers of micro- or nano-scale cells. While large numbers of cells would yield a type of continuum-limit convergence on the dynamics of shape change, the resulting transition graphs have no such convergence. The size of the transition graph goes up exponentially in the number of cells; more ominous is that the topology of the transition graph (the number of cycles) blows up as well. This is not always so with the shape complex: in several key examples, the topology of $\mathcal{S}$ converges for large systems.

Example 9. Recall the shape complex associated to the metamorphic system of N points on a graph Γ, Example 4. Consider a refinement of the underlying lattice of Γ which inserts additional vertices along edges. It follows from the techniques of [1] that the shape complex of this refined system has the same topological type (up to homotopy equivalence) after a fixed bound on the refinement ($N-2$ additional vertices per edge). Furthermore, this "stabilized" shape complex is in fact homotopic to the topological configuration space of N points of Γ (what one expects as the number of refinements goes to infinity).

Example 10. Recall the articulated robot arm of Example 3. Consider a refinement of the underlying lattice which shrinks the lattice by a factor of two (or, equivalently, which inserts an additional joint in the middle of each edge). It can be shown [2] that the shape complex of the refined system does not change homotopy type, even though the dimension of the complex goes up by a factor of two. In the limit as the number of refinements goes to infinity, the shape complex approximates the configuration space of a smooth curve of fixed length.

Not all systems obey this stabilization principle. For example, in the hex-lattice systems of Example 1, the shape complex possesses loops represented by "braiding" of pairs of hex elements about the boundary of the aggregate. Since a lattice rescaling increases the number of cells along the boundary, such loops grow in number. However, if we scale the geometry of the shape complex in accordance with the lattice scaling, these loops shrink to length zero in the continuum limit.

6 The Geometry of $\mathcal{S}$

Shape complexes are built from Euclidean (i.e., flat) cells of unit length. However, this does not imply that the complex is, as a whole, flat. Indeed, non-zero curvature can be concentrated at places where several cells meet. A simple example appears in Fig. 8: here, a surface built from flat 2-cells can be seen to have curvature which depends on the number of 2-cells incident to a vertex. Four incident cells implies zero curvature; three cells implies positive curvature; and five or more cells implies negative curvature. Such an extension of curvature to general metric spaces is made precise in Gromov's work on curved metric spaces [10] (extending the classical work of Andronov, Busemann, and others) in which triangles with geodesic edges are used to measure curvature bounds. In brief, let X be a metric space and $p \in X$ a

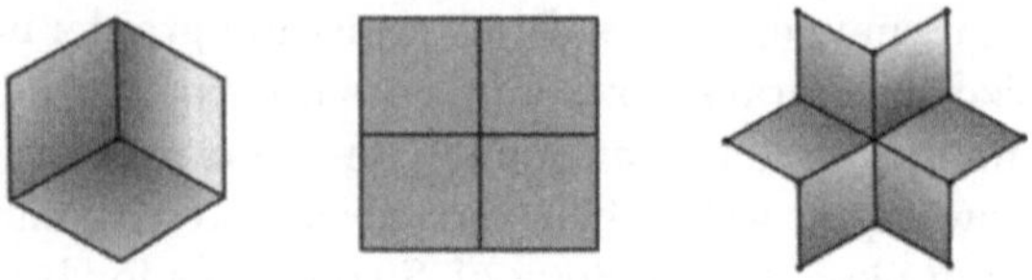

Fig. 8. Curvature about a vertex: $+, 0, -$.

point. To bound the curvature of X at p, consider a small triangle T about p with geodesic edges of length a, b, and c. Build a **comparison triangle** T' in the Euclidean plane whose sides also have length a, b, and c respectively. Choose a geodesic chord of T and measure its length d. In T', measure the length d' of the associated chord.

Definition 6. A metric space X is **nonpositively curved** (or **NPC**) if for every sufficiently small geodesic triangle T and for every chord of T, it follows that $d \leq d'$.

Fig. 9. Comparison triangles measure curvature bounds.

In other words, geodesic chords are no longer than Euclidean comparison chords. It should be stressed that the NPC property is very special and highly desirable. Indeed, being NPC implies a variety of topological consequences reminiscent of smooth nonpositively curved spaces: for example, the universal cover is contractible.

Despite the variety of (local) metamorphic systems, all shape complexes share this special geometric property.

Theorem 1 **([2]).** *The shape complex $\mathcal{S}$ of any local metamorphic system is nonpositively curved.*

The proof requires some machinery which we do not include here. The key observations required are that (1) commutativity of a set of actions is independent of the states implicated; and (2) any collection of pairwise commutative actions is totally commutative. These conditions, properly translated into the language of Gromov's theory, imply that the complex never has positive curvature.

Example 11. To see where the NPC property can fail, consider the following non-local metamorphic system. Fig. 10[left] gives a generator for a planar hexagonal pivoting system which (along with its rotations) allows for local disconnection of the aggregate. If we impose a global rule that requires the aggregate to be connected, for, say, considerations of power transmission, then we may create positive curvature. Fig. 10 shows a configuration in which three actions of local generators act to disconnect the aggregate locally but not globally. These actions commute pairwise, and any two do not disconnect globally. However, performing all three actions disconnects the space and leads to an illegal state. Therefore, the corresponding state complex for this non-local system has a "corner" of positive curvature as a local factor.[3]

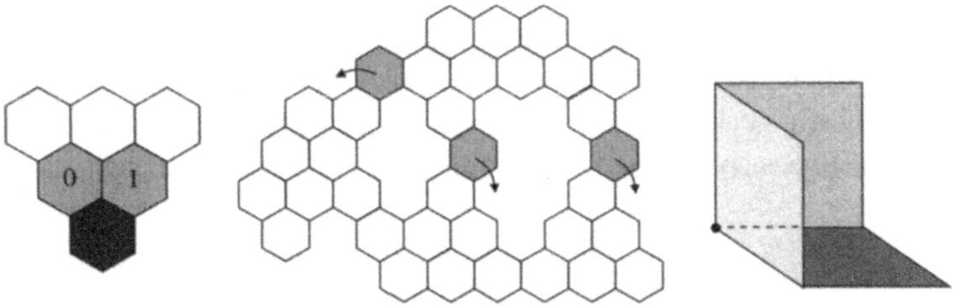

Fig. 10. A generator which allows for local disconnection [left] admits configurations [center] for which pairs of actions leave the aggregate connected, but triples do not; this non-local system has positive curvature in the shape complex at this state [right].

7 Geodesics and Time on $\mathcal{S}$

Besides displaying a unifying geometric feature, the nonpositive curvature has implications in the shape planning problem.

Corollary 2 *Geodesics on any shape complex are unique in their homotopy class.*

PROOF: Every NPC space has contractible universal cover [3]. □

Spheres, for example, do not share this property: there are many geodesics between, say, antipodal points which are homotopic. This corollary is the key ingredient in the applications of NPC geometry to path-planning on a configuration space.

One expects geodesics on $\mathcal{S}$ to coincide with optimal solutions to the shape planning problem. Corollary 2 implies that any length-decreasing algorithm

[3] The shape complex will not be two-dimensional here, but will be locally a product of this with another cubical complex. The positive curvature persists.

must converge to the minimal-length curve possible: there are no spurious local minima (or even saddles) to prevent convergence. This good news is tempered with ill: it is not necessarily easy to find the true geodesic quickly.

However, in the context of robotics applications, the goal of solving the shape-planning problem is *not* necessarily coincident with the geodesic problem on the shape complex. Fig. 11[left] illustrates the matter concisely. Consider a portion of a shape complex $\mathcal{S}$ which is planar and two-dimensional. To get from point p to point q in $\mathcal{S}$, any edge-path which is weakly monotone increasing in the horizontal and vertical directions is of minimal length in the transition graph. The true geodesic is, of course, the straight line.

Given the assumption that *each elementary move can be executed at a uniform maximum rate*, it is clear that the true geodesic on $\mathcal{S}$ is **time-minimal** in the sense that the elapsed time is minimal among all paths from p to q. However, there is an envelope of non-geodesic paths which are yet time-minimizing. Indeed, the true geodesic in Fig. 11[left] "slows down" some of the moves unnecessarily in order to maintain the constant slope.

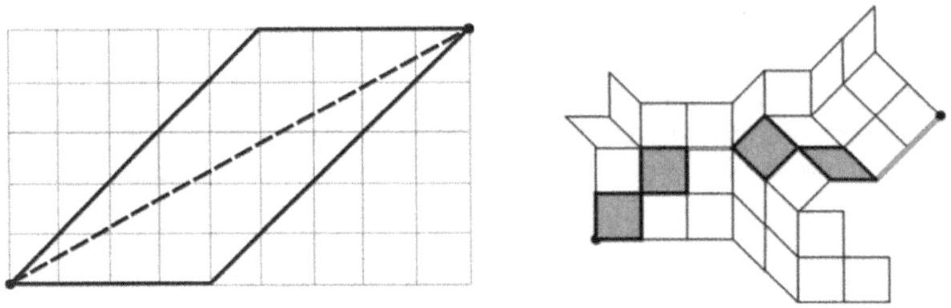

Fig. 11. [left] The true geodesic lies within an envelope of time-minimizing paths. No minimal edge-paths are time-minimizing; [right] a normal cube-path in an NPC complex (after [16]).

If, instead of the Euclidean metric on the cells of $\mathcal{S}$, one measures the ℓ^∞ norm of tangent vectors (the sup of the components), then, geodesics in this category are paths which are *time minimizing*. By using the techniques of [16], one can prove that **cube paths** are the appropriate class of generalized paths.

Definition 7. A **cube path** from vertices v_0 to v_N is an ordered sequence of closed cubes $\{C_i\}_1^N$ in $\mathcal{S}$ which satisfy (1) $C_i \cap C_{i+1} = v_i$; and (2) C_i is the smallest cell of $\mathcal{S}$ containing v_{i-1} and v_i. A cube path is said to be **normal** if in addition $\forall i$ (3) $C_{i+1} \cap \mathsf{St}(C_i) = v_i$, where $\mathsf{St}(C_i)$ is the **star** of C_i (all cells, including C_i, which have C_i as a face).

Roughly speaking, a normal cube path is one for which uses the highest dimensional cubes as early as possible in the path.

Theorem 3. *In any shape complex $\mathcal{S}$, there exists a unique normal cube path from p to q in each homotopy class. The chain of diagonals through this cube path is a globally time-minimal path over all paths in its homotopy class.*

Existence and uniqueness follows from Theorem 1 and a result of [16] (or, see the constructive algorithm of Section 8 below). Time-minimization follows from geodesic-like properties of normal cube paths: see [2] for proof details.

8 Optimizing Paths

In cases where the shape complex is sparse — local principal cells being of high dimension with few neighbors — it is possible to compress the size of the transition graph significantly. However, shape-planning via constructing all of $\mathcal{S}$ and determining geodesics is, in general, computationally infeasible: the total size of the shape complex is often exponential in the number of cells in the aggregate. We therefore assume that some shape trajectory has been determined (by a perhaps ad hoc method) and turn to the problem of optimizing this trajectory. The nonpositive curvature of $\mathcal{S}$ allows for a time-optimization which requires exploration of small regions within $\mathcal{S}$. We detail an algorithm for transforming any given edge-path to a time-optimal normal cube path.

From Definition 5, an n-dimensional cube C of $\mathcal{S}$ can be represented by a set of n commutative actions $\{\Phi_i\}$ along with an admissible state U. In the following algorithm, we suppress the state for notational convenience and consider C as the set of actions. Addition and subtraction is defined by adding or taking away admissible commutative actions to the list.

Algorithm 4 (`TimeGeodesic`) Given: a cube path $\mathcal{C} = \{C_i\}_{i=1..N}$ in $\mathcal{S}$.
1: Let $N :=$ `Length`$(\mathcal{C})$.
2: Call `ShrinkCubePath`$(\mathcal{C})$.
3: If `Length`$(\mathcal{C}) < N$ then 1: else `stop`.

Algorithm 5 (`ShrinkCubePath`) Given: a cube path $\mathcal{C} = \{C_i\}_{i=1..N}$ in $\mathcal{S}$.
1: Let $i = 1$.
2: Let $X :=$ `Commute`$(C_i; C_{i+1})$.
3: Update $C_i := C_i + X$;
4: Update $C_{i+1} := C_{i+1} - X$.
5: Call `ExciseTrivial`(C_{i+1}).
6: Call `CommonEdge`$(\mathcal{C})$.
7: If $X = \emptyset$ then $i = i + 1$.
8: If $C_{i+1} \neq \emptyset$ then 2: else `stop`.

Subroutine 6 (`Commute`) Given: a pair of cubes $C_j = \{\Phi^j_\alpha\}_{1..m}$ and $C_{j+1} = \{\Phi^{j+1}_\alpha\}_{1..n}$,

1: Let $S := \bigcup_\alpha \mathsf{sup}(\Phi^j_\alpha)$
2: Let $T := \bigcup_\alpha \mathsf{tr}(\Phi^j_\alpha)$
3: For $i = 1..n$ return Φ^{j+1}_i if
 3.1: $S \cap \mathsf{tr}(\Phi^{j+1}_i) = \emptyset$; and
 3.2: $T \cap \mathsf{sup}(\Phi^{j+1}_i) = \emptyset$.

Subroutine `ExciseTrivial` checks the cubes for empty lists, removes these from the path, and reindexes the cube path, thus reducing the length. Subroutine `CommonEdge` checks adjacent cubes in the sequence for common edges and deletes these as necessary, returning a genuine cube path (recall Definition 7).

Subroutine `Commute` takes as its argument a pair of cubes C_i and C_{i+1} and returns those elements of C_{i+1} which commute with *all* elements of C_i. The following lemma is a manipulation of the definitions.

Lemma 2. *The result of* `Commute`$(C_i; C_{i+1})$ *is precisely the set of edges in* $\mathsf{St}(C_i) \cap C_{i+1}$.

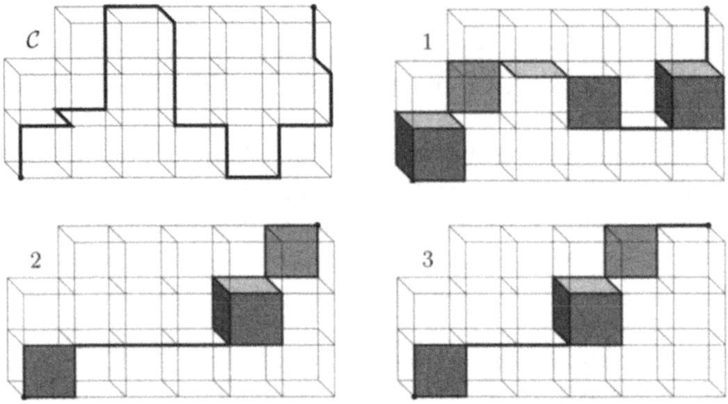

Fig. 12. Three rounds of shortening an edge path C. Cube path length stabilizes after two calls to `ShrinkCubePath`. Three more calls produces a normal cube path.

Theorem 7. *Algorithm 4 computes a globally time-optimal path in the homotopy class of* C.

PROOF: Repeated calls to Algorithm `ShrinkCubePath` eventually leaves a cube path fixed. To show this, note that the function $f(C) := \sum_i i \cdot \dim(C_i)$ decreases in each call to `ShrinkCubePath` which changes the path. Lemma 2 implies that Algorithm 5 leaves a cube path fixed if and only if it is a normal cube path. From Theorem 3, this yields a time-optimal path.

The final step is to show that once the length of a cube path in unchanged by a call to `ShrinkCubePath`, this is equal to the length of the associated normal cube path. To show this, assume that **ShrinkCubePath** completes a round without excising any trivial cubes (i.e., shortening the length). We show that further rounds merely redistribute actions earlier in the list C without deleting any permanently. Without a loss of generality, assume that the first call to **ShrinkCubePath** does not change the length of the cube path, but that in the subsequent call, the cube C_{i+1} is eliminated via C_i commuting with a portion of C_{i-1}: see the schematic of Fig. 13.

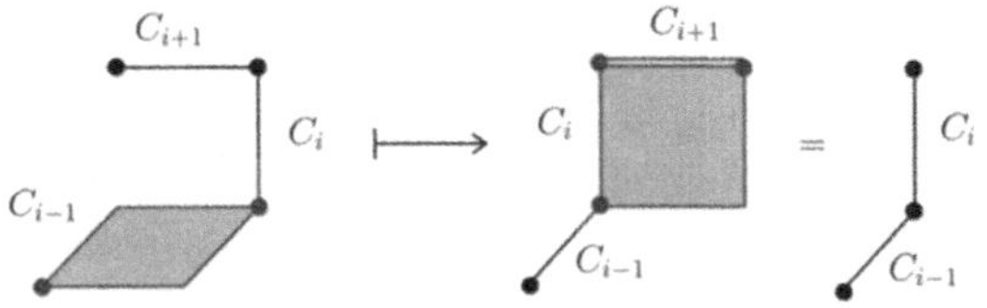

Fig. 13. Schematic of exchanging actions in sequential rounds. Each dimension in this cartoon represents a multi-dimensional cube.

Since C_{i+1} is not eliminated in the first call, there must have been actions in C_i which did not commute with those of C_{i-1} but which are subsequently pushed forward in the second call. In addition, all the actions of C_{i+1} are contained in ("parallel to" in the figure) the subset of C_{i-1} which commutes with C_i in the second call. This being the case, the commutativity of C_{i+1} with C_i would have occurred in the first call: contradiction. □

It is not surprising that Algorithm 4 converges to a *locally* time-minimal solution. The interesting implication is that the cube-path obtained is the *global* minimum for all paths obtainable from the initial: there is no way to get a quicker reconfiguration path by first lengthening the path and then shortening. Fig. 12 gives an example of paths generated by Algorithm 4.

It remains an important computational question to determine whether the homotopy class of the initial path (given, say, from a distributed algorithm) is optimal in the sense that its geodesic is the shortest among all homotopy classes.

The complexity of Algorithm 4 is best measured as a function of N, the number of edges in the initial cube path. In the worst case, Algorithm 4 calls **ShrinkCubePath** $\lceil N/2 \rceil$ times. This worst case is achieved by an initial edge path in a planar shape complex with one corner at the end of edge number $\lfloor N/2 \rfloor$. One can use a 2-stage induction on N to show this (we do not include the full argument for reasons of space). Assume that N is even and that the statement holds for paths of length up to N. For a path of length $N + 1$, each call to **ShrinkCubePath** does not register the last edge until the first N edges are processed. To keep this path a worst case, the last edge should not commute with any other cell faces of C until the very last call to **ShrinkCubePath**. Hence, the penultimate vertex remains fixed, and the worst case proceeds from induction: the shape complex is planar and the path has one corner in the middle. The final edge must lie in this plane. The case of N odd follows a slightly different argument: the extra edge is added at the *beginning* of the sequence and commutes at the very last call.

In this worst case example there are $\frac{1}{2}(N^2 - N)$ calls to `Commute`. This can almost certainly be improved with a more sophisticated algorithm. The subroutine `Commute` is an improvement over the more obvious method of computing $\mathsf{St}(C_i)$ at each step — the algorithm merely requires checking if each action of C_{i+1} commutes with every action of C_i.

The worst case is a totally flat shape complex (which is desirable in other respects). The presence of true negative curvature in a shape complex greatly increases the speed of convergence. If, as in Example 8, the shape complex has negative curvature at every vertex, then Algorithm 4 converges to the normal cube path with a *linear* number of calls to `Commute`. This transition from quadratic to linear as one passes from zero to negative curvature is a persistent phenomenon [8].

9 Discussion

Our principal contributions are:

1. *A mathematical definition* of a metamorphic system which encompass many models currently studied and suggests seemingly unrelated (and often simpler) systems. Given the difficulty of building large metamorphic systems, simpler examples possessing the same formal structure may be valuable.
2. *The shape complex*, whose naturality is manifested on the level of its topology ($\mathcal{S}$ can be homotopically simple) and its geometry (non-positive curvature is universal, rare, and highly desirable for proving theorems).

There are drawbacks to this approach. Primary among them are the dual dilemmas that shape planning is inherently complex, and that there are many types of reconfiguration possible. A shape complex approach is not meant for all systems. Indeed, it is possible to design degenerate metamorphic systems with little to no commutativity. Nevertheless, paying attention to the geometry lurking behind our higher-dimensional versions of transition graphs leads to a nontrivial result on the optimality of path-shortening.

As an interesting side-note, we observe that in the present context, the optimal reconfiguration path is not always coincident with a geodesic on the configuration space. This principal holds in more generality.

Finally, in this initial work, we have focused only on the first-order problem of shape planning. We hope that the mathematical framework here suggested finds a use in more sophisticated task-planning problems for metamorphic and reconfigurable systems.

References

1. A. Abrams, *Configuration spaces and braid groups of graphs.* Ph.D. thesis, UC Berekeley, 2000.
2. A. Abrams and R. Ghrist, *State complexes for reconfigurable systems*, in preparation.
3. M. Bridson and A. Haefliger, Metric Spaces of Nonpositive Curvature, Springer-Verlag, Berlin, 1999.
4. Z. Butler, S. Byrnes, and D. Rus, *Distributed motion planning for modular robots with unit-compressible modules*, in Proc. IROS 2001.
5. Z. Butler, K. Kotay, D. Rus, and K. Tomita, *Cellular automata for decentralized control of self-reconfigurable robots*, in Proc. IEEE ICRA Workshop on Modular Robots, 2001.
6. G. Chirikjian, *Kinematics of a metamorphic robotic system*, in Proc. IEEE ICRA, 1994.
7. G. Chirikjian and A. Pamecha, *Bounds for self-reconfiguration of metamorphic robots*, in Proc. IEEE ICRA, 1996.
8. D. Epstein et al., Word Processing in Groups. Jones & Bartlett Publishers, Boston MA, 1992.
9. R. Ghrist, *Configuration spaces and braid groups on graphs in robotics*, AMS/IP Studies in Mathematics volume 24, 29-40, 2001. ArXiv preprint math.GT/9905023.
10. M. Gromov, *Hyperbolic groups*, in Essays in Group Theory, MSRI Publ. **8**, Springer-Verlag, 1987.
11. K. Kotay and D. Rus, *The self-reconfiguring robotic molecule: design and control algorithms*, in Proc. WAFR, 1998.
12. C. McGray and D. Rus, *Self-reconfigurable molecule robots as 3-d metamorphic robots*, in Proc. Intl. Conf. Intelligent Robots & Design, 2000.
13. S. Murata, H. Kurokawa, and S. Kokaji, *Self-assembling machine*, in Proc. IEEE ICRA, 1994.
14. S. Murata, H. Kurokawa, E. Yoshida, K. Tomita, and S. Kokaji, *A 3-d self-reconfigurable structure*, in Proc. IEEE ICRA, 1998.
15. A. Nguyen, L. Guibas, and M. Yim, *Controlled module density helps reconfiguration planning*, in Proc. WAFR, 2000.
16. G. Niblo and L. Reeves, *The geometry of cube complexes and the complexity of their fundamental groups*, Topology, **37**(3), 621-633, 1998.
17. A. Pamecha, I. Ebert-Uphoff, and G. Chirikjian, *Useful metric for modular robot motion planning*, in IEEE Trans. Robotics & Automation, **13**(4), 531-545, 1997.
18. J. Walter, J. Welch, and N. Amato, *Distributed reconfiguration of metamorphic robot chains*, in Proc. ACM Symp. on Distributed Computing, 2000.
19. M. Yim, *A reconfigurable robot with many modes of locomotion*, in Proc. Intl. Conf. Adv. Mechatronics, 1993.
20. M. Yim, J. Lamping, E. Mao, and J. Chase, *Rhombic dodecahedron shape for self-assembling robots*, Xerox PARC Tech. Rept. P9710777, 1997.
21. Y. Zhang, M. Yim, J. Lamping, and E. Mao, *Distributed control for 3-d shape metamorphosis*, Aut. Robots. J., 41-56, 2001.
22. E. Yoshida, S. Murata, K. Tomita, H. Kurokawa, and S. Kokaji, *Distributed formation control of a modular mechanical system*, in Proc. Intl. Conf. Intelligent Robots & Sys., 1997.

Flexonics

John Canny, Jeremy Risner, and Vivek Subramanian

Electrical Engineering and Computer Sciences, University of California, Berkeley

Abstract. We introduce a new approach to design and fabrication of mechatronic systems called flexonics. Flexonics integrates structural, mechanical and electronic elements based on 3D printing techniques, in particular inkjet printing. This paper outlines some principles for flexonic design, which is based on flexion rather than sliding or rolling motion as in traditional mechanics. It describes our preliminary explorations of materials and processes. And it describes a kinematic approach to joint design that has so far led to one promising new design. Finally, we give a brief prospectus of future applications of flexonics.

1 Introduction

In this paper, we describe a new class of mechatronic devices, and propose methods for their fabrication and design. Our overall goal is to build fully functional mechatronic devices without assembly. These devices will integrate structural, mechanical, and electronic components during fabrication using an all inkjet printing process. In order to achieve the above goal, we intend to develop a complete 3D printer, and create a vocabulary of flexonic components (both mechanical and electronic).

Flexonics are radically different from traditional electromechanical systems in both form and manner of construction. The term "flexonics" derives from the fact that the proposed mechatronic systems achieve motion through flexion (bending). There will be no traditional sliding or rolling movements in actuators, sensors, or joints. Thus, flexonic devices will not utilize or need gears, bearings, or any sliding surface. This limits traditional design drawbacks such as friction, backlash, and mechanical wear. However, new issues arise when designing with flexures: material fatigue, inherent elastic energy storage, limited range of motion, and undesirable degrees-of-freedom require a different design strategy.

Flexonic devices will be fabricated using an inkjet printing technique. Through selective material deposition, a mechanism can be built layer by layer with a high level of complexity and integration. A necessary component of inkjet polymer printing will be a diverse set of printable materials. These include solution-based polymers and oligomers, nanoparticle suspensions, and other polymers that flow when heated. The range of current polymer-oligomer-nanoparticle capabilities includes electro-mechanical sensing and actuation, specialized structural support,

J.-D. Boissonnat et al. (Eds.): Algorithmic Foundations of Robotics V, STAR 7, pp. 203–219, 2004.

and electrical conduction. Our goal is to utilize this diversity to construct both active and passive mechanical components, as well as driver and control electronics.

This paper is organized as follows: In Sect. 2, we give some background work relevant to flexonics including 3D printing, integrated manufacturing, polymer actuation, and flexure-based design. In Sect. 3, we describe the printing technology we are currently exploring, along with several materials under study. Sect. 4 contains a discussion of our current work with organic electronics, as well as proposed work to meet the requirements imposed by flexonic design and fabrication. In Sect. 5, we describe our work so far on flexonic design: design of joints which are compositions of simple elements like rectangular plates. We include a promising new joint design we derived through this process. Later in that section, we also describe higher order mechanisms, actuators, and bulk volumes. We provide some example devices to demonstrate the design possibilities within the flexonics environment. Finally, in Sect. 6 we suggest several candidate applications for flexonics. We believe flexonic devices would be appropriate for human-machine interfaces, biomimetic robots, and functional prototyping.

2 Related Work

Various methods of 3D printing have emerged in recent years [8,9]. They are used for rapid prototyping during design and for building molds for manufacturing. Various systems use materials and layering processes ranging from fragile photo-cured resins (Stereolithography or SLA), to powders melted by a powerful laser (Selective Laser Sintering or SLS), to extruded thin thermoplastic filaments (Fused Deposition Modeling or FDM – Stratasys, Inc.). With FDM, fully functional devices can be made from either rigid ABS plastic or flexible elastomer. Complex parts with regions of overhang and irregular surfaces can be easily built using a system that integrates build material and sacrificial support material. We have built several flexure-based designs using FDM. These devices are shown in subsequent figures as a demonstration of flexonic components (Sect. 4). However, FDM has disadvantages when applied to the problem of fabricating a broad range of flexonic devices, including dielectric elastomer actuators. Few materials are available, resolution is relatively poor, part strength and properties are influenced by build orientation, and electronic components are not realizable.

Shape Deposition Manufacturing (SDM) [1] has similar goals to flexonics manufacturing: integrating functional components with a 3D manufacturing process so that complex devices can be built without assembly. The difference is that SDM employs several traditional technologies – electronics or mechanical components are built separately and then "dropped in" during the build. SDM does not employ one, but a large family of manufacturing processes. By contrast, our goal is to build complete, fully functional electromechanical systems using a 3D inkjet printing process.

Inkjet printing is currently being investigated as a tool for manufacturing in several areas including MEMS, organic displays, and biological assays [6,14]. In fact, commercially available systems exist (MicroFab Technologies, Inc.) for printing a variety of materials. We have targeted these systems to provide the basis for the proposed 3D printing station described in Sect. 3. Solders, adhesives, optical polymers, and bioactive molecules have all been printed using piezoelectric, drop-on-demand techniques. Demonstration of printed solution-based polymers is of particular interest for flexonic manufacturing.

Electroactive polymers (EAPs) are a set of materials that have received a lot of attention with regards to "artificial muscle" actuation [3]. Some EAPs, such as ionomeric polymer-metal composites (IPMC), bend when an applied voltage causes a rearrangement of ions within the material. Other types, such as dielectric elastomers, expand when a large electric field is applied across a thin sheet of the material. The latter seem to offer the most promise for flexonics. Their low weight, high energy density, and polymer construction are attractive features. They are also highly configurable: actuator designs include unimorphs, bimorphs, rolls, tubes, bellows, and diaphragms [13]. Planar actuators made from thin films of silicone and acrylic elastomers have demonstrated linear strains of 63% and 215%, respectively, and energy densities five times that of human muscle [12]. However, certain requirements must be met in order to achieve this motion. First, very large electric fields around 100 MV/m must be used: With off-the-shelf material layer thickness, voltages around 5kV are required. Second, the elastomers must be pre-strained prior to activation. Specifically, when high pre-strain is applied in one planar direction, expansion occurs primarily in the other planar direction, and at a much greater magnitude than without pre-strain. The flexonic context offers potential solutions to both difficulties. First, layer thickness of 1-10 microns are common with inkjet polymer printing, and reduce the voltage requirements with today's materials to 100 volts or less. The loss of actuator volume per layer can be made up by laminating many layers. We expect that new material research will offer even lower working voltages in the near future. The pre-strain issue is being studied through actuator housing design (see Sect. 4) and through the deposition process itself (strains developed during solvent evaporation).

Current techniques for fabricating planar dielectric elastomer actuators are quite primitive. Dielectric elastomer is first stretched across a rigid frame. Next, closed-profile structures are attached which contain rigid constraints perpendicular to the intended direction of motion (parallel to the high pre-strain direction) and flexible expanding support parallel to the direction of motion. Compliant electrodes, usually conductive carbon grease, are applied to both surfaces of the film. Electrical contact points are attached, and the actuator is then cut from the rigid frame. This process produces inconsistent and unreliable actuators. We believe flexonic design and fabrication has the potential to greatly improve performance and longevity.

Micro Electro-Mechanical Systems (MEMS) [17] have a number of common features with flexonics. First of all, they use a single material (usually silicon) that provides both electronics and mechanical actuation functions. They allow the construction of "systems on a chip" that work as sensors with matching amplifier electronics, and occasionally as actuators (inkjet heads are in fact MEMS devices). The design principles at play in MEMS are similar in several ways to flexonics. It is usually not practical to have rolling or sliding surfaces. Therefore, movement is accomplished through flexion. Secondly, MEMS devices use a lithographic process that does not penalize device complexity. It is very desirable to use previously developed modules (e.g. the "comb drive" electro-static actuator) in new designs, and very complex designs can be built from such modules. We expect flexonic designs to have similar properties. There are also many differences. Flexonics are much more appropriate for larger scale appliances that humans use directly. Polymer materials are low cost, and are well matched to natural environments (less sensitive to impurities or surface abrasions). Finally, MEMS designs tend to be largely planar, with flexure deformation remaining within the linear range. Flexonics, however, encourages full three-dimensional devices with large nonlinear deformations.

Flexure-based devices, or compliant mechanisms, represent a substantial research area on their own. Some recent work has focused on developing systematic design methods, but most 3D compliant mechanisms are designed through ad hoc methods [7]. A significant portion of compliant mechanism design is rotating joints. Initially, most joints were designed using notches or leaf springs. However, these simple designs are not adequate for high precision, high strength joints. Part of our goals within this proposal is the development of a vocabulary of flexible joints and mechanisms. Other groups are investigating new types of joints that contain more complex geometry and three-dimensional construction [5,11]. As stated above, flexonics is perfectly suited for the task of manufacturing these devices.

3 Fabrication of Flexonic Devices

3.1 Process

We are developing a novel manufacturing process requiring two steps. The first step is software based and follows a procedure similar to traditional rapid prototyping methods. Initially, an appropriate 3D CAD model of the device must be created. Because flexonics integrates all components without assembly, this model can be viewed as a single part. Depending on the orientation and design of a device, there may be regions of unsupported overhang. Therefore, within this step, locations for support material should be determined and added to the model. This support material will be sacrificial; following fabrication, it will be removed either by dissolving or breaking. Continuing within the software stage, the model would be divided into horizontal layers. Each layer will represent a cross section of the part and will contain information pertaining to component type and geometry.

The second step is the physical realization of the device. Here, our proposed 3D printer will produce the part by reading the layer information and selectively dispensing materials. We have begun work using an inkjet-based printer for organic electronics (detailed in Sect. 4). Our goal is to incorporate a similar set-up for flexonics, but with the capacity to print a multitude of mechanical and structural materials as well.

A flexonics 3D printer requires a specialized inkjet printing head, a high-resolution x-y-z stage, and corresponding drivers and control electronics. The number of individual print head jets on the printing head will depend on the polymer material set necessary to complete a flexonic device. Each jet will then have its own material supply and driver for individual control. The material set is dependent on the flexonic component types: actuators, joints, etc. MicroFab Technologies, Inc. manufactures specialized inkjet print heads for several applications ranging from printing optical polymers to solder bumps [6,14]. The individual jets use piezoelectric actuators to pump single drops of liquid onto a substrate surface. The jet orifice is on the order of 50 microns. The type of liquid that can be dispensed varies from solution-based polymers to molten metals. Two of their dispensing units are shown in Fig. 1. These heads should be ideal for our flexonic fabrication needs.

The print head is suspended above an x-y planar stage with a resolution on the order of 1 micron. Therefore, the limiting factor for resolution is droplet size. For a given layer of the device, the stage translates according to the geometry of the cross section. When a layer is complete, the stage is incremented downward, and the next layer is deposited. Assuming that drops forming a proceeding layer have dried before a successive layer is started, there should be no delay in the build process from layer to layer. The time to complete a given device will only be dependent on the deposited volume of material.

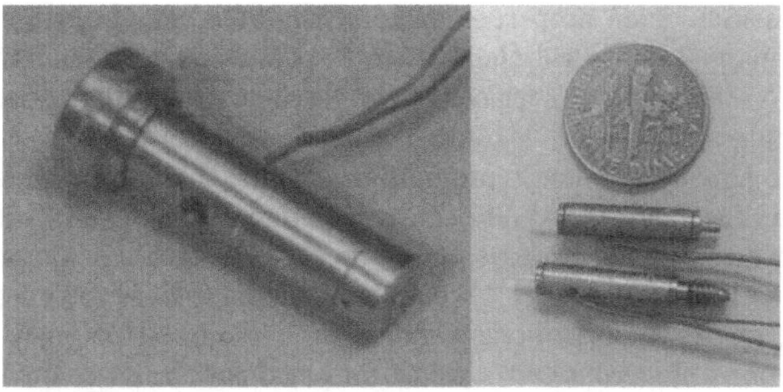

Fig. 1. MicroFab dispensers. (*left*) SolderJet high temperature dispenser. (*right*) MicroJet room temperature dispenser. Images from www.microfab.com

3.2 Materials

The 3D printing process requires materials meeting two criteria: First, the material viscosity must be low enough to allow controlled ejection from the inkjet during fabrication. Second, the material must remain a liquid for sufficient time before jetting so as not to clog the print orifice, yet solidify within a reasonable time after jetting. These requirements constrain the usable material set to those that can be dissolved into solution, or those that will flow when heated. Polymer and oligomer solutions, polymer resins, molten solders, and nanoparticle suspensions are all suitable for printing.

In Sect. 4, our current work using organic materials for electronics is discussed in greater detail. These materials include solution-based oligomers and nanoparticles. Materials for mechanical components that are being explored include thermoplastics like ABS and silicone thermoplastic and solvent-based elastomers. Flexonics integrates joints, actuators, structural volumes, and electronics. Therefore, it is necessary to match material properties to component function. Joints require both rigid and flexible regions; two polymers with high and low elastic moduli are necessary. Actuators require the properties of joints, but also need conductive and non-conductive polymers as well. Structural volumes tend to be rigid; however, the overall stiffness and anisotropy might be tuned using a selection of polymers. Flexonic manufacturing also requires a suitable sacrificial material for supporting overhanging geometry and complex forms during fabrication. Water-soluble thermoplastics are commonly used in FDM machines for sacrificial volumes and give a good starting point for our work as well.

4 Electronics

Active electronic components based on organic semiconductors are the natural choice for use as control elements in flexonics. Organic electronic devices may be fabricated using solution-based processing technologies such as inkjet deposition to produce complete electronic circuits without the use of lithography and/or vacuum processing. Such properties make organic electronic circuits ideally suited for use as integrated control elements in flexonics applications. Circuits may be fabricated on the polymer surfaces of the flexonics using inkjet-based processing facilities integrated into the flexonics manufacturing flow.

We have already initiated a program to develop inkjet-based organic semiconductor materials, device, and fabrication technology [16]. The work is targeted at developing an oligomer-based solution-processed organic semiconductor device and manufacturing technology. We are developing inkjet-based deposition protocols for formation of organic semiconductor circuits on polymer surfaces. We have developed numerous test structures to evaluate the associated materials and fabrication technologies, including FETs, conductor/semiconductor chains and organic LEDs. These may be integrated to form complete circuits for actuation and control of flexonics.

The performance of organic semiconductor-based transistors today is substantially lower than the achievable performance in inorganic semiconductors. For example, the best carrier mobility achieved in an organic semiconductor is approximately two orders of magnitude below that routinely achieved in silicon. Additionally, organic semiconductor FETs typically have substantially higher operating voltage requirements – supply voltage requirements as high as 80V have been reported [10]. Therefore, most of the research work reported to date has concentrated on reducing operating voltage requirements and improving carrier mobility to achieve higher current densities.

Broadly speaking, the work can be classified into two broad categories – polymer-based electronics and small molecule / oligomer-based electronics. Of the two, the small molecule work has shown the most promise for achieving technologically viable carrier mobilities and device performance levels. Unfortunately, the best results using small molecules have all been achieved using vacuum evaporated semiconductor layers. This process is not particularly compatible with flexonics, and hence is undesirable for this application.

The Maxwell-effect based elastomer actuators used in flexonics are actually ideal interface elements for conventional organic semiconductors FETs. These elastomers typically require very low current levels to actuate, but require high voltages. It is routinely possible to fabricate ≈ 10 micron size organic FETs with operating voltages approaching 100V, but with rather low carrier mobilities of 0.1-1 cm^2/V-s. These performance levels approach those required for control of elastomer-based actuators. Well-known techniques already used in high-voltage silicon circuits may be applied to organic FET-based circuits as well. For example, the voltage across an individual FET may be reduced by using a cascade output stage, consisting of multiple transistors biased to distribute the voltage drop. Furthermore, LDMOS-type device designs may be used to minimize the peak field within individual devices, enabling high-voltage operation of lower dielectric strength materials. Note that in many cases, much of this may be unnecessary since materials design alone may enable suitably high voltage organic semiconductors, exploiting the well-known high-voltage behavior of these materials.

Fig. 2. Inkjet printed inductor

To achieve fully integrated control circuits in the flexonics environment, it will be necessary to develop numerous associated device and circuit technologies beyond the organic semiconductor materials technology described above. We are already developing inkjet-based conductor technologies using heavily doped aniline. While this material has a relatively high resistance, it should be adequate for use in flexonics, given the low current requirements of elastomer actuators. Current work aims to use this as the gate and interconnect material in organic FET based circuits. If higher conductivity interconnects are required, we have initiated a development program for inkjet deposited metal interconnects based on deposition of metallic nanocrystals. Numerous groups have already demonstrated fabrication of all-organic FETs, including organic conductors, insulators, and semiconductors; we are currently migrating this process to an all-inkjet flow. An example inductor is shown in Fig. 2.

5 Mechanical Components

Traditional mechatronic devices require several components: structural support, mechanical joints and linkages, actuators, and control circuitry. These requirements do not change within the flexonic environment. Rather, the components take on different forms. For some, this is simply a matter of designing a suitable flexonic replacement component. For others, the discrete parts of the traditional design are replaced with one integrated, highly complex mechanism.

5.1 Simple Joints

One of the requirements for most mechanical designs is some form of rotary joint. A traditional joint might employ some rolling or sliding interaction between the two connected links in the form of a hinge or bearing assembly. The flexonic counterpart moves by deflection, where elastic energy storage provides an increasing resistance to rotation. Other issues when designing flexonic joints include range of motion, axis-drift, off-axis stiffness, and stress concentrations [11]. We have identified several joint designs for possible inclusion within flexonic mechanisms; these are shown in Figs. 3, 6, and 7.

As a design example, consider the use of flexible strips for creating rotary joints. The cross-flexure joint (also called the cross-axis flexural pivot) [7] consists of three or more simple beam flexures oriented at +/- 45 degrees (Fig. 3). Rotation occurs about the axis formed at the intersections of the flexures. This design exhibits a good approximation to pure rotary motion near the zero angle. To see this, consider a straight thin strip as shown on the left in Fig. 4. The thin strip has three principal bending degrees of freedom, which are T_y, R_x and R_z where T_d denotes translation in the direction d and R_d is rotation about d. The two rotation modes are obvious. The T_y translation mode is caused by S-shaped double bending around axes parallel to the x-axis. Other motions are possible, but bending

Fig. 3. Cross-flexure joint

stiffness is proportional to the cube of thickness, so a strip like the one in Fig. 4 is several orders of magnitude stiffer against motion about R_y and the other two translation modes.

If we represent the tangent space of the strip tip as (T_x, T_y, T_z, R_x, R_y, R_z), then the degrees of freedom are represented as the vector (*0, a, 0, b, 0, c*). Now, suppose we rotate such a strip by 45° CCW about the X-axis through its center. Then the tangent space becomes (*0, a, a, b, -c, c*) (we can ignore constant factors between degrees of freedom). Then we introduce a second strip with degrees of freedom (*0, a', 0, b', 0, c'*). We rotate the second strip by 45° CW, giving a new tangent space of (*0, a', -a', b', -c', c'*). The two strips are joined at their ends. Assume the lower join is rigid, and call it the "fixed" end. The upper join is also rigid; call it the "moving" end. The resulting structure is the simple X-joint shown on the right in Fig. 4. The degrees of freedom of the moving end are the intersection of degrees of freedom of the two strips, i.e. the solution to the following:

$$(0, a, a, b, -c, c) = (0, a', -a', b', -c', c').$$

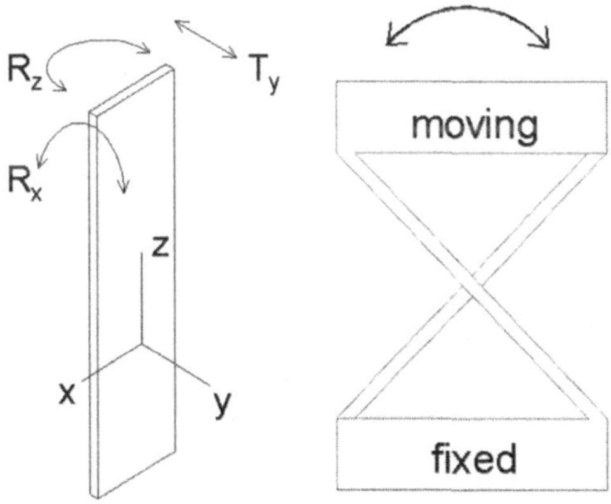

Fig. 4. Thin strip flexures. (*left*) Single strip tangent space. (*right*) Two thin strips connected at ends

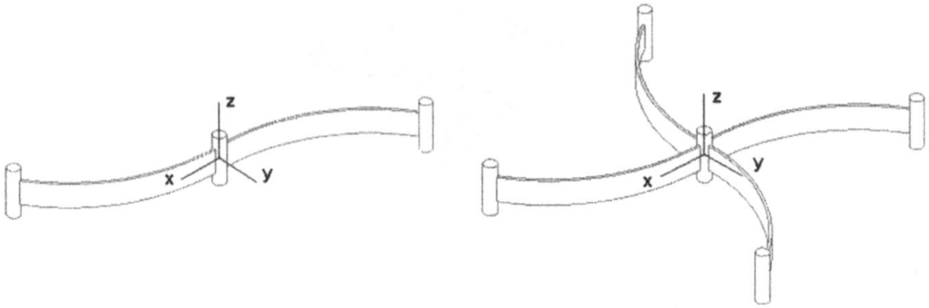

Fig. 5. Thin strip arrangement to minimize twisting during joint flexion

The solution is $a = a' = c = c' = 0$, which means no translation, and no rotation about Y- or Z-axes. The rest of the solution is $b = b'$, which means rotation about X is possible. The analysis above works when other angles (than 45°) are used to rotate the strips, so there is a family of possible X-joint designs. In summary, while a simple strip or sheet makes a poor joint, using two (or more) crossed strips gives nearly pure rotation near the zero angle.

While the basic X-joint design works well near zero flexion, at larger angles it is much less stiff in undesired directions. The problem is that while a straight strip has few degrees of freedom (three), a strip that is significantly bent and twisted is capable of motion in almost any direction. With a flexed X-joint configuration, the most problematic motion is a twisting relative to the ends of the joint. By rearranging the orientation and attachment points of the strips, it might be possible to reduce the twisting effect. Consider two flexed thin strips connected in series as shown in Fig. 5. We assume the center node is free to move in any direction, while the perimeter nodes are fixed in space. In this configuration, we can see that the center node has two rotational degrees of freedom (we will ignore the two additional translational degrees for now). The first degree of freedom is Z-axis rotation, which is the desirable rotation for this analysis. The second degree of freedom is around the X-axis. To remove this undesirable degree of freedom, we can simply add another pair of thin strips 90° from the first pair. Only Z-axis rotation is allowed in this configuration.

The combination of strips in Fig. 5 will not result in a useful joint because we have assumed the perimeter nodes are fixed. In order to rotate around the Z-axis, the perimeter nodes must get closer to the center node. The original X-joint configuration allows this behavior. We can solve the problem by using two X-joints that are oriented 180° apart. We are currently analyzing a novel alternative X-joint shown in Fig. 6. We believe this design will also have the added benefit of restricting buckling from compressive loads.

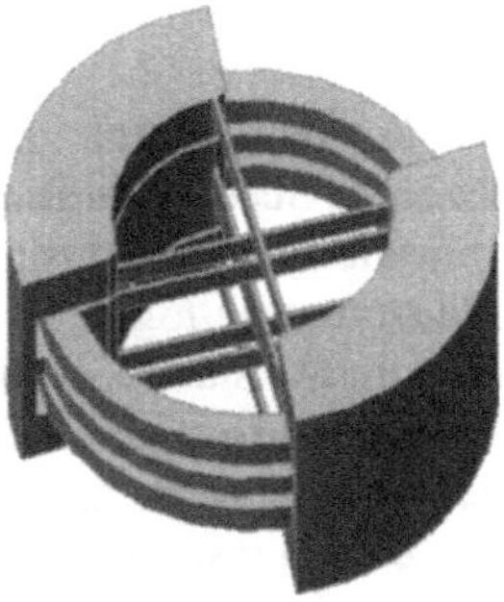

Fig. 6. Novel rotary joint using dual X-joints

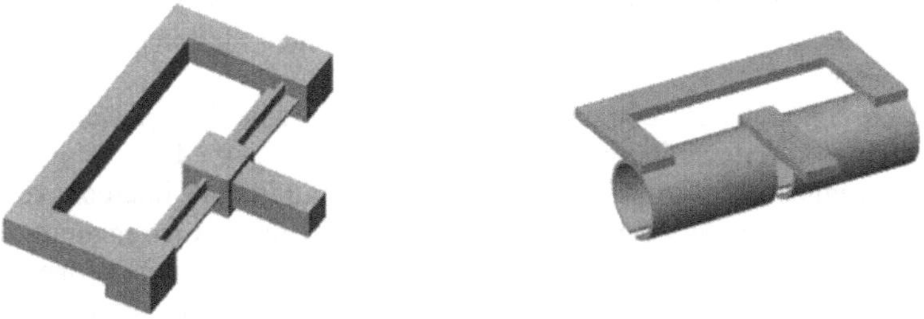

Fig. 7. Other rotary joints. (*left*) Compliant revolute (CR) joint from [11]. (*right*) Split-tube flexure from [5]

A rotary joint can also be designed by utilizing the twisting degree of freedom of a thin strip (R_z in Fig. 4). A realization of this joint is proposed and analyzed in detail in [11] and depicted in Fig. 7. Following the analysis as before, start with a thin strip oriented as shown with degrees of freedom (*0, a, 0, b, 0, c*). Next, add a second strip rotated 90 around the Z-axis with degrees of freedom (*a', 0, 0, 0, b', c'*). The intersection is simply $c = c'$, or rotation about the Z-axis. The joint utilizes perpendicular ribs along the flexure region to constrain off-axis motion. The ribs are essentially the intersection of two perpendicular thin strips. The axis of rotation is along the center of the ribbed flexure.

More complex joint designs are also feasible within the framework of flexonics. The split-tube flexure [5] is based on the principle that torsional stiffness of an open-section tube is much lower than a closed-section tube, while the remaining axial and bending stiffness remain essentially the same. The axis of rotation is along the surface of the tube, opposite from the side with the slit. Fig. 7 illustrates the split-tube design.

5.2 Complex Mechanisms

While flexonics can provide a base set of simple mechanical joints, it also encourages highly complex specialized devices. Traditional motion generation mechanisms can be redesigned within the flexonics design space. As an example, consider the Peaucellier linkage illustrated in Fig. 8. The function of the linkage is to convert rotational motion into pure translation. Under traditional design methods, the links are separate parts, connected to each other via pivots at their ends. The flexonic counterpart, however, can be one continuous piece. A partial realization of this device is shown in Fig. 8 with traditional pivots being replaced by small-length flexures (also called notch flexures or "living hinges").

Mechanisms for specific functions like gripping are also relatively easy to design. One such possibility is shown in Fig. 9. Instead of discrete regions of flexion as used in the Peaucellier linkage, we implement distributed flexion along the gripper "fingers". A simple translation of the center rigid beam brings both fingers together.

Another area where complex mechanisms are useful is power transmission. One attempt at a flexonic transmission is shown in Fig. 9. This design consists of a series connection of cross-flexure joints, with additional interconnecting beam flexures. When one cross-flexure joint is rotated, each of the remaining joints exhibits rotation as well. As before, the entire device is fabricated as one continuous piece. The transmission also demonstrates another anticipated feature of flexonics: modular construction. The flexonic transmission is essentially a repeated pattern of cross-flexure joints. Although this is only a one-dimensional pattern, we

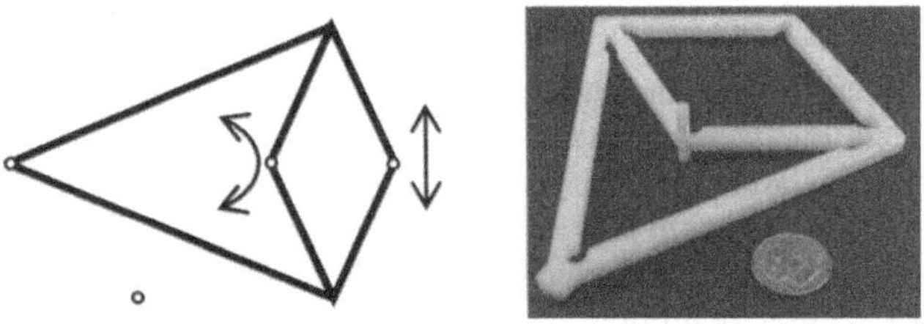

Fig. 8. Peaucellier device. (*left*) Link arrangement. (*right*) FDM partial realization

Fig. 9. Other devices. (*left*) Compliant gripper. (*right*) Power transmitting linkage

might design other devices which contain a different modular unit and employ two- or three-dimensional configurations.

5.3 Actuators

We believe dielectric elastomer actuation is the most promising technology for producing flexonic motion. In order to develop practical actuators with the performance described in Sect. 2, several key issues must be addressed. First of all, actuator design should minimize the film thickness between electrodes in order to reduce high voltage requirements. This will not pose a major challenge using 3D printing techniques. Inkjetted polymer films around 1 micron are certainly feasible. To offset the reduced volume in thin layers, we plan to use laminated stacks of actuator and electrode material. As mentioned previously, there is no penalty for this type of modular construction other than the time to print the added volume. There are some challenges in uniformity of layer thickness, but low-viscosity solvent-based materials exhibit good uniformity under gravity if drying time can be made long enough.

The overall design of the actuator frame must fill two roles. First, it must have sufficient compliance to allow the elastomer to expand in the desired direction of motion. Second, the frame should contain built-in geometry necessary to produce and maintain a pre-strain perpendicular to the direction of motion. One candidate frame is shown in Fig. 10. A complete actuator would contain two frames sandwiching a dielectric elastomer film. At the left of Fig. 10, the frame is shown prior to applying pre-strain. Secondary flexures (oval-shaped) not coupled with the dielectric elastomer can be "set" into a pre-strained state. The right of Fig. 10 shows the secondary flexures extended and locked in place. The middle portion of the dielectric material (bowtie shaped) would be stretched parallel with the secondary flexures. In addition to providing pre-strain, these flexures provide the unidirectional rigid support perpendicular to the desired direction of motion. Note that the flexible ends of the bowtie remain in the same configuration following pre-strain, thereby allowing linear expansion of the dielectric elastomer.

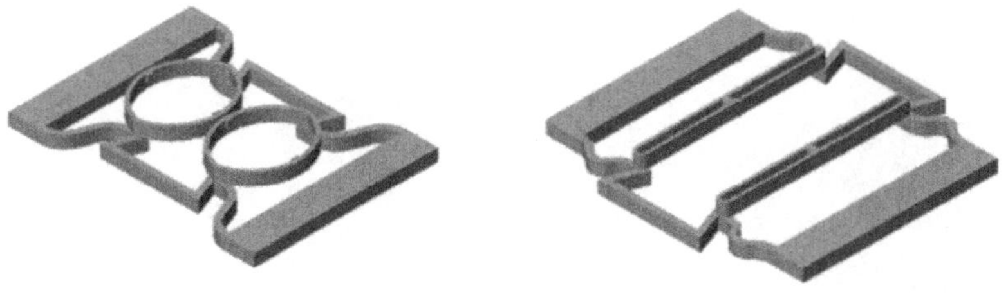

Fig. 10. Proposed actuator design. (*left*) Unstrained configuration. (*right*) Pre-strain configuration

5.4 Bulk Volumes

Mechatronic devices need passive structural volumes to support and connect moving components. In traditional designs, these bulk volumes take on the form of three-dimensional solid links or, at best, two-dimensional structures like honeycombs. High strength-to-weight performance can be achieved through one-dimensional trusses, but this is usually only feasible on a large scale or through very expensive investment casting. However, with flexonic manufacturing, we can print intricate geometry to create lightweight truss and lattice volumes. One very strong, low-density lattice is the diamond lattice shown in Fig. 11. The FDM ABS plastic structure shown has approximately 1% density and weighs just over a gram, but has a theoretical yield load of over 50 kg.

The strength of these structures raises interesting design questions. Most electro-mechanical devices are quite heavy. The determining factors seem to be the weight of components such as motors, metal shafts and bearings, and heavy electrical components such as transformers or batteries. A 100% polymer device has no such constraints. Actuators are extremely light, and their weight scales down with the weight of the entire device, since the total mass of the device determines the actuators' load. Similarly, polymer electronics are extremely light, and there is no packaging or circuit boards per se. Thus, the weight of the device is primarily dictated by its strength to withstand external loads. Because of the excellent strength-to-weight performance possible with lattice structures, the structural components of flexonic devices (and the entire device) can be made very light.

For instance, consider the design of a life-size humanoid robot. Its volume will be about 80 liters (typical for a real human). With 1% density ABS polymer construction, the robot would weigh less than 1 kilogram. Assuming 10 cm diameter limbs, it could nevertheless accept total loads of several hundred kilograms. At this time, the actuators would not be able to produce the necessary torques, but their higher energy density would allow an anthropomorphic robot to (theoretically) hold or lift loads of several kilograms. This contrasts with the present Honda robot design has a hand-load capacity of about 5 kg but a weight of several hundred.

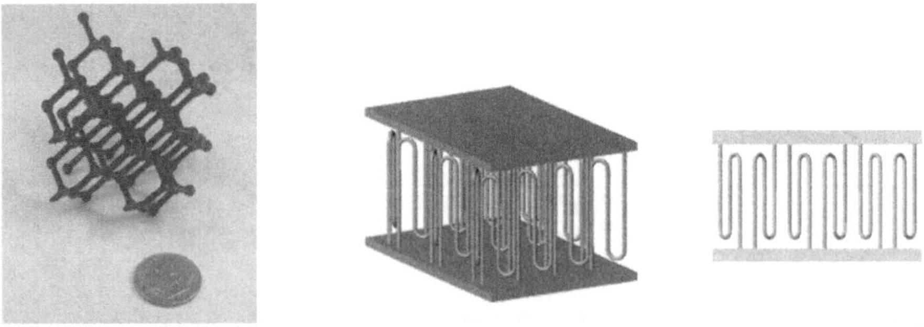

Fig. 11. Bulk volumes. (*left*) Diamond lattice. (*right*) Shearing support volume with high vertical stiffness (*isometric and side views*)

Flexonic design can produce strong, lightweight structures, with very high stiffness. However, in many cases, high stiffness may only be desirable in certain directions. The actuator frames described above are one such example of this anisotropic design. Another example would be the shearing support system shown in Fig. 11. Vertically oriented flexures resist compressive loading, but allow movement in the two planar directions.

6 Potential Application

Flexonics offers a unique set of properties not found in traditional electromechanical systems. Most notable is the ability to manufacture fully functional devices without assembly. When coupled with a diverse material set, these devices can exhibit a high level of component integration. Structures can be made lightweight, yet strong, by exploiting complex geometry and polymer construction. Flexonic devices are not intended to replace all traditional mechanical systems. For instance, systems utilizing wheels to achieve continuous and infinite travel would not be possible. However, the properties inherent to flexonic manufacturing and design are appropriate for a wide array of applications where joint range of motion is finite.

6.1 Human-Machine Interfaces

Flexonic design allows human-machine interfaces to be individualized according to the ergonomic needs of the user and the workspace constraints. In fact, the user can be inserted directly into the design loop. Due to the low cost of flexonic manufacturing (assuming access to a flexonic manufacturing station), a specific device design can be iterated until the user's requirements are satisfied. Flexonics also offers other attractive properties for interfaces in terms of user safety. For example, life-size humanoid robots with metal limbs pose a number of risks to the people around them. An ultra-light polymer design would have low mass, natural compliance, and low impact energy if an accidental contact with a person occurred. These "soft" characteristics are also beneficial in conveying life-like form and behaviors. Since there is no sliding or rolling motion in joints, motors or bearings, flexonic devices can be almost noiseless in operation. When designed for use as human-like representations (however abstract) in a collaborative setting, these flexonic traits may have advantages over traditional electromechanical designs in achieving social acceptance.

6.2 Biomimetic Robots

Flexonics also suggests advantages in the area of biomimetic design in robotics. Biology provides an abundance of robust mechanisms for articulation and locomotion. However, these mechanisms are often highly complex and necessitate a high level of functional and structural integration. For example, consider the requirements for a biomimetic actuator. In this design space, a counterpart to natural muscle is needed. One feature of muscle is an ability to function in a variety of modalities. They can operate as motors to generate power, as dampers or brakes to absorb power, as variable springs to modulate stiffness, and as rigid struts to transmit forces [4]. These functions are all contained within a relatively compact structure. In addition to pure mechanical components, muscle also integrates all the required chemical and electrical components for power and control. With proper design, polymer actuators may fill the mechanical roles of muscle. As previously discussed, organic electronics shows promise as a technology for interfacing driver and control circuitry with polymer actuators. Flexonic mechatronic systems can combine these components in one complex, integrated package.

6.3 Functional Prototypes

Even for large-volume devices that might ultimately be manufactured by some other means, such as phones, PIMs, remote controls, etc., flexonics allows fully functional prototypes to be built with extremely short turn-around. It allows more complete user testing and evaluation earlier in the design cycle. It also allows different design elements: electrical, mechanical, aesthetic, to be combined much earlier during design, because the first prototypes can incorporate all elements. It should therefore encourage shorter design times. It would also encourage shorter times between "generations" of a device, by allowing a reference design to continually evolve and be improved.

7 Conclusion

We presented a vision of a new class of mechatronic devices that integrate structural, mechanical, and electronic components in a one-piece, no assembly fabrication process. Our initial goals include the development of a complete inkjet-based 3D printing system, and the exploration of flexonic capabilities through the development of a vocabulary of mechatronic components. The basic components of an inkjet printer were described, along with the materials we are currently exploring. Methods for designing devices within the framework of flexonics were presented, along with a novel joint design. We presented several examples of joints, structures, and actuators that we have fabricated using existing 3D printing techniques. We outlined our current capabilities using organic electronics, as well as anticipated work to integrate organic electronics with a complete flexonic system. We envision flexonics as the foundation for many existing mechatronic systems, and have provided example applications including human-machine interfaces, biomimetic robots, and fully functional, prototypes.

References

1. Amon CH, Beuth JL, Merz R, Prinz FB, Weiss LE (1998) Shape deposition manufacturing with microcasting: Processing, thermal and mechanical issues. J. Manufacturing Science and Engineering 120(3):502-656
2. Ananthasuresh GK, Kota S (1995) Designing compliant mechanisms. ASME Mechanical Engineering 117(11):93-96
3. Bar-Cohen Y (2001) Transition of EAP material from novelty to practical applications - are we there yet? In Bar-Cohen Y (ed) Proc. SPIE, Smart Structures and Materials 2001: Electroactive Polymer Actuators and Devices 4329: 1-6
4. Dickinson MH, Farley CT, Full RJ, Koehl MAR, Kram R, Lehman S (2000) How animals move: An integrative view. Science 288:100-106
5. Goldfarb M, Speich JE (1999) A Well-Behaved Revolute Flexure Joint for Compliant Mechanism Design. Journal of Mechanical Design, Trans. ASME 121(3):424-429
6. Hayes DJ, Cox WR, Wallace DB (2001) Printing System for MEMS Packaging. In Proceedings, SPIE Micromachining & Microfabrication Conference, San Francisco CA, October 22-25
7. Howell LL (2001) Compliant Mechanisms. Wiley, New York
8. Jacobs PF (1995) Stereolithography and Other RP&M Technologies: From Rapid Prototyping to Rapid Tooling. ASME Press
9. Johnson J (1994) Principles of Computer Automated Fabrication. Palatino Press
10. Lin YY, Dodabalapur A, Sarpeshkar R, Bao Z, Li W, Crone B, Raju VR, Katz HE (1999) Organic semiconductor logic circuits. In 57th Annual Device Research Conference Digest:166-167
11. Moon YM, Trease BP, Kota S (2002) Design of Large-Displacement Compliant Joints. In Proceedings, ASME International Design Engineering Technical Conference, Montreal, Canada, September 29 – October 2
12. Pelrine R, Kornbluh R, Pei Q, Joseph J (2000) High-speed electrically actuated elastomers with strain greater than 100%. Science 287(5454):836-839
13. Pelrine R, Sommer-Larsen P, Kornbluh R, Heydt R, Kofod G, Pei Q, Gravesen P (2001) Applications of Dielectric Elastomer Actuators. In Bar-Cohen (ed) Proc. SPIE, Smart Structures and Materials 2001: Electroactive Polymer Actuators and Devices 4329: 335-349
14. Shah VG, Hayes DJ (2002) Trimming and Printing of Embedded Resistors Using Demand-Mode Ink-Jet Technology and Conductive Polymer. Proc., IPC Printed Circuits Expo 2002, Long Beach, CA, March 24-28
15. Shahinpoor M, Bar-Cohen Y, Simpson JO, Smith J (1998) Ionic Polymer-Metal Composites (IPMC) as Biomimetic Sensors, Actuators & Artificial Muscles - A Review. Smart Materials and Structures, 7:15-30
16. Subramanian V (2000) Applications of printing technology in organic electronics and display fabrication. In The Half Moon Bay Maskless Lithography Workshop, DARPA/SRC, Half Moon Bay
17. Trimmer W (ed) (1990) Micromechanics and MEMS: Classic and Seminal Papers to 1990. Institute of Electrical and Electronics Engineers

Coordinating Multiple Robots with Kinodynamic Constraints along Specified Paths

Jufeng Peng and Srinivas Akella

Department of Computer Science
Rensselaer Polytechnic Institute
Troy, New York 12180, USA

Abstract. This paper focuses on the collision-free coordination of multiple robots with kinodynamic constraints along specified paths. We present an approach to generate continuous velocity profiles for multiple robots that avoids collisions and minimizes the completion time. The approach, which combines techniques from optimal control and mathematical programming, consists of identifying collision segments along each robot's path, and then optimizing the robots velocities along the collision and collision-free segments. First, for each path segment for each robot, the minimum and maximum possible traversal times that satisfy the dynamics constraints are computed by solving the corresponding two-point boundary value problems. The collision avoidance constraints for pairs of robots can then be combined to formulate a mixed integer nonlinear programming (MINLP) problem. Since this nonconvex MINLP model is difficult to solve, we describe two related mixed integer linear programming (MILP) formulations that provide schedules that are lower and upper bounds on the optimum; the upper bound schedule is designed to be a continuous velocity schedule. The approach is illustrated with coordination of multiple robots, modeled as double integrators subject to velocity and acceleration constraints. Implementation results for coordination of 12 robots are described.

1 Introduction

Coordinating multiple robots with kinodynamic constraints, i.e. simultaneous kinematic and dynamics constraints ([11]), in a shared workspace without collisions has applications in manufacturing cells ([33]), AGV coordination in harbors and airports ([2]), and air traffic control ([5]). The general problem requires finding the trajectory (path and velocity profile) of each robot such that the specified objective, such as the task completion time, total time, or energy consumption, of the system is minimized.

This paper deals with the optimal coordination of multiple robots moving with kinodynamic constraints along specified paths. While prior work mostly addressed either the collision-free path coordination problem of several robots without considering dynamics constraints ([29],[26],[41]), or the search for time-optimal motions for a single robot ([7],[39]), the contribution of this paper is an approach to generate continuous velocity profiles that avoid

J.-D. Boissonnat et al. (Eds.): Algorithmic Foundations of Robotics V, STAR 7, pp. 221–237, 2004.

collisions between multiple robots and that also minimize the task completion time under the kinodynamic constraints. An example application is the coordination of AGVs along fixed paths in harbors and airports. We must satisfy kinematic constraints, such as avoiding collisions between robots and with moving obstacles, and dynamics constraints, such as velocity and acceleration bounds, on the robot motions. By identifying the collision segments along a robot's path, we can combine the collision avoidance constraints for pairs of robots to formulate a mixed integer nonlinear programming (MINLP) problem. Since the resulting nonconvex MINLP formulation is difficult to solve, we use two related mixed integer linear programming (MILP) formulations, the *improved instantaneous model*, which provides a lower bound on the optimal solution, and the *setpoint model*, which provides a continuous velocity schedule that is an upper bound on the optimal solution. In this paper, we illustrate the approach using robots modeled as double integrators subject to velocity and acceleration constraints, and briefly discuss applications to more general robot systems.

2 Related Work

Multiple Robot Coordination: Motion planning for multiple robots requires moving each robot from its initial to its goal configuration, while avoiding collisions with static obstacles or with other robots ([24]). This problem is highly underconstrained, and Hopcroft, Schwartz, and Sharir [16] showed that even a simplified two-dimensional case of the problem is PSPACE-hard. Recent efforts have focused on reducing the dimension of the configuration space by grouping robots (Aronov et al. [3]) or using probabilistic approaches. A potential field randomized path planner was applied to multiple robot planning (Barraquand, Langlois, and Latombe [4]), and probabilistic roadmap planner have been developed for multiple car-like robots (Svestka and Overmars [43]) and multiple manipulators (Sanchez and Latombe [35]).

A slightly more constrained version of the problem is obtained when all but one of the robots have specified trajectories. This is the problem of planning a path and velocity for a single robot among moving obstacles (Reif and Sharir [32], Kant and Zucker [19]). Erdmann and Lozano-Perez [12] obtain a heuristic solution for planning the motions of multiple robots by assigning priorities to robots and sequentially searching for collision-free paths for the robots in the configuration-time space.

If the problem is further constrained so that the paths of the robots are specified, one obtains a path coordination problem. O'Donnell and Lozano-Perez [29] developed a method for path coordination of two robots. LaValle and Hutchinson [26] addressed a similar problem where each robot was constrained to a specified configuration space roadmap. Simeon, Leroy, and Laumond [41] perform path coordination for a very large number of car-like robots, where robots with intersecting paths can be partitioned into smaller sets. Trajectory coordination is a closely related problem where the trajectory (path and velocity) of each robot is specified. Previous work on trajectory coordination focused almost exclusively on dual robot systems (Shin and Zheng [40], Bien and Lee [6], Chang, Chung, and Lee [9]). Akella and Hutchinson [1] recently developed an MILP formulation for the trajectory coordination of large numbers of robots by changing robot start times. Our work extends these problem classes by additionally considering dynamics constraints and generating continuous velocity profiles.

Time-optimal Trajectory Planning: There is a large body of work on the time optimal control of a single manipulator. Bobrow, Dubowsky, and Gibson [7], and Shin and McKay [39] developed algorithms to generate the time-optimal velocity profile of a manipulator moving along a specified path. Trajectory planning directly in the $2n$-dimensional state space that considers both kinematic and dynamic constraints is called *kinodynamic planning.* Sahar and Hollerbach [34] and later Shiller and Dubowsky [38] developed algorithms for global near minimum-time trajectory generation (path and velocity) for a manipulator with dynamics and actuator constraints using grid-based search spaces. Donald et al. [11] developed a polynomial time approximation algorithm to generate near time-optimal trajectories for a single robot that satisfy kinematic and dynamic constraints. Fraichard [14] describes a trajectory planner for a car-like robot with dynamics constraints moving along a given path among moving obstacles. Recent work has focused on randomized kinodynamic planning, including the use of rapidly exploring random trees (Lavalle and Kuffner [27]) and probabilistic roadmaps (Hsu et al. [17]).

Air Traffic Control: Conflict resolution among multiple aircraft in a shared airspace is closely related to multiple robot coordination. Tomlin, Pappas, and Sastry [42] synthesized provably safe conflict resolution maneuvers for two aircraft using speed and heading changes. Kosecka et al. [21] used potential field planners to generate conflict resolution maneuvers. Bicchi and Pallottino [5] modeled aircraft with constant velocity and curvature bounds and generated minimum total time collision-free paths using given waypoints for three aircraft. Schouwenaars et al. [37] used a discretized system model to develop an MILP formulation for fuel-optimal path planning of multiple vehicles, including moving obstacles. Pallottino, Feron, and Bicchi [30] generate conflict-free paths to minimize the total flight time for the cases when either instantaneous velocity changes or heading angle changes are allowed.

3 Problem Overview

Given a set of n robots $\mathcal{A}_1, \ldots, \mathcal{A}_n$ with specified paths and dynamics constraints, the goal is to find the control inputs along the specified paths so that the completion time of the set of robots is minimized and their motions are collision free. We assume that the start and goal configurations of each robot are collision-free. This assumption guarantees the existence of a solution since a feasible schedule is for one robot at a time to move along its path, for some arbitrary sequencing of the robots. The only assumptions about the specified paths are that they are free of static obstacles, and can be traversed by the robots without violating kinematic constraints. We further assume that each robot moves only forward along its path.

3.1 Paths and Collision Zones

Each robot $\mathcal{A}_i$ is given a path γ_i, which is a continuous mapping $[0,1] \rightarrow \mathcal{C}_i^{free}$. Let $\mathcal{S}_i = [0,1]$ denote the set of parameter values s_i that place the robot along the path γ_i. The *coordination space* for n robots is defined as $\mathcal{S} = \mathcal{S}_1 \times \mathcal{S}_2 \times \cdots \times \mathcal{S}_n$. A feasible coordination is a schedule $\psi(t) : \mathcal{R}^+ \rightarrow \mathcal{S}$ in which $s_{init} = (0, 0, \ldots, 0)$ and $s_{goal} = (1, 1, \ldots, 1)$ and the robots do not collide. Note that there is a 1-to-1 mapping between s and the path length. Based on the topology of the coordination space, we define *collision zones.*

A *collision pair* $\mathcal{CP}_{ij}(s_i, s_j)$, where s_i and $s_j \in [0,1]$, is defined as a pair of configurations $(\gamma_i(s_i), \gamma_j(s_j))$ where robot $\mathcal{A}_i$ and robot $\mathcal{A}_j$ collide, i.e., $\mathcal{A}_i(\gamma_i(s_i)) \cap \mathcal{A}_j(\gamma_j(s_j)) \neq \emptyset$. A *collision segment* for robot $\mathcal{A}_i$ is a contiguous interval $[s_i^{start}, s_i^{end}]$ over which $\mathcal{A}_i$ collides with some $\mathcal{A}_j$. That is, $\forall s_i \in [s_i^{start}, s_i^{end}], \exists s_j$ such that $\mathcal{A}(\gamma_i(s_i)) \cap \mathcal{A}(\gamma_j(s_j)) \neq \emptyset$.

An ordered pair of maximal contiguous intervals $([s_i^{start}, s_i^{end}], [s_j^{start}, s_j^{end}])$ in the coordination space $\mathcal{S}$ constitute a *collision zone* $\mathcal{CZ}_{ij}$ if and only if any point in one interval results in a collision with at least one point in the other interval (Figure 1). That is, $\forall s_i \in [s_i^{start}, s_i^{end}], \exists s_j \in [s_j^{start}, s_j^{end}]$ such that $\mathcal{A}(\gamma_i(s_i)) \cap \mathcal{A}(\gamma_j(s_j)) \neq \emptyset$, and $\forall s_j \in [s_j^{start}, s_j^{end}], \exists s_i \in [s_i^{start}, s_i^{end}]$ such that $\mathcal{A}(\gamma_i(s_i)) \cap \mathcal{A}(\gamma_j(s_j)) \neq \emptyset$.

In Figure 1, the collision zones are $([a_1, a_2], [b_3, b_4])$, and $([a_3, a_4], [b_1, b_2])$. A maximal interval that is not within any collision zone is called a *collision-free segment* . Each robot's path is decomposed into one or more collision segments and collision-free segments.

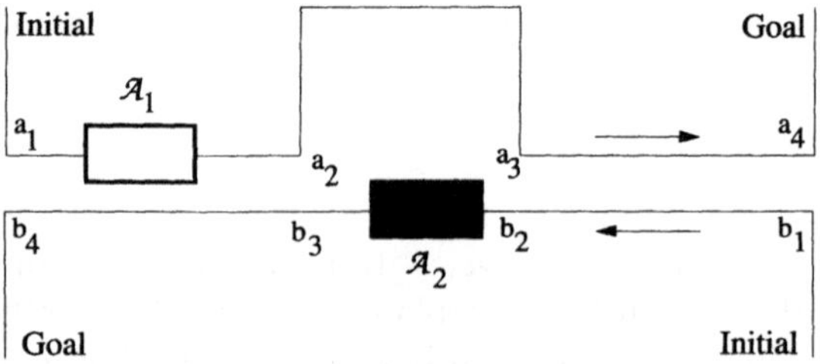

Fig. 1. Example with two translating robots with two collision zones.

3.2 Optimal Control Problem for a Single Robot

For a single robot $\mathcal{A}$ moving along a path segment, let $\mathbf{x}(t)$ represent the state, $\mathbf{u}(t)$ be the control, γ be the path of the robot, $J(\mathbf{x}, \mathbf{u})$ be the objective function, and $g(\mathbf{x})$ and $q(\mathbf{u})$ be the inequality constraints on the state variables and controls respectively. Then the optimal control problem, to compute the minimum and maximum time taken by the robot to traverse the segment subject to its dynamics and path constraints, can be written as:

$$\begin{aligned} \text{Minimize} \quad & J(\mathbf{x}, \mathbf{u}) \\ \text{subject to:} \quad & \\ \dot{\mathbf{x}} &= f(\mathbf{x}, \mathbf{u}) \\ g(\mathbf{x}) &\leq 0 \\ q(\mathbf{u}) &\leq 0 \\ \mathbf{x}(0) &= \mathbf{x}_{start} \\ \mathbf{x}(\Delta T) &= \mathbf{x}_{end} \\ \mathbf{x} &\in \gamma \end{aligned}$$

The minimum time control problem has $J(\mathbf{x}, \mathbf{u}) = \Delta T$, and the maximum time control problem has $J(\mathbf{x}, \mathbf{u}) = -\Delta T$ where $\Delta T = \int_0^{\Delta T} 1 dt$ is the time to traverse the segment. Feasible robot motions that give a minimum and a maximum of the objective over each segment are obtained by solving two TPBVPs (two-point boundary value problems) for each segment.

3.3 Coordination of Multiple Robots

Now consider the multiple robot system in which each robot has a specified path and dynamics constraints. The goal is to coordinate these robots to minimize a specified objective; in this paper it is the global completion time. The path of each robot is decomposed into collision segments and collision-free segments. The coordination of multiple robots can then be modeled as a mixed integer nonlinear programming (MINLP) problem, with each robot satisfying the traversal time and collision avoidance constraints over each of its segments. Since this MINLP problem with nonconvex constraints is difficult to solve, we obtain schedules that provide a lower bound and an upper bound on the optimal solution by solving two related mixed integer linear programming (MILP) problems. We illustrate this approach using the double integrator model from optimal control ([8]).

4 Instantaneous Model

We first consider a simplified model where each robot always moves using its highest speed v_{max}, and is permitted to instantaneously change its velocity.

That is, each robot has infinite acceleration, and can instantaneously accelerate to v_{max} or instantaneously decelerate to zero velocity from v_{max}. We will refer to this as the *instantaneous model*, since it provides a schedule with instantaneous starts and stops.

4.1 Mixed Integer Linear Programming (MILP) Formulation

We now present an MILP formulation for the instantaneous model. Let t_{ik} be the time when robot $\mathcal{A}_i$ begins moving along its kth segment and τ_{ik} be the traversal time for $\mathcal{A}_i$ to pass through segment k. Let ΔT_{ik}^{min} and ΔT_{ik}^{max} represent the minimum and maximum traversal time for $\mathcal{A}_i$ between the start point of segment k and the start point of segment $k+1$, giving the *traversal time constraints* $\Delta T_{ik}^{max} \geq \tau_{ik} \geq \Delta T_{ik}^{min}$. For the instantaneous model, $\Delta T_{ik}^{max} = \infty$. The minimum time for $\mathcal{A}_i$ to traverse a segment of length S_{ik} at its maximum velocity $v_{i,max}$ is $\Delta T_{ik}^{min} = S_{ik}/v_{i,max}$. The motion completion time C_{max} for the set of robots is greater than or equal to the completion time of each robot, leading to the *completion time constraint* $C_{max} \geq t_{i,last} + \tau_{i,last}$ for each robot $\mathcal{A}_i$. Consider robots $\mathcal{A}_i$ and $\mathcal{A}_j$ with a shared collision zone where k and h are their respective collision segments. A sufficient condition for collision avoidance is that $\mathcal{A}_i$ and $\mathcal{A}_j$ are not simultaneously in their shared collision zone. That is, $t_{jh} \geq t_{i(k+1)}$ (when $\mathcal{A}_i$ exits segment k before $\mathcal{A}_j$ enters segment h) or $t_{ik} \geq t_{j(h+1)}$ (when $\mathcal{A}_j$ exits segment h before $\mathcal{A}_i$ enters segment k). These disjunctive constraints are converted to standard conjunctive form ([28]) by introducing δ_{ijkh}, a binary variable that is 1 if robot $\mathcal{A}_i$ goes first along its kth segment and 0 if robot $\mathcal{A}_j$ goes first along its hth segment, and M, a sufficiently large positive number. The resulting *collision avoidance constraints* to ensure the two robots $\mathcal{A}_i$ and $\mathcal{A}_j$ are not simultaneously in their shared collision zone are:

$$
\begin{aligned}
&t_{jh} - t_{i(k+1)} + M(1 - \delta_{ijkh}) \geq 0 \\
&t_{ik} - t_{j(h+1)} + M\delta_{ijkh} \geq 0
\end{aligned}
$$

The constraints for all robots are combined to form the instantaneous MILP formulation:

$$
\begin{aligned}
&\text{Minimize } C_{max} \\
&\text{subject to:} \\
&C_{max} \geq t_{i,last} + \tau_{i,last} \qquad \text{for } i = 1, \ldots, n \\
&t_{ik} \geq 0 \\
&t_{i(k+1)} = t_{ik} + \tau_{ik} \\
&\Delta T_{ik}^{max} \geq \tau_{ik} \geq \Delta T_{ik}^{min} \\
&t_{jh} - t_{i(k+1)} + M(1 - \delta_{ijkh}) \geq 0 \\
&t_{ik} - t_{j(h+1)} + M\delta_{ijkh} \geq 0 \\
&\delta_{ijkh} \in \{0, 1\}
\end{aligned}
$$

The above model is conservative in requiring that two robots should not simultaneously be in their shared collision zone, and can lead to solutions that are not truly optimal. When a robot has overlapping collision zones with more than one robot, we subdivide its overlapping collision segments into several subsegments. The relevant pairs of subdivided collision zones are used to generate collision avoidance constraints as before. This increases the number of binary variables in the formulation.

The coordination of multiple robots under the instantaneous model can be viewed as a *job shop scheduling problem* (JSP). Each *job*, composed of several operations, is a robot's motion along its path. Each *operation* is the motion along a segment. Each *machine* is a collision zone or a collision-free zone. The job shop scheduling problem is NP-hard ([31]), and by reduction, the instantaneous model for robot coordination is NP-hard.

5 Continuous Velocity Model

We now consider generating a schedule with continuous velocity profiles for the robots that is consistent with their maximum velocity and acceleration bounds. To find the minimum and maximum times taken by a robot to traverse a segment, we solve two TPBVPs (two-point boundary value problems) over the segment. We illustrate this procedure using the *double integrator* model from classical optimal control ([8]).

5.1 Single Robot on a Segment

A single robot moving along a path segment can be modeled as a double integrator ([8]) with inequality constraints on the control input (acceleration) and the velocity state variable. Let $x(t)$, $v(t) = \frac{dx(t)}{dt}$, and $a(t) = \frac{dv(t)}{dt}$ be the position, velocity, and acceleration of the robot at time t, S be the length of the segment, and ΔT be the time taken to traverse the segment. Computing the minimum and maximum time taken by the robot to traverse the segment, subject to constraints on its velocities v_{start} and v_{end} at the segment endpoints and inequality constraints on its velocity and acceleration, can be solved as a TPBVP. The double integrator system can be written as:

$$\text{Minimize or Maximize } \Delta T = \int_0^{\Delta T} 1 dt$$

subject to:

$$\begin{pmatrix} \dot{x} \\ \dot{v} \end{pmatrix} = \begin{pmatrix} 0 & 1 \\ 0 & 0 \end{pmatrix} \begin{pmatrix} x \\ v \end{pmatrix} + \begin{pmatrix} 0 \\ 1 \end{pmatrix} a(t)$$

$$x(0) = 0 \quad x(\Delta T) = S$$

$$v(0) = v_{start} \quad v(\Delta T) = v_{end}$$

$$0 \leq v \leq v_{max}$$

$$-a_{max} \leq a \leq a_{max}$$

5.2 Minimum and Maximum Time Control for a Single Robot over a Segment

The minimum time control of the double integrator model is well known ([8]). We have extended this to obtain the maximum time control using the restricted maximum principle ([20]). Basically these are TPBVPs, and the solutions have a bang-bang or bang-off-bang control structure. The minimum ΔT and maximum ΔT each have two different cases, depending on whether S is sufficiently long for the robot to reach v_{max} (zero) for the minimum (maximum) time case. Note that if $\frac{|v_{end}^2 - v_{start}^2|}{2a_{max}} > S$, there is no feasible velocity profile since the distance is too short.

1. Minimum ΔT (Figure 2):

   (a) If $S \geq \frac{1}{2}(\frac{(v_{max}^2 - v_{start}^2)}{a_{max}} + \frac{(v_{max}^2 - v_{end}^2)}{a_{max}})$,
   $\Delta T^{min} = \frac{S}{v_{max}} - \frac{((v_{max}^2 - v_{start}^2) + (v_{max}^2 - v_{end}^2))}{2a_{max} \cdot v_{max}} + \frac{v_{max} - v_{start}}{a_{max}} + \frac{v_{max} - v_{end}}{a_{max}}$

   (b) If $\frac{1}{2}(\frac{(v_{max}^2 - v_{start}^2)}{a_{max}} + \frac{(v_{max}^2 - v_{end}^2)}{a_{max}}) > S \geq \frac{1}{2}\frac{|v_{end}^2 - v_{start}^2|}{a_{max}}$,
   $\Delta T^{min} = \frac{(v_{middle} - v_{start})}{a_{max}} + \frac{(v_{middle} - v_{end})}{a_{max}}$
   where $v_{middle} = \frac{1}{2}(2v_{start}^2 + 2v_{end}^2 + 4Sa_{max})^{\frac{1}{2}}$

2. Maximum ΔT (Figure 3):

   (a) If $S \geq \frac{1}{2}\frac{(v_{start}^2 + v_{end}^2)}{a_{max}}$, $\Delta T^{max} = \infty$.

   (b) If $\frac{1}{2}\frac{(v_{start}^2 + v_{end}^2)}{a_{max}} > S \geq \frac{1}{2}\frac{|(v_{end}^2 - v_{start}^2)|}{a_{max}}$,
   $\Delta T^{max} = \frac{(v_{start} - v_{middle})}{a_{max}} + \frac{(v_{end} - v_{middle})}{a_{max}}$
   where $v_{middle} = \frac{1}{2}(2v_{start}^2 + 2v_{end}^2 - 4Sa_{max})^{\frac{1}{2}}$

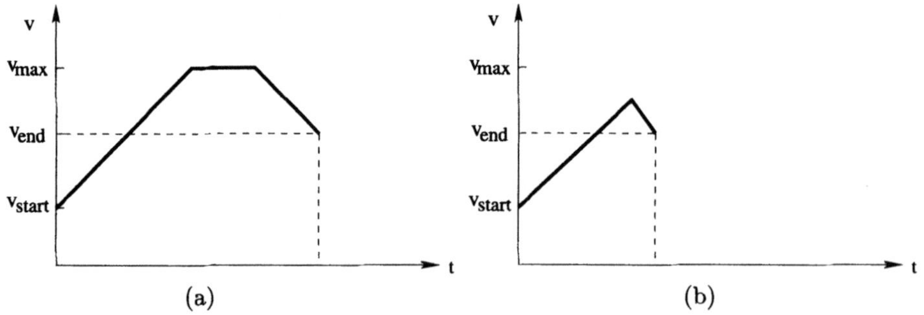

Fig. 2. Minimum ΔT. Case (a): Velocity reaches v_{max}. Case (b): Velocity cannot reach v_{max}.

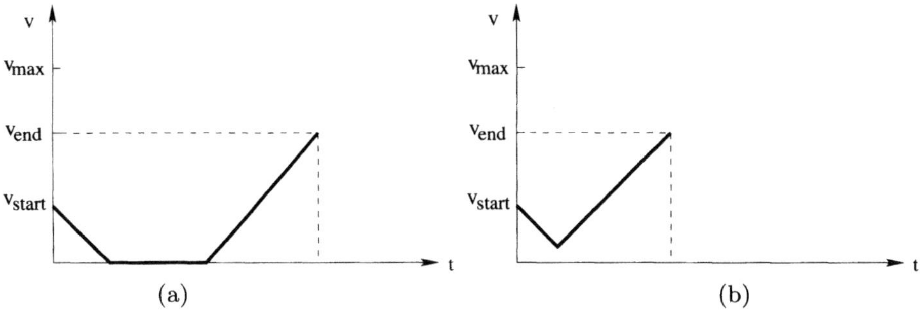

Fig. 3. Maximum ΔT. Case (a): Velocity can decrease to zero. Case (b): Velocity cannot decrease to zero.

5.3 Continuous Velocity MINLP Formulation

Since the robot velocities are variables in the minimum and maximum time control for a robot over a segment, they introduce nonlinear constraints. We therefore formulate a mixed integer nonlinear programming problem (MINLP) for the multiple robot coordination problem. We have the usual completion time constraints and collision avoidance constraints. The traversal time constraints are more complicated since the segment endpoint velocities are variables. Let $a_{i,max}$ be the maximum acceleration of robot $\mathcal{A}_i$ and let v_{ik} represent the velocity of $\mathcal{A}_i$ at the start of segment k. Let ΔT_{ik}^{min} and ΔT_{ik}^{max} be the minimum and maximum traversal times for $\mathcal{A}_i$ along the segment k. Let $\Delta T_{ik,1}^{min}(\Delta T_{ik,1}^{max})$ and $\Delta T_{ik,2}^{min}(\Delta T_{ik,2}^{max})$ represent the two possible minimum (maximum) traversal time values (as described in Section 5.2). The binary variables $y_{ik,1}$ and $y_{ik,2}$ ($z_{ik,1}$ and $z_{ik,2}$) depend on whether or not the values of v_{ik} and $v_{i(k+1)}$ permit the robot to reach v_{max} (zero) in S_{ik}, and are used to select the feasible value of $\Delta T_{ik}^{min}(\Delta T_{ik}^{max})$. The MINLP formulation for the optimal continuous velocity schedule is:

$$
\begin{aligned}
&\text{Minimize } C_{max} \\
&\text{subject to:} \\
&C_{max} \geq t_{i,last} + \tau_{i,last} \qquad \text{for } i = 1, \ldots, n \\
&t_{i(k+1)} - t_{ik} = \tau_{ik} \\
&t_{jh} - t_{i(k+1)} + M(1 - \delta_{ijkh}) \geq 0 \\
&t_{ik} - t_{j(h+1)} + M\delta_{ijkh} \geq 0 \\
&S_{ik} \geq \frac{(v_{i(k+1)}^2 - v_{ik}^2)}{2a_{max}} \geq -S_{ik} \\
&(S_{ik} - \frac{(v_{i,max}^2 - v_{ik}^2) + (v_{i,max}^2 - v_{i(k+1)}^2)}{2a_{i,max}}) - My_{ik,1} \leq 0 \\
&(S_{ik} - \frac{(v_{i,max}^2 - v_{ik}^2) + (v_{i,max}^2 - v_{i(k+1)}^2)}{2a_{i,max}}) + My_{ik,2} \geq 0
\end{aligned}
$$

$$
\begin{aligned}
&(S_{ik} - \frac{v_{ik}^2 + v_{i(k+1)}^2}{2a_{i,max}}) - M z_{ik,1} \leq 0 \\
&(S_{ik} - \frac{v_{ik}^2 + v_{i(k+1)}^2}{2a_{i,max}}) + M z_{ik,2} \geq 0 \\
&\Delta T_{ik,1}^{min} = \frac{S_{ik}}{v_{i,max}} - \frac{(v_{i,max}^2 - v_{ik}^2 + v_{i,max}^2 - v_{i(k+1)}^2)}{2a_{i,max}v_{i,max}} \\
&\qquad + \frac{v_{i,max} - v_{ik}}{a_{i,max}} + \frac{v_{i,max} - v_{i(k+1)}}{a_{i,max}} \\
&\Delta T_{ik,2}^{min} = \frac{(v_{middle,ik}^{min} - v_{ik})}{a_{i,max}} + \frac{(v_{middle,ik}^{min} - v_{i(k+1)})}{a_{i,max}} \\
&(v_{middle,ik}^{min})^2 = \frac{1}{4}(2v_{ik}^2 + 2v_{i(k+1)}^2 + 4S_{ik}a_{i,max}) \\
&\Delta T_{ik,1}^{max} = \infty \\
&\Delta T_{ik,2}^{max} = \frac{(v_{ik} - v_{middle,ik}^{max})}{a_{i,max}} + \frac{(v_{i(k+1)} - v_{middle,ik}^{max})}{a_{i,max}} \\
&(v_{middle,ik}^{max})^2 = \frac{1}{4}(2v_{ik}^2 + 2v_{i(k+1)}^2 - 4S_{ik}a_{i,max}) \\
&\Delta T_{ik}^{min} = y_{ik,1} \cdot \Delta T_{ik,1}^{min} + y_{ik,2} \cdot \Delta T_{ik,2}^{min} \\
&\Delta T_{ik}^{max} = z_{ik,1} \cdot \Delta T_{ik,1}^{max} + z_{ik,2} \cdot \Delta T_{ik,2}^{max} \\
&y_{ik,1} + y_{ik,2} = 1 \qquad y_{ik,1}, y_{ik,2} \in \{0,1\} \\
&z_{ik,1} + z_{ik,2} = 1 \qquad z_{ik,1}, z_{ik,2} \in \{0,1\} \\
&\Delta T_{ik}^{max} \geq \tau_{ik} \geq \Delta T_{ik}^{min} \\
&v_{i,max} \geq v_{ik} \geq 0 \\
&v_{i,initial} = v_{i,goal} = 0 \\
&t_{ik} \geq 0 \\
&\delta_{ijkh} \in \{0,1\}
\end{aligned}
$$

This MINLP problem describing the optimal solution has difficult nonconvex constraints. Existing techniques to solve MINLPs either require convexity or are not guaranteed to find the optimal solution for large problem sizes.

We therefore solve two MILPs, which differ only in their ΔT^{max} values, to obtain good lower and upper bounds on the optimal solution. For simplicity, the velocities at the initial and goal configurations, $v_{initial}$ and v_{goal}, are assumed zero for each robot. Initially also assume that the first segment is sufficiently long for the robot to reach $v_{i,max}$ by its end, and that the last segment is sufficiently long for the robot to decelerate from $v_{i,max}$ to zero.

1. Lower bound problem: A lower bound for the MINLP problem can clearly be obtained by solving the MILP for the instantaneous model, assuming infinite acceleration. We obtain a tighter lower bound by formulating an *improved instantaneous model* that considers the time to accelerate and decelerate over the first and last segments for each robot. Since the segments are sufficiently long for the robot to go from zero to v_{max} (and

vice versa), the minimum traversal times for the first and last segments are $\Delta T^{min} = S/v_{max} + v_{max}/2a_{max}$. Solving the MILP for this improved instantaneous model gives a lower bound for the MINLP problem.

2. Upper bound problem: Here the original MINLP model is transformed into an MILP problem by setting the velocities at the endpoints of each segment (except the initial and goal configurations) to v_{max}, the highest feasible velocity given the segment lengths. Solving the MILP problem for this *setpoint model* (described in the next section) gives a feasible continuous velocity schedule that is an upper bound for the MINLP problem.

Note that under the assumption that the initial and goal configurations are collision-free, a valid solution always exists for both problems since a feasible schedule is for one robot at a time to move along its path, for some arbitrary sequencing of the robots.

6 Setpoint Model

We now describe the *setpoint model* to generate a continuous velocity schedule for a set of robots with maximum velocity $v_{i,max}$ and maximum acceleration $a_{i,max}$. We generate a continuous velocity profile for each robot by making the following additional assumption: *Each robot travels at its maximum velocity at the endpoints of each of its collision segments.* The intuition is that by setting the velocity v_{ik} at the endpoints of the collision zones to be the maximum velocity $v_{i,max}$, the robots are biased to move through their collision zones in the least time. Since any continuous velocity schedule is an upper bound on the optimal continuous velocity schedule, this solution is guaranteed to be an upper bound on the optimal solution. Further, this reduces the MINLP formulation to an MILP formulation.

6.1 MILP Formulation

Since the segment endpoint velocities are $v_{i,max}$ by the setpoint assumption, the minimum and maximum times to traverse a segment of length S_{ik} are:

$$\Delta T_{ik}^{min} = \begin{cases} S_{ik}/v_{i,max} & \text{if } k \text{ an interior segment} \\ S_{ik}/v_{i,max} + v_{i,max}/2a_{i,max} & \text{if the first or last segment} \end{cases}$$

$$\Delta T_{ik}^{max} = \begin{cases} \infty & \text{if } S_{ik} \geq v_{i,max}^2/a_{i,max} \\ \frac{2v_{i,max} - 2(v_{i,max}^2 - a_{i,max}S_{ik})^{\frac{1}{2}}}{a_{i,max}} & \text{if } S_{ik} < v_{i,max}^2/a_{i,max} \end{cases}$$

The MILP formulation for the setpoint model is identical to the MILP formulation for the improved instantaneous model, except for the difference in the ΔT^{max} values. The solution to the setpoint MILP is a continuous velocity schedule that is guaranteed to be an upper bound on the optimal continuous velocity schedule. When the segment traversal time τ_{ik} generated by the MILP does not correspond to either minimum time or maximum time trajectories over the segment, we generate a velocity profile, based on a case analysis for the double integrator, so the robot can traverse the segment in the given amount of time subject to its velocity and acceleration constraints. Note that the setpoint model is also NP-hard.

6.2 Relaxing Segment Length Constraints

We now relax the requirement that the first and last segments are sufficiently long for the robot velocity to reach v_{max} and zero respectively, by assuming that each path is long enough for the robot to reach its highest speed v_{max} at some point along the path. We then set the velocity at each segment endpoint to be the highest velocity consistent with the initial and goal velocities.

The velocity v_{ik} at the start of the kth segment, assuming robot $\mathcal{A}_i$ begins with an initial velocity of zero and ends with a goal velocity of zero, is:

$$v_{ik} = \begin{cases} \sqrt{2a_{i,max} \sum_{j=1}^{k-1} S_{ij}} & \text{if } \sum_{j=1}^{k-1} S_{ij} \leq \frac{v_{i,max}^2}{2a_{i,max}} \\ \sqrt{2a_{i,max} \sum_{j \geq k} S_{ij}} & \text{if } \sum_{j \geq k} S_{ij} \leq \frac{v_{i,max}^2}{2a_{i,max}} \\ v_{i,max} & \text{otherwise.} \end{cases}$$

Note that the velocity v_{ik} at the start of each segment is uniquely determined since the path is sufficiently long for the robot to reach its highest speed $v_{i,max}$ at some point along the path.

The corresponding bounds on the segment traversal times ΔT_{ik}^{min} and ΔT_{ik}^{max} for the setpoint model are computed from $\Delta T^{min}(v_{ik}, v_{i(k+1)})$ and $\Delta T^{max}(v_{ik}, v_{i(k+1)})$ as described in Section 5.2.

We correspondingly update the improved instantaneous model so that the velocity v_{ik} at each segment endpoint is the highest velocity physically possible, which is the corresponding setpoint velocity. For the improved instantaneous model, ΔT_{ik}^{min} takes the same values as in the setpoint model, while $\Delta T_{ik}^{max} = \infty$ since the robot can pause its motion instantaneously.

7 General Robot Models

We can apply the approach for coordinating multiple robots in this paper to general robots with linear or nonlinear dynamics if we have analytical or numerical solutions for the extremal TPBVPs for each robot over a segment. Additionally, we should be able to compute the collision zones for the robots and express the collision avoidance constraints in the MILP formulation. We briefly discuss some illustrative cases of interest.

Moving obstacles: We have applied both the instantaneous and setpoint formulations to include multiple moving obstacles with known trajectories.

Each moving obstacle is treated like a robot with a specified trajectory. The collision constraints for each obstacle are computed from its known velocity profile, and are easily added to the MILP formulations.

Car-like robots: Our approach can be applied to car-like nonholonomic robots with continuous curvature paths. Paths that satisfy the nonholonomic constraints ([25]) typically require the robot to stop whenever there is a discontinuity in curvature (to change the steering direction) or there is a cusp point (to reverse the robot direction of motion). Therefore we use continuous curvature paths for a forward moving robot (Scheuer and Fraichard [36]) so that the robot does not have to stop at curvature discontinuities. When the class of continuous curvature paths includes cusp points ([15], [22]), we add the constraint that the robot velocity must be zero at every such point.

Aircraft: To apply this approach to aircraft, we can use a simplified planar aircraft model ([5]). Since the path is specified, the linear velocity is a variable with magnitude bounds and the acceleration is the control input. As with car-like robots, we require the paths to have continuous curvature.

Articulated robots: To apply this approach to articulated robots, we must compute appropriate collision zones and solve the TPBVPs for the dynamic model of the robots. Due to the difficulty of solving the TPBVPs analytically, numerical methods can be used to approximate the optimal controls. Algorithms for minimum time control along a given path exist ([7],[39],[10]), and we expect they can be modified for maximum time control.

8 Implementation

We have implemented software in C++ to coordinate the motions of polyhedral robots with specified paths (Figure 4). We compute the collision zones using the PQP collision detection package (Larsen et al. [23]) by sampling uniformly along each robot's path. We generate the MILP formulations from the computed collision zones and solve them using the AMPL [13] and CPLEX [18] optimization packages. Since the setpoint formulation with its tighter constraints is usually solved much more quickly than the improved instantaneous formulation, we use the setpoint solution as an upper bound constraint for the improved instantaneous formulation. See Table 1 for running times measured on a Sun Ultra 60. The problem complexity depends primarily on the number of collision zones, and to a lesser extent on the number of robots. For a particularly difficult problem (for example, the symmetric radial case with a bottleneck at the center) or for a sufficiently large number of collision zones, the MILP time will dominate the running time.

In these experiments, the gap between the objective function values returned by the instantaneous and setpoint formulations was zero in all the cases, indicating that an optimal solution had been found. Example animations may be seen at `www.cs.rpi.edu/~sakella/multikino/`.

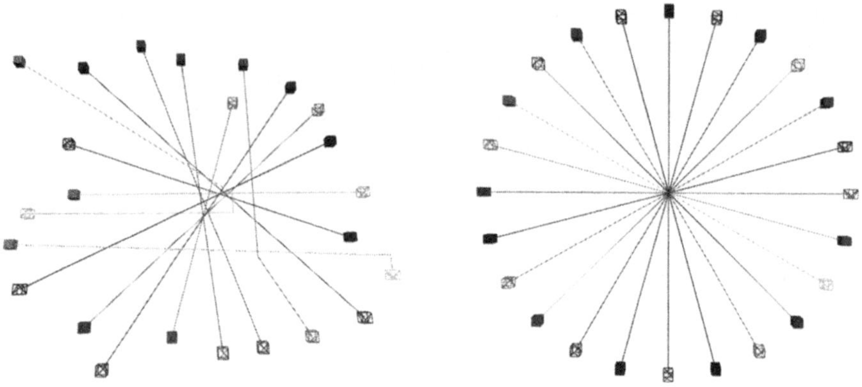

(a) (b)

Fig. 4. Overhead view of example paths for 12 robots: (a) Typical paths (b) Radial paths with symmetry, with a bottleneck at the center. Goal configurations are indicated by solid cubes.

Table 1. Sample run times for setpoint formulation (MILP-S) and improved instantaneous formulation (MILP-I). (The improved instantaneous formulation used the setpoint solutions as upper bounds.) Collision checks were performed at 200 points along each path.

Num. of robots	Num. of collision zones	Collision time (secs)	Num. of binary variables	MILP-S time (secs)	Num. of binary variables	MILP-I time (secs)
5	13	18.67	20	0.04	14	0
8	42	55.67	64	0.13	62	0.08
10	71	88.26	102	0.53	100	0.17
12	82	115.81	124	0.61	123	0.25
8 (radial, unsymmetric)	32	54.76	54	0.167	54	0.095
12 (radial, unsymmetric)	94	170.15	128	2.2	128	0.49
8 (radial, symmetric)	29	30.53	54	3.87	54	0.095
12 (radial, symmetric)	86	70.53	128	160	128	60.67

9 Conclusion

We have developed an approach to generate continuous velocity profiles for (near) minimum time collision-free coordination of multiple robots with kinodynamic constraints along specified paths. The principal advantage of our MILP formulations is that they permit the collision-free coordination of a large number of robots. The MILP formulations for coordination of multiple robots are NP-hard, and their complexity increases directly with the number of potential collisions. However efficient collision detection software and integer programming solvers make this approach practical for reasonable problem sizes.

There are several directions for future work. Analytically characterizing the gap between the improved instantaneous model and setpoint model would indicate the optimality of the generated solutions. Developing polynomial time approximation or randomization algorithms to schedule robots can be useful in generating bounds for the MILP formulations. Developing computational procedures to solve the continuous velocity MINLP formulation is important. Demonstrating the approach on robots with more complex dynamics, such as car-like mobile robots, aircraft, and articulated robots, is an interesting next step since the complex constraints on the state and control variables make solving the TPBVPs for individual segments challenging. Automatic modification of robot paths to reduce completion time would be a useful extension. Another interesting direction is online coordination of multiple robots using sensor-based estimates of robot positions and velocities.

Acknowledgments.
This work was supported in part by RPI and by NSF under CAREER Award No. IIS-0093233. Andrew Andkjar implemented animation software that interfaced with the PQP software from the University of North Carolina and helped generate examples. Thanks to Seth Hutchinson for early discussions, to Prasad Akella for suggesting the problem, and John Mitchell for advice. Thanks also to the reviewers for helpful suggestions.

References

1. S. Akella and S. Hutchinson. Coordinating the motions of multiple robots with specified trajectories. In *IEEE International Conference on Robotics and Automation*, pages 624–631, Washington, DC, May 2002.
2. R. Alami, S. Fleury, M. Herrb, F. Ingrand, and F. Robert. Multi-robot cooperation in the MARTHA project. *IEEE Robotics and Automation Magazine*, 5(1):36–47, Mar. 1998.
3. B. Aronov, M. de Berg, A. F. van der Stappen, P. Svestka, and J. Vleugels. Motion planning for multiple robots. *Discrete and Computational Geometry*, 22:505–525, 1999.
4. J. Barraquand, B. Langlois, and J.-C. Latombe. Numerical potential field techniques for robot path planning. *IEEE Transactions on Systems, Man, and Cybernetics*, 22(2):224–241, 1992.
5. A. Bicchi and L. Pallottino. On optimal cooperative conflict resolution for air traffic management systems. *IEEE Transaction on Intelligent Transportation Systems*, 1(4):221–231, Dec. 2000.
6. Z. Bien and J. Lee. A minimum time trajectory planning method for two robots. *IEEE Transactions on Robotics and Automation*, 8(3):414–418, June 1992.

7. J. E. Bobrow, S. Dubowsky, and J. S. Gibson. Time-optimal control of robotic manipulators along specified paths. *International Journal of Robotics Research*, 4(3):3–17, Fall 1985.
8. A. E. Bryson and Y.-C. Ho. *Applied Optimal Control.* Hemisphere, Washington D.C., 1975.
9. C. Chang, M. J. Chung, and B. H. Lee. Collision avoidance of two robot manipulators by minimum delay time. *IEEE Transactions on Systems, Man, and Cybernetics*, 24(3):517–522, Mar. 1994.
10. Y. Chen and A. A. Desrochers. A proof of the structure of the minimum-time control law of robotic manipulators using a Hamiltonian formulation. *IEEE Transactions on Robotics and Automation*, 6(3):388–393, June 1990.
11. B. R. Donald, P. Xavier, J. Canny, and J. Reif. Kinodynamic motion planning. *Journal of the ACM*, 40(5):1048–1066, Nov. 1993.
12. M. Erdmann and T. Lozano-Perez. On multiple moving objects. *Algorithmica*, 2(4):477–521, 1987.
13. R. Fourer, D. M. Gay, and B. W. Kernighan. *AMPL: A Modeling Language for Mathematical Programming.* Duxbury Press, 1993.
14. T. Fraichard. Trajectory planning in dynamic workspace: a 'state-time space' approach. *Advanced Robotics*, 13(1):75–94, 1999.
15. T. Fraichard, A. Scheuer, and R. Desvigne. From Reeds and Shepp's to continuous-curvature paths. In *Proceedings of the International Conference on Advanced Robotics*, pages 585–590, Tokyo, Japan, Oct. 1999.
16. J. E. Hopcroft, J. T. Schwartz, and M. Sharir. On the complexity of motion planning for multiple independent objects: PSPACE-hardness of the "warehouseman's problem". *International Journal of Robotics Research*, 3(4):76–88, 1984.
17. D. Hsu, R. Kindel, J.-C. Latombe, and S. Rock. Randomized kinodynamic motion planning with moving obstacles. In B. R. Donald, K. M. Lynch, and D. Rus, editors, *Algorithmic and Computational Robotics: New Directions*, pages 247–264. A. K. Peters, Natick, Massachusetts, 2001.
18. ILOG, Inc., Incline Village, NV. *CPLEX 6.0 Documentation Supplement*, 1998.
19. K. Kant and S. W. Zucker. Toward efficient trajectory planning: The path-velocity decomposition. *International Journal of Robotics Research*, 5(3):72–89, Fall 1986.
20. G. Knowles. *An Introduction to Applied Optimal Control.* Mathematics in Science and Engineering, Volume 159. Academic Press, New York, 1981.
21. J. Kosecka, C. Tomlin, G. Pappas, and S. Sastry. Generation of conflict resolution maneuvers for air traffic management. In *IEEE/RSJ International Conference on Robots and Systems*, pages 1598–1603, Grenoble, France, 1997.
22. F. Lamiraux and J.-P. Laumond. Smooth motion planning for car-like vehicles. *IEEE Transactions on Robotics and Automation*, 17(4):498–502, Aug. 2001.
23. E. Larsen, S. Gottschalk, M. Lin, and D. Manocha. Fast distance queries using rectangular swept sphere volumes. In *IEEE International Conference on Robotics and Automation*, pages 3719–3726, San Francisco, CA, Apr. 2000.
24. J.-C. Latombe. *Robot Motion Planning.* Kluwer Academic Publishers, Norwell, MA, 1991.
25. J.-P. Laumond, editor. *Robot Motion Planning and Control.* Lecture Notes in Control and Information Sciences, No. 229. Springer, 1998.

26. S. M. LaValle and S. A. Hutchinson. Optimal motion planning for multiple robots having independent goals. *IEEE Transactions on Robotics and Automation*, 14(6):912–925, Dec. 1998.
27. S. M. LaValle and J. J. Kuffner. Randomized kinodynamic planning. *International Journal of Robotics Research*, 20(5):378–401, May 2001.
28. G. L. Nemhauser and L. A. Wolsey. *Integer and Combinatorial Optimization.* John Wiley and Sons, New York, 1988.
29. P. A. O'Donnell and T. Lozano-Perez. Deadlock-free and collision-free coordination of two robot manipulators. In *IEEE International Conference on Robotics and Automation*, pages 484–489, Scottsdale, AZ, May 1989.
30. L. Pallottino, E. Feron, and A. Bicchi. Conflict resolution problems for air traffic management systems solved with mixed integer programming. *IEEE Transactions on Intelligent Transportation Systems*, 3(1):3–11, Mar. 2002.
31. M. Pinedo. *Scheduling: Theory, Algorithms, and Systems.* Prentice-Hall, Englewood Cliffs, NJ, 1995.
32. J. Reif and M. Sharir. Motion planning in the presence of moving obstacles. In *Proceedings of the 26th Annual Symposium on the Foundations of Computer Science*, pages 144–154, Portland, Oregon, Oct. 1985.
33. A. A. Rizzi, J. Gowdy, and R. L. Hollis. Distributed coordination in modular precision assembly systems. *International Journal of Robotics Research*, 20(10):819–838, Oct. 2001.
34. G. Sahar and J. M. Hollerbach. Planning of minimum-time trajectories for robot arms. *International Journal of Robotics Research*, 5(3):97–140, Fall 1986.
35. G. Sanchez and J. Latombe. On delaying collision checking in PRM planning — application to multi-robot coordination. *International Journal of Robotics Research*, 21(1):5–26, Jan. 2002.
36. A. Scheuer and T. Fraichard. Continuous-curvature path planning for car-like vehicles. In *Proc. of the IEEE-RSJ Int. Conf. on Intelligent Robots and Systems, volume 2*, pages 997–1003, Grenoble, France, Sept. 1997.
37. T. Schouwenaars, B. D. Moor, E. Feron, and J. How. Mixed integer programming for multi-vehicle path planning. In *European Control Conference 2001*, Porto, Portugal, 2001.
38. Z. Shiller and S. Dubowsky. On computing the global time-optimal motion of robotic manipulators in the presence of obstacles. *IEEE Transactions on Robotics and Automation*, 7(6):785–797, Dec. 1991.
39. K. G. Shin and N. D. McKay. Minimum-time control of robotic manipulators with geometric path constraints. *IEEE Transactions on Automatic Control*, AC-30(6):531–541, June 1985.
40. K. G. Shin and Q. Zheng. Minimum-time collision-free trajectory planning for dual robot systems. *IEEE Transactions on Robotics and Automation*, 8(5):641–644, Oct. 1992.
41. T. Simeon, S. Leroy, and J.-P. Laumond. Path coordination for multiple mobile robots: A resolution-complete algorithm. *IEEE Transactions on Robotics and Automation*, 18(1):42–49, Feb. 2002.
42. C. Tomlin, G. J. Pappas, and S. Sastry. Conflict resolution for air traffic management: A study in multi-agent hybrid systems. *IEEE Transactions on Automatic Control*, 43(4):509–521, Apr. 1998.
43. P. Švestka and M. H. Overmars. Coordinated motion planning for multiple car-like robots using probabilistic roadmaps. In *IEEE International Conference on Robotics and Automation*, pages 1631–1636, Nagoya, Japan, May 1995.

Hybrid Motion Planning: Coordinating Two Discs Moving among Polygonal Obstacles in the Plane*

Shai Hirsch and Dan Halperin

School of Computer Science, Tel Aviv University, Tel Aviv, Israel 69978
{shaihi,danha}@cs.tau.ac.il

Abstract. The basic motion-planning problem is to plan a collision-free motion for an object moving among obstacles between free initial and goal positions, or to determine that no such motion exists. The basic problem as well as numerous variants of it have been intensively studied over the past two decades yielding a wealth of results and techniques, both theoretical and practical. In this paper, we propose a novel approach to motion planning, *hybrid motion planning*, in which we integrate complete solutions along with probabilistic roadmap (PRM) methods in order to combine their strengths and offset their weaknesses. We incorporate robust tools, that have not been available before, in order to implement the complete solutions. We exemplify our approach in the case of two discs moving among polygonal obstacles in the plane. The planner we present easily solves problems where a narrow passage in the workspace can be arbitrarily small. Our planner is also capable of providing correct nontrivial "no" answers, namely it can, for some queries, detect the situation where no solution exists. We envision our planner not as a total solution but rather as a new tool that cooperates with existing planners. We demonstrate the advantages and shortcomings of our planner with experimental results.

1 Introduction

In this paper we study the motion-planning problem for two discs moving among polygonal obstacles in the plane. This is one of the many extensions to the basic motion planning problem in which only one robot is concerned. Given an *initial* and *goal* configurations, the planner has to plan a *collision-free path* between these configurations, or to determine that no such motion exists. We follow the standard practice to study the motion-planning problem in terms of the *configuration space (C-space)*, the space of all possible configurations of the robot. Workspace obstacles transform into obstacles in the

* Work reported in this paper has been supported in part by the IST Programme of the EU as a Shared-cost RTD (FET Open) Project under Contract No IST-2000-26473 (ECG – Effective Computational Geometry for Curves and Surfaces), by The Israel Science Foundation founded by the Israel Academy of Sciences and Humanities (Center for Geometric Computing and its Applications), and by the Hermann Minkowski–Minerva Center for Geometry at Tel Aviv University.

J.-D. Boissonnat et al. (Eds.): Algorithmic Foundations of Robotics V, STAR 7, pp. 239–255, 2004.

configuration space, or *C-obstacles*. Inter-robot collisions induce additional C-obstacles. We refer to inter-robot collisions as *coordination collisions* to differentiate them from obstacle-robot collisions. A motion of the robot system corresponds to a curve in C-space. A motion is collision-free if the corresponding curve does not intersect any C-obstacle, that is, it lies completely within the free part of C-space. Motion planning has been proven a hard problem (see, e.g., [21]) and there is strong evidence that its solution requires exponential time in the number of dofs (degrees of freedom, the dimension of C-space). See [18] for more information on motion planning.

1.1 Complete Solutions: Theory and Practice

In robot motion planning we consider an algorithm to be *complete* for a problem if it is guaranteed, for all instances of the problem, to find a solution when one exists and to return failure otherwise. General complete solutions are exponential in the number k of dofs [6]. A lot of effort was invested in giving worst-case near-optimal solutions to motion-planning problems with a small number of dofs [23].

Until recently, complete algorithms have rarely been implemented and some were implemented over a decade ago (e.g., [2]). The task is far from trivial even for motion-planning problems with a small number of dofs. Notice however, that a genuinely *complete implementation*, namely an implementation that can correctly cope with *all* instances of the problem, requires the usage of special arithmetic in order to deal with arbitrary input and in particular to handle tight or narrow passages for the robot in the workspace. We will refer to a complete implementation of a complete motion-planning algorithm as a *complete solution*. The difficulties in implementing complete solutions arise in the implementation of geometric algorithms in many other areas as well. This has led to a major effort by several research groups in computational geometry to develop the infrastructure for implementation of such algorithms. One such effort is the CGAL library of geometric data structures and algorithms [7].

Within CGAL there are means to construct and manipulate *arrangements* of curves in the plane [11], namely the subdivision of the plane into vertices, edges, and two-dimensional faces induced by a collection of curves. Arrangements can serve as a key ingredient in complete solutions to motion-planning problems with two dofs (this connection as well as the underlying software support for it are described in [13]). On top of the arrangement package we have built a complete solution to the polygonal translation problem [10] and to the problem of a disc moving among polygonal obstacles [15]. We shall be using the latter in the novel hybrid solution that we present in Sect. 3 and we shall borrow ideas from the complete solution for two discs [24].

1.2 Probabilistic Roadmaps

In the face of the prohibitive complexity of complete general planners and the difficulty in implementing them other research directions were sought. Among these, the introduction of probabilistic solutions [3,17] proved as an important advance in the field of motion planning. Probabilistic solutions trade a limited amount of completeness against a gain in computing efficiency. In addition, they are much simpler than complete solutions, easier to implement, and easily adaptable to different settings, including problems with many dofs. The form of completeness that describes this family of solutions is referred to as *probabilistic completeness*: if a solution path exists, the planner finds one in bounded time with high probability.

A class of probabilistic solutions that has been intensively studied and applied is *probabilistic roadmaps (PRMs)*, which consists of sampling the configuration space at random, retaining the free samples as graph nodes, and connecting these nodes by a "local" planner to form a *connectivity graph* (the roadmap). In PRMs, the prohibitive computation of an explicit representation of the free configuration space is abandoned altogether. Instead, the free configuration space is implicitly provided by some local function, e.g., collision detection or distance measurement. The main drawback of PRMs is their difficulty to solve cases where the configuration space contains narrow passages. Many variants of the basic PRM scheme have been proposed in order to overcome this hurdle [1,5,16]. LaValle and Branicky have developed deterministic variants of the PRM that use quasi-randomized sampling [19]. They show that these variants offer performance advantages in comparison with common PRMs. All advanced variants work well on certain types of problems but unfortunately all fail on others. A synthesis is proposed in [8], where a *meta-planner* breaks down the original problem into subproblems, and matches appropriate different motion planner to each of them. The reader is referred to a comparative study of PRMs [12]. We remark that no matter how successful PRMs are, when such a planner fails to find a path, it is not certain whether no path exists, or that the planner simply has not adequately sampled the free configurations space. It is also possible that the time limit to which a PRM is being confined is insufficient.

1.3 Multi-robot Motion Planning

A natural extension of the basic motion-planning problem is the multi-robot variant, which ranges from the two-robot case to scenarios with giant fleets of moving objects. In *centralized* motion planners, planning is performed concurrently for all the robots in a composite configuration space, offering complete solutions, which are however exponential in the dimension of the composite configuration space. *Decoupled* solutions avoid this high complexity by first computing a path for each of the robots independently. Then, the different paths are coordinated in an attempt to avoid mutual collisions among the robots. Since deadlock situations are possible such methods are inherently incomplete. See [18] for details.

Sharir and Sifrony [24] described a complete $O(n^2)$ solution for two independent robots moving among polygonal obstacles in the plane that is a special case of the complete algorithm to the general motion planning problem described in [22]. Our solution is based in part on ideas from this algorithm, ideas that are briefly presented in Sect. 2.

Švestka and Overmars describe a roadmap-coordination method [26], i.e., of robot coordination along independent roadmaps rather than paths. They go a step further and reduce that size of the graph that represents the coordination of the robots, by distinguishing between situations where the robots are located near each other, and situations where this is not the case. In this paper we apply similar ideas. Our planner identifies subregions of the free configuration space where mutual collision are guaranteed not to exist, and coordination is hence not required, even prior to sampling.

1.4 Our Results

Even as PRMs become more sophisticated it is clear that between these methods and complete solutions there is still a wide gap. This suggests the possibility of a further synergy between the two approaches. Such a combination has been hard to achieve until recently because, as explained in Sect. 1.1, complete solutions require special arithmetic along with robust geometric tools for them to be widely used in practice. As such tools are now available, e.g., as part of CGAL [7], we believe that there is a place for a new type of planners that try to integrate complete solutions with PRM techniques. In other words, there is a whole spectrum of potential solutions that incorporate the seemingly two extreme approaches, solutions that would combine the strengths of both approaches and offset their weaknesses.

The hybrid approach we suggest is to break an existing problem into simpler subproblems (e.g., with less dofs or with simpler geometry), such that a composition of solutions to the subproblems is a solution to the original one. We present a specific example of this hybrid approach in the form of a planner for the coordination of two discs moving among polygonal obstacles in the place. We refer to the hybrid motion planner as *HyMP*. Instead of having an explicit representation of $\mathcal{C}_{\text{free}}$ or none at all, HyMP computes two explicit representations, $\mathcal{C}^-_{\text{free}}$ *for a subset* of $\mathcal{C}_{\text{free}}$ and $\mathcal{C}^+_{\text{free}}$ *for a superset* thereof (which is smaller than the entire configuration space $\mathcal{C}$). These explicit partial representations, as we refer to them, provide HyMP with a complete solution for an important subset of the possible queries. We emphasize that we do not give a complete solution of the two-disc problem, as we avoid the most complex stage of complete solutions, namely computing an explicit representation of $\mathcal{C}_{\text{free}}$. In order to answer the full set of queries HyMP incorporates a PRM within $\mathcal{C}^+_{\text{free}} \setminus \mathcal{C}^-_{\text{free}}$. This PRM has to cope with a subproblem that is smaller and simpler than the original one. Furthermore, it is aided by information derived from the explicit partial representations. The probabilistic part of HyMP is interesting in its own right and, for clarity, we will denote it by hPRM. Unlike PRMs, a complete solution for a subset of the queries enables our planner to result with correct nontrivial "no" answers (in fact, *disconnection proofs*, a term we borrow from [4]) rather than "failure" for a subset of the queries. We also show experimental results that demonstrate how HyMP easily solves previously hard problems.

HyMP is not a total solution that replaces existing ones. Beyond its evident advantages, our solution has disadvantages. Both are described in Sect. 4. We consider HyMP as a novel tool in the accumulating motion-planning toolbox and we envision it as part of a meta-planner framework, like the one presented in [8].

The rest of the paper is organized as follows. We begin by presenting the setting of our study in Sect. 2. In Sect. 3, we give a taxonomy of configurations that leads to the introduction of the general scheme of the hybrid planner, and we mention some implementation details. The advantages and disadvantages of the planner along with experimental results are described in Sect. 4. Suggestions for future research directions are given in Sect. 5.

2 Preliminaries

2.1 The Setting

The moving objects (robots) considered in this paper are two planar discs $\mathcal{D}_1$ and $\mathcal{D}_2$ of radii r_1 and r_2, respectively. The discs are free to move in a planar rectangular workspace $\mathcal{W}$, which contains N polygonal obstacles, $\mathcal{P}_1, \ldots, \mathcal{P}_N$, with a total of n edges, as long as the discs *do not* collide with the obstacles (the complement of the bounding rectangle of $\mathcal{W}$ is regarded as another obstacle) or with each other. A configuration for which a robot interior overlaps the interior of an obstacle, or disc interiors overlap is *forbidden*. All non-forbidden configurations are *semi-free*, but only semi-free configurations in which no robot touches an obstacles and the robots do not touch one another are considered *free*. Semi-free configurations that are not free are *boundary* configurations. In this paper we do not assume general position and, as a rule, allow boundary configurations save for a small fraction thereof. We exclude boundary configurations for which at least one robot touches more than one obstacle edge or vertex simultaneously. In other words, we avoid obstacle-robot tight passages. This constraint serves to simplify both the description of the planner as well as its implementation. The tools we used themselves allow the inclusion of tight passages (as implemented for one disc in [15]). We represent configurations using four-tuples (x_1, y_1, x_2, y_2), where $p_i = (x_i, y_i)$, $i = 1, 2$ is the placement of the center of the disc $\mathcal{D}_i$ in $\mathcal{W}$.

The configuration space $\mathcal{C}$ is four-dimensional. We let $\mathcal{C}_{\text{free}}$ denote the (closed) subset of (collision-)free configurations and let $\mathcal{C}_{\text{forb}}$ denote the subset of forbidden configurations.

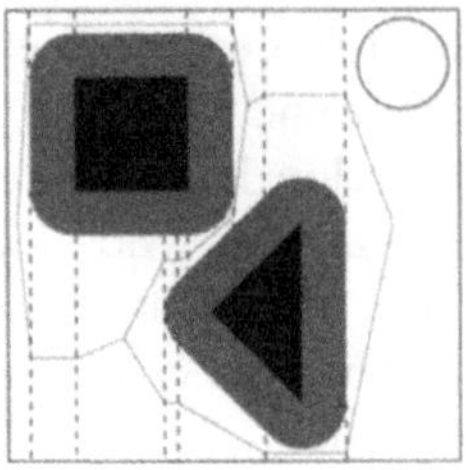

Fig. 1. Motion planning for one disc: roadmap embedding in the trapezoidal decomposition of a configurations space; for clarity the obstacles (black polygons) and the disc are shown.

2.2 Complete Solution for One or Two Discs

Our solution relies on a complete solution for the problem of one disc moving among polygonal obstacles, and borrows ideas from a complete solution for the coordination of two discs. We briefly sketch each of these solutions.

Complete Solution for One Disc A motion planner for a polygonal robot among polygonal obstacles is presented in [9]. We use an adaptation of this $O(n \log n)$ algorithm for a disc robot, which is presented in [15].

The planner begins with computing a representation of the $2D$ free configuration space $\mathcal{C}_{\text{free}}$ of the robot. The boundary of $\mathcal{C}_{\text{free}}$ consists of line segments and circular arcs. The planner continues by vertically decomposing $\mathcal{C}_{\text{free}}$. The result is a collection of (pairwise openly disjoint) simple connected cells. Each cell has a trapezoid-like shape with two x-monotone bases, any of which may be either a straight-line segment or a circular arc. We refer to these cells as *trapezoids.* If the upper or lower base of a trapezoid is a circular arc then the trapezoid is non-convex, because the convex side of a circular arc is always part of a C-obstacle. One of the vertical attachments of a trapezoid may collapse to a point. We say that two trapezoids are *adjacent* if they meet along a vertical attachment.

Next, the planner constructs a roadmap which reflects the connectivity of $\mathcal{C}_{\text{free}}$. Roughly, the roadmap nodes are the trapezoids and nodes of adjacent trapezoids are connected with an edge. This roadmap is embedded in the plane so that it represents free paths for the robots. Additional intermediate nodes may be added to the roadmap so that its embedding would consist of line segments alone. One of the required adaptations of the planner to the case of the disc robot is to ensure that this rule holds for non-convex trapezoids [15]. Given a pair of query configurations, there is a path connecting them in $\mathcal{C}_{\text{free}}$ if and only if the nodes of their containing trapezoids belong to the same connected component of the roadmap.

Complete Solution for Two Discs An efficient complete algorithm for two discs running in $O(n^2)$ time is described in [24]. The algorithm is based on decomposing $\mathcal{C}_{\text{free}}$ into a collection of simple connected $4D$ cells, such that each cell has $O(1)$ complexity. Then, each cell is repartitioned by $O(1)$ *critical facets.* By determining adjacency of pairs of these subcells, a discrete combinatorial representation of $\mathcal{C}_{\text{free}}$ in the form of a connectivity graph is obtained.

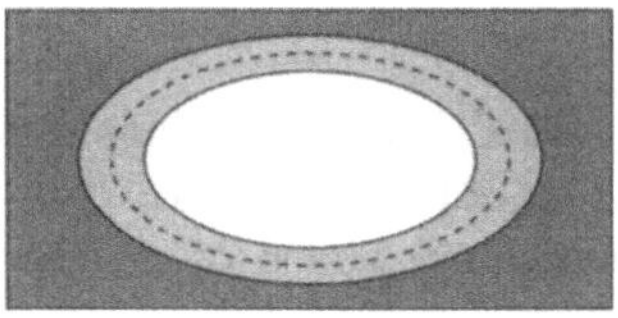

Fig. 2. Configuration space partition: The interior of the dashed ellipse is free space. The dark gray area is $\mathcal{C}^-_{\text{forb}}$ and the white area is $\mathcal{C}^-_{\text{free}}$, both whose representation is explicit. The light gray annulus, which stands for $\mathcal{C}^\Delta_{\text{free}} \cup \mathcal{C}^\Delta_{\text{forb}}$, is the domain of the probabilistic planner.

The initial cell decomposition is based on the vertical decomposition of the free $2D$ configuration space of each disc. As mentioned above, software tools are now available to maintain the arrangements representing these underlying $2D$ configuration spaces. In this paper we perform only the initial cell decomposition but avoid the costly computation of critical facets. The initial decomposition supplies our planner with useful albeit partial information regarding the connectivity of $\mathcal{C}_{\text{free}}$.

3 The Hybrid Planner

3.1 Taxonomy of Configurations

We differentiate between *obstacle collisions*, where one of the robots collides with some obstacle and *coordination collisions*, where each robot is disjoint from the obstacles but the robots overlap. We refer to configurations that represent such collisions as *obstacle-forbidden* and *coordination-forbidden*, respectively. (Obstacle-forbidden configurations may represent cases where the robots overlap, in addition to the obstacle collision, but we do not consider these cases as coordination collisions.) Let $\mathcal{C}^-_{\text{forb}}$ and $\mathcal{C}^\Delta_{\text{forb}}$ denote the sets of obstacle-forbidden configurations and coordination-forbidden configurations, respectively.

Let us partition $\mathcal{C}_{\text{free}}$ as well. Consider $\overline{\mathcal{C}^-_{\text{forb}}}$, the complement of $\mathcal{C}^-_{\text{forb}}$. $\overline{\mathcal{C}^-_{\text{forb}}}$ equals $\mathcal{C}_{\text{free}} \cup \mathcal{C}^\Delta_{\text{forb}}$, i.e, it includes configurations that are either free or coordination-forbidden. $\overline{\mathcal{C}^-_{\text{forb}}}$ is a superset of $\mathcal{C}_{\text{free}}$ and we use the shorthand $\mathcal{C}^+_{\text{free}}$ to refer to it. Let $C_1, \ldots, C_M$ be a disjoint finite partition of $\mathcal{C}^+_{\text{free}}$ into connected cells. (We will shortly propose such a partition.) Some of these cells may contain only free configurations, while others may contain coordination-forbidden configurations as well. We denote by $\mathcal{C}^-_{\text{free}}$ the union of all the cells C_i that contain only free configurations. If q_{init} and q_{goal} are both in the same connected component of $\mathcal{C}^-_{\text{free}}$, then it is evident there is a solution to the motion-planning problem, although solutions may be possible even if the query configurations are in different connected component of $\mathcal{C}^-_{\text{free}}$. We denote by $\mathcal{C}^\Delta_{\text{free}}$ the rest of the cells C_i, namely $\mathcal{C}_{\text{free}} \backslash \mathcal{C}^-_{\text{free}}$. The result of the partitioning is that the full configuration space $\mathcal{C}$ of our problem is the disjoint union $\mathcal{C}^-_{\text{free}} \cup \mathcal{C}^\Delta_{\text{free}} \cup \mathcal{C}^-_{\text{forb}} \cup \mathcal{C}^\Delta_{\text{forb}}$. Fig. 2 schematically illustrates this partition.

3.2 General Scheme

HyMP computes explicit representations of $\mathcal{C}^+_{\text{free}}$ and $\mathcal{C}^-_{\text{free}}$, but it does not compute an explicit representation of $\mathcal{C}_{\text{free}}$. The planner begins with computing $\mathcal{C}^1_{\text{free}}$ and $\mathcal{C}^2_{\text{free}}$, where $\mathcal{C}^i_{\text{free}}$ is the two-dimensional space of all free placements of $\mathcal{D}_i$, ignoring the presence of the other disc. Let $\{c^1_1,\ldots,c^1_{m_1}\}$ and $\{c^2_1,\ldots,c^2_{m_2}\}$ be the trapezoids in the vertical decompositions for $\mathcal{C}^1_{\text{free}}$ and $\mathcal{C}^2_{\text{free}}$, respectively. For each $1 \le i \le m_1$ and $1 \le j \le m_2$, let $C_{i,j}$ denote the Cartesian product $c^1_i \times c^2_j$. Obviously, $C_{i,j}$ is the free configuration space of the combined system $\mathcal{D}_1$ and $\mathcal{D}_2$ when moving in the presence of only $O(1)$ obstacles, or, more precisely, subject to only $O(1)$ collision constraints, *ignoring coordination collisions*. Each $C_{i,j}$ is a connected four-dimensional region that we call a 4*D cell*, or *cell* for short. By definition, we obtain the following lemma.

Lemma 1. $\mathcal{C}^+_{\text{free}} = \bigcup_{i=1}^{m_1} \bigcup_{j=1}^{m_2} C_{i,j}$.

$\mathcal{C}^+_{\text{free}}$ is the result of "gluing" all these 4D cells together. A pair of 4D cells $C_{i,j}$ and $C_{k,l}$ are *adjacent* if and only if one of the following conditions hold: (**1**) $c^1_i = c^1_k$ and c^2_j and c^2_l are adjacent. (**2**) $c^2_j = c^2_l$ and c^1_i and c^1_k are adjacent. (**3**) both c^1_i and c^1_k and c^2_j and c^2_l are adjacent in their respective 2D configuration space. The first two entries describe cases where $\mathcal{D}_1$ (resp. $\mathcal{D}_2$) moves within the same trapezoid, while $\mathcal{D}_2$ (resp. $\mathcal{D}_1$) crosses over a vertical attachment from one trapezoid to an adjacent trapezoid. The third entry describes the case where both discs cross over from one trapezoid to an adjacent one.

Let us look at a 4D cell $C_{i,j}$, that describes all possible configurations where $\mathcal{D}_1$ (resp. $\mathcal{D}_2$) is in c^1_i (resp. c^2_j). Assume that we embed the two-dimensional configuration space of each disc in the workspace in the obvious manner (i.e., a configuration is mapped to the location of the center of the disc). Let $\mathcal{X} \oplus \mathcal{Y}$ denote the Minkowski sum of the sets $\mathcal{X}$ and $\mathcal{Y}$, that is $\mathcal{X} \oplus \mathcal{Y} = \{x + y \mid x \in \mathcal{X},\ y \in \mathcal{Y}\}$. The following lemma gives a sufficient condition for a 4D cell to contain only free configurations.

Lemma 2. *$C_{i,j}$ contains only free configurations if $c^1_i \oplus \mathcal{D}_1$ and $c^2_j \oplus \mathcal{D}_2$ are disjoint.*

We define $\mathcal{C}^-_{\text{free}}$ to be the union of 4D cells that fulfill the condition of Lemma 2 and hence contain only free configurations.

HyMP computes two (undirected) connectivity graphs. The connectivity graph $\mathcal{G}^+$ is built to reflect the connectivity of $\mathcal{C}^+_{\text{free}}$. The nodes of $\mathcal{G}^+$ are the 4D cells. The graph contains an edge connecting a pair of 4D cells if these cells are adjacent. In addition, a roadmap $\mathcal{G}$ is constructed in two parts. Within $\mathcal{C}^-_{\text{free}}$ no coordination collisions are possible, so we apply the exact solution for one disc within each of the appropriate regions of the underlying 2D configuration spaces. This results in two roadmaps for each of the robots independently. We then perform what is, in general, a Cartesian product of the two independent roadmaps. hPRM, the PRM within HyMP completes the roadmap $\mathcal{G}$ by sampling and connecting configurations within $\mathcal{C}^{\Delta}_{\text{free}} \cup \mathcal{C}^{\Delta}_{\text{forb}}$. In the course of the algorithm the two parts of $\mathcal{G}$ are glued to one another. We next present hPRM and this gluing process.

Sampling Scheme The default sampling scheme is simple: We sample a constant number k_1 of configurations in every $4D$ cell of $\mathcal{C}^{\Delta}_{\text{free}} \cup \mathcal{C}^{\Delta}_{\text{forb}}$. Clearly, it is not possible for a sampled configuration to be obstacle-forbidden.

Collision Detection and Simple Local Planning Given a configuration q in $\mathcal{C}^{\Delta}_{\text{free}} \cup \mathcal{C}^{\Delta}_{\text{forb}}$, collision detection amounts to an invocation of a predicate, that answers true if and only if the discs overlap. Given a pair of configurations p, q in $\mathcal{C}^{\Delta}_{\text{free}} \cup \mathcal{C}^{\Delta}_{\text{forb}}$, the simplest local planner tries to connect these configurations along a straight line $\overline{pq}$. Within hPRM, we invoke this primitive only for pairs p, q contained in the same $4D$ cell. More complex connection attempts are reduced to a set of consecutive invocations of this primitive. Since a $4D$ cell has $O(1)$ complexity, and since its shape is simple and well characterized, we gain an instantaneous connection primitive. In turn, this enables us to invoke this primitive many times without sacrificing efficiency.

In the case of two discs the straight-line local planner may prove too crude. We examine the different types of configurations to refine it. The path $\overline{pq}$ may be invalid because it contains obstacle-forbidden or coordination-forbidden configurations. The explicit representations of $\mathcal{C}^{1}_{\text{free}}$ and $\mathcal{C}^{2}_{\text{free}}$ allow us to break the connection attempt between p and q into three questions via two simpler predicates. For each disc we independently check if the movement is free of obstacles. In addition, we verify that $\mathcal{D}_1$ and $\mathcal{D}_2$ do not overlap along the path. Thanks to the explicit representations of $\mathcal{C}^{1}_{\text{free}}$ and $\mathcal{C}^{2}_{\text{free}}$, a single computation is sufficient to answer whether the path of each of the discs is free in its $2D$ configuration space in comparison with the need to check many consecutive configuration at some resolution along the path in common PRMs. If there are no obstacle-forbidden configurations along $\overline{pq}$ but the discs overlap along it, we further check if it is possible to coordinate the robots so that one waits to the other and only then moves.

Roadmap Connection Similar to OBPRM [1], there are three general stages in our PRM connection strategy, where each stage is composed of sub-stages of its own. The first stage (In-Cell Connection) attempts to connect configurations sampled within a $4D$ cell into a local connectivity graph. The second stage (Inter-Cell Connection) attempts to connect local graphs of adjacent $4D$ cells. The third stage (Stitching) improves the connectivity of the roadmap by attempting to join different connected components of the initial roadmap.

1. In-Cell Connection: This step (along with the sampling stage) is a set of simplified motion-planning problems within every $4D$ cell. Each cell is of $O(1)$ complexity, it contains only free or coordination-forbidden configurations, and it has a simple and well characterized shape. Therefore, a modest

sampling scheme and a local planner are sufficient to grasp the connectivity of a cell in most cases. This is backed up by our experimental results. Let $\mathcal{G}_{i,j}$ denote the connectivity graph of the cell $\mathcal{C}_{i,j}$. (The planner can be further enhanced by applying sophisticated sampling schemes within $4D$ cells, e.g., OBPRM-like sampling [1], or quasi-randomized sampling [19].)

2. INTER-CELL CONNECTION: The second stage goes over all pairs of adjacent $4D$ cells $\mathcal{C}_{i,j}$ and $\mathcal{C}_{k,l}$ and attempts to connect the connected components of $\mathcal{G}_{i,j}$ with the connected components of $\mathcal{G}_{k,l}$. As in Stage 1, the problem at this stage is also simple. The additional difficulty is due to the fact that pairs of adjacent $4D$ cells can form more complicated shapes than one cell. Given two configurations $p \in \mathcal{C}_{i,j}$ and $q \in \mathcal{C}_{k,l}$, we find an intermediate free configuration r on the common boundary (facet) between the two cells and we try to connect both p and q to r. Each of the connecting paths in these attempts lie entirely within the closure of a $4D$ cell and therefore we can use the same simple local planning primitive as in in-cell connection. If one of the $4D$ cells $\mathcal{C}_{i,j}$ is in $\mathcal{C}^-_{\text{free}}$ then connecting $\mathcal{G}_{i,j}$ to the configuration on the common boundary is easily achieved deterministically by applying the solution of one robot for each of the discs.

3. STITCHING: At each iteration of this stage, the planner takes pairs of connected components $\mathcal{V}_i$ and $\mathcal{V}_j$ of the roadmap constructed so far and tries to "stitch" them into a single connected component in order to improve the overall connectivity of the roadmap. The motivation is to find discrepancies between the connectivity of $\mathcal{C}_{\text{free}}$ and the possible motions as represented in $\mathcal{G}$. Having no explicit representation of $\mathcal{C}_{\text{free}}$, the planner takes advantage of $\mathcal{C}^+_{\text{free}}$ to predict such possible discrepancies. For each pair of connected components the planner picks up different pairs of representative configurations $p \in \mathcal{V}_i$, $q \in \mathcal{V}_j$. Next, the planner finds the $4D$ cell $\mathcal{C}_p$ (resp. $\mathcal{C}_q$) that contains p (resp. q). The planner tries to find a path in $\mathcal{G}^+$ that connects $\mathcal{C}_p$ and $\mathcal{C}_q$ and corresponds to a "channel" in $\mathcal{C}^+_{\text{free}}$ with p and q on its opposite ends. The existence of such a path (there can be one, many, or none) does not guarantee that a free motion is possible between p and q along this "channel," since it may contain coordination-forbidden configurations, which may block robot movements. It may still however be possible that other paths between the cells of the same pair of representatives or paths that connect cells of other pairs of representatives, will enable a free motion from one connected components to the other. (We leave the study of sophisticated strategies of finding qualitatively different paths in configuration spaces to future research.)

Given a path in $\mathcal{G}^+$, its cells are traversed in order to find pairs of representatives of $\mathcal{V}_i$ and $\mathcal{V}_j$ such that these representatives are both in the same cell or in consecutive cells. If such a pair is found then a complex probabilistic local planner is invoked in an attempt to find a path that connects these representatives. The traversal terminates upon connection success or when no more pairs of representatives are found. The planner skips cells and pairs of cells that are part of $\mathcal{C}^-_{\text{free}}$ since for them connections were made determinis-

Alg. 1 Preprocessing

Construct $\mathcal{C}^i_{\text{free}}$ exactly, $i = 1, 2$
Vertically decompose $\mathcal{C}^i_{\text{free}}$ into $c^i_1 \dots c^i_{m_i}$, $i = 1, 2$
Let $\mathcal{C}_{i,j} := c^1_i \times c^2_j$, $i = 1 \dots m_1$, $j = 1 \dots m_2$
for all $4D$ cells $\mathcal{C}_{i,j}$ **do**
    Construct $\mathcal{G}_{i,j}$
for all pairs of adjacent $4D$ cells $\mathcal{C}_{i,j}$, $\mathcal{C}_{k,l}$ **do**
    Connect connected components of $\mathcal{G}_{i,j}$, $\mathcal{G}_{k,l}$
Let $\mathcal{V}_1 \dots \mathcal{V}_m$ be the connected components of the initial roadmap $\mathcal{G}$
for all pairs $\mathcal{V}_i, \mathcal{V}_j, i \neq j$ **do**
    Stitch $\mathcal{V}_i, \mathcal{V}_j$

tically and are certain not to cause any discrepancy. One failed traversal may not prevent the relevant connected components from getting transitively connected as they may belong to a sequence of connected components, for which each and every consecutive pair along the sequence was directly connected. We summarize the preprocessing stage of HyMP in Algorithm 1.

Query Processing Given a motion-planning query, i.e., a pair of free configurations q_{init} and q_{goal}, the $4D$ cells $\mathcal{C}_{\text{init}}$ and $\mathcal{C}_{\text{goal}}$ that contain q_{init} and q_{goal}, respectively, are found. If the $\mathcal{G}^+$ nodes of $\mathcal{C}_{\text{init}}$ and $\mathcal{C}_{\text{goal}}$ are in different connected components of $\mathcal{G}^+$ then it is clear that no solution is possible and HyMP stops with a correct "no" answer (a disconnection proof). Otherwise, HyMP adds roadmap nodes for q_{init} and q_{goal} and attempts to connect these nodes to the same connected component of $\mathcal{G}$. For each connected component of $\mathcal{G}$ the planner looks for representative nodes whose underlying configurations lie within $\mathcal{C}_{\text{init}}$ (resp. $\mathcal{C}_{\text{goal}}$) and applies a local planner in an attempt to connect q_{init} (resp. q_{goal}) and to these underlying configurations. If a connection is found, then a simple graph-searching algorithm is applied to retrieve a path between the nodes of q_{init} and q_{goal}. Otherwise the planner responses with "failure".

3.3 Implementation Details

HyMP was implemented in C++. We used the CGAL library [7] as the geometrical engine of our implementation. The CGAL kernel provided the set of primitive objects (points, line segments, circles, etc.) and predicates required. The planar map and arrangement package was used for the representation and manipulation of the $2D$ configuration spaces and proved easy to use in order to perform complex operations such as trapezoidal decomposition and Minkowski sums [11,14]. $4D$ cells of $\mathcal{C}^+_{\text{free}}$ were represented with pairs of trapezoids from $\mathcal{C}^1_{\text{free}}$ and $\mathcal{C}^2_{\text{free}}$. Our work relied on our previously-developed tools to compute Minkowski sums of polygons and discs as well as to solve the complete problem of one disc among polygonal obstacles in the plane [10,15]. Handling line segments and circular arcs simultaneously relies mainly on our

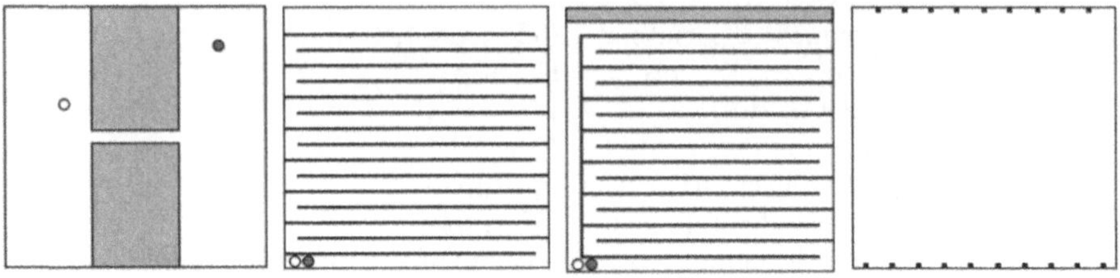

(a) H shape (b) Maze swap (c) Annulus (d) Jagged boundary

Fig. 3. Workspaces are the unit square $[0, 1]^2$, each with a different set of obstacles. Hence, the corresponding C-spaces are confined to the $4D$ unit cube $[0, 1]^4$.

ability to manipulate conic curves *exactly* [27]. This requires an exact number type that can represent algebraic numbers (irrationals in particular). We used the LEDA `real` number type for this purpose [20]. Finally, we used the graph package of LEDA to represent, manipulate, and query connectivity graphs. As additional levels of filtering, which are crucial to expedite exact computations, we were using Sturm sequences and keeping track of the history of computations. See the full version of the paper for more details.

4 Features of HyMP and Experimental Results

HyMP stops short of computing an explicit representation of $\mathcal{C}_{\text{free}}$ (doing so would provide us with a complete solution) but it does sufficient preprocessing so that despite some disadvantages it enjoys distinctive advantages. In this section we present these features and provide experimental results that demonstrate them. A larger set of experiments is available in the full paper. Both advantages and disadvantages stem from three major traits of the planner and the combination thereof. These are (**1**) the breaking of the original problem into subproblems, (**2**) the application of complete solutions to some of the subproblems, and (**3**) the cell decomposition scheme. These advantages allow HyMP to rely on deterministic computations whenever possible and leave as little work as possible to a specialized probabilistic part and as a result lead to faster convergence of the planner toward correct solutions.

The complete solutions of each of the free $2D$ configuration spaces, $\mathcal{C}^1_{\text{free}}$ and $\mathcal{C}^2_{\text{free}}$, provide the planner with the explicit representation of $\mathcal{C}^+_{\text{free}}$. The cell decomposition of $\mathcal{C}^1_{\text{free}}$ and $\mathcal{C}^2_{\text{free}}$, which is used to define and construct the cell decomposition of $\mathcal{C}^+_{\text{free}}$, also provides the planner with the explicit representation of $\mathcal{C}^-_{\text{free}}$. This is achieved by identifying $4D$ cells that fulfill the condition of Lemma 2. In what follows we show how this information leads to a specialized PRM. We then show and demonstrate other advantages that derive from this information. We conclude with a disadvantage along with the means to overcome it.

We remark that, as mentioned in the introduction, complete solutions per se can indeed have many disadvantages. They may have prohibitive complexity and may be hard to implement. This is not the case here. The complete solution incorporated (a disc moving among polygonal obstacles in the plane) is very efficient. In addition, the explicit representations of $\mathcal{C}^1_{\text{free}}$ and $\mathcal{C}^2_{\text{free}}$ indeed demand an intricate implementation that requires exact and rather involved data structures in comparison with the straightforward implementations of PRMs, but the means to construct and manipulate the required data structures (mainly arrangements of curves in the plane) are now available and can be used as black boxes. Moreover, relevant $2D$ algorithms and the cell decomposition are far less complex than the complete solution of the original two-disc problem [24].

Specialized PRM hPRM is a specialized PRM and for various reasons. First, its sampling domain, $\mathcal{C}^{\Delta}_{\text{free}} \cup \mathcal{C}^{\Delta}_{\text{forb}}$ is smaller than the entire configuration space $\mathcal{C}$, often much smaller. This allows hPRM to sample (far) fewer configurations in order to obtain a threshold of free configurations than it should have if it were to sample all of $\mathcal{C}$. In turn, this allows hPRM to concentrate on the difficult areas. The probability for hPRM to adequately sample its domain, in general, and narrow passages, in particular, increases. Second, this subspace contains only free or coordination-forbidden configurations so collision detection amounts to an overlapping predicate of the discs which is trivial. Third, because of the cell decomposition within this domain, the robots locally operate within simple well-characterized $4D$ cells and the connectivity of these cells is known from their adjacency relation. As a result, the local planner we apply can be very simple and efficient, as presented in Sec. 3.2. Fourth, cell decomposition gives additional weight to configurations within narrow passages in the process of sampling. Narrow passages tend to result in smaller $4D$ cells because their width is smaller along at least one axis, but the same number of configuration is sampled within these cells as in bigger cells and more if needed. Finally, the explicit representation of $\mathcal{C}^+_{\text{free}}$ along with the cell decomposition provides a good prediction of the connectivity of $\mathcal{C}_{\text{free}}$ since $\mathcal{C}^+_{\text{free}}$ equals $\mathcal{C}_{\text{free}} \cup \mathcal{C}^{\Delta}_{\text{forb}}$. This enables the planner to look for possible discrepancies in the connectivity of the roadmap in comparison of the connectivity of $\mathcal{C}_{\text{free}}$ and apply the stitching stage to attempt and overcome such discrepancies. An experiment described in the full version of the paper demonstrates the contribution of the stitching stage, which often proves critical to the final connectivity of the roadmap. We show that stitching can overcome scenarios where we (artificially) force the planner to avoid any prior connection attempts.

Disconnection Proofs The explicit representation of $\mathcal{C}^+_{\text{free}}$ enables HyMP to recognize queries where the initial configuration is in a different connected component of $\mathcal{C}^+_{\text{free}}$ from the goal configuration. As $\mathcal{C}_{\text{free}} \subseteq \mathcal{C}^+_{\text{free}}$, these configurations are in different connected components of $\mathcal{C}_{\text{free}}$. For such queries

HyMP issues correct nontrivial "no" answers, which are in fact disconnection proofs, rather than "failure"

Workspace Narrow Passages It is clear that complete solutions are in principal indifferent to narrow passages. $4D$ cells of $\mathcal{C}^-_{\text{free}}$ that represent narrow passages do not contain coordination-forbidden configuration. These regions are deterministically represented in the final roadmap because within them HyMP constructs the roadmap using the $2D$ complete solutions applied for each robot. This makes HyMP almost insensitive to narrow passages where no coordination is involved.

Sensitivity to the Complexity of $\mathcal{C}_{\text{free}}$ HyMP as described above has the following obvious disadvantage which is we can easily overcome. If the complexity of the obstacles is n, then there are $\Theta(n^2)$ $4D$ cells in $\mathcal{C}^+_{\text{free}}$ and a total of $\Theta(n^2)$ pairs of adjacent $4D$ cells. The hybrid planner constructs a connectivity graph for every one of the $\Theta(n^2)$ $4D$ cells and tries to connect connected components between every one of the $\Theta(n^2)$ pairs of adjacent $4D$ cells. In simple motion-planning problems, such as depicted in Fig. 3(**d**), this may prove an effort far beyond what is actually needed. Fortunately, this pitfall is easy to overcome as most PRMs are quick to answer queries where the solution is easy. We envision HyMP as part of meta-planner framework [8] rather than a total solution on its own. Such a framework is expected to start with simpler planning methods, that can quickly answer easy-to-solve queries, and reserve complex solutions for complex problems. An experiment presented in the full paper demonstrates this sensitivity.

Experiments The workspace used are illustrated in Fig. 3 and the results are presented in Table 1 along with additional details. Despite the use of exact arithmatic of algebraic numbers (an overkill for polygonal robots), preprocessing took no longer than a minute in the experiments presented. The first two experiments demonstrate the low sensitivity of the hybrid planner to the width of workspace narrow passages (in the first) and their quantity (in the second). For the first experiment ("H_1") we use an H-shaped workspace environment where each of the vertical slabs of the shape ("slabs") spans a third of the width of the workspace (Fig. 3(**a**)). This provides the robots with sufficient space to coordinate their movements. The query configurations are chosen so that both robots have to succeed in crossing the narrow passage for the planning to be successful. The height of the corridor is 0.05 and we set the robot radii to 0.0249 each. Consequently, the corresponding corridors in $\mathcal{C}^i_{\text{free}}$ are the width of 0.0002. Ten configurations were sampled in each $4D$ cell of $\mathcal{C}^+_{\text{free}} \backslash \mathcal{C}^-_{\text{free}}$. The results show that there is no problem for the hybrid planner to sample configurations that represent robot placements within the corridor, to connect these configurations to one another, and to connect them to configurations in adjacent cells. All ten runs were successful and all ended with a roadmap of one big connected component.

For the second experiment ("maze_1") we used a maze-like workspace (Fig. 3(**b**)). All the corridors, the long horizontal ones and the short vertical ones, have a width of 0.05, and the radii are both 0.02, resulting in configuration-space corridors whose width is 0.01. This time we also added a coordination challenge of medium difficulty by confining the area where the robots can bypass one another to the the uppermost horizontal bar, whose width is 0.1. The robots are both placed at the bottom-left corner of the maze and have to swap positions. The initial and goal configurations are in the farthest regions in $\mathcal{C}_{\text{free}}$ as far as the shortest path distance is considered. Topologically, $\mathcal{C}_{\text{free}}$ is like a "channel" with these regions on its opposite ends. In order for the robots to succeed, they have to go from one end of the "channel" to the other, i.e., each robot has to go through *all* possible narrow passages twice and they have to succeed in bypassing one another as well. Additional difficulty is the need to connect the long sequence of narrow passages. This can only be possible if the configurations sampled within a short vertical passage have visibility that covers both incident long horizontal passages. There were seven successful runs out of ten. The reason for failure in these three cases is that the planner did not find a way for the robots to swap in the top part of the workspace. We believe that a more sophisticated local planner should overcome this difficulty.

The third experiment ("annulus$_1$") presents how material the explicit representation of $\mathcal{C}^{-}_{\text{free}}$ can be to the representation of the connectivity of the $\mathcal{C}_{\text{free}}$. The workspace environment is similar to the maze but differs in two major aspects (Fig. 3(**c**)). First, nowhere in the workspace can the robots locally swap positions. Second, topologically, the free configuration space is an annulus. In order for the robots to swap positions (like in the maze) one of the robots should remain in place, while the other should go all the way through the corridors and reach the lower left corner from the other side of the stationary robot. All ten runs were successful. Notice that most of the roadmap, i.e. 85% of its vertices and 95% of its edges, were determined deterministically. To take this point further we ran the planner again on this input, but preventing it from sampling so the planner could use only the roadmap portion that is computed for $\mathcal{C}^{-}_{\text{free}}$ cells ("annulus$_2$"). We made sure the initial and goal configuration are positioned in coordination-free cells so that the planner could connect them to the roadmap. This experiment stresses the fact that in many scenarios we can avoid sampling and still be able to solve difficult and interesting problems.

5 Future Research

Obvious future application of the hybrid-motion-planning approach is in cases of more than two bodies in the plane but also in 3-Space, where, e.g., several polyhedra are translating among polyhedral obstacles. We already have the key ingredients of a complete solution for the problem of translating a single polyhedron among polyhedra in 3-space [25] (*polyhedral translation problem* for short). We believe however, that a hybrid approach can also be

Table 1. Experimental results. The table presents the number of nodes (V), edges (E), and connected components ($\#cc$) in the final roadmap. It presents what is the part of the final roadmap which was constructed deterministically for cells in $\mathcal{C}^{-}_{\mathrm{free}}$ in terms of vertices (V^{-}) and edges (E^{-}). Information about hPRM includes the number of sampled configurations (sc) and the free portion thereof (fc), as well as the number of successful connection attempts (sca) out of total attempts (ca). The data on the first three rows of the table are the average over 10 runs.

	vertices, sampling						edges, local planner						
exper.	V	V^-	V^-/V	fc	sc	fc/sc	E	E^-	E^-/E	sca	ca	sca/ca	#cc
H_1	176	32	18%	2309	3712	60%	303	108	36%	195	441	45%	1
$maze_1$	16027	13560	85%	3363	5633	59%	56889	108	94%	3273	4357	75%	1.2
$annulus_1$	18695	15952	85%	2743	4873	56%	67135	63448	95%	3687	4810	77%	1
$annulus_2$	15952	15952	100%	0	0	–	63448	63448	100%	0	0	–	1

helpful in obtaining improved solutions to problems other than coordinated motion planning. For example, we would like to use the complete solution for the polyhedral translation problem in order to solve the full rigid motion-planning problem for a polyhedron among polyhedra (that is, allowing translation *and rotation*). One way to do that is to use a "slicing" method, where we build a coarse grid in the three-dimensional rotation space. For each vertex in this grid (which fixes the rotational dofs of the robot) we construct an explicit representation of the free space (we call these representations *complete cross-sections*). We then use PRM techniques to connect between the complete cross-sections. How to effectively make these connections is a non-trivial challenge. Along different lines, one could use a PRM to solve a motion-planning problem and resort to complete solutions in only selected regions of the configuration space where it has failed to make a connection and there is some indication that a connection could be found.

References

1. N. Amato, O. Bayazit, L. Dale, C. Jones, and D. Vallejo. OBPRM: An obstacle-based PRM for 3D workspaces. In *Proc. Int. Workshop on Alg. Found. of Rob.*, pages 155–168, 1998.
2. J. Bañon. Implementation and extension of the ladder algorithm. In *Proc. IEEE Conf. on Rob. and Auto.*, pages 1548–1553, 1990.
3. J. Barraquand, L. E. Kavraki, J.-C. Latombe, T.-Y. Li, R. Motwani, and P. Raghavan. A random sampling scheme for path planning. *Internat. J. Robot. Res.*, 16(6):759–774, 1997.
4. J. Basch, L. Guibas, D. Hsu, and A. Nguyen. Disconnection proofs for motion planning. In *Proc. IEEE Int. Conf. on Rob. and Auto.*, pages 1765–1772, 2001.
5. V. Boor, M. Overmars, and A. van der Stappen. The Gaussian sampling strategy for probabilistic roadmap planners. In *Proc. IEEE Int. Conf. on Rob. and Auto.*, pages 1018–1023, 1999.
6. J. Canny. *The Complexity of Robot Motion Planning.* ACM – MIT Press Doctoral Dissertation Award Series. MIT Press, Cambridge, MA, 1987.
7. CGAL. The Computational Geometry Algorithms Library. www.cgal.org.

8. L. Dale and N. Amato. Probabilistic roadmaps - putting it all together. In *Proc. IEEE Int. Conf. on Rob. and Auto.*, 2001.
9. M. de Berg, M. van Kreveld, M. Overmars, and O. Schwarzkopf. *Computational Geometry: Algorithms and Applications.* Springer-Verlag, Berlin, Germany, 2nd edition, 2000.
10. E. Flato. Robust and efficient construction of planar Minkowski sums. Master's thesis, Dept. Comput. Sci., Tel-Aviv Univ., 2000.
11. E. Flato, D. Halperin, I. Hanniel, O. Nechushtan, and E. Ezra. The design and implementation of planar maps in CGAL. *The ACM Journal of Exper. Alg.*, 5, 2000. Also in LNCS Vol. 1668 (WAE '99), Springer, pp. 154–168.
12. R. Geraerts and M. Overmars. A comparative study of probabilistic roadmap planners. These proceedings.
13. D. Halperin. Robust geometric computing in motion. *International Journal of Robotics Research*, 21(3):219–232, 2002.
14. I. Hanniel and D. Halperin. Two-dimensional arrangements in cgal and adaptive point location for parametric curves. In *Proc. 4th Workshop on Alg. Eng.*, 2000.
15. S. Hirsch and E. Leiserowitz. Exact construction of Minkowski sums of polygons and a disc with application to motion planning. Technical Report ECG-TR-181205-01, Tel-Aviv University, 2002.
16. D. Hsu, L. E. Kavraki, J.-C. Latombe, R. Motwani, and S. Sorkin. On finding narrow passages with probabilistic roadmap planners. In *Proc. Int. Workshop on Alg. Found. Rob.*, Wellesley, MA, 1998. A. K. Peters.
17. L. E. Kavraki, P. Švestka, J.-C. Latombe, and M. Overmars. Probabilistic roadmaps for path planning in high dimensional configuration spaces. *IEEE Trans. Robot. Autom.*, 12:566–580, 1996.
18. J.-C. Latombe. *Robot Motion Planning.* Kluwer Academic Publishers, Boston, 1991.
19. S. LaValle and M. Branicky. On the relationship between classical grid search and probabilistic roadmaps. These proceedings.
20. K. Mehlhorn and S. Näher. *The LEDA Platform of Combinatorial and Geometric Computing.* Cambridge University Press, 1999.
21. J. H. Reif. Complexity of the mover's problem and generalizations. In *Proc. 20th Annu. IEEE Sympos. Found. Comput. Sci.*, pages 421–427, 1979.
22. J. T. Schwartz and M. Sharir. On the "piano movers" problem II: General techniques for computing topological properties of real algebraic manifolds. *Adv. Appl. Math.*, 4:298–351, 1983.
23. M. Sharir. Algorithmic motion planning. In J. E. Goodman and J. O'Rourke, editors, *Handbook of Discrete and Comput. Geom.*, chapter 40, pages 733–754. CRC Press LLC, Boca Raton, FL, 1997.
24. M. Sharir and S. Sifrony. Coordinated motion planning for two independent robots. *Ann. Math. Artif. Intell.*, 3:107–130, 1991.
25. H. Shaul and D. Halperin. Improved construction of vertical decompositions of three dimensional arrangements. In *Proc. 18th Annu. ACM Sympos. Comput. Geom.*, pages 283–292, 2002.
26. P. Švestka and M. Overmars. Coordinated path planning for multiple robots. *Robotics and Autonomous Systems*, 23:125–152, 1998.
27. R. Wein. High level filtering for arrangements of conic arcs. In *Proc. 10th European Symposium on Algorithms*, volume 2461 of *Lecture Notes in Computer Science*, pages 884–895. Springer-Verlag, Rome, 2002.

Decentralized Algorithms for Multirobot Manipulation via Caging

Guilherme A. S. Pereira[1,2], Vijay Kumar[1], and Mario F. M. Campos[2]

[1] GRASP Lab., University of Pennsylvania, Philadelphia, PA, USA.
[2] VERLab, Universidade Federal de Minas Gerais, Belo Horizonte, MG, Brazil.

Abstract. This paper addresses the problem of transporting objects with multiple mobile robots using the concept of *object closure*. In contrast to other manipulation techniques that are typically derived from form or force closure constraints, object closure requires the less stringent condition that the object be trapped or caged by the robots. Our basic goal in this paper is to develop decentralized control policies for a group of robots to achieve a condition of object closure, and then, move toward a goal position while maintaining this condition. We present experimental results that show car-like robots controlled using visual feedback, transporting an object in an obstacle free environment toward a prescribed goal.

1 Introduction

Object manipulation with mobile robots have been extensively discussed in the literature. Most approaches use the notions of force and form closure to perform the manipulation of relative large objects [1–4]. *Force closure* is a condition that implies that the grasp can resist any external force applied to the object [5]. *Form closure* can be viewed as the condition guaranteeing force closure, without requiring the contacts to be frictional. In general, robots are the agents that induce contacts with the object, and are the only source of grasp forces. But, when external forces acting on the object, such as gravity and friction, are used together with contact forces to produce force closure, we have a situation of *conditional force closure*. Several research groups have used conditional closure to transport an object by pushing it from an initial position to a goal [6,7].

In contrast to these approaches, as shown in Fig. 1, object closure requires the less stringent condition that the object be trapped or caged by the robots. (Our use of the concept of caging is slightly different from the definition in [8], and hence the object closure.) In other words, although the object may have some freedom to move, it cannot be completely removed [9,10]. Because a caging operation requires a relatively low degree of precision in relative positions and

J.-D. Boissonnat et al. (Eds.): Algorithmic Foundations of Robotics V, STAR 7, pp. 257–273, 2004.

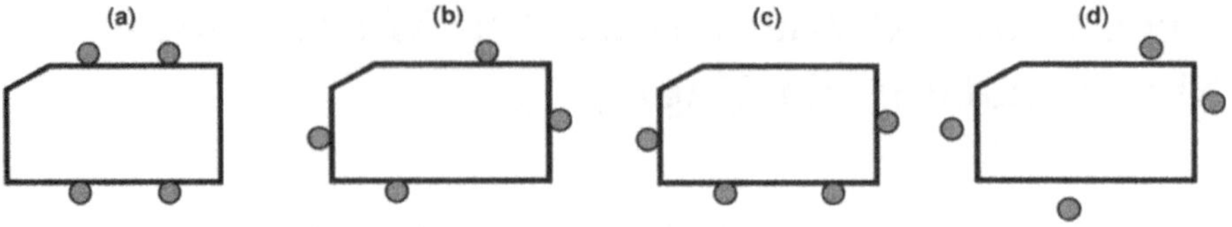

Fig. 1. Four technics of manipulation: (a) force closure (robots pressing the object); (b) form closure; (c) conditional closure (robots pushing the object up); (d) object closure or caging.

orientations, manipulation strategies based on caging operations are potentially more robust than, for example, approaches relying on force closure.

Caging was first introduced by Rimon and Blake in [8] for non-convex objects and two fingered gripers. Other papers addressing variation on this basic theme are [9–13]. Broadly speaking, our work may be considered closest to the work by Sudsang and Ponce [13]. They develop a centralized algorithm for moving three robots with circular geometry in an object manipulation task.

Our basic goal in this paper is to develop decentralized control policies for a group of robots to achieve a condition of object closure, and then, move toward a goal position while maintaining the object closure condition. Unlike previous work [13,10], we do not require the robots to be circular. However, we do introduce a number of simplifying assumptions to enable real-time implementation. Further, our interest is in transporting the object from an initial position toward a goal position in $\mathbb{R}^2$. We do not address the problem of precisely positioning and orienting the object in the plane.

The remaining of this paper is organized as follows. We first discuss the mathematical modeling of the object and the robots. Section 3 presents our definition of object closure, and several key necessary conditions for establishing and maintaining object closure. Section 4 describes our approach to cooperative control. We briefly present results from experiments in Section 6. Finally, the main points of the paper and directions for future work are presented in Section 7.

2 Mathematical Modeling

Consider a planar world, $\mathcal{W} = \mathbb{R}^2$, occupied by a convex, polygonal object $\mathcal{O}$, and a group of n convex, polygonal robots. The ith robot R_i is described by the convex set $\mathcal{A}_i(q_i) \in \mathcal{W}$, where $q_i = (x_i, y_i, \theta_i)$

denotes the configuration of R_i. The configuration of the object is described by the coordinates $q = (x, y, \theta)$. We will use $\mathcal{C}_{\mathcal{R}_i}$ to denote the configuration space of a robot, while $\mathcal{C}$ will denote the configuration space for the object $\mathcal{O}$.

Convex robots and objects are represented by an intersection of m half planes derived from the equations for each edge. The edge from (x_j, y_j) to (x_{j+1}, y_{j+1})[1] is given by $f_j(x, y) = a_j x + b_j y + c_j$ and $f_j(x, y) < 0$ for all points in the interior of the polygon.

If robot positions and orientations are held fixed, the region in the configuration space that corresponds to an interpenetration between the object $\mathcal{O}$ and the robot i is:

$$\mathcal{C}_{obj\_i} = \{q \in \mathcal{C} \mid interior(\mathcal{A}_i(q_i) \cap \mathcal{O}(q) \neq \emptyset)\} , \tag{1}$$

where $\mathcal{O}(q)$ is the representation of $\mathcal{O}$ in the configuration q.

3 Object Closure

3.1 Definition

Before we proceed further, we will make three assumptions in this section.

Assumption A1. All robots are holonomic and identical in terms of geometry, and in terms of capabilities and constraints related to sensing, control, and mobility.

Assumption A2. All robots are point robots – $\mathcal{A}_i(q_i) = q_i = (x_i, y_i)$.

Assumption A3. The manipulated object cannot rotate – the coordinates of the object are given by $q = (x, y)$ and $\mathcal{C} \subset \mathbb{R}^2$.

Assumptions *A2* and *A3* make it easier to explain the basic ideas and will be relaxed in the next sections.

Figure 2 shows the boundary of $\mathcal{C}_{obj\_i}$ for a five-sided polygonal object and the point robot, R_i.

The union of $\mathcal{C}_{obj\_i}$ for $1 \leq i \leq n$ determines the region in $\mathcal{C}$ which cannot be occupied by the object. Then,

$$\mathcal{C}_{obj} = \bigcup_{i=1}^{n} \mathcal{C}_{obj\_i} . \tag{2}$$

[1] $j + 1$ is replaced by 1 for $j = m$ and $j - 1$ by m for $j = 1$.

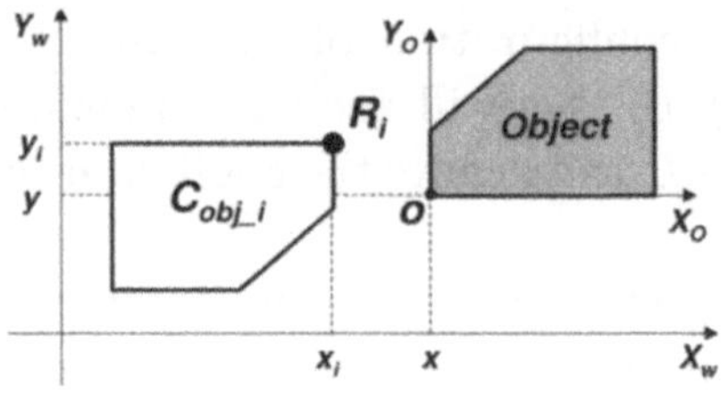

Fig. 2. $\mathcal{C}_{obj\_i}$ for a point robot R_i considering only object translations. By sliding the object around the robot, the origin, o, of the object-fixed reference frame traces out the boundaries of $\mathcal{C}_{obj\_i}$.

Let the complement of $\mathcal{C}_{obj}$ in $\mathcal{C}$ be $\bar{\mathcal{C}}_{obj}$. When $\bar{\mathcal{C}}_{obj}$ consists of two (or more) disjoint sets, we use the term *object closure* to refer to the condition when one of these sets is compact and contains the object configuration, q. This is shown for four robots in Fig. 3, where the compact set, which we refer to as the *closure configuration space* and denote by $\mathcal{C}_{cls}$, is shown shaded. Observe that the object is trapped or caged (in the terminology of [8]) when its origin is in $\mathcal{C}_{cls}$.

We can easily relax Assumption *A3* to accommodate the more general case with translations and rotations. In this case, (2) remains the same, but $\mathcal{C}_{obj\_i}$ in (1) is a three-dimensional solid whose cross-section, for a given angular orientation, is similar to the picture in Fig. 2, and the compact subset $\mathcal{C}_{cls}$ consists of one or more three dimensional solids whose cross-section (slice) is similar to the one shown in Fig. 3 [10].

We now define a *non essential robot* with the help of Fig. 4. In contrast to Fig. 3 in which all four $\mathcal{C}_{obj\_i}$ (and therefore all four robots) are essential to construct the boundary for the closure configuration space, R_3 is not essential for object closure in Fig. 4. In a group of robots maintaining object closure, a non essential robot, R_x, is a robot whose removal (and consequently the absence of the constraint due to $\mathcal{C}_{obj\_x}$) does not violate the state of object closure.

We now introduce a fourth assumption:

Assumption A4. We will only consider groups of robots with no non essential robots.

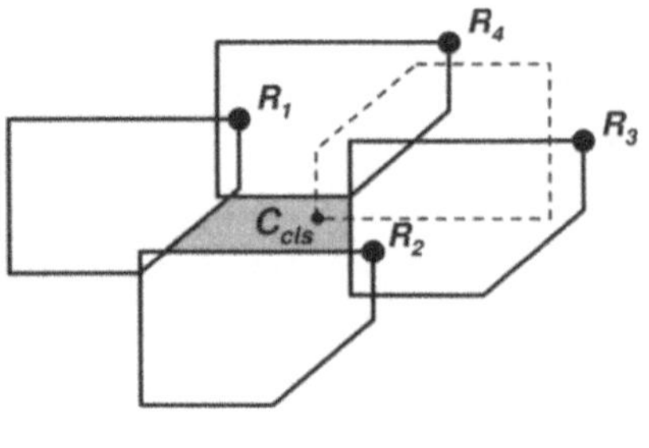

Fig. 3. Object closure: The interior (shaded gray) represents the *closure configuration space*, $\mathcal{C}_{cls}$, for a team of 4 robots. The dashed polygon represents the object. Notice that the origin of the object's reference frame is inside $\mathcal{C}_{cls}$, a compact set, indicating a condition of object closure.

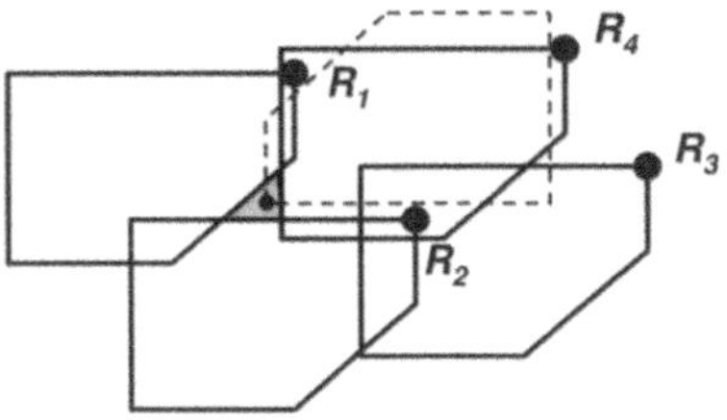

Fig. 4. Essential Robots: Even with the removal of R_3 the closure properties of the group is preserved and so, R_3 is a non essential robot.

3.2 A Test for Object Closure

Checking the object closure condition involves two steps: (a) Establishing the existence of $\mathcal{C}_{cls}$; and (b) Verifying $q \in \mathcal{C}_{cls}$. Step (a) requires obtaining state information from all robots and Step (b) requires obtaining position (pose, in the more general case) of the object.

The key idea comes from Figure 3 where robots are numbered R_1 through R_n in a counter clockwise fashion. A necessary condition for object closure with non essential robots is that the ith robot's position satisfies: $\mathcal{C}_{obj\_i-1} \cap \mathcal{C}_{obj\_i} \neq \emptyset$; and $\mathcal{C}_{obj\_i} \cap \mathcal{C}_{obj\_i+1} \neq \emptyset$. This condition is not sufficient. The sufficient condition involves verifying $q \in \mathcal{C}_{cls}$. However, this condition is necessary and sufficient for the existence of $\mathcal{C}_{cls}$ (step (a) above), and for maintaining object closure. Hence we can state the following:

Proposition 1. *If an object is in a state of object closure with a group of robots with no non essential robots, a sufficient condition for maintaining object closure is* $\mathcal{C}_{obj\_i-1} \cap \mathcal{C}_{obj\_i} \neq \emptyset$; *and* $\mathcal{C}_{obj\_i} \cap \mathcal{C}_{obj\_i+1} \neq \emptyset$.

We now explain how to derive the algebraic equations for object closure. Define $\mathcal{I}_i$ to be a set of the robots configuration space that represents the intersection between $\mathcal{C}_{obj\_i}$ and $\mathcal{C}_{obj\_k}$:

$$\mathcal{I}_i = \{q_k \in \mathcal{C}_{R_k} \mid \mathcal{C}_{obj\_i}(q_i) \cap \mathcal{C}_{obj\_k}(q_k) \neq \emptyset\},$$

Note that $\mathcal{C}_{obj\_i}(q_i)$ and $\mathcal{C}_{obj\_k}(q_k)$ are identical polygons, which introduces a symmetry in the form of $\mathcal{I}_i$.

It can be observed that:

$$\mathcal{C}_{obj\_i} \cap \mathcal{C}_{obj\_k} \neq \emptyset \Leftrightarrow (q_k \in \mathcal{I}_i \wedge q_i \in \mathcal{I}_k).$$

Thus, the object closure conditions for each robot, which can be rewritten as $q_i \in \mathcal{I}_{i-1}$ and $q_i \in \mathcal{I}_{i+1}$ (see Fig. 5(a)) are represented as a set of inequality constraints of the form $g_j(q_{i-1}, q_i) \leq 0$ or $g_j(q_i, q_{i+1}) \leq 0$,

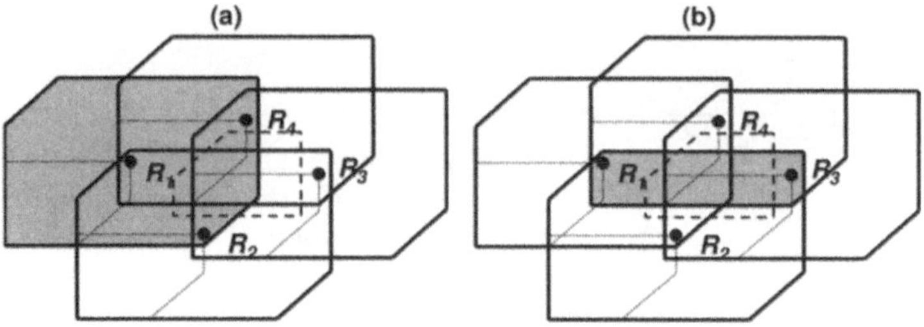

Fig. 5. Object closure is achieved if each robot i is inside Γ_i. The shaded areas represent (a) $\mathcal{I}_1$ and (b) Γ_1.

where g_j are the functions that delimit $\mathcal{I}_{i-1}$ or $\mathcal{I}_{i+1}$ respectively. $\mathcal{I}_{i-1}$ ($\mathcal{I}_{i+1}$) is a $2m$-sided polygon defined by $2m$ algebraic constraints , each linear in q_{i-1} and q_i (q_i and q_{i+1}). Since each polygon has up to $2m$ sides, the number of constraints for each robot is $4m$. For the situation we are considering, where the robots are points and the object cannot rotate, the boundary of $\mathcal{C}_{obj\_i}$ is formed by the same edges of $\mathcal{O}$ but ordered in a different way (see Fig. 2). Then, each $\mathcal{I}_i$, which depends on $\mathcal{C}_{obj\_i-1}$ and $\mathcal{C}_{obj\_i+1}$, is bounded by two sets of the object's edges (refer to the algorithm presented in [14] for proofs). Consequently, $\mathcal{I}_{i-1}$ is given by functions $g_j(q_{i-1}, q_i)$, while $\mathcal{I}_{i+1}$ is given by another set of functions $g_j(q_i, q_{i+1})$. Each function is directed derived from the functions $f_i(x, y)$ used to describe the object.

We can now rewrite Proposition 1 as follows:

Proposition 2. *If an object is in a state of object closure with a group of robots with no non essential robots, a sufficient condition for maintaining object closure is* $q_i \in \Gamma_i$, *where* $\Gamma_i = \mathcal{I}_{i-1} \cap \mathcal{I}_{i+1}$.

Note Γ_i is bounded by a subset of the constraints $g_j(q_i) \leq 0$, $1 \leq j \leq 4m$. A example of Γ_i can be seen in Fig. 5(b).

3.3 Introducing Rotations

Thus far, we have ignored rotations. In reality, since the robots will collide with and bump against the object, the object can rotate. Even if object closure is guaranteed for a given object orientation, a small rotation followed by a translation may cause the object to escape from the robot formation.

Our approach to incorporate rotations is to establish guarantees for object closure under the worst case rotation. Because the object

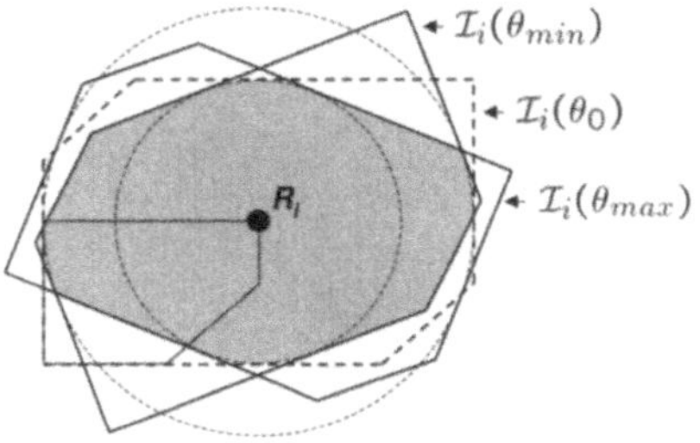

Fig. 6. Closure region for a maximum rotation of 20°.

has no actuators, its maximum velocity is limited by the maximum velocity of the robots. Thus, if the object orientation at any instant is estimated to be θ_o, the orientation in the ensuing interval ΔT must be in the interval, $[\theta_{min}, \theta_{max}]$, where $\theta_{min} = \theta_o - \Delta T\,\omega_{max}$, $\theta_{max} = \theta_o + \Delta T\,\omega_{max}$, and ω_{max} is the (estimated) maximum object's angular velocity. Let $\mathcal{J}_i$ be defined as:

$$\mathcal{J}_i = \bigcap_{\theta=\theta_{min}}^{\theta_{max}} \mathcal{I}_i(\theta)\,,$$

where $\mathcal{I}_i(\theta)$ is $\mathcal{I}_i$ computed for an object orientation θ. Following the previous methodology, the conditions that guarantee object closure for all $\theta \in [\theta_{min}, \theta_{max}]$ are: $q_i \in \mathcal{J}_{i-1}$ and $q_i \in \mathcal{J}_{i+1}$.

Since $\mathcal{C}_{obj\_i}$ is represented by the same polygon for every robot, the shape of $\mathcal{I}_i(\theta)$ is independent of the object orientation. As θ changes, $\mathcal{I}_i(\theta)$ is obtained by simply rotating $\mathcal{I}_i(\theta_o)$ around R_i. The intersection set $\mathcal{J}_i$, can be constructed as shown in Fig. 6. The shaded area represents the configuration space where q_i must be in order to guarantee object closure for object orientations between θ_{min} and θ_{max}. It is bounded by circular arcs and the sides of $\mathcal{I}_i(\theta_{min})$ and $\mathcal{I}_i(\theta_{max})$. Notice that we continue with a set of inequalities constraints but now two of them are nonlinear. The set $\mathcal{J}_i$ is still convex. From a practical standpoint, this set-valued approach for modeling the uncertainty in orientation allows us to be robust to errors in pose estimation.

3.4 Working with Polygonal Robots

The main difficulty of working with polygonal robots is the computation of $\mathcal{C}_{obj\_i}$ in real time. Although an efficient algorithm exists, it must run every time step since changes in robot orientation alters not only the orientation but also the shape of $\mathcal{C}_{obj\_i}$. Furthermore, differences on the shapes of $\mathcal{C}_{obj\_i}$ and its neighbors make the form of closure

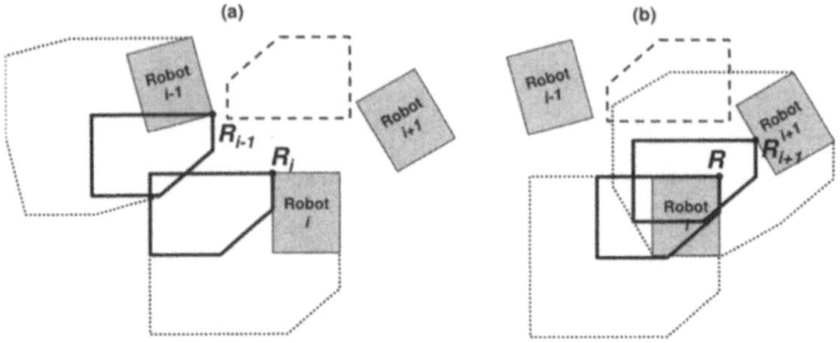

Fig. 7. Robot i checks closure (a) using the imaginary point robots, R_i and R_{i-1} (left) (b) using a different set of point robots, R_i and R_{i+1} (right). The dotted polygons are the actual object configuration space.

constraints very complicated. However, since $\mathcal{C}_{obj\_i}$ can be constructed by the union of the $\mathcal{C}_{obj\_i}$ of the points in the convex polygon that form the robot, we can easily establish a sufficient condition that guarantees closure. If the intersection between $\mathcal{C}_{obj\_i}$ of two virtual point robots located at the closest pair of points between robot k and robot $k+1$ is non zero, $\mathcal{C}_{obj\_k} \cap \mathcal{C}_{obj\_k+1} \neq \emptyset$. Since $\mathcal{C}_{obj\_i}$ of a point robot can be computed off-line, the online computation is limited to the translation of this set to the location of the virtual point robots. This computation is illustrated in Fig 7.

For each slice of the configuration space (a specific orientation), the object configuration space for the polygonal robot k can be written as:

$$\mathcal{C}_{OBJ\_k} = \mathcal{A}_k(q_k) \oplus \mathcal{C}_{obj\_i}(0) \ ,$$

where $\mathcal{C}_{obj\_i}(0)$ is the the object configuration space for the point robot located in the origin of the world reference frame and $\oplus$ is the Minkowski sum operator. We can observe also that:

$$\mathcal{C}_{obj\_i}(q) = \{q\} \oplus \mathcal{C}_{obj\_i}(0) \ .$$

Thus, it is not too hard to verify:

Proposition 3. *If $\mathcal{C}_{obj\_i}(q_a) \cap \mathcal{C}_{obj\_i}(q_b) \neq \emptyset$, and $q_a \in \mathcal{A}_k(q_k)$ and $q_b \in \mathcal{A}_l(q_l)$ are the closest pair of points between the two convex robots, then:*

$$(\mathcal{A}_k(q_k) \oplus \mathcal{C}_{obj\_i}(0)) \cap (\mathcal{A}_l(q_l) \oplus \mathcal{C}_{obj\_i}(0)) \neq \emptyset \ .$$

Thus, using the closest pair of points as reference for our computations leads us to a conservative but simple test for object closure for polygonal robots.

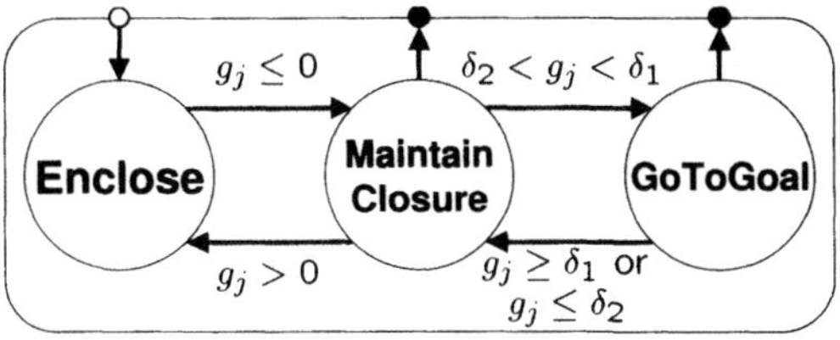

Fig. 8. The switched control system with 3 modes for multirobot manipulation. δ_1 and δ_2 are thresholds for activating the transition between modes.

4 Control

Our control system is decentralized and implemented using a set of reactive controllers. Each robot switches between the controllers as shown in Fig. 8. The switches are governed by the activation of constraints that depend on the relative positioning of a robot with respect to its neighbors and the robots' estimate of the object orientation.

Recall from Proposition 2, that object closure constraints for R_i are defined by inequalities, $g_j(q_{i-1}, q_i) \leq 0$ or $g_j(q_i, q_{i+1}) \leq 0$. We consider the jth constraint to be active when $g_j = \delta_1$, where δ_1 is a small negative number that can be thought of as a threshold. In addition to ensuring $q_i \in \Gamma_i$, it is also necessary to ensure the robots do not try to cluster together thus crushing the object. From a practical standpoint, although the object may be rigid and immune to damage, this will cause large contact forces and jamming due to friction. To avoid this, we introduce a new set of constraints that prevent the robot from being very close of its neighbors: $g_j \geq \delta_2$, where $\delta_2 < \delta_1 < 0$. This defines a "safe" configuration space for each q_i where the object is caged but jamming is avoided. Practically, the set of constraints $\delta_2 \leq g_j \leq \delta_1$ define two polygons with the same shape as Γ_i but with different sizes. The reactive controllers and the sequential composition of these controllers are shown in Fig. 8.

We consider a simple kinematic model for each robot. For the ith robot, the dynamical model is given by:

$$\dot{q}_i = u_i \,,$$

where $q_i = (x_i, y_i)$.

We will denote the constraints due to R_{i-1}, which have the form $g_j(q_{i-1}, q_i) \leq 0$, by g_j^l and those due to R_{i+1}, which have the form $g_j(q_i, q_{i+1}) \leq 0$ by g_j^r. In the *Enclose* mode each robot tries to initially achieve object closure. The control input in this mode is:

$$u_i = -k_1 \left[a\nabla g_j^r + b\nabla g_j^l \right] , \tag{3}$$

where ∇g_j^x is the gradient of the constraint defined as $\partial g_j^x / \partial q_i$. ∇g_j^l is due to robot $i-1$ and ∇g_j^r is due to robot $i+1$. We assume that these vectors are normalized and are of unit length. The variables a and b can each be -1, 0, or 1. When $g_j \leq \delta_2$ the value -1 is assigned. When $\delta_1 > g_j > \delta_2$, the value 0 is assigned. And when $g_j \geq \delta_1$, the value 1 is assigned.

In the *MaintainClosure* mode a robot tries to maintain object closure while navigating towards the goal. The control input for this state is:

$$u_i = -k_1 \left[a\nabla g_j^r + b\nabla g_j^l \right] + k_2 \, u_T , \tag{4}$$

where u_T is a common input to the team of robots that leads them to the goal. It can be, for example, a proportional control law that is a function of the object position, $q - q_{goal}$. Alternatively, u_T can be derived from a potential function that also serves as a Lyapunov function. We will assume that

$$u_T = -\nabla \phi(q) , \tag{5}$$

where $\phi(q)$ is a potential function with a single minimum at q_{goal}.

The *GoToGoal* mode has the following input:

$$u_i = k_2 \, u_T . \tag{6}$$

It is clear that we can always design u_T in (5) to guarantee that each robot will go to the specified target. Although, if one of the constraints $g_j \geq \delta_1$ or $g_j \leq \delta_2$ is violated, the controller switches to the *MaintainClosure* mode in (4). We will now prove that this controller guarantees that the condition of object closure is maintained.

Before we do that, we will need to make an important observation about the constraints $g_j(q_i, q_k)$. Let us consider, for illustration purposes, the special case of object translation. The set of constraints, Γ_i, for the point robot i can be subdivided in two sets of linear inequalities. The equations of each set are of the form $a_j(x_i+x_k)+b_j(y_i+y_k)+c_j \leq 0$, where $k = i-1$ (robot on the left) for the first set and $k = i+1$ (robot on the right) for the second set. Observe that each constraint

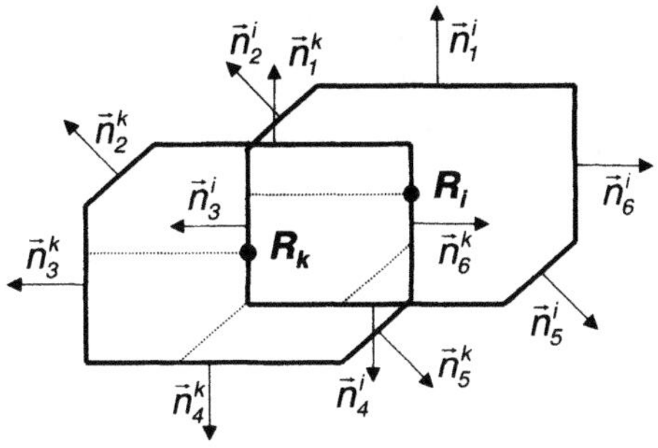

Fig. 9. An active constraint for robot i ($g_6^k(q_i, q_k) = 0$) indicates the activation of a identical constraint with opposite sign for one of its neighbors ($g_3^i(q_i, q_k) = 0$). In this picture, $\delta_1 = 0$. Notice that the normal vector of the active constraint for robot i, $\boldsymbol{n}_3^i$, is equal to $-\boldsymbol{n}_6^k$, the negative of the normal vector of the active constraint for robot k.

describes a line in the world reference frame translated by the position of one of the neighbors. Since $\mathcal{I}_i$ has the same form for all robots, when one constraint, $g_j(q_i, q_k)$, is active for one robot, there is an identical constraint with opposite sign, $-g_j(q_i, q_k)$, active for one of its neighbors. Figure 9 shows a typical situation when one constraint is active for the ith robot and a identical constraint, with opposite sign, is active for one of its neighbors. This is also the case when $g_j(q_i, q_k)$ is not a linear function (as is the case when rotations are considered). In other words,

$$\frac{\partial g_j}{\partial q_k} = -\frac{\partial g_j}{\partial q_i} \tag{7}$$

We use this observation to prove that, once the robots have captured the object, the controller (4) guarantees object closure is maintained.

Proposition 4. *Once the robots achieve a situation of object closure the switched control system shown in Fig. 8 guarantees object closure.*

Proof. We need to prove that every time a generic constraint, $g_j(q_i, q_k) \leq \delta_1$, is active the control input makes $\dot{g}_j(q_i, q_k) \leq 0$. The time derivative of $g_j(q_i, q_k)$ is given by:

$$\dot{g}_j(q_i, q_k) = \frac{\partial g_j}{\partial q_i}\dot{q}_i + \frac{\partial g_j}{\partial q_k}\dot{q}_k \tag{8}$$

Denote $\partial g_j/\partial q_i$ by ∇g_j^k (induced by R_k), and $\partial g_j/\partial q_k$ by ∇g_j^i. Because of (7), $\nabla g_j^i = -\nabla g_j^k$. Assuming that for the ith robot $g_j(q_i, q_k)$, whose gradient is ∇g_j^k, is active, then for the kth robot, $g_p(q_i, q_k)$, whose gradient is $\nabla g_p^i = -\nabla g_j^k$, is also active. If, for example, R_k is the left neighbor of R_i then, in (4), $\nabla g_j^l = \nabla g_j^k$ for R_i and $\nabla g_j^r = \nabla g_p^i = -\nabla g_j^k$

for R_k. Substituting $\dot{q}_i$ and $\dot{q}_k$ in (8) by the control inputs in (4) (call ∇g_α the constraint induced by the other neighbor of R_i and ∇g_β the constraint induced by the other neighbor of R_k) we rewrite the time derivative of $g_j(q_i, q_k)$ as:

$$\begin{aligned}
\dot{g}_j(q_i, q_k) = &- k_1 \left[\nabla g_j^k \nabla g_j^k + a \nabla g_j^k \nabla g_\alpha\right] + k_2 \nabla g_j^k u_T + \\
&- k_1 \left[(-\nabla g_j^k)(-\nabla g_j^k) + b(-\nabla g_j^k)\nabla g_\beta\right] + \\
&+ k_2 \,(-\nabla g_j^k) u_T \\
= &- k_1 \left[2 \nabla g_j^k \nabla g_j^k + a \nabla g_j^k \nabla g_\alpha - b \nabla g_j^k \nabla g_\beta\right] \\
= &- k_1 \Big[2 \|\nabla g_j^k\|^2 + a \|\nabla g_j^k\| \, \|\nabla g_\alpha\| \cos\theta_a + \\
&- b \|\nabla g_j^k\| \, \|\nabla g_\beta\| \cos\theta_b\Big] \\
= &- k_1 \,[2 + a\cos\theta_a - b\cos\theta_b] \leq 0 \,.
\end{aligned}$$

Since $-1 \leq a\cos\theta \leq 1$, for all $a \in \{-1, 0, 1\}$, $\dot{g}_j(q_i, q_k) \leq 0$. Therefore, given the initial conditions, $g_j(q_i, q_k) \leq 0$, for all $1 \leq i \leq n$, $k \in \{q-1, q+1\}$ and $1 \leq j \leq 4m$. □

Using the same methodology presented above we can also prove that when a constraint $g_j(q_i, q_k) = \delta_2$, the control law in the *MaintainClosure* mode maintains the condition $g_j(q_i, q_k) \geq 0$. Thus, if the robots are in the *MaintainClosure* mode, they either stay in this mode while moving toward the goal (the u_T component guarantees this [16]), or they switch to the *GoToGoal* mode.

It is more difficult to prove that the control law (3) in the *Enclose* mode leads to a condition of object closure. The main difficulties come from the assumption related to non essential robots and the book-keeping associated with numbering the robots. However, simple potential field controllers, like the one presented in [15] have the attractive property of symmetrically distributing the robots around the object and producing initial conditions that are favorable for the *Enclose* mode.

It is also natural to ask if the kinematic model can be extended to non-holonomic robots. For non-holonomic robots, u_i, which is a two dimensional vector, can be used as a set-point for controllers that take in account the non-holonomic constraints. An example of such an approach is shown in [16]. This is a direction of future research.

5 Computational Complexity

Considering that $\mathcal{I}_i$ (or $\mathcal{J}_i$ when rotations are considered) can be computed off-line, the decentralized manipulation algorithm consists in (a) determining the object orientation θ and the two neighbors' position; (b) computing $\mathcal{I}_i(\theta)$ by rotating $\mathcal{I}_i$; (c) translating $\mathcal{I}_i(\theta)$ to q_{i-1} and q_{i+1} and computing $\mathcal{I}_{i-1}$ and $\mathcal{I}_{i+1}$ respectively; (d) verifying if there are active constraints; and (e) computing the control signals.

The estimation of θ, q_{i-1} and q_{i+1} is not addressed here but it is important to mention that in case of polygonal robots an $O(l)$ [17] algorithm needs to be used to determine the closest pair of points between the robots. Since $\mathcal{I}_i$ ($\mathcal{J}_i$) is defined by up to $2m$ functions, each robot needs to compute up to $2m$ rotations and $4m$ translations in order to compute $\mathcal{I}_{i-1}$ and $\mathcal{I}_{i+1}$ ($\mathcal{J}_{i-1}$ and $\mathcal{J}_{i+1}$). The determination of the active constraints, if there are any, can be done by evaluating the $4m$ inequalities that define Γ_i. Observe however, that Γ_i don't need to be computed explicitly since the cost of computing this region is higher than evaluating all constraints for $\mathcal{I}_{i-1}$ and $\mathcal{I}_{i+1}$. The computation of the control laws can be done in constant time. Therefore, the cost of the algorithm is $O(m+l)$, and is independent of the number of robots in the group.

6 Experiments

Our mobile robots are car-like robot equipped with an omnidirectional camera as its only sensor. Although we have performed experiments with teams of up to five mobile robots, we report here experiments with three robots. The communication among the robots relies on IEEE 802.11 networking. To facilitate the visual processing, each team member and the goal position are marked with different colors. Because each robot has only one camera we use communication between robots and cooperative sensing for (a) localization with respect to each other; and (b) estimating the pose of the object [18]. The communication is essentially used for multi-eyed stereo algorithms but is not used for control or decision making. Ground truth information is obtained from a calibrated overhead camera.

Figure 10 illustrates the test for object closure performed by Robot 1 (R_1). R_1 estimates the position of its neighbor R_2, as well as the orientation of the object. It then computes $\mathcal{C}_{obj\_1}$ and $\mathcal{C}_{obj\_2}$ based on its estimate of the pair of closest points, one on R_1 and one on R_2. As the

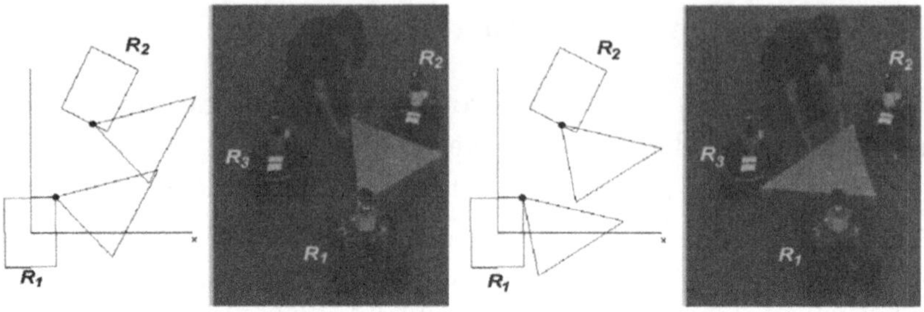

Fig. 10. Three robots caging a triangular object. R_1's computation of $\mathcal{C}_{obj\_1}$ and $\mathcal{C}_{obj\_2}$ for the imaginary point robots located at the closest pair of points are shown. The overlap (left) indicates the object is constrained for this specific orientation, and the lack of overlap (right) shows that object closure is not maintained for this slice of the configuration space.

figure shows, the snapshot on the left shows overlap and therefore a positive test for object closure. The snapshot on the right shows that the object can actually escape. A similar test (not shown in the figure) needs to be performed with robot R_3 .

Data collected from the overhead camera for typical experimental runs are shown in Figures 11 and 12. Robots R_1, R_2, and R_3 transport a triangular box toward a goal position. Figure 11 shows a situation where robots R_2 and R_3 change their control behaviors in order to perform the task. In Fig. 12, the actual $\mathcal{C}_{OBJ\_i}$ for the rectangular robot geometry is overlaid on the experimental data. Note, however, that the robots do not use the $\mathcal{C}_{OBJ\_i}$ for maintaining object closure, but instead they work with the virtual point robots as explained in Sect. 3.4. The object is caged in the three snap-shots shown.

7 Concluding Remarks

We presented algorithms for manipulating objects with multiple mobile robots combining the paradigms of pushing and caging. We defined the concept of object closure, a condition that ensures the objects are caged during manipulation. The main contributions of the paper are: (a) an algorithm that enable each robot to independently verify the condition of object closure; and (b) a decentralized control algorithm that enables each robot to move while maintaining object closure.

There are two main advantages of our approach. The decentralized algorithms only rely on the robots' ability to estimate the positions of their neighbors. Because robots are easily instrumented (in our case,

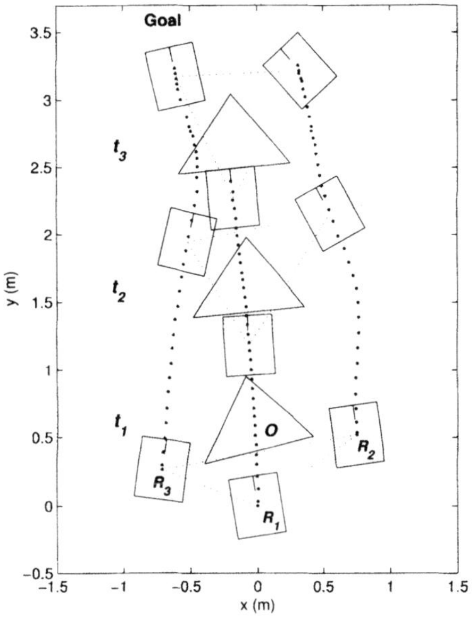

Fig. 11. Object transportation: t_1 – R_2 and R_3 are in the *Enclose* mode (see Fig. 8) trying to achieve object closure; t_2 – Object closure constraints are satisfied, R_2 and R_3 are in the *MaintainClosure* mode; t_3 – The robots are in the *GoToGoal* mode. R_1 is in the *GoToGoal* mode in all three snapshots.

this is done by tagging them with colored collars), this is relative easy even in an unstructured environment. Therefore, our methodology is potentially scalable for larger groups of robots operating in unstructured environments. Second, our algorithms do not rely on exact estimates of the position and orientation of the manipulated object. Therefore they are robust to errors in estimates of position and orientation of the manipulated object.

The main limitations of the algorithms used here include (a) the assumption of convex shapes; (b) the over approximation that is involved in verifying object closure when rotations are present; and (c) the use of the virtual point robots which result in sufficient conditions for maintaining object closure. All these result in conservative results with associated degradation in performance. For example, ensuring object closure with concave objects is often simpler than than is the case for convex objects. However, these assumptions and over approximations enable real-time performance and decentralized decision making with *guarantees*, and are important from a practical standpoint.

There are important directions of future work. First, it is necessary to explicitly model the nonholonomic behavior of the robots. The work in [16] provides a starting point in this direction. Second, we do not specifically consider algorithms for acquiring the object, establishing object closure (the *Enclose* mode) here. The papers [10,15]

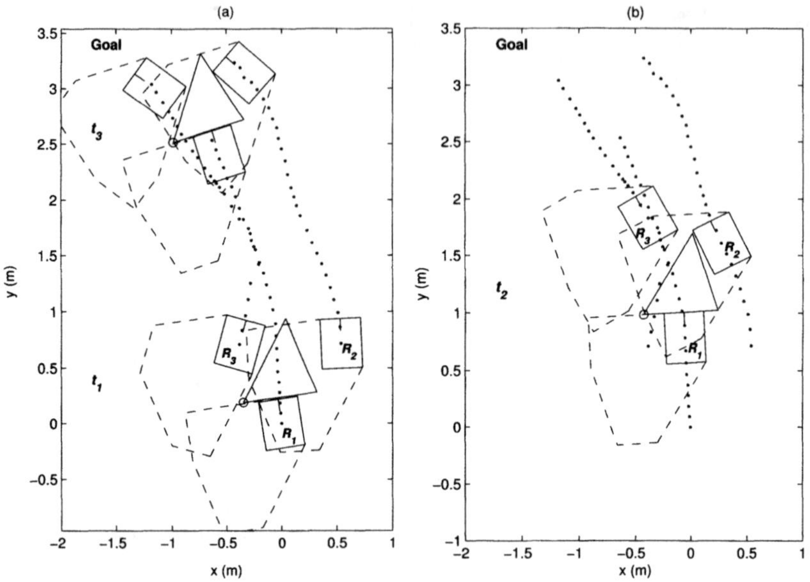

Fig. 12. The actual $\mathcal{C}_{OBJ\_i}$ (*dashed polygons*) for each robot. The origin of the object (∘) is always inside $\mathcal{C}_{cls}$ (the compact set delimited by the three $\mathcal{C}_{obj\_i}$) indicating a object closure condition. (a) – initial and final configurations; (b) – an intermediate configuration.

provide some approaches to this, with guarantees for small (3-4) teams of robots. There are challenges in designing decentralized policies that scale up to large numbers of robots. One of the key steps here is to remove the assumption related to non essential robots. Finally, we do not address the precise positioning and orienting of the object. By varying the threshold δ_2, we can get tighter tolerances on the object position relative to the robots. However, it is also essential to plan trajectories for the individual robots, instead of simply prescribing a common feedforward control signal u_T. The work in [13] provides a starting point in this direction.

Acknowledments.

We gratefully acknowledge the support of AFOSR grant no. F49620-01-1-0382, NSF grant no. CDS-97-03220, NSF grant no. IIS-0083420, DARPA ITO MARS Program, grant no. 130-1303-4-534328-xxxx-2000-0000, and CNPq-Brazil grants 200765/01-9 and 300212/99-2.

References

1. Ota J., Miyata N., Arai T. (1995) Transferring and regrasping a large object by cooperation of multiple mobile robots. In: *IEEE/RJS Int'l. Conf. on Intelligent Robots and Systems*, 543–548.
2. Kosuge K., Oosumi T. (1996) Decentralized control of multiple robots handling an object. In: *IEEE/RJS Int'l. Conf. on Intel. Robots and Systems*, 318–323.
3. Rus D. (1997) Coordinated manipulation of objects in a plane. *Algorithmica*, 19(1/2):129–147.
4. Sugar T., Kumar V. (1998) Decentralized control of cooperating mobile manipulators. In: *IEEE Int'l. Conf. on Robotics and Automation*, 2916–2921.
5. Rimon E., Burdick J. (1996) On force and form closure for multiple finger grasps. In: *IEEE Int'l. Conf. on Robotics and Automation*, 1795–1800.
6. Mataric M., Nilsson M., and Simsarian K. (1995) Cooperative multi-robot box-pushing. In: *IEEE/RJS Int'l. Conf. on Intelligent Robots and Systems*, 556–561.
7. Lynch K. M. (1996) Stable pushing: Mechanics, controllability, and planning. *Int'l. Journal of Robotics Research*, 15(6):533–556.
8. Rimon E., Blake A. (1996) Caging 2D bodies by one parameter, two-fingered gripping systems. In: *IEEE Int'l. Conf. on Robotics and Automation*, 1458–1464.
9. Davidson C., Blake A. (1998) Caging planar objects with a three-finger one-parameter gripper. In: *IEEE Int'l. Conf. on Robotics and Automation*, 2722–2727.
10. Wang Z., Kumar V. (2002) Object closure and manipulation by multiple cooperative mobile robots. In: *IEEE Int'l. Conf. on Robotics and Automation*, 394–399.
11. Sudsang A., Ponce J. (1998) On grasping and manipulating polygonal objects with disc-shaped robots in the plane. In: *IEEE Int'l. Conf. on Robotics and Automation*, 2740–2746.
12. Sudsang A., Ponce J., Hyman M., Kriegman D. J. (1999) On manipulating polygonal objects three 2-DOF robots in the plane. In: *IEEE Int'l. Conf. on Robotics and Automation*, 2227–2234.
13. Sudsang A., Ponce J. (2000) A new approach to motion planning for disc-shaped robots manipulating a polygonal object in the plane. In: *IEEE Int'l. Conf. on Robotics and Automation*, 1068–1075.
14. Latombe J-C. (1991) *Robot Motion Planning.* Kluwer Academic Publishers, Boston, MA.
15. Song P., Kumar V. (2002) A potential field based approach to multi-robot manipulation. In: *IEEE Int'l. Conf. on Robotics and Automation*, 1217–1222.
16. Esposito J. M., Kumar V. (2002) A method for modifying closed-loop motion plans to satisfy unpredictable dynamic constraints at runtime. In: *IEEE Int'l. Conf. on Robotics and Automation*, 1691–1696.
17. Ponamgi M. K., Manocha D., Lin M. C. (1997) Incremental algorithms for collision detection between polygonal models. *IEEE Trans. on Visualization and Computer Graphics*, 3(1):51–64.
18. Pereira G. A. S., Kumar V., Spletzer J., Taylor C. J., Campos M. F. M. (2002) Cooperative transport of planar objects by multiple mobile robots using object closure. In: Siciliano B., Dario P. (Eds.) *Experimental Robotics VIII, 8th Int'l Symp. on Experimental Robotics*, Springer, 275–284.

References

Communication Complexity of Multi-robot Systems

Eric Klavins*

California Institute of Technology, Pasadena, CA

Abstract. We examine the scalability of multi-robot algorithms. In particular, we attempt to capture the idea that the less coordination a multi-robot system requires, the better it should scale to large numbers of robots. To that end, we introduce a notion of *communication complexity* of multi-robot (or more generally, distributed control) systems as a surrogate for coordination. We describe a formalism, called CCL, for specifying multi-robot systems and algorithms for which the definition of communication complexity arises naturally. We then analyze the communication complexity of several, in some cases novel, multi-robot communication schemes each representative of one of several natural complexity classes.

1 Introduction

Research in multi-robot systems has matured to the point where systems with tens, hundreds or even thousands of robots are being proposed. Usually, to achieve a given task, the robots must share information — about what they are sensing, for example. Because more sharing requires more resources (time, sensory effort and communication bandwidth), the amount of information that must be shared determines how *coordinated* a task is.

The adverse effects of coordination are well known to researchers in parallel processing: Often, increasing the number of processors results in a corresponding decrease in the amount of time needed to perform a given computation only up to a point, after which the time spent coordinating inter-process communication outweighs the benefits of having more processors.

In contrast, multiple robots are not usually used to increase the speed of a computation. Rather, such systems are designed to cover a large area with sensors; or achieve fault tolerance by using large numbers of cheap, expendable units; or manipulate an object in parallel with vast numbers of tiny, relatively independent actuators. Nevertheless roboticists are faced with problems similar to those of the parallel algorithm designer: If the task charged to a multi-robot system by necessity requires a great deal of coordination, then the performance of the system will necessarily degrade as more robots are employed.

Thus a measure of *coordination complexity* is needed to evaluate proposed algorithms. Ideally such a measure should be a function of the minimal *information flow* required to perform a given task and be independent of how

* This research is supported in part by the DARPA SEC program under grant number F33615-98-C-3613 and by AFOSR grant number F49620-01-1-0361.

J.-D. Boissonnat et al. (Eds.): Algorithmic Foundations of Robotics V, STAR 7, pp. 275–291, 2004.

that flow is mediated (by sensors, communication networks, etc.). It should, furthermore, account for how that information changes (becomes obsolete) as a result of the dynamics of the environment. We do not yet have a way to characterize arbitrary information flow, however, so we use the more basic notion of communication complexity (Definition 4) to capture coordination complexity. The model we assume is that each robot knows its own state and can communicate directly with any other other robot via some (unspecified) communication system.[1] The cost of a communication event, in terms of bandwidth used, latency or time spent waiting, is summarized into a single, abstract cost (denoted by γ in Definition 4). At every step of a multi-robot algorithm, we sum the costs of the communication events that occurred in that step and then take the average over all the (infinitely many) steps in the control algorithm.

With this definition, it turns out that the worst case complexity of any reasonable algorithm is $O(n^2)$: all robots must communicate constantly with all other robots. With $O(n^2)$ complexity, the bandwidth of the communication system must increase with the square of the number of robots. For this reason, an $O(n^2)$ complexity algorithm or task is not considered scalable. In an $O(n)$ complexity algorithm, on the other hand, bandwidth need only scale linearly with the number of robots. In this case we imagine that, for example, for every 100 robots, we add one new communication node, and do not suffer a decrease in performance.

Ultimately, the question we would like to answer is: Given a task what is the minimum amount of communication required to achieve the task — for any number of robots? Such results are usually difficult to obtain in complexity theory. Thus, for now we examine algorithms we believe to be representative of various natural complexity classes, leaving the above question to future work. In particular, we explore communication schemes that depend on some minimal information about the motion of the robots – for example, how the distances between robots change. These schemes are intended to be used as components of higher level control algorithms, which we do not address here.

The communication schemes we investigate range from full communication to no communication. Two of the schemes are new. The first, *Distance Modulated Communication (DMC)*, uses the idea that a robot might need a very accurate estimate of the position of nearby robots (to avoid colliding with them, for example) while needing only a coarse estimate of more distant robots. Thus, the scheme has any two robots communicating at a frequency proportional to the distance between them. We show that DMC has communication complexity $O(n \log n)$ or $O(n^{1.5})$ depending on assumptions about the dynamics of the robots. The second, the *Wandering Communication Scheme (WCS)*, defines a protocol by which only a (small) constant number of "wandering" robots are allowed to move and communicate, while

[1] TCP/IP is an example of a protocol that, at a certain level of abstraction, can be supposed to provide point to point communication.

others must remain immobile. A wanderer may transfer its right to move, and its information about the world, to an immobile robot in a short burst of communication. It is shown that WCS has communication complexity $O(n)$ or $O(1)$ depending on assumptions about how certain higher level decisions are made by the robots.

In this paper, systems are specified in a variant of the UNITY language [1] which we call CCL (for Computation and Control Language) [5]. UNITY is a popular language for describing parallel algorithms in a way amenable to analysis. CCL is quite similar to UNITY, but modified for real time systems. In CCL, it is easy to both model the environment (e.g. the physics of the robotic systems involved) and represent the algorithms that operate on the environment as defined in Section 2.1. The definition of communication complexity (Definition 4), which is similar to that used in parallel algorithms, arises quite naturally in CCL. However, the definition can certainly be adapted to other formalisms. The reader is advised, in this paper, to consider CCL as merely a convenient and formal specification language.

Related Work We are primarily interested in scalable algorithms for formation forming [3] and other cooperative tasks for multi-vehicle systems [2] as well as distributed and self assembly [6]. The specification language CCL that we describe in Section 2.1 is very similar to the UNITY language [1]. Our main goal with CCL (not addressed here) is to develop a modeling, synthesis and verification tool for multi-robot and decentralized control systems, as described in [5]. Our notion of communication complexity borrows heavily from notions of communication complexity defined in the analysis of parallel algorithms as in, for example, [8] — the difference being that in the control systems we design we are concerned with maintaining some invariant indefinitely, as opposed to performing some finite computation. The idea of examining communication complexity was inspired by, but only superficially related to, the sort of communication complexity developed in [7]. Bandwidth-aware control and communication schemes are receiving increased attention in the controls literature [9], but scalability as addressed in this paper has not been addressed elsewhere.

2 Definitions

2.1 CCL: The Computation and Control Language

We describe CCL briefly here, details can be found in [5]. We suppose we have a set V of **variable symbols** and define a **state** s to be a function from V into some universe U of values: $s : V \to U$. We let S denote the set of all states.

Definition 1. A **clause** is a pair $g : r$ where $g : S \to \texttt{Bool}$ is called the **guard** and $r : S \times S \to \texttt{Bool}$ is called the **rule**. If c is a clause, then the guard of c is denoted $c.g$ and the rule is denoted $c.r$.

Guards are expressed syntactically as logical expressions over V and rules are expressed syntactically as logical expressions over $V \cup V'$ where $V' = \{v' : v \in V\}$ is the set of *primed* variable symbols from V. A primed variable symbol refers to the next value of a variable while an unprimed variable symbol refers to the present value. For example, if c is the clause

$$x > 0 : y' = z \wedge x' < 0$$

then $c(s_1, s_2)$ essentially means

$$(s_1(x) > 0 \rightarrow s_2(y) = s_1(z) \wedge s_2(x) > 0) \;\wedge\; (s_1(x) \leq 0 \rightarrow s_2 = s_1).$$

A variable $v \in V$ is said to **occur** in a clause $g : r$ if v appears in the expression for either g or r. It is said to **occur primed** in $g : r$ if v' occurs in r. If $c.g(s) = \mathit{true}$ we say that c is **applicable** in state s. A specification is essentially a collection of clauses, along with an initial condition, a special clause that describes the dynamics of the environment, and a communication cost function.

Definition 2. A **specification** is a quadruple $\Pi = (\mathcal{I}, \mathcal{C}, \Delta, \gamma)$ where

1. $\mathcal{I} : S \rightarrow \texttt{Bool}$ is the **initial condition**
2. $\mathcal{C}$ is a set of clauses
3. Δ is a clause called the **dynamics** clause
4. $\gamma : \mathcal{C} \rightarrow \mathbb{R}^+ \cup \{0\}$ gives the **communication cost** of each clause.

Definition 3. Let $\Pi = (\mathcal{I}, \mathcal{C}, \Delta, \gamma)$ be a specification. An **execution** of Π is a sequence $\{s_k\}$ for $k \in \mathbb{N}$ such that $\mathcal{I}(s_0)$ and for all $k \geq 0$

1. for all $c \in \mathcal{C}$, if $c.g(s_k)$ then $c.r(s_k, s_{k+1})$
2. $\Delta.r(s_k, s_{k+1})$
3. if $v \in V$ does not occur primed in any clause applicable in state s_k, then $s_{k+1}(v) = s_k(v)$.

We denote the set of all executions of Π by $\mathcal{E}(\Pi)$.

2.2 Communication Complexity of Specifications

We have already defined the cost of a clause. We next define the *communication complexity* of a step in an execution, of an execution, and finally of a specification. For each of these notions of complexity we use the polymorphic symbol $cc(\cdot)$, the exact meaning of which should always be clear from context.

Definition 4. Assume a fixed specification $\Pi = (\mathcal{I}, \mathcal{C}, \Delta, \gamma)$ is given. The **communication complexity** of a *state* $s \in S$ is

$$cc(s) \triangleq \sum_{c \in \mathcal{C} \wedge c.g(s)} \gamma(c).$$

The **communication complexity** of an *execution* $\{s_k\} \in \mathcal{E}(\Pi)$ is

$$cc(\{s_k\}) = \lim_{T\to\infty} \frac{1}{T}\sum_{k=1}^{T} cc(s_k).$$

The (worst case) **communication complexity** of the *specification* Π is

$$cc(\Pi) = \max_{\{s_k\}\in\mathcal{E}(\Pi)} cc(\{s_k\}).$$

2.3 Basic Properties and Proof Methods

We next supply some basic results that can aid in determining the communication complexity of a particular specification. Generally, to prove that a given specification has a certain communication complexity, we determine the frequency with which each clause is applicable in any execution. This may be done deterministically (Lemma 1) or probabilistically (Lemmas 2 and 3). The proofs of these properties are given in the Appendix to this paper. The first lemma allows us to easily compute the cost of a specification that has periodic executions.

Lemma 1. *Let $\Pi = (\mathcal{I}, \mathcal{C}, \Delta, \gamma)$ be a specification and suppose that each clause $c_i \in \mathcal{C}$ is applicable every r_i steps in any execution of Π. Then*

$$cc(\Pi) = \sum_{i=1}^{|\mathcal{C}|} \frac{\gamma(c_i)}{r_i}.$$

Next suppose that the cost of each step k of any execution is modeled by a non-negative real valued random variable C_k with expectation $E[C_k]$. In the proofs of Theorems 3 and 4, C_k is essentially a function of a probabilistic model of the frequencies with which clauses in the corresponding specifications are applicable. The probabilistic models arise from either assumptions about the relative locations of the robots (e.g. their locations may be uniformly distributed in their workspace) or assumptions on the frequency with which certain guards are true (e.g. a guard may be true at some step if a flip of a biased coin comes up heads at that step). Given C_k, we would like to compute the expected cost of the specification which, in general, is

$$E[cc(\Pi)] = \lim_{T\to\infty} \frac{1}{T}\sum_{k=1}^{T} E[C_k].$$

We begin with the case where each C_k is independent and identically distributed.

Lemma 2. *Let $\Pi = (\mathcal{I}, \mathcal{C}, \Delta, \gamma)$ be a specification and suppose the cost of executing each step k is modeled by the random variable C. Then $E[cc(\Pi)] = E[C]$.*

The next lemma allows us to compute the communication complexity of a specification if we know what the limit of the step cost is. It will be useful when we consider specifications modeled by Markov chains (as in Theorem 4). Of course, Lemma 2 is a special case of Lemma 3.

Lemma 3. *Let* $\Pi = (\mathcal{I}, \mathcal{C}, \Delta, \gamma)$ *be a specification and suppose the cost of each step* k *in any execution is* C_k. *If* $\lim_{k\to\infty} C_k = \kappa$ *then* $cc(\Pi) = \kappa$. *Furthermore, if each* C_k *is a random variable and* $\lim_{k\to\infty} E[C_k] = \kappa$, *then* $E[cc(\Pi)] = \kappa$.

Remark: Observe that the definition of communication complexity is an infinite average. A finite, initial period of $O(n^2)$ communication will not affect an ultimately less complex algorithm. Thus, in the specifications in this paper, we may assume that the global state is known initially since this could have been obtained by a brief period of total communication.

2.4 Specifications of Multi-robot Systems

In the sequel we consider systems with n robots with locations $(x_i, y_i) \in \mathbb{R}^2$ for each $i \in \{1, ..., n\}$. We suppose (implicitly) that certain variables in V and certain clauses in $\mathcal{C}$ are *owned* by particular robots. Variables owned by robot i are denoted as such via an initial subscript i, as in x_i or $e_{i,j}$. The cost $\gamma(c)$ of a clause c is generally the result of an explicit reference to a variable v_j in a clause owned by robot i where $i \neq j$. Variables not sub-scripted are not owned (i.e. they are global) and the dynamics clause has no cost, being executed, as it is, by the environment (which by definition includes all of the robots).

In the dynamics clause we will generally be non-committal about the details of the system specified. For example, we will make use of the clause $\Delta(n)$ defined by

$$\mathit{true} : t' = t + \delta \wedge \forall i. \, (||x_i' - x_i|| < \delta v_{max}) \tag{1}$$

that simply states that time goes forward by some constant $0 < \delta \ll 1$ and the velocity of any robot is bounded by some constant $v_{max} > 0$.[2] Note that we have defined a family of dynamics clauses, one for each $n \in \mathbb{N}$ and have used the parameter n implicitly in the rule for $\Delta(n)$ by assuming that i is quantified over $\{1, ..., n\}$.

3 Multiple Robot Communication Schemes

In this section we present multi-robot communication schemes for several natural complexity classes: $O(n^2)$, $O(n^{1.5})$, $O(n \log n)$, $O(n)$ and $O(1)$. We

[2] This constraint on the velocity can be considered as a crude discretization of the actual continuous dynamics. We assume that $1/\delta$ is greater than the frequency of the actual dynamics so that aliasing problems do not occur.

usually assume the dynamics of the robots are given by (1) or some variant of it. A discussion of how these schemes may be used, augmented and extended appears in Section 4. The scheme we omit is the empty communication scheme where no communication occurs at all. This scheme clearly has communication complexity $O(0)$.

3.1 The Worst Case: $\mathbf{O(n^2)}$ Complexity

Many simple multi-robot algorithms treat a group of robots as a single (albeit disconnected) robot. Assuming there is no leader (a fragility one might wish to avoid), such algorithms usually require that a good estimate of the global state of the system be known by each robot. Thus, each robot i must maintain, for each j, an estimate of x_j which we denote $e_{i,j}$. Furthermore we require that the property

$$||e_{i,j} - x_j|| < \varepsilon \tag{2}$$

hold at any time, for some constant $\varepsilon > \delta v_{max}$ which may be, for example, required for the stability of whatever control algorithm the system is using. If we assume the dynamics are given by (1), the best that can be achieved is for each robot i to include the clause $c_{i,j}$ defined by

$$(t + \delta - l_{i,j})v_{max} > \varepsilon \ : \ e'_{i,j} = x_j \wedge l'_{i,j} = t \tag{3}$$

for $i, j \in \{1, ..., n\}$ and $i \neq j$. We set $\gamma(c_{i,j}) = 1$. In (3), $l_{i,j}$ represents the time at which the last update to $e_{i,j}$ took place. The inequality in the guard assumes the worst case movement (a velocity of exactly v_{max}) by robot j — that is, $(t + \delta - l_{i,j})v_{max}$ represents the worst case value for $||e_{i,j} - x_j||$ in the next step (at time $t + \delta$). If this exceeds ε, then the estimate $e_{i,j}$ and the time of last update $l_{i,j}$ are updated.

Supposing that the guard in $c_{i,j}$ is an equality and solving for $t - l_{i,j}$ gives that the clause will be applied every $r_{i,j} \triangleq \frac{\varepsilon - \delta v_{max}}{v_{max}}$ steps. Using Lemma 1 we obtain the simple result

Theorem 1. *Let* $\mathcal{I}(n) = \forall i, j \ . \ e_{i,j} = x_j \wedge l_{i,j} = 0$ *and* $\mathcal{C} = \{c_{i,j} \mid 1 \leq i \neq j \leq n\}$ *and put*

$$\Pi_{global}(n) = (\mathcal{I}(n), \mathcal{C}(n), \Delta(n), \gamma)$$

where $c_{i,j}$ *is as in (3) and* Δ *is as in (1). Then* $cc(\Pi_{global}(n)) = \mathrm{O}(n^2)$.

We use the fact that $|\mathcal{C}| = n(n-1)$. Note that the result is independent of the constants δ, v_{max} and ε which essentially determine (the constant) $r_{i,j}$. Thus, even techniques which increase the period $r_{i,j}$ of communication significantly, by having each robot use a sophisticated estimator of all other robot's positions, for example [9], ultimately do not scale in the sense introduced in this paper.

Define $c_{i,j}$ by

$$(t - l_{i,j})v_{max} \geq k\,(||e_{i,j} - x_i|| - (t - l_{i,j})v_{max}) : \\ e'_{i,j} = x_j \wedge l'_{i,j} = t$$

$$\gamma(c_{i,j}) = 1$$

$$\mathcal{I}(n) \equiv \forall i,j \in \{1,...,n\}\ (\ e_{i,j} = x_j \wedge l_{i,j} = 0 \wedge t = 0\ )$$

$$\mathcal{C}(n) = \{c_{i,j} \mid i,j \in \{1,...,n\} \wedge i \neq j\}$$

Then $\Pi_{DMC}(n) = (\mathcal{I}(n), \mathcal{C}(n), \Delta(n), \gamma)$

Fig.1: The specification of the distance modulated communication scheme Π_{DMC}. $\Delta(n)$ refers to the dynamics clause in equation (1).

3.2 Sub-O(n^2) Complexity

In this section we introduce *distance modulated communication* (DMC). The specification, shown in Figure 1, is based on the idea that closer robots should communicate their positions more frequently than distant ones. The goal is to maintain the invariant

$$||e_{i,j} - x_j|| \leq k||x_i - x_j|| \tag{4}$$

for each $i \neq j$. Here $e_{i,j}$ is the ith robot's estimate of the jth robot's position and $k > 0$ is a constant. The behavior of each robot i is specified by the clause $c_{i,j}$, for $j \in \{1,...,n\} - \{i\}$. The variable $l_{i,j}$ is used by robot i to keep track of the last time robot i updated its estimate $e_{i,j}$, thereby accruing 1 unit of communication cost. The quantity $(t - l_{i,j})v_{max}$ in the guard represents the maximum possible difference between $e_{i,j}$ and x_j. The quantity $||e_{i,j} - x_i|| - (t - l_{i,j})v_{max}$ represents the minimum possible distance between the two robots. That is

$$||e_{i,j} - x_j|| \leq (t - l_{i,j})v_{max} \text{ and } ||e_{i,j} - x_i|| - (t - l_{i,j})v_{max} \leq ||x_i - x_j||.$$

Each of these bounds can be computed by agent i without reference to x_j and so require no communication. When the bound on the estimate error exceeds k times the bound on the distance, the guard becomes true and $e_{i,j}$ and $l_{i,j}$ are updated.

The communication complexity of Π_{DMC}, of course, depends on the relative locations of the robots while the specification is enforced. We will show that under two different reasonable assumptions about the robot locations, that $cc(\Pi_{DMC}(n))$ is better than $O(n^2)$. First we suppose that the robots are arranged in a line. We specify this in the initial condition and by using a trivial dynamics clause. Of course, the theorem holds for any spacing and any dynamics clause that (even approximately) preserves it, and the extremely simple dynamics clause in the statement of the theorem below is used merely for simplicity in presentation.

Theorem 2. *Let $\mathcal{J}(n) = \forall i \in \{1, ..., n\}(x_i = (i, 0))$ and $\Delta = true : t' = t+\delta$ and put*

$$\Pi_{DMC}(n) = (\mathcal{I}(n) \wedge \mathcal{J}(n), \mathcal{C}(n), \Delta, \gamma)$$

where $\mathcal{I}(n)$, $\mathcal{C}(n)$ and γ are as in Figure 1. Then $cc(\Pi_{DMC}(n)) = O(n \log n)$.

Proof: Let $d_{i,j} \triangleq ||x_i - x_j|| = |i - j|$. Solving $tv_{max} = k(d_{i,j} - tv_{max})$ for t shows that clause $c_{i,j}$ will be applicable every

$$r_{i,j} \triangleq \frac{kd_{i,j}}{(1+k)\delta v_{max}} = \alpha|i-j|$$

steps where $\alpha \triangleq \frac{k}{(1+k)\delta v_{max}}$. By Lemma 1, the communication complexity is thus

$$cc(\Pi_{DMC}(n)) = \sum_{1 \le i \ne j \le n} \frac{1}{\alpha|i-j|} = \frac{2}{\alpha} \sum_{i=1}^{n} \sum_{j=i+1}^{n} \frac{1}{j-i}.$$

It is straightforward to show that this sum evaluates to $\frac{2}{\alpha}(nH(n) - n) = O(n \log n)$ where $H(n)$ is the nth harmonic number. □

We next explore the somewhat more natural assumption that the robots are randomly distributed in a square on the plane with an average density of ρ robots per square meter and determine the *expected value* of the communication complexity of $\Pi_{DMC}(n)$. We therefore suppose that the position of robot i is given by a pair of independent identically distributed random variables, (X_i, Y_i), with the common density function

$$f_m(x) \triangleq \begin{cases} \frac{1}{2m} \text{ if } x \in [-m, m] \\ 0 \text{ otherwise.} \end{cases}$$

For some $m \in \mathbb{R}$ yet to be determined. We then have

Theorem 3. *Let $\mathcal{J}(n) = \forall i \in \{1, ..., n\}((x_i, y_i) \in [-m, m]^2)$ where $m = \sqrt{\frac{n}{4\rho}}$ and let Δ be as in Theorem 2 and let*

$$\Pi_{DMC}(n) = (\mathcal{I}(n) \wedge \mathcal{J}(n), \mathcal{C}(n), \Delta(n), \gamma)$$

where $\mathcal{I}(n)$, $\mathcal{C}(n)$ and γ are as in Figure 1. Then $E[cc(\Pi_{DMC}(n))] = O(n^{1.5})$.

This result suggests that a uniform distribution of robots in a square of density ρ is less sparse than an equal spacing of the robots in a line. In fact, one expects that the complexity of Π_{DMC} approaches $O(n^2)$ as the dimension of the workspace grows.

Proof: For each $i \neq j$, define the random variable $D_{i,j} = ||X_i - X_j||$. Then clause $c_{i,j}$ will be applicable every $R_{i,j} = \alpha D_{i,j}$ steps, where α is the same as in the proof of Theorem 2. The cost of a single step is then

$$C \triangleq \sum_{1 \le i \neq j \le n} \frac{1}{R_{i,j}}$$

units. Using Lemma 2 we have that $E[cc(\Pi_{DMC}(n))] = E[C]$. Next we make use of the following lemma (proved in the Appendix).

Lemma 4. *The expected cost at each step is* $E[C] = \frac{\beta}{\alpha}\frac{1}{16m}n(n-1)$ *where* $\beta = \frac{8}{3}(4 - 4\sqrt{2} - 9\ln(\sqrt{2}-1) + 3\ln(\sqrt{2}+1))$.

The area of the square in which the robots are placed is $4m^2$. Taking $m = \sqrt{\frac{n}{4\rho}}$ gives an area of n/ρ so that there are ρ robots per square meter. We then have

$$E[cc(\Pi_{DMC}(n))] = E[C] = 2\frac{\beta\sqrt{\rho}}{16\alpha}\sqrt{n}(n-1) = O(n^{1.5})$$

as desired. □

The conditions under which the DMC protocol can be used are much broader than the two situations we have analyzed in Theorems 2 and 3. However, we believe these two cases to be typical. The first represents a common *formation* and other formations likely produce similar complexity results. The second represents the reasonable assumption that many multi-robot algorithms would tend to keep the robots more or less evenly distributed in the workspace: i.e. that the workspace would grow if the number of robots were increased.

3.3 O(n) Complexity

In [8] it is shown (in a slightly somewhat different model than that presently considered) that the communication complexity of systems with n processors arranged in a graph with constant degree is O(n). We can easily obtain a similar result here with respect to multi-robot systems. In particular, suppose we have a graph $G = (\{1, ..., n\}, E)$ where $E \subseteq V \times V$ and that we wish to maintain the property $(i,j) \in E \Rightarrow ||e_{i,j} - x_j|| < \varepsilon$ in place of (2). Then the obvious corollary can be shown:

Corollary 1. *Let* $\mathcal{I}(n) = \forall i,j \,.\, e_{i,j} = x_j \wedge l_{i,j} = 0$ *and* $\mathcal{C} = \{c_{i,j} \,|(i,j) \in E\}$ *and put*

$$\Pi_{nbr}(n) = (\mathcal{I}(n), \mathcal{C}(n), \Delta(n), \gamma)$$

where $\Delta(n)$ *is as in (1) and* $c_{i,j}$ *is as in (3). Then* $cc(\Pi_{nbr}(n)) = O(n)$.

Another commonly used scheme for controlling a group of robots is to have each robot transmit its location to a leader robot (or control computer) with some frequency. The leader then sends to each robot a command based on the collected global state. The communication complexity of this scheme is also $O(n)$, assuming the size of the commands does not depend on n. This scheme (in its simplest form) is not robust to failures in the leader, however.

3.4 Better than O(n) Complexity

The obvious way to get better than O(n) complexity is to have no communication at all. The subject of how to do anything useful without communication has been investigated by others and is beyond the scope of this paper. Instead, in this section, we examine a task similar to that explored in Section 3.1: to maintain the invariant $||e_{i,j} - x_j|| < \varepsilon$ for some constant ε for any *currently moving* robot. This is possible with lower than O(n^2) communication complexity if we restrict the motions of the robots, allowing only certain robots to move at any given time (the dynamics clause in Figure 2, for example, enforces this constraint). In particular, we present a communication and movement scheme that has either O(n) or O(1) communication complexity for any number of robots, depending on our assumptions. The scheme takes advantage of the presumption that any robot that is not moving does not need to know any other robot's states. In particular, we choose a constant κ and only allow, for any number of robots n, that κ of them can move at any time. A robot can give up its right to move by stopping and uploading its estimates of the other robot locations to a newly chosen and formerly idle robot.

The specification of the scheme, called the *Wandering Communication Scheme* (WCS), is shown in Figure 2. The meanings of the variables are as follows. The variable q_i denotes the state of the robots as either wandering ($q_i = 1$), uploading ($q_i = 2$) or idle ($q_i = 3$). The variable $e_{i,j}$ is the ith robot's estimate of x_j. Setting the "reservation" variable r_i to j specifies that robot j is uploading to robot i and its appearance in $c_{i,3}$ prevents two robots from uploading to a same new robot.

Clause $c_{i,1}$ is used to keep a wandering robot's estimate of the other moving robots accurate. Clause $c_{i,2}$ is used to decide whether to switch from wandering to idle. It uses the random number generator $coin(p, t)$, which is true at time t with probability p. Clause $c_{i,3}$ is used by robot i to choose a new robot j, to whom it will hand over its "right" to move: uploading its state estimates to j and switching j's state to wandering and its own state to idle.

For a given n and κ we model an execution of $\Pi_{WCS}(n)$ as a Markov Chain whose states X_m correspond to the possible number of robots in each specification state $q_i \in \{1, 2, 3\}$ at step m. The number of robots that are in state 3 is always $n - \kappa$. Thus X_m is just the number of robots in state 1. The number of robots in state 2 is just κ minus the number of robots in state 1.

For $i \in \{1, ..., n\}$ define clauses $c_{i,1}$, $c_{i,2}$ and $c_{i,3}$ by

$$
\begin{aligned}
&q_i = 1 : \forall j (q_j \neq 3 \rightarrow e'_{i,j} = x_j) \\
&q_i = 1 \wedge coin(p,t) : q'_i = 2 \\
&q_i = 2 : q'_i = 3 \wedge r'_i = \bot \wedge \\
&\qquad \exists j \left[q_j = 3 \wedge q'_j = 1 \wedge r'_j = i \wedge \forall k (e'_{j,k} = e_{i,k}) \right]
\end{aligned}
$$

respectively. Then set

$$
\begin{aligned}
&\gamma(c_{i,1}) = \kappa - 1 \\
&\gamma(c_{i,2}) = 0 \\
&\gamma(c_{i,3}) = n
\end{aligned}
$$

$$
\begin{aligned}
\mathcal{I}(n) \equiv \forall i [& (i \leq \kappa \rightarrow q_i = 1) \wedge (i > \kappa \rightarrow q_i = 3) \\
& \wedge r_i = \bot \wedge \forall j (e_{i,j} = x_j)]
\end{aligned}
$$

$$\mathcal{C}(n) = \bigcup_{i=1}^{n} \{c_{i,1}, c_{i,2}, c_{i,3}\}$$

$$
\begin{aligned}
&\Delta(n) = true : t' = t + \delta \wedge \\
&\quad \forall i [(q_i = 1 \rightarrow ||x'_i - x_i|| \leq \delta v_{max} \\
&\quad \wedge (q_i \neq 1 \rightarrow x'_i = x_i)]
\end{aligned}
$$

$$\Pi_{WCS}(n) = (\mathcal{I}(n), \mathcal{C}(n), \Delta(n), \gamma)$$

Fig.2: The specification of the Wandering Communication Scheme (WCS). $coin(p,t)$ is a random number generator that is true at time t with probability p. We assume quantification over $\{1, ..., n\}$ in the notation.

The transition probabilities are, for each i and j,

$$
P[X_{m+1} = \kappa - j \mid X_m = \kappa - i] = \begin{cases} \binom{\kappa - i}{j} p^j (1-p)^{\kappa - i - j} & \text{if } 1 \leq j \leq \kappa - i \\ 0 \text{ otherwise.} \end{cases} \tag{5}
$$

We define the $\kappa + 1$ dimensional transition matrix P to be the matrix whose jth row and ith column is given by (5). Next, we determine the cost of each step X_m. The communication cost of a robot in state 1 is $\kappa - 1$ and the the communication cost of a robot in state 2 is n. In state 3, a robot does not communicate. Thus, the cost at step m is

$$C_m \triangleq (\kappa - 1) X_m + n(\kappa - X_m).$$

And the expected total cost at step m is

$$E[C_m] = (1 \underbrace{0 \ldots 0}_{\kappa \text{ times}})P^m r \tag{6}$$

where r is the cost vector

$$r \triangleq ((\kappa-1)\kappa + 0n,\ (\kappa-1)(\kappa-1)+1n,\ \ldots,\ (\kappa-1)0 + n\kappa)^T .$$

We show, by straightforward analysis (as described in many texts [4] on the subject) in the Appendix that

Lemma 5. *Let x be the $\kappa+1$ dimensional vector whose ith element is $\binom{\kappa}{i} p^{i-\kappa}$ for $i \in \{0, \ldots, \kappa\}$. Then as m goes to infinity, P^m goes to the matrix whose rows are all given by*

$$\tilde{x}^T \triangleq \left(\frac{p}{p+1}\right)^\kappa x^T .$$

Using this result we can show that

Theorem 4. *If the value of p in $\Pi_{WCS}(n)$ is constant, then $E[cc(\Pi_{WCS}(n))] = \mathrm{O}(n)$ and if $p \triangleq 1/n$ then $E[cc(\Pi_{WCS}(n))] = \mathrm{O}(1)$.*

Proof: The limit of the expected value of the cost (6) as $m \to \infty$ is

$$\tilde{x}^T r = \left(\frac{p}{1+p}\right)^\kappa \sum_{j=0}^{\kappa} [(\kappa-1)j + (\kappa-j)n] \binom{\kappa}{j} p^{j-k} = \frac{\kappa}{1+p}(\kappa - 1 + np).$$

We use Lemma 3 to conclude that if p is constant, that $E[cc(\Pi_{WCS}(n))] = \mathrm{O}(n)$. If, on the other hand, $p \triangleq 1/n$, then

$$\lim_{m\to\infty} E[C_m(n)] = \frac{\kappa}{1+\frac{1}{n}}\kappa < \kappa^2 = \mathrm{O}(1).$$

Once again using Lemma 3 yields the desired result. □

4 Discussion

Control Pairing a communication scheme with a high level controller depends on the requirements of the controller and the guarantees of the scheme. Using DMC, for example, a reactive planner for each robot may be constructed as follows. Consider two robots with indices i and j and suppose that i has last communicated with j at time $l_{i,j}$. It can easily be determined, from the guard in clause $c_{i,j}$ in Figure 1, at what time $l_{i,j} + \Delta t$ it will communicate with j again. Thus, vehicle i can simply avoid the region

$$R_j = \{\, x \,:\, ||e_{i,j} - x|| < \Delta t \, v_{max} \,\}$$

while planning its route through the workspace. Furthermore, to make the path planning problem easier, other distance metrics can be used in DMC such as the distance between robot j and the line segment connecting robot i with its goal position.

Assumptions The analyses in Section 3 make assumptions that are certainly not as general as they could be. For example, Theorem 2 should hold for any formation that increases in size linearly with the number of robots used. Theorem 2 could be extended to allow motions that are guaranteed to preserve a certain expected distance between robots. The WCS has many variants, each of which affect the expected communication complexity. For example, constant communication complexity can be obtained by requiring that all idle robots be positioned in one of some large number of (sparsely placed) "island regions". The wandering robots wouldn't necessarily know where an idle robot was, but they would know at least that it was located in *some* island region. The transfers of position estimates in clause $c_{i,3}$ in Figure 2 would then have cost κ resulting in O(1) complexity overall.

Sensing Although sensing is not really mentioned in Section 3, it is modeled. For example, the assumption that robots know their own locations implies a certain amount of sensing. Also, clause $c_{i,3}$ where it states "$\exists j [q_j = 3...$" assumes that robot i can find another robot whose state is 3 (idle). This information can not be obtained from the scheme itself and adding the communication required to make the discrete states of the robots global would increase the communication complexity of the algorithm. The original intent, however, was to have wanderers *find* idle robots using their local sensors and then upload their estimates to them. More generally, however, incorporating sensing assumptions into CCL specifications remains an art and no clear path toward investigating, say, sensing vs. communication trade-offs has yet emerged.

Variants There are many possible variants of Definition 4. For example, suppose that the cost of communication γ were proportional to the distance between the communicating robots, as is the case with low power sensor nets, or "smart dust". With this definition, communicating indirectly via hops in a network instead of directly may give a lower overall complexity. A *power aware* communication scheme could be devised that determines the lowest energy route from one robot to another based on current estimates (possibly obtained by a variant of DMC or WCS) of the vehicle locations.

5 Conclusion

We introduced a notion of communication complexity as a means of investigating the scalability of multi-robot algorithms in terms of how much coordination they require and presented several communication schemes, two of which are (to the best of our knowledge) new, that maintain estimates of the global state of varying accuracies in the face of simply-modeled possible changes in the environment. By analyzing their complexities, we show that these schemes cover several natural communication complexity classes from $O(n^2)$ communication to $O(1)$ communication. It is our hope that techniques similar to these will be used to check the scalability of newly proposed multi-robot or distributed control algorithms.

We have not included a model of possible sensing modalities and costs. Nor have we described the actual control algorithms and higher level planners that could be used in conjunction with the schemes we have presented. More importantly, we have only analyzed instances of algorithms that achieve a given task but have not answered the question: For a given task, what is the minimum communication complexity of *any* algorithm that achieves the task? It is simple to show that the algorithms we supply for maintaining (2) and (4) respectively are the best possible, but for more complex tasks the question appears to be more difficult and we plan to attempt address it in a future paper.

Acknowledgments. This work grew out of discussions with Richard Murray and Jason Hickey. Abhishek Tiwari provided many helpful comments on a draft of this paper. This research is supported in part by the DARPA SEC program under grant number F33615-98-C-3613 and by AFOSR grant number F49620-01-1-0361.

A Appendix

A.1 Proof of Lemma 1

In T steps, clause i is executed $\frac{T}{r_i}$ times. Thus,

$$\sum_{k=1}^{T} cc(s_k) = T \sum_{i=1}^{|C|} \frac{\gamma(c_i)}{r_i}.$$

Multiplying by $1/T$ and taking the limit yields the desired result. □

A.2 Proof of Lemma 2

Suppose the cost of each step in any execution of Π can be modeled as a random variable C and that the costs at each step are independent. Then

$$E[cc(\Pi)] = \lim_{T\to\infty} \frac{1}{T} \sum_{k=1}^{T} E[C] = \lim_{T\to\infty} \frac{1}{T} T E[C] = E[C].$$ □

A.3 Proof of Lemma 3

Suppose $\lim_{k\to\infty} C_k = \kappa$. Then for any $\epsilon > 0$, a positive integer T_0 can be found so that $\kappa - \varepsilon < C_k < \kappa + \varepsilon$ for all $k > T_0$. Thus,

$$\frac{1}{T}\sum_{k=1}^{T} C_k = \frac{1}{T}\left(\sum_{k=1}^{T_0} C_k + \sum_{k=T_0+1}^{T} C_k\right) < \frac{1}{T}\left[\alpha + (T - T_0 - 1)(\kappa + \varepsilon)\right]$$

where $\alpha = \sum_{k=1}^{T_0} C_k$. Similarly,

$$\frac{1}{T}\sum_{k=1}^{T} C_k > \frac{1}{T}\left[\alpha + (T - T_0 - 1)(\kappa - \varepsilon)\right].$$

Taking limits, we arrive at

$$\kappa - \varepsilon < \lim_{T\to\infty} \frac{1}{T}\sum_{k=1}^{T} C_k < \kappa + \varepsilon.$$

This argument holds for any ε and thus we may conclude, by the *squeeze theorem*, that

$$\lim_{T\to\infty} \frac{1}{T}\sum_{k=1}^{T} C_k = \kappa.$$

The rest of the lemma follows. □

A.4 Proof of Lemma 4

Let $g(x_1, y_1, x_2, y_2) = \left((x_2 - x_1)^2 + (y_2 - y_1)^2\right)^{-\frac{1}{2}}$. Straightforward integration shows that

$$\int_{-m}^{m}\int_{-m}^{m}\int_{-m}^{m}\int_{-m}^{m} g(x_1, y_1, x_2, y_2)\, dx_1 dy_1 dx_2 dy_2$$

$$= \frac{8}{3}m^3\left(4 - 4\sqrt{2} - 9\ln(\sqrt{2} - 1) + 3\ln(\sqrt{2} + 1)\right) \triangleq \beta m^3.$$

The expected value $E[C]$ is given by

$$\frac{1}{\alpha}\left(\frac{1}{2m}\right)^{2n} \underbrace{\int_{-m}^{m} \cdots \int_{-m}^{m}}_{n(n-1)\text{ times}} \sum_{1\le i\ne j\le n} g(x_i, y_i, x_j, y_j)\, dx_1 ... dy_n = \frac{\beta}{\alpha}\frac{1}{16m}n(n-1). \quad \square$$

A.5 Proof of Lemma 5

It is known that the matrix P^T has 1 as an eigenvalue and that the steady state of the system $y^T P^m$ (for any given $y \neq 0$) is given by (any nonzero multiple of) the eigenvector corresponding to 1 [4]. We first show that $x^T P = x^T$ is one such eigenvector,where x is as in the statement of the lemma. By the definition of matrix multiplication, the ith element of $x^T P$ is

$$\sum_{j=0}^{\kappa - i} \binom{k}{j} p^{j-\kappa} \binom{\kappa - j}{i} p^i (1-p)^{\kappa - i - j}.$$

Define $s \triangleq \kappa - i$. Then the sum becomes

$$p^{-s} \sum_{j=0}^{s} \binom{k}{j} \binom{\kappa - j}{\kappa - s} p^j (1-p)^{s-j}.$$

It is straightforward to show that sum part of this expression is $\binom{\kappa}{s}$. Thus, the ith element of $x^T P$ is

$$p^{-s} \binom{\kappa}{s} = \binom{\kappa}{\kappa - i} p^{i-\kappa} = \binom{\kappa}{i} p^{i-\kappa}$$

as desired.We now have that for any $y \neq 0$ that $y^T P^m \to \alpha x$ for some $\alpha \neq 0$. In particular, $(1\ 0\ \ldots\ 0) \lim_{m\to\infty} P^m = \alpha x$ so that the first row of the limit of P^m is given by αx. Since each row of P^m must sum to 1, we obtain that

$$\alpha = \frac{1}{||x||} = \left(\sum_{i=0}^{\kappa} \binom{\kappa}{i} p^{i-\kappa} \right)^{-1} = \left(\frac{p}{p+1} \right)^{\kappa}$$

using the binomial theorem. □

References

1. K. M. Chandy and J. Misra. *Parallel Program Design: A Foundation.* Addision-Wesley, 1988.
2. L. Cremean, B. Dunbar, D. van Gogh, J. Hickey, E. Klavins, J. Meltzer, and R. M. Murray. The Caltech multi-vehicle wireless testbed. In *Conference on Decision and Control*, 2002. Submitted for Review.
3. J. A. Fax and R. M. Murray. Information flow and cooperative control of vehicle formations. In *IFAC World Congress*, 2002.
4. W. Feller. *An Introduction to Probability Theory and Its Applications*, volume I. Wiley, 1950.
5. E. Klavins. A formal language approach to embedded systems modeling and verification. In preparation.
6. E. Klavins. Automatically synthesized controllers for distributed assembly: Partial correctness. In S. Butenko, R. Murphey, and P. M. Pardalos, editors, *Cooperative Control: Models, Applications and Algorithms.* Kluwer, 2002.
7. E. Kushilevitz and N. Nisan. *Communication Complexity.* Cambridge University Press, 1997.
8. J. N. Tsitsiklis and G. D. Stamoulis. On the average communications complexity of asynchronous distributed algorithms. *Journal of the Association for Computing Machinery*, 42(2):382–400, March 1995.
9. J. K. Yook, D. M. Tilbury, and N. R. Soparkar. Trading computation for bandwidth: Reducing communication in distributed control systems using state estimators. *IEEE Transactions on Control Systems Technology*, 10(5), July 2002.

Reconstructing the Shape and Motion of Unknown Objects with Active Tactile Sensors*

Mark Moll[1] and Michael A. Erdmann[2]

[1] Rice University, Houston TX 77005, USA
[2] Carnegie Mellon University, Pittsburgh PA 15213, USA

Abstract. We present a method to simultaneously reconstruct the shape and motion of an unknown smooth convex object. The object is manipulated by planar palms covered with tactile elements. The shape and dynamics of the object can be expressed as a function of the sensor values and the motion of the palms. We present a brief review of previous results for the planar case. In this paper we show that the 3D case is fundamentally different from the planar case, due to increased tangent dimensionality. The main contribution of this paper is a shape-dynamics analysis in 3D, and the synthesis of shape approximation methods via reconstructed contact point curves.

1 Introduction

Robotic manipulation of objects of unknown shape and weight is very difficult. To manipulate an object reliably a robot typically requires precise information about the object's shape and mass properties. Humans, on the other hand, seem to have few problems with manipulating objects of unknown shape and weight, even without visual feedback. During the manipulation of an unknown object the tactile sensors in the human hand give enough information to find the pose and shape of that object. At the same time humans can infer some mass properties of the object to determine a good grasp. Manipulation and sensing continuously interact with each other. These observations are an important motivation for our research. Our long-term goal is to develop combined manipulation and sensing strategies that allow robots to interact more robustly with unknown or uncertain environments. In this paper we present a model that integrates manipulation and tactile sensing. We derive equations for the shape and motion of an unknown object as a function of the motion of the manipulators and the sensor readings.

Figure 1 illustrates the basic idea for planar shapes. There are two palms that each have one rotational degree of freedom at the point where they connect, allowing the robot to change the angle between palm 1 and palm 2 and between the palms and the global frame. As the robot changes the palm angles it keeps track of the contact points through tactile elements on the palms.

* This work was supported in part by the National Science Foundation under grant IIS-9820180. The research was performed while the first author was with Carnegie Mellon University.

J.-D. Boissonnat et al. (Eds.): Algorithmic Foundations of Robotics V, STAR 7, pp. 293–309, 2004.

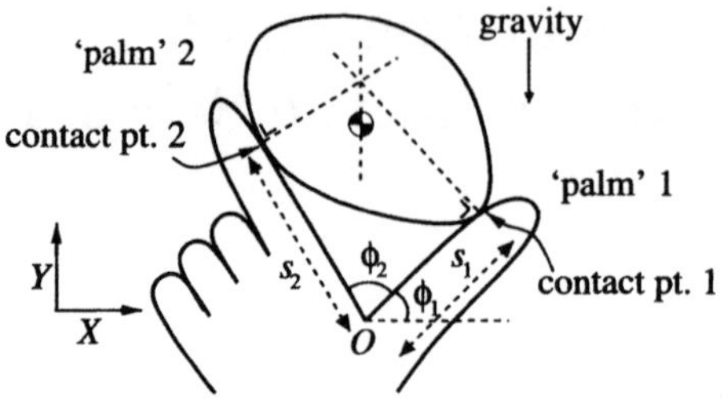

Fig. 1. A smooth convex object resting on two fingers that are covered with tactile sensors.

In the next section we will give an overview of the related work in shape reconstruction using tactile sensors. In section 3 we will briefly review planar shape reconstruction, which has been described in more detail in previous work [18, 19]. In sections 4 through 6 we will derive expressions for the shape and motion of an unknown 3D object being manipulated by three palms. In section 7 we will present some simulation results. Section 8 describes how we can use the curves traced out by the contact points to construct bounds and approximations of the entire shape. In section 9 we discuss how the model can be extended to arbitrary palm shapes. Finally, in section 10 we conclude and outline directions for future research.

2 Related Work

With tactile exploration the goal is to build up an accurate model of the shape of an unknown object. One early paper by Goldberg and Bajcsy [11] described a system requiring very little information to reconstruct an unknown shape. The system consisted of a cylindrical finger covered with 133 tactile elements. The finger could translate and tap different parts of an object.

Often the unknown shape is assumed to be a member of a parametrized class of shapes, so one could argue that this is really just shape recognition. Nevertheless, with some parametrized shape models, a large variety of shapes can still be characterized. In [10], for instance, results are given for recovering generalized cylinders. Allen and Roberts [2] model objects as superquadrics. Roberts [23] proposed a tactile exploration method for polyhedra. In [6] tactile data are fit to a general quadratic form. Finally, Liu and Hasegawa [15] use a network of triangular B-spline patches.

Allen and Michelman [1] presented methods for exploring shapes in three stages, from coarse to fine: grasping by containment, planar surface exploring and surface contour following. Montana [20] described a method to estimate curvature based on a number of probes. Montana also presented a control law for contour following. Charlebois et al. [4, 5] introduced two different tactile exploration methods. The first method is based on rolling a finger around the object to estimate the curvature using Montana's contact equations. Charlebois et al. analyze the sensitivity of this method to noise. With the second method a B-spline surface is fitted to the contact points and normals obtained by sliding multiple fingers along an unknown object.

Marigo et al. [16] showed how to manipulate a known polyhedral part by rolling it between the two palms of a parallel-jaw gripper. Bicchi et al. [3] extended these results to tactile exploration of unknown objects with a parallel-jaw gripper equipped with tactile sensors. The two palms of the gripper roll the object without slipping and track the contact points. Using tools from regularization theory they produce spline-like models that best fit the sensor data. The work by Bicchi and colleagues is different from most other work on tactile shape reconstruction in that the object being sensed is not immobilized. With our approach the object is not immobilized either, but whereas Bicchi and colleagues assumed pure rolling we assume pure sliding.

A different approach is taken by Kaneko and Tsuji [13], who try to recover the shape by pulling a finger over the surface. With this finger they can also probe concavities. In [22, 21] the emphasis is on detecting fine surface features such as bumps and ridges. Sensing is done by rolling a finger around the object. Okamura et al. [21] show how one can measure friction by dragging a block over a surface at different velocities, measure the forces and solve for the unknowns.

Much of our work builds forth on [9]. There, the shape of planar objects is recognized by three palms; two palms are at a fixed angle, the third palm can translate compliantly, ensuring that the object touches all three palms. Erdmann [9] derives the shape of an unknown object with an unknown motion as a function of the sensor values. In our work we no longer assume that the motion of the object is completely arbitrary. Instead, we model the dynamics of the object as it is manipulated by the palms. Only gravity and the contact forces are acting on the object. As a result we can recover the shape with fewer sensors. By modeling the dynamics we need one palm less in 2D. In this paper we will show that 3D objects can be reconstructed with three palms. In the long term we plan to develop a unified framework for reconstructing the shape and motion of unknown objects with varying contact modes.

3 Planar Shape Reconstruction

In this section we describe a method for reconstructing the shape and motion of an unknown planar object. Let the unknown shape be parametrized by the function $\boldsymbol{x} : [0, 2\pi] \to \mathbb{R}^2$, such that $\boldsymbol{x}(\theta)$ is the vector from the center of mass to the point on the surface where the outward-pointing normal $\boldsymbol{n}(\theta)$ is equal to $(\cos\theta, \sin\theta)^T$. Let the tangent $\boldsymbol{t}(\theta)$ be equal to $(\sin\theta, -\cos\theta)^T$ so that $[\boldsymbol{t}, \boldsymbol{n}]$ constitutes a right-handed frame. The projections of $\boldsymbol{x}(\theta)$ onto the normal and tangent are written as $r(\theta)$ and $d(\theta)$, respectively. The function r is called the *radius function*. To reconstruct $\boldsymbol{x}$ it is sufficient to reconstruct r: given the values of r at the contact points, the palm configurations, and palm angles we can solve for the values of the d function. Moreover, we have that $r'(\theta) = \boldsymbol{x}'(\theta) \cdot \boldsymbol{n}(\theta) - x(\theta) \cdot \boldsymbol{t}(\theta) = -d(\theta)$.

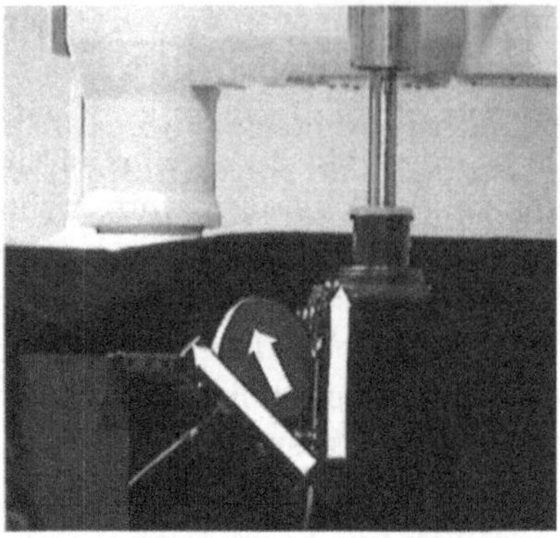

(a)

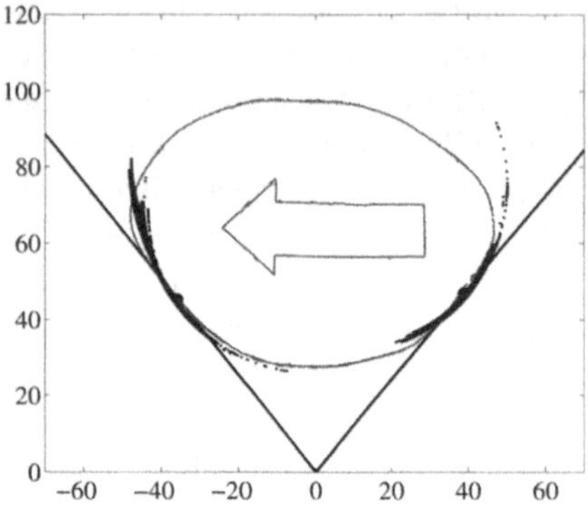

(b)

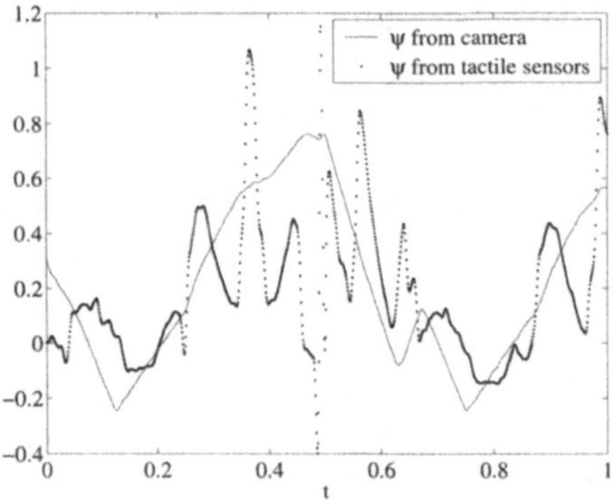

(c)

Fig. 2. Experimental Results. (**a**) Experimental setup. (**b**) Partially reconstructed shape. (**c**) Orientation measured by the vision system and the observed orientation

The rate of change of a sensor value is due to the relative motion of the object and the change of the contact point on the surface of the object. So if we would know the motion of the object, we could solve for the change of the contact point on the surface of the object. We can obtain the motion of the object in two different ways. One way is to assume the dynamics are quasistatic. At any point in time the object will be in force/torque balance. This constraint has the following geometric interpretation: the center of mass lies on a vertical line that passes through the lines of force at the contact points. This allows us to solve for the X-coordinate of the center of mass of the object. Assuming the object does not break contact, the object has only one degree of freedom, and thus we can solve for the motion of the object. This approach is described in more detail in [18]. Figure 2 shows some experimental results of shape reconstruction based on this model for the dynamics.

Another way to solve for the motion of the object is by analyzing the full dynamics. Based on the wrenches exerted by the palms on the object we can solve for the acceleration and angular acceleration of the object. Let $\boldsymbol{q}$ be a vector describing the state of the system. It contains the values of the

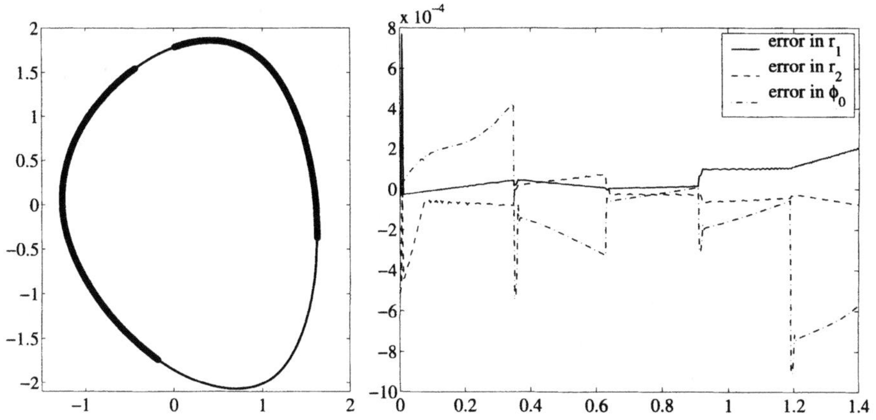

(a) (b)

Fig. 3. Shape reconstruction with an observer based on Newton's method. (**a**) The reconstructed shape. (**b**) The absolute error in the radius function values, r_1 and r_2, and the orientation of the object, ϕ_0

radius function at the contact points, the rotational velocity of the object, the sensor values, and palm angles. Then we can write the solution for the shape and motion as a system of differential equations of the following form [19]:

$$\dot{\boldsymbol{q}} = \boldsymbol{f}(\boldsymbol{q}) + \tau_1 \boldsymbol{g}_1(\boldsymbol{q}) + \tau_2 \boldsymbol{g}_2(\boldsymbol{q}), \tag{1}$$

$$\boldsymbol{y} = \boldsymbol{h}(\boldsymbol{q}), \tag{2}$$

where $\boldsymbol{f}$, $\boldsymbol{g}_1$ and $\boldsymbol{g}_2$ are vector fields, and $\boldsymbol{h}$ is called the *output function.* The output function returns the sensor readings. The vector fields $\boldsymbol{g}_1$ and $\boldsymbol{g}_2$ are called the *input vector fields* and describe the rate of change of our system as torques are being applied on palm 1 and palm 2, respectively, at their point of intersection. The vector field $\boldsymbol{f}$ is called the *drift vector field.* It includes the effects of gravity. In [19] we describe this system in more detail and prove observability. If a system is observable, it is possible to construct a state estimator (called an observer) that, given an initial estimate near the true state, quickly converges to the true state as time progresses. In [17] we describe an observer that uses Newton's method to minimize the error in the estimate. This observer is due to [25]. The error is defined as the integral over a certain time interval of the squared norm of the difference in predicted and real output function values. Figure 3 shows some simulation results obtained using this observer. In this particular case the angle between the palms was fixed and the motion of the palms was simply a counter-clockwise rotation.

4 Three-Dimensional Shape Reconstruction

Naturally we would like to extend the results from the planar case to three dimensions. Although the same general approach can also be used for the 3D case, there are also some fundamental differences. First of all, the difference is not just one extra dimension, but three extra dimensions: a planar object has three degrees of freedom and a 3D object has six. Analogous to the planar case we can derive a constraint on the position of the center of mass if we assume quasistatic dynamics. This constraint gives us the X and Y coordinate of the center of mass. However, assuming quasistatic dynamics is not sufficient to solve for the motion of an unknown object: if the object stays in contact with the palms it has *three* degrees of freedom, but the quasistatic dynamics give us only *two* constraints. Another difference from the planar case is that in 3D we cannot *completely* recover an unknown shape in finite time, since the contact points trace out only curves on the surface of the shape.

In 3D the surface of an arbitrary smooth convex object can be parameterized with spherical coordinates $\boldsymbol{\theta} = (\theta_1, \theta_2) \in [0, 2\pi) \times [-\frac{\pi}{2}, \frac{\pi}{2}]$. For a given $\boldsymbol{\theta}$ we define the following right-handed coordinate frame

$$\boldsymbol{t}_1(\boldsymbol{\theta}) = \begin{pmatrix} -\sin\theta_1 \\ \cos\theta_1 \\ 0 \end{pmatrix}, \quad \boldsymbol{t}_2(\boldsymbol{\theta}) = \begin{pmatrix} -\cos\theta_1 \sin\theta_2 \\ -\sin\theta_1 \sin\theta_2 \\ \cos\theta_2 \end{pmatrix}, \quad \boldsymbol{n}(\boldsymbol{\theta}) = \begin{pmatrix} \cos\theta_1\cos\theta_2 \\ \sin\theta_1\cos\theta_2 \\ \sin\theta_2 \end{pmatrix}. \tag{3}$$

The function $\boldsymbol{x} : S^2 \to \mathbb{R}^3$ describing a smooth shape can then be defined as follows. The vector $\boldsymbol{x}(\boldsymbol{\theta})$ is defined as the vector from the center of mass to the point on the shape where the surface normal is equal to $\boldsymbol{n}(\boldsymbol{\theta})$.

The *contact support function* $(r(\boldsymbol{\theta}), d(\boldsymbol{\theta}), e(\boldsymbol{\theta}))$ is defined as

$$r(\boldsymbol{\theta}) = \boldsymbol{x}(\boldsymbol{\theta}) \cdot \boldsymbol{n}(\boldsymbol{\theta}), \qquad d(\boldsymbol{\theta}) = \boldsymbol{x}(\boldsymbol{\theta}) \cdot \boldsymbol{t}_1(\boldsymbol{\theta}), \qquad e(\boldsymbol{\theta}) = \boldsymbol{x}(\boldsymbol{\theta}) \cdot \boldsymbol{t}_2(\boldsymbol{\theta})$$

The function $r(\boldsymbol{\theta})$ is called the *radius function.* The radius function completely describes the surface. The other two components of the contact support function appear in the derivatives of the radius function:

$$\frac{\partial r}{\partial \theta_1} = \frac{\partial \boldsymbol{x}}{\partial \theta_1} \cdot \boldsymbol{n} + \boldsymbol{x} \cdot \frac{\partial \boldsymbol{n}}{\partial \theta_1} = 0 + (\boldsymbol{x} \cdot \boldsymbol{t}_1)\cos\theta_2 = d\cos\theta_2, \tag{4}$$

$$\frac{\partial r}{\partial \theta_2} = \frac{\partial \boldsymbol{x}}{\partial \theta_2} \cdot \boldsymbol{n} + \boldsymbol{x} \cdot \frac{\partial \boldsymbol{n}}{\partial \theta_2} = 0 + (\boldsymbol{x} \cdot \boldsymbol{t}_2) = e. \tag{5}$$

By definition, the partial derivatives of $\boldsymbol{x}$ lie in the tangent plane, so the dot product of the partials with the normal is equal to 0. Let $\boldsymbol{\theta}_i = (\theta_{i1}, \theta_{i2})^T$ denote the surface parameters for contact point i. Below we will drop the argument $\boldsymbol{\theta}_i$ and replace it with a subscript i where it does not lead to confusion. For instance, we will write $\boldsymbol{n}_i$ for $\boldsymbol{n}(\boldsymbol{\theta}_i)$, the surface normal at contact point i in body coordinates.

The palms are modeled as three planes. The point of intersection of these planes is the origin of the world frame. Let us assume we can rotate each plane around the line of intersection with the horizontal plane through the origin. For each palm we can define a right-handed frame as follows. Let $\bar{\boldsymbol{n}}_i$ be the normal to palm i (in world coordinates) pointing toward the object, let $\bar{\boldsymbol{t}}_{i2}$ be the tangent normal to the axis of rotation and pointing in the positive

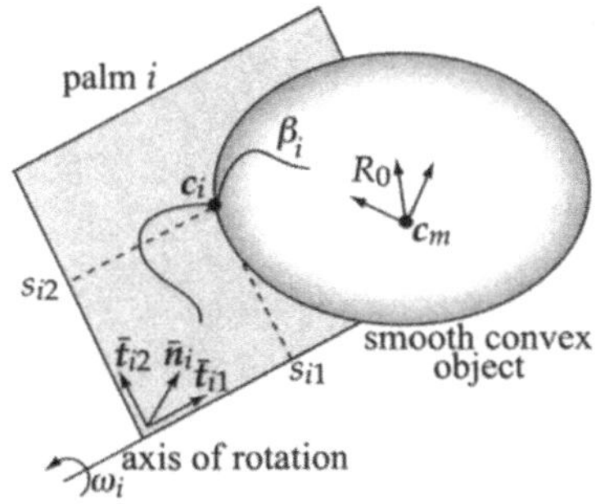

Fig. 4. Illustration of the notation.

Z direction and let $\bar{\boldsymbol{t}}_{i1}$ be $\bar{\boldsymbol{t}}_{i2} \times \bar{\boldsymbol{n}}_i$. Then $R_i = [\bar{\boldsymbol{t}}_{i1}, \bar{\boldsymbol{t}}_{i2}, \bar{\boldsymbol{n}}_i]$ is a right-handed frame for palm i. The configuration of the palms is completely described by the three rotation matrices R_1, R_2, and R_3. Let $\boldsymbol{s}_i$ denote the coordinates in palm frame i of contact point $\boldsymbol{c}_i$, so that $R_i \boldsymbol{s}_i = \boldsymbol{c}_i$. Note that the third component of $\boldsymbol{s}_i$ is always zero; by definition the distance of the contact point along the normal is zero. See Fig. 4 for an illustration.

The position and orientation of the unknown object are described by $\boldsymbol{c}_m$ and R_0, respectively. The center of mass is located at $\boldsymbol{c}_m$. The object coordinate frame defined by R_0 is chosen such that it coincides with the principal axes of inertia. The inertia matrix $\mathcal{I}$ can then be written as

$$\mathcal{I} = m \begin{pmatrix} \varrho_x^2 & 0 & 0 \\ 0 & \varrho_y^2 & 0 \\ 0 & 0 & \varrho_z^2 \end{pmatrix},$$

where m is the mass of the object, and the ϱ's correspond to the radii of gyration. We will write $\boldsymbol{\beta}_i$ for the curve traced out on the surface of the object by contact point i. So $\boldsymbol{\beta}_i(t) = \boldsymbol{x}(\boldsymbol{\theta}_i(t))$.

5 Local Shape

The derivation of the shape and motion proceeds as follows. In this section we derive expressions for the contact point velocities in body coordinates. These expressions depend on the sensor values, the motions of the palms, and the motion of the object. In the next section we will solve for the motion of the object by analyzing the dynamics. We integrate the contact point velocities to obtain curves that describe the local shape of the object. We can recover the contact point velocities by considering the distances between the contact points and the rates at which they change. We derive additional constraints on the contact point velocities and curvature by considering the acceleration constraints induced by the position constraints.

We can write the constraint that the object maintains contact with each palm as

$$\boldsymbol{c}_i = \boldsymbol{c}_m + R_0 \boldsymbol{\beta}_i, \qquad i = 1, 2, 3. \tag{6}$$

The velocity of contact point i is therefore

$$\dot{\boldsymbol{c}}_i = \dot{\boldsymbol{c}}_m + \boldsymbol{\omega}_0 \times R_0\boldsymbol{\beta}_i + R_0\dot{\boldsymbol{\beta}}_i. \tag{7}$$

The difference between two contact point velocities is

$$\dot{\boldsymbol{c}}_i - \dot{\boldsymbol{c}}_j = \boldsymbol{\omega}_0 \times R_0(\boldsymbol{\beta}_i - \boldsymbol{\beta}_j) + R_0(\dot{\boldsymbol{\beta}}_i - \dot{\boldsymbol{\beta}}_j) \tag{8}$$

$$= \boldsymbol{\omega}_0 \times (\boldsymbol{c}_i - \boldsymbol{c}_j) + R_0(\dot{\boldsymbol{\beta}}_i - \dot{\boldsymbol{\beta}}_j). \tag{9}$$

Since we assume the object is smooth, we have that $\boldsymbol{n}_i \cdot \dot{\boldsymbol{\beta}}_i = 0$. Furthermore, the palm normals and object are related by the object orientation matrix:

$$\bar{\boldsymbol{n}}_i = -R_0\boldsymbol{n}_i, \tag{10}$$

since $\bar{\boldsymbol{n}}_i$ is in world coordinates and $\boldsymbol{n}_i$ is in object coordinates. We can combine these constraints to solve for $\dot{\boldsymbol{\beta}}_i$:

$$\bar{\boldsymbol{n}}_i \cdot R_0\dot{\boldsymbol{\beta}}_i = 0 \tag{11}$$

$$\bar{\boldsymbol{n}}_j \cdot R_0\dot{\boldsymbol{\beta}}_i = \bar{\boldsymbol{n}}_j \cdot \big(\dot{\boldsymbol{c}}_i - \dot{\boldsymbol{c}}_j - \boldsymbol{\omega}_0 \times (\boldsymbol{c}_i - \boldsymbol{c}_j)\big) \tag{12}$$

$$\bar{\boldsymbol{n}}_k \cdot R_0\dot{\boldsymbol{\beta}}_i = \bar{\boldsymbol{n}}_k \cdot \big(\dot{\boldsymbol{c}}_i - \dot{\boldsymbol{c}}_k - \boldsymbol{\omega}_0 \times (\boldsymbol{c}_i - \boldsymbol{c}_k)\big), \tag{13}$$

such that i, j, and k are distinct. Let Q be the defined as the 3×3 matrix with entries q_{ji}:

$$q_{ji} = \bar{\boldsymbol{n}}_j \cdot \big(\dot{\boldsymbol{c}}_i - \dot{\boldsymbol{c}}_j - \boldsymbol{\omega}_0 \times (\boldsymbol{c}_i - \boldsymbol{c}_j)\big). \tag{14}$$

Then we can write the solution for $\dot{\boldsymbol{\beta}}_1$, $\dot{\boldsymbol{\beta}}_2$, and $\dot{\boldsymbol{\beta}}_3$ more compactly as

$$\boxed{\big(\dot{\boldsymbol{\beta}}_1\ \dot{\boldsymbol{\beta}}_2\ \dot{\boldsymbol{\beta}}_3\big) = B^{-1}Q,} \tag{15}$$

where $B = \big(\bar{\boldsymbol{n}}_1\ \bar{\boldsymbol{n}}_2\ \bar{\boldsymbol{n}}_3\big)^T R_0$. As long as the palms are in general position, B will be invertible. Equation (15) describes the curves traced out by the contact points on the surface of the object (in body coordinates) as a function of the motion of the palms, the sensor values and the motion of the object. Note that these curves are not independent of each other. We know the configurations of the palms and the sensor values. If we also know one of the curves and the motion of the object, we can reconstruct the other curves. Below we will show that we can reconstruct all three curves by solving for the values of the radius function along the curves. Using (4) and (5), the derivative with respect to time of the radius function at contact point i is

$$\dot{r}_i = \frac{\partial r}{\partial \theta_{i1}}\dot{\theta}_{i1} + \frac{\partial r}{\partial \theta_{i2}}\dot{\theta}_{i2} = \begin{pmatrix} d_i \cos\theta_{i2} \\ e_i \end{pmatrix} \cdot \dot{\boldsymbol{\theta}}_i. \tag{16}$$

We will rewrite the right-hand side of this equation as a function of the motion of the palms and the object, and the values of the radius function at the contact points. Using the position constraints we can rewrite d_i and e_i as a function of the configuration of the palms and r_i. We can write $\boldsymbol{\beta}_i$ as

$$\boldsymbol{\beta}_i = r_i \boldsymbol{n}_i + d_i \boldsymbol{t}_{i1} + e \boldsymbol{t}_{i2}. \tag{17}$$

The vector between contact point i and contact point j is then

$$\begin{aligned} \boldsymbol{c}_i - \boldsymbol{c}_j &= R_0(\boldsymbol{\beta}_i - \boldsymbol{\beta}_j) \\ &= R_0(r_i \boldsymbol{n}_i + d_i \boldsymbol{t}_{i1} + e \boldsymbol{t}_{i2} - r_j \boldsymbol{n}_j + d_j \boldsymbol{t}_{j1} + e \boldsymbol{t}_{j2}). \end{aligned}$$

By rearranging terms we can obtain the following solution for the d's and e's:

$$\begin{pmatrix} d_1 \\ e_1 \\ d_2 \\ e_2 \\ d_3 \\ e_3 \end{pmatrix} = \begin{pmatrix} & & & & 0 & 0 \\ \boldsymbol{t}_{11} & \boldsymbol{t}_{12} & -\boldsymbol{t}_{21} & -\boldsymbol{t}_{22} & 0 & 0 \\ & & & & 0 & 0 \\ & & 0 & 0 & & \\ \boldsymbol{t}_{11} & \boldsymbol{t}_{12} & 0 & 0 & -\boldsymbol{t}_{31} & -\boldsymbol{t}_{32} \\ & & 0 & 0 & & \end{pmatrix}^{-1} \begin{pmatrix} R_0^T(\boldsymbol{c}_1 - \boldsymbol{c}_2) - (r_1 \boldsymbol{n}_1 - r_2 \boldsymbol{n}_2) \\ R_0^T(\boldsymbol{c}_1 - \boldsymbol{c}_3) - (r_1 \boldsymbol{n}_1 - r_3 \boldsymbol{n}_3) \end{pmatrix}. \tag{18}$$

'Hidden' in the tangent vectors are the $\boldsymbol{\theta}_i$'s. Using (3) we can write $\boldsymbol{\theta}_i$ as a function of the palm surface normal $\boldsymbol{n}_i$:

$$\boldsymbol{\theta}_i = \begin{pmatrix} \arctan(n_{i2}, n_{i1}) \\ \arcsin n_{i3} \end{pmatrix}. \tag{19}$$

The relationship between the normal $\boldsymbol{n}_i$ and the orientations of the palms and object is given by (10). The expression for $\dot{r}_i$ also contains $\dot{\boldsymbol{\theta}}_i$. By considering the derivative of the normal $\boldsymbol{n}_i$ we can obtain simple expressions for $\dot{\boldsymbol{\theta}}_i$. On the one hand we have that

$$\dot{\boldsymbol{n}}_i = \frac{\partial \boldsymbol{n}}{\partial \theta_{i1}} \dot{\theta}_{i1} + \frac{\partial \boldsymbol{n}}{\partial \theta_{i2}} \dot{\theta}_{i2} = \dot{\theta}_{i1} \cos \theta_2 \boldsymbol{t}_{i1} + \dot{\theta}_{i2} \boldsymbol{t}_{i2}. \tag{20}$$

But we can also obtain $\dot{\boldsymbol{n}}_i$ by differentiating (10):

$$\boldsymbol{\omega}_i \times \bar{\boldsymbol{n}}_i = \dot{\bar{\boldsymbol{n}}}_i = -\boldsymbol{\omega}_0 \times R_0 \boldsymbol{n}_i - R_0 \dot{\boldsymbol{n}}_i \Rightarrow \quad \dot{\boldsymbol{n}}_i = R_0^T\big((\boldsymbol{\omega}_0 - \boldsymbol{\omega}_i) \times \bar{\boldsymbol{n}}_i\big). \tag{21}$$

Here $\boldsymbol{\omega}_i$ is the rotational velocity of palm i. Combining these two expressions for $\dot{\boldsymbol{n}}_i$ we can write $\dot{\boldsymbol{\theta}}_i$ as

$$\dot{\boldsymbol{\theta}}_i = \big(\boldsymbol{t}_{i1} / \cos \theta_{i2}\ \boldsymbol{t}_{i2}\big)^T R_0^T\big((\boldsymbol{\omega}_0 - \boldsymbol{\omega}_i) \times \bar{\boldsymbol{n}}_i\big). \tag{22}$$

Let us now consider the constraints on the acceleration of the object induced by the three point contact assumption. This will provide us with an additional constraint on $\boldsymbol{\beta}_i$ and will give us some more insight into how the 3D case is fundamentally different from the planar case. By differentiating (7) we obtain the following constraint on the acceleration:

$$\ddot{\boldsymbol{c}}_i = \boldsymbol{a}_0 + \boldsymbol{\alpha}_0 \times R_0 \boldsymbol{\beta}_i + \boldsymbol{\omega}_0 \times (\boldsymbol{\omega}_0 \times R_0 \boldsymbol{\beta}_i + 2 R_0 \dot{\boldsymbol{\beta}}_i) + R_0 \ddot{\boldsymbol{\beta}}_i, \tag{23}$$

where $\boldsymbol{a}_0$ and $\boldsymbol{\alpha}_0$ are the acceleration and angular acceleration of the object. (We will solve for $\boldsymbol{a}_0$ and $\boldsymbol{\alpha}_0$ in the next section by analyzing the dynamics.) Observe that from differentiation of the smoothness constraint $\dot{\boldsymbol{\beta}}_i \cdot \boldsymbol{n}_i = 0$ it follows that $\ddot{\boldsymbol{\beta}}_i \cdot \boldsymbol{n}_i = -\dot{\boldsymbol{\beta}}_i \cdot \dot{\boldsymbol{n}}_i$. We can therefore rewrite the acceleration constraint in the normal direction as a constraint on $\dot{\boldsymbol{\beta}}_i$. First, we rewrite the terms containing $\dot{\boldsymbol{\beta}}_i$ and $\ddot{\boldsymbol{\beta}}_i$:

$$
\begin{aligned}
&\bar{\boldsymbol{n}}_i \cdot (\boldsymbol{\omega}_0 \times 2R_0\dot{\boldsymbol{\beta}}_i) + \bar{\boldsymbol{n}}_i \cdot R_0\ddot{\boldsymbol{\beta}}_i \\
&\quad = 2(\bar{\boldsymbol{n}}_i \times \boldsymbol{\omega}_0) \cdot R_0\dot{\boldsymbol{\beta}}_i + R_0(\dot{\boldsymbol{n}}_i \cdot \dot{\boldsymbol{\beta}}_i) \\
&\quad = 2(\bar{\boldsymbol{n}}_i \times \boldsymbol{\omega}_0) \cdot R_0\dot{\boldsymbol{\beta}}_i + ((\boldsymbol{\omega}_0 - \boldsymbol{\omega}_i) \times \bar{\boldsymbol{n}}_i) \cdot R_0\dot{\boldsymbol{\beta}}_i \\
&\quad = (\bar{\boldsymbol{n}}_i \times (\boldsymbol{\omega}_0 + \boldsymbol{\omega}_i)) \cdot R_0\dot{\boldsymbol{\beta}}_i.
\end{aligned}
$$

The constraint on $\dot{\boldsymbol{\beta}}_i$ is therefore

$$
\begin{aligned}
&(\bar{\boldsymbol{n}}_i \times (\boldsymbol{\omega}_0 + \boldsymbol{\omega}_i)) \cdot R_0\dot{\boldsymbol{\beta}}_i = \\
&\qquad \bar{\boldsymbol{n}}_i \cdot \left(\ddot{\boldsymbol{c}}_i - \boldsymbol{a}_0 - \boldsymbol{\alpha}_0 \times R_0\boldsymbol{\beta}_i - \boldsymbol{\omega}_0 \times (\boldsymbol{\omega}_0 \times R_0\boldsymbol{\beta}_i)\right). \qquad (24)
\end{aligned}
$$

Let us now consider what how the acceleration constraint describes certain time-independent shape properties of the contact curves. The *Darboux frame field* [24] can be used to describe curves on a surface. For the curve $\boldsymbol{\beta}_i$ the Darboux frame field is defined by the unit tangent T of $\boldsymbol{\beta}_i$, the surface normal U restricted to $\boldsymbol{\beta}_i$, and $V = U \times T$. The normal U coincides with $\boldsymbol{n}_i$. Note that the normal of the *curve* does not necessarily coincide with the normal of the *surface*. Similar to the Frenet frame field, the derivatives of T, V, and U can be expressed in terms of T, V, and U:

$$\dot{T} = v(\qquad \kappa_g V + \kappa_n U) \qquad (25)$$

$$\dot{V} = v(-\kappa_g T \qquad + \tau_g U) \qquad (26)$$

$$\dot{U} = v(-\kappa_n T - \tau_g V \qquad) \qquad (27)$$

Here $v = \|\dot{\boldsymbol{\beta}}_i\|$ is the velocity of the curve, κ_g the *geodesic curvature*, κ_n the *normal curvature*, and τ_g the *geodesic torsion*. The geodesic curvature at a point describes the 'bending' of the curve in the tangent plane of the surface at that point. The normal curvature at a point describes the 'bending' of the curve in the surface normal direction. Using this frame field we can write $\ddot{\boldsymbol{\beta}}_i$ as

$$\ddot{\boldsymbol{\beta}}_i = \dot{v}T + v\dot{T} = \dot{v}T + v^2(\kappa_g V + \kappa_n U). \qquad (28)$$

So by taking the dot product with the normal on both sides of the acceleration constraint we can obtain a constraint on the normal curvature of the curve. In the planar case the velocity of the curve is equal to the radius of curvature, and the acceleration constraint determines the (normal) curvature at the contact points. In the 3D case, the acceleration constraint puts a constraint on the normal curvature at the contact points. But now we have two extra curve shape parameters, κ_g and τ_g, which are equal to zero in the planar case. In other words, in 3D the contact point curves are less constrained than in 2D.

6 Dynamics

The solution for the shape of the object depends on the motion of the object. We can solve for the motion by writing out the dynamics equations for the system formed by the palms and the object. If the object remains in contact with the palms, the object has only three degrees of freedom. It is thus sufficient to solve for the angular acceleration. We obtain the rotational velocity by integration. Let I_i be the moment of inertia of palm i around its axis of rotation, α_i the angular acceleration around that axis, τ_i the torque produced by palm i's motor at the axis of rotation, and f_i the magnitude of the contact force. Then the motion of palm i is described by

$$I_i \alpha_i = \tau_i - f_i s_{i2}, \tag{29}$$

where s_{i2} is the second component of $\boldsymbol{s}_i$. From the definition of the palm frame it follows that s_{i2} measures the distance to the axis of rotation (see also Fig. 4). The net force and net torque on the object are given by Newton's and Euler's equations:

$$\boldsymbol{F}_0 = m\boldsymbol{a}_0 = \boldsymbol{F}_g + \textstyle\sum_{i=1}^{3} f_i \bar{\boldsymbol{n}}_i \tag{30}$$

$$\boldsymbol{\tau}_0 = \mathcal{I}'\boldsymbol{\alpha}_0 + \boldsymbol{\omega}_0 \times \mathcal{I}'\boldsymbol{\omega}_0 = \textstyle\sum_{i=1}^{3} \boldsymbol{\tau}_{c_i} \tag{31}$$

where

$$\mathcal{I}' = R_0 \mathcal{I} R_0^T \tag{32}$$

$$\boldsymbol{\tau}_{c_i} = (R_0\boldsymbol{\beta}_i) \times (f_i \bar{\boldsymbol{n}}_i) = -f_i R_0(\boldsymbol{\beta}_i \times \boldsymbol{n}_i). \tag{33}$$

From these equations we can solve for the angular acceleration of the object, $\boldsymbol{\alpha}_0$:

$$\boldsymbol{\alpha}_0 = -\mathcal{I}'^{-1}\big(\boldsymbol{\omega}_0 \times \mathcal{I}'\boldsymbol{\omega}_0 + \textstyle\sum_{i=1}^{3} \frac{\tau_i - I_i\alpha_i}{s_{i2}} R_0(\boldsymbol{\beta}_i \times \boldsymbol{n}_i)\big). \tag{34}$$

Let us assume we can control the palms to move at a constant rotational velocity. The angular acceleration terms α_i will then disappear. We can summarize the simultaneous solution for the shape and motion of an unknown smooth convex object manipulated by three flat palms with the following system of differential equations

$$\dot{r}_i = (d_i \cos\theta_{i2}, e_i)^T \cdot \dot{\boldsymbol{\theta}}_i, \qquad i = 1,2,3 \tag{35}$$

$$\dot{R}_0 = \hat{\omega}_0 R_0 \tag{36}$$

$$\dot{\boldsymbol{\omega}}_0 = -\mathcal{I}'^{-1}\big(\boldsymbol{\omega}_0 \times \mathcal{I}'\boldsymbol{\omega}_0 + \textstyle\sum_{i=1}^{3} \frac{\tau_i}{s_{i2}} R_0(\boldsymbol{\beta}_i \times \boldsymbol{n}_i)\big). \tag{37}$$

Here $\hat{\omega}_0$ is the matrix form of the cross product, i.e., the 3×3 matrix such that $\hat{\omega}_0 \boldsymbol{p} = \boldsymbol{\omega}_0 \times \boldsymbol{p}$ for any vector $\boldsymbol{p}$. Equation (35) describes the shape of the object at the contact points. Equations (36) and (37) describe the dynamics of the object. We can replace the variables d_i, e_i, θ_{i2}, $\dot{\boldsymbol{\theta}}_i$, and $\boldsymbol{\beta}_i$ with the solutions given in (18), (19), (22) so that the system of differential equations only depends on the values of the radius function at the contact points, palm configurations and sensor values. This allows us to integrate the system of differential equations given the palm configurations and sensor values.

7 Simulation Results

We have written a program to simulate the motion of an arbitrary smooth convex object supported by three planes. To reconstruct the shape we need to integrate out the system of differential equations given by (35)–(37). There are several reasons why straightforward integration is not likely to produce good results. We can improve the results in the following two ways. First, we should use quaternions [7] for orientations to improve numerical stability and avoid singularities of three-parameter representations of orientations. Second, we should try to enforce *all* the constraints on the contact point curves given by (15) and (24). We use a prediction-correction numerical integration method. We use the difference between prediction and correction as an error measure when searching for the initial conditions of the system of differential equations.

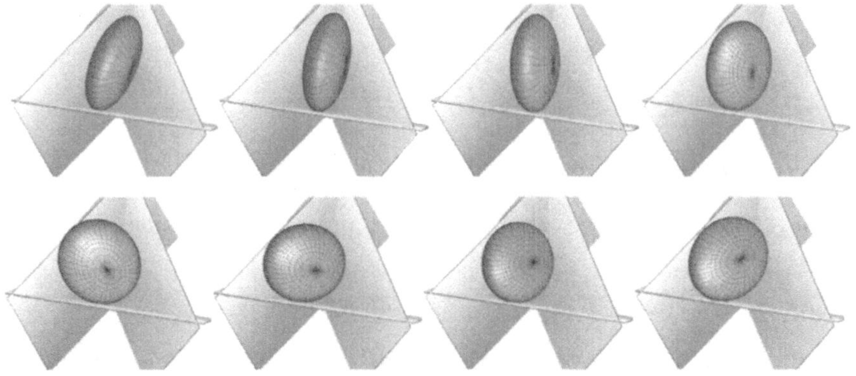

Fig. 5. An object rolling and sliding on immobile palms with gravity and contact forces acting on it. The object is given some initial rotational velocity. It is shown at $t = 0, 0.6, \ldots, 4.2$.

Figure 5 shows the motion of an ellipsoid supported by three palms. The palms are at 60 degree angles with the horizontal plane. The rotational axes are at 60 degree angles with each other. The radii of the axes of the ellipsoid are 2.5, 2, and 1. One problem with simulating this system without friction is that if the palms are moving they continuously increase the kinetic energy of

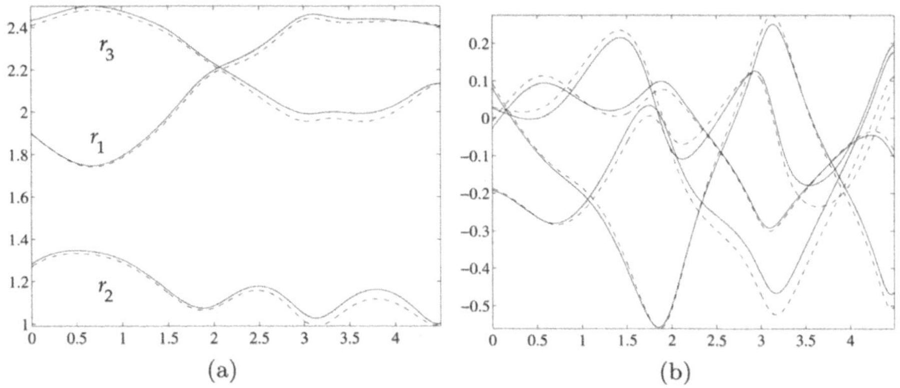

Fig. 6. Differences between real and observed shape and motion. Real values are plotted as solid lines, observed values as dashed lines. (**a**) The radius function values at the contact points. (**b**) The four quaternion components of the rotational velocity

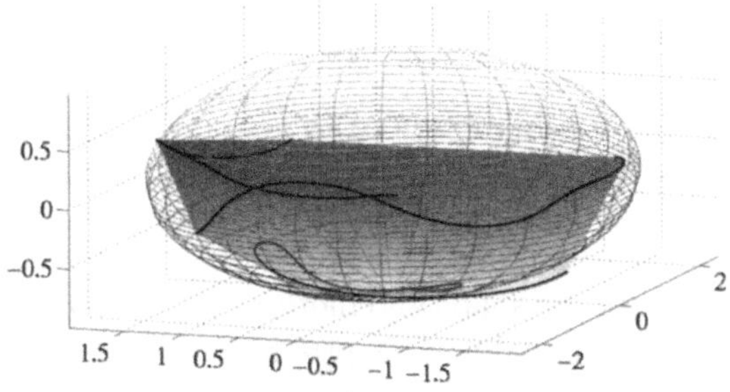

Fig. 7. The convex hull of the contact curves gives a lower bound on the volume occupied by the object. The true shape of the object is shown as a wire-frame model.

the object. In our simulations the ellipsoid breaks contact before a significant part of the object is recovered. To get around this problem, we keep the palms in the same position and give the ellipsoid some initial velocity. In Fig. 5 the object has an initial rotational velocity of $(0, 0.3, 0.3)^T$. In the search for initial conditions we assume the initial rotational velocity is known. This is, of course, not very realistic. If we would model friction, then moving the palms would become feasible and we could start the system with the object at rest. In that case the rotational velocity would be zero. The reconstructed shape and motion are shown in Fig. 6.

8 Shape Approximations

Given the recovered motion and contact curves we can also give a lower and an upper bound on the volume occupied by the entire shape. Since we assume the object is convex, the convex hull of the contact curves in the object frame is a lower bound on the shape. Figure 7 shows the lower bound on the shape obtained from the example in the previous section. We can obtain an upper bound by observing that at each contact point the corresponding palm plane introduces a half-space constraint: the object has to lie entirely on one side of the plane. Clearly, the intersection of half-spaces along the contact curves in the object frame forms an upper bound on the shape. Note that we can obtain slightly tighter upper and lower bounds if we are given bounds on the curvature of the object.

These bounds could form the basis for a manipulation strategy. Having bounds makes it possible to speculate about the outcomes of actions. Suppose we would like to get an estimate of the entire shape. One objective for a planner could be to minimize the difference between the upper bound and lower bound. A simple 'greedy' planner would always try to move the contact points toward the largest distance between upper and lower bound. A smarter planner would also take into account the path length, such that the palms would minimize nearby differences between the bounds before trying to minimize faraway differences. This assumes we have some reasonable metric to measure distance between palm/object configurations.

9 Arbitrary Palm Shapes

In this paper we have written the shape and motion of an unknown object as a function of the contact point positions $\boldsymbol{c}_i$ and the contact point velocities $\dot{\boldsymbol{c}}_i$, all in world coordinates. We wrote out these positions and velocities in terms of poses of the palms and the sensor values. Effectively, we have used world coordinates to decouple the solution for the shape and motion of the unknown object from the shape and motion of the palm. This procedure allows us to use differently shaped palms without changing the solution for the shape and motion of the unknown object. Suppose we were to use spherical palms. For example, Fig. 8 shows two spherical palms holding an object. The contact point on palm i can be written as

$$\boldsymbol{c}_i = b_i R_i \begin{pmatrix} \cos s_{i1} \cos s_{i2} \\ \sin s_{i1} \cos s_{i2} \\ \sin s_{i2} \end{pmatrix}, \tag{38}$$

where b_i is the radius of sphere/palm i, and (s_{i1}, s_{i2}) are the spherical sensor coordinates. We can obtain expressions for $\dot{\boldsymbol{c}}_i$ by differentiating the right-hand side of equation 38 with respect to R_i and $\boldsymbol{s}_i$. The solution for the dynamics of the object would change slightly with spherical palms, but the solution for the shape and motion of the object would remain structurally

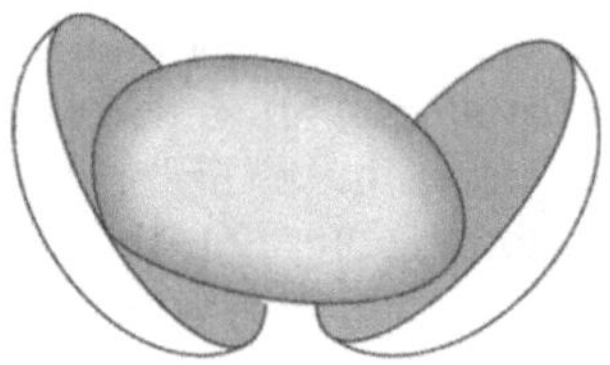

Fig. 8. Two spherical palms holding an object.

the same. Equations 14 and 15 show clearly how the solution for the shape of the object depends on the motion of the contact points.

Changing the palm shapes raises many new questions. Intuitively, it seems that spherical palms would be 'better' than flat palms, but can we quantify this? We can use the shape of the palms to get tighter upper bounds on the shape of the unknown object. Instead of taking intersections of half-spaces, we can now use intersections of spheres as an upper bound. It also seems that with spherical palms we would have to move the palms less to reach a point on the surface.

If we allow some of the palms to be convex, we could actually recover concavities on the object provided the radius of curvature along the concavities is larger than the radius of the palm. Allowing the object to have concavities would invalidate our bounds from the previous section. We can define a new upper bound as the intersection of half-spaces at those points on the contact curves that are also on the boundary of the convex hull of the curves. We can also define other approximate representations of the shape of the object that may be closer to the true shape, such as spline interpolations, α-shapes [8] or principal surfaces [12, 14].

10 Discussion

We have described a method for reconstructing the shape and motion of unknown smooth convex objects with tactile sensors. We required no immobilization of the object or even control of its exact motion. Instead, we used nonprehensile manipulation with planar palms covered with tactile elements to simultaneously manipulate and sense an object. The motion and shape of the object can be written as a function of the motion of the palms and the sensor values. This work will have applications in grasping (partially) unknown objects and tactile exploration of unknown environments.

In this paper we assumed there was no friction and that the unknown object being manipulated by the palms was smooth and convex. In future work we are planning to remove these assumptions. In the presence of friction the normal forces will no longer be along the normal, but at an angle with the normal. If the contact point is moving this angle is proportional to the coefficient of friction. If we know the coefficient of friction, we can again solve for the motion of the object. If a contact point is not moving (i.e., the contact force is inside the friction cone), but one of the other contact points is, we may be able to write the shape and motion of that contact point relative to the fixed contact point. Non-smoothness of the object results in some cases in additional constraints or at least non-smoothness in the sensor values, that can be used to detect corners and edges.

Removing the assumptions makes the system more complex, but the basic formulation remains the same. One question that arises if we allow various contact modes is whether we can detect which contact mode the system is in. We can run observers for each contact mode in parallel and have a higher level algorithm decide which observer is (most likely) to have the smallest error in the state.

The contact kinematics lead naturally to a description of the shape of an object in terms of the curvature at the contact points. One disadvantage of this approach is that the shape has an infinite number of degrees of freedom. Although this allows us to reconstruct *any* smooth convex shape, in many cases it may be sufficient to have a 'reasonable' approximation. With future work we would like to find finite dimensional models for shape that admit a wide range of shapes and can be constructed efficiently in an incremental way. We may be able to apply some of the techniques developed for reconstructing the shape of immobilized objects.

Finally, we would like to construct a state-estimation method for the 3D case with observer-like properties. Since we currently do not have a state-space description of the system, constructing an observer is not possible. However, we derived more constraints on the shape and motion of the unknown than there are unknowns. These extra constraints can be used to adjust our state estimate. Borrowing ideas from reduced-order observers, we may be able construct something similar for our system. Alternatively, we can use the three additional constraints to estimate the moments of inertia, which we had previously assumed to be known.

References

1. Allen, P. K. and Michelman, P. (1990). Acquisition and interpretation of 3-D sensor data from touch. *IEEE Trans. on Robotics and Automation*, 6(4):397–404.
2. Allen, P. K. and Roberts, K. S. (1989). Haptic object recognition using a multi-fingered dextrous hand. In *Proc. 1989 IEEE Intl. Conf. on Robotics and Automation*, pp. 342–347, Scottsdale, AZ.
3. Bicchi, A., Marigo, A., and Prattichizzo, D. (1999). Dexterity through rolling: Manipulation of unknown objects. In *Proc. 1999 IEEE Intl. Conf. on Robotics and Automation*, pp. 1583–1588, Detroit, Michigan.
4. Charlebois, M., Gupta, K., and Payandeh, S. (1996). Curvature based shape estimation using tactile sensing. In *Proc. 1996 IEEE Intl. Conf. on Robotics and Automation*, pp. 3502–3507.
5. Charlebois, M., Gupta, K., and Payandeh, S. (1997). Shape description of general, curved surfaces using tactile sensing and surface normal information. In *Proc. 1997 IEEE Intl. Conf. on Robotics and Automation*, pp. 2819–2824.
6. Chen, N., Rink, R., and Zhang, H. (1996). Local object shape from tactile sensing. In *Proc. 1996 IEEE Intl. Conf. on Robotics and Automation*, pp. 3496–3501.
7. Chou, J. C. (1992). Quaternion kinematic and dynamic differential equations. *IEEE Trans. on Robotics and Automation*, 8(1):53–64.

8. Edelsbrunner, H. and Mücke, E. P. (1994). Three-dimensional alpha shapes. *ACM Transactions on Graphics (TOG)*, 13(1):43–72.
9. Erdmann, M. A. (1999). Shape recovery from passive locally dense tactile data. In *Robotics: The Algorithmic Perspective*. A.K. Peters.
10. Fearing, R. S. (1990). Tactile sensing for shape interpretation. In *Dexterous Robot Hands*, chapter 10, pp. 209–238. Springer Verlag, Berlin; Heidelberg; New York.
11. Goldberg, K. Y. and Bajcsy, R. (1984). Active touch and robot perception. *Cognition and Brain Theory*, 7(2):199–214.
12. Hastie, T. and Stuetzle, W. (1989). Principal curves. *Journal of the American Statistical Association*, 84(406):502–516.
13. Kaneko, M. and Tsuji, T. (2001). Pulling motion based tactile sensing. In *Algorithmic and Computational Robotics: New Directions*, pp. 157–167. A. K. Peters.
14. LeBlanc, M. and Tibshirani, R. (1994). Adaptive principal surfaces. *Journal of the American Statistical Association*, 89(425):53–64.
15. Liu, F. and Hasegawa, T. (2001). Reconstruction of curved surfaces using active tactile sensing and surface normal information. In *Proc. 2001 IEEE Intl. Conf. on Robotics and Automation*, pp. 4029–4034.
16. Marigo, A., Chitour, Y., and Bicchi, A. (1997). Manipulation of polyhedral parts by rolling. In *Proc. 1997 IEEE Intl. Conf. on Robotics and Automation*, pp. 2992–2997.
17. Moll, M. (2002). *Shape Reconstruction Using Active Tactile Sensors*. PhD thesis, Computer Science Department, Carnegie Mellon University, Pittsburgh, PA.
18. Moll, M. and Erdmann, M. A. (2001). Reconstructing shape from motion using tactile sensors. In *Proc. 2001 IEEE/RSJ Intl. Conf. on Intelligent Robots and Systems*, pp. 691–700, Maui, HI.
19. Moll, M. and Erdmann, M. A. (2002). Dynamic shape reconstruction using tactile sensors. In *Proc. 2002 IEEE Intl. Conf. on Robotics and Automation*, pp. 1636–1641.
20. Montana, D. J. (1988). The kinematics of contact and grasp. *Intl. J. of Robotics Research*, 7(3):17–32.
21. Okamura, A. M., Costa, M. A., Turner, M. L., Richard, C., and Cutkosky, M. R. (1999). Haptic surface exploration. In *Experimental Robotics VI*, Sydney, Australia. Springer Verlag.
22. Okamura, A. M. and Cutkosky, M. R. (2001). Feature detection for haptic exploration with robotic fingers. *Intl. J. of Robotics Research*, 20(12):925–938.
23. Roberts, K. S. (1990). Robot active touch exploration: Constraints and strategies. In *Proc. 1990 IEEE Intl. Conf. on Robotics and Automation*, pp. 980–985.
24. Spivak, M. (1999). *A Comprehensive Introduction to Differential Geometry*, volume 3. Publish or Perish, Houston, Texas, third edition.
25. Zimmer, G. (1994). State observation by on-line minimization. *Intl. J. of Control*, 60(4):595–606.

A General Manipulation Task Planner

Thierry Siméon, Juan Cortés, Anis Sahbani, and Jean-Paul Laumond

LAAS-CNRS, Toulouse, France

Abstract. This paper addresses the manipulation planning problem which deals with motion planning for robots manipulating movable objects among static obstacles. We propose a general manipulation planning approach capable to deal with continuous sets for modeling both the possible grasps and the stable placements of the movable object, rather than discrete sets generally assumed by the existing planners. The algorithm relies on a topological property that characterizes the existence of solutions in the subspace of configurations where the robot grasps the object placed at a stable position. This property leads to reduce the problem by structuring the search-space. It allows us to devise a manipulation planner that directly captures in a probabilistic roadmap the connectivity of sub-dimensional manifolds of the composite configuration space. Experiments conducted with the planner demonstrate its efficacy to solve complex manipulation problems.

Several applications such as robot programming, manufacturing or animation of digital actors require some real or virtual artifact to move and manipulate objects within an environment composed of obstacles.

Manipulation planning concerns the automatic generation of the sequence of robot motions allowing to manipulate *movable objects* among obstacles. The presence of movable objects, i.e. objects that can only move when grasped by a robot, leads to a more general and computationally complex version of the classical motion planning problem [14]. Indeed, the robot has the ability to modify the structure of its configuration space depending on how the movable object is grasped and where it is released in the environment. Motion planning in this context appears as a constrained instance of the coordinated motion planning problem. Indeed, movable objects can not move by themselves; either they are transported by robots or they must rest at some stable placement. The solution of a manipulation planning problem consists in a sequence of sub-paths satisfying these motion restrictions. In related literature (e.g. [2,14]), motions of the robot holding the object at a fixed grasp are called *transfer paths*, and motions of the robot while the object stays at a stable placement are called *transit paths*.

Consider the manipulation planning example illustrated by Figure 1. The manipulator arm has to get a movable object (the bar) out of the cage, and to place it on the other side of the environment. Solving this problem requires to automatically produce the sequence of transfer/transit paths separated by grasps/ungrasps operations, allowing to get one extremity of the bar out of the cage; the manipulator can then re-grasp the object by the extremity that was made accessible by the previous motions, perform a transfer path to ex-

J.-D. Boissonnat et al. (Eds.): Algorithmic Foundations of Robotics V, STAR 7, pp. 311–327, 2004.

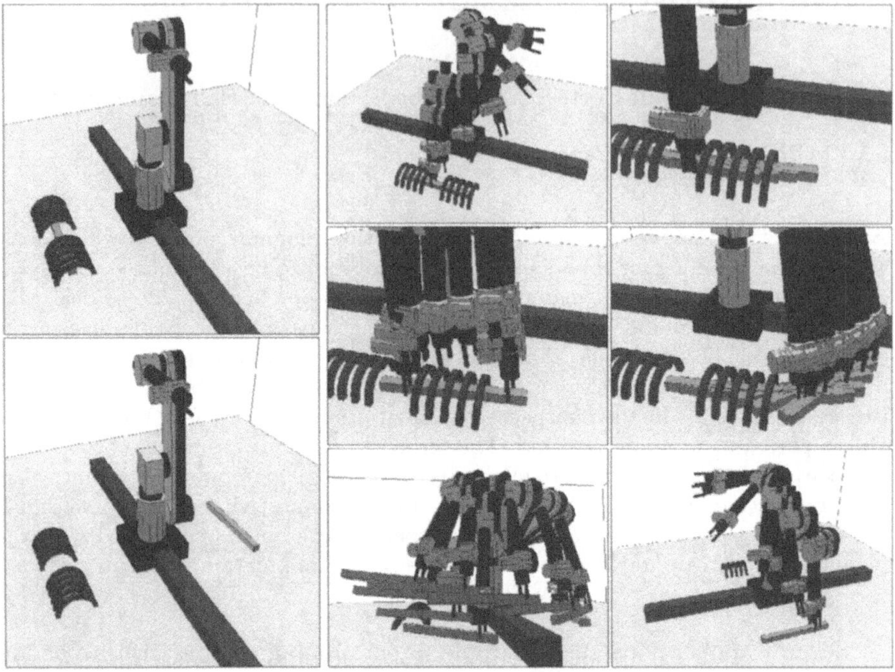

Fig. 1. A manipulation planning problem (left) and its computed solution (right)

tract the bar from the cage, and finally reach the specified goal position. This example shows that a manipulation task possibly leads to a complex sequence of motions including several re-grasping operations. A challenging aspect of manipulation planning is to consider the automatic task decomposition into such elementary collisions-free motions.

Most of existing algorithms (e.g. [1,2,5,12,18]) assume that finite sets of stable placements and possible grasps of the movable object are given in the definition of the problem. Consequently, a part of the task decomposition is thus resolved by the user since the initial knowledge provided with these finite sets has to contain the grasps and the intermediate placements required to solve the problem. Referring back to the example, getting the bar out of the cage would require a large number of precise grasps and placements to be given as input data.

In this paper, we describe a more general approach based onto recent results presented in [22,23]. The proposed approach is capable to deal with a continuous setting of the manipulation problem. It allows us to design a manipulation planner that automatically generates among continuous sets the grasps and the intermediate placements required to solve complicated manipulation problems like the one illustrated onto Figure 1. The approach relies on a topological property first established in [3] and recalled in Section 1. This property allows us to reduce the problem by characterizing the existence of a solution in the lower dimensional subspace of configurations where the robot *grasps* the movable object *placed* at a stable position. Section 2 describes the proposed two-stage approach. First, Section 2.1 shows how the connected components of this sub-space can be captured in a probabilistic roadmap computed for a virtual closed-chain system. Using the visibility technique [20] extended to deal with such closed systems [7], this first stage of the approach captures the connectivity of the search space into a small roadmap generally composed of a low number of connected components. Connections between these components using transit or transfer motions are then computed in a second stage (Section 2.2) by solving a limited number of point-to-point path planning problems. The details of an implemented planner interleaving both stages in an efficient way are described in Section 3. Finally, we discuss in Section 4 some experiments and the performance of the planner.

1 Problem Statement

Let us consider a 3-dimensional workspace with a robot $\mathcal{R}$ and a movable object $\mathcal{M}$ moving among static obstacles[1] . The robot has n degrees of freedom and $\mathcal{M}$ is a rigid object with 6 degrees of freedom that can only move when it is grasped by the robot. Let C_{rob} and C_{obj} be the configuration spaces of the robot and the object, respectively. The composite configuration space of the system is $CS = C_{rob} \times C_{obj}$ and we call CS_{free} the subset in CS of all admissible configurations, i.e. configurations where the moving bodies do not intersect together or with the static obstacles.

Manipulation Constraints: A solution to a manipulation planning problem corresponds to a constrained path in CS_{free}. Such a solution path is an alternate sequence of two types of sub-paths verifying the specific constraints of the manipulation problem, and separated by grasp/ungrasp operations:

- *Transit Paths* where the robot moves alone while the object $\mathcal{M}$ stays stationary in a stable position. The configuration parameters of $\mathcal{M}$ remain constant along a transit path. Such motions allow to place the robot at a configuration where it can grasp the object. They are also involved when changing the grasp of the object.

- *Transfer Paths* where the robot moves while holding $\mathcal{M}$ with the same grasp. Along a transfer path, the configuration parameters q_{obj} of $\mathcal{M}$ change according to the grasp mapping induced by the forward kinematics of the robot: $q_{obj} = g(q_{rob})$.

Let P (resp. G) denote the set of stable placements (resp. grasps) of $\mathcal{M}$. Both types of paths lie in lower dimensional sub-spaces of CS_{free}. These sub-spaces are defined as follows:

[1] See [2,14] for a more formal and general description of the manipulation problem

- The *Placement* space CP is the sub-space of CS_{free} defined as the set of free configurations where the mobile object is placed at a stable position, i.e. a position $p \in P$ at which $\mathcal{M}$ can rest when it is not grasped by the robot. Each transit path lies in CP. However, note that a path in CP is not generally a transit path since such path has to belong to the sub-manifold corresponding to a given placement p.

- The *Grasp* space CG is the sub-space of CS_{free} defined as the set of free configurations under all possible grasps $g \in G$ of the object $\mathcal{M}$. A transfer path belongs to the sub-manifold defined by a particular grasp kept constant along the path.

Problem: Given the two sets (discrete or continuous) P and G defining the stable placements and feasible grasps, the manipulation planning problem is to find a manipulation path (i.e. an alternate sequence of transit and transfer paths) connecting two given configurations q_i and q_f of $CG \cup CP$. Manipulation planning then consists in searching for transit and transfer paths in a collection of sub-manifolds corresponding to particular grasps or stable placements of the movable object. Note that the intersection $CG \cap CP$ between the sub-manifold defines the places where transit paths and transfer paths can be connected. The manipulation planning problem appears as a constrained path planning problem inside (and between) the various connected components of $CG \cap CP$.

1.1 Discrete Case

In the case of a discrete number of grasps and stable placements, the intersection $CG \cap CP$ consists of a finite set [2] of configurations. In this case, it is possible to build a *Manipulation Graph* (MG) where each node corresponds to a configuration of $CG \cap CP$, and edges are constructed by searching for transfer (or transit) paths between nodes sharing the same grasp (or placement) of the mobile object(s).

Most implemented manipulation planners consider that such discrete sets P and G are given as input to the planner. The general framework proposed in [2] for one robot and several moving objects with discrete grasps and placements was first implemented for two simple robots: a translating polygon [2] and a $3dof$ planar manipulator [15]. This framework was extended in [11,12] to multi-arm manipulation planning where several robots cooperate to carry a single movable object amidst obstacles. The planner proposed in [12] first plans the motions of the movable object using a *RPP* technique [4], and then finds the transfer/transit motions of the arms to move the object along the computed path. This heuristic approach led to impressive

[2] Note however that the case of a redundant robot considered in [1] also leads to an infinite cardinality of the set $CG \cap CP$.

results for solving realistic problems. Another heuristic planner proposed in [5] iteratively deforms a coordinated path first generated in the composite configuration space using a variational dynamic programming technique that progressively enforces the manipulation constraints.

More recently, two other contributions extended existing planning techniques to manipulation planning. In [1], the *Ariane's Clew* algorithm [6] is applied to a redundant robot manipulating a single object in a 3D workspace. The method assumes discrete grasps of the movable object; it is however capable to deal in realistic situations with redundant manipulators for which each grasp possibly corresponds to an infinite number of robot configurations (i.e. the cardinality of $CG \cap CP$ is infinite). Finally, [18] proposed another practical manipulation planner based onto the extension of the PRM framework [10,19]. The planner builds as in [2] a manipulation graph between discrete configurations of $CG \cap CP$, but connections are computed using a *Fuzzy PRM* planner that builds a roadmap with edges annotated by a probability of collision-freeness. Such roadmaps improve the efficiency of the planner for solving the possibly high number of path planning queries in changing environments required to compute the connections.

1.2 Continuous Case

The structure of the search space is more complex when dealing with continuous sets P and G. For example, one may describe the set of stable placements by constraining the movable object to be placed on top of some horizontal faces of the static obstacles; such placement constraints would define P as a domain in a 3-dimensional manifold of CS_{obj} (two translations in the horizontal place and one rotation around the vertical axis). Also, one may consider G as a set of continuous domains such that the jaws of a parallel gripper have a sufficient contact with two given faces of $\mathcal{M}$. With such grasp constraints, G also corresponds to a 3-dimensional manifold (two translations parallel to the grasped faces and one rotation around the axis perpendicular to the faces).

In such cases, $CG \cap CP$ is defined by continuous configuration domains. Extending the notion of manipulation graph to continuous domains, first requires that the connectivity of $CG \cap CP$ adequally reflects the existence of a manipulation path. This is not *a priori* obvious since paths computed in $CG \cap CP$ correspond to a continuous change of the grasp, while transfer paths need to stay in the sub-manifold defined by a constant grasp.

Reduction Property: It is shown in [3] [3] that two configurations which are in a same connected component of $CG \cap CP$ can be connected by a manipulation path. This property means that any path in $CG \cap CP$ can be transformed into a finite sequence of transit and transfer paths. It is then sufficient to

[3] Note that this property only holds for a single movable object under the hypothesis that the robot does not touch the static obstacles

study the connectivity of the various components of $CG \cap CP$ by transit and transfer paths.

Manipulation Graph: The reduction property allows to extend the Manipulation Graph to the continuous case: $MG's$ nodes correspond to the connected components of $CG \cap CP$ and an edge between two nodes indicates the existence of a transit/transfer path connecting two configurations of different connected components. This property was used in [3] to derive, for the specific case of a translating polygonal robot, a cell decomposition algorithm capable to manipulate one movable polygon with an infinite set of grasps.

2 A General Approach to Manipulation Planning

We now describe a more general approach for solving manipulation problems under continuous manipulation constraints. The main idea is to exploit the reduction property of Section 1 to decompose the construction of the manipulation graph into the two following steps:

- compute the connected components of $CG \cap CP$. Section 2.1 describes a method that directly captures the topology of $CG \cap CP$ inside a probabilistic roadmap [10,19].

- determine the connectivity of these connected components using transit-paths and transfer-paths. Section 2.2 explains the method used to compute such connections.

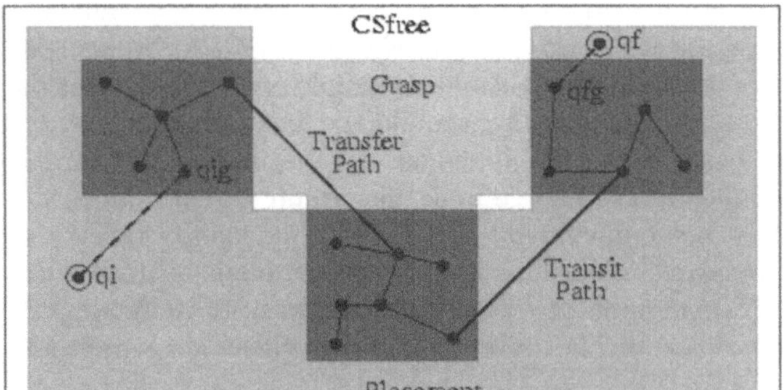

Fig. 2. Structure of the Manipulation Roadmap

Figure 2 illustrates the structure of a manipulation roadmap computed using this approach. The roadmap is structured into a small number of nodes (the connected components of $CG \cap CP$) connected with transit or transfer paths. Once the manipulation roadmap is computed, queries are solved by searching a path inside MG. Note that the obtained solution possibly alternates transfer/transit paths (when traversing edges of MG) with other elementary paths computed in $CG \cap CP$ (inside nodes of MG) which do not correspond to feasible paths. However, thanks to the reduction property, any such $CG \cap CP$ paths can be transformed in a post-processing stage into a finite sequence of transit and transfer paths.

2.1 Capturing $CG \cap CP$ Topology via Closed-Chain Systems

The main critical issue is to capture into a probabilistic roadmap the topology of $CG \cap CP$ which is a sub-manifold of the global configuration space with a lower dimension. The difficulty in using probabilistic roadmap approaches is to face sub-dimensional manifolds. Indeed the probability to choose a configuration at random on a given sub-dimensional manifold is null.

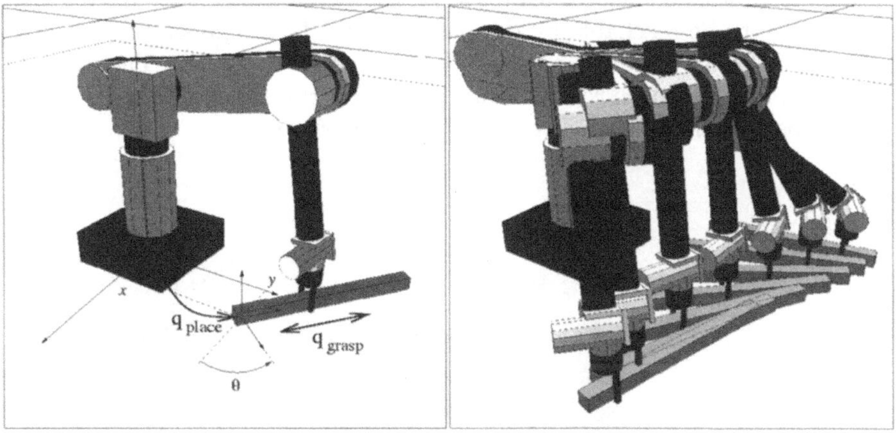

Fig. 3. Virtual closed-chain system and a feasible $CG \cap CP$ motion (the bar moves on the floor while sliding into the gripper's jaws)

The idea here is to explore $CG \cap CP$ as such. For this, we consider that $CG \cap CP$ is the configuration space of a single system consisting of the robot together with the movable object *placed* at a stable position. Maintaining the stable placement while the object is grasped by the robot induces a closed chain for the global system.

Figure 3 shows the example of a closed chain system formed by a $6dof$ arm manipulating a long bar. The bar moves in contact with the floor while sliding within the gripper. The sliding motion of the gripper results from the additional degrees of freedom q_{grasp} introduced in the system to characterize the infinite set of grasps. In this example q_{grasp} is chosen to allow a translation of the parallel jaw gripper along the bar. Similarly, the set P of stable placements corresponds to the planar motions parameterized by a vector $q_{place} = (x, y, \theta)$ (two horizontal translations and a vertical rotation), that maintain the contact of the bar with the floor. The motion shown in the Figure is a feasible motion in $CG \cap CP$. It is not admissible from the manipulation problem point of view. However, it can be transformed into a finite sequence of feasible transit and transfer paths.

We then propose to apply planning techniques for closed chain systems to capture the topology of $CG \cap CP$. Indeed, several recent contributions extended the PRM framework [10,19] to deal with such closed chain mechanisms [17,9,7]. In particular, we use the algorithm [7] that demonstrated good performance onto complex 3D closed chains involving tenths of degrees of freedom. As initially proposed in [9] the loop is broken into two open sub-chains, but the random node generation combines a sampling technique called *Random Loop Generator* (*RLG*) with forward kinematics for the *active* chain and inverse kinematics for the remaining *passive* part of the loop in order to force the closure. The advantage of the *RLG* algorithm is that it that it produces random samples for the active chain that have a high probability to be reachable by the passive part. This significantly decreases the cost of computing and connecting closure configurations. The practical efficacy of our approach results from the good performance reached today by these closed-chain extensions of the *PRM* framework.

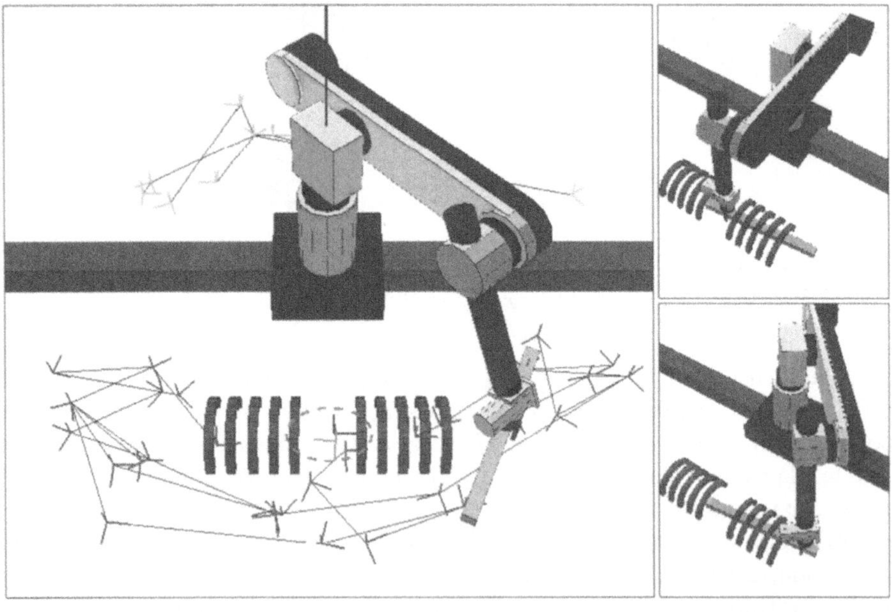

Fig. 4. A Visibility Roadmap computed in $CG \cap CP$ (left) and two placements of the system inside two different connected components of $CG \cap CP$ (right)

The $CG \cap CP$ roadmap is computed using the *Visibility-PRM* technique [20,16]. This technique keeps the roadmap as small as possible by only adding two types of useful samples: *guards* that correspond to samples not already "seen" by the current roadmap, and *connectors* allowing to merge several connected components. Its interest is first to control the quality of the roadmap in term of coverage, and also to capture the connectivity of possibly complex spaces into a small data structure. We believe that the small size of the visibility roadmaps, combined with the proposed structuring of $CG \cap CP$ contributes to the overall efficiency of our approach by limiting the number of costly path-planning queries to be performed during the second stage, when searching the connections with collision-free transfer or transit paths.

Figure 4 shows the visibility roadmap computed in $CG \cap CP$ by the algorithm for the introduction example shown in Figure 1. While the collision-free configuration space of the arm alone is connected, $CG \cap CP$ is not. The computed roadmap has four connected components: two main components separated by the long static obstacle, and two other small components that correspond to placements of the movable object inside the cage obstacle while it is grasped by the arm through the open passage in the middle of the cage. These two small components (inside the dashed circle of the left image) correspond to a similar position of the bar, rotated or not by 180 degrees. One associated placement of the system is shown onto the top right image. The bottom right image corresponds to a node of the main component with the bar placed at the same position, but using a different grasp. Connecting this node to the small component is not possible because of the cage obstacle that limits the continuous change of grasp. Such re-grasping requires the computation of collision-free paths outside $CG \cap CP$ as explained below.

2.2 Connections with Transit and Transfer Paths

Consider two configurations in $CG \cap CP$ that can not be directly connected by a collision-free path in $CG \cap CP$ (i.e. configurations that do not belong to the same connected component of MG). These configurations correspond to fixed grasps and placements of the movable object, noted $(g_i, p_i)_{i=1,2}$. The problem is now to determine whether both configurations can be connected by collision-free transit/transfer paths searched inside the corresponding leaves of CP and CG. Such a connection is possible if there exists:

- a transfer path from (g_1, p_1) to (g_1, p_2) followed by a transit path from (g_1, p_2) to (g_2, p_2).
- or a transit path from (g_1, p_1) to (g_2, p_1) followed by a transfer path from (g_2, p_1) to (g_2, p_2).

We next explain the two-stage method used by the planner to solve these path-planning queries. Note that each query has to be performed in a partially modified environment since it corresponds to a particular grasp or placement of the movable object. The motivation is simply to amortize the cost of dealing with such partial changes by re-using at each query some of the paths precomputed during the first stage by disregarding the movable object.

First, we compute a roadmap for the robot and the static obstacles, without considering the presence of the movable object. Then, before to solve a given (transit or transfer) path query, the roadmap is updated by checking whether each edge is collision-free with respect to the current position of the movable object. Colliding edges are labeled as *blocked* in the roadmap.

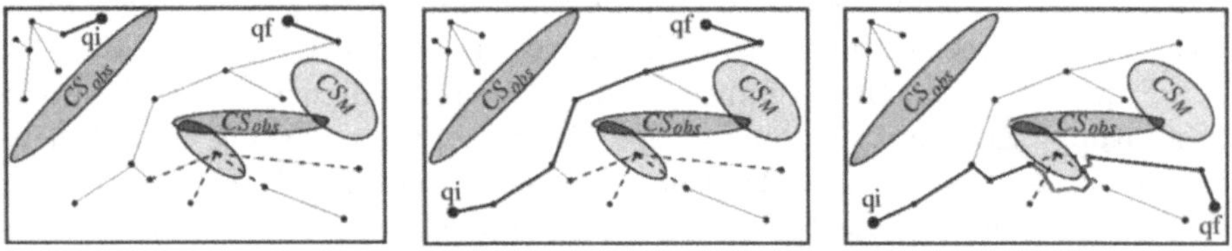

Fig. 5. Search into the *labeled* roadmap: **a/** no solution exists **b/** a solution is found **c/** local update of the *blocked* edges

The search for a given path is then performed within the *labeled* roadmap. As illustrated by Figure 5, three cases possibly occur. When the search fails, this means that no path exists even in the absence of the movable object; the problem has no solution. Similarly, when the computed path does not contain any blocked edge (dashed edges onto the Figure) then a solution is found. Now let us consider the intermediate situation where the a path necessarily contains blocked edges. In such case, the algorithm tries to solve the problem locally by using a Rapid Random Tree (*RRT*) algorithm [13] to connect the endpoints of the blocked edges. The main interest of *RRT* is to perform well locally. Its complexity depends on the length of the solution path. This means that the approach quickly finds easy solutions. It may be viewed as a dynamic updating of the roadmaps.

3 Manipulation Planning Algorithm

We now describe in more details the algorithm that was developed to implement the approach presented above. The algorithm takes as input several classes of possible continuous grasps G_i. Each G_i is defined by a transformation matrix T_{g_i} and a set of parameters, noted q_{grasp}, varying in a given interval. Similarly, several continuous regions of placement P_j can be defined, each characterized by a transformation matrix T_{p_j} that defines a stable situation of the object, and a vector $q_{place} = (x, y, \theta)$ representing two horizontal translations inside a rectangular domain and a rotation around an axis perpendicular to this plane. For each couple of sets (G_i,P_j), the resulting virtual closed-chain system is considered when analyzing the connectivity of $CG \cap CP$ sub-spaces corresponding to the couple.

3.1 Building the Manipulation Graph

The algorithm incrementally constructs the manipulation graph MG by interleaving the two steps of the approach: computing $CG \cap CP$ connected components and linking them. Following the principle of the *Visibility-PRM* technique, the algorithm stops when it is not able to expand the graph after a given number of tries (MAX_NTRY). This number of failures is related to an estimated coverage of the composite configuration-space by MG.

EXPAND_GRAPH(MG)
$ntry \leftarrow 0$
$expanded \leftarrow$ FALSE
while ($ntry <$ MAX_NTRY(MP)) and ($\neg expanded$)
  $N \leftarrow$ NEW_NODE(MP)
  $expanded \leftarrow$EXPAND_G∩P(MG,N)
  if ($\neg expanded$) and (TRY_OTHER_LINKS($ntry$)) **then**
    $expanded \leftarrow$CONNECT_OUTSIDE_G∩P(MG,N)
  $ntry = ntry + 1$
return *expanded*

The function EXPAND_GRAPH performs one expansion step of MG. Candidate nodes are first sampled in the $CG \cap CP$ manifold and the different types of connections to the graph are then tested.

A candidate node N is generated as follows by the function NEW_NODE: it first randomly selects one of the P_j sets given as input to MP and a stable configuration $p \in P_j$ of the movable object is then chosen by randomly sampling q_{place}. Similarly, one of the grasp classes G_i and the corresponding parameters q_{grasp} are also randomly chosen. The candidate node N is generated when the sampled grasps and placements are collision-free and feasible for the virtual closed system.

Following the discussion in Section 2.1, it is more efficient to explore $CG \cap CP$ than to check connections outside. Therefore, the sample N is first used by EXPAND_IN_G∩P which checks connections to the graph components using $CG \cap CP$ paths. This function keeps N to expand the graph when the node creates a new component or merges existing components of the graph. In the other case, the choice of testing other connections with CONNECT_OUTSIDE_G∩P is made by the boolean function TRY_OTHER_LINKS, based onto a biased random choice {True, False} that depends on the evolution of the size of MG: the first expansion steps starts with a low probability to call CONNECT_OUTSIDE_G∩P; when the roadmap grows, this probability increases as the percentage of the coverage estimated by the fraction $(1-\frac{1}{ntry})$ (see [20] for details).

Expansion in $CG \cap CP$. The function EXPAND_G∩P carries out the exploration of the $CG \cap CP$ manifold using LINKED_IN_G∩P to test the link

```
EXPAND_G∩P(MG,N)
n_linked comp. ← 0
for i = 1 to N_COMP(MG) do
  if LINKED_IN_G∩P(N,C_i) then
    n_linked comp. = n_linked comp. + 1
if n_linked comp. ≠ 1 then
  ADD_NODE(N, MG)
  UPDATE_GRAPH(MG)
  return TRUE
else
  return FALSE
```

between N and each connected component of MG by means of paths in $CG \cap CP$. Note that such connections can be achieved only with the nodes computed from the same couple $(G_i,\ P_j)$ as the one used to generate N. For each component C_i, nodes with such characteristics are tested until a connection is feasible or all those nodes have been checked. The connections are determined from local paths computed in $CG \cap CP$ for the virtual closed system induced by $(G_i,\ P_j)$. Following the visibility principle, the node N is added to the graph only if it was linked to none or to more than one connected component. In the second case, the linked components are merged.

Connections with transit and transfer paths. The function EXPAND_G∩P fails when N can be linked by $CG \cap CP$ paths to only one connected component of MG. In this case, the role of CONNECT_OUTSIDE_G∩P is to test the link to the other connected components using the composition of a transfer and a transit path, as explained in Section 2.2. Function LINKED_OUTSIDE_G∩P tests the connection between N and each node of C_i using such paths. This function stops checking a component when a valid connection is found. In such case the component of node N is merged with the linked C_i.

```
CONNECT_OUTSIDE_G∩P(MG,N)
n_linked comp. ← 0
for i=1 to (N_COMP(MG) | C_i ≠COMP(N)) do
  if LINKED_OUTSIDE_G∩P(N,C_i) then
    n_linked comp. = n_linked comp. + 1
if n_linked comp. ≠ 0 then
  UPDATE_GRAPH(MG)
  return TRUE
else
  return FALSE
```

3.2 Solving Manipulation Queries

Once the manipulation roadmap is built, queries can be performed using the three following steps. First, the start and goal configurations are connected to MG using the LINKED_OUTSIDE_G∩P function described above and the manipulation graph is searched for a path between both configurations. The second step is necessary to transform $CG \cap CP$ portions of the solution path into a finite sequence of transfer/transit paths. This is done by a simple procedure that iteratively splits the $CG \cap CP$ paths into pieces whose endpoints can be connected by a composition of two transit/transfer motions. Finally, the solution is smoothed by a procedure that eliminates unnecessary motions.

4 Experimental Results

The manipulation planner was implemented within the software platform *Move3D* [21] currently developed at LAAS. Several environments have been used as test-bed of the planner. In this section, we present the results obtained onto three of them. The computation times correspond to experiments conducted on a 330MHz Sparc Ultra 10 workstation. The first example corresponds to the problem of Figure 1. We refer to it as the *Cage* example. Two other scenes are shown in Figure 6: the left image illustrates a problem (*MobM*) involving a manipulator arm mounted onto a mobile platform. The right example (*RoBr*) corresponds to an industrial logistics problem treated in the framework of the European Project *MOLOG* [4] . The manipulation device is a rolling bridge.

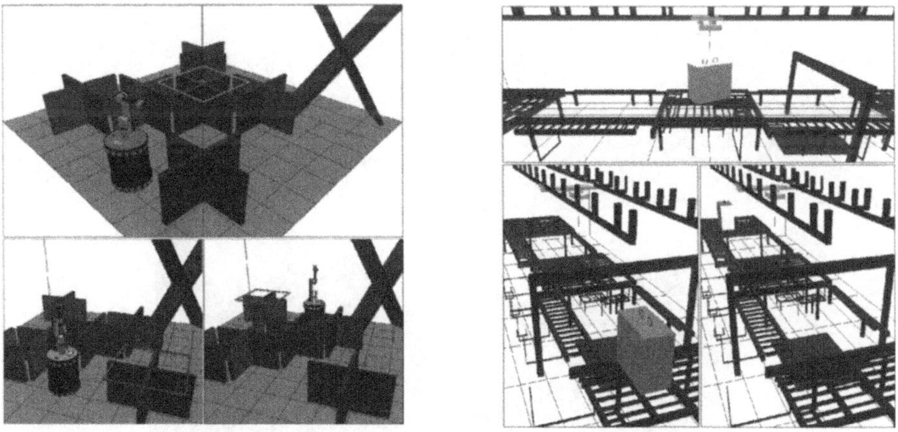

Fig. 6. Scenes and start/goal positions for the *MobM* and *RoBr* examples

[4] see `http://www.laas.fr/molog`.

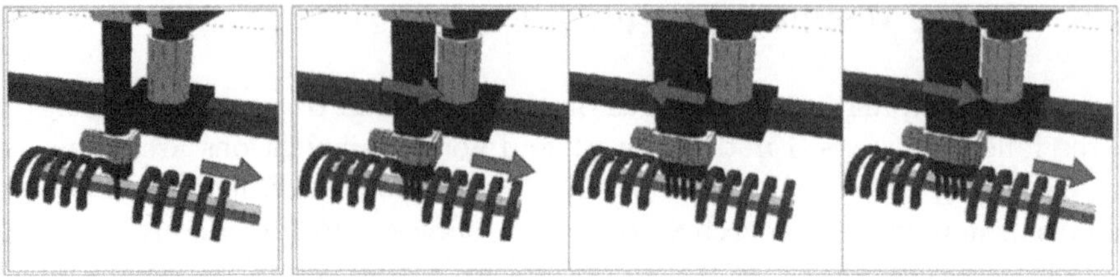

Fig. 7. A $CG \cap CP$ path with a sliding motion of the bar (left) transformed into a sequence of three feasible *transfer/transit/transfer* manipulation paths (right)

The difficulty with the *Cage* example is the complexity of the manipulation task. Several consecutive re-graspings motions through the middle of *cage* obstacle are necessary to move the bar to a position where it can be regrasped by its extremity. The planner automatically computes the required configurations from only one continuous placement domain (the floor) and one grasping zone all along the bar. The path to get the bar out of the cage is found in the $CG \cap CP$ manifold, and then transformed during the post-processing step in a sequence of transit and transfer paths (see Figure 7). The final path contains 20 elementary paths with 8 re-graspings of the movable object. This difficult manipulation problem was solved in less than two minutes, which demonstrates the efficacy of the proposed approach.

In the example *MobM*, the mobile manipulator ($9dof$) can only pass from one side of the scene to the other side through the passage under the X-shaped obstacle. However this passage is too narrow for the movable object

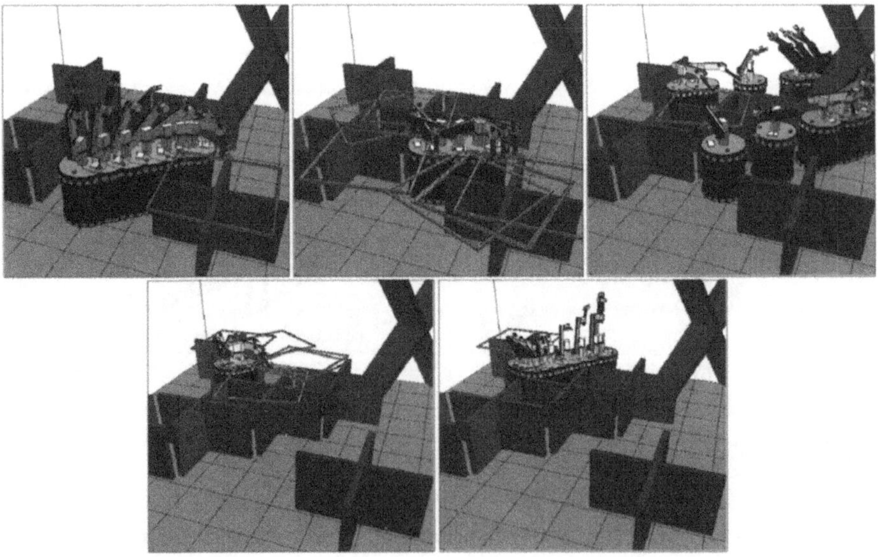

Fig. 8. Manipulation path computed for the *MobM* problem

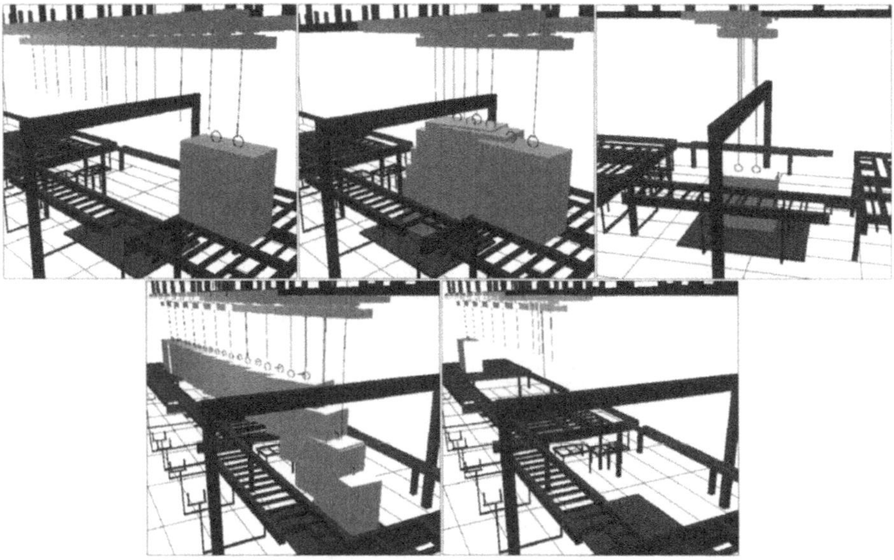

Fig. 9. Manipulation path computed for the *RoBr* problem

(the square frame). A continuous grasping set is defined all around this object. The frame can be placed on the central obstacles. Figure 8 shows the manipulation solution computed by the planner. Here, the manipulation task is simpler compared to the *Cage* example; less re-graspings are needed to solve the problem. The difficulty illustrated by the example is to deal with a redundant system. An infinite set of solutions exists to achieve the same grasp. Redundancy is a challenge when treating closed chain mechanisms. The exploration of the $CG \cap CP$ manifold for such systems is efficiently performed by using the approach in [7].

In the example *RoBr*, transporting the load between the extreme sides of the scene requires to generate an intermediate placement under the "arc" obstacle such that the rolling bridge can reach one of the rings at the top of the load from each side of the obstacle. Figure 9 shows the manipulation solution computed by the planner. Here, the rolling bridge a simple system ($4dof$). However, the geometric complexity of the static scene ($\approx$ 10.000 facets) and the large size of the movable object (which results in many *RRT* calls during the second stage) increase the overall cost of this connection stage.

Table 1 shows for the three examples numerical results of the performance of the algorithm. All examples were solved after less than five minutes of computation. One can also note that most of the computation time is spent for checking connections with transit and transfer paths; this shows the interest of the proposed approach which limits the number of such connection tests by first computing connected components inside $CG \cap CP$. Also, the use of the visibility technique is the reason for the small size of the manipulation roadmaps; such small roadmaps also reduces the number of connections to be tested.

Table 1. Numerical results

Examples:	**Cage**	**MobM**	**RoBr**
Total Time	96 s	293 s	146 s
Computing $CG \cap CP$	23 s	6 s	1 s
Transit/Transfer paths	70 s	284 s	143 s
N. manip.nodes	32	45	10
N. manip. paths	12	14	6

5 Conclusion

We have presented a new approach to manipulation planning. The power of the approach lies in the fact that it can deal with continuous grasps and placements. It relies on a structuring of the search space allowing to directly capture into a probabilistic roadmap the connectivity of the sub-manifolds that correspond to the places where transit paths and transfer path can be connected. This structuring allows us to design a manipulation planner that automatically generates inside continuous domains, the particular grasps and placements that make the problem solvable. Simulations results show the effectiveness of the approach for solving complex manipulation problems.

There remain several possible improvements, in particular in the way to interleave the construction of $CG \cap CP$ with the connection of its connected components via transit or transfer paths. Also, our manipulation planner is currently restricted to a single movable object manipulated by a single robot. Considering the case of multiple movable objects and robots first requires studying the conditions under which the reduction property can be extended to such situations. We also begin to investigate an alternative approach [8] combining a symbolic task planning level with our planner for solving a higher level of problems complexity in presence of multiple objects and robots.

References

1. J.M. Ahuactzin, K.K. Gupta and E. Mazer. Manipulation planning for redundant robots: a practical approach. In *Practical Motion Planning in Robotics*, K.K. Gupta and A.P. Del Pobil (Eds), J. Wiley, 1998.
2. R. Alami, T. Siméon, J.P. Laumond, A geometrical Approach to planning Manipulation Tasks. The case of discrete placements and grasps." In *Fifth International Symposium on Robotics Research*, 1989.
3. R. Alami, J.P. Laumond and T. Siméon. Two manipulation planning algorithms. In *Algorithmic Foundations of Robotics (WAFR94)*, K. Goldberg et al (Eds), AK Peters, 1995.

4. J. Barraquand and J.C. Latombe. Robot motion planning: a distributed representation approach. In *International Journal of Robotics Research*, 10(6), 1991.
5. J. Barraquand and P. Ferbach. A penalty function method for constrained motion planning. In *IEEE Int. Conf. on Robotics and Automation*, 1994.
6. P. Bessiere, J. Ahuactzin, T, El-Ghazali and E. Mazer. The "Ariane's clew" algorithm: Global planning with local methods. In *IEEE Int. Conf. on Robots and Systems*, 1993.
7. J. Cortés, T. Siméon and J.P. Laumond. A Random Loop Generator for planning the motions of closed kinematic chains with PRM methods. In *IEEE Int. Conf. on Robotics and Automation*, 2002.
8. F. Gravot, R. Alami and T. Siméon. Playing with several roadmaps to solve manipulation problems. In *IEEE Int. Conf. on Intelligent Robots and Systems*, 2002.
9. L. Han and N. Amato. A kinematics-based probabilistic roadmap method for closed kinematic chains. In *Algorithmic and Computational Robotics (WAFR00)*, B.R. Donald et al (Eds), AK Peters, 2001.
10. L. Kavraki and J.C. Latombe. Randomized preprocessing of configuration space for fast path planning. In *IEEE Int. Conf. on Robotics and Automation*, 1994.
11. Y. Koga and J.C. Latombe. Experiments in dual-arm manipulation planning. In *IEEE Int. Conf. on Robotics and Automation*, 1992.
12. Y. Koga and J.C. Latombe. On multi-arm manipulation planning. In *IEEE Int. Conf. on Robotics and Automation*, 1994.
13. J. Kuffner and S. Lavalle. RRT-Connect: an efficient approach to single-query path planning. In *IEEE Int. Conf. on Robotics and Automation.*, 2000.
14. J.C. Latombe. *Robot Motion Planning*, Kluwer, 1991.
15. J.P. Laumond and R. Alami. A geometrical Approach to planning Manipulation Tasks in robotics. In *LAAS Technical Report n° 89261*, 1989.
16. JP. Laumond, T. Siméon. Notes on visibility roadmaps for motion planning. In *Algorithmic and Computational Robotics (WAFR00)*, B.R. Donald et al (Eds), AK Peters, 2001.
17. S. LaValle, J.H. Yakey, and L. Kavraki. A probabilistic roadmap approach for systems with closed kinematic chains. In *IEEE Int. Conf. on Robotics and Automation*, 1999.
18. Ch. Nielsen, L. Kavraki. A two-level fuzzy PRM for manipulation planning. In *IEEE Int. Conf. on Intelligent Robots and Systems*, 2000.
19. M. Overmars and P. Švestka. A Probabilistic learning approach to motion planning. In *Algorithmic Foundations of Robotics (WAFR94)*, K. Goldberg et al (Eds), AK Peters, 1995.
20. T. Siméon, J.P. Laumond and C. Nissoux. Visibility-based probabilistic roadmaps for motion planning. In *Advanced Robotics Journal* 14(6), 2000 (a short version appeared in IEEE IROS, 1999).
21. T. Siméon, JP. Laumond, C. van Geem and J. Cortés. Computer Aided Motion: Move3D within MOLOG. In *IEEE Int. Conf. on Robotics and Automation*, 2001.
22. T. Siméon, J. Cortés, A. Sahbani and J.P. Laumond. A Manipulation Planner for Pick and Place Operations under Continuous Grasps and Placements. In *IEEE Int. Conf. on Robotics and Automation*, 2002.
23. A. Sahbani, J. Cortés, T. Siméon. A probabilistic algorithm for manipulation planning under continuous grasps and placements. In *IEEE Int. Conf. on Intelligent Robots and Systems*, 2002.

Computation on Parametric Curves with Applications in Localization and Grasping

Yan-Bin Jia

Iowa State University, Ames, IA 50011-1040, USA

Summary. This paper investigates efficient curve processing that originates from the localization and grasping of 2D curved objects.

The first algorithm locates the boundary section of an object traced out by a rolling finger based on the length and total curvature information obtained with the finger's tactile sensor. The algorithm slides an imaginary segment along the object boundary by alternatively marching its two endpoints forward, stretching or contracting the segment if necessary.

The second algorithm computes all pairs of antipodal points on an object. Two fingers placed at such a pair of points can achieve a force-closure grasp in the presence of friction. Dissecting the boundary into segments everywhere convex or everywhere concave, the algorithm marches simultaneously on every two segments with opposing normals and alternates marching with numerical bisection recursively. It builds on a procedure that constructs common tangents of two curves with quadratic local convergence.

Completeness (up to numerical resolution) of the above algorithms is established by applying curvature-based analyses. Dissection and the coupling of marching with bisection introduced in this paper are potentially applicable to optimization problems involving curves and curved surfaces.

1 Introduction

Geometry often plays an essential role in robot tasks such as sensing, grasping, pushing, path planning, dextrous manipulation, and control. Sensing, manipulation, and planning strategies for polygonal and polyhedral objects have been studied extensively in the past. Many of these strategies are designed using combinatorial techniques that take advantage of the absence of local (differential) geometry.

Curved shapes are frequent subjects of maneuvers by the human hand and they share the differential nature with contact kinematics and dynamics which are responsible for such maneuvers [2]. This paper presents efficient algorithms that process plane curves for tasks including parts localization and grasping.

Section 2 studies how a disk-like finger with tactile ability localizes itself while rolling on a stationary curved object. Section 3 describes an algorithm that constructs all common tangents of two curve segments. This algorithm is part of the preprocessing in Section 5 that prepares for the computation of antipodal grasps in Section 4.

The curve $\boldsymbol{\alpha}(s)$ considered in this paper is simple, regular, and twice continuously differentiable. In case no ambiguity arises, the parameter s also

J.-D. Boissonnat et al. (Eds.): Algorithmic Foundations of Robotics V, STAR7, pp. 329–345, 2004.

refers to the point $\alpha(s)$ on the curve. The curvature $\kappa(s)$ of α is zero at only a finite number of points. A point z on α is a ***simple inflection*** if $\kappa(z) = 0$ but $\kappa'(z) \neq 0$. The ***total curvature*** of a segment of α over $[a, b]$ is given by $\Phi(a, b) = \int_a^b \kappa \|\alpha'(u)\| \, du$. This integral measures the amount of rotation of the tangent as a point moves from a to b along the curve[1].

1.1 Related Work

Hong *et al.* [12] proved the existence of two pairs of antipodal points on a closed, simple, and smooth convex curve or surface. Chen and Burdick [6] computed antipodal points on 2D and 3D shapes through minimizing a grasping energy function.

Two vertices of a polygon are antipodal if they admit parallel supporting lines. The polygon's diameter is equal to the distance between the two furthest vertices (which must be antipodal). For n points in three dimensions, the number of antipodal pairs can be $O(n^2)$ while the diameter can be achieved by $2n - 2$ pairs of points [11]. Preparata and Shamos [27] described a linear-time algorithm that finds the diameter of a convex polygon with n vertices in one traversal. Clarkson and Shor [7] presented a randomized algorithm to compute the diameter in $O(n \log n)$ expected time. Near-linear-time deterministic algorithms were developed through derandomizing their algorithm in [5, 19]. Yao [32] offered a different approach that led to an $o(n^2)$-time algorithm for computing the diameter of a point set in any dimensions.

Goodman [10] gave an upper bound on the number of inflection points on parametric spline curves. Sakai [29] obtained the distribution of inflection points and cusps on a parametric rational cubic curve. Manocha and Canny [17] employed the Sturm sequence method to find inflection points on rational curves. Mokhtarian and Mackworth [21] used inflection points for plane curve description and matching.

Preparata and Hong [26] developed a "walking" strategy to construct common supporting lines of two convex polygons in linear time. The procedure `Common-Tangent` in Section 3.2 can be viewed as its continuous counterpart where tangent lines (in place of supporting lines) are used to determine the next stops.

Allen and Roberts [1] deployed robot fingers to obtain a number of contact points around an object and used fitting to reconstruct the object's shape. Fischler [9] described an algorithm that locates points with extreme curvatures on planar curves and reconstructs the original curves based on these points. Erdmann [8] showed that the local geometry of a curved object with known angular velocity can be recovered by two passive linear tactile sensors. Extending this work, Moll and Erdmann [22] applied quasi-static dynamics

---

[1] The total curvature over $[a, b]$ has a closed form if it is within $(0, 2\pi]$ and can be fast computed otherwise.

to reconstruct the shape of a convex object and estimate its motion from tactile readings on two palms in frictionless contact with the object.

Mishra *et al.* [20] gave upper bounds on the numbers of frictionless fingers that are sufficient for equilibrium and force-closure grasps, respectively, on objects with piecewise smooth boundaries. Markenscoff *et al.* [18] determined the number of fingers to immobilize 2-D and 3-D objects with piecewise smooth boundaries. Ponce *et al.* [25] employed parallel cell decomposition to compute pairs of maximal-length segments on a piecewise smooth curved 2D object that guarantee force closure with friction.

2 Locating Segments on a Curve

We begin with the problem of determining the location of a finger rolling on a stationary object bounded by $\boldsymbol{\alpha}$. A tactile sensor is mounted on the finger.

To simplify our analysis, the finger is assumed to be a disk described by $r(\cos u, \sin u)$, where u locates the contact on the disk and is determined from the tactile data. The disk is rolling with angular velocity ω. Let s be the location of contact on $\boldsymbol{\alpha}$, as shown in Figure 2.1(a).

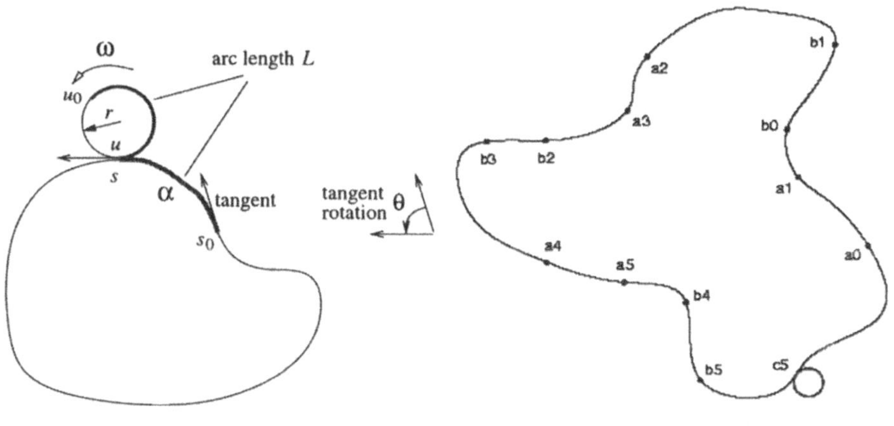

(a) (b)

Figure 2.1. (a) A disk rolling on a stationary curved object; (b) a disk localizing itself by rolling from a_5 to b_5 and then to c_5. The localization algorithm finds six segments over $[a_i, b_i]$, $0 \leq i \leq 5$, respectively, which have the length and total curvature computed from tactile data. These ambiguities are eliminated using the curve length and total curvature over $[b_5, c_5]$.

Contact kinematics [4, 23] yield the velocities of the contact point on the disk and the object, respectively:

$$\dot{u} = -\frac{\omega}{1 + r \cdot \kappa(s)}, \tag{2.1}$$

$$\dot{s} = \frac{\omega}{\|\boldsymbol{\alpha}'(s)\| \Big(1/r + \kappa(s)\Big)}. \tag{2.2}$$

From (2.1) and (2.2), we obtain

$$\begin{aligned} \dot{s}\, \|\boldsymbol{\alpha}'(s)\| &= -r\dot{u}, \\ \dot{s}\, \kappa(s)\, \|\boldsymbol{\alpha}'(s)\| &= \dot{u} + \omega. \end{aligned}$$

Integrate these two equations over the time period $[0, \tau]$:

$$\int_{s_0}^{s_1} \|\boldsymbol{\alpha}'(s)\|\, ds = -r(u_1 - u_0) = L, \tag{2.3}$$

$$\int_{s_0}^{s_1} \kappa(s)\, \|\boldsymbol{\alpha}'(s)\|\, ds = u_1 - u_0 + \int_0^{\tau} \omega(\xi)\, d\xi = \theta, \tag{2.4}$$

where $s_0 = s(0)$, $s_1 = s(\tau)$, $u_0 = u(0)$, and $u_1 = u(\tau)$.

The length L of the object boundary traced out by the contact can be determined from tactile readings u_0 and u_1. The amount of disk rotation $\int_0^{\tau} \omega(\xi) d\xi$ is also known. Thus the total curvature θ over $[s_0, s_1]$ is known by (2.4). Localization thus reduces to *finding a curve segment on $\boldsymbol{\alpha}$ with length L and total curvature θ.*

When multiple segments are found to have length L and total curvature θ, the disk eliminates the ambiguities by continuing the rolling and using the length and total curvature of the next segment it traces out (see Figure 2.1(b)).

Equations (2.3) and (2.4) need to be solved numerically. In general, arc length $\ell(a, b) = \int_a^b \|\boldsymbol{\alpha}'(s)\|\, ds$ has no closed form. We first consider the case that the boundary curve $\boldsymbol{\alpha}$ is convex and present a marching algorithm.

The algorithm marches two endpoints s and t of a hypothesized segment counterclockwise along $\boldsymbol{\alpha}$. It starts at location $s_0 = 0$ and advances repeatedly by a step size until it reaches the point q where $\Phi(s_0, q) = \theta$. If $\ell(s_0, q) > L$, at step $i \geq 0$ advance from t_{i-1} to t_i (set $t_0 \leftarrow q$) where $\Phi(s_i, t_i) = \theta$ and then advance from s_i to s_{i+1} where $\ell(s_{i+1}, t_i) = L$. If $\ell(s_0, q) < L$, at step i advance from t_{i-1} (or from q if $i = 0$) to t_i where $\ell(s_i, t_i) = L$ and then advance from s_i to s_{i+1} where $\Phi(s_{i+1}, t_i) = \theta$. Figure 2.2 illustrates the working of the algorithm in these two cases.

It can be shown by induction [15] that the two sequences $s_0, s_1, \ldots$ and $t_0, t_1, \ldots$ generated above are monotone and will converge to the endpoints of the first segment on $\boldsymbol{\alpha}$ that satisfies equations (2.3) and (2.4). Furthermore, the convergence rate [31, p. 264] is linear and given by the ratio $\min\{\kappa(a)/\kappa(b), \kappa(b)/\kappa(a)\}$. To find the next segment on $\boldsymbol{\alpha}$, we reset s_0 to be slightly greater than s_i for large enough i, and repeat the same procedure.

The correctness of the above march relies on that the total curvature $\Phi(s, t)$ has partial derivatives $\partial\Phi/\partial s < 0$ and $\partial\Phi/\partial t > 0$ for all $s < t$. But

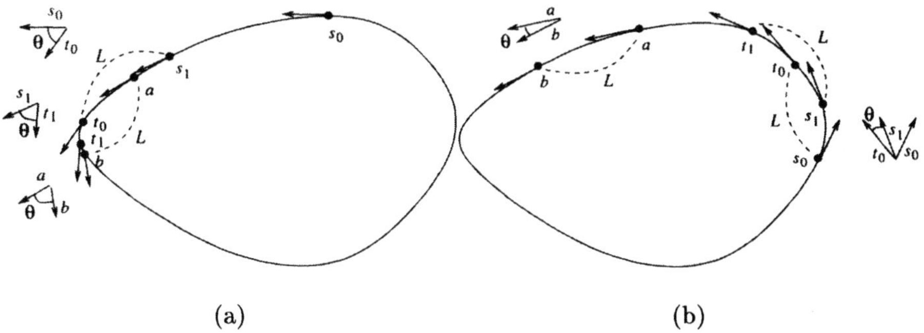

Figure 2.2. Two cases of marching s and t: **(a)** $\Phi(s_i, t_i) = \theta$ but $\ell(s_i, t_i) > L$; **(b)** $\ell(s_i, t_i) = L$ but $\Phi(s_i, t_i) > \theta$. Both marches converge to (a, b) where $\ell(a, b) = L$ and $\Phi(a, b) = \theta$.

this is no longer true when $\boldsymbol{\alpha}$ has concavities. For example, $\partial\Phi/\partial s > 0$ when $\kappa(s) < 0$.

When $\boldsymbol{\alpha}$ is not convex, marching has four basic modes: *convex-convex* ($\kappa(s) \geq 0$ and $\kappa(t) \geq 0$), *concave-concave* ($\kappa(s) \leq 0$ and $\kappa(t) \leq 0$), *convex-concave* ($\kappa(s) \geq 0$ and $\kappa(t) \leq 0$), and *concave-convex* ($\kappa(s) \leq 0$ and $\kappa(t) \geq 0$). Within each mode, the hypothesized curve segment over (s, t) slides along $\boldsymbol{\alpha}$ until a simple inflection point is reached by either s or t so that the mode changes. Location(s) of the desired curve segment, if exists, is also found during the advancement. Sliding is done by increasing only one of s and t at a time while simultaneously keeping track of where the other should be (without actually increasing it).

For a detailed description of all operations, we refer to [15] where the following result is established:

Theorem 2.1. *All segments with length L and total curvature θ can be found on a closed simple curve over domain $[0, D]$ up to numerical resolution in at most $5D/h$ steps, where h is the step size.*

The algorithm runs asymptotically as fast as evaluating the length of the entire curve through numerical integration.

3 Common Tangent Construction

For clarity of presentation, in Sections 3 and 4 we consider that the curve $\boldsymbol{\alpha}$ is *unit-speed*, that is, $\|\boldsymbol{\alpha}'(s)\| = 1$. All algorithms to be presented have been extended straightforwardly to and implemented on arbitrary-speed curves.

Two segments $\mathcal{S}$ and $\mathcal{T}$ of $\boldsymbol{\alpha}$ are defined on subdomains (s_a, s_b) and (t_a, t_b), respectively, where $(s_a, s_b) \cap (t_a, t_b) = \emptyset$. Here $s_a < s_b$ always holds but $t_a > t_b$ is allowed, in which case (t_a, t_b) refers to the interval (t_b, t_a). Denote by $T(s) = \boldsymbol{\alpha}'(s)$ the unit tangent of $\boldsymbol{\alpha}$ and denote by $N(s)$ the unit normal such that $T(s) \times N(s) = 1$. The following conditions are satisfied by $\mathcal{S}$ and $\mathcal{T}$:

(i) $\kappa > 0$ everywhere or $\kappa < 0$ everywhere on $\mathcal{S}$ and $\mathcal{T}$, with $\kappa = 0$ possible only at s_a, s_b, t_a, t_b.
(ii) $N(s_a) + N(t_a) = 0$ and $N(s_b) + N(t_b) = 0$.
(iii) $-\pi \leq \Phi(s_a, s_b) = -\Phi(t_a, t_b) \leq \pi$.

For now we present algorithms working on $\mathcal{S}$ and $\mathcal{T}$. Section 5 will describe how to preprocess $\boldsymbol{\alpha}$ to generate pairs of segments satisfying (i)–(iii).

Under these conditions, a one-to-one correspondence exists between a point s on $\mathcal{S}$ and a point t on $\mathcal{T}$:

$$N(s) + N(t) = 0, \qquad \text{or equivalently,} \tag{3.1}$$

$$T(s) + T(t) = 0. \tag{3.2}$$

Let $g(s,t) = N(s) \times N(t)$. Since $\frac{\partial g}{\partial t} = N(s) \times (-\kappa(t)T(t)) = -\kappa(t) \neq 0$ under (3.1),[2] by the Implicit Function Theorem, the equation $g(s,t) = 0$ defines t locally as a function of s. From now on, we refer to t as the *opposite point* of s. Differentiate (3.2) with respect to s and then substitute (3.1) in. We obtain that $\kappa(s) - \kappa(t)\,\frac{dt}{ds} = 0$ and

$$\frac{dt}{ds} = \frac{\kappa(s)}{\kappa(t)}. \tag{3.3}$$

3.1 Classification of Common Tangents

Theorem 3.1. *The segments $\mathcal{S}$ and $\mathcal{T}$ under conditions (i)–(iii) have at most two common tangents. The segments are always on the same side of every common tangent or always on different sides of every common tangent.*

A proof of Theorem 3.1 is given in [14]. Based on the theorem, we classify four configurations (shown in Figure 3.1) in which at least one common tangent of $\mathcal{S}$ and $\mathcal{T}$ exists.

In configurations (a) and (c) the two curve segments bend in the same direction while in (b) and (d) they bend in opposite directions. The bending direction of the curve $\boldsymbol{\alpha}$ at a point u is indicated by $\kappa(u)N(u)$. So we check the dot product $\kappa(s_c)N(s_c) \cdot \kappa(t_c)N(t_c) = -\kappa(s_c)\kappa(t_c)$, where $s_c = \frac{s_a+s_b}{2}$ and $N(t_c) = -N(s_c)$. If the dot product is positive, then the two segments $\mathcal{S}$ and $\mathcal{T}$ bend in the same direction; otherwise, they bend in opposite directions.

Define the function $d(s)$ as the translational distance along the normal $N(s)$ from the tangent line of $\mathcal{S}$ at s to the tangent line of $\mathcal{T}$ at its opposite point t. This function is continuous since the curve $\boldsymbol{\alpha}$ is continuously differentiable. Obviously, $d(s) = 0$ if and only if the tangent line through s

---

[2] The Frenet formulas [24, pp. 56–58] for unit-speed plane curves state that $T'(s) = \kappa(s)N(s)$ and $N'(s) = -\kappa(s)T(s)$.

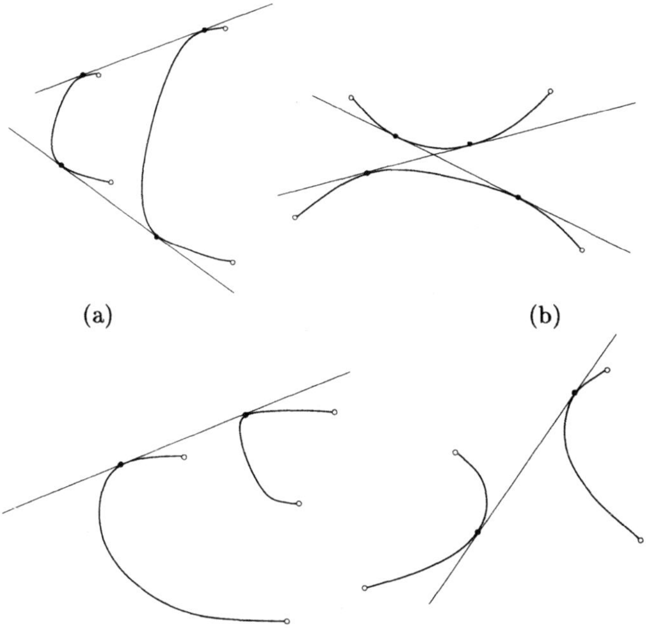

Figure 3.1. Four configurations where two segments satisfying conditions (i)–(iii) have at least one common tangent.

is a common tangent. It follows from condition (i) that $d'(s) \neq 0$ whenever $d(s) = 0$.

When $d(s_a)$ and $d(s_b)$ have different signs, there exists some $c \in (s_a, s_b)$ such that $d(c) = 0$. Namely, the line passing through c and its opposite point on $\mathcal{T}$ is a common tangent. Because there are at most two common tangents and d has only simple zeros, we can infer that this is the unique common tangent of $\mathcal{S}$ and $\mathcal{T}$ and the configuration is either (c) or (d) in Figure 3.1. These two configurations can be further distinguished.

When $d(s_a)$ and $d(s_b)$ have the same sign, $\mathcal{S}$ and $\mathcal{T}$ may have two common tangents as in configurations (a) and (b) or they may not have a common tangent at all. We will need to look at whether the two segments bend in the same direction or in opposite directions and make use of the regions partitioned by the segments as well as the four tangent lines at their endpoints. A procedure is given in [14] to determine whether two or zero common tangent exists.

3.2 An Iterative Method

Below we offer an iterative method to construct a common tangent. The method makes the following two assumptions that will be removed in the end of the section.

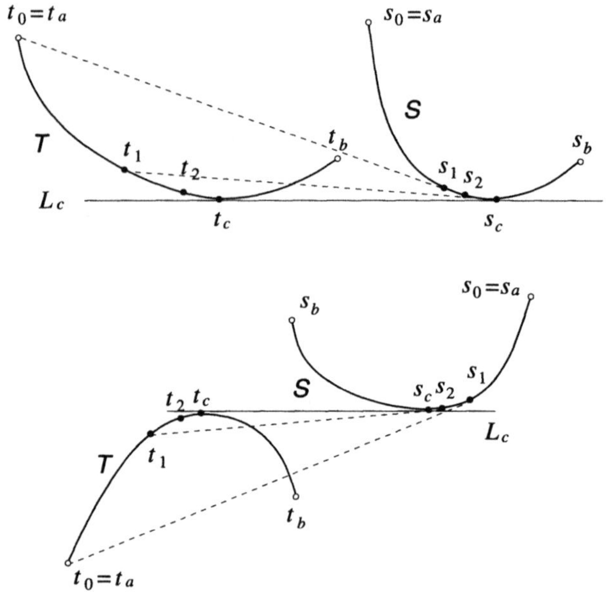

Figure 3.2. Constructing a common tangent iteratively.

1. There exists exactly one common tangent L_c of $\mathcal{S}$ and $\mathcal{T}$ incident on s_c and t_c, respectively.
2. A point traversing $\mathcal{T}$ from t_a to t_b is moving towards s_c at the time it reaches t_c. So the two segments and L_c are in one of the two configurations shown in Figure 3.2.

The iteration starts at $s_0 = s_a$ and $t_0 = t_a$ and maintains two invariants:

$$\Big(\boldsymbol{\alpha}(s_{i+1}) - \boldsymbol{\alpha}(t_i)\Big) \times T(s_{i+1}) = 0; \qquad (3.4)$$

$$N(s_{i+1}) + N(t_{i+1}) = 0. \qquad (3.5)$$

The point $s_{i+1} \in (s_i, s_b)$ is found by a primitive `point-of-tangency`. Such a point of tangency always exists under assumptions 1 and 2. The procedure call for finding the two points of tangency s_c and t_c is `Common-Tangent`(s_a, s_b, t_a, t_b).

We have established [14] that the two sequences $\{s_i\}$ and $\{t_i\}$ generated according to (3.4) and (3.5) are strictly increasing (decreasing) and bounded by s_c and t_c. They must converge to two points that satisfy (3.4), i.e., s_c and t_c.

Next, we determine the order of local convergence of $\{s_i\}$ and $\{t_i\}$. Let g be the iteration function such that $s_{i+1} = g(s_i)$. Differentiate equation (3.4) with respect to s_i, simplify, and substitute (3.3) in:

$$-T(t_i) \times T(s_{i+1}) \frac{\kappa(s_i)}{\kappa(t_i)} + \|\boldsymbol{\alpha}(s_{i+1}) - \boldsymbol{\alpha}(t_i)\| \, \kappa(s_{i+1}) \frac{ds_{i+1}}{ds_i} = 0.$$

From above we obtain the derivative of the iteration function $g'(s_i)$, in particular,

$$g'(s_c) = \frac{T(t_c) \times T(s_c)}{\|\boldsymbol{\alpha}(s_c) - \boldsymbol{\alpha}(t_c)\|\kappa(s_c)} \frac{\kappa(s_c)}{\kappa(t_c)} = 0,$$

because $T(t_c) \times T(s_c) = 0$. Meanwhile, we also obtain

$$g''(s_c) = \frac{\kappa(s_c)}{\kappa(t_c)\|\boldsymbol{\alpha}(s_c) - \boldsymbol{\alpha}(t_c)\|} \neq 0.$$

Hence the two sequences $\{s_i\}$ and $\{t_i\}$ have *quadratic* convergence rate.

To remove assumptions 1 and 2 made earlier, we need to determine the order of the four endpoints passed on as arguments for calling `Common-Tangent`. This order determines which of the two pairs of endpoints to start the iteration at, and, within the pair, which endpoint to update first. This would not be an issue for configuration (b) in Figure 3.1, where the two common tangents can be found using the procedure calls

`Common-Tangent`(s_a, s_b, t_a, t_b) and `Common-Tangent`(s_b, s_a, t_b, t_a)

regardless of the vertex labeling.

When the configuration is one of (a), (c), (d), we look at the parallel tangent lines L_a and L'_a of $\boldsymbol{\alpha}$ at s_a and t_a, respectively. For instance, if the translation from L_a to L'_a is along the bending direction of $\mathcal{S}$ then we start the iteration by updating from $t_0 = t_a$ to t_1, the point of tangency of a tangent line to $\mathcal{T}$ through s_a, then from $s_0 = s_a$ to s_1, where $N(s_1) + N(t_1) = 0$, and so on.

4 Computation of Antipodal Points

From now on we require that $\boldsymbol{\alpha}$ be also closed. Two points a and b on $\boldsymbol{\alpha}$ are *antipodal* if their normals are collinear and pointing at each other, or more precisely,

$$\begin{aligned} N(a) + N(b) &= 0, \\ N(a) \times \Big(\boldsymbol{\alpha}(b) - \boldsymbol{\alpha}(a)\Big) &= 0, \\ N(a) \cdot \Big(\boldsymbol{\alpha}(b) - \boldsymbol{\alpha}(a)\Big) &> 0. \end{aligned}$$

We need only consider that the curvature of $\boldsymbol{\alpha}$ is not constant[3]. For example, the closed curve in Figure 4.1(a) has eight pairs of antipodal points (numbered the same within each pair). Two fingers placed at any of these eight pairs of points will form a force-closure grasp that can resist any arbitrary external

[3] This excludes a circle on which any two points determining a diameter are antipodal.

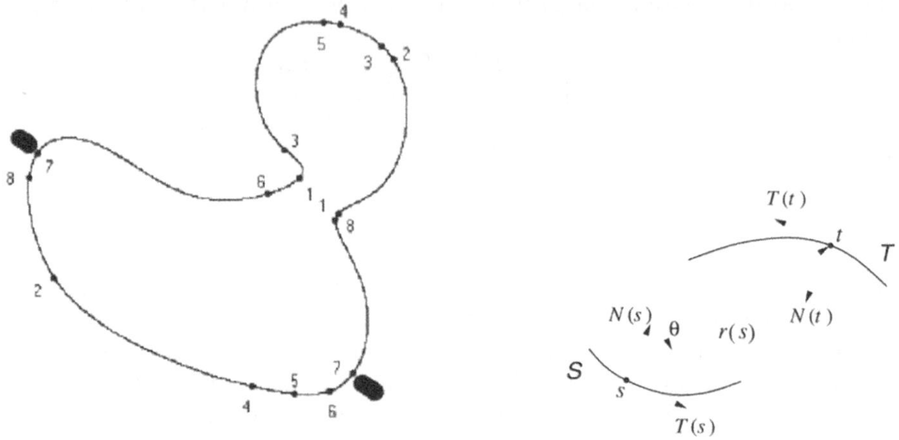

(a) (b)

Figure 4.1. (a) Eight pairs of antipodal points on a curved shape. The grasp at pair 7 (or at any other pair) is force-closure. (b) Antipodal angle θ.

force and torque in the presence of contact friction. Such a grasp is referred to as an *antipodal grasp.*

In this section, we present an algorithm that *finds all antipodal points on a pair of segments* $\mathcal{S}$ *and* $\mathcal{T}$ *of* $\boldsymbol{\alpha}$ that meet conditions (i)–(iii) in Section 3 as well as the conditions below:

(iv) Neither s_a and t_a nor s_b and t_b are antipodal.

(v) $N(s) \cdot \Big(\boldsymbol{\alpha}(t) - \boldsymbol{\alpha}(s)\Big) > 0$ for all $s \in (s_a, s_b)$.

Under condition (iii), no two points on $\mathcal{S}$ (which does not include s_a or t_a) are antipodal and the same holds for $\mathcal{T}$. Condition (v) addresses that two opposite points may be antipodal only if their normals do not point away from each other.

Pairs of segments on $\boldsymbol{\alpha}$ that meet conditions (i)–(v) will be generated in the preprocessing step described in Section 5.

4.1 Antipodal Angle

Define the *antipodal angle*[4] $\theta(s)$ as the rotation angle from the normal $N(s)$ to the vector $\boldsymbol{r}(s) = \boldsymbol{\alpha}(t) - \boldsymbol{\alpha}(s)$, where t is the opposite point of s (see Figure 4.1(b)). Under condition (v), $\theta \in (-\frac{\pi}{2}, \frac{\pi}{2})$. By definition, s and t are antipodal *if and only if* $\theta(s) = 0$.

To determine the change rate of θ, we first calculate the derivative:

$$\frac{d}{ds}\|\boldsymbol{r}(s)\| \quad = \quad \frac{d}{ds}\sqrt{\boldsymbol{r}(s)\cdot\boldsymbol{r}(s)} \quad = \quad \left(\frac{\kappa(s)}{\kappa(t)}+1\right)\sin\theta, \qquad \text{by (3.3).}$$

[4] In [3], it is referred to as the friction angle.

From Figure 4.1(b) we see that $\sin\theta = N(s) \times \boldsymbol{r}(s)/\|\boldsymbol{r}(s)\|$. Differentiate both sides of this equation and substitute the above expression for $\frac{d}{ds}\|\boldsymbol{r}(s)\|$ in. After a few more steps, we obtain

$$\theta'(s) \quad = \quad -\kappa(s) + \frac{\cos\theta}{\|\boldsymbol{r}(s)\|}\left(\frac{\kappa(s)}{\kappa(t)} + 1\right). \tag{4.1}$$

Suppose $\boldsymbol{\alpha}$ is at least $k + 1 \geq 2$ times continuously differentiable. Two points s^* on $\mathcal{S}$ and t^* on $\mathcal{T}$ are called *antipodal points of order k* if $\theta(s^*) = \theta'(s^*) = \cdots = \theta^{(k-1)}(s^*) = 0$ but $\theta^{(k)}(s^*) \neq 0$. When $k = 1$, s^* and t^* are *simple* antipodal points. When $k = 2$, equation (4.1) reduces to $\|\boldsymbol{r}(s^*)\| = \frac{1}{\kappa(s^*)} + \frac{1}{\kappa(t^*)}$; that is, the osculating circles at s^* and t^* are *concentric*.

We are primarily interested in finding simple antipodal points. So from now on we assume that *higher order antipodal points do not exist*. Our algorithm employs different strategies according as whether the signs of the antipodal angles $\theta(s_a)$ and $\theta(s_b)$ at the endpoints of $\mathcal{S}$ are the same.

4.2 Opposite Angle Signs at Endpoints

When $\theta(s_a)$ and $\theta(s_b)$ have different signs, at least one pair of antipodal points exists on $\mathcal{S}$ and $\mathcal{T}$. Bisection method [28, pp. 261–263] guarantees to find one such pair.

In the case where both segments are concave, $\kappa(s) < 0$ and $\kappa(t) < 0$, we have $\theta'(s) > 0$ by (4.1). The antipodal angle θ increases monotonically from s_a to s_b. A unique pair of antipodal points exists if $\theta(s_a) < 0$ and $\theta(s_b) > 0$. Otherwise, no antipodal points exist.

4.3 Same Angle Sign at Endpoints

In this case, at least one of $\mathcal{S}$ and $\mathcal{T}$ must be convex in order to have antipodal points. Multiple pairs of antipodal points may exist.

4.3.1 Two Convex Segments. When $\mathcal{S}$ and $\mathcal{T}$ are both convex, we march on both segments using the fact that the vector $\boldsymbol{r}(s)$ rotates counterclockwise as s increases from s_a to s_b. This is because $\frac{d\boldsymbol{r}}{ds} \times \boldsymbol{r} < 0$. To see the inequality, note that the vector $\frac{d\boldsymbol{r}}{ds} = T(t)\left(1 + \frac{\kappa(s)}{\kappa(t)}\right)$ is in the direction of $T(t)$ since $\kappa(s), \kappa(t) > 0$. But $T(t) \times \boldsymbol{r}(s) = -T(s) \times \boldsymbol{r}(s) < 0$ under condition (v).

Figure 4.2 illustrates the march when $\theta(s_a) < 0$ and $\theta(s_b) < 0$. It starts with s and t at $s_0 = s_b$ and $t_0 = t_b$, respectively. As s moves toward s_a, the vector $\boldsymbol{r}(s)$ rotates clockwise. At the ith iteration step move s from s_i to s_{i+1} at which the normal is parallel to $\boldsymbol{r}(s_i)$. If no such point s_{i+1} exists, stop. Otherwise, move t from t_i to t_{i+1} where $N(t_{i+1}) + N(s_{i+1}) = 0$. The iteration continues until s_i and t_i converge to a pair of antipodal points, as in the figure, or it reaches s_a and t_a, in which case no antipodal points.

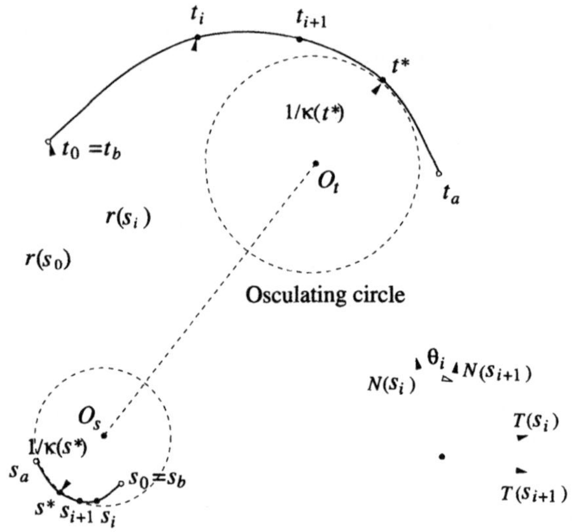

Figure 4.2. Geometry at a pair of antipodal points s^* and t^*: $\theta'(s^*) < 0$ and $\|\boldsymbol{r}(s^*)\| > \frac{1}{\kappa(s^*)} + \frac{1}{\kappa(t^*)}$.

It can be shown by induction [14] that the sequence $s_0, s_1, \ldots$, generated above according to $N(s_{i+1}) \times \boldsymbol{r}(s_i) = 0$ is *strictly decreasing* and no antipodal point exists on $[s_i, s_b) = \cup_{k=1}^{i}[s_k, s_{k-1})$ for all $i > 0$. The monotone sequence $\{s_i\}$ is bounded from below by s^*, the closest antipodal point to s_b if it exists. So it must converge to some $\xi \in (s_a, s_b)$ where $N(\xi) \times \boldsymbol{r}(\xi) = 0$. Hence $\xi = s^*$.

Next, we show that the local convergence rate of the sequence is *linear*. It suffices to establish that $0 < \frac{ds_{i+1}}{ds}(s^*) = f'(s^*) < 1$. As shown in Figure 4.2, $\sin\theta_i = \sin(\theta(s_i)) = N(s_i) \times N(s_{i+1})$. We can derive $f'(s_i)$ by differentiating both sides of this equation with respect to s_i, and in particular,

$$f'(s^*) = \frac{\theta'(s^*)}{\kappa(s^*)} + 1 = \frac{\kappa(s^*) + \kappa(t^*)}{\kappa(s^*)\kappa(t^*)\|\boldsymbol{r}(s^*)\|} > 0,$$

where t^* on $\mathcal{T}$ is the opposite point of s^*. The march starts at s_b where $\theta(s_b) < 0$ and never passes s^*. So $\theta'(s^*) < 0$ must hold. Hence $f'(s^*) < 1$.

When $\theta(s_a) > 0$ and $\theta(s_b) > 0$, the march starts at s_a and t_a and moves toward s_b and t_b. The result on convergence still holds.

4.3.2 A Convex Segment and a Concave Segment. We only consider the case that $\mathcal{S}$ is convex and $\mathcal{T}$ is concave as the other case is symmetric. We first determine if one of the rays extending the normals $N(t_a)$ and $N(t_b)$ intersects $\mathcal{S}$. Under conditions (i)–(v), this can be done by checking cross products. If neither ray intersects $\mathcal{S}$, then no antipodal points exist (a proof is given in [14]).

When either the ray of $N(t_a)$ or the ray of $N(t_b)$ intersects $\mathcal{S}$, we carry out the following march. Start at $s_0 = s_a$ and $t_0 = t_a$ if the ray of $N(t_a)$

intersects $\mathcal{S}$, or at $s_0 = s_b$ and $t_0 = t_b$ if the ray of $N(t_b)$ intersects $\mathcal{S}$. In each round, s_{i+1} is generated as the intersection of the ray of $N(t_i)$ and $\mathcal{S}$ and t_{i+1} is generated as its opposite point.

The sequence $\{s_i\}$ is shown to be *strictly increasing* and will converge at *linear* rate to the first antipodal point s^* from s_a if it exists.

Now we combine the results from Sections 4.3.1 and 4.3.2:

Theorem 4.1. *Let $\mathcal{S}$ and $\mathcal{T}$ be two curve segments satisfying conditions (i)-(v). Suppose the two antipodal angles $\theta(s_a)$ and $\theta(s_b)$ at the endpoints s_a and s_b of $\mathcal{S}$ have the same sign. Then the two iterative strategies described in Sections 4.2 and 4.3 satisfy the following:*

1. *When no antipodal points exist, the iteration starting at a pair of opposite endpoints of $\mathcal{S}$ and $\mathcal{T}$ will terminate at the other pair of endpoints.*
2. *Otherwise, the iteration will converge at* linear rate *to a pair of antipodal points s^* and t^* that is the closest to the two endpoints at which the iteration starts. Furthermore, $\theta'(s^*) < 0$ if $\mathcal{S}$ and $\mathcal{T}$ are both convex and $\theta'(s^*) > 0$ if one of them is concave.*

4.4 Finding All Pairs of Antipodal Points

Once the first pair of antipodal points s^* and t^* has been found, other pairs, if exist, can be found by alternating marching with bisection recursively.

To illustrate, suppose $\theta(s_a), \theta(s_b) < 0$ and $\mathcal{S}$ and $\mathcal{T}$ are both convex. Then s^* and t^* have been found by a march from s_b. So we already know that no antipodal points exist in (s^*, s_b). That $\theta'(s^*) < 0$ and $\theta(s^*) = 0$ implies $\theta(s^* - \epsilon) > 0$ for small enough $\epsilon > 0$. Therefore the interval $(s_a, s^* - \epsilon)$ contains at least one antipodal point, which can be found using bisection.

Suppose $\theta(s_a)$ and $\theta(s_b)$ have different signs. Then s^* and t^* have been found through bisection. So $\theta(s^* - \epsilon)$ has the sign of $\theta(s_a)$ while $\theta(s^* + \epsilon)$ has the sign of $\theta(s_b)$. Then we march over the intervals $(s_a, s^* - \epsilon)$ and $(s^* + \epsilon, s_b)$ separately to search for possible antipodal points.

Figure 4.3 illustrates the recursive procedure as applied on an ellipse.

5 Preprocessing, Generalization, and Implementation

Curve $\boldsymbol{\alpha}$ is preprocessed in four steps to generate pairs of segments that satisfy conditions (i)–(v). These steps are illustrated on an example in Figure 5.1. To locate antipodal points on $\boldsymbol{\alpha}$, we perform all four steps of preprocessing to generate segment pairs that satisfy conditions (i)–(v). Then find antipodal points on each pair[5] as described in Section 4..

We have implemented all algorithms in C++ for arbitrary-speed curves. The eight pairs of antipodal points in Figure 4.1(a) has all three combinations

[5] For each pair, we also report opposing endpoints that are antipodal.

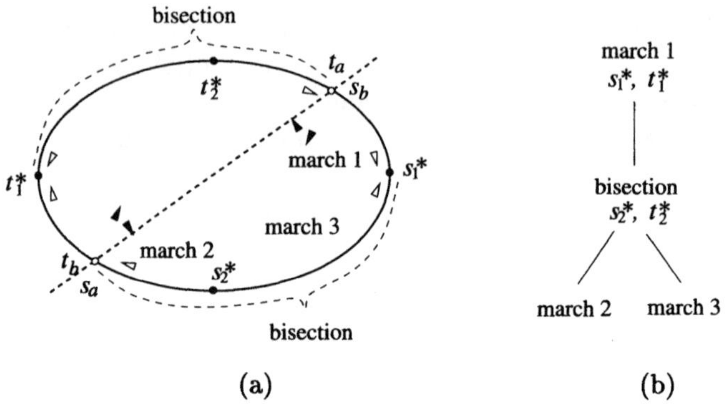

Figure 4.3. (a) Two pairs of antipodal points (s_1^*, t_1^*) and (s_2^*, t_2^*) on an ellipse; and (b) the recursion tree in search for them. Here $\boldsymbol{\alpha}(s_b) = \boldsymbol{\alpha}(t_a)$ and $N(s_a) = -N(t_a)$. The recursion always ends at marching.

of curvature signs. Figure 5.2 displays the antipodal points found on three other shapes [30]: (a) an elliptic lemniscate $\rho = \sqrt{6^2 \cos^2 \phi + 3^2 \sin^2 \phi}$; (b) a limaçon $\rho = 4 + \frac{5}{2} \cos \phi$; and (c) a curve with convexities $\rho = 9/(2 + \frac{2}{5} \cos 4\phi)$.

Let n be the number of inflection points and m the number of pairs of antipodal points. There are $O(n^2)$ pairs of segments after the preprocessing. The total number of calls to the marching and bisection procedures in Section 4 is $O(n^2 + m)$. This bound is indeed tight since a quadratic number of pairs of segments dissected at inflection points can have at least one pair of antipodal points. An example is the curve $\rho(\phi) = \frac{p}{1+\epsilon \cos m\phi}$ with $p > 0$, $0 < \epsilon < 1$, and integer $m > 1$. In Figure 5.2(c), $p = \frac{9}{2}$, $m = 4$, and $\epsilon = \frac{1}{5}$.

6 Discussion

We have presented a collection of algorithms that compute geometric substructures on a simple, closed, and twice continuously differentiable curve. These substructures include curve segments of specified length and total curvature, common tangents, and antipodal points. Based on the monotonicity of dissected curve segments and local geometry, various marching strategies (interleaved with bisection) are employed.

A conventional nonlinear programming approach, inherently local, would rely heavily on initial guesses of antipodal positions. It would be slow and not guarantee to always find antipodal points, not to mention all of them. Our results demonstrate that computational efficiency and completeness can be achieved by taking advantage of both global and differential geometry.

The algorithms can be extended in a straightforward way to curves that are piecewise twice continuously differentiable. Extension to a curved surface in 3D will likely hinge on how efficiently the surface can be partitioned into

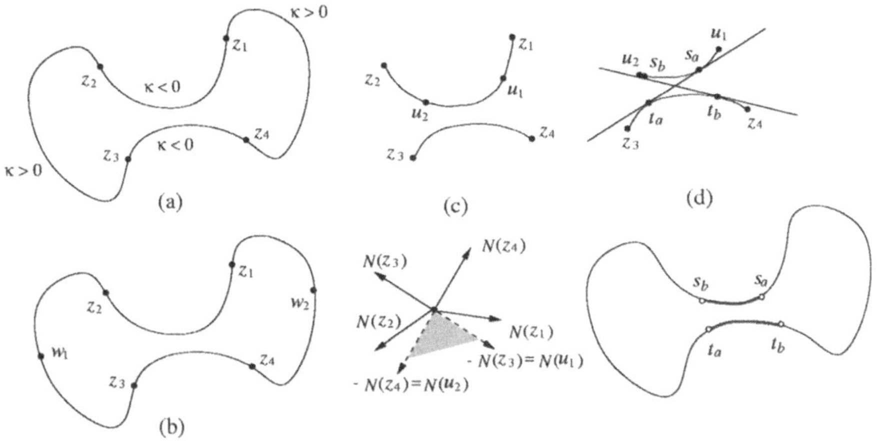

Figure 5.1. Four preprocessing steps illustrated on a closed curve. (a) Find all four inflection points z_1, z_2, z_3, and z_4. (b) Split the segments $[z_2, z_3]$ and $[z_4, z_1]$ with total curvatures beyond $[-\pi, \pi]$ at the points w_1 and w_2, respectively. Carry out the next two steps on every pair of segments, for instance, $[z_1, z_2]$ and $[z_3, z_4]$. (c) Intersect the cone of inward normals over $[z_1, z_2]$ with the cone of outward normals over $[z_3, z_4]$. Now (u_1, u_2) and (z_3, z_4) satisfy conditions (i)-(iv) but not (v). (d) Extract the portions defined by the tangency points s_a, t_a, s_b, and t_b of their two common tangents.

patches on which a search or an objective function assumes similar monotonicity.

Both the localization algorithm and the algorithm for computing antipodal points step over a found segment for more segments or a pair of antipodal points for more pairs. The completeness is thus subject to the numerical resolution. Marching in localization is also subject to the resolution although marching in antipodal point computation is not.

The design of these algorithms based on dissection suggests a measure of the "combinatorial size" of a curve by the number of inflections. The algorithm for finding antipodal points employs several specialized numerical primitives to complete a basic operation such as moving a point on the curve or determining if a local geometric condition holds. Such a routine call usually completes in tens to hundreds of iteration steps. It seems reasonable to analyze the asymptotic running time of a curve processing algorithm by the number of calls to such numerical primitives.

Acknowledgement Support for this research was provided in part by Iowa State University, and in part by the National Science Foundation through a CAREER award IIS-0133681. The author would like to thank David Persky for helping with the implementation, David Kriegman for pointing out the possibility of globally solving low-dimensional problems without resorting to optimization, and Ravi Janardan for suggesting to analyze the presented

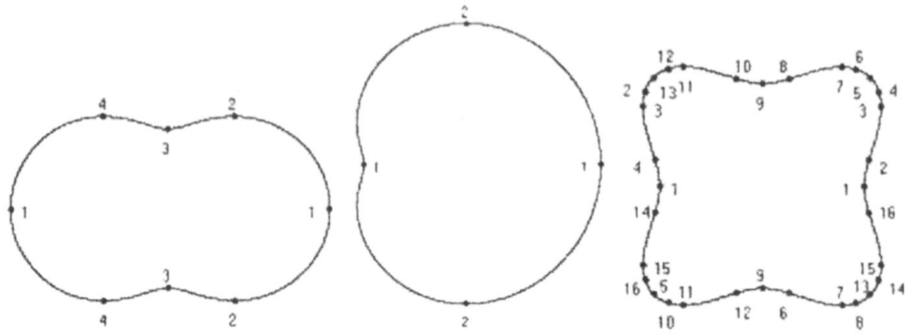

(a) (b) (c)

Figure 5.2. Antipodal points on three parametric curves.

algorithms. He is also very grateful to the anonymous reviewers of the earlier results [13, 16] for providing valuable feedback that has helped improve the structure and presentation of the paper.

References

1. P. K. Allen and K. S. Roberts. Haptic object recognition using a multi-fingered dextrous hand. *Proc. IEEE Intl. Conf. Robot. and Automat.* pp. 342–347, 1989.
2. A. Bicchi. Hands for dextrous manipulation and robust grasping: a difficult road toward simplicity. *IEEE Trans. Robot. and Automation*, 16(6):652-662, 2000.
3. A. Blake and M. Taylor. Planning planar grasps of smooth contours. *Proc. IEEE Intl. Conf. Robot. and Automation*, pp. 834–839, 1993.
4. C. Cai and B. Roth. On the spatial motion of a rigid body with point contact. *Proc. IEEE Intl. Conf. Robot. and Automat.*, pp. 686–695, 1987.
5. B. Chazelle, H. Edelsbrunner, L. J. Guibas, and M. Sharir. Diamester, width, closest line pair and parametric searching. *Disc. & Comp. Geometry*, 10:183–196, 1993.
6. I-M. Chen and J. W. Burdick. Finding antipodal point grasps on irregularly shaped objects. *Proc. IEEE Intl. Conf. Robot. and Automation*, pp. 2278–2283, 1992.
7. K. L. Clarkson and P. W. Shor. Applications of random sampling in computational geometry, ii. *Disc. & Comp. Geometry*, 4:387–421, 1989.
8. M. Erdmann. Shape recovery from passive locally dense tactile data. In P. K. Agarwal *et al.*, ed., *Robotics: The Algorithmic Perspective*, pp. 119–132. A. K. Peters, Boston, MA, 1998.
9. M. A. Fischler and H. C. Wolf. Locating perceptually salient points on planar curves. *IEEE Trans. Pattern Analysis and Machine Intell.*, 16(2):113–129, 1994.
10. T. N. T. Goodman. Inflections on curves in two and three dimensions. *Comp. Aided Geom. Design*, 8(1):37–50, 1991.
11. B. Grunbaum. A proof of Vazsonyi's conjecture. *Bulletin of Research Council Israel*, 6(A):77–78, 1956.

12. J. Hong, G. Lafferriere, B. Mishra, and X. Tan. Fine manipulation with multifinger hands. *Proc. IEEE Intl. Conf. Robot. and Automation*, pp. 1568–1573, 1990.
13. Y.-B. Jia. Localization on curved objects using tactile information. *Proc. IEEE/RSJ Intl. Conf. Robots and Systems*, pp. 701-706, 2001.
14. Y.-B. Jia. Geometry and computation of antipodal points on plane curves. Tech. Report ISU-CS-01-04, Computer Science Department, Iowa State University, Ames, IA, 2001. **http://www.cs.iastate.edu/~jia/papers/isu-cs-01-04.pdf**.
15. Y.-B. Jia. Localization and grasping of curved objects using tactile information. Tech. Report ISU-CS-01-08, Computer Science Department, Iowa State University, Ames, IA, 2001. **http://www.cs.iastate.edu/~jia/papers/isu-cs-01-08.pdf**.
16. Y.-B. Jia. Curvature-based computation of antipodal grasps. In *Proc. IEEE Intl. Conf. Robot. and Automation*, pp. 1571-1577, 2002.
17. D. Manocha and J. F. Canny. Detecting cusps and inflection points in curves. *Comp. Aided Geom. Design*, 9(1):1–24, 1992.
18. X. Markenscoff, L. Ni, and C. H. Papadimitriou. The geometry of grasping. *Intl. J. Robot. Res.*, 9(1):61–74, 1990.
19. J. Matoušek and O. Schwarzkopf. A deterministic algorithm for the three-dimensional diameter problem. *Comp. Geometry: Theory and Applications*, 6:253–362, 1996.
20. B. Mishra, J. T. Schwartz, and M. Sharir. On the existence and synthesis of multifinger positive grips. *Algorithmica*, 2(4):541–558, 1987.
21. F. Mokhtarian and A. Mackworth. Scale-based description and recognition of planar curves and two-dimensional shapes. *IEEE Trans. Pattern Analysis and Machine Intell.*, 8(1):34–43, 1986.
22. M. Moll and M. A. Erdmann. Reconstructing shape from motion using tactile sensors. *Proc. IEEE/RSJ Intl. Conf. Intell. Robots and Systems*, pp. 692–700, 2001.
23. D. J. Montana. The kinematics of contact and grasp. *Intl. J. Robot. Res.*, 7(3):17–32, 1988.
24. B. O'Neill. *Elementary Differential Geometry*. Academic Press, Inc., 1966.
25. J. Ponce, D. Stam, and B. Faverjon. On computing two-finger force-closure grasps of curved 2D objects. *Intl. J. Robot. Res.*, 12(3):263–273, 1993.
26. F. P. Preparata and S. J. Hong. Convex hulls of finite sets of points in two and three dimensions. *Communications of the ACM*, 20(2):87–93, 1977.
27. F. P. Preparata and M. I. Shamos. *Computational Geometry: an Introduction*. Springer-Verlag, 1985.
28. W. H. Press, B. P. Flannery, S. A. Teukolsky, and W. T.Vetterling. *Numerical Recipes in C*. Cambridge Univ. Press, Inc., 1988.
29. M. Sakai. Inflection points and singularities on planar rational cubic curve segments. *Comp. Aided Geom. Design*, 16:149–156, 1999.
30. E. V. Shikin. *Handbook and Atlas of Curves*. CRC Press, Inc., 1995.
31. J. Stoer and R. Bulirsch. *Introduction to Numerical Analysis*. Springer-Verlag, 2nd edition, 1993.
32. A. C. Yao. On constructing minimum spanning trees in k-dimensional space and related problems. *SIAM Journal of Computing*, 11(4):721–736, 1982.

A Near-Quadratic Algorithm for Fence Design*

Pankaj K. Agarwal[1], Robert-Paul Berretty[2], and Anne D. Collins[3]

[1] Duke University, Box 90129, Durham, NC 27708-0129. pankaj@cs.duke.edu
[2] Philips Research Laboratories, Building WDC 1-053, Prof. Holstlaan 4, 5656 AA Eindhoven, The Netherlands. robert-paul.berretty@philips.com
[3] Stanford University, Bldg. 380, Stanford, CA 94305-2125 collins@math.stanford.edu

Abstract. A part feeder is a mechanism that receives a stream of identical parts in arbitrary orientations and outputs them oriented the same way. Various sensorless part feeders have been proposed in the literature. The feeder we consider consists of a sequence of fences that extend partway across a conveyor belt; a polygonal part P carried by the belt is reoriented by each fence it encounters. We present an $O(m + n^2 \log^3 n)$-time algorithm to compute a sequence of fences that uniquely orients P, if one exists, where m is the total number of vertices and n is the number of stable edges of P. As in [3], we reduce the problem to searching for a path in a state graph that has $O(n^3)$ edges. By exploiting various geometric properties of the state graph, we show that it can be represented implicitly and a desired path can be found in $O(m + n^2 \log^3 n)$ time. Our technique is quite general and is applicable to other part manipulation problems.

1 Introduction

Robotic manipulation deals with various part handling problems in industrial automation [11]. One such problem, which arises in automated assembly, is the so-called *part feeding* or *orienting* problem. Many automated manufacturing processes require parts to be oriented prior to assembly. A part feeder receives a stream of identical parts in arbitrary orientations and outputs them oriented the same way. Although many part feeders use some kind of sensors, we are interested in *sensorless* orientation of parts, for which the initial orientation of the part is unknown and parts are oriented using passive mechanical compliance. A variety of sensorless part feeders have been proposed. For example, parts on a conveyor belt can be oriented using a sequence of stationary fences or a single moving fence. Parts can be toppled

* Research by P.A. is supported by the NSF under grants CCR-00-86013 EIA-98-70724, EIA-01-31905, ITR–333–1050, and CCR-97-32787, and by a grant from the U.S.–Israel Binational Science Foundation. Research by R.B. was supported by the Dutch Organization for Scientific Research (N.W.O.). Research by A.C. was supported by the NSF under grants ITR–333–1050, CCR–97-32787, DMS–0107621, DMS–9983320, and DMS-0101364. Part of this work was done while R.B. was visiting Department of Computer Science, University of North Carolina, Chapel Hill.

J.-D. Boissonnat et al. (Eds.): Algorithmic Foundations of Robotics V, STAR 7, pp. 347–362, 2004.

using horizontal pins suspended above the belt or using a collection of conveyor belts at varying heights. Parts can be pushed with fences, squeezed or rolled between parallel-jaw grippers, pulled from the inside-out, dropped through traps, tilted on a table, or subjected to vibrating plates with programmable vector fields. See [2,14,11] and the references therein.

Part feeders are typically created on a case-by-case basis, and it can take a long time to design a feeder for a single part. Only recently, researchers have begun to focus on automating the design process itself. In this paper we focus on designing a system of fences that uniquely orients a 2-dimensional part moving on a conveyor belt.

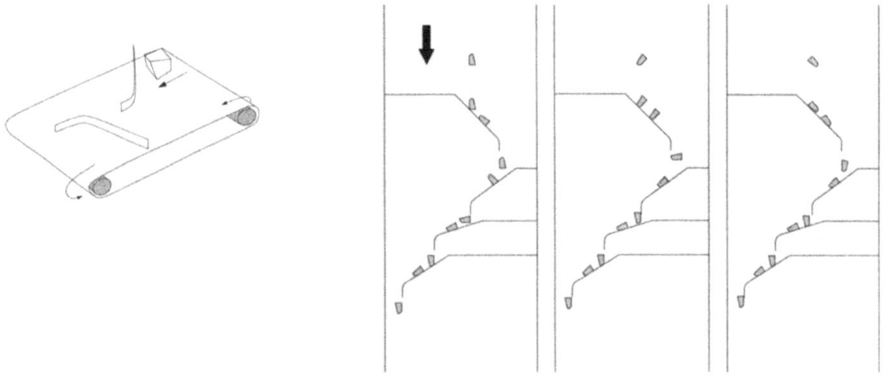

Fig. 1. A polygonal part is moved by a conveyor belt past a series of rigid fences. The fences passively reorient the part, which leaves the last fence in a unique orientation regardless of its initial orientation.

Our model. Our feeder consists of a conveyor belt equipped with a series of fences designed to passively reorient a part as it moves by. The fences are rigid, frictionless bars attached to walls on either side of the belt that extend partway across the belt at some fixed angle. See Figure 1 for an example.

We assume the part to be a planar polygon P, which lies on the conveyor belt and is translated past each fence. Since a fence is only in contact with the boundary of its convex hull, we can assume P to be a convex m-gon. When the part encounters a fence, it simultaneously rotates and slides until one of its stable edges aligns with the fence; we assume that the fences are long enough to allow ample time for this reorientation.[1] Once aligned, the part slides compliantly along the fence until it reaches the end.

In order to avoid any uncertainty in the orientation of the part as it leaves a fence, we add a carefully curved tail, as in [4]. This ensures that when the

[1] We assume that the motion of P follows the pushing model proposed by Mason [10]. See the original paper and [2] for details, which we omit from here.

part leaves the fence, the aligned edge is parallel to and facing the wall from which the fence emanates.

We work in a frame of reference in which the belt moves downward. A fence is specified by its *fence angle* ϕ, the direction of the ray normal to the fence with positive vertical component. If the fence is attached on the left side of the belt, then $\phi \in (0, \pi/2)$, while $\phi \in (\pi/2, \pi)$ for a right fence (Figure 2). Note that these are open intervals; a part cannot pass a fence with $\phi = \pi/2$, and the vertical fences at $\phi = 0$ and π have no effect.

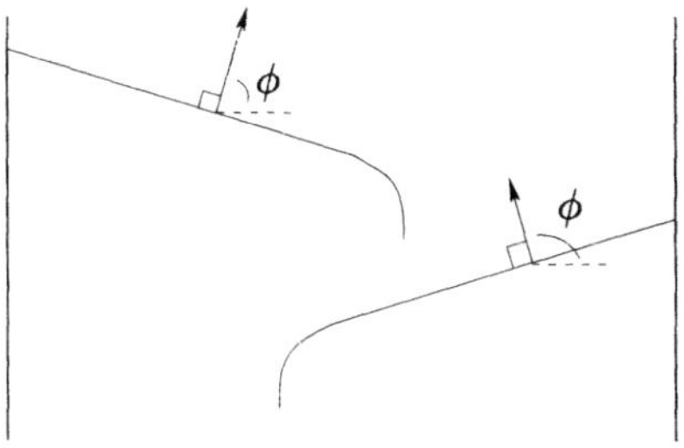

Fig. 2. The fence angle ϕ is the direction of the upward normal to the fence.

Given a polygonal part P, the *fence-design problem* is to construct a sequence $\Phi = \langle \phi_1, \ldots, \phi_k \rangle$ of fence angles so that, regardless of its initial orientation, P always leaves the last fence in a unique final orientation.

Related work. Goldberg [9] showed that it is possible to reorient any polygon in the plane from an unknown initial orientation to a unique final one by a fixed sequence of normal pushes with a straight fence. Each push is in a direction orthogonal to the length of the fence, and the reorientation of the fence between pushes is independent of the orientation of the part. His algorithm requires $\Omega(m^2)$ pushes in the worst-case. Chen and Ierardi [5] constructed a linear sequence of such pushes.

The fence-design problem is closely related to orienting a part by pushing. The main difference is that the fences can no longer be reoriented arbitrarily. If the belt moves downward, then any fence encountered by a part will affect a push with positive vertical component; thus, only half of the possible push directions are available at a given time. This restriction makes the problem significantly more complicated, and requires a very different approach.

The fence-design problem was first considered by Peshkin and Sanderson [13]. They introduce a graph representation of the state space, in which each node represents a set of orientations in which the part can be at a given moment and in which certain paths represent successful fence designs. They discretize the set of allowable fence angles, so their solution is not complete, in the sense that when their algorithm fails to design a feeder, there may still exist a solution that requires an angle not in their set. Later, Brokowski et al. [4] constrained the part to one of $O(m)$ orientations by adding a curved tail to each fence. This led to the complete algorithm of Wiegley et al. [15], which allows for all possible fence angles and is guaranteed to find a sequence of fences to orient a part if one exists. They conjecture that a polynomial-time fence-design algorithm exists for any polygonal part.

For parts with a certain asymmetry, Berretty et al. [3] present an $O(m)$ time algorithm, similar to Chen and Ierardi's [5]; unfortunately, it does not extend to arbitrary parts. Application of a general result by Eppstein [7] yields an $O(m^4)$-time algorithm for any part; successful fence designs again correspond to certain paths in a state graph that requires only $O(m^2)$ of the $O(2^m)$ possible states of the part. Berretty et al. [3] construct a related graph of size $O(m^3)$, thereby improving the design time to $O(m^3)$. Actually, the running time of their algorithm is $O(m + n^3)$, where n is the number of "stable" edges in P (see Section 2), which can be much smaller than m.

Our results. The main result of this paper is an $O(m+n^2 \log^3 n)$ algorithm for the fence-design problem, where n is the number of stable edges of the convex m-gon P. By exploiting the geometry of the state graph, we show that the graph can be represented implicitly, as the union of a family of complete bipartite subgraphs, and a desired path can be found in this implicit representation. Although similar techniques have been used in the past in other contexts [1,8], we believe this is the first application of this approach for a manipulation problem.

We believe that our approach is versatile and will find many other applications in those part manipulation problems that can be formulated as searching in a graph. For example, the same technique can give an improved $O(n^2 \log^3 n)$ algorithm for the problem of orienting micro-scale parts using squeeze and roll primitives, studied by Moll et al. [12].

2 Fence Function and Fence Graph

We first describe the push mechanism and the motion of the part when it encounters a fence. Then we explain how the fence-design problem can be formulated as a graph-searching problem.

Radius and push functions. Let c be the center of mass of P. We attach a local frame of reference to P with c as its origin. The *radius function* $\rho : \mathbb{S}^1 \to \mathbb{R}^+$ of P is defined as follows: Let ℓ be the line tangent to P whose inward (towards P) normal points in the direction α, as measured in the frame attached to P. Then $\rho(\alpha)$ is the (shortest) distance from c to ℓ; see Figure 3.

The radius function is piecewise sinusoidal, and its local maxima and minima occur when a ray originating from a contact point and normal to ℓ passes through c. Let γ_i and ε_i be the angles where ρ attains its local

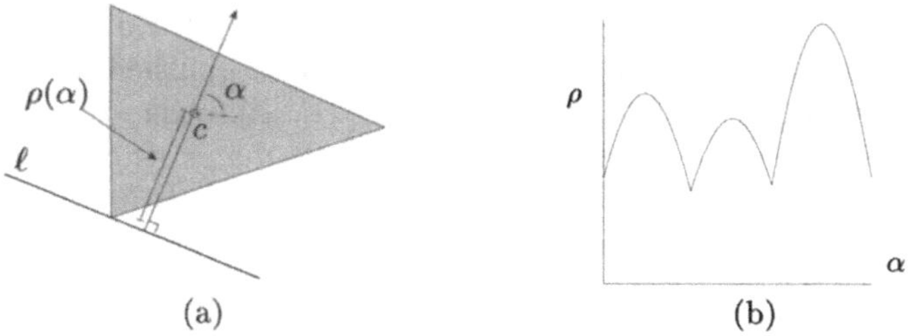

Fig. 3. The radius function $\rho : \mathbb{S}^1 \to \mathbb{R}^+$. (a) α is the direction of the ray normal to ℓ that passes through c. (b) ρ is a piecewise-sinusoidal function of α.

maxima and minima, respectively, ordered in the counter-clockwise direction with $\gamma_0 < \varepsilon_0 < \gamma_1 < \cdots < \varepsilon_{n-1}$. An edge e of P is called ***stable*** if the normal direction ε of e is a local minimum of ρ; if e is stable, then the ray from c normal to e intersects e. Let $\langle e_0, \ldots, e_{n-1} \rangle$ be the sequence of n stable edges of P; they can be computed in time $O(m)$.

Now, suppose that we push P with a fence, in the direction normal to the fence. As described in [10], during the push, P rotates in the direction that decreases ρ until a stable edge of P aligns with the fence, and then P translates in the direction normal to the fence. The ***push function*** $p : \mathbb{S}^1 \to \mathbb{S}^1$ for P is defined as follows: If a fence applies a normal push to P in the direction α, then $p(\alpha)$ is the final orientation of the fence, where both α and $p(\alpha)$ are measured in P's coordinate frame; Figure 4. That is,

$$p(\alpha) = \begin{cases} \varepsilon_i & \text{if } \gamma_i < \alpha < \gamma_{i+1}, \\ \gamma_i & \text{if } \alpha = \gamma_i. \end{cases}$$

We will avoid pushing in the directions γ_i, as these are unstable equilibria.

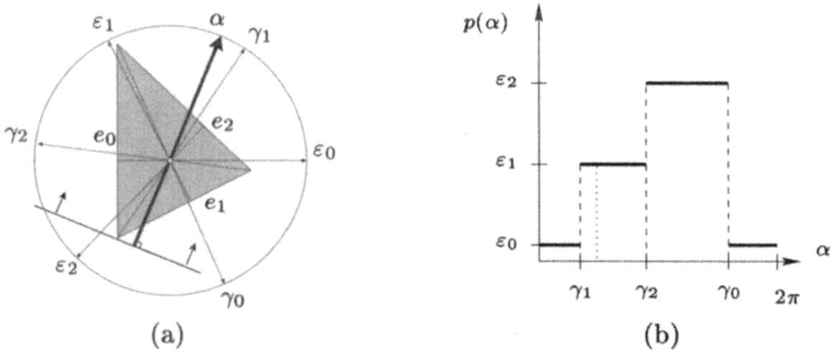

Fig. 4. The push function $p : \mathbb{S}^1 \to \mathbb{S}^1$. (a) A push in direction α with $p(\alpha) = \varepsilon_1$. (b) $p(\alpha)$ is a step function. ε_i and γ_i are the local minima and maxima of ρ.

Note that the belt does not translate the part in a direction normal to the fence; in general, the motion of the part when pushed in such a fashion is quite complicated, even unpredictable [11]. However, the component of the push force along the length of the fence is due to friction alone, so our assumption that the contact is frictionless implies that the force of the push felt by the part is always orthogonal to the fence. Thus the push function p correctly predicts the resulting orientation of P when it encounters a fence on the belt.

Fence function. Finally, we turn to the fence function, which describes the action of a fence, with fence angle ϕ, on the orientation of the part. Recall that once P is aligned with a fence, the curved tail reorients P so that the aligned edge faces the left side of the belt if $\phi \in (0, \pi/2)$, or the right side of the belt if $\phi \in (\pi/2, \pi)$. This ensures that P is in one of only $2n$ orientations as it travels between fences. We will denote by (i, L) and (i, R) the orientation of P with stable edge e_i parallel to and facing the left and right walls of the belt, respectively.

Suppose P is initially in orientation (i, L). When it encounters a fence with fence angle ϕ, P feels a push in direction $\varepsilon_i + \phi$ in its own frame. If $\gamma_j < \varepsilon_i + \phi < \gamma_{j+1}$, then the push function $p(\varepsilon_i + \phi) = \varepsilon_j$ dictates that edge e_j aligns with the fence. If, on the other hand, P is initially aligned to the right, say in orientation (i, R), then it feels a push in the $\varepsilon_i + \phi + \pi$ direction, and the final orientation depends on $p(\varepsilon_i + \phi + \pi)$. See Figure 5.

The equivalent action in planar pushing is to reorient the fence by θ. The difference here is that, although values of θ can in general lie anywhere in $[0, 2\pi)$, we are restricted to $\theta = \phi$ or $\theta = \phi + \pi$ for any particular push, where $\phi \in (0, \pi)$. Only half of the possible reorientations are available at a given time, which implies that some orientations are unattainable from others. It is precisely this fact that makes fence design harder than pushing.

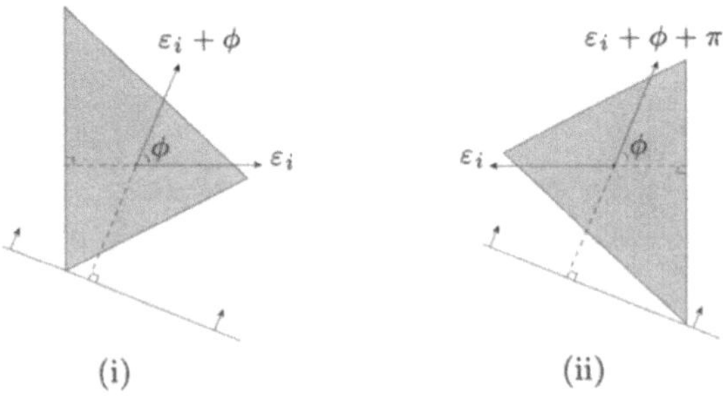

Fig. 5. The action of a fence: (i) If P is initially in orientation (i, L), it feels a push in direction $\alpha = \varepsilon_i + \phi$. (ii) If P is initially in orientation (i, R), it feels a push in direction $\alpha = \varepsilon_i + \phi + \pi$.

Definition 1 Given polygon P with n stable edges, define its *fence function*

$$\mathcal{F} : \{0, \ldots, n-1\} \times \{L, R\} \times (0, \pi) \to \{0, \ldots, n-1\} \times \{L, R\}$$

so that if P is initially in position (i, s) when it encounters a fence with fence angle $\phi \in (0, \pi)$, the resulting orientation is $\mathcal{F}(i, s, \phi) = (j, t)$, where

$$\varepsilon_j = \begin{cases} p(\varepsilon_i + \phi) & \text{if } s = L, \\ p(\varepsilon_i + \phi + \pi) & \text{if } s = R, \end{cases}$$

and

$$t = \begin{cases} L & \text{if } \phi \in (0, \pi/2), \\ R & \text{if } \phi \in (\pi/2, \pi). \end{cases}$$

As observed in [3], $\mathcal{F}$ is *monotonic* in the sense that $i_1 \leq i_2 \leq i_3$ implies that $\mathcal{F}(i_1, s, \phi) \leq \mathcal{F}(i_2, s, \phi) \leq \mathcal{F}(i_3, s, \phi)$, for fixed s and ϕ. This monotonicity was crucial for their algorithm, and it will be crucial for ours as well.

A convenient representation of the fence function is by a collection of n circles. For each stable edge e_i of P, circle C_i is rotated clockwise by ε_i, and marked with the intervals $\mathcal{X}_{ij} = (\gamma_j - \varepsilon_i, \gamma_{j+1} - \varepsilon_i)$, for $0 \leq j < n$; Figure 6. The location of $\mathcal{X}_{ij}$ on C_i indicates all triples s, t, ϕ for which $\mathcal{F}(i, s, \phi) = (j, t)$.

Suppose that $\theta \in \mathcal{X}_{ij} \cap (0, \pi)$. Then $\gamma_j < \varepsilon_i + \theta < \gamma_{j+1}$, which implies that $p(\varepsilon_i + \theta) = \varepsilon_j$; thus, the fence at angle θ takes the left orientation (i, L) to one of the orientations (j, t). Specifically, if $\theta \in (0, \pi/2)$, then $\mathcal{F}(i, L, \theta) = (j, L)$, while if $\theta \in (\pi/2, \pi)$, then $\mathcal{F}(i, L, \theta) = (j, R)$.

Similarly, if $\mathcal{X}_{ij}$ intersects the lower half of C_i, then there exists a fence which takes the right orientation (i, R) to one of the orientations (j, t). Suppose $\theta \in \mathcal{X}_{ij} \cap (\pi, 2\pi)$. Of course, θ itself is not a valid fence angle, but $\phi = \theta - \pi$ is, and $p(\varepsilon_i + \phi + \pi) = \varepsilon_j$. Therefore, $\mathcal{F}(i, R, \theta - \pi) = (j, t)$, where $t = L$ if $\theta \in (\pi, 3\pi/2)$, and $t = R$ if $\theta \in (3\pi/2, 2\pi)$.

Consider the circle in Figure 6. $\mathcal{X}_{ij}$ intersects quadrants I and II, so there exist fence angles ϕ_1 and ϕ_2 with $\mathcal{F}(i, L, \phi_1) = (j, L)$ and $\mathcal{F}(i, L, \phi_2) = (j, R)$. Since $\mathcal{X}_{ij}$ does not intersect the lower semicircle, $\mathcal{F}(i, R, \phi) \neq (j, t)$ for any $\phi \in (0, \pi)$ or $t \in \{L, R\}$. On the other hand, $\mathcal{X}_{ij-1}$ does intersect quadrant IV, so $\mathcal{F}(i, R, \theta - \pi) = (j-1, R)$ for some $\theta \in \mathcal{X}_{ij-1} \cap (3\pi/4, 2\pi)$.

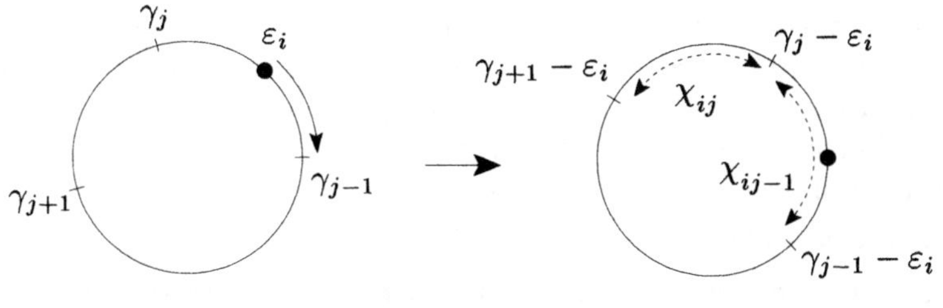

Fig. 6. The fence circle C_i is obtained by rotating the circle clockwise by ε_i.

Note that when choosing ϕ for our push plans, we will avoid the discrete set of values $\gamma_j - \varepsilon_i, \gamma_j + \pi - \varepsilon_i$, for $0 \le i, j < n$, since each can lead to an unstable equilibrium.

Fence graph. Next, we show how the fence-design problem can be formulated as a path search in a graph. Note that the graph defined here is slightly different from the one constructed in [3]. While the modified definition provides an almost identical $O(m + n^3)$ solution to the fence design problem, it also allows us to *compress* the graph to obtain a faster $O(m + n^2 \log^3 n)$ algorithm.

Definition 2 Given a polygon P with n stable edges and fence function $\mathcal{F}$, we define the *fence graph* G as follows:

- The nodes of G are subsets of the form $[i, j]^s = \{(i, s), (i+1, s), \ldots, (j, s)\}$, where $0 \le i, j < n$, and $s = L$ or R. Note that $[i, j]^s \ne [j, i]^s$.

- For $i' \ne j'$, there is an edge from $[i, j]^s$ to $[i', j']^t$ whenever there exists a fence angle ϕ such that both $\mathcal{F}(i, s, \phi) = (i', t)$ and $\mathcal{F}(j, s, \phi) = (j', t)$. We refer to these edges as *non-sink* edges.

- There is an edge from $[i, j]^s$ to $[i', i']^t$ if there exists a fence angle ϕ with $\mathcal{F}(k, s, \phi) = (i', t)$ for *all* $i \le k \le j$. We refer to these edges as *sink* edges.

Lemma 1. *Every node in G has out-degree $O(n)$.*

Proof: The neighbors of $[i, j]^s$ can be determined by overlaying the fence circles C_i and C_j. If $[i', j']^t$ is a neighbor, then $\mathcal{X}_{ii'} \cap \mathcal{X}_{jj'} \ne \emptyset$. The overlay of C_i and C_j is partitioned into $2n$ intervals by the endpoints of the $\mathcal{X}$'s, and each of these intervals corresponds to at most one edge of G. □

Lemma 2. *Every successful fence design corresponds to a path in G from a node of the form $[i, i-1]^s$ to a singleton node $[j, j]^t$, and vice versa.*

Proof: The monoticity of $\mathcal{F}$ implies that G has an edge $[i, j]^s \to [i', j']^t$ if and only if there exists a fence angle ϕ so that $\mathcal{F}(k, s, \phi) \in [i', j']^t$ for all $(k, s) \in [i, j]^s$. Then given a path in G from $[i-1, i]^s$ to $[j, j]^t$, we construct a fence design by choosing one such fence angle for each edge in the path.

The proof of the converse, that every succesful fence design corresponds to such a path, is a bit more subtle, but follows from a more general result by Eppstein [7] relating feeder design problems to finite monotonic automata. Strictly speaking, the monotonicity of $\mathcal{F}$ is not as strong as that required by Eppstein, but it is not difficult to adapt his result to our problem. We omit the details from this version of the paper. □

By Lemmas 1 and 2,

Theorem 3. *Given a convex polygonal part P with m vertices and n stable edges, a fence design can be constructed in $O(m + n^3)$ time.*

Sink edges. Before proceeding with the fence graph compression, we pause to consider in more detail the sink edges, i.e. edges of the form $([i,j]^s, [i',i']^t)$. Suppose $\mathcal{F}(i,s,\phi) = (i',t)$ and $\mathcal{F}(j,s,\phi) = (j',t)$. If $i' \neq j'$, the monotonicity of $\mathcal{F}$ ensures that $\mathcal{F}(k,s,\phi) \in [i',j']^t$ for all $(k,s) \in [i,j]^s$, so we always add an edge to G. However, this is not guaranteed when $i' = j'$, and it is unclear whether $[i',i']^t$ should represent the set of all orientations or just the singleton node $\{(i',t)\}$. We therefore need to take extra care with the sink edges when we compress G in the next section.

To begin, we note that the solution is trivial for certain parts.

Lemma 3. *If $\varepsilon_{j+1} - \varepsilon_j > \gamma_i - \gamma_{i+1}$ for some i and j, then two fences are sufficient to orient P.*

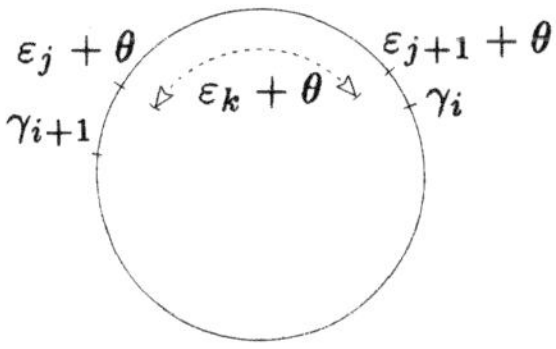

Fig. 7. For $\varepsilon_{j+1} - \varepsilon_j > \gamma_i - \gamma_{i+1}$, there exists θ s.t. $p(\varepsilon_k + \theta) = \varepsilon_i$ for all k.

Proof: Since $\varepsilon_{j+1} - \varepsilon_j > \gamma_i - \gamma_{i+1}$, we have $\gamma_i < \varepsilon_{j+1} + \theta < \varepsilon_j + \theta < \gamma_{i+1}$ for some $\theta \in [0, 2\pi)$. Since $\varepsilon_k \in (\varepsilon_{j+1}, \varepsilon_j)$ for all k, we have $p(\varepsilon_k + \theta) = \varepsilon_i$ for all k (see Figure 7). A reorientation by θ can be accomplished with exactly two fences, as follows. If $\theta \in (0,\pi)$, then we set $\phi_1 = \pi/4$ and $\phi_2 = \theta$; if $\theta \in (\pi, 2\pi)$, then we set $\phi_1 = 3\pi/4$ and $\phi_2 = \theta - \pi$. The first fence aligns the part with the left (resp. right) side of the belt if $\theta \in (0,\pi)$ (resp. $\theta \in (\pi, 2\pi)$), and the second fence orients the part in direction ε_i. □

For the rest of this paper, we assume that P has the following property:

$$(*) \quad \varepsilon_{j+1} - \varepsilon_j \leq \gamma_i - \gamma_{i+1} \text{ for all } 0 \leq i,j < n.$$

Lemma 4. *Suppose that P satisfies $(*)$, $\mathcal{F}(i,s,\phi) = \mathcal{F}(j,s,\phi) = (i',t)$ for some $\phi \in (0,\pi)$, and $\gamma_{i'+1} - \gamma_{i'} \leq \pi$. Then there is a sink edge $[i,j]^s \to [i',i']^t$ in G if and only if $\varepsilon_j - \varepsilon_i \leq \pi$.*

Proof: Set $\theta = \phi$ if $s = L$ and $\theta = \phi + \pi$ if $s = R$. Suppose first that $\varepsilon_j - \varepsilon_i \leq \pi$. Since $\varepsilon_i + \theta, \varepsilon_j + \theta \in (\gamma_{i'}, \gamma_{i'+1})$ and $\gamma_{i'+1} - \gamma_{i'} \leq \pi$, we must have $\gamma_{i'} < \varepsilon_i + \theta < \varepsilon_j + \theta < \gamma_{i'+1}$; Figure 8(a). This implies that $\mathcal{F}(k,s,\phi) = (i',t)$ for all $i \leq k \leq j$, so G has a sink edge $[i,j]^s \to [i',i']^t$.

On the other hand, if $\varepsilon_j - \varepsilon_i > \pi$, then $\varepsilon_i + \theta < \gamma_{i'+1} < \gamma_{i'} < \varepsilon_j + \theta$; Figure 8(b). If $[i,j]^s \to [i',i']^t$ is an edge of G, there must be some k with $i \leq k < j$ such that $\varepsilon_k + \theta < \gamma_{i'+1} < \gamma_{i'} < \varepsilon_{k+1} + \theta$; but this implies that $\varepsilon_{k+1} - \varepsilon_k > \gamma_{i'} - \gamma_{i'+1}$, violating $(*)$. □

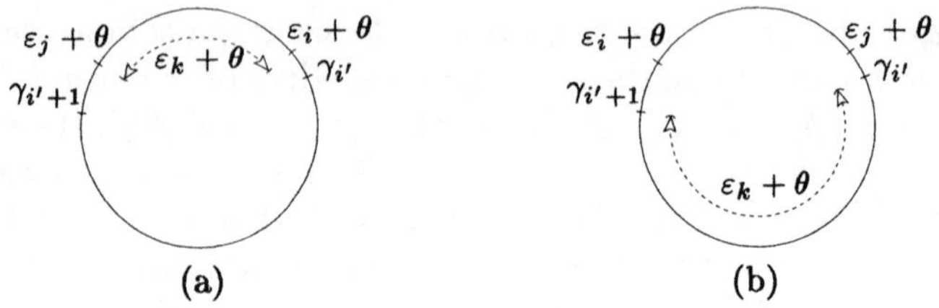

Fig. 8. For most parts, $[i,j]^s \to [i',i']^t$ exists if and only if $\varepsilon_j - \varepsilon_i \leq \pi$. (a) If $\varepsilon_j - \varepsilon_i \leq \pi$, then $\varepsilon_k + \theta \in (\gamma_{i'}, \gamma_{i'+1})$ for all $i \leq k \leq j$, (b) but not if $\varepsilon_j - \varepsilon_i > \pi$.

3 Fence Graph Compression

In this section we describe how to compute a compressed representation of the fence graph for a polygon that satisfies $(*)$. The compressed representation needs considerably less space and still allows us to quickly compute a path between two vertices. We first explain the compression scheme in general and then apply it to the fence graph.

3.1 Bipartite Clique Covers

Given a directed graph $G = (V, E)$, a *bipartite clique cover* of G is a collection

$$\mathcal{G} = \{(A_1, B_1), \ldots, (A_k, B_k)\}$$

such that

(i) $A_i, B_i \subseteq V$, (iii) $E = \bigcup_i E_i$,

(ii) $E_i = A_i \times B_i \subseteq E$, (iv) $E_i \cap E_j = \emptyset$, if $i \neq j$.

Conditions (i) and (ii) imply that each clique $(A_i \cup B_i, E_i)$ is a complete bipartite subgraph of G, while (iii) and (iv) imply that every edge of E is represented exactly once. The *size* of $\mathcal{G}$ is $|\mathcal{G}| = \sum_{i=1}^{k} |A_i| + |B_i|$.

Given a bipartite clique cover $\mathcal{G} = \{(A_1, B_1), \ldots, (A_k, B_k)\}$ of G, we can generate a *compressed representation* $\widetilde{G} = (\widetilde{V}, \widetilde{E})$ of G by setting

$$\widetilde{V} = V \cup \{1, \ldots, k\}, \text{ and}$$
$$\widetilde{E} = \bigcup_{i=1}^{k} \{(a,i) \mid a \in A_i\} \cup \{(i,b) \mid b \in B_i\}.$$

See Figure 9.

There is an edge $(a, b) \in A_i \times B_i$ in G if and only if $(a,i), (i,b) \in \widetilde{E}$, which immediately implies the following.

Lemma 5. *There is a path of length l in G from a vertex μ to another vertex ν if and only if there is a path of length $2l$ from μ to ν in $\widetilde{G}$.*

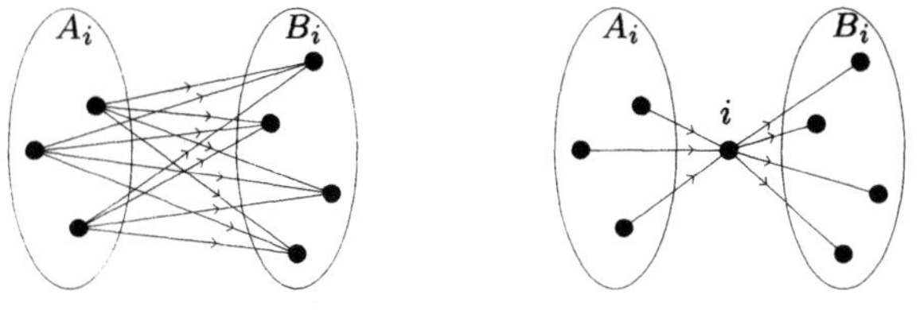

Fig. 9. G and $\widetilde{G}$ store the same connectivity information, but $\widetilde{G}$ has fewer edges.

We now present a method for generating a bipartite clique cover for $G = (V, E)$. Suppose that we have a family $\mathcal{N} = \{B_1, \ldots, B_k\}$ of subsets $B_i \subseteq V$, with the property that the set of neighbors $N(\mu) = \{\nu \in V \mid (\mu, \nu) \in E\}$ of a vertex μ can be expressed as a disjoint union of subsets of $\mathcal{N}$. That is, for each $\mu \in V$, there is a subfamily $\mathcal{N}_\mu \subseteq \mathcal{N}$ such that

$$\text{(i) } N(\mu) = \bigcup_{B_i \in \mathcal{N}_\mu} B_i, \text{ and}$$

$$\text{(ii) } B_i \cap B_j = \emptyset \quad \text{if} \quad B_i, B_j \in \mathcal{N}_\mu \text{ and } i \neq j.$$

We refer to $\mathcal{N}$ as a family of *canonical* subsets. For each $1 \le i \le k$, define

$$A_i = \{\mu \in V \mid B_i \in \mathcal{N}_\mu\},$$

and set $\mathcal{G} = \{(A_1, B_1), \ldots, (A_k, B_k)\}$.

Lemma 6. *$\mathcal{G}$ is a bipartite clique cover for G.*

Proof: It is clear that (A_i, B_i) is a complete bipartite subgraph of G. Indeed, $A_i, B_i \subseteq V$, and $(\mu, \nu) \in A_i \times B_i$ implies that $\nu \in B_i \subseteq N(\mu)$, i.e., $(\mu, \nu) \in E$. Conversely, if $(\mu, \nu) \in E$, then there exists a subset $B_i \in \mathcal{N}_\mu$ such that $\nu \in B_i$, implying $(\mu, \nu) \in A_i \times B_i$. Finally, if there are two indices i, j such that $(\mu, \nu) \in A_i \times B_i \cap A_j \times B_j$, then $\nu \in B_i \cap B_j$ and both $B_i, B_j \in \mathcal{N}_\mu$, contradicting the assumption (ii) of $\mathcal{N}_\mu$. Hence, (i)–(iv) are satisfied, and $\mathcal{G}$ is a bipartite clique cover of G. □

3.2 Compressing the Fence Graph

We now describe how we compress the fence graph G for a polygon that satisfies $(*)$, i.e., $\varepsilon_{j+1} - \varepsilon_j \le \gamma_i - \gamma_{i+1}$ for all i, j. Recall that the nodes of G are of the form $[i, j]^s = \{(i, s), \ldots, (j, s)\}$, where (i, s) is the orientation of the part with stable edge e_i aligned with side s of the belt, and if there is an edge $[i, j]^s \to [i', j']^t$, then $\mathcal{F}(i, s, \phi) = (i', t)$ and $\mathcal{F}(j, s, \phi) = (j', t)$ for some $\phi \in (0, \pi)$. Recall from the discussion in Section 2 that the sink edges need to be handled carefully. If there is an index i' such that $\gamma_{i'+1} - \gamma_{i'} > \pi$ (there is at most one such i'), we check in $O(1)$ time for each vertex $[i, j]^s$ of G and for each pair $s, t \in \{L, R\}$ whether $[i,j]^s \to [i',i']^t$. If the answer is yes, we add the pair $([i,j]^s, [i',i']^t)$ to $\mathcal{G}$. There are only $O(n^2)$ such pairs and we spend $O(n^2)$ time on them. Therefore, from now on we are interested in computing a compressed representation of only those neighbors of a vertex that correspond to either a non-sink edge or a sink edge $[i',i']$ with $\gamma_{i'+1} - \gamma_{i'} \leq \pi$.

We first describe how we compute the family of canonical subsets and then show how to represent these neighbors of a vertex as a disjoint union of canonical subsets.

Constructing canonical subsets. We first give an overview of the construction and postpone description of the necessary data structures until Section 4. Let $[i:j]$ denote the set of integers $\{i, i+1, \ldots, j\}$, mod n.

The construction of $\mathcal{N}$ proceeds in two phases. In the first phase, we construct a family $\mathcal{I}$ of $O(n)$ canonical subsets of $[0:n-1]$ with the following properties:

(F.1) Each $I_a \in \mathcal{I}$ is of the form $I_a = [l_a : r_a]$.

(F.2) $|\mathcal{I}| = \sum_{I_a} n_a = O(n \log n)$, where $n_a = r_a - l_a + 1$.

(F.3) For any interval $A = (\alpha_1, \alpha_2)$ of push angles, there is a subfamily $\mathcal{I}_A \subseteq \mathcal{I}$ of $O(\log n)$ disjoint subsets such that $p(A) = \bigcup_{I_a \in \mathcal{I}_A} \bigcup_{i \in I_a} \varepsilon_i$.

We show in Section 4 how we compute in $O(n \log n)$ time the family $\mathcal{I}$, and in $O(\log n)$ time the subfamily $\mathcal{I}_A$ for an arbitrary angular interval $A \subseteq \mathbb{S}^1$.

For each pair of canonical subsets $I_a, I_b \in \mathcal{I}$, we define the function $V_{ab} : \mathbb{S}^1 \to 2^{I_a \times I_b}$ as follows: For a given *shift* angle δ, $V_{ab}(\delta)$ is the set of pairs $[i,j] \in I_a \times I_b$ such that the angular intervals (γ_i, γ_{i+1}) and $(\gamma_j + \delta, \gamma_{j+1} + \delta)$ intersect on $\mathbb{S}^1$. We also define another function $\bar{V}_{ab}$, which is the same as V_{ab} except that we remove all nodes that would represent sink edges; that is, $\bar{V}_{ab}(\delta) = V_{ab}(\delta) \setminus \{[i,i] \mid 1 \leq i < n\}$.

The second phase of the algorithm constructs a family $\mathcal{N}^{ab}$ of canonical subsets of $I_a \times I_b$, for each pair $I_a, I_b \in \mathcal{I}$, with the following properties:

(S.1) $|\mathcal{N}^{ab}| = \sum_{N \in \mathcal{N}^{ab}} |N| = O(n_a n_b \log n)$, and

(S.2) Given a shift angle δ, there are two subfamilies $\mathcal{N}^{ab}_\delta, \bar{\mathcal{N}}^{ab}_\delta \subseteq \mathcal{N}^{ab}$ of $O(\log n)$ disjoint canonical subsets each such that

$$V_{ab}(\delta) = \bigcup_{N \in \mathcal{N}^{ab}_\delta} N \quad \text{and} \quad \bar{V}_{ab}(\delta) = \bigcup_{N \in \bar{\mathcal{N}}^{ab}_\delta} N.$$

Again, we describe in Section 4 how to compute the families $\mathcal{N}^{ab}$.

Computing the neighbors. With these families of canonical subsets in hand, we are now ready to compute the neighbor sets of each vertex in the fence graph G. For a vertex $[i,j]^L \in G$, let $N_L([i,j]^L)$ be the set of neighbors $[i',j']^L$ of $[i,j]^L$ such that either $i' \neq j'$, or $i' = j'$ and $\gamma_{i'+1} - \gamma_{i'} \leq \pi$. We refer to N_L as the set of *left neighbors*. Recall that if there is an edge $[i,j]^L \to [i',j']^L$ in E, then we must have $\mathcal{F}(i,L,\phi) = (i',L)$ and $\mathcal{F}(j,L,\phi) = (j',L)$ for some $\phi \in (0,\pi/2)$. Furthermore, $[i,j]^L \to [i',i']^L$ with $\gamma_{i'+1} - \gamma_{i'} \leq \pi$ is a sink edge if and only if $\mathcal{F}(i,L,\phi) = \mathcal{F}(j,L,\phi) = (i',L)$ and $\varepsilon_j - \varepsilon_i \leq \pi$.

We compute $N_L([i,j]^L)$, the set of left neighbors of $[i,j]^L$, as follows. Set $A_i = (\varepsilon_i, \varepsilon_i + \pi/2)$, $A_j = (\varepsilon_j, \varepsilon_j + \pi/2)$, and $\delta_{ij} = \varepsilon_i - \varepsilon_j$. Using the family $\mathcal{I}$ of canonical subsets, we compute subfamilies $\mathcal{I}_{A_i}$ and $\mathcal{I}_{A_j}$. For each pair $I_a \in \mathcal{I}_{A_i}$, $I_b \in \mathcal{I}_{A_j}$, we compute the subfamily $\mathcal{N}^{ab}_\delta \subseteq \mathcal{N}^{ab}$, representing the pairs $[i',j'] \in V_{ab}(\delta_{ij})$ if $\varepsilon_j - \varepsilon_i \leq \pi$; and the subfamily $\bar{\mathcal{N}}^{ab}_\delta \subseteq \mathcal{N}^{ab}$, representing the non-sink pairs $[i',j'] \in \bar{V}_{ab}(\delta_{ij})$ if $\varepsilon_j - \varepsilon_i > \pi$.

Lemma 7. *The set of left neighbors of a left node $[i,j]^L \in V$ is*

$$N_L([i,j]^L) = \begin{cases} \bigcup_{I_a \in \mathcal{I}_{A_i}} \bigcup_{I_b \in \mathcal{I}_{A_j}} \bigcup_{N \in \mathcal{N}^{ab}_\delta} N & \text{if } \varepsilon_j - \varepsilon_i \leq \pi, \\ \bigcup_{I_a \in \mathcal{I}_{A_i}} \bigcup_{I_b \in \mathcal{I}_{A_j}} \bigcup_{N \in \bar{\mathcal{N}}^{ab}_\delta} N & \text{if } \varepsilon_j - \varepsilon_i > \pi. \end{cases}$$

Proof: Let us assume that $\varepsilon_j - \varepsilon_i \leq \pi$. Since P satisfies $(*)$, by Lemma 4, the left-neighbor set of $[i,j]^L$ is the set of all $[i',j']^L$ such that $X_{ii'} \cap X_{jj'} \cap [0,\pi/2] \neq \emptyset$ in the overlay of the fence circles C_i and C_j.

The family $\mathcal{I}_{A_i}$ partitions the first quadrant of C_i into $O(\log n)$ intervals, and similarly for C_j. For each pair $I_a \in \mathcal{I}_{A_i}$ and $I_b \in \mathcal{I}_{A_j}$, we want to list the pairs $[i',j']$ such that $(\gamma_{i'} - \varepsilon_i, \gamma_{i'+1} - \varepsilon_i)$ and $(\gamma_{j'} - \varepsilon_j, \gamma_{j'+1} - \varepsilon_j)$ intersect for some $i' \in I_a$ and $j' \in I_b$. This is equivalent to determining $V_{ab}(\delta)$, where $\delta = \varepsilon_i - \varepsilon_j$, which we can express as the disjoint union of $O(\log(n_a n_b))$ canonical subsets in the subfamily $\mathcal{N}^{ab}_\delta \subseteq \mathcal{N}^{ab}$. The argument for the case $\varepsilon_j - \varepsilon_i > \pi$ is the same, except that $[i,j]$ does not have any outgoing sink edges, and we use $\bar{V}_{ab}$ and $\bar{\mathcal{N}}^{ab}_\delta$. □

In the preceding argument, we focused on edges $[i,j]^s \to [i',j']^t$ with $s = t = L$. The three other cases can be handled similarly. For $s = L, t = R$, we set $A_i = (\varepsilon_i + \pi/2, \varepsilon_i + \pi)$; for $s = R, t = L$, we set $A_i = (\varepsilon_i + \pi, \varepsilon_i + 3\pi/2)$; and for $s = R, t = R$, we set $A_i = (\varepsilon_i + 3\pi/2, \varepsilon_i + 2\pi)$. We do the same for A_j in each case, and then follow the above procedure.

Theorem 4. *Given a convex m-gon P with n stable edges, we can compute a fence design in $O(m + n^2 \log^3 n)$ time.*

Proof: We first compute the stable edges of P in $O(m)$ time. We then compute the families $\mathcal{I}$ and $\mathcal{N}^{ab}$ of canonical subsets of nodes of the fence graph G, as described above. Next, we define the bipartite clique cover

$\mathcal{G} = \{(A_1, B_1), \ldots, (A_k, B_k)\}$, where the B_i are the canonical subsets that make up the $\mathcal{N}^{ab}$. Then the compressed graph $\widetilde{G}$, described in Section 3.1, has $|\widetilde{V}| = |V| + k$ nodes and $|\widetilde{E}| = \sum_i |A_i| + |B_i|$ edges.

We initially added $O(n^2)$ pairs in $\mathcal{G}$ corresponding to the sink edges $[i,j]^s \to [i', i']^t$ such that $\gamma_{i'+1} - \gamma_{i'} > \pi$. Next, for each pair I_a, I_b, there are $O(n_a n_b)$ subsets in $\mathcal{N}^{ab}$, so the total number of canonical subsets is

$$k = O(n^2) + \sum_{I_a, I_b \in \mathcal{I}} O(n_a n_b) = \sum_{I_a \in \mathcal{I}} O(n_a n \log n) = O\left(n^2 \log^2 n\right).$$

Hence $|\widetilde{V}| = O(n^2 \log^2 n)$.

Recall that $\mu \in A_i$ if and only if $B_i \in \mathcal{N}_\mu$, so $\sum_i |A_i| = \sum_{\mu \in V} |\mathcal{N}_\mu|$. For $\mu = [i,j]$, we first compute $\mathcal{I}_{A_i}$ and $\mathcal{I}_{A_j}$. Each contains $O(\log n)$ canonical subsets I_a, but only $O(\log n)$ pairs will overlap at a given shift. For each overlapping pair, we collect $O(\log n)$ canonical subsets from $\mathcal{N}^{ab}$, so $\mathcal{N}_\mu$ contains $O(\log^2 n)$ subsets total. Therefore

$$\sum_{i=1}^{k} |A_i| = \sum_{\mu \in V} |\mathcal{N}_\mu| = O(n^2 \log^2 n).$$

Moreover,

$$\sum_i |B_i| = \sum_{a,b} \sum_{N \in \mathcal{N}^{ab}} |N| = \sum_{a,b} O(n_a n_b \log(n_a n_b)) = O(n^2 \log^3 n).$$

Hence $|\widetilde{E}| = O(n^2 \log^3 n)$ and $\mathcal{G}$ can be computed in time $O(m + n^2 \log^3 n)$. Finally, once $\widetilde{G}$ is computed, the time to find a path in $\widetilde{G}$ from one of the nodes $[i, i-1]^s$ to a singleton node $[j,j]^t$ is $O(|\widetilde{V}| + |\widetilde{E}|)$, using a simple breadth-first search. □

4 Data Structures

In this section, we present the data structures needed to generate the families of canonical subsets. In particular, we introduce two tree data structures, called *fence tree* and *shift tree*, which generate the canonical families $\mathcal{I}$ and $\mathcal{N}^{ab}$, respectively.

Fence tree. We describe the construction of a family $\mathcal{I}$ of $O(n)$ canonical subsets that satisfies properties (F.1)–(F.3). Let T be a minimum-height binary tree whose leaves store $\{\gamma_i \mid i \in [0 : n-1]\}$. Each internal node of T stores the leftmost leaf of its right subtree to guide the search. We associate the *canonical* interval $I_a = [l_a : r_a]$ with each node a of T, where $\gamma_{l_a}, \ldots, \gamma_{r_a}$ are the leaves in the subtree rooted at a. We set $\mathcal{I} = \{I_a \mid a \in T\}$. Since each index is stored in exactly one I_a at each level of T and the height of T is $O(\log n)$, we have $|\mathcal{I}| = \sum_a |I_a| = O(n \log n)$. Hence (F.1) and (F.2) hold.

Let $A = (\alpha_1, \alpha_2)$ be a push-angle interval. If $0 \in (\alpha_1, \alpha_2) \pmod{2\pi}$, we split A into two intervals $(\alpha_1, 2\pi]$ and $[0, \alpha_2)$, and work with each of them separately, so let us assume that $0 \notin (\alpha_1, \alpha_2)$. Recall that $p(\alpha) = \varepsilon_i$ if and only if $\alpha \in (\gamma_i, \gamma_{i+1})$. Then $p(A) = \{\varepsilon_l, \dots, \varepsilon_r\}$, where l and r are the maximum values for which $\gamma_l \le \alpha_1$ and $\gamma_r < \alpha_2$.

In order to compute $\mathcal{I}_A$, we regard T as a 1-dimensional range tree [6], and query T with α_1 and α_2 to determine l and r. We then report the set of indices $[l : r]$ as a disjoint union of $O(\log n)$ canonical intervals as follows: Let u be the least common ancestor in T of γ_l and γ_r. For any node on the path from u to γ_l (resp. γ_r), if its right (resp. left) child v is not on the path, we add $[l_v : r_v]$ to $\mathcal{I}_A$.

Shift trees. For each pair I_a, I_b of canonical subsets in $\mathcal{I}$, we show how to construct a family $\mathcal{N}^{ab}$ of $O(n_a n_b)$ subsets of $I_a \times I_b$ with properties (S.1) and (S.2). Notice that the angular intervals (γ_i, γ_{i+1}) and $(\gamma_j + \delta, \gamma_{j+1} + \delta)$ intersect if and only if the shift angle δ is in the interval $\Delta_{ij} = (\gamma_i - \gamma_{j+1}, \gamma_{i+1} - \gamma_j)$. Set $\mathcal{D}_{ab} = \{\Delta_{ij} \mid i \in I_a, j \in I_b\}$. The following lemma is straightforward.

Lemma 8. *Let $i \in I_a$ and $j \in I_b$. Then $[i, j] \in V_{ab}(\delta)$ if and only if $\delta \in \Delta_{ij}$.*

Thus, computing $V_{ab}(\delta)$ is reduced to the following 1-dimensional *stabbing query*: determine which of the intervals in $\mathcal{D}_{ab}$ contain the query point $\delta \in \mathbb{S}^1$.

To this end, we build a segment tree T_{ab} on the set $\mathcal{D}_{ab}$ of intervals; see [6]. As with the fence tree, if Δ_{ij} contains 0, it is split into two intervals and each of them is inserted separately. We associate two subsets $N_c^{ab}, \bar{N}_c^{ab}$ with each node c of T_{ab}. Suppose the interval Δ_{ij} is stored at a node c of T_{ab}. If $i \ne j$, then we add $[i, j]$ to both N_c^{ab} and $\bar{N}_c^{ab}$. If $i = j$ and $\gamma_{i+1} - \gamma_i \le \pi$, we add $[i, i]$ to N_c^{ab}. We set $\mathcal{N}^{ab} = \{N_c^{ab}, \bar{N}_c^{ab} \mid c \in T_{ab}\}$. There are $n_a n_b$ intervals in $\mathcal{D}_{ab}$, so there are $O(n_a n_b)$ nodes in T_{ab}, and thus $O(n_a n_b)$ canonical subsets $N_c^{ab}, \bar{N}_c^{ab} \in \mathcal{N}^{ab}$. Since each $[i, j] \in I_a \times I_b$ is stored in $O(\log n)$ canonical subsets, $|\mathcal{N}^{ab}| = O(n_a n_b \log n)$, as claimed in (S.1).

To compute $V_{ab}(\delta)$ (resp. $\bar{V}_{ab}(\delta)$) for a query shift δ, we follow the path from the root to the leaf whose associated interval contains δ and report N_c^{ab} (resp. $\bar{N}_c^{ab}$) for all $O(\log n)$ nodes c on this path. The correctness of the procedure follows from Lemma 8. Hence, (S.2) is obviously satisfied.

5 Conclusion

In this paper, we presented a new technique for computing a fence design in near-quadratic $O(m + n^2 \log^3 n)$ time, thus improving the previous $O(m + n^3)$ algorithm. Here, the polygonal part to be oriented has m vertices and n stable edges. The approach, based on a graph compression scheme, is quite general. There are a number of other part-feeding problems which can be solved by searching for certain paths in a discrete state space, and our compression technique applies to these as well.

References

1. P. K. Agarwal and K. R. Varadarajan. Efficient algorithms for approximating polygonal chains. *Discrete & Computational Geometry*, 23:273–291, 2000.
2. R.-P. Berretty. *Geometric Design of Part Feeders.* PhD thesis, Utrecht University, Utrecht, The Netherlands, 2000.
3. R.-P. Berretty, K. Goldberg, M. Overmars, and A. F. van der Stappen. Algorithms for fence design. In *Robotics:The Algorithmic Perspective*, (P. K. Agarwal, L. E. Kavraki, M. T. Mason, eds.), pages 279–296. AK Peters, Natick, MA, 1998.
4. M. E. Brokowski, M. A. Peshkin, and K. Goldberg. Optimal curved fences for part alignment on a belt. *ASME Journal of Mechanical Design*, 117(1):27–34, 1995.
5. Y.-B. Chen and D. Ierardi. The complexity of oblivious plans for orienting and distinguishing polygonal parts. *Algorithmica*, 14(5):367–397, 1995.
6. M. de Berg, M. van Kreveld, M. Overmars, and O. Schwarzkopf. *Computational Geometry: Algorithms and Applications.* Springer Verlag, Berlin, 1997.
7. D. Eppstein. Reset sequences for monotonic automata. *SIAM J. Comput.*, 19(3):500–510, 1990.
8. T. Feder and R. Motwani. Clique partitions, graph compression, and speeding-up algorithms. In *Proc. 23rd ACM Symp. on Theory of Computing*, pages 123–133, 1991.
9. K. Y. Goldberg. Orienting polygonal parts without sensors. *Algorithmica*, 10:210–225, 1993.
10. M. T. Mason. *Manipulator Grasping and Pushing Operations.* PhD thesis, MIT, 1982. Published in Robot Hands and the Mechanics of Manipulation, MIT Press, Cambridge, 1985.
11. M. T. Mason. *Mechanics of Robotic Manipulation.* MIT Press, August 2001. Intelligent Robotics and Autonomous Agents Series.
12. M. Moll, K. Goldberg, M. Erdmann, and R. Fearing. Orienting micro-scale parts with squeeze and roll primitives. In *Proc. IEEE International Conference on Robotics and Automation*, pages 1931–1936, May 2002.
13. M. A. Peshkin and A. C. Sanderson. Planning robotic manipulation strategies for workpieces that slide. *IEEE Journal of Robotics and Automation*, 4(5):524–531, 1988.
14. A. F. van der Stappen, R.-P. Berretty, K. Goldberg, and M. H. Overmars. Geometry and part feeding. In *Sensor Based Intelligent Robot Systems*, (G. D. Hager, H. I. Christensen, H. Bunke, and R. Klein, eds.) Lecture Notes in Computer Science 2238, pages 259–281. Springer Verlag, Berlin, 2002.
15. J. Wiegley, K. Goldberg, M. Peshkin, and M. Brokowski. A complete algorithm for designing passive fences to orient parts. *Assembly Automation*, 17(2):129–136, 1997.

Simultaneous Mapping and Localization with Sparse Extended Information Filters: Theory and Initial Results

Sebastian Thrun[1], Daphne Koller[2], Zoubin Ghahramani[3], Hugh Durrant-Whyte[4], and Andrew Y. Ng[2]

[1] Carnegie Mellon University, Pittsburgh, PA, USA
[2] Stanford University, Stanford, CA, USA
[3] Gatsby Computational Neuroscience Unit, University College London, UK
[4] University of Sydney, Sydney, Australia

Abstract. This paper describes a scalable algorithm for the simultaneous mapping and localization (SLAM) problem. SLAM is the problem of determining the location of environmental features with a roving robot. Many of today's popular techniques are based on extended Kalman filters (EKFs), which require update time quadratic in the number of features in the map. This paper develops the notion of sparse extended information filters (SEIFs), as a new method for solving the SLAM problem. SEIFs exploit structure inherent in the SLAM problem, representing maps through local, Web-like networks of features. By doing so, updates can be performed in constant time, irrespective of the number of features in the map. This paper presents several original constant-time results of SEIFs, and provides simulation results that show the high accuracy of the resulting maps in comparison to the computationally more cumbersome EKF solution.

1 Introduction

The simultaneous localization and mapping (SLAM) problem is the problem of acquiring a map of an unknown environment with a moving robot, while simultaneously localizing the robot relative to this map [6,12]. The SLAM problem addresses situations where the robot lacks a global positioning sensor, and instead has to rely on a sensor of incremental ego-motion for robot position estimation (e.g., odometry, inertial navigation). Such sensors accumulate error over time, making the problem of acquiring an accurate map a challenging one. Within mobile robotics, the SLAM problem is often referred to as one of the most challenging ones [28].

In recent years, the SLAM problem has received considerable attention by the scientifi c community, and a flurry of new algorithms and techniques has emerged, as attested, for example, by a recent workshop on this topic [11]. Existing algorithms can be subdivided into batch and online techniques. The former provide sophisticated techniques to cope with perceptual ambiguities [2,24,31], but they can only generate maps after extensive batch processing. Online techniques are specifi cally suited to acquire maps as the robot navigates [6,27], which is of great practical importance in many navigation and exploration problems [25]. Today's most widely used online algorithms are based on extended Kalman fi lters (EKFs), based on a seminal series of papers [18,27,26]. EKFs calculate Gaussian posteriors over the locations of environmental features and the robot itself.

J.-D. Boissonnat et al. (Eds.): Algorithmic Foundations of Robotics V, STAR 7, pp. 363–380, 2004.

A key bottleneck of EKFs—which has been subject to intense research—is their computational complexity. The standard EKF approach requires time quadratic in the number of features in the map, for each incremental update. This computational burden restricts EKFs to relatively sparse maps with no more than a few hundred features. Recently, several researchers have developed hierarchical techniques that decompose maps into collections of smaller, more manageable submaps [1,8,32]. While in principle, hierarchical techniques can solve this problem in linear time, most of these techniques still require quadratic time per update. However, they do so with a much reduced constant factor, enabling them to manage significantly more features. One recent technique updates the estimate in constant time [13], but with a loss of global consistency that can lead to inaccurate maps when closing loops in cyclic environments [9]. A different line of research has relied on particle filters for efficient mapping [7]. The FastSLAM algorithm [17] and related mapping algorithms [19] require time logarithmic in the number of features in the map, but they depend linearly on a particle-filter specific parameter (the number of particles), whose scaling with environmental size is still poorly understood. None of these approaches, however, offer constant time updating while simultaneously maintaining global consistency of the map.

This paper proposes a new SLAM algorithm whose updates require constant time, independent of the number of features in the map. Our approach is based on the well-known *information form* of the EKF, also known as the extended information filter (EIF) [22]. To achieve constant time updating, we develop an approximate EIF which maintains a *sparse* representation of environmental dependencies. Empirical simulation results provide evidence that the resulting maps are comparable in accuracy to the computationally much more cumbersome EKF solution, which is still at the core of most work in the field.

Our approach is best motivated by investigating the workings of the EKF. Figure 1 shows the result of EKF mapping in an environment with 50 landmarks. The left panel shows a moving robot, along with its Gaussian estimates of the location of all 50 point features. The central information maintained by the EKF solution is a covariance matrix of these different estimates. The normalized covariance, i.e., the correlation, is visualized in the center panel of this figure. Each of the two axes lists the robot pose (x-y location and orientation) followed by the x-y-locations of the 50 landmarks. Dark entries indicate strong correlations. It is known that in the limit of SLAM, all x-coordinates and all y-coordinates become fully correlated [6]. The checkerboard appearance of the correlation matrix illustrates this fact. Maintaining these cross-correlations—of which there are quadratically many in the number of features in the map—are essential to the SLAM problem. This observation has given rise to the (false) suspicion that online SLAM is inherently quadratic in the number of features in the map.

The key insight that motivates our approach is shown in the right panel of Figure 1. Shown there is the *inverse* covariance matrix (also known as *information matrix* [16,22]), normalized just like the correlation matrix. Elements in this normalized information matrix can be thought of as constraints, or links, between the locations of different features: The darker an entry in the display, the stronger the link. As this depiction suggests, the normalized information matrix appears to be

Figure 1. Typical snapshots of EKFs applied to the SLAM problem: Shown here is a map (left panel), a correlation (center panel), and a normalized information matrix (right panel). Notice that the normalized information matrix is naturally almost sparse, motivating our approach of using sparse information matrices in SLAM.

naturally sparse: it is dominated by a small number of strong links, and possesses a large number of links whose values, when normalized, are near zero. Furthermore, link strength is related to distance of features: Strong links are found only between geometrically nearby features. The more distant two landmarks, the weaker their link. This observation suggest that the EKF solution to SLAM possesses important structure that can be exploited for more effi cient solutions. While any two features are fully correlated in the limit, the correlation arises mainly through a network of local links, which only connect nearby landmarks.

Our approach exploits this structure by maintaining a *sparse* information matrix, in which only nearby features are linked through a non-zero element. The resulting network structure is illustrated in the right panel of Figure 2, where disks corresponds to point features and dashed arcs to links, as specifi ed in the information matrix visualized on the left. Shown also is the robot, which is linked to a small subset of all features only, called active features and drawn in black. Storing a sparse information matrix requires linear space. More importantly, updates can be performed in constant time, regardless of the number of features in the map. The resulting fi lter is a *sparse extended information filter*, or SEIF. We show empirically that the SEIFs tightly approximate conventional extended information fi lters, which previously applied to SLAM problems in [20,22] and which are functionally equivalent to the popular EKF solution.

Our technique is probably most closely related to work on SLAM fi lters that represent relative distances, such as Newman's geometric projection fi lter [23] and extensions [5], and Csorba's relative fi lter [4]. Neither of these alternative approaches permits constant time updating in SLAM, though it appears that these techniques could be developed into constant time algorithms, using approximations similar to the ones described here. Our work is also related to the rich body of literature on topological mapping [3,10,15,33], which typically does not represent about dependencies and correlations in the representation of uncertainty.

2 Extended Information Filters

This section reviews the extended information fi lter (EIF), which forms the basis of our work. EIFs are computationally equivalent to extended Kalman fi lters (EKFs),

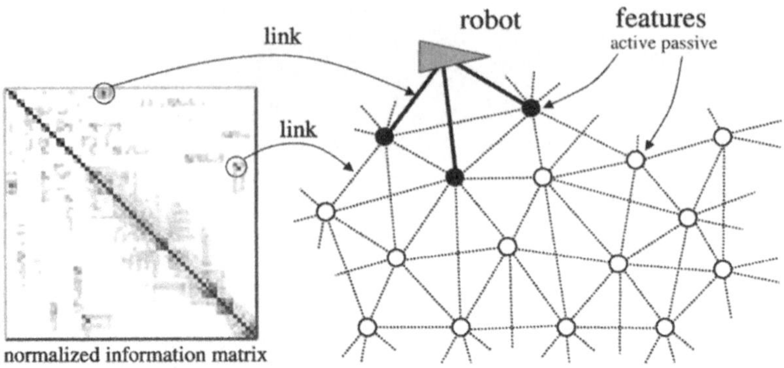

Figure 2. Illustration of the network of landmarks generated by our approach. Shown on the left is a sparse information matrix, and on the right a map in which entities are linked whose information matrix element is non-zero. As argued in the paper, the fact that not all landmarks are connected is a key structural element of the SLAM problem, and at the heart of our constant time solution.

but they represent information differently: instead of maintaining a covariance matrix, the EIF maintains an inverse covariance matrix, also known as information matrix. EIFs have previously been applied to the SLAM problem, most notably by Nettleton and colleagues [20,22], but they are much less common than the EKF approach.

Most of the material in this section applies equally to linear and non-linear filters. We have chosen to present all material in the extended, non-linear form, since robots are inherently non-linear.

2.1 Information Form of the SLAM Problem

Let x_t denote the pose of the robot at time t. For rigid mobile robots operating in a planar environment, the pose is given by its two Cartesian coordinates and the robot's heading direction. Let N denote the number of features (e.g., landmarks) in the environment. The variable y_n with $1 \leq n \leq N$ denotes the pose of the n-th feature. For example, for point landmarks in the plane, y_n may comprise the two-dimensional Cartesian coordinates of this landmark. In SLAM, it is usually assumed that features do not change their pose (or location) over time.

The robot pose x_t and the set of all feature locations Y together constitute the *state* of the environment. It will be denoted by the vector $\xi_t = \begin{pmatrix} x_t \; y_1 \; \ldots \; y_N \end{pmatrix}^T$, where the superscript T refers to the transpose of a vector.

In the SLAM problem, it is impossible to sense the state ξ_t directly—otherwise there would be no mapping problem. Instead, the robot seeks to recover a probabilistic estimate of ξ_t. Written in a Bayesian form, our goal shall be to calculate a posterior distribution over the state ξ_t. This posterior $p(\xi_t \mid z^t, u^t)$ is conditioned on past sensor measurements $z^t = z_1, \ldots, z_t$ and past controls $u^t = u_1, \ldots, u_t$. Sensor measurements z_t might, for example, specify the approximate range and bearing to nearby features. Controls u_t specify the robot motion command asserted in the time interval $(t-1; t]$.

Following the rich EKF tradition in the SLAM literature, our approach represents the posterior $p(\xi_t \mid z^t, u^t)$ by a multivariate Gaussian distribution over the state ξ_t. The mean of this distribution will be denoted μ_t, and covariance matrix Σ_t:

$$p(\xi_t \mid z^t, u^t) \propto \exp\left\{-\tfrac{1}{2}(\xi_t - \mu_t)^T \Sigma_t^{-1}(\xi_t - \mu_t)\right\} \tag{1}$$

The proportionality sign replaces a constant normalizer that is easily recovered from the covariance Σ_t. The representation of the posterior via the mean μ_t and the covariance matrix Σ_t is the basis of the EKF solution to the SLAM problem (and to EKFs in general).

Information fi lters represent the same posterior through a so-called *information matrix* H_t and an *information vector* b_t—instead of μ_t and Σ_t. These are obtained by multiplying out the exponent of (1):

$$\begin{aligned} &= \exp\left\{-\tfrac{1}{2}\left[\xi_t^T \Sigma_t^{-1}\xi_t - 2\mu_t^T \Sigma_t^{-1}\xi_t + \mu_t^T \Sigma_t^{-1}\mu_t\right]\right\} \\ &= \exp\left\{-\tfrac{1}{2}\xi_t^T \Sigma_t^{-1}\xi_t + \mu_t^T \Sigma_t^{-1}\xi_t - \tfrac{1}{2}\mu_t^T \Sigma_t^{-1}\mu_t\right\} \end{aligned} \tag{2}$$

We now observe that the last term in the exponent, $-\frac{1}{2}\mu_t^T \Sigma_t^{-1}\mu_t$ does not contain the free variable ξ_t and hence can be subsumed into the constant normalizer. This gives us the form:

$$\propto \exp\{-\tfrac{1}{2}\xi_t^T \underbrace{\Sigma_t^{-1}}_{=:H_t} \xi_t + \underbrace{\mu_t^T \Sigma_t^{-1}}_{=:b_t} \xi_t\} \tag{3}$$

The information matrix H_t and the information vector b_t are now defi ned as indicated:

$$H_t = \Sigma_t^{-1} \qquad \text{and} \qquad b_t = \mu_t^T H_t \tag{4}$$

Using these notations, the desired posterior can now be represented in what is commonly known as the *information form* of the Kalman fi lter:

$$p(\xi_t \mid z^t, u^t) \propto \exp\left\{-\tfrac{1}{2}\xi_t^T H_t \xi_t + b_t \xi_t\right\} \tag{5}$$

As the reader may easily notice, both representations of the multi-variate Gaussian posterior are functionally equivalent (with the exception of certain degenerate cases): The EKF representation of the mean μ_t and covariance Σ_t, and the EIF representation of the information vector b_t and the information matrix H_t. In particular, the EKF representation can be 'recovered' from the information form via the following algebra:

$$\Sigma_t = H_t^{-1} \qquad \text{and} \qquad \mu_t = H_t^{-1} b_t^T = \Sigma_t b_t^T \tag{6}$$

The advantage of the EIF over the EKF will become apparent further below, when the concept of sparse EIFs will be introduced.

Of particular interest will be the geometry of the information matrix. This matrix is symmetric and positive-defi nite:

$$H_t = \begin{pmatrix} H_{x_t,x_t} & H_{x_t,y_1} & \cdots & H_{x_t,y_N} \\ H_{y_1,x_t} & H_{y_1,y_1} & \cdots & H_{y_1,y_N} \\ \vdots & \vdots & \ddots & \vdots \\ H_{y_N,x_t} & H_{y_N,y_1} & \cdots & H_{y_N,y_N} \end{pmatrix} \tag{7}$$

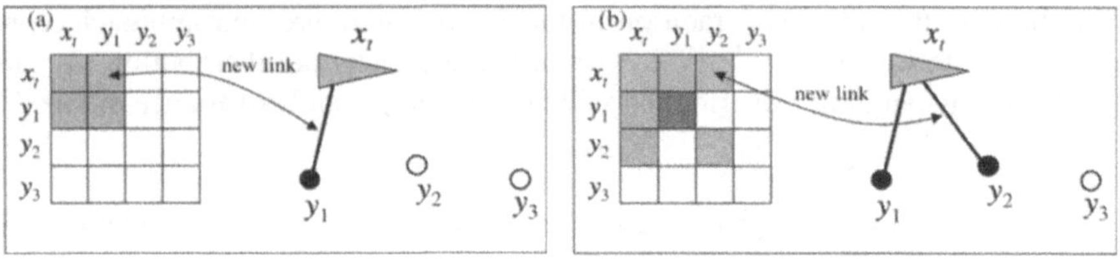

Figure 3. The effect of measurements on the information matrix and the associated network of features: (a) Observing y_1 results in a modification of the information matrix elements H_{x_t,y_1}. (b) Similarly, observing y_2 affects H_{x_t,y_2}. Both updates can be carried out in constant time.

Each element in the information matrix constraints one (on the main diagonal) or two (off the main diagonal) elements in the state vector. We will refer to the off-diagonal elements as *links*: the matrices H_{x_t,y_n} link together the robot pose estimate and the location estimate of a specifi c feature, and the matrices $H_{y_n,y_{n'}}$ for $n \neq n'$ link together two feature locations y_n and $y_{n'}$. Although rarely made explicit, the manipulation of these links is the very essence of Gaussian solutions to the SLAM problem. It will be an analysis of these links that ultimately leads to a constant-time solution to the SLAM problem.

2.2 Measurement Updates

In SLAM, measurements z_t carry spatial information on the relation of the robot's pose and the location of a feature. For example, z_t might be the approximate range and bearing to a nearby landmark. Without loss of generality, we will assume that each measurement z_t corresponds to exactly one feature in the map. Sightings of multiple features at the same time may easily be processed one-after-another.

Figure 3 illustrates the effect of measurements on the information matrix H_t. Suppose the robot measures the approximate range and bearing to the feature y_1, as illustrated in Figure 3a. This observation links the robot pose x_t to the location of y_1. The strength of the link is given by the level of noise in the measurement. Updating EIFs based on this measurement involves the manipulation of the off-diagonal elements $H_{x_t,y}$ and their symmetric counterparts H_{y,x_t} that link together x_t and y. Additionally, the on-diagonal elements H_{x_t,x_t} and H_{y_1,y_1} are also updated. These updates are additive: Each observation of a feature y increases the strength of the total link between the robot pose and this very feature, and with it the total information in the fi lter. Figure 3b shows the incorporation of a second measurement of a different feature, y_2. In response to this measurement, the EIF updates the links $H_{x_t,y_2} = H^T_{y_2,x_t}$ (and H_{x_t,x_t} and H_{y_2,y_2}). As this example suggests, measurements introduce links only between the robot pose x_t and observed features. Measurements never generate links between pairs of landmarks, or between the robot and unobserved landmarks.

For a mathematical derivation of the update rule, we observe that Bayes rule enables us to factor the desired posterior into the following product:

$$\begin{aligned} p(\xi_t \mid z^t, u^t) &\propto p(z_t \mid \xi_t, z^{t-1}, u^t)\, p(\xi_t \mid z^{t-1}, u^t) \\ &= p(z_t \mid \xi_t)\, p(\xi_t \mid z^{t-1}, u^t) \end{aligned} \tag{8}$$

The second step of this derivation exploited common (and obvious) independences in SLAM problems [30]. For the time being, we assume that $p(\xi_t \mid z^{t-1}, u^t)$ is represented by $\bar{H}_t$ and $\bar{b}_t$. Those will be discussed in the next section, where robot motion will be addressed. The key question addressed in this section, thus, concerns the representation of the probability distribution $p(z_t \mid \xi_t)$ and the mechanics of carrying out the multiplication above. In the 'extended' family of filters, a common model of robot perception is one in which measurements are governed via a deterministic non-linear measurement function h with added Gaussian noise:

$$z_t = h(\xi_t) + \varepsilon_t \tag{9}$$

Here ε_t is an independent noise variable with zero mean, whose covariance will be denoted Z. Put into probabilistic terms, (9) specifies a Gaussian distribution over the measurement space of the form

$$p(z_t \mid \xi_t) \propto \exp\left\{-\tfrac{1}{2}(z_t - h(\xi_t))^T Z^{-1}(z_t - h(\xi_t))\right\} \tag{10}$$

Following the rich literature of EKFs, EIFs approximate this Gaussian by linearizing the measurement function h. More specifically, a Taylor series expansion of h gives us

$$h(\xi_t) \approx h(\mu_t) + \nabla_\xi h(\mu_t)[\xi_t - \mu_t] \tag{11}$$

where $\nabla_\xi h(\mu_t)$ is the first derivative (Jacobian) of h with respect to the state variable ξ, taken $\xi = \mu_t$. For brevity, we will write $\hat{z}_t = h(\mu_t)$ to indicate that this is a prediction given our state estimate μ_t. The transpose of the Jacobian matrix $\nabla_\xi h(\mu_t)$ and will be denoted C_t. With these definitions, Equation (11) reads as follows:

$$h(\xi_t) \approx \hat{z}_t + C_t^T(\xi_t - \mu_t) \tag{12}$$

This approximation leads to the following Gaussian approximation of the measurement density (10):

$$p(z_t \mid \xi_t) \propto \exp\left\{-\tfrac{1}{2}(z_t - \hat{z}_t - C_t^T\xi_t + C_t^T\mu_t)^T Z^{-1}(z_t - \hat{z}_t - C_t^T\xi_t + C_t^T\mu_t)\right\} \tag{13}$$

Multiplying out the exponent and regrouping the resulting terms gives us

$$\begin{aligned} &= \exp\left\{-\tfrac{1}{2}\xi_t^T C_t Z^{-1} C_t^T \xi_t + (z_t - \hat{z}_t + C_t^T\mu_t)^T Z^{-1} C_t^T \xi_t \right. \\ &\quad \left. -\tfrac{1}{2}(z_t - \hat{z}_t + C_t^T\mu_t)^T Z^{-1}(z_t - \hat{z}_t + C_t^T\mu_t)\right\} \end{aligned} \tag{14}$$

As before, the final term in the exponent does not depend on the variable ξ and hence can be subsumed into the proportionality factor:

$$\propto \exp\left\{-\tfrac{1}{2}\xi_t^T C_t Z^{-1} C_t^T \xi_t + (z_t - \hat{z}_t + C_t^T\mu_t)^T Z^{-1} C_t^T \xi_t\right\} \tag{15}$$

We are now in the position to state the measurement update equation, which implement the probabilistic law (8).

$$\begin{aligned} p(\xi_t \mid z^t, u^t) \;&\propto\; \exp\left\{-\tfrac{1}{2}\xi_t^T \bar{H}_t \xi_t + \bar{b}_t \xi_t\right\} \\ &\cdot \exp\left\{-\tfrac{1}{2}\xi_t^T C_t Z^{-1} C_t^T \xi_t + (z_t - \hat{z}_t + C_t^T\mu_t)^T Z^{-1} C_t^T \xi_t\right\} \\ &= \exp\{-\tfrac{1}{2}\xi_t^T \underbrace{(\bar{H}_t + C_t Z^{-1} C_t^T)}_{H_t}\xi_t + \underbrace{(\bar{b}_t + (z_t - \hat{z}_t + C_t^T\mu_t)^T Z^{-1} C_t^T)}_{b_t}\xi_t\} \end{aligned} \tag{16}$$

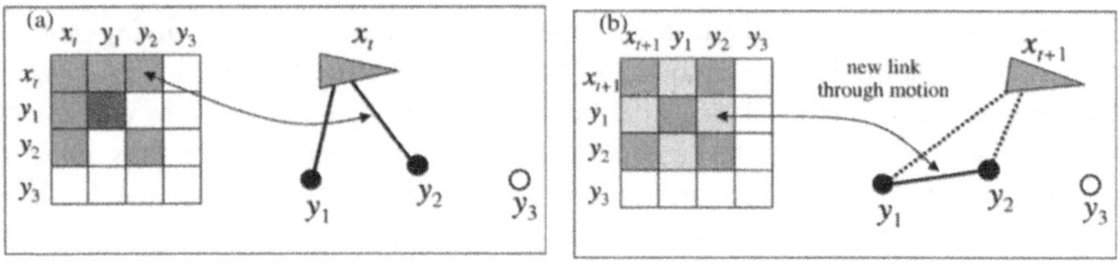

Figure 4. The effect of motion on the information matrix and the associated network of features: (a) before motion, and (b) after motion. If motion is non-deterministic, motion updates introduce new links (or reinforce existing links) between any two active features, while weakening the links between the robot and those features. This step introduces links between pairs of landmarks.

Thus, the measurement update of the EIF is given by the following additive rule:

$$H_t = \bar{H}_t + C_t Z^{-1} C_t^T \tag{17}$$

$$b_t = \bar{b_t} + (z_t - \hat{z}_t + C_t^T \mu_t)^T Z^{-1} C_t^T \tag{18}$$

In the general case, these updates may modify the entire information matrix H_t and vector b_t, respectively. A key observation of all SLAM problems is that the Jacobian C_t is *sparse*. In particular, C_t is zero except for the elements that correspond to the robot pose x_t and the feature y_t observed at time t.

$$C_t = \left(\frac{\partial h}{\partial x_t} \; 0 \cdots 0 \; \frac{\partial h}{\partial y_t} \; 0 \cdots 0 \right)^T \tag{19}$$

This sparseness is due to the fact that measurements z_t are only a function of the relative distance and orientation of the robot to the observed feature. As a pleasing consequence, the update $C_t Z^{-1} C_t^T$ to the information matrix in (17) is only non-zero in four places: the off-diagonal elements that link the robot pose x_t with the observed feature y_t, and the main-diagonal elements that correspond to x_t and y_t. Thus, the update equations (17) and (18) are well in tune with our intuitive description given in the beginning of this section, where we argued that measurements only strengthen the links between the robot pose and observed features, in the information matrix.

To compare this to the EKF solution, we notice that even though the change of the information matrix is local, the resulting covariance usually changes in non-local ways. put differently, the difference between the old covariance $\bar{\Sigma}_t = \bar{H}_t^{-1}$ and the new covariance matrix $\Sigma_t = H_t^{-1}$ is usually non-zero everywhere.

2.3 Motion Updates

The second important step of SLAM concerns the update of the fi lter in accordance to robot motion. In the standard SLAM problem, only the robot pose changes over time. The environment is static.

The effect of robot motion on the information matrix H_t are slightly more complicated than that of measurements. Figure 4a illustrates an information matrix and the associated network before the robot moves, in which the robot is linked to two (previously observed) landmarks. If robot motion was free of noise, this link structure would not be affected by robot motion. However, the noise in robot actuation weakens the link between the robot and all active features. Hence H_{x_t,y_1} and H_{x_t,y_2} are decreased by a certain amount. This decrease reflects the fact that the noise in motion induces a loss of information of the relative location of the features to the robot. Not all of this information is lost, however. Some of it is shifted into between-landmark links H_{y_1,y_2}, as illustrated in Figure 4b. This reflects the fact that even though the motion induced a loss of information of the robot relative to the features, no information was lost between individual features. Robot motion, thus, has the effect that features that were indirectly linked through the robot pose become linked directly.

To derive the update rule, we begin with a Bayesian description of robot motion. Updating a fi lter based on robot motion motion involves the calculation of the following posterior:

$$p(\xi_t \mid z^{t-1}, u^t) = \int p(\xi_t \mid \xi_{t-1}, z^{t-1}, u^t)\, p(\xi_{t-1} \mid z^{t-1}, u^t)\, d\xi_{t-1} \tag{20}$$

Exploiting the common SLAM independences [30] leads to

$$= \int p(\xi_t \mid \xi_{t-1}, u_t)\, p(\xi_{t-1} \mid z^{t-1}, u^{t-1})\, d\xi_{t-1} \tag{21}$$

The term $p(\xi_{t-1} \mid z^{t-1}, u^{t-1})$ is the posterior at time $t-1$, represented by H_{t-1} and b_{t-1}. Our concern will therefore be with the remaining term $p(\xi_t \mid \xi_{t-1}, u_t)$, which characterizes robot motion in probabilistic terms.

Similar to the measurement model above, it is common practice to model robot motion by a non-linear function with added independent Gaussian noise:

$$\xi_t = \xi_{t-1} + \Delta_t \qquad \text{with} \qquad \Delta_t = g(\xi_{t-1}, u_t) + S_x \delta_t \tag{22}$$

Here g is the motion model, a vector-valued function which is non-zero only for the robot pose coordinates, as feature locations are static in SLAM. The term labeled Δ_t constitutes the state change at time t. The stochastic part of this change is modeled by δ_t, a Gaussian random variable with zero mean and covariance U_t. This Gaussian variable is a low-dimensional variable defi ned for the robot pose only. Here S_x is a projection matrix of the form $S_x = (\ I\ \ 0\ \ldots\ 0\ )^T$, where I is an identity matrix of the same dimension as the robot pose vector x_t and as of δ_t. Each 0 in this matrix refers to a null matrix, of which there are N in S_x. The product $S_x\delta_t$, hence, give the following generalized noise variable, enlarged to the dimension of the full state vector ξ: $S_x\delta_t = (\ \delta_t\ \ 0\ \ldots\ 0\ )^T$. In EIFs, the function g in (22) is approximated by its fi rst degree Taylor series expansion:

$$\begin{aligned} g(\xi_{t-1}, u_t) &\approx g(\mu_{t-1}, u_t) + \nabla_\xi g(\mu_{t-1}, u_t)[\xi_{t-1} - \mu_{t-1}] \\ &= \hat{\Delta}_t + A_t\xi_{t-1} - A_t\mu_{t-1} \end{aligned} \tag{23}$$

Here $A_t = \nabla_\xi g(\mu_{t-1}, u_t)$ is the derivative of g with respect to ξ at $\xi = \mu_{t-1}$ and u_t. The symbol $\hat{\Delta}_t$ is short for the predicted motion effect, $g(\mu_{t-1}, u_t)$. Plugging this approximation into (22) leads to an approximation of ξ_t, the state at time t:

$$\xi_t \approx (I + A_t)\xi_{t-1} + \hat{\Delta}_t - A_t\mu_{t-1} + S_x\delta_t \tag{24}$$

Hence, under this approximation the random variable ξ_t is again Gaussian distributed. Its mean is obtained by replacing ξ_t and δ_t in (24) by their respective means:

$$\bar{\mu}_t = (I + A_t)\mu_{t-1} + \hat{\Delta}_t - A_t\mu_{t-1} + S_x 0 = \mu_{t-1} + \hat{\Delta}_t \tag{25}$$

The covariance of ξ_t is simply obtained by scaled and adding the covariance of the Gaussian variables on the right-hand side of (24):

$$\begin{aligned} \bar{\Sigma}_t &= (I + A_t)\Sigma_{t-1}(I + A_t)^T + 0 - 0 + S_x U_t S_x^T \\ &= (I + A_t)\Sigma_{t-1}(I + A_t)^T + S_x U_t S_x^T \end{aligned} \tag{26}$$

Update equations (25) and (26) are in the EKF form, that is, they are defi ned over means and covariances. The information form is now easily recovered from the defi nition of the information form in (4) and its inverse in (6). In particular, we have

$$\begin{aligned} \bar{H}_t &= \bar{\Sigma}_t^{-1} = \left[(I + A_t)\Sigma_{t-1}(I + A_t)^T + S_x U_t S_x^T\right]^{-1} \\ &= \left[(I + A_t)H_{t-1}^{-1}(I + A_t)^T + S_x U_t S_x^T\right]^{-1} \end{aligned} \tag{27}$$

$$\begin{aligned} \bar{b}_t &= \bar{\mu}_t^T \bar{H}_t = \left[\mu_{t-1} + \hat{\Delta}_t\right]^T \bar{H}_t = \left[H_{t-1}^{-1} b_{t-1}^T + \hat{\Delta}_t\right]^T \bar{H}_t \\ &= \left[b_{t-1} H_{t-1}^{-1} + \hat{\Delta}_t^T\right] \bar{H}_t \end{aligned} \tag{28}$$

These equations appear computationally involved, in that they require the inversion of large matrices. In the general case, the complexity of the EIF is therefore cubic in the size of the state space. In the next section, we provide the surprising result that both $\bar{H}_t$ and $\bar{b}_t$ can be computed in constant time if H_{t-1} is sparse.

3 Sparse Extended Information Filters

The central, new algorithm presented in this paper is the *Sparse Extended Information Filter*, or *SEIF*. SEIF differ from the extended information fi lter described in the previous section in that is maintains a *sparse* information matrix. An information matrix H_t is considered *sparse* if the number of links to the robot and to each feature in the map is bounded by a constant that is independent of the number of features in the map. The bound for the number of links between the robot pose and other features in the map will be denoted θ_x; the bound on the number of links for each feature (not counting the link to the robot) will be denoted θ_y. The motivation for maintaining a sparse information matrix was already given above: In SLAM, the normalized information matrix is already *almost* sparse. This suggests that by enforcing sparseness, the induced approximation error is small.

3.1 Constant Time Results

We begin by proving three important constant time results, which form the backbone of SEIFs. All proofs can be found in [29].

Lemma 1: *The measurement update in Section (2.2) requires constant time, irrespective of the number of features in the map.*

This lemma ensures that measurements can be incorporated in constant time. Notice that this lemma does *not* require sparseness of the information matrix; rather, it is a well-known property of information filters in SLAM.

Less trivial is the following lemma:

Lemma 2: *If the information matrix is sparse and $A_t = 0$, the motion update in Section (2.3) requires constant time. The constant-time update equations are given by:*

$$\begin{aligned} L_t &= S_x[U_t^{-1} + S_x^T H_{t-1} S_x]^{-1} S_x^T H_{t-1} \\ \bar{H}_t &= H_{t-1} - H_{t-1} L_t \\ \bar{b}_t &= b_{t-1} + \hat{\Delta}_t^T H_{t-1} - b_{t-1} L_t + \hat{\Delta}_t^T H_{t-1} L_t \end{aligned} \tag{29}$$

This result addresses the important special case $A_t = 0$, that is, the Jacobian of pose change with respect to the absolute robot pose is zero. This is the case for robots with linear mechanics, and with non-linear mechanics where there is no 'cross-talk' between absolute coordinates and the additive change due to motion.

In general, $A_t \neq 0$, since the x-y update depends on the robot orientation. This case is addressed by the next lemma:

Lemma 3: *If the information matrix is sparse, the motion update in Section (2.3) requires constant time if the mean μ_t is available for the robot pose and all active landmarks. The constant-time update equations are given by:*

$$\begin{aligned} \Psi_t &= I - S_x(I + [S_x^T A_t S_x]^{-1})^{-1} S_x^T \\ H'_{t-1} &= \Psi_t^T H_{t-1} \Psi_t \\ \Delta H_t &= H'_{t-1} S_x [U_t^{-1} + S_x^T H'_{t-1} S_x]^{-1} S_x^T H'_{t-1} \\ \bar{H}_t &= H'_{t-1} - \Delta H_t \\ \bar{b}_t &= b_{t-1} - \mu_{t-1}^T (\Delta H_t - H_{t-1} + H'_{t-1}) + \hat{\Delta}_t^T \bar{H}_t \end{aligned} \tag{30}$$

For $A_t \neq 0$, a constant time update requires knowledge of the mean μ_{t-1} before the motion command, for the robot pose and all active landmarks (but not the passive features). This information is *not* maintained by the standard information filter, and extracting it in the straightforward way (via Equation (6)) requires more than constant time. A constant-time solution to this problem will now be presented.

3.2 Amortized Approximated Map Recovery

Before deriving an algorithm for recovering the state estimate μ_t from the information form, let us briefly consider what parts of μ_t are needed in SEIFs, and when. SEIFs need the state estimate μ_t of the robot pose and the active features in the map. These estimates are needed at three different occasions: (1) the linearization of the non-linear measurement and motion model, (2) the motion update according to Lemma 3, and (3) the sparsification technique described further below. For linear systems, the means are only needed for the sparsification (third point above). We also note that we only need constantly many of the values in μ_t, namely the estimate of the robot pose and of the locations of active features.

As stated in (6), the mean vector μ_t is a function of H_t and b_t:

$$\mu_t = H_t^{-1} b_t^T \; = \; \Sigma_t b_t^T \tag{31}$$

Unfortunately, calculating (31) directly involves inverting a large matrix, which would requires more than constant time.

The sparseness of the matrix H_t allows us to recover the state incrementally. In particular, we can do so on-line, as the data is being gathered and the estimates b and H are being constructed. To do so, it will prove convenient to pose (31) as an optimization problem:

Lemma 4: *The state μ_t is the mode $\hat{\nu}_t := \text{argmax}_{\nu_t}\, p(\nu_t)$ of the Gaussian distribution, defined over the variable ν_t:*

$$p(\nu_t) = \text{const.} \cdot \exp\left\{-\tfrac{1}{2}\nu_t^T H_t \nu_t + b_t^T \nu_t\right\} \tag{32}$$

Here ν_t is a vector of the same form and dimensionality as μ_t. This lemma suggests that recovering μ_t is equivalent to fi nding the mode of (32). Thus, it transforms a matrix inversion problem into an optimization problem. For this optimization problem, we will now describe an iterative hill climbing algorithm which, thanks to the sparseness of the information matrix, requires only constant time per optimization update.

Our approach is an instantiation of coordinate descent. For simplicity, we state it here for a single coordinate only; our implementation iterates a constant number K of such optimizations after each measurement update step. The mode $\hat{\nu}_t$ of (32) is attained at:

$$\begin{aligned} \hat{\nu}_t &= \underset{\nu_t}{\text{argmax}}\ p(\nu_t) \ = \ \underset{\nu_t}{\text{argmax}}\ \exp\left\{-\tfrac{1}{2}\nu_t^T H_t \nu_t + b_t^T \nu_t\right\} \\ &= \underset{\nu_t}{\text{argmin}}\ \tfrac{1}{2}\nu_t^T H_t \nu_t - b_t^T \nu_t \end{aligned} \tag{33}$$

We note that the argument of the min-operator in (33) can be written in a form that makes the individual coordinate variables $\nu_{i,t}$ (for the i-th coordinate of ν_t) explicit:

$$\tfrac{1}{2}\nu_t^T H_t \nu_t - b_t^T \nu_t = \tfrac{1}{2}\sum_i \sum_j \nu_{i,t}^T H_{i,j,t} \nu_{j,t} - \sum_i b_{i,t}^T \nu_{i,t} \tag{34}$$

where $H_{i,j,t}$ is the element with coordinates (i,j) in H_t, and $b_{i,t}$ if the i-th component of the vector b_t. Taking the derivative of this expression with respect to an arbitrary coordinate variable $\nu_{i,t}$ gives us

$$\frac{\partial}{\partial \nu_{i,t}} \left\{ \tfrac{1}{2}\sum_i \sum_j \nu_{i,t}^T H_{i,j,t} \nu_{j,t} - \sum_i b_{i,t}^T \nu_{i,t} \right\} = \sum_j H_{i,j,t} \nu_{j,t} - b_{i,t}^T \tag{35}$$

Setting this to zero leads to the optimum of the i-th coordinate variable $\nu_{i,t}$ given all other estimates $\nu_{j,t}$:

$$\nu_{i,t}^{[k+1]} = H_{i,i,t}^{-1} \left[b_{i,t}^T - \sum_{j \neq i} H_{i,j,t} \nu_{j,t}^{[k]} \right] \tag{36}$$

The same expression can conveniently be written in matrix notation, were S_i is a projection matrix for extracting the i-th component from the matrix H_t:

$$\nu_{i,t}^{[k+1]} = (S_i^T H_t S_i)^{-1} S_i^T \left[b_t - H_t \nu_t^{[k]} + H_t S_i S_i^T \nu_t^{[k]} \right] \tag{37}$$

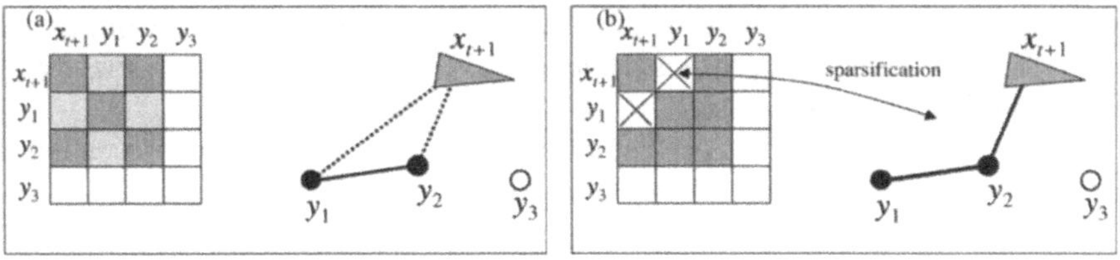

Figure 5. Sparsification: A feature is deactivated by eliminating its link to the robot. To compensate for this change in information state, links between active features and/or the robot are also updated. The entire operation can be performed in constant time.

All other estimates $\nu_{i',t}$ with $i' \neq i$ remain unchanged in this update step, that is, $\nu_{i',t}^{[k+1]} = \nu_{i',t}^{[k]}$.

As is easily seen, the number of elements in the summation in (36), and hence the vector multiplication in (37), is constant if H_t is sparse. Hence, each update requires constant time. To maintain the constant-time property of our SLAM algorithm, we can afford a constant number of updates K per time step. This will generally not lead to convergence, but the relaxation process takes place over multiple time steps, resulting in small errors in the overall estimate.

3.3 Sparsification

The fi nal step in SEIFs concerns the sparsifi cation of the information matrix H_t. Sparsifi cation is necessarily an approximative step, since information matrices in SLAM are naturally not sparse—even though normalized information matrices tend to be almost sparse. In the context of SLAM, it suffi ces to remove links (deactivate) between the robot pose and individual features in the map; if done correctly, this also limits the number of links between pairs of features.

To see, let us briefly consider the two circumstances under which a new link may be introduced. First, observing a passive feature activates this feature, that is, introduces a new link between the robot pose and the very feature. Thus, measurement updates potentially violate the bound θ_x. Second, motion introduces links between any two active features, and hence lead to violations of the bound θ_y. This consideration suggests that controlling the number of active features can avoid violation of both sparseness bounds.

Our sparsifi cation technique is illustrated in Figure 5. Shown there is the situation before and after sparsifi cation. The removal of a link in the network corresponds to setting an element in the information matrix to zero; however, this requires the manipulation of other links between the robot and other active landmarks. The resulting network is only an approximation to the original one, whose quality depends on the magnitude of the link before removal.

We will now present a constant-time sparsifi cation technique. To do so, it will prove useful to partition the set of all features into three subsets:

$$Y = Y^+ \uplus Y^0 \uplus Y^- \tag{38}$$

where Y^+ is the set of all active features that shall remain active. Y^0 are one or more active features that we seek to deactivate (remove the link to the robot). Finally, Y^- are all currently passive features.

The sparsification is best derived from first principles. If $Y^+ \uplus Y^0$ contains all currently active features, the posterior can be factored as follows:

$$\begin{aligned} p(x_t, Y \mid z^t, u^t) &= p(x_t, Y^0, Y^+, Y^- \mid z^t, u^t) \\ &= p(x_t \mid Y^0, Y^+, Y^-, z^t, u^t)\, p(Y^0, Y^+, Y^- \mid z^t, u^t) \\ &= p(x_t \mid Y^0, Y^+, Y^- = 0, z^t, u^t)\, p(Y^0, Y^+, Y^- \mid z^t, u^t) \end{aligned} \tag{39}$$

In the last step we exploited the fact that if we know the active features Y^0 and Y^+, the variable x_t does not depend on the passive features Y^-. We can hence set Y^- to an arbitrary value without affecting the conditional posterior over x_t, $p(x_t \mid Y^0, Y^+, Y^-, z^t, u^t)$. Here we simply chose $Y^- = 0$.

To sparsify the information matrix, the posterior is approximated by the following distribution, in which we simply drop the dependence on Y^0 in the first term. It is easily shown that this distribution minimizes the KL divergence to the exact, non-sparse distribution:

$$\begin{aligned} \tilde{p}(x_t, Y \mid z^t, u^t) &= p(x_t \mid Y^+, Y^- = 0, z^t, u^t)\, p(Y^0, Y^+, Y^- \mid z^t, u^t) \\ &= \frac{p(x_t, Y^+ \mid Y^- = 0, z^t, u^t)}{p(Y^+ \mid Y^- = 0, z^t, u^t)}\, p(Y^0, Y^+, Y^- \mid z^t, u^t) \end{aligned} \tag{40}$$

This posterior is calculated in constant time. In particular, we begin by calculating the information matrix for the distribution $p(x_t, Y^0, Y^+ \mid Y^- = 0)$ of all variables but Y^-, and conditioned on $Y^- = 0$. This is obtained by extracting the submatrix of all state variables but Y^-:

$$H_t' = S_{x,Y^+,Y^0} S_{x,Y^+,Y^0}^T H_t S_{x,Y^+,Y^0} S_{x,Y^+,Y^0}^T \tag{41}$$

With that, the inversion lemma leads to the following information matrices for the terms $p(x_t, Y^+ \mid Y^- = 0, z^t, u^t)$ and $p(Y^+ \mid Y^- = 0, z^t, u^t)$, denoted H_t^1 and H_t^2, respectively:

$$\begin{aligned} H_t^1 &= H_t' - H_t' S_{Y_0} (S_{Y_0}^T H_t' S_{Y_0})^{-1} S_{Y_0}^T H_t' \\ H_t^2 &= H_t' - H_t' S_{x,Y_0} (S_{x,Y_0}^T H_t' S_{x,Y_0})^{-1} S_{x,Y_0}^T H_t' \end{aligned} \tag{42}$$

Here the various S-matrices are projection matrices, analogous to the matrix S_x defined above. The final term in our approximation (40), $p(Y^0, Y^+, Y^- \mid z^t, u^t)$, has the following information matrix:

$$H_t^3 = H_t - H_t S_{x_t} (S_{x_t}^T H_t S_{x_t})^{-1} S_{x_t}^T H_t \tag{43}$$

Putting these expressions together according to Equation (40) yields the following information matrix, in which the landmark Y^0 is now indeed deactivated:

$$\begin{aligned} \tilde{H}_t = H_t^1 - H_t^2 + H_t^3 \;=\; & H_t - H_t' S_{Y_0} (S_{Y_0}^T H_t' S_{Y_0})^{-1} S_{Y_0}^T H_t' \\ & + H_t' S_{x,Y_0} (S_{x,Y_0}^T H_t' S_{x,Y_0})^{-1} S_{x,Y_0}^T H_t' - H_t S_{x_t} (S_{x_t}^T H_t S_{x_t})^{-1} S_{x_t}^T H_t \end{aligned} \tag{44}$$

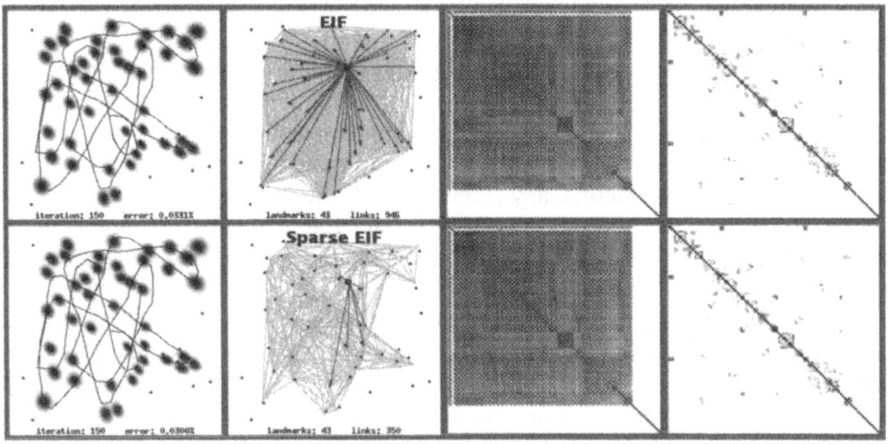

Figure 6. Comparison of EKFs with SEIFs using a simulation with $N = 50$ landmarks. In both diagrams, the left panels show the final filter result, which indicates higher certainties for our approach due to the approximations involved in maintaining a sparse information matrix. The center panels show the links (red: between the robot and landmarks; green: between landmarks). The right panels show the resulting covariance and normalized information matrices for both approaches. Notice the similarity!

The resulting information vector is now obtained by the following simple consideration:

$$\begin{aligned}\tilde{b}_t = \mu_t^T \tilde{H}_t &= \mu_t^T (H_t - H_t + \tilde{H}_t) \\ = \mu_t^T H_t + \mu_t^T (\tilde{H}_t - H_t) &= b_t + \mu_t^T (\tilde{H}_t - H_t)\end{aligned} \tag{45}$$

All equations can be computed in constant time. The effect of this approximation is the deactivation of the features Y^0, while introducing only new links between active features. The sparsifi cation rule requires knowledge of the mean vector μ_t for all active features, which is obtained via the approximation technique described in the previous section. From (45), it is obvious that the sparsifi cation does not affect the mean μ_t, that is, $H_t^{-1} b_t^T = [\tilde{H}_t]^{-1} [\tilde{b}_t]^T$. Furthermore, our approximation minimizes the KL divergence to the correct posterior. These property is essential for the consistency of our approximation.

The sparsifi cation is executed whenever a measurement update of a motion update would violate a sparseness constraint. Active features are chosen for deactivation in reverse order of the magnitude of their link. This strategy tends to deactivate features whose last sighting is furthest away in time. Empirically, it induces approximation errors that are negligible for appropriately chosen sparseness constraints θ_x and θ_y.

4 Experimental Results

Our present experiments are preliminary: They only rely on simulated data, and they require known data associations. Our primary goal was to compare SEIFs to the computationally more cumbersome EKF solution that is currently in widespread use.

An example situation comparing EKFs with our new fi lter can be found in Figure 6. This result is typical and was obtained using a sparse information matrix with $\theta_x = 6$, $\theta_x = 10$, and a constant time implementation of coordinate descent that updates $K = 10$ random landmark estimates in addition to the landmark estimates connected to the robot at any given time. The key observation is the apparent similarity between the EKF and the SEIF result. Both estimates are almost indistinguishable, despite the fact that EKFs use quadratic update time whereas SEIF require only constant time.

We also performed systematic comparisons of three algorithms: EKFs, SEIFs, and a variant of SEIFs in which the exact state estimate μ_t is available. The latter was implemented using matrix inversion (hence does not run in constant time). It allowed us to tease apart the error introduced by the amortized mean recovery step, from the error induced through sparsifi cation. The following table depicts results for $N = 50$ landmarks, after 500 update cycles, at which point all three approaches are near convergence.

	# experiments (so far)	final error (with 95% conf. interval)	final # of links (with 95% conf. interval)	computation (per update)
EKF	1,000	$(5.54 \pm 0.67) \cdot 10^{-3}$	1,275	$O(N^2)$
SEIF with exact μ_t	1,000	$(4.75 \pm 0.67) \cdot 10^{-3}$	549 ± 1.60	$O(N^3)$
SEIF (constant time)	1,00	$(6.35 \pm 0.67) \cdot 10^{-3}$	549 ± 1.59	$O(1)$

As these results suggest, our approach approximates EKF very tightly. The residual map error of our approach is with $6.35 \cdot 10^{-3}$ approximately 14.6% higher than that of the extended Kalman fi lter. This error appears to be largely caused by the coordinate descent procedure, and is possibly inflated by the fact that $K = 10$ is a small value given the size of the map. Enforcing the sparseness constraint seems not to have any negative effect on the overall error of the resulting map, as the results for our sparse fi lter implementation suggest. Experimental results using a real-world data set can be found in [14].

5 Discussion

This paper proposed a constant time algorithm for the SLAM problem. Our approach adopted the information form of the EKF to represent all estimates. Based on the empirical observation that in the information form, most elements in the normalized information matrix are near-zero, we developed a *sparse* extended information fi lter, or SEIF. This fi lter enforces a sparse information matrix, which can be updated in constant time. In the *linear* SLAM case, all updates can be performed in constant time; in the *non-linear case*, additional state estimates are needed that are not part of the regular information form of the EKF. We proposed a amortized constant-time coordinate descent algorithm for recovering these state estimates from the information form.

The approach has been fully implemented and compared to the EKF solution. Overall, we found that SEIFs produce results that differ only marginally from that of the EKFs. Given the computational advantages of SEIFs over EKFs, we believe that SEIFs should be a viable alternative to EKF solutions when building high-dimensional maps.

Our approach puts a new perspective on the rich literature on hierarchical mapping, briefly outlined in the introduction to this paper. Like SEIF, these techniques focus updates on a subset of all features, to gain computational effi ciency. SEIFs, however, composes submaps dynamically, whereas past work relied on the defi nition of static submaps. We conjecture that our sparse network structures capture the natural dependencies in SLAM problems much better than static submap decompositions, and in turn lead to more accurate results. They also avoid problems that frequently occur at the boundary of submaps, where the estimation can become unstable. However, the verifi cation of these claims will be subject to future research. A related paper discusses the application of constant time techniques to information exchange problems in multi-robot SLAM [21].

Acknowledgment

The authors would like to acknowledge invaluable contributions by the following researchers: Wolfram Burgard, Geoffrey Gordon, Kevin Murphy, Eric Nettleton, Michael Stevens, and Ben Wegbreit. This research has been sponsored by DARPA's MARS Program (contracts N66001-01-C-6018 and NBCH1020014), DARPA's CoABS Program (contract F30602-98-2-0137), and DARPA's MICA Program (contract F30602-01-C-0219), all of which is gratefully acknowledged. The authors furthermore acknowledge support provided by the National Science Foundation (CAREER grant number IIS-9876136 and regular grant number IIS-9877033).

References

1. M. Bosse, J. Leonard, and S. Teller. Large-scale CML using a network of multiple local maps. In [11].
2. W. Burgard, D. Fox, H. Jans, C. Matenar, and S. Thrun. Sonar-based mapping of large-scale mobile robot environments using EM. *Proc. OCML-99.*
3. H. Choset. *Sensor Based Motion Planning: The Hierarchical Generalized Voronoi Graph.* PhD thesis, Caltech, 1996.
4. M. Csorba. *Simultaneous Localization and Map Building.* PhD thesis, Univ. of Oxford, 1997.
5. M. Deans and M. Hebert. Invariant filtering for simultaneous localization and mapping. *Proc. ICRA-00.*
6. G. Dissanayake, P. Newman, S. Clark, H.F. Durrant-Whyte, and M. Csorba. A solution to the simultaneous localisation and map building (SLAM) problem. *Transactions of Robotics and Automation*, 2001.
7. A. Doucet, J.F.G. de Freitas, and N.J. Gordon, editors. *Sequential Monte Carlo Methods In Practice.* Springer, 2001.
8. J. Guivant and E. Nebot. Optimization of the simultaneous localization and map building algorithm for real time implementation. *Transactions of Robotics and Automation*, 2001.
9. J.-S. Gutmann and K. Konolige. Incremental mapping of large cyclic environments. *Proc. ICRA-00.*
10. B. Kuipers and Y.-T. Byun. A robot exploration and mapping strategy based on a semantic hierarchy of spatial representations. *Journal of Robotics and Autonomous Systems*, 8, 1991.

11. J. Leonard, J.D. Tardós, S. Thrun, and H. Choset, editors. *ICRA Workshop Notes (W4)*, 2002.
12. J. J. Leonard and H. F. Durrant-Whyte. *Directed Sonar Sensing for Mobile Robot Navigation*. Kluwer, 1992.
13. J.J. Leonard and H.J.S. Feder. A computationally efficient method for large-scale concurrent mapping and localization.
14. Y. Liu and S. Thrun. Results for outdoor-SLAM using sparse extended information filters. Submitted to ICRA-03.
15. M. J. Matarić. A distributed model for mobile robot environment-learning and navigation. MIT AITR-1228.
16. P. Maybeck. *Stochastic Models, Estimation, and Control, Volume 1*. Academic Press, 1979.
17. M. Montemerlo, S. Thrun, D. Koller, and B. Wegbreit. FastSLAM: A factored solution to the simultaneous localization and mapping problem. *Proc. AAAI-02*.
18. P. Moutarlier and R. Chatila. An experimental system for incremental environment modeling by an autonomous mobile robot. *Proc. ISER-89*.
19. K. Murphy. Bayesian map learning in dynamic environments. *Proc. NIPS-99*.
20. E. Nettleton, H. Durrant-Whyte, P. Gibbens, and A. Goktoğan. Multiple platform localisation and map building. *Sensor Fusion and Decentralised Control in Robotic Stystems III*: 4196, 2000.
21. E. Nettleton, S. Thrun, and H. Durrant-Whyte. A constant time communications algorithm for decentralised SLAM. Submitted.
22. E.W. Nettleton, P.W. Gibbens, and H.F. Durrant-Whyte. Closed form solutions to the multiple platform simultaneous localisation and map building (slam) problem. *Sensor Fusion: Architectures, Algorithms, and Applications IV*: 4051, 2000.
23. P. Newman. *On the Structure and Solution of the Simultaneous Localisation and Map Building Problem*. PhD thesis, University of Sydney, 2000.
24. H Shatkay and L. Kaelbling. Learning topological maps with weak local odometric information. *Proc. IJCAI-97*.
25. R. Simmons, D. Apfelbaum, W. Burgard, M. Fox, D. an Moors, S. Thrun, and H. Younes. Coordination for multi-robot exploration and mapping. *Proc. AAAI-00*.
26. R. Smith, M. Self, and P. Cheeseman. Estimating uncertain spatial relationships in robotics. *Autonomous Robot Vehicles*, Springer, 1990.
27. R. C. Smith and P. Cheeseman. On the representation and estimation of spatial uncertainty. TR 4760 & 7239, SRI, 1985.
28. C. Thorpe and H. Durrant-Whyte. Field robots. *Proc. ISRR-01*.
29. S. Thrun, D. Koller, Z. Ghahramani, H. Durrant-Whyte, and A.Y. Ng. Simultaneous mapping and localization with sparse extended information filters:theory and initial results. TR CMU-CS-02-112, CMU, 2002.
30. S. Thrun. Robotic mapping: A survey. *Exploring Artificial Intelligence in the New Millenium*, Morgan Kaufmann, 2002.
31. S. Thrun, D. Fox, and W. Burgard. A probabilistic approach to concurrent mapping and localization for mobile robots. *Machine Learning*, 31, 1998.
32. S. Williams and G. Dissanayake. Efficient simultaneous localisation and mapping using local submaps. In [11].
33. U.R. Zimmer. Robust world-modeling and navigation in a real world. *Neurocomputing*, 13, 1996.

Expected Shortest Paths for Landmark-Based Robot Navigation

Amy J. Briggs[1], Carrick Detweiler[1], Daniel Scharstein[1], and Alexander Vandenberg-Rodes[2]

[1] Middlebury College, Middlebury VT 05753, USA
[2] Princeton University, Princeton NJ 08544, USA

Abstract. In this paper we address the problem of planning reliable landmark-based robot navigation strategies in the presence of significant sensor uncertainty. The navigation environments are modeled with directed weighted graphs in which edges can be traversed with given probabilities. To construct robust and efficient navigation plans, we compute *expected shortest paths* in such graphs. We formulate the expected shortest paths problem as a Markov decision process and provide two algorithms for its solution. We demonstrate the practicality of our approach using an extensive experimental analysis using graphs with varying sizes and parameters.

1 Introduction

Reliable strategies for mobile robot navigation using visual landmarks must be robust in the face of significant uncertainty in landmark detection. Varying lighting conditions, temporary occlusion, and unreliability of landmark recognition are all factors that contribute to such uncertainty, whether the visual landmarks are artificial or extracted from the environment. This paper presents algorithms and experimental results for computing expected shortest paths for use in navigation between landmarks. The paper builds on our real-time landmark detection system [16] and on our vision-based navigation framework [3,4]. Here we extend our prior work with significant contributions to the understanding of probabilistic navigation. First, we present two algorithms for the solution to the expected shortest paths problem and prove their convergence. Second, we present a careful analysis of the properties of the two algorithms, with an extensive empirical evaluation of their performance in practice. We first review the natural graph formulation of the problem, and then show how the problem can be concisely described as a Markov decision process.

The specific problem we address is the following. We assume an unmapped indoor environment (such as a factory floor or hospital) in which a mobile robot must repeatedly navigate. To enable the robot to localize itself quickly and then navigate in this space, the environment is enhanced with artificial visual landmarks. During an initial exploratory phase the robot systematically visits all landmarks and constructs their visibility graph. This graph is continually updated during subsequent navigation. All motion paths are

J.-D. Boissonnat et al. (Eds.): Algorithmic Foundations of Robotics V, STAR 7, pp. 381–398, 2004.

planned from landmark to landmark, moving along the visibility edges of the graph. Because landmark detection is unreliable, each edge is annotated not only with its length, but also with an estimate of the ***probability*** that the target landmark can be detected. These estimates are based on the history of all observations made by the robot, and are updated as the robot navigates the environment. Given such a graph, planning an optimal path to a goal landmark involves computing the ***expected shortest path*** (ESP), i.e., the path with expected shortest length. Using probabilistic techniques, our navigation system constructs reliable and efficient motion paths while compensating for occlusion, unreliability of landmark detection, and variations in visibility due to changes in the environment over time.

The ESP algorithms can be applied to important problems in other domains as well. For example, probabilistic graphs can be used to model communication networks in which individual links can fail. The ESP algorithms could then be used for optimal package routing given continually changing models of link reliability. Another application is that of traffic planning in a road network where some roads may be closed due to poor weather, accidents, congestion, road construction, or other temporary interruptions (e.g., draw bridges or railway crossings). The ESP formulation can also be applied to other robot path planning problems, for example to configuration-space planning in uncertain or changing environments.

After reviewing related work in Section 2, we define the expected shortest path problem and give two algorithms for its solution in Section 3. We then present the Markov decision process formulation of the problem and analyze the properties of our algorithms in Section 4. We report on our experimental results evaluating the performance of the algorithms on large sets of graphs in Section 5. We conclude with ideas for future work in Section 6.

2 Related Work

Techniques for mobile robot navigation based on landmarks include those planned within a geometric environment map enhanced with perceptual landmarks [7,10], and those based on a topological description of landmark locations without a global map [12,17,23]. Many research groups have used artificial landmarks that can be easily and unobtrusively added to the environment for localization and navigation tasks [1,8,20].

Several different approaches to probabilistic path planning have been developed. Kavraki and Latombe [6] propose a randomized method for configuration space preprocessing that generates a network of collision-free configurations in a known environment. Overmars and Švestka [11] describe a similar probabilistic learning approach that extends to a number of motion planning problems. Several researchers have used partially observable MDPs for robot localization and navigation [18,19,21,23]. These approaches model the robot's state with a probability distribution, as opposed to our method, which assumes a known state. However, POMDPs are computationally intractable and require approximate solutions such as a discretization of the state space [19] or "coastal navigation" [21].

Probabilistic graphs in which each edge is passable with a given probability have also been considered [9,13,14,22]. Most existing work, however, assumes a static graph, i.e., that unpassable edges do not become passable and vice versa. A typical problem in this scenario is to compute the probability that the graph is connected [14,22]. Mani et al. [9] study the problem of finding shortest paths in such static graphs, and focus on series-parallel graphs. Besides length and probability, they also associate a detection cost with each edge. Finally, Blei and Kaelbling [2] describe Markov decision processes for finding shortest paths in stochastic graphs with partially unknown topologies. Their work differs from ours in that they assume that an edge is either passable or not, but that the state of each edge is only known with a certain probability.

3 The Expected Shortest Paths Problem

We assume an environment augmented with N visual landmarks $a, b, c, \dots$ that can be detected by the robot, albeit unreliably. We assume an edge from landmark a to landmark b has associated probability $p_{ab} \in [0, 1]$ and length $l_{ab} > 0$. The probability p_{ab} represents the likelihood that landmark b can be detected from landmark a; the length l_{ab} represents a measure of cost to traverse the edge (e.g., time of travel). We assume that the robot can only travel to landmarks that are currently visible. Thus, the probabilities p_{ab} more generally represent the likelihoods that edges are passable.

The robot, at any given landmark (or node) n, must choose among the currently visible landmarks (i.e., passable edges) where to travel to next. It also has the option of staying at the current node, and waiting for new landmarks to become visible. If no edge is currently passable, this is the only option. We represent the possibility of staying with a self-edge at each node n. Since staying is always an option, the associated probability is $p_{nn} = 1$. To prevent the robot from staying at a landmark indefinitely, we associate with the self-edge a non-zero cost $l_{nn} > 0$, for example, the time it takes to acquire a new image and to search it for landmarks.

A navigation task in such a *probabilistic graph* consists of a designated goal node g that the robot wants to reach from its current node s. In a non-probabilistic graph, the optimal path could be computed easily, for example using Dijkstra's shortest-path algorithm. In our case, however, we have to change the notion of a shortest path to that of a *path with shortest expected length*, or *expected shortest path* (ESP).

In previous work we have described a system in which a robot explores and navigates an unknown environment augmented with artificial visual landmarks, and constructs and updates a probabilistic graph in the process [3].

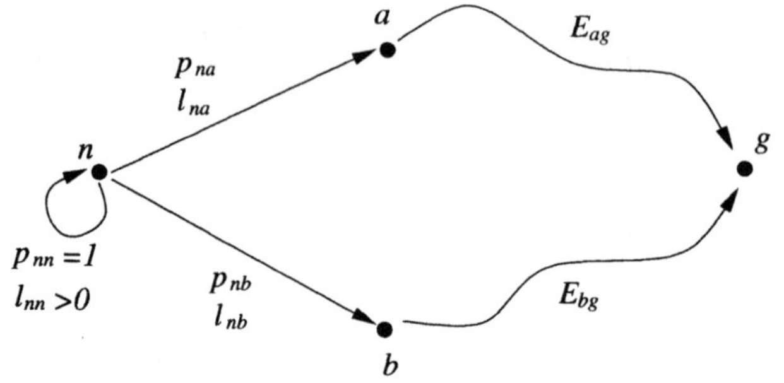

Fig. 1. A node n with self-edge and two outgoing edges to nodes a and b. All edges have lengths l_{ij} and probabilities of being passable p_{ij}. We wish to compute the expected lengths E_{ig} of the shortest paths from each node i to goal g. The four possible visibility scenarios at node n are reflected in Equation 2

Edge probabilities are estimated based on the history of observations made at each landmark.

Here we focus on the ESP problem and on the theoretical and practical properties of two algorithms for its solution. As mentioned earlier, the ESP problem and algorithms have applications beyond that of visual robot navigation and are important in their own right. We thus assume that a probabilistic graph as described above is given as input.

3.1 The ESP Equations

Given a goal node g, the ESP problem is to compute for each node n the expected length E_{ng} of the shortest path from n to g. Clearly,

$$E_{gg} = 0. \tag{1}$$

We now derive equations that recursively relate the $N-1$ unknowns E_{ng}, $n \neq g$.

Consider first a node n with two outgoing edges $n \to a$ and $n \to b$ (see Figure 1). The robot, upon arriving at n, will be faced with one of four *visibility scenarios*, depending on which of a and b can be detected. Suppose for example that a is visible but b is not. This scenario occurs with probability $p_{na}\,\overline{p_{nb}}$, where $\overline{p}$ denotes $(1-p)$. Given this situation, the robot has two options: to go to a, or to stay at n. (The latter may be desirable if b is usually visible and yields a much shorter path.) The expected total length of the path through node a is $l_{na} + E_{ag}$, while the expected total length of staying at node n is $l_{nn} + E_{ng}$. For notational convenience, let

$$L^i_{ng} = l_{ni} + E_{ig}$$

denote the total expected length of the path from n to g through i. The length of the shortest expected path (with b not visible) is then $\min(L^a_{ng}, L^n_{ng})$, i.e., the smaller of the two candidate lengths. We can now write the complete equation for E_{ng} by weighting the shortest candidate path with its corresponding probability in each of the four visibility scenarios:

$$\begin{aligned} E_{ng} = &\quad \overline{p_{na}}\,\overline{p_{nb}}\,L^n_{ng} \\ &+ p_{na}\,\overline{p_{nb}}\,\min(L^a_{ng},\,L^n_{ng}) \\ &+ \overline{p_{na}}\,p_{nb}\,\min(L^b_{ng},\,L^n_{ng}) \\ &+ p_{na}\,p_{nb}\,\min(L^a_{ng},\,L^b_{ng},\,L^n_{ng}). \end{aligned} \tag{2}$$

It is easy to see how this equation generalizes to nodes with more than two outgoing edges. In particular, a node with k outgoing edges $n \to a$, $n \to b$, $\ldots$, $n \to z$ yields an equation with 2^k terms:

$$\begin{aligned} E_{ng} = &\quad \overline{p_{na}}\,\overline{p_{nb}}\cdots\overline{p_{nz}}\,L^n_{ng} \\ &+ p_{na}\,\overline{p_{nb}}\cdots\overline{p_{nz}}\,\min(L^a_{ng},\,L^n_{ng}) \\ &+ \overline{p_{na}}\,p_{nb}\cdots\overline{p_{nz}}\,\min(L^b_{ng},\,L^n_{ng}) \\ &+ p_{na}\,p_{nb}\cdots\overline{p_{nz}}\,\min(L^a_{ng},\,L^b_{ng},\,L^n_{ng}) \\ &+ \cdots \\ &+ p_{na}\,p_{nb}\cdots p_{nz}\,\min(L^a_{ng},\,L^b_{ng},\,\ldots,L^z_{ng},\,L^n_{ng}). \end{aligned} \tag{3}$$

We thus have a system of $N-1$ equations that recursively relate the $N-1$ unknowns E_{ng}, $n \neq g$. Because of the minimum expressions the system cannot be solved directly. Note, however, that Equation 3 could be written as a linear equation with only k terms if the *ordering* among the candidate lengths L^i_{ng} were known. This observation plays an important role in our algorithms, so we will elaborate a bit:

Going back to the simpler Equation 2, consider the possible orderings among L^a_{ng}, L^b_{ng}, and L^n_{ng}. While one can deduce that L^n_{ng} cannot be the smallest of the three, each of the remaining four orderings is possible:

$$\begin{aligned} &L^a_{ng} \le L^b_{ng} \le L^n_{ng}, \qquad && L^a_{ng} \le L^n_{ng} \le L^b_{ng}, \\ &L^b_{ng} \le L^a_{ng} \le L^n_{ng}, && L^b_{ng} \le L^n_{ng} \le L^a_{ng}. \end{aligned} \tag{4}$$

Assume for example that the first of these orderings holds: $L^a_{ng} \le L^b_{ng} \le L^n_{ng}$. Such an ordering translates literally into a navigation strategy for a robot at landmark n: go to landmark a if possible (visible); if not, try to go to b if possible; else remain at n. Given this ordering, Equation 2 simplifies to

$$E_{ng} = p_{na}\,L^a_{ng} + \overline{p_{na}}\,p_{nb}\,L^b_{ng} + \overline{p_{na}}\cdot\overline{p_{nb}}\,L^n_{ng}. \tag{5}$$

A final observation: if L^n_{ng} is not last in the ordering, the subsequent terms in the equation drop out since $p_{nn} = 1$ (the self-edge can always be taken) and thus $\overline{p_{nn}} = 0$. For example, given the ordering $L^b_{ng} \le L^n_{ng} \le L^a_{ng}$, Equation 2 simplifies to:

$$E_{ng} = p_{nb}\,L^b_{ng} + \overline{p_{nb}}\,L^n_{ng}. \tag{6}$$

3.2 Algorithms for the ESP Problem

To solve the ESP problem, we have developed variants of two algorithms commonly used to solve Markov decision processes (MDPs) (see Section 4 below). Borrowing the names from the MDP literature, these are *value iteration* (VI) and *policy iteration* (PI). Here we describe our algorithms in terms of the above notation. We then analyze some of their properties in the context of an MDP formulation in the next section.

Value iteration Recall that our goal is to solve the system of $N-1$ equations, each of the form of Equation 3. Collecting the $N-1$ unknowns E_{ng}, $n \neq g$, into a vector v we can write the entire system of equations concisely:

$$v = F(v). \tag{7}$$

Thus, the desired solution v of this system of equations is a *fixed point* of function F. The value iteration algorithm finds this fixed point by treating Equation 7 as an iterative process

$$v^{(n+1)} = F(v^{(n)}). \tag{8}$$

In [3] we have shown that this process converges to a unique fixed point v^* given an initial value $v^{(0)} = 0$, if there exists a path with non-zero probabilities from each node to the goal. Convergence is geometric once the current value is in the vicinity of the fixed point.

For an efficient implementation, we use the observation about orderings made in the previous section. That is, to evaluate Equation 3 for a node n with out-degree k, we first sort the candidate lengths L^i_{ng} in $O(k \log k)$ time, and then evaluate the resulting linear equation in $O(k)$ time. The total time for one iteration of Equation 8 is thus almost linear in the total number of edges in the graph E.

Policy iteration The idea of the second algorithm is to hypothesize edge orderings (as in Equation 4) at all nodes, and to *solve* the resulting linear system of Equation 7. The process is then repeated, using the previous solution to determine the new edge orderings. The iteration terminates when the solution to the current system of equations yields the same orderings as the previous solution. The name of the algorithm, policy iteration, reflects that instead of iterating over the *values* of the expected lengths, we iterate over the *ordering* of the expected lengths of the outgoing paths at each node. Recall that these orderings represent strategies (or policies) for the robot when faced with different visibility scenarios at a node.

Note that while each iteration of the VI algorithm requires evaluating $N-1$ linear equations (which together with the sorting step takes $O(N^2 \log N)$ time for dense graphs), each iteration of the PI algorithm requires *solving* this linear system, requiring a total time of $O(N^3)$ for dense graphs. As we show in Section 5, however, the PI algorithm is competitive since it converges in very few iterations.

Analyzing the convergence properties of the policy iteration algorithm is easiest in the context of a Markov decision process (MDP) formulation, which we present next.

4 MDP Formulation

The ESP problem can be formulated nicely as a Markov decision process (MDP). In brief, a MDP consists of a set of states S, and a set of allowable actions A_s for each state $s \in S$. Each action $\alpha \in A_s$ taken in state s yields a reward $r(s, \alpha)$, and results in a new (random) state s' according to a transition probability distribution $p(\cdot|s, \alpha)$. The objective is to devise a ***policy*** with a stationary *decision rule* $\delta : S \to A_s$ that selects a certain (fixed) action in each state so as to optimize a function of the total reward. This brief discussion ignores many common variations of MDPs, including time-dependent or discounted rewards, and non-stationary policies. For a comprehensive introduction to Markov decision processes, see the book by Puterman [15].

The ESP problem specifically translates into a non-discounted negative expected total-reward MDP. This means that each reward is interpreted as cost or penalty, and that the objective is to minimize the total expected cost. Upon reaching the goal g, no further cost is incurred.

In our case, the set of states S is the collection of landmarks. The concept of an allowable action at a landmark is slightly more complicated: it is not simply a destination landmark the robot should go to (which may not always be visible), but rather a ***strategy*** telling the robot what to do in any visibility scenario. Of course, such strategies correspond to the familiar edge orderings from Equation 4. Thus, the set of allowable actions A_s at landmark s is the set of all orderings among the outgoing edges of s.

For example, suppose the robot is at a node s with three outgoing edges $s \to a, s \to b, s \to c$. Then an example of $\alpha \in A_s$ would be: go to landmark c if possible; otherwise, go to landmark a if possible; otherwise, remain at s (take the self-edge) and try again. We notate this as $\alpha = s : c, a, s$. A robot following this action (strategy) will end up at c, a, or s, depending on the visibility scenario. The transition probabilities are thus

$$\begin{aligned} p(c|s, \alpha) &= p_{sc} \\ p(a|s, \alpha) &= \overline{p_{sc}}\, p_{sa} \\ p(s|s, \alpha) &= \overline{p_{sc}}\, \overline{p_{sa}}. \end{aligned} \tag{9}$$

The expected cost for α is simply the expected length for this action

$$r(s, \alpha) = p_{sc}\, l_{sc} + \overline{p_{sc}}\, p_{sa}\, l_{sa} + \overline{p_{sc}}\, \overline{p_{sa}}\, l_{ss}. \tag{10}$$

Note that $p(i|s,\alpha) = 0$ for any other landmark i since the self-edge can always be taken. We can therefore write the expected cost for action α as

$$r(s,\alpha) = \sum_{i \in S} l_{si}\, p(i|s,\alpha). \tag{11}$$

A decision rule (or policy) δ assigns to each landmark one action (strategy) α. Under a given policy, the robot traverses the graph until it reaches the goal landmark. We assume the action at goal g is always $\alpha = g : g$, and $r(g,\alpha) = 0$. Since we wish to find the expected shortest path, we seek a policy that minimizes the total expected cost, i.e., the sum of the costs accrued by the actions leading to the goal.

We now derive the total expected cost for a given decision rule δ. Let us denote the landmarks $x_1, \ldots, x_N$. Let r_δ be the vector containing the expected cost for the action given by δ at each of the N landmarks. That is, entry s of r_δ is

$$r_\delta[s] = r(s, \delta(x_s)). \tag{12}$$

Let P_δ be the $N \times N$ transition probability matrix under the decision rule δ, i.e., the entry at row s and column d is

$$P_\delta[s,d] = p(x_d | x_s, \delta(x_s)). \tag{13}$$

Now define the j-step transition probability matrix ${}_jP_\delta$ such that ${}_jP_\delta[s,d]$ is the probability of arriving at x_d after j transitions from x_s. The probabilities after $j+1$ steps are then

$$\begin{aligned} {}_{j+1}P_\delta[s,d] &= \sum_{i=1}^{N} {}_jP_\delta[s,i]\, p(x_d|x_i, \delta(x_i)) \\ &= \sum_{i=1}^{N} {}_jP_\delta[s,i]\, P_\delta[i,d], \end{aligned} \tag{14}$$

and therefore ${}_{j+1}P_\delta = {}_jP_\delta\, P_\delta$. Since ${}_1P_\delta = P_\delta$, we get by induction

$${}_jP_\delta = P_\delta^j. \tag{15}$$

The vector of expected costs for the first transition is just r_δ. It is easy to see that the costs for the second transition are $P_\delta\, r_\delta$; for the third, $P_\delta^2\, r_\delta$, etc. The vector v_δ of total expected costs for δ is thus the infinite sum

$$v_\delta = \sum_{i=0}^{\infty} P_\delta^i\, r_\delta. \tag{16}$$

Expanding this sum we get

$$\begin{aligned} v_\delta &= r_\delta + P_\delta\, r_\delta + P_\delta^2\, r_\delta + P_\delta^3\, r_\delta + \ldots \\ &= r_\delta + P_\delta(r_\delta + P_\delta\, r_\delta + P_\delta^2\, r_\delta + \ldots), \end{aligned}$$

and thus

$$v_\delta = r_\delta + P_\delta\, v_\delta. \tag{17}$$

So, given a decision rule δ we can compute the expected lengths by solving

$$(I - P_\delta)\, v_\delta = r_\delta. \tag{18}$$

We now ask the question "which policy results in the shortest total expected lengths?" and present our two algorithms, value iteration and policy iteration, in this MDP formulation. The difference between the two is that VI iterates over the expected lengths v, while PI iterates over the policies δ.

4.1 Value Iteration

The value iteration algorithm computes a sequence of values $v^{(0)}, v^{(1)}, \ldots,$ as follows:

1. Choose an initial $v^{(0)}$, and set $n = 0$.
2. Compute

$$v^{(n+1)}[k] = \min_{\alpha \in A_{x_k}} \left(r(x_k, \alpha) + \sum_{i=1}^{N} p(x_i | x_k, \alpha)\, v^{(n)}[i] \right). \tag{19}$$

3. If $v^{(n+1)}$ is very close to $v^{(n)}$, then terminate and return the policy δ consisting of the minimizing actions α for all nodes. Else, increment n by 1, and go to step 2.

When written in matrix form, Equation 19 becomes

$$v^{(n+1)} = \min_{\delta}(r_\delta + P_\delta v^{(n)}), \tag{20}$$

which parallels Equation 8 in Section 3.2. Our convergence result from [3] guarantees that Equation 20 has a unique fixed point v^*, and that the sequence $v^{(n)}$ converges monotonically to v^* when $v^{(0)} = 0$.

4.2 Policy Iteration

The policy iteration algorithm computes a sequence of policies $\delta^{(0)}, \delta^{(1)}, \ldots,$ as follows:

1. Choose an initial policy $\delta^{(0)}$, and set $n = 0$.
2. Solve $(I - P_{\delta^{(n)}})\, v^{(n)} = r_{\delta^{(n)}}$ for $v^{(n)}$.
3. If $\delta^{(n)}$ is in the set

$$D = \arg\min_{\delta}(r_\delta + P_\delta v^{(n)}),$$

then terminate and return $\delta^{(n)}$. Else choose some $\delta^{(n+1)} \in D$, increment n by 1, and go to step 2.

We now turn to the convergence proof. It turns out that for general non-discounted negative expected reward MDPs, the policy iteration algorithm does *not* always converge to the optimal policy [15]. Given the conditions in our model, however, it does.

We will first prove the following lemma, analog to Proposition 7.3.13 in the text by Puterman [15].

Lemma 1 *For each iteration n in policy iteration,* $v^{(n+1)} \leq v^{(n)}$.

Proof: By step 3 in the policy iteration algorithm, if $\delta^{(n)}$ is in the set D, then $v^{(n+1)} = v^{(n)}$. If $\delta^{(n)}$ is not in D, then

$$r_{\delta(n+1)} + P_{\delta(n+1)} v^{(n)} \leq r_{\delta(n)} + P_{\delta(n)} v^{(n)} = v^{(n)}. \tag{21}$$

Because all entries of $P^k_{\delta(n)}$ are non-negative for every $k > 0$, we can substitute $r_{\delta(n+1)} + P_{\delta(n+1)} v^{(n)}$ for $v^{(n)}$ in the above inequality, giving:

$$r_{\delta(n+1)} + P_{\delta(n+1)}(r_{\delta(n+1)} + P_{\delta(n+1)} v^{(n)}) \tag{22}$$
$$= (I + P_{\delta(n+1)}) r_{\delta(n+1)} + P^2_{\delta(n+1)} v^{(n)} \leq v^{(n)}. \tag{23}$$

So by induction,

$$(\sum_{k=0}^{K-1} P^k_{\delta(n+1)} r_{\delta(n+1)}) + P^K_{\delta(n+1)} v^{(n)} \leq v^{(n)} \tag{24}$$

for any positive K. Since all entries of $v^{(n)}$ are non-negative,

$$\sum_{k=0}^{K-1} P^k_{\delta(n+1)} r_{\delta(n+1)} \leq v^{(n)} \tag{25}$$

for $K > 0$. Taking the limit as K goes to infinity gives

$$\sum_{k=0}^{\infty} P^k_{\delta(n+1)} r_{\delta(n+1)} \leq v^{(n)}. \tag{26}$$

Therefore by Equation 16, $v^{(n+1)} \leq v^{(n)}$. □

We are now prepared to prove our main theorem:

Theorem 1. *Given the above model and an initial policy* $\delta^{(0)}$ *such that* $v_{\delta^{(0)}}$ *is finite, policy iteration converges to a policy* δ *with minimal* v_δ.

Proof: Suppose we start with some $\delta^{(0)}$ such that $v^{(0)}$ is finite. By Lemma 1, for each iteration n, $v^{(n+1)} \leq v^{(n)}$. There are finitely many edges in the graph, so there are only finitely many decision rules (since each decision rule is an ordering of the edges). Since the sequence $v^{(n)}$ is non-increasing and no decision rule appears twice, the algorithm must terminate, because there exists an optimal decision rule. When it does at the n-th iteration, $\delta^{(n)}$ is in $\arg\min(r_\delta + P_\delta v^{(n)})$, so $v^{(n)} = \min(r_\delta + P_\delta v^{(n)})$. Our fixed-point result from

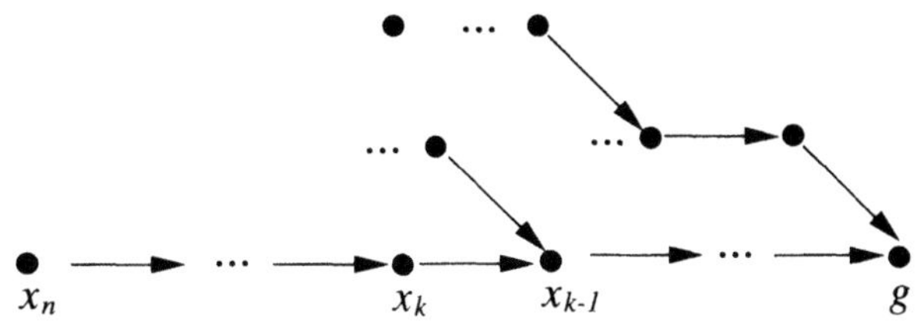

Fig. 2. Every landmark x_n is part of a chain of landmarks that leads to goal g

the VI algorithm then guarantees that we have the optimal decision rule δ. □

The remaining problem is to find an initial policy $\delta^{(0)}$ that yields finite expected costs $v_{\delta^{(0)}}$. Note that there are many policies that yield infinite expected costs. Intuitively, such policies can "trap" the robot within a subgraph from which it cannot reach the goal. The simplest example is a policy containing an action $\alpha = s\colon s$ that commands the robot to always stay at a node s. Circular traps such $s\colon t, s$ and $t\colon s, t$ are also possible.

A policy δ with finite v_δ can easily be constructed, however, using a breadth-first search that starts at the goal node g and follows the edges in reverse direction. Initially, the action $\alpha = s\colon g, s$ is assigned to each node s that has a direct edge to the goal. Recursively, actions $\alpha = s : b(s), s$ are then assigned to other nodes s, where $b(s)$ denotes the node's predecessor in the search (i.e., a node one step closer to the goal).

The above algorithm performs a backwards search through every edge, and thus takes $O(E)$ time, where E is the number of edges in the graph. It assigns actions that impose a (non-circular) tree structure on the graph, with goal g forming the root of the tree (see Figure 2). In particular, each node x_n is part of a chain of nodes leading to the goal

$$x_n \rightarrow \ldots \rightarrow x_k \rightarrow x_{k-1} \rightarrow \ldots \rightarrow g.$$

Assume x_k and x_{k-1} are two successive nodes along that chain. By Equation 17,

$$v[k] = r(x_k, \alpha) + p_{x_k x_{k-1}} v[k-1] + \overline{p_{x_k x_{k-1}}}\, v[k], \tag{27}$$

so

$$v[k] = \frac{r(x_k, \alpha)}{p_{x_k x_{k-1}}} + v[k-1]. \tag{28}$$

Since the probabilities for each landmark are non-zero, and the costs are finite, by induction from the goal landmark g, each $v[k]$ through $v[n]$ is finite. Therefore the above breadth-first search algorithm gives a suitable starting policy $\delta^{(0)}$ for the policy iteration algorithm.

We now have convergence proofs for both VI and PI algorithms, and also an algorithm for constructing an initial policy for PI. To our knowledge, no theoretical results exist about the convergence speed of either algorithm. Thus, in order to assess the practicality of both VI and PI algorithms, we now turn to our empirical evaluation.

5 Experiments

We have implemented both the value iteration and policy iteration algorithms, and have performed an extensive experimental evaluation to assess their respective performance. Our implementation is in C, and we use Matlab for solving the linear system of equations in the policy iteration algorithm.

As a first step in both algorithms we compute the actual shortest paths (ignoring the probabilities) using Dijkstra's shortest path algorithm. This is done for several reasons: First, in a graph that is being constructed by a robot exploring the environment, the goal may not be reachable from all nodes (e.g., from nodes that the robot has only seen but not yet visited). Running Dijkstra's algorithm will identify these nodes, which can then be removed before solving the ESP problem. Second, the lengths of the actual shortest paths are lower bounds on the lengths of the expected shortest paths, and can be used as a better initial estimate for the unknowns E_{ng} for the value iteration algorithm. (The VI convergence proof is based on bounded increasing sequences and extends to any start value that is component-wise smaller than the solution.) Finally, for policy iteration, it turns out that the actual shortest paths provide an alternate way of finding an initial policy that yields finite expected lengths, and thus a solvable system. While the breadth-first search algorithm from Section 4.2 for finding an initial policy is asymptotically faster, extra memory and time are needed to establish back-pointers along all edges before the actual search, which are not required by Dijkstra's algorithm. In practice, we have found that the two methods have very similar running times.

For testing purposes we have generated more than 50,000 graphs with different properties, including number of nodes N, number of edges E, and range of probability values. Graphs used have up to 3000 nodes; graph densities range from sparse ($E \propto N$) to dense ($E \propto N^2$); and probability values are characterized as very low (0.0001–0.001), low (0.0001–0.5), or full (0.0001–1). We have also experimented with different graph structures, including large diameter "corridor" graphs, and "multi-room" graphs (sparsely connected collections of highly connected subgraphs, simulating the visibility graphs of multi-room buildings).

Here we report on a subset of our results. The plots shown in Figure 3 measure the number of iterations necessary for VI and PI as a function of the number of nodes N in the graph. Each plot shows the results of 1250 individual experiments on random graphs with up to 2500 nodes. We have performed such experiments for many different types of graphs and parameters. The plots in Figure 3 contrast sparse graphs ($E \approx N$) with dense graphs

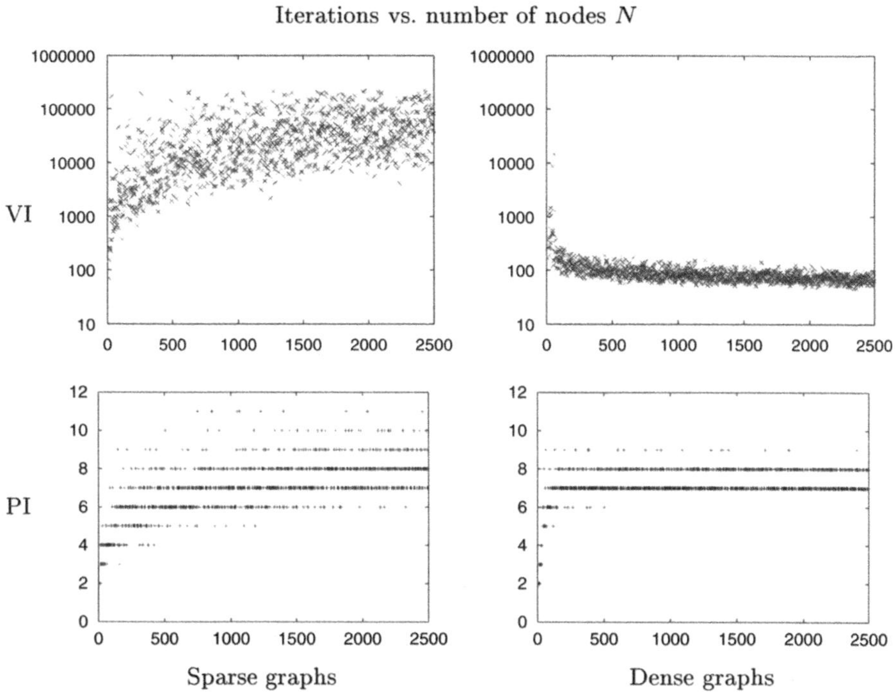

Fig. 3. Number of iterations for VI (first row) and PI (second row) algorithms as a function of the number of nodes in the graph. The plots on the left show results for sparse graphs ($E \approx N$), the plots on the right show results for dense graphs ($E \approx N^2/15$)

($E \approx N^2/15$), both with full random probabilities (ranging from 0.0001 to 1). Examining the plots in the first row of Figure 3, it can be seen that the value iteration algorithm requires large numbers of iterations (often upwards of 10,000) on sparse graphs, while it converges much faster on dense graphs (typically within a few hundred iterations). The likely reason for this is that changes in the variables can propagate much faster in densely connected graphs. Note that the initial "peak" in the second plot is due to our definition of "dense" as $E \approx N^2/15$, which for small N still results in fairly sparse graphs. Interestingly, the average number of iterations for dense graphs does not increase with larger values of N — if anything, there is a slight decrease.

Results for the policy iteration algorithm (shown in the second row of Figure 3) are vastly different. In all cases, PI only takes very few iterations to converge. While the number of iterations increases with the number of nodes, the curves flatten out quickly, and the distributions of iterations for $N = 1000$ are very similar to those for $N = 2500$. Even for the largest N, most sparse graphs take no more than 6–9 iterations to converge; the dense graphs rarely take more than 7 or 8 iterations. In fact, among the more than

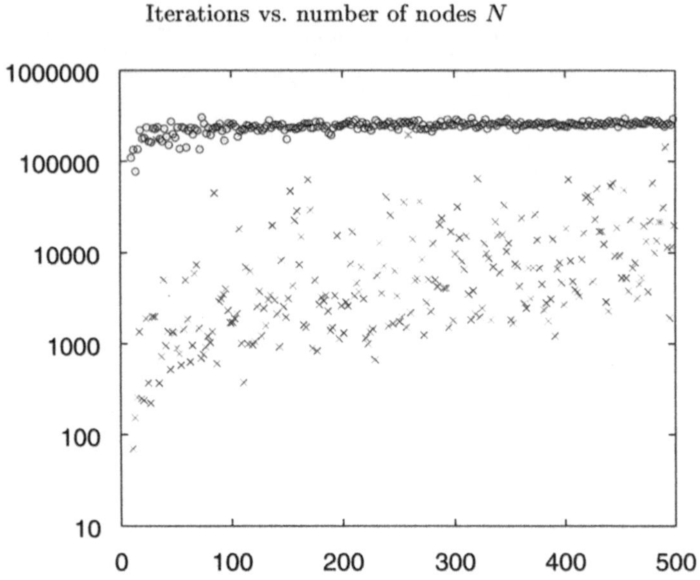

Fig. 4. Number of iterations for VI for normal (red ×) and very low (black ∘) edge probabilities on sparse graphs

50,000 graphs we have generated, only a handful require 11 iterations, only 3 require 12 iterations, and we have yet to find a graph that requires more than 12 iterations. As is the case for VI, the distribution of iterations for dense graphs is narrower and has fewer outliers.

We have also investigated the effect of different probability ranges. The number of iterations used by the PI algorithm remains roughly the same, although a low probability range (0.0001 to 0.5) often requires one extra iteration. The VI algorithm, however, is strongly affected by the probability ranges. This is demonstrated in Figure 4, which shows the number of iterations of the VI algorithm on sparse graphs for two different probability ranges: full (0.0001–1), and very low (0.0001–0.001).

The results of other experiments we have performed are consistent with the ones reported here. In particular, for PI, the curve defined by the number of iterations vs. graph size always resembles a logarithmic shape, indicating that few iterations are necessary even for very large graphs. Also, we have been unable to construct "pathological cases" that require a certain minimum number of iterations for PI. Value iteration, on the other hand, can be made to converge arbitrarily slowly by constructing graphs whose edges all have very low probabilities as shown in Figure 4. Although such graphs may not be common in practical applications, this is further evidence for the unpredictability of the convergence speed of VI that is also exhibited in the top left plot in Figure 3.

We now turn to the actual running times of the two algorithms. Figure 5 shows plots that compare the running times of VI and PI as a function of N on the same sets of graphs as in Figure 3. All experiments were performed under Linux on a 1.53 GHz Athlon machine with 512 MB of memory. It can be seen that policy iteration clearly outperforms value iteration, in particular on sparse graphs, where PI is faster by several orders of magnitude. Much of this performance gain is due to the fact that we use Matlab's sparse matrix representation and equation solver. (Using a regular solver for sparse graphs is slower by at least a factor of 10.) On dense graphs, the margin between PI and VI is much smaller, since VI takes fewer iterations than it does on sparse graphs, but PI's equation solving step requires more work. Even in this case, however, PI is the winner, running at least a factor of 2 faster than VI. It should be noted that the reported running times are process times. When we measure total time, the running times of PI and VI on dense graphs become almost identical. The likely reason for this is that the time for PI's memory allocation steps are not measured in the process time.

Graphs encountered in real environments span the range of the two types of graphs discussed here. In practical applications, sparse graphs will be more common, and PI, although slightly more difficult to implement, will be the better choice.

A practical issue is the memory requirements of the two algorithms. VI, which only needs to *evaluate* the system of equations, requires only enough memory to store the graph. PI, which needs to *solve* the system, needs up to $O(N^2)$ space to store the matrix representing this system. Given a large enough graph, the matrix will be too large to be allocated, making PI impossible to use. For our implementation and hardware, this happens for dense graphs upwards of about 2800 nodes. In such situations, VI would be the only choice. Since we are using Matlab's sparse matrix representation, however, the memory is sufficient for much larger sparse graphs. For example, a sparse graph with 15,000 nodes and about 25,000 edges can still be solved using PI in less than 1 second.

6 Conclusion

In this paper we have defined the ***expected shortest path*** (ESP) problem, which arises in landmark-based robot navigation where the reliability of landmark detection is modeled explicitly. We have presented two algorithms for its solution, *value iteration* (VI) and *policy iteration* (PI), and have proven their convergence properties in the context of a MDP formulation. Using an extensive experimental analysis, we have demonstrated the practicality of both algorithms, and have shown that PI usually outperforms VI. In particular, on sparse graphs, PI is orders of magnitudes faster. On very large dense graphs, however, PI cannot be used due to its larger memory requirements.

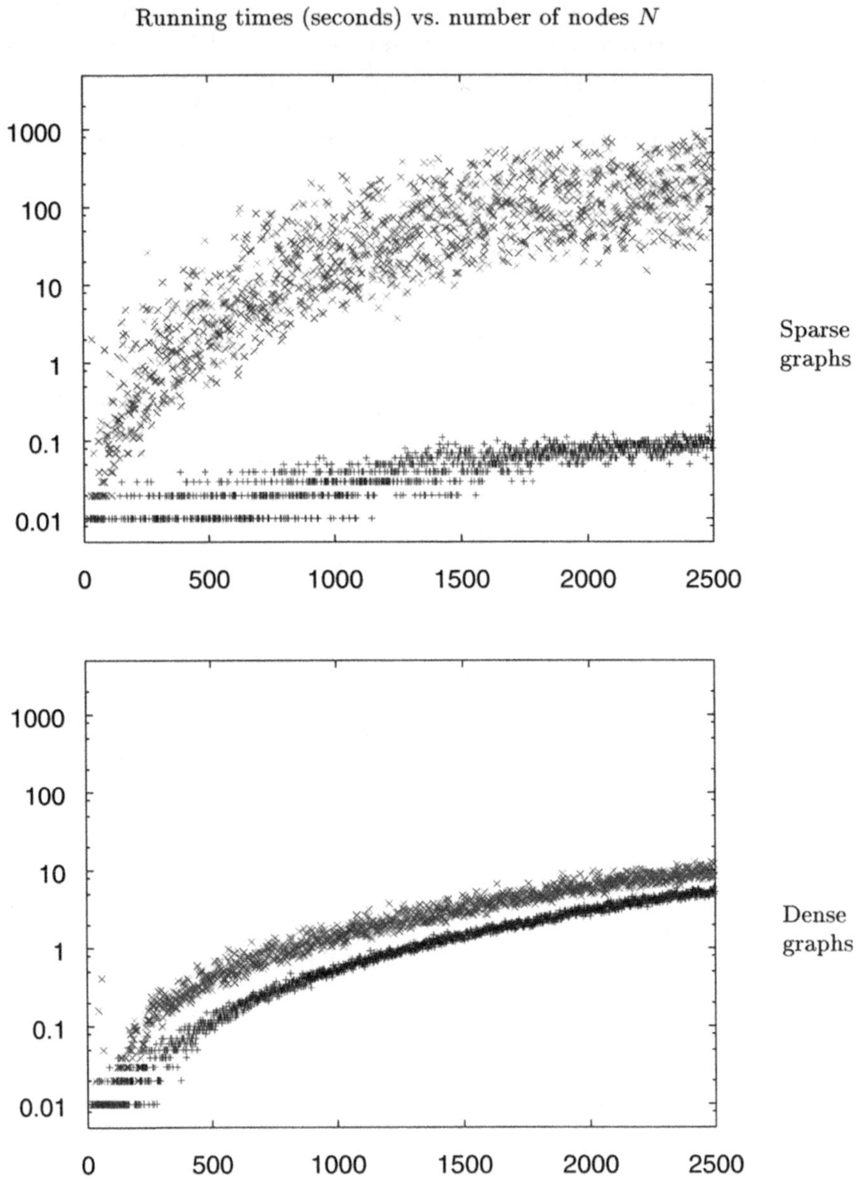

Fig. 5. Running times in seconds for VI (red ×) and PI (blue +) algorithms as a function of the number of nodes in the graph. The first plot shows results for sparse graphs ($E \approx N$); the second plot shows results for dense graphs ($E \approx N^2/15$)

In future work we plan to investigate the application of our algorithms to navigation using natural landmarks extracted from the environment. Building on related work in this area [5,10,12,17], we aim to extend our probabilistic framework to obtain more reliable and efficient path planners for vision-based robot navigation in arbitrary environments.

Acknowledgements.
The authors are grateful for the insights of Steve Abbott and Deniz Sarioz that contributed to this work, and for the assistance of Darius Braziunas, Victor Dan, Cristian Dima, Huan Ding, David Ehringer, Dan Knights, Jeff Lanza, Fafa Paku, and Peter Wall in the implementation of the navigation framework presented here.

Support for this work was provided in part by the National Science Foundation under grants IIS-0118892, CCR-9902032, CAREER grant 9984485, POWRE grant EIA-9806108, by Middlebury College, by the Howard Hughes Medical Institute, and by the Council on Undergraduate Research.

References

1. C. Becker, J. Salas, K. Tokusei, and J.-C. Latombe. Reliable navigation using landmarks. In *Proceedings of IEEE International Conference on Robotics and Automation*, pages 401–406, June 1995.
2. D. M. Blei and L. P. Kaelbling. Shortest paths in a dynamic uncertain domain. In *Proceedings of the IJCAI Workshop on Adaptive Spatial Representations of Dynamic Environments*, 1999.
3. A. Briggs, D. Scharstein, and S. Abbott. Reliable mobile robot navigation from unreliable visual cues. In Donald, Lynch, and Rus, editors, *Algorithmic and Computational Robotics: New Directions, A. K. Peters*, pages 349–362, 2001.
4. A. Briggs, D. Scharstein, D. Braziunas, C. Dima, and P. Wall. Mobile robot navigation using self-similar landmarks. In *Proceedings of the IEEE International Conference on Robotics and Automation (ICRA 2000)*, pages 1428–1434, April 2000.
5. C. Fennema, A. Hanson, E. Riseman, J. R. Beveride, and R. Kumar. Model-directed mobile robot navigation. *IEEE Transactions on Systems, Man, and Cybernetics*, 20(6):1352–1369, 1990.
6. L. Kavraki and J.-C. Latombe. Randomized preprocessing of configuration space for fast path planning. In *Proceedings of IEEE International Conference on Robotics and Automation*, pages 2138–2145, May 1994.
7. A. Lazanas and J.-C. Latombe. Landmark-based robot navigation. *Algorithmica*, 13(5):472–501, May 1995.
8. C. Lin and R. Tummala. Mobile robot navigation using artificial landmarks. *Journal of Robotic Systems*, 14(2):93–106, 1997.
9. M. Mani, A. Zelikovsky, G. Bhatia, and A. Kahng. Traversing probabilistic graphs. Technical Report 990010, UCLA, 1999.

10. B. Nickerson, P. Jasiobedzki, D. Wilkes, M. Jenkin, E. Milios, J. Tsotsos, A. Jepson, and O. N. Bains. The ARK project: Autonomous mobile robots for known industrial environments. *Robotics and Autonomous Systems*, 25:83–104, 1998.
11. M. H. Overmars and P. Švestka. A probabilistic learning approach to motion planning. In Goldberg, Halperin, Latombe, and Wilson, editors, *1994 Workshop on the Algorithmic Foundations of Robotics, A. K. Peters*, pages 19–37, 1995.
12. C. Owen and U. Nehmzow. Landmark-based navigation for a mobile robot. In *Proceedings of Simulation of Adaptive Behaviour.* MIT Press, 1998.
13. C. H. Papadimitriou and M. Yannakakis. Shortest paths without a map. *Theoretical Computer Science*, 84:127–150, 1991.
14. J. S. Provan and M. O. Ball. The complexity of counting cuts and of computing the probability that a graph is connected. *SIAM Journal of Computing*, 12(4), November 1983.
15. M. Puterman. *Markov Decision Processes: Discrete Stochastic Dynamic Programming.* John Wiley & Sons, New York, NY, 1994.
16. D. Scharstein and A. Briggs. Real-time recognition of self-similar landmarks. *Image and Vision Computing*, 19(11):763–772, September 2001.
17. R. Sim and G. Dudek. Mobile robot localization from learned landmarks. In *Proceedings of IEEE/RSJ Conference on Intelligent Robots and Systems (IROS), Victoria, BC*, October 1998.
18. R. Simmons, J. Fernandez, R. Goodwin, S. Koenig, and J. O'Sullivan. Xavier: An autonomous mobile robot on the web. *IEEE Robotics and Automation Magazine*, 1999.
19. R. Simmons and S. Koenig. Probabilistic navigation in partially observable environments. In *Proceedings of the International Joint Conference on Artificial Intelligence*, pages 1080–1087, 1995.
20. C. J. Taylor and D. J. Kriegman. Vision-based motion planning and exploration algorithms for mobile robots. In Goldberg, Halperin, Latombe, and Wilson, editors, *1994 Workshop on the Algorithmic Foundations of Robotics, A. K. Peters*, pages 69–83, 1995.
21. S. Thrun, M. Beetz, M. Bennewitz, W. Burgard, A. B. Cremers, F. Dellaert, D. Fox, D. Hahnel, C. Rosenberg, N. Roy, J. Schulte, and D. Schulz. Probabilistic algorithms and the interactive museum tour-guide robot Minerva. *International Journal of Robotics Research*, 19(11), 2000.
22. L. G. Valiant. The complexity of enumeration and reliability problems. *SIAM Journal of Computing*, 8(3), August 1979.
23. F. Zanichelli. Topological maps and robust localization for autonomous navigation. In *Proceedings of the International Joint Conference on Artificial Intelligence, Workshop ROB-2*, 1999.

Sensor-Based Coverage: Incremental Construction of Cellular Decompositions

Ercan U. Acar and Howie Choset

Carnegie Mellon University, Pittsburgh PA 15213, USA

Abstract. Coverage path planning determines a path that passes a robot, a detector, or some type of effector over all points in an environment. Previous work in coverage used cellular decompositions to pass a robot over all points in a target space. We defined a class of cellular decompositions termed *Morse Decompositions* which were defined by Morse functions; varying the Morse function changes the pattern by which the space was covered. In this paper, we prescribe a provably complete sensor-based coverage path planner that can accommodate detectors with variable effective detecting ranges that go beyond the robot's periohery. In vast open spaces, the robot can use the full range of its detector, so we cover these spaces as if the robot were as big as the detector. In narrow and cluttered spaces, we obstacles lie within the detector range and thus the detection range "fills" the surrounding area; in this case, the robot simply follows the generalized Voronoi diagram in the cluttered spaces. Therefore, we introduce a new hierarchical decomposition that combines Morse decompositions and generalized Voronoi diagrams. We show how to use this hierarchical decomposition to cover unknown spaces.

1 Introduction

Coverage path planning determines a path that directs a robot to pass over all points in its free space. Integrating the robot's footprint (detector range) along the coverage path yields an area identical to that of the target region. We achieve coverage by introducing new types of *exact cellular decompositions.* Exact cellular decompositions [8], [18] represent the free space by dividing it into non-overlapping regions called *cells* such that adjacent cells share a common boundary, the interior of each cell intersects no other cell, and the union of all of the cells fills the free space. An adjacency graph encodes the topology of the cell decomposition where nodes represent the cells and edges connect nodes corresponding to adjacent cells. Traditionally, the problem of determining a path between two points, i.e., navigation, can be solved in two steps: first identify the cells containing the start and the goal, and then search the adjacency graph for a sequence of cells that connect the start cell to the goal cell.

We can use cellular decompositions to produce a path that *covers* the free space. We define cells such that each cell can be covered via a

J.-D. Boissonnat et al. (Eds.): Algorithmic Foundations of Robotics V, STAR 7, pp. 399–415, 2004.

simple pattern, say simple back-and-forth motions. A planner achieves complete coverage simply by ensuring that the robot identifies and visits each cell in the region, i.e., finds an exhaustive walk through the adjacency graph.

In previous work, we define a family of cellular decompositions termed *Morse Decompositions* whose cells are defined by critical points of a Morse function. Our work is motivated by Canny's roadmap work [5] where he uses critical points to ensure the connectivity of his roadmap. The critical points are useful because topologically meaningful events occur at them. Recall that Morse functions are real-valued functions with non-degenerate critical points. The Morse decompositions allowed us to design sensor-based coverage algorithms that use simple motions such as back and forth, for coverage with a robot-size detector. Varying the Morse function that defines the decomposition changes the pattern by which coverage is acheived.

We can consider two ends of the coverage spectrum: coverage with detectors the same size of the robot and infinite range detectors. The Morse decompositions are suitable for coverage with detectors that are the same size of the robot; we call these *robot-size* detectors. If the robot has an infinite range detector, then it need not use Morse decompositions. Instead, the robot can follow the generalized Voronoi diagram (GVD) (sets of points equidistant to two obstacles [10,21]) to completely cover a bounded free space.

In this paper, we address a problem in the middle of the extremes. This paper describes a new method to cover unknown spaces with detectors whose ranges are larger than the robot, but still less than infinite. We term these *extended-range detectors.* In a sense, an extended-range detector has a "variable" effector size: in a vast open space, the maximum range of the detector is the effector size. Covering a vast open space with back and forth motions, as if the robot were as big as its maximum detector range, produces an "good" coverage path when the objective is to minimize path length with respect to area covered. In narrow spaces, however, the detector's range is limited by the presence of obstacles. As suggested in previous work of Hert*et al.* [15] and then Lumelsky *et al.* [19], performing simple back and forth motions, again as if the robot were as big as its maximum detector range, does not produce an efficient path (Fig. 2). When the robot is in a cluttered or narrow space, its detector is effectively infinite and thus can cover the space as if it had an infinite detector. See Fig. 1 for an overview of

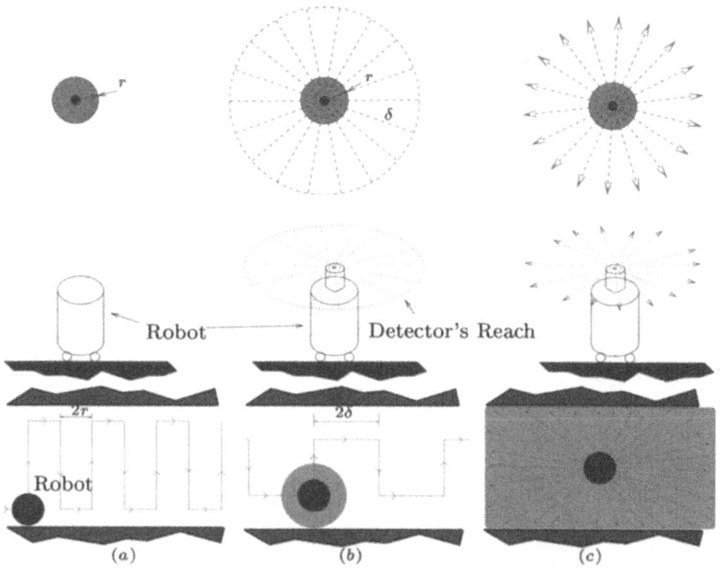

Fig. 1. (*a*) Robot with a detector as wide as the robot's width. (*b*) Robot with a detector that has extended-range δ. (*c*) Robot with an infinite-range detector. The top row shows the top views of the robot and the detector range. The bottom row shows coverage paths with respective detector ranges for the same area. Note that in (*c*) the robot does not need to move to "detect" everything in this example.

three modes of coverage: robot-size detector, extended-range detector, and infinite range detector.

We introduce a hierarchical decomposition for coverage with extended-range detector. This decomposition combines Morse decompositions and generalized Voronoi diagrams. We divide the robot's free space into two regions: vast regions where obstacles are outside of the detector range of the robot and narrow regions where obstacles are within the detector range of the robot. For the vast regions, we use Morse decomposition of the space but "pretend" that the size of the robot is the detector range of the robot. In the narrow regions, we simply have the robot follow the generalized Voronoi diagram to cover the unknown space. We also introduce a method that uses sensor data to interchange coverage modes from vast to narrow. This method allows the robot to locate vast/narrow regions while it is covering narrow/vast regions in an unknown space.

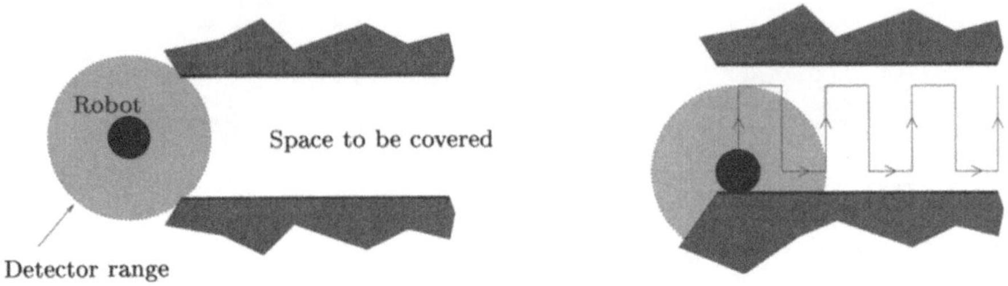

Fig. 2. The robot is required to pass the extended-range detector over all the points in the corridor. Even though the robot can perform back and forth motions to cover the space, these motions will not make efficient use of the available detector range.

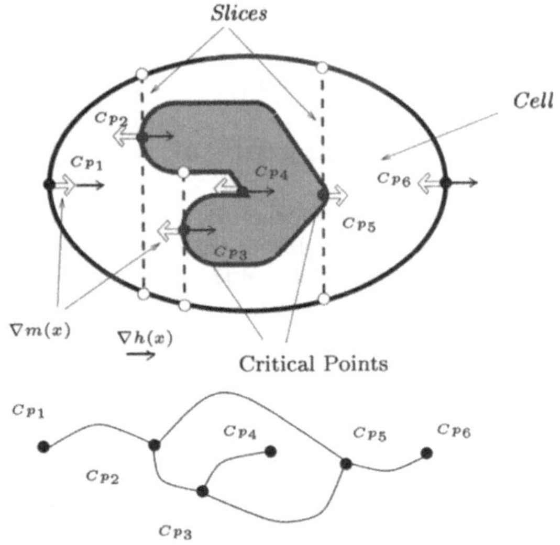

Fig. 3. Exact cellular decomposition of a space and its Reeb graph. At the critical points of $h|_{\partial\mathcal{C}\mathcal{C}}$, the gradient of h $\nabla h(x)$ and the surface normals $\nabla m(x)$ are parallel. The cell boundaries have two parts: slices that contain critical points and portions of the obstacle boundaries. The Reeb graph with nodes as critical points and edges as cells represents the topology of the decomposition and enables as to reduce the sensor-based coverage to an incremental construction procedure.

2 Coverage with a Robot-Size Detector: Prior Work in Farming Patterns

We believe that cellular decompositions can be used to describe previous coverage work with robot-size detectors. Cao *et al.* [7] implicitely use a Morse decomposition but they assume all obstacles are convex. Butler *et al.* [4] present a coverage algorithm that uses decompositions

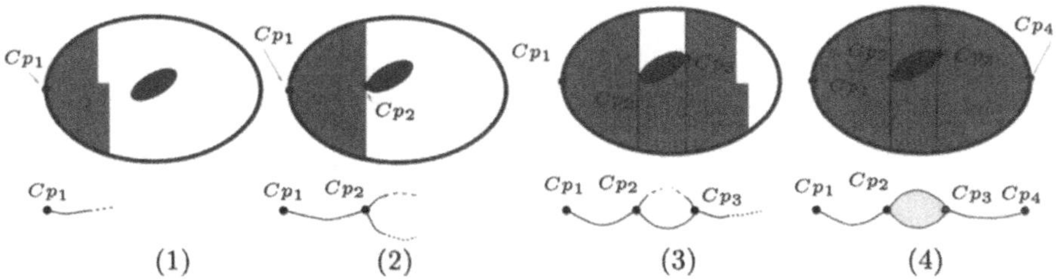

Fig. 4. Incremental construction of the graph while the robot is covering the space.

of rectilinear spaces where all critical points are degenerate. In our work, we only consider non-degenerate spaces. Grid-based methods can be thought of cellular decompositions with very fine cells. Such a method is presented by Zelinsky *et al.* [24] where they generate a coverage path using a path transform which can be regarded as a numeric potential field.

We achieve coverage with a robot-size detector using a cellular decomposition called the *boustrophedon cellular decomposition* [12], where coverage in each cell can be achieved by performing back and forth motions with an inter-lap spacing equal to robot's diameter. Since coverage in each cell is trivial, visiting each cell ensures complete coverage. Essentially, the boustrophedon decomposition provides the bookkeeping to ensure back-and-forth motions indeed fully cover the free space.

We define the cells by passing a slice (a line segment), the pre-image of a real-valued function $h(x) = x_1$, termed a *slice function*, i.e., $\mathcal{CS}_\lambda = \{x \in \mathcal{CS} | h(x) = \lambda, \lambda \in \Re\}$ ($\mathcal{CS}$ denotes the configuration space[1] of a disk robot) through the free space. This approach relies on Canny's slicing method [5], [6] where he uses the connectivity changes of the slice to connect his roadmap. In our work, we use connectivity changes to locate the cell boundaries. The "left-most" and "right-most" boundaries of the cells occur at slices where the connectivity of the slice in the free space changes. These connectivity changes occur at the *critical points*[2] of $h|_{\partial\mathcal{C}\mathcal{C}}$, the restriction of the slice function to the obstacle boundaries, $\partial\mathcal{C}\mathcal{C}$. We assume that all the critical points of the function $h|_{\partial\mathcal{C}\mathcal{C}}$ are non-degenerate, and thus is a Morse function [20]. At

[1] Even though we present results in the configuration space of a disk robot, our results are valid in the workspace too. Essentially we achieve coverage in the configuration space using the workspace distance measurements.

[2] At the critical points of a function, its first derivative vanishes.

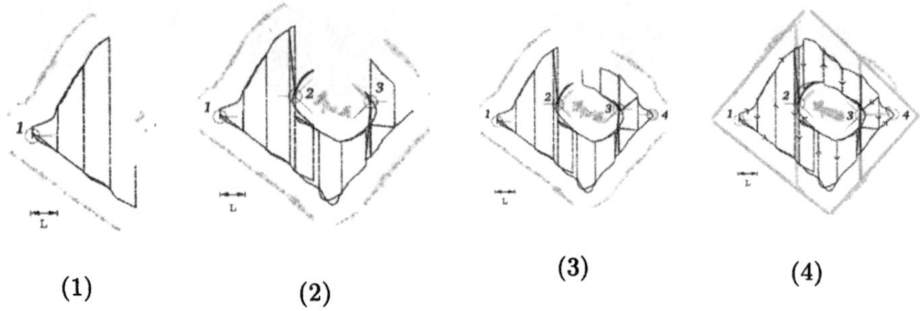

Fig. 5. Four stages of coverage in an unknown environment with a robot-size detector. The coverage path followed by the robot is shown by dotted black lines. We depict the critical points as circles with lines emanating from them. The lines represent the directions of the corresponding adjacent cells. The robot incrementally constructs the graph representation by sensing the critical points $1, 2, 3, 4, 3, 2$ (in the order of appearance) while covering the space. In the final stage (4), since all the critical points have explored edges, the robot concludes that it has completely covered the space. For the sake of discussion, we outlined the boundaries of the obstacles and cells in (4). $L = 0.53[m]$.

the critical points of $h|_{\partial\mathcal{C}\mathcal{C}}$, the gradient of h, $\nabla h(x)$ and the surface normals $\nabla m(x)$ of the obstacles are parallel [3,9].

We use the critical points to form cells and a graph representation that encodes the topology of the cellular decomposition (Fig. 3). This graph representation, that has nodes as critical points and edges as cells, enables us to reduce the sensor-based coverage with a robot-size detector to an incremental construction procedure. To realize this procedure with range sensors, we developed a critical point sensing method and a coverage algorithm that guarantees that the robot will encounter all critical points in an unknown space (hence achieve complete coverage) [1,2].

In Fig. 4-(1), the robot starts to cover the space at the critical point Cp_1 and it instantiates an edge with only one node. When the robot is done covering the cell between Cp_1 and Cp_2, it joins their corresponding nodes with an edge in the graph representation (Fig. 4-(2)). Now the robot has two new uncovered cells. Since the space is *a priori* unknown, the robot arbitrarily chooses the lower cell to cover. When the robot reaches Cp_3, nodes of Cp_2 and Cp_3 become connected with an edge and the lower cell is completed (Fig. 4-(3)). At Cp_3, the robot decides to cover the cell to the right of Cp_3. When the robot senses Cp_4, it goes back to Cp_3 and starts to cover the upper cell. When it comes back to Cp_2, it determines that all the edges of all the nodes (critical points) have been explored (Fig. 4-(4)). Thus the robot concludes that it has completely covered the space. This incremental construction method serves a basis for the sensor-based coverage algorithm with a robot-size detector. Fig. 5 shows different stages of this incremental construction in *a priori* unknown $2.75 \times 3.65[m^2]$ room with a Nomad mobile robot that has a sonar ring.

3 Coverage with Robot-Size Detector: Different Patterns

All of the figures and examples in the previous section assumed a slice function $h(x, y) = x$, which emits the boustrophedon decomposition and its simple farming coverage pattern, back and forth motions. Without causing too much of a distraction, we wanted to point out that by considering different "shapes" of slices, a planner can produce different decompositions, which in turn yeilds different patterns to cover a space. In this section, we simply show how different functions give rise to different patterns and then return to covering spaces with extended range detectors in the subsequent sections.

The function $h(x) = \sqrt{\sum_i (x_i)^2}$ emits a pattern of concentric circles in the plane. The critical points occur at points where the circle is tangent to an obstacle. Critical points are then used to form annular-arc shaped cells (Fig. 6). Since the slice, here, is not a straight line segment but rather a circle or subset of a circle, the planner initially directs the robot to circumnavigate a circle, move the inter-lap distance along the radius of the circle and circumnavigate a circle of a larger radius again. If the robot encounters an obstacle while circumnavigating a circle, the planner simply directs the robot to follow the obstacle boundary until the robot has moved an inter-lap distance in the radial direction of the decomposition. Instead of following a circle and stepping outward, the robot can follow a spiral pattern until it encounters critical points (Fig. 6). The spiral pattern bypasses the need to step along the radial direction. This yields a path that maximizes the area covered per unit distance traveled in regions sparsely populated with obstacles. The function $h(x) = |x|_1$ produces cells that look like rotated squares or diamonds (Fig. 7). We can achieve coverage here by driving in concentric squares or following square-like spirals which we term *squarels*.

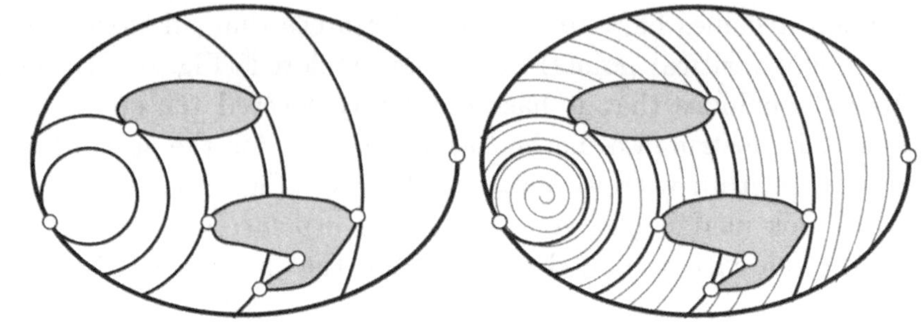

Fig. 6. Cellular decomposition for $h(x) = \sqrt{(x_1)^2 + (x_2)^2}$ and an associated spiral coverage pattern. The slices are the circles that are the pre-images of h. At the critical points, demarked with little circles (not to be confused with the slices), the circle-shaped slices become tangent to the obstacles. Rather than moving along circular paths and stepping outward, the robot follows a spiral pattern.

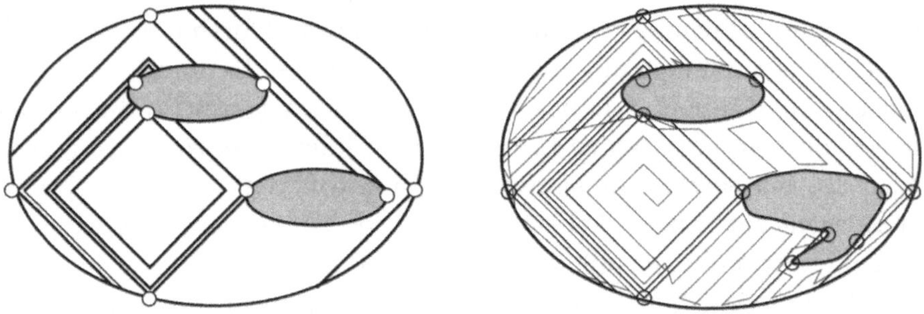

Fig. 7. Decomposition for $h(x) = |x|_1$ and a coverage pattern. Squares are the slices and at the critical points the corner of a square touches an obstacle and/or side of it becomes tangent to an obstacle. Since it is easier for the robot to move along straight lines rather than circles, this pattern can be used to approximate the spiral pattern.

4 Coverage with an Infinite-Range Detector: Generalized Voronoi Diagrams

To cover an unknown space with an infinite-range detector, we use generalized Voronoi diagram. The GVD comprises GVD edges and nodes termed *meet points*. Each GVD edge is the set of points equidistant to two obstacles, i.e., $\{x \in \Re^2 | d_i(x) = d_j(x) \leq d_h(x)\ \forall h,\ \nabla d_i(x) \neq \nabla d_j(x)\}$, where $d_i(x)$ measures the distance to the closest point on obstacle C_i and $\nabla d_i(x)$ is its gradient. At meet points where GVD edges "meet", there are more than two equidistant obstacles. In the context of conventional motion planning, the robot can use the GVD to plan a path from a start to a goal by first planning a path from the start to the GVD, then along the GVD edge to the vicinity of the goal, and then from the GVD to the goal. See Fig. 8 for the generalized Voronoi diagram of a simple space and a sample path between start and goal locations.

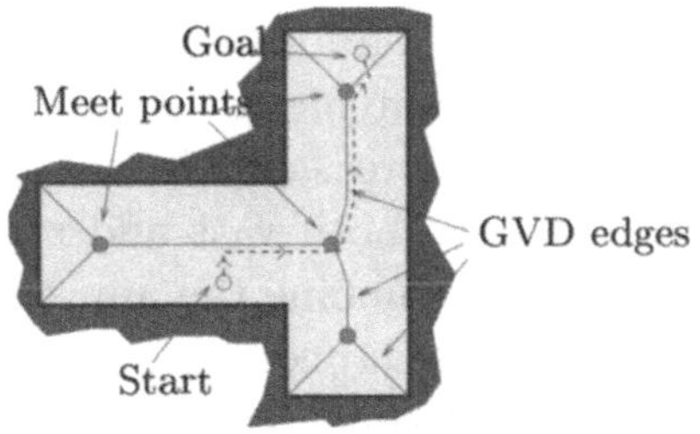

(a)

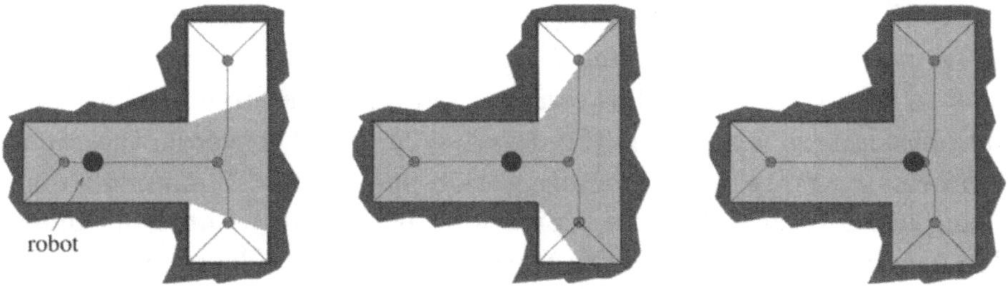

(1) (2) (3)

Fig. 8. (*a*) The robot uses the GVD to plan a path from a start to a goal by first planning a path from start to the GVD, then along the GVD edge to the vicinity of goal, and then from the GVD to goal. (1) The robot starts to cover the space with an infinite-range detector by following the GVD edges from right hand-side. In (2) almost two thirds of the space is covered. In (3), the free space is completely covered.

In the context of coverage, if the robot follows all the GVD edges, then it is ensured to pass an infinite-range detector over all points in the free space. Therefore, covering an unknown region with an infinite-range detector can be achieved by incrementally constructing the GVD ([10,11]). Fig. 8 depict stages of coverage with an infinite-range detector in a simple T-shaped room.

5 Hierarchical Decomposition in Terms of Generalized Voronoi Diagrams and Morse Decompositions

By following the GVD in narrow sub-regions of a space, the robot guarantees that it will pass an extended-range detector over all points in the narrow region. This is because in narrow regions obstacles lie within the detector range. However, in vast open regions, the robot cannot ensure complete coverage by following the GVD edges because there can be obstacles located out of the detector range. On the other hand, the robot can ensure complete coverage of the vast open regions by executing the algorithms presented in the previous chapters. To cover environments that have both vast and narrow regions with an extended-range detector, we combine the approaches described in the previous section and chapters. Our combined approach treats the space as the union of two types of cells: VAST and NARROW. A VAST-cell is the set of all points such that the distance between any point in the cell and the obstacles is greater than the detector range, i.e., $\{x \in \Re^2 | D(x) > \delta\}$ where $D(x)$ is the minimum distance between point x and the obstacles and δ is the extended detector range. Note that in VAST-cells the robot can perform back and forth motions as if its physical size is as big as the detector 2δ. In other words, the robot performs coverage in free configuration space of a 2δ diameter disk using our coverage algorithms presented before but assumes an inter-lap spacing of δ as opposed to r (setting $r = \delta$) via lapping and wall following motions (Fig. 9). A NARROW-cell is the union of balls with δ radius and centers, y, located on the GVD such that the distance from y to obstacles is less than δ, i.e., $\bigcup B_\delta(y)$ such that $y \in GVD$ and $D(y) \leq \delta$. When the robot is located in a NARROW-cell, obstacles lies within the detector range δ. Therefore, the detector acts like an infinite-range detector and we can simply have the robot follow the GVD edges for complete coverage.

In unknown spaces, the robot starts to cover the space in a NARROW or VAST cell without knowing the existence of other cells. Therefore, while the robot is covering the space, it has to identify features of the cells that indicate the existence of new VAST or NARROW cells. For this purpose, we use *cusp points*. At cusp points, the surface normal to the closest obstacle in the free configuration space, $\nabla m(x)$ is ill defined, either because the obstacle is not smooth, not convex, or there is a sudden change in the closest obstacle. (Fig. 10). Recall that the robot covers a VAST-cell by performing boundary following and

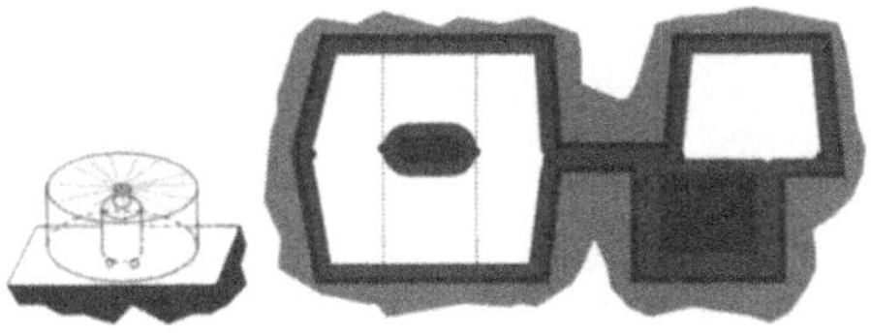

Fig. 9. The robot is assumed to be as big as the extended-range detector to cover vast open spaces. The configuration space of the detector size robot is disconnected even though the configuration space of the robot itself is not.

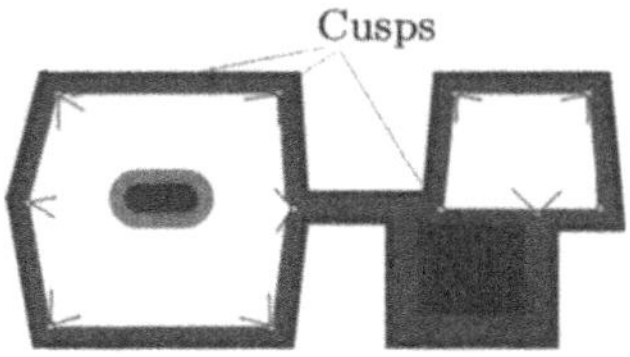

Fig. 10. Configuration space for a 2δ diameter disk detector. The space becomes disconnected and the corridors are not accessible to the 2δ disk. At cusp points, the boundary of the configuration space is non-smooth.

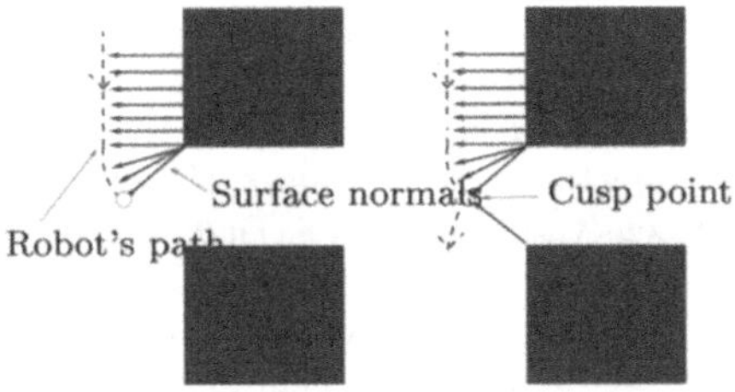

Fig. 11. The robot determines the location of the cusp point by sensing the jump in the surface normal while it is following the boundary of the obstacle.

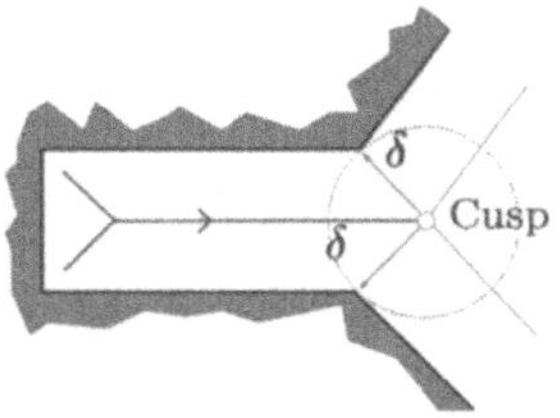

Fig. 12. The robot determines the location of the cusp point while it is following the GVD edge by sensing the increase in the minimum distance to the obstacles ($d_i(x) > \delta$).

lapping motions. In this work, we restrict the robot to look for cusp points while it is following the obstacle boundaries in VAST-cells or GVD edges in NARROW-cells.

During boundary following motions, the robot looks for cusp points that are characterized by jumps in the surface normal of the obstacles, a discontinuity in $\nabla m(x)$. Since we can determine surface normals using range sensors, the robot can easily determine the jumps in the surface normal using its sensors (Fig. 11). To determine cusp points while covering narrow regions, the robot uses the magnitude of the distance measurements. Recall that the robot covers narrow regions by following the GVD edges and thus the minimum distance to the obstacles is always less than the detector range. Therefore, in a NARROW-cell the robot looks for an increase in the minimum distance. When the minimum distance to the obstacles is greater than the detector range, the robot locates the cusp point (Fig. 12).

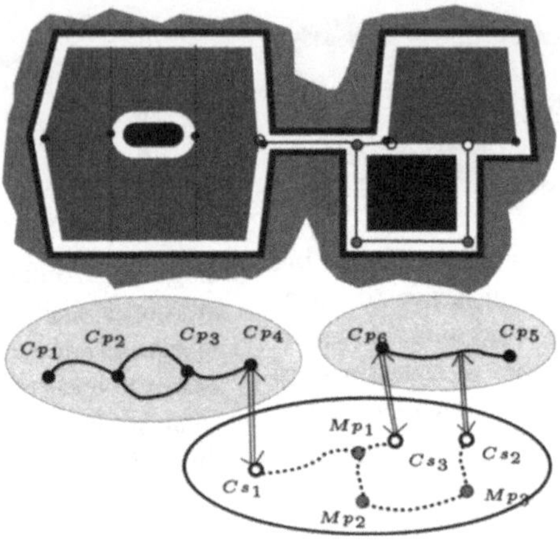

Fig. 13. In the graph, gray ellipses depict the VAST-cells that contain VAST-subcells represented as solid edges. Each VAST-subcell has two associated critical points of $h|_{\partial cc}$ represented as black dots. NARROW-cell is represented by the white ellipse and it contains the NARROW-subcells depicted as dashed edges. Hollow dots correspond to cusp points and gray dots represent the meet points. Double arrows show the links between the NARROW-cells and their neighboring VAST-cells. In this particular example, some of the NARROW-cells such as the ones at the corners, are very short, therefore for the sake of clarity, we do not show them in the graph.

We encode the entire information that the robot needs for coverage with extended-range detectors in a hierarchical graph representation. Just like before, with this graph representation in hand, we not only simplify the data representation but also reduce sensor-based coverage to an incremental construction procedure. The hierarchical graph representation has mainly two elements; gray ellipses that represent VAST-cells that contain VAST-subcells and white ellipses that correspond to NARROW-cells that contain NARROW-subcells (Fig. 13). We represent cusp points that connect VAST-cells to NARROW-cells with hollow dots and use double arrow lines to depict the connection. Each VAST-cell contains VAST-cells represented by the Reeb graph. Recall that in the Reeb graph, the cells correspond to edges and nodes depict the critical points. Note that within each gray ellipse the Reeb graph is connected. Each NARROW-cell contains NARROW-subcells depicted by dashed edges. NARROW-subcells are connected to each other via meet points where there are more than two equidistant obstacles. Each meet point is represented by a gray dot.

6 Incremental Construction of the Hierarchical Decomposition

We demonstrate incremental construction of the hierarchical decomposition while the robot is covering its free space using an example depicted in Fig. 14. The robot starts to cover the space in a VAST-cell by executing our prior coverage algorithm from left-most corner that corresponds to the critical point Cp_1[3].

When the robot reaches the critical point Cp_2 that is on the left-most side of the middle obstacle, it finishes covering the VAST-subcell between Cp_1 and Cp_2 (Fig. 14-1). At this point, since the space is unknown *a priori*, the robot arbitrarily chooses the lower VAST-subcell among the two new VAST-subcells to cover. The robot finishes the new VAST-subcell when it senses the critical point Cp_3 that is on the right-most side of the obstacle and starts to cover the VAST-subcell to the right side of the obstacle (Fig. 14-2).

While the robot is covering this subcell, it locates the cusp point Cs_1 and instantiates a NARROW-cell that contains a NARROW-subcell (Fig. 14-3). The VAST-cell (gray ellipse) and the NARROW-cell (white ellipse) are connected via the cusp point represented by the double arrow line. Note that in this example the cusp point and the critical point Cp_4 occur at the same location but in general this is not the case. The robot finishes left room by covering the final VAST-subcell on the top of the obstacle and coming back to the cusp point Cs_1 (Fig. 14-4). Starting at Cs_1, the robot begins to follow the GVD edge along the corridor.

While the robot is following the GVD edge in the corridor, it encounters meet point Mp_1, shown as a gray dot in the graph that identifies two new NARROW-subcells. The robot chooses the lower NARROW-subcell and follows the GVD edge in the corridor until it reaches the other end of the corridor which is the cusp point Cs_2[4].

[3] Note that Cp_1 is also a cusp point. Therefore the robot instantiates a NARROW-cell. However in this particular example, since the NARROW-cell is very short, for the sake of explanation, we do not show it in the graph, but the robot definitely covers the NARROW-cell. The robot also covers other short NARROW-cells associated with cusp points on the outer boundary whenever it senses them, but again for the sake of discussion, we do not depict them.

[4] Note that at meet points Mp_2 and Mp_3, the robot instantiates very short new NARROW-subcells and covers them on the spot. However we do not depict them for the sake of explanation.

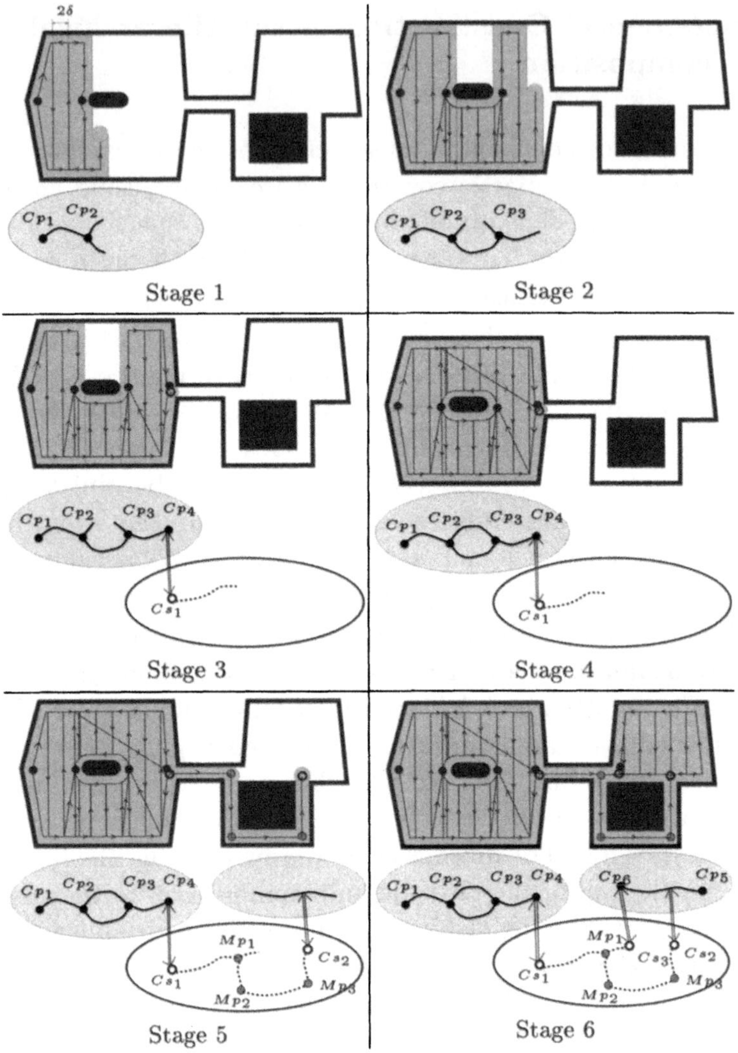

Fig. 14. Depiction of the stages of the incremental construction of the hierarchical decomposition while the robot is covering the space.

At the end of the corridor the robot senses the cusp point Cs_2 and determines the existence of a new VAST-cell (Fig. 14-5). In the graph, it instantiates a gray ellipse for this new VAST-cell and a solid edge without nodes for the new VAST-subcell. Again this new VAST-cell is connected to the NARROW-cell via the cusp point. The robot starts to follow the outer boundary toward right starting from the cusp point Cs_2 until it senses the critical point Cp_5. Then the robot starts to cover the VAST-subcell (toward left). When the robot senses the critical point Cp_6 that is also the cusp point Cs_3, the robot finishes covering the right room and starts to follow the corridor toward left. When the robot comes back to Mp_1, the robot determines that it has visited all branches of the graph and hence has passed the extended-range detector over all points in the space.

7 Conclusion

We have been tasked by the Navy to develop inexpensive de-mining robots that require minimal computational power. The methods presented in this paper are all designed for this purpose. We presented decompositions of planar spaces that are suitable for coverage of unknown spaces with variable size detectors. First, we introduced a family of exact cellular decompositions, termed *Morse decompositions*, that are suitable for coverage with detectors that are as wide as the robot. We started to formulate the Morse decompositions with conventional slice algorithms where the slice is defined by the pre-image of a real-valued function whose restriction to obstacle boundaries does not have any degenerate critical points, i.e., it is Morse. We used the critical points to form the cell boundaries such that the structure of each cell enables a planner to use simple motions such as back-and-forth for coverage. We furthered this approach by altering the Morse function, and hence the slice geometry, to obtain a variety of coverage patterns.

Coverage tasks such as looking for people with a thermal camera require a coverage algorithm that can handle variable range detectors. We combined coverage with robot-size and infinite-range detectors to prescribe a hierarchical decomposition to achieve coverage with extended-range detectors. We formed two types of cells: VAST and NARROW. Our algorithms ensured coverage of each type of cell. We linked these cells using cusp points that lie on the boundaries of both types of cells. These cusp points are easy to sense with range sensors such as sonars. Therefore, a mobile robot equipped with a range sensing system such as a sonar ring can cover the space completely with extended-range detectors. By using Morse decompositions and Voronoi diagrams for coverage with extended-range detectors, we were able to demonstrate that one can take advantage of the strong sides of two different structures. Voronoi diagrams are more useful in narrow spaces than in vast open spaces. Whereas in vast open spaces Morse decompositions are more useful than the Voronoi diagrams for coverage. By introducing switching methods, we are able to combine these two structures for coverage with extended-range detectors. In the future, we are going to investigate optimality issues related to coverage path planning problem. In this paper we only considered coverage with omni-directional detectors. As a part of future work, we are planning to develop coverage algorithms for detectors with limited angle of view.

Finally, we would like to note that this paper mainly focuses on complete, but not optimal coverage of unknown environments. For reference on optimal coverage algorithms that work in discrete or known spaces see [22], [23], [14], [16].

References

1. E.U. Acar and H. Choset. Critical point sensing in unknown environments. In *Proc. of IEEE ICRA'00, International Conference on Robotics and Automation*, pages 3803–3810, San Francisco, CA, 2000.
2. E.U. Acar and H. Choset. Sensor-based coverage of unknown environments: Incremental construction of morse decompositions. *To appear in The International Journal of Robotics Research*, 2002.
3. E.U. Acar, H. Choset, A.A. Rizzi, P.N. Atkar, and D. Hull. Morse decompositions for coverage tasks. *To appear in The International Journal of Robotics Research*, 2002.
4. Z. Butler, A. A. Rizzi, and R. L. Hollis. Contact-sensor based coverage of rectilinear environments. In *Proc. of IEEE Int'l Symposium on Intelligent Control*, Sept. 1998.
5. J.F. Canny. *The Complexity of Robot Motion Planning.* MIT Press, Cambridge, MA, 1988.
6. J.F. Canny and M. Lin. An opportunistic global path planner. In *Proc. IEEE Int. Conf. on Robotics and Automation*, pages 1554–1561, 1990.
7. Z. L. Cao, Y. Huang, and E. Hall. Region filling operations with random obstacle avoidance for mobile robots. *Journal of Robotic Systems*, pages 87–102, February 1988.
8. B. Chazelle. Convex partition of polyhedra: A lower bound and worst-case optimal algorithm. *SIAM Journal on Computing*, 13(3):488–507, 1984.
9. H. Choset, E. Acar, A. Rizzi, and J. Luntz. Exact cellular decompositions in terms of critical points of Morse functions. In *Proc. of IEEE ICRA'00, Int. Conf. on Robotics and Automation*, pages 2270–2277, San Francisco, CA, 2000.
10. H. Choset and J. Burdick. Sensor based motion planning: Incremental construction of the hierarchical generalized voronoi graph. *International Journal of Robotics Research*, 19(2):126–148, February 2000.
11. H. Choset and J. Burdick. Sensor based motion planning: The hierarchical generalized voronoi graph. *International Journal of Robotics Research*, 19(2):96–125, February 2000.

12. H. Choset and P. Pignon. Coverage path planning: The boustrophedon decomposition. In *Proceedings of the International Conference on Field and Service Robotics*, Canberra, Australia, December 1997.
13. F. H. Clarke. *Optimization and Nonsmooth Analysis*. Society of Industrial and Applied Mathematics, Philadelphia, PA, 1990.
14. Y. Gabriely and E. Rimon. Spanning-tree based coverage of continous areas by a mobile robot. *Annals of Mathematics and Artificial Intelligence*, Accepted, 2000.
15. S. Hert, S. Tiwari, and V. Lumelsky. A Terrain-Covering Algorithm for an AUV. *Autonomous Robots*, 3:91–119, 1996.
16. W.H. Huang. Optimal line-sweep-based decompositions for coverage algorithms. In *IEEE Int'l. Conf. on Robotics and Automation*, volume 1, pages 27–32, Seoul,Korea, 2001.
17. R.A. Jarvis. Collision free trajectory planning using distance transforms. *Mech Eng Trans of the IE Aust*, ME10:197–191, 1985.
18. J.C. Latombe. *Robot Motion Planning*. Kluwer Academic Publishers, Boston, MA, 1991.
19. V. J. Lumelsky, Snehasis Mukhopadhyay, and Kang Sun. Dynamic path planning in sensor-based terrain acquisition. *IEEE Transactions on Robotics and Automation*, 6(4):462–472, August 1990.
20. J. Milnor. *Morse Theory*. Princeton University Press, Princeton, New Jersey, 1963.
21. C. Ó'Dúnlaing and C.K. Yap. A "Retraction" Method for Planning the Motion of a Disc. *Algorithmica*, 6:104–111, 1985.
22. V.S. Spires and S.Y. Goldsmith. Exhaustive geographic search with mobile robots along space-filling curves. In *Proc. of First International Workshop on Collective Robotics*, pages 1–12, Paris, France, July 1998.
23. I.A. Wagner and Bruckstein A.M. Cooperative cleaners: A study in ant-robotics. Technical Report CIS-9512, Center for Intelligent Systems, Technion, Haifa, 1995.
24. A. Zelinsky, R.A. Jarvis, J.C. Byrne, and S. Yuta. Planning Paths of Complete Coverage of an Unstructured Environment by a Mobile Robot. In *Proceedings of International Conference on Advanced Robotics*, pages pp533–538, Tokyo, Japan, November 1993.

Online Scan Coverage of Grid Environments by a Mobile Robot

Yoav Gabriely[1] and Elon Rimon[1]

Technion, Israel Institute of Technology
meeryg@tx.technion.ac.il, elon@robby.technion.ac.il

Abstract. *We describe an on-line sensor based algorithm for covering planar areas with a uniform scan pattern, where the covering is executed by a* $D \times D$ *square-shaped tool attached to a mobile robot. The algorithm, called* Scan-STC, *incrementally subdivides the planar area into a grid of* D*-size cells, while following a spanning tree of a graph whose nodes are* $2D$*-size cells. The algorithm covers any planar grid with a scan path whose total length is at most* $(n+m)D$*, where* n *is the number of* D*-size cells and* $m \leq n$ *is the number of* boundary cells, *defined as cells that share at least one point with the grid boundary.* Scan-STC *additionally strives to minimize the total length of path segments orthogonal to the scan direction, and we report a preliminary bound on the length of these segments. We also demonstrate that any on-line coverage algorithm generates a covering path whose length is at least* $(2-\epsilon)l_{opt}$ *in worst case, where* l_{opt} *is the length of the optimal off-line covering path. Since* $(n+m)D \leq 2l_{opt}$*, the bound is tight and* Scan-STC *is worst-case optimal. Moreover, in practical environments* $m << n$*, and* Scan-STC *generates close-to-optimal covering paths in such environments.*

1 Introduction

In the *mobile robot covering problem*, a tool of a specific planar shape is attached to a mobile robot. Given a continuous planar area bounded by obstacles, the mobile robot has to move the tool along a path such that every point of the area is eventually covered by the tool. In the on-line version of the problem the robot must collect information about obstacles in the environment using its sensors while covering the planar area. Note that on-line coverage is *distinct* from on-line mapping or area exploration by a mobile robot. In the latter problem an autonomous robot has to completely scan an unknown environment with its sensors [5,15], while area coverage requires physical sweep of a tool over every point of a given work-area. The mobile robot covering problem has several important applications such as floor cleaning and coating, hazardous waste cleaning, field demining, and lawn mowings.

An optimal solution of the off-line covering problem generalizes the Traveling Salesperson Problem (TSP) to continuous domains, and is NP-hard [2]. It is therefore natural to seek competitive polynomial-time algorithms for the on-line version of the problem. (An on-line algorithm is *competitive* if its solution to every problem instance is a constant times the optimal solution to

J.-D. Boissonnat et al. (Eds.): Algorithmic Foundations of Robotics V, STAR 7, pp. 417–433, 2004.

the problem with full information available [12].) The on-line covering problem was first considered by Kalyanasundaram and Pruhs [14], in the context of on-line TSP tours on weighted planar graphs. They have shown that the problem has a competitive ratio of sixteen. In the special case of graphs representing grid environments the edges have uniform weights. The trivial DFS algorithm covers such graphs with a path whose length is at most twice the length of the optimal path. Our Scan-STC algorithm runs on general grid environments and has the same competitive ratio as DFS. However, Scan-STC generates a covering path whose length is bounded by $(n+m)D$ (n is the total number of grid cells, $m \leq n$ is the number of boundary cells defined below, and D is the tool size), while DFS generates a path whose length is always $2nD$. In practice $m << n$, and Scan-STC significantly improves on DFS.

In the context of on-line mobile robot coverage, the first competitive covering algorithms used environmental markers such as pebbles [6,7] or pheromone-like traces [16] to aid in the robot's coverage process. More recently, two covering algorithms that rely solely on the robot's on-board recourses are the URC covering strategy proposed by Icking et al. [11], and our Spiral-STC algorithm [9]. The two methods are similar in the sense that both have a competitive ratio of two but they significantly improve on DFS. However, both algorithms (as well as DFS) make no attempt to control the shape of the covering path. All of these algorithms generate a covering path whose eventual shape is determined by the obstacles encountered during the coverage process. In practical covering scenarios it is highly desirable to cover the planar work-area with a predictable scan pattern. Scan patterns are also amenable to Morse Theory analysis [4], a property that we use in the analysis of the algorithm. Moreover, the transition between scans usually involves costly maneuvers both in terms of time and wheel slippage. Hence it is desirable not only to minimize the total length of the covering path, but also to minimize the total number of scans. Arkin et al. [1] recently proposed an off-line scan coverage algorithm for pocket machining, that minimizes a weighted sum of path length and number of scans. Our Scan-STC algorithm is the first on-line algorithm that generates competitive scan coverage. The algorithm is 2-competitive as the on-line covering algorithms mentioned above, and it strives to minimize the total number of path-segments orthogonal to the desired scan direction.

We make the following assumptions. First, we assume that the tool is a square of size D. Second, the robot is allowed to move the tool only in directions orthogonal to the tool's four sides, without rotating the tool during this motion. Third, we focus on the path taken by the covering tool rather than the path taken by the robot. However, the covering tool is commonly mounted under the mobile robot's platform, and in this case the robot and covering tool follow the same path. Fourth, we approximate the continuous work-area by a discrete grid of D-size cells as shown in Figure 1. Note that the Hamiltonian path problem on planar grids (i.e. the construction of a path

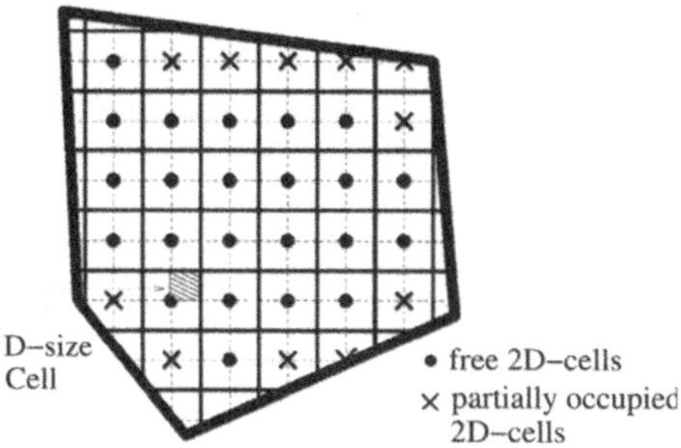

Fig. 1. Grid approximation of a given work-area.

that visits every node of the grid graph precisely once) is NP-complete [13]. The optimal covering of planar grids, being a more general problem, is NP-hard. Finally, we assume that the environment is populated by stationary obstacles unknown to the robot. Rather, the robot must use its sensors to detect obstacles while covering the work-area grid.

The paper makes the following two contributions. The first is description and analysis of the sensor-based covering algorithm Scan-STC. The algorithm is presented in Section 2 and analyzed in Section 3. The main result is that Scan-STC covers any planar grid in $O(n)$ time using a path whose length is at most $(n+m)D$. In this bound n is the number of D-size cells and $m \leq n$ is the number of *boundary cells*, defined as cells that share at least one point with the grid boundary. Another result is a preliminary bound on the number of path-segments orthogonal to the scanning direction generated by Scan-STC. The second contribution, reported in Section 4, is a universal lower bound on the competitiveness of on-line coverage algorithms. We demonstrate that any on-line coverage algorithm generates a covering path whose length is at least $(2-\epsilon)l_{opt}$ in worst case, where l_{opt} is the length of the shortest covering path and ϵ is an arbitrarily small positive parameter. Interestingly, Icking et al. independently obtained the same lower bound [11]. Since $l_{opt} \geq nD$, the path generated by Scan-STC satisfies $(n+m)D \leq 2l_{opt}$. Consequently, the $(2-\epsilon)l_{opt}$ bound is *tight*, and Scan-STC is optimal in worst case. Finally, simulation results demonstrate the effectiveness of the algorithm in office-like environments.

2 The Scan-STC Algorithm

We first describe a preliminary version of the algorithm called 2D-Scan-STC. The preliminary algorithm covers only $2D$-size cells which are completely free of obstacles. The full algorithm covers general grids of D-size cells by treating $2D$-size cells which are partially occupied by obstacles as special nodes of the spanning tree. Both versions of Scan-STC use the following two sensors. The first is a *position-and-orientation sensor* that allows the robot to locally recognize the cells comprising the work-area. The second is a *range sensor* capable of identifying obstacles in a neighborhood of the robot's current $2D$-cell. The size of this neighborhood is key to the on-line generation of a scan pattern. All previous on-line covering algorithms inspect the *four* cells neighboring the current cell. Scan-STC requires detection of obstacles in a larger neighborhood, consisting of the ring of *eight* $2D$-cells surrounding the current cell. For notational simplicity $2D$-size cells are simply called *cells*, while D-size cells are called *subcells*.

2.1 The Preliminary 2D-Scan-STC Algorithm

The input to the 2D-Scan-STC algorithm is a desired scanning direction, which we assume to be the *vertical direction*. Using the robot's orientation sensor, the algorithm aligns the incrementally constructed work-area grid with the vertical scanning direction. The preliminary algorithm incrementally subdivides the work-area into $2D$-size cells, and discards cells which are partially occupied by obstacles. The free cells induce a graph structure whose nodes are the center points of the cells, and whose edges are the line segments connecting centers of adjacent cells (Figure 1). The algorithm incrementally constructs a spanning tree for this graph, and uses the spanning tree to generate a coverage path as follows. During the spanning tree construction, the robot subdivides every cell it encounters into four identical subcells of size D, each being identical to the covering tool in size and shape. The covering tool follows a path of subcells that circumnavigates the incrementally constructed spanning tree, until the entire collection of free cells is covered.

The algorithm generates a vertical scan pattern by making sure that the spanning tree consists almost entirely of vertical edges. To do that, the algorithm inspects cells in the ring surrounding the current cell. These cells are denoted as follows. Given a neighbor cell y, the cells $y + 45$ and $y + 90$ are the cells that form 45° and 90° angles with the line from x to y in counter-clockwise order (Figure 2). In the following description of the algorithm, a free cell is *new* when its four subcells have not yet been covered, otherwise the cell is *old*.

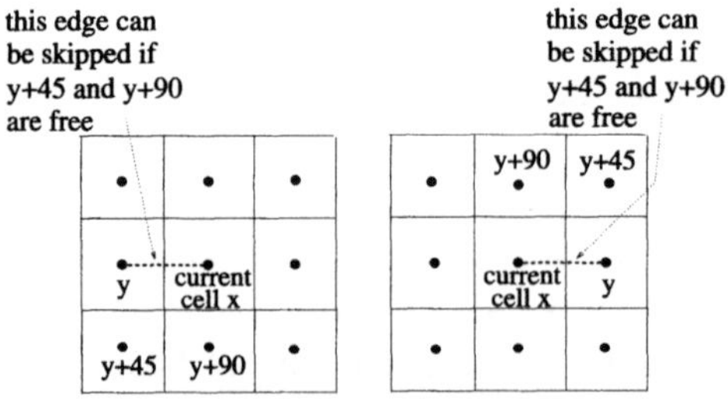

Fig. 2. The cells inspected by 2D-Scan-STC when considering whether to skip the construction of a horizontal spanning-tree edge.

2D-Scan-STC Algorithm:

Sensors: A position and orientation sensor. An 8-cells obstacle detection sensor.

Input: A starting cell S, but no apriori knowledge of the environment.

Recursive function: $STC1(w, x)$, where x is the current cell and w the parent cell in the spanning tree.

Initialization: Call $STC1(Null, S)$, where S is the starting cell.

$STC1(w, x)$:

1. Mark the current cell x as an *old* cell.
2. While x has a *new* obstacle-free neighboring cell which has not been inspected:

2.1 Scan in counterclockwise order for the first new neighbor of x which has not been inspected yet, starting with the parent w. Call this neighbor y.

2.2 If y is a *horizontal* neighbor of x and the cells y+45 and y+90 are free: *Skip* the construction of a spanning-tree edge from x to y. Goto Step 2.

2.3 Construct a spanning-tree edge from x to y.

2.4 Move to a subcell of y by following the right-side of the spanning tree edges.

2.5 Execute $STC1(x, y)$.

End of while loop.

3. If $x \neq S$, move back from x to a subcell of w along the right-side of the spanning tree edges.
4. Return. (End of $STC1(w, x)$)

We now discuss several details of the preliminary algorithm. First, the counterclockwise scanning of neighbors specified in step 2.1 ensures that the covering tool circumnavigates the incrementally constructed spanning tree in counterclockwise order (Figure 3(a)). Also note that the starting cell S has no parent cell, and in step 2.1 any neighbor of S can be designated as its parent. Second, in step 2.3 the covering tool is located in a subcell of x and has to move into a new cell y. By construction there is already a spanning-tree edge from x to y. The covering tool moves from its current subcell in x to a subcell of y by following the right-side of the spanning tree edges, measured with respect to the tool's direction of motion. When the covering tool returns to a parent cell w in step 3, it again moves through subcells that lie on the right-side of the spanning-tree edge connecting x with w. Consider now step 2.2, where the algorithm skips the construction of spanning-tree edges to horizontal neighbors. Let us verify that the spanning tree still reaches all the free cells accessible from S. First, when the covering tool enters a new vertical column of cells, the entire column is covered since the algorithm never skips vertical grid-edges. Second, the covering tool skips the construction of a horizontal spanning-tree edge to a neighbor y *only when it knows that y has a neighboring free cell in its own column,* the cell y+45, that can be accessed from the current column at a later stage of the covering process, from the cell $y+90$. Otherwise the algorithm constructs a horizontal spanning-tree edge that takes the covering tool into y.

An execution example of the algorithm is illustrated in Figure 3(a). The figure shows the subcell path taken by the covering tool. It can be seen that the robot moves the covering tool through subcells that lie on the right-side of the spanning-tree edges, measured with respect to the tool's direction of motion. The figure also shows the spanning tree constructed on-line by the algorithm. It can be seen that the spanning-tree edges are vertically aligned, except for a single horizontal edge between every pair of adjacent columns. The algorithm in general strives to minimize the number of horizontal edges, and this number is quantified below. For clarity, the tool size D is shown unrealistically large with respect to the work-area size, and the tool path is shown curved while it is rectilinear according to the algorithm.

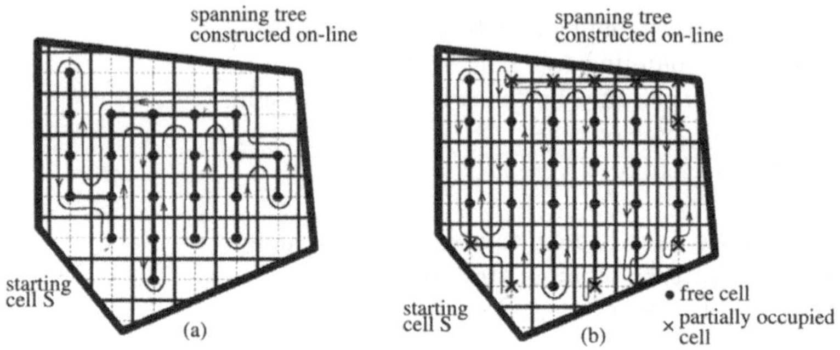

Fig. 3. An execution example of (a) the preliminary 2D-Scan-STC algorithm, and (b) the full Scan-STC algorithm.

2.2 The Full Scan-STC Algorithm

The full algorithm augments the preliminary algorithm with coverage of *partially occupied* 2D*-cells,* defined as cells that contain at least one obstacle-free D-size subcell. The full algorithm uses the following augmented graph structure. The nodes of the augmented graph are the center points of all free and partially occupied cells. The edges of this graph connect center points of adjacent cells that contain free subcells with a common boundary. This construction gives rise to two types of edges. The first type, called *double-sided edges,* possess only free subcells on both sides (Figure 4(a)). The other type, called *single-sided edges,* possess at least one occupied subcell on either side (Figure 4(b)). Scan-STC incrementally constructs a spanning tree for the augmented grid graph, while guiding the covering tool along a subcell path that circumnavigates the spanning tree. However, the circumnavigation of a single-sided edge incurs repetitive coverage of certain subcells associated with this edge. Another new structure in the augmented grid graph involves *node doubling.* A partially occupied cell with diagonally opposite free subcells is locally disconnected. Since the covering tool may be able to reach the free subcells from neighboring cells, we represent such a cell with two nodes, each having edges to adjacent cells from which a free subcell can be accessed (Figure 4(c)).

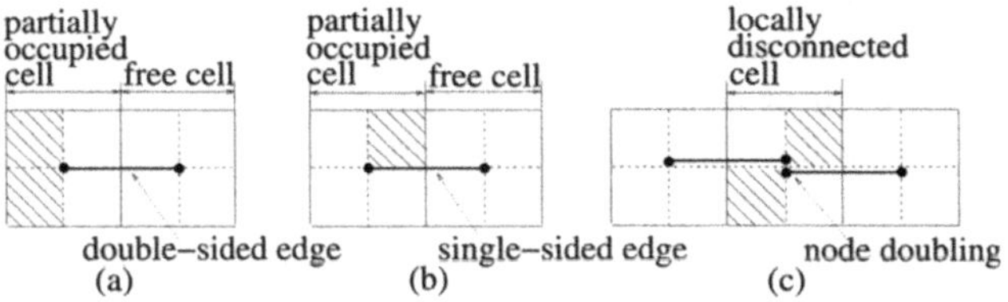

Fig. 4. (a) A single-sided, and (b) double-sided edge. (c) Node doubling at a disconnected cell.

For practical reasons, we wish to ensure that the full algorithm inspects only subcells which are directly visible from the covering-tool's current location. The algorithm thus inspects only the following subcells. Let p be the point where the cells x, y, $y+45$, and $y+90$ meet. Then the full algorithm inspects the four subcells surrounding p. These subcells are denoted $x_p, y_p, (y+45)_p, (y+90)_p$ and are depicted in Figure 5. The full algorithm has the same structure as the preliminary algorithm. Hence we only describe the recursive function, called $STC2(w, x)$, which is modified in order to account for the two types of edges that occur in the augmented grid graph.

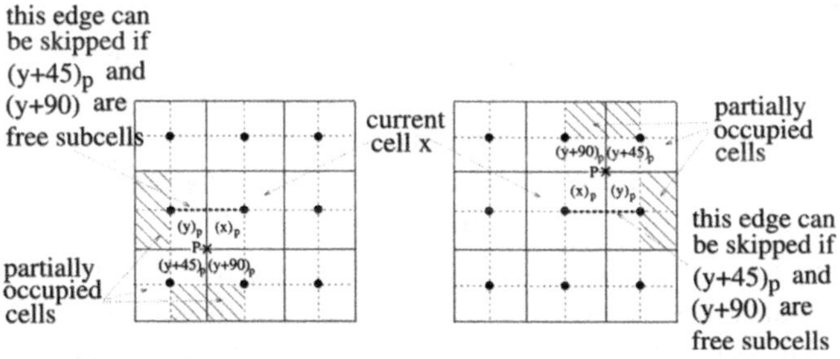

Fig. 5. The subcells inspected by the full Scan-STC when considering whether to skip the construction of a horizontal spanning-tree edge.

Full Scan-STC Algorithm:

Initialization: Call $STC2(Null, S)$, where S is the starting cell.

$STC2(w, x)$:

1. Mark the current cell x as an *old* cell.

2. While x has a *new* free or partially occupied neighboring cell which has not been inspected:
2.1 Scan in counterclockwise order for the first new neighbor of x which has not been inspected yet, starting with the parent w. Call this neighbor y.
2.2 If y is a *horizontal* neighbor of x and the four subcells x_p, y_p, $(y+45)_p$ and $(y+90)_p$ are obstacle-free: *Skip* the construction of the spanning-tree edge from x to y. Goto Step 2.
2.3 Construct a spanning-tree edge from x to y.
2.4 Move to a subcell of y along the spanning tree edges, using a path determined by the type of edge from x to y as described below.
2.4 Execute $STC2(x, y)$.
End of while loop.
3. If $x \neq S$, move back from x to a subcell of w along a path determined by the type of edge from x to w as described below.
4. Return. (End of $STC2(w, x)$.)

Let us describe the subcell path taken by the covering tool along a single-sided edge in steps 2.4 and 3. Consider for concreteness the partially occupied cell depicted in Figure 6(a). This cell is a leaf node in the spanning tree, and the spanning-tree edge entering this cell is single-sided. Let α be the subcell path that would have been taken by the covering tool if the cell were completely free. Then the actual path taken by the covering tool, denoted β, is obtained by *deforming* α *away from the occupied subcell without changing the path's sense of direction, until every point of* β *lies at a distance of* $D/2$ *from the occupied subcell.* Figure 6(a) shows the original path α as well as the deformed path β. Note that we use a maximum metric[1] in the path deformation, to ensure that β takes the covering tool only through obstacle-free subcells. Note, too, that the deformed path β crosses the spanning-tree edge. The deformed path may cross other spanning-tree edges emanating from a partially occupied cell, as shown in Figure 6(b).

Next consider step 2.2, where the algorithm skips the construction of a horizontal spanning-tree edge. The rule specified in this step is the subcell-version of the corresponding rule in the preliminary algorithm. Given a horizontal neighbor y, the cell $y+45$ is a vertical neighbor of y. The algorithm skips the construction of a horizontal spanning-tree edge from x to y when it knows that y can be accessed from the subcell $(y+45)_p$ at a later stage of the covering process from the subcell $(y+90)_p$. The four subcells involved in this test are visible from any subcell of the current cell x, for the following reason. If the subcell x_p is occupied by an obstacle, the test fails and there

[1] The maximum metric is $d(p, q) = \max\{|p_x - q_x|, |p_y - q_y|\}$.

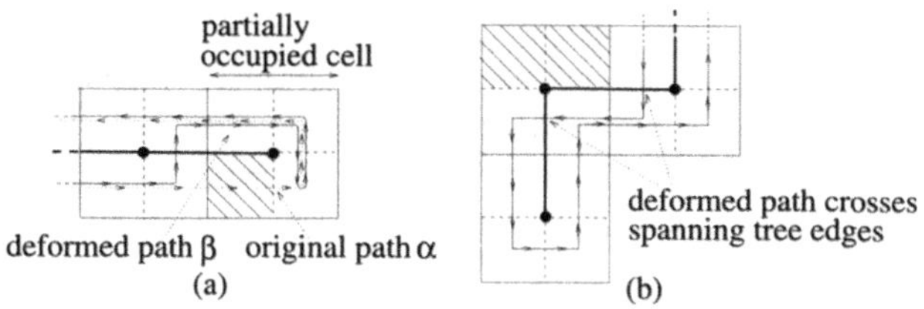

Fig. 6. (a) Path deformation along a single-sided edge. (b) Crossing of spanning-tree edges.

is no need to inspect the other three subcells. Otherwise x_p is free and the other three subcells are directly visible from any subcell of x. Finally, the full algorithm need only inspect the ring of subcells surrounding the current cell, rather than the ring of cells required by the preliminary algorithm.

An execution example of the full algorithm is illustrated in Figure 3(b). Note that repetitive coverage occurs at places where subcells on the right-side of the spanning-tree edges are occupied by obstacles. As discussed below, the total number of repetitive coverages is bounded by the number of boundary subcells in the work-area grid. Simulations of the full Scan-STC algorithm are presented in Section 5.

3 Analysis of Scan-STC

This section lists several properties of Scan-STC. The two main results are a bound on the length of the covering path (Theorem 1), and a bound on the total length of horizontal path segments (Proposition 3.4).

Lemma 3.1 (completeness) *Scan-STC covers every free subcell accessible from the starting cell S.*

Let us sketch the proof. The algorithm constructs a spanning tree that reaches every free and partially occupied cell in the accessible work-area grid. Consequently, every free subcell touches the spanning tree either at a point (a leaf node), or along a segment (part of a spanning-tree edge). It follows that the subcell path generated by circumnavigating the spanning tree with suitable deformations away from occupied subcells passes through every free subcell which is accessible from S. The next lemma gives the run-time and memory requirement of the algorithm.

Lemma 3.2 *Let n be the number of free subcells accessible from S. Then* Scan-STC *covers these subcells in $O(n)$ time using $O(n)$ memory.*

Let us sketch the derivation of these bounds. In each step the covering tool enters either a new subcell or a previously visited subcell. According to Theorem 1 below, the total number of repetitive visits is bounded by m, where $m \leq n$ is the number of boundary subcells. Hence there are $O(n)$ covering steps. The $O(n)$ memory requirement allows the algorithm to identify which cells of the grid have already been visited by the covering tool. This memory requirement imposes a strict limitation on the size of the work-areas that can be covered by a bounded-memory robot. In Ref. [8] we describe a spanning-tree based algorithm that uses cell markers to identify previously visited subcells. This algorithm requires only $O(1)$ memory.

The following key theorem provides a bound on the length of the covering path generated by Scan-STC. Recall that *boundary subcells* are obstacle-free subcells that share either a point or a segment with the boundary of the work-area grid.

Theorem 1 *Let n be the number of free subcells in the accessible work-area grid. Let $m \leq n$ be the number of boundary subcells. Then* Scan-STC *covers the work-area grid using a path of total length $l \leq (n+m)D$, where D is the tool size.*

A proof of the theorem appears in Ref. [10]. The $(m+n)D$ bound does not reflect an important property of the algorithm, that its repetitive coverage is localized around partially occupied cells (Figure 10). The following corollary provides a tighter bound that emphasizes this property. Consider the collection of occupied subcells contained in the partially occupied cells. Let k be the total number of free subcells that share either a point or a segment with these occupied subcells. (Note that k is a property of the work-area grid.) Then $k \leq m$, and the length of the covering path is bounded in terms of k as follows.

Corollary 3.3 *Let n be the number of free subcells in the accessible work-area grid. Let $k \leq m$ be the number of free subcells that touch occupied subcells contained in partially occupied cells. Then* Scan-STC *covers the work-area grid using a path of total length $l \leq (n+k)D$, where D is the tool size.*

The next result concerns the covering pattern generated by Scan-STC. The objective of Scan-STC is to cover the work-area grid with a vertical-scanning pattern that minimizes the number of horizontal path segments. We present an upper bound on the number of horizontal path segments using the following terminology. A *column segment* is a contiguous sequence of free cells arranged vertically and bounded by the boundaries of the work-area grid. Consider now the possible connectivity changes that can occur in a column-segment as we move horizontally, say from left to right. A column-segment can split into two disjoint column-segments (Figure 7(a)), or two column-segments can join into a single connected column-segment (Figure 7(b)).
In both cases we say that the column-segment experiences a *connectivity change*. For clarity, Figure 7(c) shows a case where a column-segment experiences no connectivity change. The following proposition provides an upper bound on the number of horizontal path segments in terms of the number of

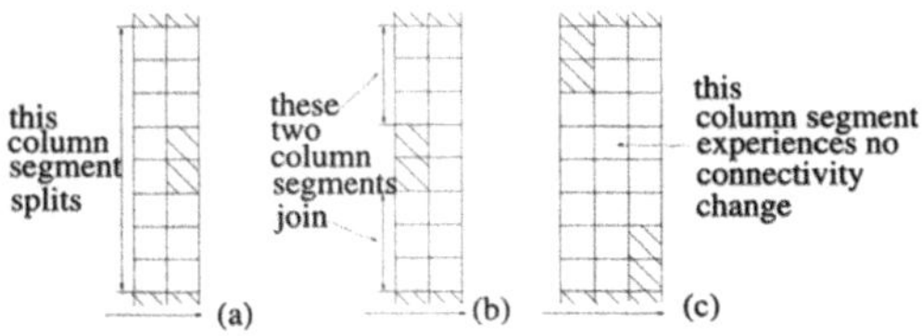

Fig. 7. (a)-(b) Split and join connectivity changes in a column-segment. (c) A column-segment can deform without experiencing any connectivity change.

horizontal spanning-tree edges constructed by the algorithm. The bound is stated for the simpler 2D-Scan-STC algorithm.

Proposition 3.4 *The number of horizontal spanning-tree edges constructed by* 2D-Scan-STC*, denoted* h*, is bounded by* $h \leq h^* + p + 1$*, where* h^* *is the off-line optimal number of horizontal spanning-tree edges*[2]*, and* p *is the number of connectivity changes in the column-segments of the grid.*

Let us discuss some aspects of the proposition, leaving its proof to Ref. [10]. First, the total length of the horizontal segments along the covering path is $2Dh$, since the covering tool moves on both sides of each spanning-tree edge. Second, the unity term in the bound for h corresponds to a wasteful horizontal spanning-tree edge that can occur in the column-segment containing the starting cell. This term can be eliminated if the covering tool always starts on the grid boundary. Last, it seems that any on-line covering algorithm must "pay" for the connectivity changes. For instance, let a column-segment C experience a join connectivity change. Then C can be accessed from two column-segments on its left side. Since only local sensory information is available to the robot, it may lead the covering tool into C from both routes. Not knowing that it has arrived to the same column-segment, the robot may independently cover several portions of C, generating a wasteful covering pattern.

Finally consider the competitive ratio of Scan-STC, defined as l/l_{opt} where l_{opt} is the length of the optimal coverage path. The total area of the grid is $A = nD^2$, and the area occupied by the boundary subcells is $\partial A = mD^2$. The optimal coverage path satisfies $l_{opt} \geq A/D$, and the length of the coverage path generated by Scan-STC satisfies $l \leq (n+m)D = A/D + \partial A/D$. It follows that the competitive ratio of Scan-STC satisfies $l/l_{opt} \leq 1 + \partial A/A$. In practice $\partial A << A$, and Scan-STC achieves a close-to-optimal coverage of practical work-area grids.

4 A Universal On-Line Coverage Bound

In this section we derive a lower bound on the path-length of any on-line covering algorithm for planar environments. First let us demonstrate the bound in a simple corridor whose width is identical to the tool size D. Suppose

[2] h^* is equal to the number of column-segments in the grid minus one.

Fig. 8. A corridor environment with an on-line covering path depicted.

the robot executing the algorithm has a range sensor with a detection range of $1-\delta$, where δ is a small positive parameter. We initially place the robot in an infinite corridor and wait until it covers a length-L portion of the corridor, where $L >> 1$. At that instant we truncate the corridor at a unit distance from both sides of the length-L portion, as shown in Figure 8. Since the robot has no way of knowing about this change in the environment, its on-line behavior would be identical if it starts at the same point in the truncated corridor. The length of the on-line covering path is $l \geq 2L + 3$, since the covering tool must visit both ends of the corridor before coverage is complete. The shortest off-line covering path, achieved by sweeping the corridor from end to end, has length $l_{opt} = L + 2$. Thus $l/l_{opt} \geq 2 - \epsilon$, and this is the lower bound we wish to establish. However, *the corridor environment does not support the bound for covering tours where the robot must return to its starting point.* Another caveat with the corridor is that the shortest off-line covering path does not start at the same point as the on-line algorithm.

The more sophisticated environment presented in Figure 9 is circular, and the lower bound is established in this environment both for covering paths and covering tours. Moreover, the circular environment gives no advantage to any particular initial point. Hence the off-line covering path might as well start at the same point as the on-line algorithm.

Theorem 2 *Every on-line covering algorithm of a bounded planar environment generates in worst case a covering path or tour of length at least* $(2-\epsilon)l_{opt}$*, where* l_{opt} *is the length of the shortest off-line covering path, and* ϵ *is an arbitrarily small positive parameter.*

Moreover, the bound is ϵ*-tight, since there exist on-line algorithms (such as Scan-STC) that generate covering paths whose length is at most* $2l_{opt}$.

Proof: In the proof we identify the robot with the covering tool. Furthermore, we consider for simplicity a point robot covering a graph. We loose no generality by this simplification, as a thickening of the graph to rectangular corridors of width D and lengths that are integer multiples of D would imply the same bound in grid as well as continuous planar environments. The robot is able to detect nodes of the graph by "looking" along edges of the graph, and it has a detection range of $1-\delta$ where δ is a small positive parameter. Given an on-line coverage algorithm, we execute the algorithm in the *double-ring environment* depicted in Figure 9(a). The ring consists of $2k$ edges arranged in parallel pairs, where $k >> 1$. Each edge has length L such that $L >> 1$. Adjacent edges are connected by an X-shaped node that allows the robot to freely move between the four edges meeting at the node. During the execution of the algorithm, we perform k *local incisions* that remove portions of the edges. However, the removed portions always lie beyond the robot's detection range.

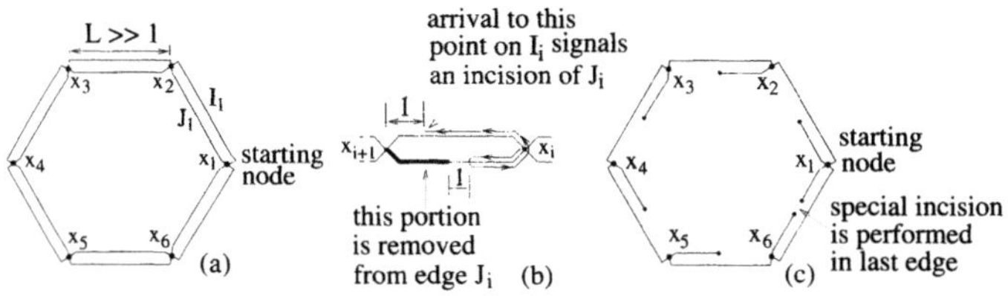

Fig. 9. (a) A double-ring environment. (b) The local incision performed on the edge J_i. (c) The environment resulting from k incisions.

The local incision procedure is as follows. Let $x_1, ..., x_k$ be the nodes of the ring, arranged in counterclockwise order. Let I_i and J_i be the parallel edges connecting x_i and x_{i+1}, where indexing is modulo k. Let the starting point be at the node x_1. The robot initially covers portions of the four edges incident to x_1. The first incision occurs at the instant when the robot reaches within a distance of *one unit* from one of the nodes adjacent to x_1. Suppose this event occurs first on the edge I_1 which leads to x_2. We remove from the parallel edge, J_1, the portion that lies at a distance of *one unit* beyond the covered portion of J_1 (Figure 9(b)). There is no problem with this incision, since $L >> 1$ and the covered portion of J_1 is more than one unit away from x_2. Note that the portion removed from J_1 has not been detected by the robot. The robot next continues with its covering path, and at a certain instant it would reach a new node. We may assume that the robot visits the nodes in counterclockwise order, until it reaches the last node x_k. (The robot may visit the nodes in any other order, but the covering path would only be longer.) The same local incision is repeated in every edge J_i for $i = 1, ..., k-1$, while the edges I_i are left intact. The last edges I_k and J_k have been already partially covered by the robot, since these edges are incident to x_1. In these edges only, we wait until the robot approaches within *one unit* distance from the covered portion of either I_k or J_k. Suppose this event occurs first in I_k. Then we remove the middle portion of J_k which lies *one unit* away from the covered portions of J_k. The environment resulting from the k incisions is depicted in Figure 9(c).

The modified environment consists of a loop of k edges $I_1, ..., I_k$ each having a length L. Every node x_i also has a truncated edge emanating from it (x_1 has two truncated edges). For purpose of path-length computation, let H_i be the length of the portion of J_i that has been covered up to the instant of incision in J_i. (The piece of each edge J_i has a total length of $H_i + 1$.) Also, let $H = \sum_{i=1}^{k} H_i$. A key property of the modified environment is that *the on-line algorithm has covered a length-H_i portion of the piece of J_i emanating from x_i when it reaches within one-unit distance from x_{i+1} along the edge I_i.* In the best scenario the robot first completes the coverage of the edges $I_1, ..., I_k$, then performs a second loop while pausing at each node to cover the truncated edge (two truncated edges in the case of x_1) emanating from it. The total length of the motion along the edges I_i is $2kL$. During the first circumnavigation the robot covers a length-H_i portion of the piece of J_i emanating from x_i, then retreats through x_i into the parallel edge I_i. Hence the total length of the motion along the pieces of the edges J_i is $2H$ during the first round, and an additional $2(H + k)$ during the second round. The total length of the on-line coverage path is therefore $l \geq 2kL+4H+2k$. Note that the assumption $k >> 1$ allows us to ignore the small difference in the length of a covering path and a covering tour for this particular environment.

Let us now compute the off-line optimal covering path. Starting from x_1, the optimal covering path circumnavigates the modified environment once, stopping at each node to cover the edge-piece (two edge-pieces in the case of x_1) emanating from it. The total length of the optimal covering path is $l_{opt} = kL + 2(H + k)$, since the robot must enter and retreat from every truncated edge J_i. Comparing l with l_{opt}, we obtain that $l \geq (2 - \epsilon)l_{opt}$, where $\epsilon = 2k/(kL + 2(H + k)) << 1$ since $L >> 1$. □

5 Simulation Results

We present simulations of Scan-STC in an environment that resembles several rooms populated by pieces of furniture (Figure 10). In Figure 10(a) the covering tool follows a vertical-scanning pattern that extends leftward from the starting cell. In Figure 10(b) the covering tool enters the lower-left room, with the spanning tree constructed during this stage is overlayed. It can be seen that repetitive coverage, shown in dark gray, occurs each time subcells on the right-side of the spanning-tree edges are occupied by obstacles. In Figure 10(c) the covering tool completes the covering of the lower-left room and retreats along the wall into the initial room. Intuitively speaking, Scan-STC exhibits a desired behavior, whereby it covers entire rooms except for an escape route which is left along the walls. This feature is not shared by previous on-line covering algorithms, and its practical significance is discussed in the concluding section. Figure 10(d) shows the covering of the lower-right room, and Figure 10(e) shows the completion of coverage. In this example 17% of the subcells were repetitively covered, making the covering path at most 17% longer than the globally optimal path. Note that the spanning tree constructed by the algorithm almost minimizes the number of horizontal spanning-tree edges. As predicted by Proposition 3.4 for the preliminary algorithm, the spanning tree contains at most one wasteful horizontal edge in each column segment that experiences a connectivity change.

Finally, the average amount of repetitive coverage generated by the full Scan-STC algorithm can be characterized as follows. Recall that repetitive

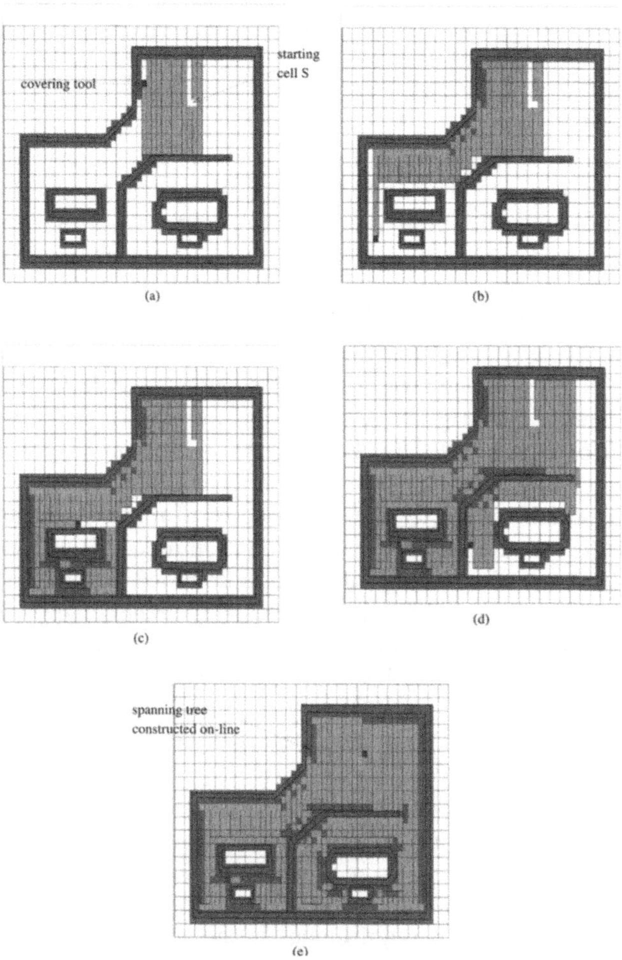

Fig. 10. Five covering stages of the full *Scan-STC*, with repetitive coverage depicted in dark gray.

coverage occurs at places where subcells on the right-side of the spanning-tree edges are occupied by obstacles. In particular, all the repetitive coverages occur in subcells that share a point or an edge with the grid boundaries. On average, half of these boundary subcells lie on the left-side of the spanning-tree edges, and the covering tool must pass through these subcells twice during the spanning tree circumnavigation. On average slightly more than 50% of the boundary subcells are repetitively covered, due to additional repetitive coverage incurred each time the covering tool crosses the spanning-tree edges. We have not attempted to rigorously measure the average amount of repetitive coverage in simulations. In the example described here, Scan-STC repetitively covered 52% of the boundary subcells (Figure 10(e)).

6 Conclusion

Scan-STC is a sensor based algorithm for covering general grid environments with a uniform scan pattern. The algorithm discretizes the work-area into a grid of D-size cells, then follows a path that circumnavigates a spanning tree constructed for a coarser grid of $2D$-size cells. The algorithm treats partially occupied $2D$-cells as special nodes of the spanning tree that incur repetitive coverage. It covers any planar grid of D-size cells in $O(n)$ time using a covering path whose length is at most $(n+m)D$, where n is the number of D-size cells and $m \leq n$ is the number of D-size cells along the grid boundary. The algorithm satisfies an even tighter path-length bound of $(n+k)D$, where $k \leq m$ is the number of D-size cells that lie along the grid boundary in the vicinity of partially occupied $2D$-cells. Furthermore, Scan-STC generates a uniform scan coverage which strives to minimize the total number of path-segments orthogonal to the desired scan direction. The number of path-segments orthogonal to the desired scanning direction is bounded by h^*+p+1, where h^* is the off-line minimum number of path-segments orthogonal to the scanning direction, and p is the number of connectivity changes in the columns of the grid aligned with the scanning direction. Finally, we showed that any on-line coverage algorithm generates a covering path whose length is at least $(2-\epsilon)l_{opt}$ in worst case. Since $(n+m)D \leq 2l_{opt}$, the bound is tight and Scan-STC is optimal in worst case.

Let us mention several open issues concerning on-line mobile robot coverage. The first issue is coverage within guaranteed return-time bounds. Given a cell in the interior of the grid, it is desirable to complete the coverage of the cell as well as its surrounding cells within a guaranteed time bound. The practical motivation for this requirement is the accumulative error incurred by the robot's position sensors. If the robot does not complete the coverage of any particular area in a reasonably short time, it will have to compensate for its position uncertainty by employing wasteful overlap in the covering pattern. Scan-STC has the desired property of covering contiguous areas while leaving only cells along the grid boundary for the return path. Unfortunately Scan-STC does not guarantee any return-time bound, and to our knowledge no other on-line covering algorithm guarantees such a bound.

A second issue is cooperative coverage by several mobile robots. Existing results on cooperative coverage employ cell markers to communicate the state-of-coverage between the group members [16]. It is more desirable to achieve cooperative coverage using local communication between the robots. A fundamental question in this context is to identify the minimal amount of communication that still allows a reasonable speedup of the cooperative coverage. A third issue is on-line coverage of curved terrains by a mobile robot. In this case a representation of the terrain's surface by a uniform grid may not be well justified, since the surface's area-form varies along the surface while the covering-tool's shape is constant. This issue becomes even more acute when a robot arm has to cover a three-dimensional structure [3].

References

1. E. M. Arkin, M. A. Bender, E. D. Demaine, S. P. Fekete, J. S. B. Mitchell, and S. Sethia. Optimal covering tours with turn costs. In *12'th ACM-SIAM Symp. on Discrete Algorithms (SODA)*, 138–147, 2001.
2. E. M. Arkin, S. P. Fekete, and J. S. Mitchell. Approximation algorithms for lawn mowing and milling. *Computational Geometry: Theory and Applications*, 17:25–50, 2000.
3. P. N. Atkar, H. Choset, A. A. Rizzi, and E. U. Acar. Exact cellular decomposition of closed orientable surfaces embedded in $\mathbb{R}^3$. In *IEEE Int. Conf. on Robotics and Automation*, 699–704, 2001.
4. H. Choset, E. Acar, A. A. Rizzi, and J. Luntz. Exact cellular decomposition in terms of critical points of Morse functions. In *IEEE Int. Conf. on Robotics and Automation*, May 2000.
5. X. Deng, Kameda, and C. Papadimitriou. How to learn an unknown environment. In *31th IEEE Symp. on Foundations of Computer Science*, 298–303, 1991.
6. X. Deng and A. Mirzaian. Competitive robot mapping with homogeneous markers. *IEEE Trans. on Robotics and Automation*, 12(4):532–542, 1996.
7. G. Dudek, M. Jenkin, E. Milios, and D. Wilkes. Robotic exploration as graph construction. *IEEE Trans. on Robotics and Automation*, 7(6):859–865, 1991.
8. Y. Gabriely and E. Rimon. Spanning-tree based coverage of continuous areas by a mobile robot. *Annals of Math. and Artificial Intelligence*, 31:77–98, 2001.
9. Y. Gabriely and E. Rimon. Spiral-STC: An on-line coverage algorithm of grid environments by a mobile robot. In *IEEE Int. Conf. on Robotics and Automation*, 954–960, 2002.
10. Y. Gabriely and E. Rimon. Competitive on-line coverage of grid environments by a mobile robot. *Computational Geometry: Theory and Applications (to appear)*, http://www.technion.ac.il/~robots.
11. C. Icking, T. Kamphans, R. Klein, and E. Langetepe. On the competitive complexity of navigation tasks. In *Sensor Based Intelligent Robots, Lecture Notes in Computer Science*, volume 2238, 245–258. Springer Verlag, 2002.
12. C. Icking and R. Klein. Competitive strategies for autonomous systems. In H. Bunke et al., editors, *Modelling and Planning for Sensor Based Intelligent Robot Systems*, 23–40. World Scientific, Singapore, 1995.
13. A. Itai, C. Papadimitrio, and J. Szwarcfiter. Hamiltonian paths in grid graphs. *SIAM J. of Computation*, 11:676–686, 1982.
14. B. Kalyanasundaram and K. R. Pruhs. Constructing competitive tours from local information. In *Lecture Notes in Computer Science*, vol. 700, 102–113. Springer Verlag, 1993.
15. N. S. V. Rao and S. S. Iyengar. Autonomous robot navigation in unknown terrains: visibility graph based methods. *IEEE Trans. on Systems, Man and Cybernetics*, 20(6):1443–1449, 1990.
16. I. A. Wagner, M. Lindenbaum, and A. M. Bruckstein. Distributed covering by ant-robots using evaporating traces. *IEEE Trans. on Robotics and Automation*, 15(5):918–933, 1999.

References

Distributed Motion Planning for 3D Modular Robots with Unit-Compressible Modules

Zack Butler and Daniela Rus

Dept. of Computer Science, Dartmouth College, Hanover NH, USA

Abstract. Self-reconfigurable robots are versatile systems consisting of large numbers of independent modules. Effective use of these systems requires parallel actuation and planning, both for efficiency and independence from a central controller. This paper presents the PacMan algorithm, a technique for distributed actuation and planning. This algorithm works on systems with two- or three-dimensional unit-compressible modules. We give a simplified version of the algorithm along with extensions that handle a larger class of parallel actuation. For both algorithms, we present correctness analysis that show the classes of reconfigurations that can be guaranteed to be achieved. For the extensions, we give proofs of parallel actuation capability that describe how several modules can move simultaneously without synchronization while retaining correctness. We have successfully instantiated the basic algorithm onto the Crystal, a self-reconfigurable robot system, and present hardware experiments.

1 Introduction

Self-reconfiguring modular robots are systems that adapt to the task at hand by changing their overall shape through the use of many independently actuating modules. A self-reconfiguring robot could, for example, form into a legged robot for rapid locomotion, then reform into a snake to navigate a tunnel, while using some of its modules to form a gripper to manipulate an object. Several self-reconfiguring systems have been proposed [3,4,7,8,10,12,13]. In all of these systems, the actuation is distributed. However, if the control is centralized, the overall efficiency of the system may be quite low. We seek to enable decentralized approaches to path planning and actuation for self-reconfiguring systems. These algorithms lead to more efficient systems by supporting parallelism, taking the burden of low-level motion planning and control off of a central controller.

Previous work in self-reconfiguring robotics has included hardware development as well as centralized and distributed reconfiguration planners. Several centralized schemes have been developed, often for complex modules where distributed approaches are more difficult [5,9,11]. Distributed self-reconfiguration planning includes Tomita *et al.*'s [14] method for shape formation based on attaining the correct sets of connections between modules, and Lee and Sanderson's [6] distributed control for the Tetrabot system. Vassilvitskii *et al.* [15] present a planning scheme for meta-modules made up of

J.-D. Boissonnat et al. (Eds.): Algorithmic Foundations of Robotics V, STAR 7, pp. 435–451, 2004.

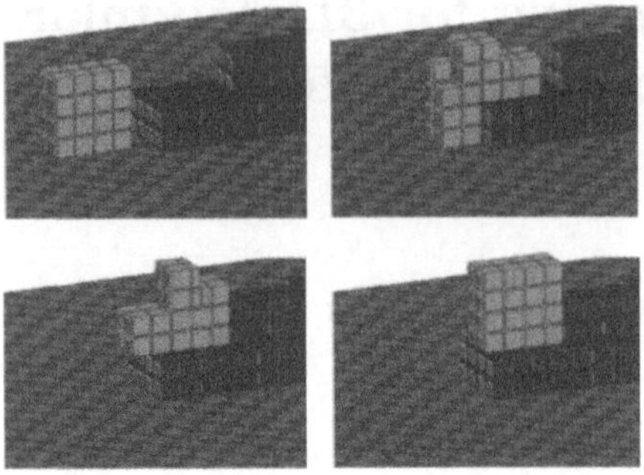

Fig. 1. In this simulation, the light gray modules represent a unit-compressible robot that uses morphing to climb stairs.

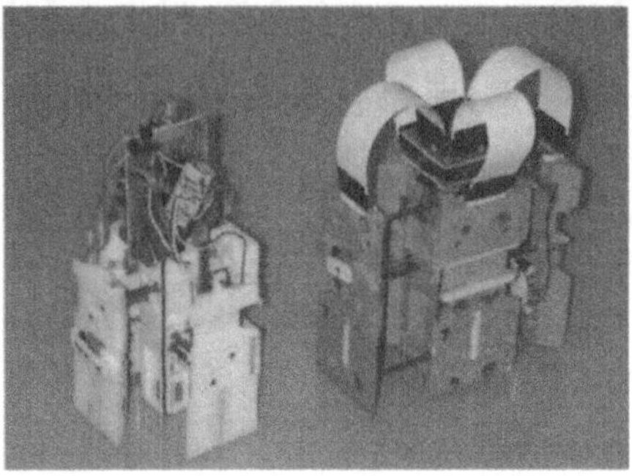

Fig. 2. A module (atom) of the first version of the crystalline robot (left), and a module of the second version (right), which is contracted along one axis and expanded along the other.

3D unit-compressible modules. In this work, groups of eight modules (2×2×2) are treated as units for planning, and arbitrary reconfiguration between any shapes made of meta-modules can be achieved. Their work focuses on planning, and does not address parallel actuation or synchronization among the modules within each meta-module.

The contribution of this paper is the PacMan algorithm, a simple scheme for distributed planning and actuation for self-reconfiguring robots with unit-compressible modules, in which actuation takes place through expansion and contraction. We also present analysis of parallel motion capabilities. The natural parallelism of the PacMan algorithm allows for increased performance of the reconfiguration, making it potentially much faster than a central controller for systems with large numbers of modules. In addition, our algorithm does not rely on meta-modules. The PacMan algorithm consists of a distributed planner that develops paths for individual modules and an actuation protocol that the modules execute in parallel. We characterize the class of self-reconfigurations achievable by PacMan and discuss extensions to the actuation protocol. A key part of our analysis shows that the robot remains connected during actuation. This work builds on an earlier 2-D version of PacMan which was presented in [1]. In this paper, we extend the algorithm and analysis from 2-D systems to 3-D systems, and also develop algorithms and guarantees for parallel path actuation which were not present in the original version.

An outline of the PacMan concept is given in Sec. 3, followed by the distributed planning and basic actuation algorithms in Sec. 4 and 5. Analysis of these algorithms is given in Sec. 6. Extensions that allow for better parallel actuation are described in Sec. 7, along with analysis of these extensions. Software and hardware implementations are discussed in Sec. 8.

2 The Crystalline Atomic Robot

The distributed algorithms described in this paper are applicable to robots with unit-compressible modules [12,13]. One specific instantiation is the *Crystalline Atomic* robot, or *Crystal*, developed in our lab [12]. It is made up of a collection of *Atoms*, which are square modules that can connect to each other and can expand and contract by a factor of two. Figure 2 shows one Atom of the first version of the Crystal along with one Atom of the second version. In the second version, a degree of freedom has been added, so that each atom can expand its North/South axis independently of its East/West axis. Module-to-module communication has been added and broadcast communication eliminated, so that any algorithm for the Crystal must be distributed. The unit-compressible module concept extends directly to the three-dimensional case, as do the algorithms presented here. The Telecube system developed at PARC [13] is a 3D unit-compressible system with six DOF which currently uses off-board power and computation.

In any unit-compressible system, a single module cannot move relative to the group on its own. The expansion and contraction of neighboring modules are required to cause it to move with respect to its neighbors, while attaching and detaching from neighbors allows for more complex global reconfigurations. A sample reconfiguration is shown in Fig. 1. An important aspect of controlling unit-compressible robots is to ensure that the overall group does not fragment during actuation due to module disconnections.

Like most self-reconfiguring modular robots, the Crystal is homogeneous, meaning that all atoms are identical. Any module can therefore occupy any position in the goal configuration. Given a goal shape, assigning atoms to specific locations can be done at run-time. This effect enables the PacMan technique, and can lead to very efficient reconfiguration algorithms, as described previously [12].

3 The PacMan Concept

PacMan is a distributed algorithm for self-reconfiguration in which the individual modules have the ability to plan and actuate a given reconfiguration in parallel using only local information and local communication. A planner run by the atoms develops paths for each module that will move. The paths are then actuated by the atoms in a parallel distributed fashion.

The PacMan algorithm was inspired by the video game of the same name, in which the main character eats "pellets" as it moves around the board. The algorithm uses data structures called pellets as a way of marking the path that each module should follow to perform its part of the reconfiguration. This representation is well suited to unit-compressible actuation, since motion of the modules takes place in the interior of the structure. However, a single physical module cannot move through the entire structure to follow its path.

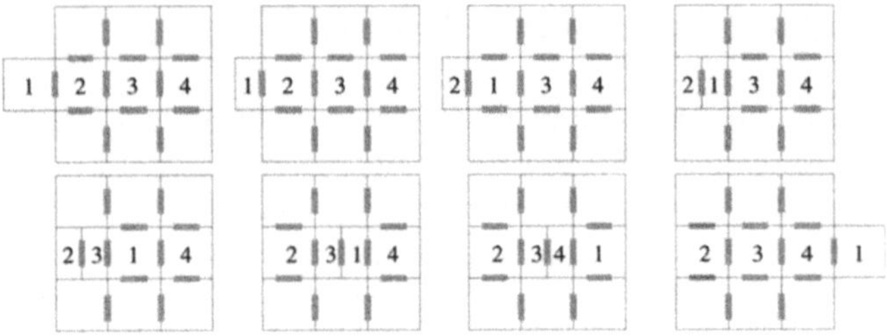

Fig. 3. A simple planar PacMan reconfiguration. The numbers represent module IDs, not physical modules. Gray bars represent connections between modules.

Instead, the module will virtually travel along the path marked by its pellets. To do this, it exchanges its identity with other modules along the path, while the physical modules move only locally. A simple PacMan reconfiguration is shown in Fig. 3. Additionally, by marking each pellet with the identity of the module that is to "eat" it, paths for several modules can coexist in the Crystal. The existence of multiple paths in turn allows for several modules to perform simultaneous reconfigurations.

More specifically, the PacMan process is two-fold. First, a path is planned for each module in a distributed fashion, using the planner described in Sec. 4. The result of planning is a set of pellets distributed through the atoms of the robot. Once the pellets are in place, the actuation will happen asynchronously, directed by the algorithm presented in Sec. 5. Each atom looks for pellets and "eats" them without adhering to a strict schedule. This means that the intermediate structure of the Crystal will be undetermined, but the final structure will be as specified. The distinction between planning and actuation suggests that a centralized planner could be used to create all necessary pellets. The modules could then perform PacMan actuation in parallel. This is reasonable for a small system, since the constraints on paths can be complicated and it may be easier to generate paths in a centralized way. For larger systems, a distributed algorithm would likely be preferred. We will now describe a distributed planner, followed by the actuation protocol.

4 Distributed Planning

To plan a reconfiguration in a distributed fashion, the algorithm operates in two stages. In the first stage, the modules determine the local difference between the current shape and the desired one. Then, the modules plan paths for individual modules to move into goal locations. These stages may happen a number of times, interleaved with actuation, resulting in a series of waves of reconfiguration. For a given reconfiguration, we define *target* atoms as those adjacent to unfilled goal locations, and *spare* atoms as those not currently part of the goal shape. We define *mobile* atoms as spare atoms that are on the edge of the system and can be moved without disconnecting the structure.

4.1 Initialization

To determine how the current configuration relates to the goal configuration, we represent the desired shape with a binary matrix S_d. One atom is chosen as the initiator of the process, and this atom is given a description of S_d along with its location in S_d[1]. It then marks its location as visited and if its location is not filled in S_d, notes that it is a spare atom. This atom then checks each of its neighbor locations (in S_d and in hardware). For each location, if a neighbor is present, S_d is passed to that neighbor. If a neighbor is not present but that location in S_d is set to 1, the current atom knows that it is a target atom. This process ensures that all spare and target atoms will be identified.

4.2 Path Planning

In the second phase of planning, each target atom will begin a search through the robot for a mobile atom to become its neighbor. The search is done using *plan-pellets*, which represent one step of a potential path. They are propagated through the atoms of the Crystal, performing a depth-first search[2] for a mobile atom in a distributed fashion as detailed in Algorithm 1. When a mobile atom is found, the plan-pellets (now forming a chain from mobile to target atom) are turned into *path-pellets*, which the mobile atom will "eat" to virtually move to the goal location. However, an atom cannot follow arbitrary paths through the Crystal under PacMan, and the planning process must take these restrictions into account.

There are two constraints that the actuation forces on the planning. First, when a path turns a corner, there is a minimum amount of surrounding structure that must be present. This includes a number of modules and a minimum path length to and from the corner. In order to keep track of segment lengths, each path-pellet contains the last direction in which it was propagated (d), and four counters T for the number of atoms adjacent to the path in each of the four directions perpendicular to the current segment. The second constraint on planning is that when a path travels along the edge of the Crystal, any *dangling* atoms (see Fig. 4) will become disconnected as the atoms along the path actuate. The planner forces the dangling atoms to move off of that edge by creating appropriate pellets for the dangler when the original path is instantiated. The original path is then put on hold until the dangler has moved from its original location.

To correctly perform the depth-first search, we define a canonical ordering of the six cardinal directions. When a path cannot continue straight, the path is propagated in the next available direction (lines 8-11 of Algorithm 1), and when a path is rejected, the canonical ordering also allows us to determine

[1] It should also be possible to give the modules only S_d and have them dynamically determine the best alignment to the current shape.

[2] Vasillvitskii *et al* [15] note that iterative deepening guarantees a shortest path, although we may instead wish to minimize turns with a similar technique.

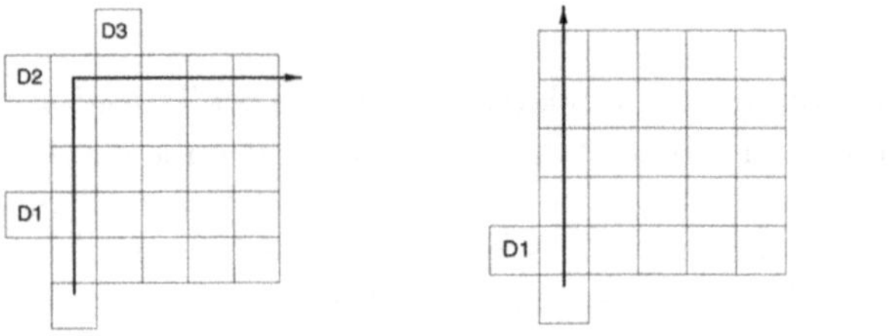

(a) (b)

Fig. 4. Dangling atoms, atoms that lie by themselves adjacent to the edge of the robot, that (a) can and (b) cannot be correctly handled under PacMan.

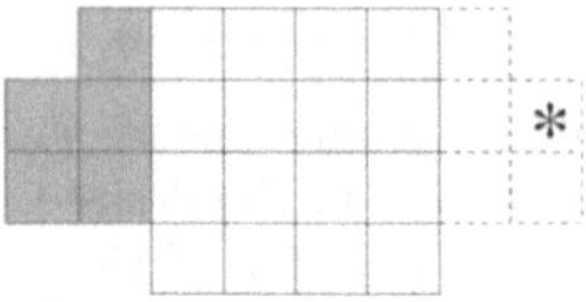

Fig. 5. An example of a reconfiguration requiring multiple waves of planning and actuation: unoccupied goal locations are given by dashed lines and spare atoms are shaded.

Algorithm 1 PacMan path planner

1: **if** Pellet received with direction d **then**
2:    **if** (I am mobile) and ($\max(T) \geq 2$) **then**
3:       set pellet ID to my ID and return to sender
4:    **if** neighbor exists in d **then**
5:       Increment T based on existence of neighbors in each direction
6:       propagate pellet in d
7:    **else**
8:       **for** each direction pd perpendicular to d **do**
9:          **if** $t_{pd} \geq 2$ and there is a neighbor in pd **then**
10:             set $d = pd$ and set $T = 0$
11:             propagate the plan-pellet to the neighbor in pd
12:       return rejection
13: **if** Rejection received (from dir rd) **then**
14:    Set d to direction of pellet that matches rejection
15:    **for** each perpendicular dir pd canonically after rd **do**
16:       **if** $t_{pd} \geq 2$ and there is a neighbor in pd **then**
17:          set $d = pd$ and set $T = 0$
18:          propagate the plan-pellet to the neighbor in pd
19:    return rejection in opp(d).

the next successor direction using only the information in the pellet and the direction of the rejection message (lines 15-18). Finally, when a path-pellet is received by a mobile atom, the mobile atom accepts the path request and sends a message back to the atom that gave it the plan-pellet. This message causes the plan-pellet to be turned into a path-pellet with the mobile atom's ID, and the message is propagated along the sequence of plan-pellets.

This procedure is guaranteed to find a path if one exists (as shown in Sec. 6), while respecting the actuation constraints inherent in the Crystal under PacMan. To guarantee that several path searches can simultaneously be completed, atoms will not begin their actuation while searches are in progress. This is ensured through a distributed semaphore that is incremented by each atom starting a search and decremented when the search ends in success or failure. No atom will actuate while this semaphore is greater than zero.

4.3 Multiple-Wave Reconfiguration

For some reconfigurations, the start and goal configurations will have large areas which do not overlap, such as in Fig. 5. In these instances, there are locations in the goal shape (such as the starred cell in Fig. 5) that are not next to any atom in the initial configuration. To perform these reconfigurations, the PacMan technique uses multiple *waves* of planning and actuation. In each wave, a layer of atoms is moved from the "back" of the group (the side that is not needed in the goal shape) to the "front." If multiple paths are executed during a single wave, it is necessary for all of them to complete before the next wave is planned so that communication between all pairs of neighbors is possible. To keep the structure synchronized from one wave to the next, we use a distributed semaphore that represents the number of currently active paths in the group. When an atom fulfills a path request, it sends out a message to the group to increment the semaphore. Then, when the atom completes its path, it sends out a decrement message. Atoms adjacent to empty target locations initiate new plan searches in the next wave when the semaphore reaches zero.

5 Basic Distributed Actuation

In this section we describe how the pellets are used to create motion in a parallel distributed context. This is the heart of the PacMan algorithm — while the pellets could be determined and placed by a variety of mechanisms, the use of the pellets is fixed to the algorithm given here (or its extension given in Sec. 7). The atom will virtually move along the path given by its pellets, exchanging its identity with the neighbor that has its pellet at each step. In addition, the atom will also contract toward the neighbor holding its pellet. The result is an inchworm motion that produces physical motion toward the goal configuration.

Since many atoms may be actuated simultaneously using PacMan actuation, deadlock, fragmentation and conflict at path intersections may all occur. To prevent these occurrences, two flags, *busy* and *dibs*, are used to ensure that the atoms complete their PacMan paths. Another constraint appears in the special case of turning a corner in a PacMan path. When turning a corner, a hole is opened up, so a pair of flags are introduced that ensure that the hole does not grow large enough to cause fragmentation.

The main part of the PacMan algorithm is given as Algorithm 2, while its supporting message handler is listed in Algorithm 3. The eating of pellets and identity trading happens in the first part of Algorithm 2 (lines 1-13). The atom first asks its neighbor for "dibs" on a trade, then contracts toward

Algorithm 2 PacMan actuation algorithm

```
1: if not(busy or dibs or waiting_for_corner) then
2:    if neighbor(dir) has pellet with my ID then
3:       while "OK" message not received do
4:          Send("request transfer") in (dir)
5:       Busy = 1
6:       if !contracted(dir) then
7:          disconnect(perp(dir)); contract(dir)
8:       Send("trading ID",ID) in (dir)
9:       Send("trading pellets",pellets) in (prev_dir)
10:      if dir ≠ prev_dir then
11:         corner_neighbor = 1
12: if at_goal then
13:    Disconnect(perp(dir)); busy = 0
14: if busy and contracted then
15:    if !(Is_nbr(prev_ID)) or at_goal(prev_ID) then
16:       Expand(prev_dir)
17:       Connect(perp(prev_dir))
18:       busy = 0
19:    if corner_neighbor then
20:       try_to_connect(corner_dir); corner_nbr = 0
```

Algorithm 3 PacMan message handler

```
Switch (message type):
if "Request transfer" then
   if not(busy or dibs) then
      Dibs = sender's ID
      Send("OK") to sender
   else
      Send("Not OK") to sender
if "Trading ID" then
   Prev_ID = ID, ID = received ID.
   Dibs = ∅
   Remove path-pellet with received ID.
   Send("Return ID",Prev_ID,prev_dir)
   to sender.
   Prev_dir = dir to sender
if "Return ID" then
   Prev_ID = ID, ID = received ID
   Prev_dir = received prev_dir
```

it, and finally exchanges identities. The busy flag is set to indicate that it is undergoing an identity transfer. The second half of Algorithm 2 (lines 15-23) takes care of a contracted atom that has traded identities with another atom. The contracted atom will wait for the atom it traded with to move on, and then expand and reconnect to its neighbors to complete the PacMan motion. If the contracted atom is next to a corner in the path, it will instead wait for the hole at the corner to be filled before expanding.

6 Basic Algorithm Analysis

In this section we analyze the correctness of the basic PacMan algorithm and define a class of reconfigurations that are possible in the case where only one PacMan path is executed at a time. In Section 7 we extend this analysis and describe a class of paths that can be executed in parallel (that is, at any arbitrary relative time). To build our arguments, we define a class of systems called *stem cells*, and restrict our analysis to this class, although reconfigurations of other shapes is possible. A stem cell is a group of modules consisting of a *core*, which is a square or cube of a given size, along with arbitrary projections from the edges/faces of the core.

We first show that PacMan actuation is sufficient for the class of stem cell robots. In both 2D and 3D we use a set of canonical path segments to show that a path can exist for any relocation, although the arguments are slightly different in 2D and 3D. In the 2D case, there is less surrounding structure, and so the analysis is more complex. We therefore present the 3D analysis here and refer readers interested in the 2D case to [1]. Note that all results other than Theorem 1 apply both to 2D and 3D.

Theorem 1. *[Actuation] For any three-dimensional stem cell of size 4 consisting of a core and a single extra atom, a PacMan path exists for the extra atom to move to any other location on the periphery.*

Proof. Consider a cube of size four with an additional atom. For the purposes of this discussion, we assume that neither the start nor goal location are on the +Z or -Z faces of the cube. (Rotation of the overall structure does not affect the possible shapes of the paths.) We start a path concatenation argument by noting that if a Z segment of a path reaches a corner of the XY plane containing its goal, it can turn and subsequently reach any peripheral location in that plane. To follow a Z segment that reaches any layer, it is sufficient to reach any ±Z pair of corners of the cube from an X or Y segment. This ability can be shown for any start location by looking at (1) locations in the top and bottom layers and (2) locations in the middle layers. Here we discuss a start location off of the -Y face, but by symmetry, all locations follow the same argument. In the first case, the farthest corner in the same layer can be reached by a path that goes in +Y and then ±X, while the corresponding corner in the far layer can be reached by a path in +Y, ±Z, ±X. For the second case, the far corner in the farther layer (two layers away) can be reached as in the first case, but the corresponding corner in the layer adjacent to the start layer requires a path in +Y, ∓Z, -Y, ±Z, ±X, +Y.

Note that any stem cell of size greater than 4 can be analyzed in a similar way to show that the extra atom can be arbitrarily relocated around the periphery. Also, a valid path can always be created by the techniques in Theorem 1, but in many cases a shorter path can be found. Our argument here simply shows that there will always be at least one valid path.

Theorem 2. *[Planning] For any relocation as described above, the distributed planner will find a valid PacMan path and leave appropriate pellets for the atom to use.*

Proof. Theorem 1 shows that a single continuous *valid* (i.e. obeying the actuation constraints) path exists for any relocation. Since the search for a moving atom is a complete DFS of the cells, a path from start to goal will be found. In addition, the search respects the PacMan turning constraint, so that any path found will be valid. The search process places pellets (incrementally) with the ID of the start atom when a path is found. Thus, the pellets created can be followed to perform the relocation.

We now continue our analysis by showing what happens when there are dangling atoms present adjacent to the stem cell, as shown in Fig. 4. In many cases (such as D1 in Fig. 4a), these atoms can simply translate across the core before the path is traveled and back to their original location once the path is completed. In other instances, a dangling atom might have to travel past other dangling atoms (see D2 and D3 in Fig. 4a). This is possible as long as the two atoms do not share a vertex such as in Fig. 4b. We define any stem cell that does not contain such pairs of dangling atoms as *well-behaved.*

Theorem 3. *[Dangler handling] Any relocation as described in Theorem 1 can be performed in any well-behaved stem cell.*

Proof. For any planar reconfiguration, there will always be at least one edge that is not adjacent to the path, and therefore "safe" for dangling atoms. The existence of a free edge can be shown on a case-by-case basis. For any non-planar reconfiguration, the presence of a cubic core gives even more safe locations. By the analysis of Theorem 1, at most eight linear path segments are required for any path. Even if all these segments are on the surface of the core, at most 25 modules can be involved in the path for a core of size four. There are 52 total surface modules with adjacent free space (i.e. not counting those on the floor), so there will always be sufficient places for dangling atoms. Since a path can be found from any peripheral location to any other, the planner will always be able to generate a path for any dangler.

Thus, a path through a single stem cell can always be performed successfully. We can extend this analysis to a collection of well-behaved stem cells. This analysis is detailed in 2-D in [1], and extends directly to the 3-D case. One additional consideration for 3-D reconfigurations is the presence of gravity. The nature of the unit-compressible actuation makes it possible to ensure stability for most shapes. In particular, the motion of a module during reconfiguration never leaves the boundaries of the initial shape. Assuming the initial and final shapes are stable, this is generally sufficient to ensure that the reconfiguring shape is stable, since the moving module leaves the edge of the group, moves through the center, and out to the perimeter. For some paths, it is possible for a module to go across the center of mass only during the middle of the path, in which case the initial and final shapes cannot be marginally stable — they must be stable with the addition of one module mass anywhere in the system.

7 Enabling Parallelism

The basic PacMan algorithm, while compact and easy to understand, has some limitations in the ability to support many intersecting paths without undergoing disconnection of the Crystal, both in 2-D and in 3-D. In [1], we showed that the basic PacMan algorithm will not produce deadlocks during parallel actuation, but the class of paths that could be performed in parallel without disconnection of the group was extremely limited. In this section, we introduce two additions to the basic PacMan algorithm which can greatly increase the class of simultaneous paths that can be assured to complete successfully. We first present and analyze an extension under which any set of paths that does not include a cycle can be executed at arbitrary relative times without fragmenting the system. A set of paths is said to contain a cycle if the directed graph formed by the paths (with vertices at each atom) contains a cycle. The second extension guarantees that PacMan will still never deadlock while avoiding fragmentation whenever possible.

7.1 Using State to Avoid Fragmentation

The first addition to the PacMan algorithm addresses how to limit the disconnections adjacent to path intersections in the Crystal. Under the original algorithm, when a path approaches a module that a different path is already using, the atom executing the second path will not disconnect. Thus, the disconnections in the Crystal will remain separate (see Fig. 6a). However, once the first path is through the intersection, the second can use the intersection immediately, as in Fig. 6b, and cause the disconnections to link up, potentially fragmenting the robot. If the first path then turns as in Fig. 6c, its atoms can be disconnected regardless of the size of the Crystal.

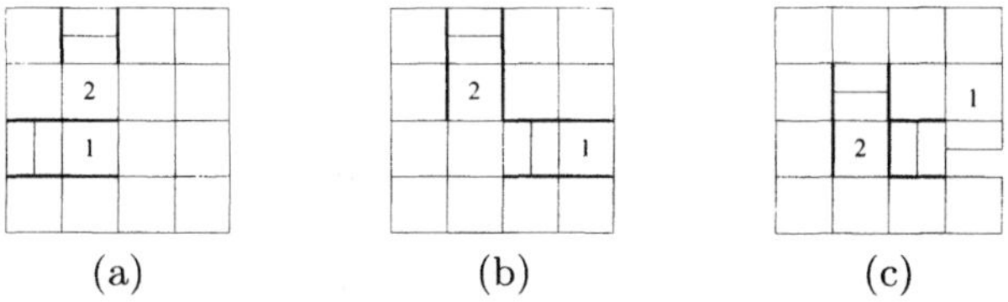

Fig. 6. Two intersecting paths that lead to fragmentation under basic PacMan.

To fix this problem, we introduce a boolean state variable and an additional message type. This will allow us to have the disconnections around the intersection be symmetric and not link up. The added state is a boolean *nbr_wait* that signifies whether an atom has recently been used by a path and has a contracted neighbor. This is set to 1 when a pellet is received after an ID return message (signifying an intersection, as this additional pellet belongs to a path yet to use this atom). It will remain set until a "Neighbor OK" message comes from the atom that it swapped with, which will happen as detailed in Algorithm 4.

Algorithm 4 Extension to PacMan

```
if busy and contracted then
   if !(Is_nbr(prev_ID)) or at_goal(prev_ID) then
      Expand(prev_dir)
      Connect(perp(prev_dir))
      busy = 0
      Send("Neighbor OK") in opp(prev_dir)
```

As long as the *nbr_wait* variable is set, the atom will refuse to give dibs to the atom executing the second path. With this additional restriction, the intersection will evolve over time as shown in Fig. 7.

Analysis The additional state and message described in Section 7.1 enables correct planning and actuation for any set of paths not containing a cycle. To prove this, we look at the immediate neighborhood of each path intersection and show that the atom(s) that occupy the location of the intersection will remain attached to all other atoms in the Crystal throughout the actuation of the two paths. We can then induce from this result that no set of intersections can disconnect the Crystal given the displacement restriction.

We show the correctness of a single intersection by analyzing the connectedness of various components of the Crystal during actuation. We use the term *mover* to refer to a virtual atom that is executing a path, that is, an atom with an ID for which pellets exist in the Crystal.

Lemma 1. *Any Crystal with two intersecting paths will remain connected during actuation when using PacMan with the extension in Algorithm 4.*

Proof. Consider two intersecting paths as shown in Figure 7a. The only possible point of fragmentation during the parallel execution of the two paths is around the intersection point marked in Figure 7a by X. Suppose path 1 obtains dibs from X first, we will have the situation shown in Fig. 7. At each of these positions, note that the structure remains globally connected despite local disconnections. Also, between each step, connection is also maintained. If path 2 obtained dibs first a similar process would take place, also retaining global connectivity.

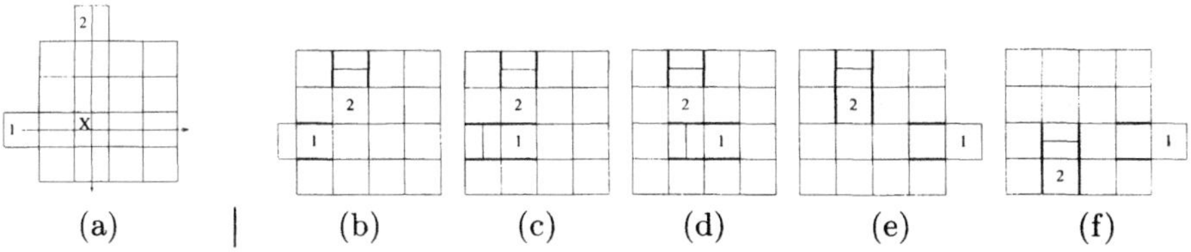

Fig. 7. (a) Two intersecting paths and (b-f) one of two actuation sequences for these paths under Algorithm 2 with the addition of Algorithm 4.

Finally, if one of these paths turns near the intersection, the actuation still succeeds. If the corner is before the intersection, then the path would disconnect near the intersection only when it got dibs on the intersection, and its disconnections due to the corner would not touch the intersection otherwise. If the corner is after the intersection, then the "Neighbor OK" message would not be sent until after the corner had been filled and the nearby atoms expanded, and so the intersection would behave the same way.

Theorem 4. *Any set of paths will be executed by the atoms without fragmenting the Crystal provided a mover is never displaced.*

Proof. For ease of exposition, we first examine the case when all the paths are straight by an inductive case analysis. Lemma 1 guarantees no fragmentation for two intersecting paths. Consider the effect of a third intersecting path. The same process will take place around the added intersection point. Note that all disconnections in the immediate neighborhood of a single intersection are only in one direction at any given time. The disconnections due to the two intersections can only adjoin if they belong to the common path or if the two intersections are adjacent and the active paths parallel (which is enforced by the minimum segment length restrictions). This inductive argument applies for each additional path and intersection in the absence of cycles. As long as the mover does not trade identities with a neighbor, it will wait to use the intersection until it can do so safely.

For paths with corners, there will be disconnections that are perpendicular and touching adjacent to the corner. However, as long as each path segment is of the required length, these disconnections alone cannot fragment the Crystal. Any additional path intersecting the turning path will be subject to the restriction above, so the disconnection due to the corner cannot extend beyond the neighborhood of the corner and cause fragmentation.

7.2 Using State to Handle Cycles and Deadlocks

The restriction in the previous theorem points out an inherent compromise in the original PacMan algorithm. In the original algorithm, deadlock is avoided by allowing one mover to displace another to break a cycle of waiting paths. However, this can cause a group of atoms to become disconnected

Algorithm 5 Augmented PacMan message handler

```
Switch (message type):
if "Request transfer" then
    if not(busy or dibs or nbr_wait) and (not mover or (waiting_for = sender)) then
        Dibs = sender's ID
        Send("OK") to sender
    else if mover then
        if waiting_for < ID then
            Send("Wait for", ID) to sender
        else
            Send("Wait for", waiting_for) to sender
    else
        Send("Not OK") to sender
if "Wait for" then
    waiting_for = received ID.
if "Trading ID" then
    Function from Algorithm 3
    Mover = 1
if "Return ID" then
    Function from Algorithm 3
    Mover = received mover
```

from the main group. A simple alternative is to disallow cycles completely (or equivalently, to allow deadlock) by not allowing a mover to give dibs[3]. This will guarantee that the Crystal stays connected by Theorem 4, but at the expense of potential deadlock. The extension presented here allows both cycles and guaranteed connection in the absence of cycles, but at the expense of additional messages and atom state.

The additional information and computation serves two purposes: (1) to allow each atom to know explicitly whether it is a mover or not, and (2) to detect cycles in the active paths. With this information, we allow the default behavior to be for a mover to not give dibs to another atom, to avoid fragmentation as detailed in Theorem 4. However, to avoid deadlock, each mover that is denied dibs will determine the identity of the mover at the head of the chain. If that mover is itself, it can override the rejection and get dibs anyway (as it would be able to immediately in the basic algorithm). The changes are only in the message handler, which is augmented as shown in Algorithm 5.

Analysis

Theorem 5. *The augmented PacMan algorithm will prevent fragmentation in the absence of cycles, and deadlock in all cases.*

Proof. The augmented algorithm centers on the detection of cycles in the current paths. When a cycle occurs, each atom will be denied dibs by the one

[3] This can be done without any additional state by simply changing the times when the "busy" flag is set and cleared.

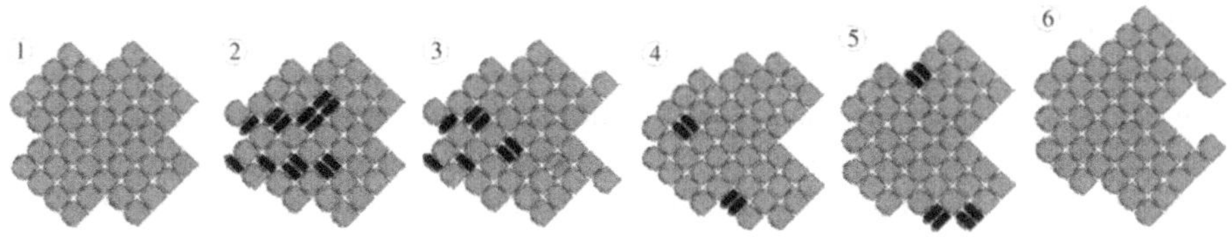

Fig. 8. A sequence of screenshots of a PacMan simulation. The black modules are contracted and are currently involved in the actuation process.

ahead of it, as all will be movers and none will be able to move. Since the atoms ask repeatedly for dibs, even if the cycle is entered at multiple points, eventually all of them will be waiting for the atom with the largest ID. This atom will then assert priority and trade IDs with its neighbor, alleviating deadlock. On the other hand, if a cycle is not present, there will always be one atom capable of advancing along its path. Any atom that must wait for that mover (or wait for one waiting for that mover, etc) will not be able to wait for itself and therefore assert priority. This will enforce serialization of paths around intersections and prevent movers from trading IDs, thus preventing fragmentation.

8 Experiments

8.1 Simulation

The PacMan primitive has been implemented in simulation, and can perform 2-D reconfigurations such as the one shown in Fig. 8 as well as 3-D reconfigurations. In the current simulation, a separate process is run for each atom, to obtain reasonable fidelity to the hardware. An additional process is run to simulate physics and facilitate communication between simulated atoms.

8.2 Hardware

In the second generation of the Crystal (seen in Fig.9), of which 18 modules have been constructed, each module is completely independent, having computation, communication and power all onboard. In this system, all communication and actuation is necessarily asynchronous, and since all communication is between neighboring modules, there is no capability for broadcast messages. Detailed descriptions of the hardware and low-level communication protocol (all developed in our lab) are presented in [2]. In support of this work, we have implemented the basic PacMan algorithm on the Crystal. Previous algorithms for the Crystal have used message loops as their basic architecture, and so it was straightforward to use the same infrastructure for PacMan. Message handlers were written as described in Sec. 5, with the PacMan actuation protocol a separate function called by the message handlers.

Our implementation was run on an 11-module robot, and was able to both plan and execute various single paths as well as intersecting straight-line paths, showing that the algorithm (including traffic control) can perform correctly in a physical distributed asynchronous system. Snapshots of one experiment are given in Fig. 9. The algorithms worked correctly, although some slight mechanical assistance was required to allow the modules to reach their intended goal configurations. These failures are solely due to compliance in the module faces and connectors, and we are currently working to address this hardware problem in a new version of the system.

Fig. 9. Crystal hardware performing a PacMan experiment. The reconfiguration being executed is the same as the one shown schematically in Fig. 3.

9 Conclusions

This paper has presented the PacMan algorithm, with which unit-compressible modular robot systems can plan and execute self-reconfigurations in a parallel and distributed fashion. Analysis shows that nearly any self-reconfiguration can be achieved with this technique. We have also presented analysis that shows that with additional messaging in the actuation, PacMan can successfully perform in parallel any set of paths that does not contain a cycle, and will not deadlock in cases where a cycle exists.

As a further extension to PacMan, we have developed a related algorithm for systems in which modules move over the surface of the group. In this work, we use an generic model of a self-reconfiguring module, and show that very similar path-planning and actuation algorithms can apply despite the different actuation. In the future we hope to instantiate this generic algorithm to particular systems, both algorithmically and in hardware.

We have also presented initial implementation on the Crystal robot hardware, and we hope to continue this work by implementing the extensions to the actuation as well as improving the hardware itself. Finally, it also seems likely that PacMan could be used for locomotion without too much difficulty, especially for surface-moving systems, and this will also be investigated.

Acknowledgments.
Support for this work was provided through the NSF CAREER award IRI-9624286, NSF awards IRI-9714332, EIA-9901589, IIS-9818299, IIS-9912193, EIA-0202789 and 0225446, and DARPA award F30602002-0585. Murphy Stein worked on the simulation and developed the reconfiguration shown in Fig. 1.

References

1. Z. Butler, S. Byrnes, and D. Rus. Distributed motion planning for modular robots with unit-compressible modules. In *Proc. of the Int'l Conf. on Intelligent Robots and Systems*, 2001.
2. Z. Butler, R. Fitch, and D. Rus. Experiments in distributed locomotion with a unit-compressible robot. In *Proc. of the Int'l Conf. on Intelligent Robots and Systems*, pages 2813–8, 2002.
3. T. Fukuda and Y. Kawakuchi. Cellular robotic system (CEBOT) as one of the realization of self-organizing intelligent universal manipulator. In *Proc. of IEEE ICRA*, pages 662–7, 1990.
4. K. Kotay and D. Rus. Locomotion versatility through self-reconfiguration. *Robotics and Autonomous Systems*, 26:217–32, 1999.
5. K. Kotay and D. Rus. Algorithms for self-reconfiguring molecule motion planning. In *Proc. of the Int'l Conf. on Intelligent Robots and Systems*, 2000.
6. W. H. Lee and A. Sanderson. Dynamic analysis and distributed control of the tetrabot modular reconfigurable robot system. *Autonomous Robots*, 10(1):67–82, 2001.
7. S. Murata, H. Kurokawa, E. Yoshida, K. Tomita, and S. Kokaji. A 3-D self-reconfigurable structure. In *Proc. of IEEE ICRA*, pages 432–9, May 1998.
8. S. Murata, E. Yoshida, K. Tomita, H. Kurokawa, A. Kamimura, and S. Kokaji. Hardware design of modular robotic system. In *Proc. of the Int'l Conf. on Intelligent Robots and Systems*, pages 2210–7, 2000.
9. A. Nguyen, L. Guibas, and M. Yim. Controlled module density helps reconfiguration planning. In *Algorithmic and Computational Robotics: Proceedings of WAFR 2000*, pages 23–35, 2000.
10. A. Pamecha, C-J. Chiang, D. Stein, and G. Chirikjian. Design and implementation of metamorphic robots. In *Proc. of the 1996 ASME Design Engineering Technical Conf. and Computers in Engineering Conf.*, 1996.
11. A. Pamecha, I. Ebert-Uphoff, and G. Chirikjian. Useful metrics for modular robot motion planning. *IEEE Trans. on Robotics and Automation*, 13(4):531–45, 1997.
12. D. Rus and M. Vona. Crystalline robots: Self-reconfiguration with unit-compressible modules. *Autonomous Robots*, 10(1):107–24, 2001.
13. J. Suh, S. Homans, and M. Yim. Telecubes: Mechanical design of a module for self-reconfigurable robotics. In *Proc. of IEEE ICRA*, 2002.
14. K. Tomita, S. Murata, H. Kurokawa, E. Yoshida, and S. Kokaji. Self-assembly and self-repair method for a distributed mechanical system. *IEEE Trans. on Robotics and Automation*, 15(6):1035–45, Dec. 1999.
15. S. Vassilvitskii, M. Yim, and J. Suh. A complete, local and parallel reconfiguration algorithm for cube style modular robots. In *Proc. of IEEE ICRA*, 2002.

Distributed Manipulation with Rolling Contact

Mark Bedillion and William Messner

Department of Mechanical Engineering, Carnegie Mellon University
capn@andrew.cmu.edu bmessner@andrew.cmu.edu

Summary. The Modular Distributed Manipulator System (MDMS) is an array of actuators that is capable of manipulating objects in the plane. This paper presents a controller designed to transport objects to the origin of a coordinate system under rolling contact. A stability proof using the multivariable circle criterion directly accounts for discontinuities due to changing actuator sets. Simulations verify the controller's effectiveness and show stability even when the contact switches between viscous friction and full rolling contact.

1 Introduction

The Modular Distributed Manipulator System (MDMS) was first introduced in 1995 [1] as a system for positioning and orienting objects in the plane. The MDMS consists of many small actuators in a fixed array that cooperate to manipulate objects that lie above them. Each unit of the array, or cell, consists of an actuator capable of approximating a vector force, a sensor capable of detecting an object, and a microcontroller that communicates with neighboring cells and controls the actuator. The system is capable of operating at a completely distributed level such that sensing, planning, and actuation are accomplished without a centralized controller.

While much of the previous work on the MDMS has focused on object dynamics under a viscous friction contact assumption [2, 8], this paper assumes no-slip rolling contact between the object and the actuator wheels. This no-slip contact makes more precise manipulation possible because the dynamics of the object are found directly from knowledge of actuator dynamics. It also reduces system power consumption and wear by turning on motors only when their wheels are in contact with the object and only at a rate consistent with the object's velocity. However, some potential problems with rolling contact are the tight coordination required between cells and strict velocity limits needed to ensure that slipping does not occur.

This paper is organized as follows. Section 2 describes previous research into actuator arrays and prior work on the MDMS. Section 3 presents the con-

J.-D. Boissonnat et al. (Eds.): Algorithmic Foundations of Robotics V, STAR 7, pp. 453–468, 2004.

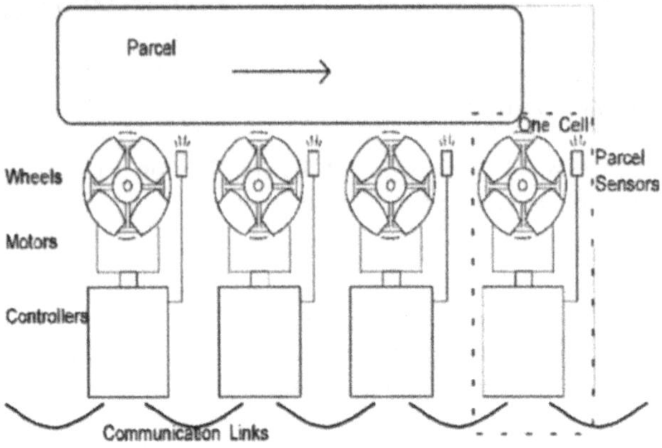

Fig. 1. MDMS

trol strategy. Section 4 converts the problem for analysis via the multivariable circle criterion. Section 5 completes a stability proof. Section 6 presents simulation results for the controller operating in both full rolling and stick-slip contact. Finally, Section 7 presents conclusions and discusses open issues.

2 Prior Work

Prior work on actuator arrays has largely focused on MEMS scale systems whose action can be described by continuous force fields. This research area has applied the sensorless manipulation approach of Mason and Erdman [3, 4] and Goldberg's use of a sequence of squeezes to orient parts [5] to actuator arrays. Bohringer and Donald [6] created a system with cilia-like actuators and analyzed object motion using continuous field approximations. Their work with Kavraki [7] developed fields that could position and orient any non-symmetric part. The modeling of the MDMS has differed from the models of [6] and [7] because spacing between actuators is not considered negligible relative to object size, and thus continuous force field approximations are inappropriate.

Luntz et. al. [2] derived the equations of motion for the MDMS assuming a single point of contact at each cell. These dynamics were exploited to develop open-loop and closed-loop control laws for positioning and orienting objects on the array [8]. Bedillion et. al. [10] derived the array dynamics assuming full rolling contact at all wheels and proposed a controller for open-loop trajectory following. This paper employs the same dynamics model in the stability proof.

Murphey and Burdick examined contact modes and open-loop instability under elliptic fields for actuator arrays [9]. Their work showed that orientation could be stabilized through simple proportional feedback. This paper differs from Murphey and Burdick's work in that it uses a dynamic rather than quasistatic model, considers full rolling contact, and explicitly considers changes in support set. Furthermore, this paper employs somewhat more standard nonlinear analysis tools than what they used.

3 A Regulating Controller

Figure 2 shows an object supported by several actuators of the MDMS. The

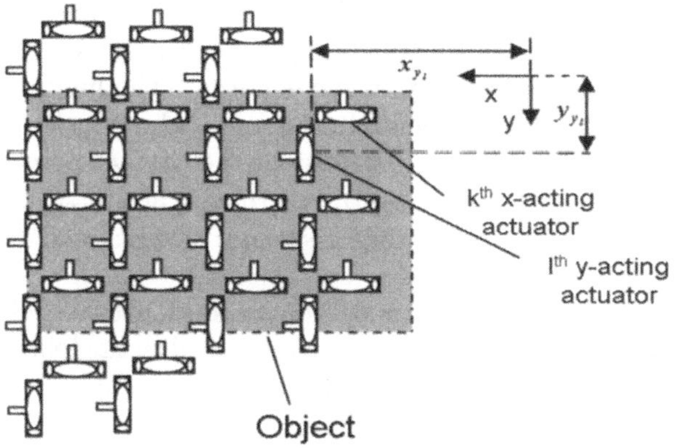

Fig. 2. Top View of Actuator Array with Coordinate System

dynamics of an object manipulated by the array are derived in [10] and are summarized here. The relationship between the force applied by an actuator, F_j, its motor current, i_j, and its rotational speed, ω_j, is given by

$$J_m\dot{\omega}_j + \beta_m\omega_j = \frac{F_j r}{N_g} + k_t i_j, \tag{1}$$

where J_m is the motor inertia, β_m is the motor damping, k_t is the torque constant, r is the wheel radius, and N_g is the gear ratio. A standard electrical model of each motor including resistance R, inductance L, applied voltage u_j, and back emf relates the electrical properties of the motor to its mechanical properties by

$$L\frac{di_j}{dt} + Ri_j = u_j - k_t\omega_j. \tag{2}$$

Assuming that the inertia, damping, and inductance of the motor are small enough to be neglected in the analysis, the preceding differential equations reduce to the algebraic equations

$$\frac{F_j r}{N_g} + k_t i_j = 0 \tag{3}$$

$$Ri_j = u_j - k_t\omega_j. \tag{4}$$

Combining equations to eliminate the motor current and using the relation that $\omega_j = \frac{v_j N_g}{r}$ where v_j is the velocity at the wheel perimeter, the force acting on each wheel is given by

$$F_j = \frac{k_t N_g}{Rr}(\frac{k_t N_g}{r}v_j - u_j) = \alpha v_j - \beta u_j, \tag{5}$$

where $\beta = \frac{k_t N_g}{Rr}$ and $\alpha = R\beta^2$.

Because there is no relative motion between the wheel and the object under rolling contact, the object velocity at the point of contact may be substituted for the wheel velocity in the equation. The linear velocity of the object at the point of contact is equal to the sum of the velocity of the center of mass, $(\dot{x}_{cm}, \dot{y}_{cm})$, and the relative velocity at the contact point due to object rotation, $(\dot{\theta}(y_{x_k} - y_{cm}), \dot{\theta}(x_{y_l} - x_{cm}))$, where $\dot{x}_{cm}$ and $\dot{y}_{cm}$ are the x and y components of the velocity of the object's center of mass, and $\dot{\theta}$ is the object's angular velocity.

Substitution into Equation 5 and the assumption that cell positions are measured relative to the object's center of mass lead to the following equations for the force on the wheel in terms of the object motion and the input voltage

$$F_{x_k} = \alpha\dot{x}_{cm} - \alpha\dot{\theta}y_{x_k} - \beta u_{x_k}, k = 1, \ldots, m \tag{6}$$

$$F_{y_l} = \alpha\dot{y}_{cm} + \alpha\dot{\theta}x_{y_l} - \beta u_{y_l}, l = 1, \ldots, n. \tag{7}$$

By summing the contributions from all supporting cells the following equations of motion of an object on the array are arrived at [10]

$$M\ddot{x}_{cm} + \alpha m\dot{x}_{cm} - \alpha\dot{\theta}\sum_{k=1}^{m} y_{x_k} - \beta\sum_{k=1}^{m} u_k = 0 \tag{8}$$

$$M\ddot{y}_{cm} + \alpha n\dot{y}_{cm} + \alpha\dot{\theta}\sum_{l=1}^{n} x_{y_l} - \beta\sum_{l=1}^{n} u_l = 0 \tag{9}$$

$$J\ddot{\theta} + \alpha\sum_{l=1}^{n}[x_{y_l}\dot{y}_{cm} + \dot{\theta}y_{y_l}^2] - \alpha\sum_{k=1}^{m}[y_{x_k}\dot{x}_{cm} - \dot{\theta}x_{x_k}^2] - \beta\sum_{l=1}^{n} x_{y_l}u_{y_l} + \beta\sum_{k=1}^{m} y_{x_k}u_{x_k} = 0. \tag{10}$$

where M and J are the object's mass and inertia and m and n are the number of supporting cells oriented in the x and y directions. These equations of motion exhibit strong coupling between translational and rotational motion, thus complicating the control problem. Note that m, n, and the terms involving the positions of supporting cells change discontinuously as the object changes sets of supporting cells.

The control problem consists of finding a controller that can transport an object from an arbitrary initial configuration to the origin of a coordinate system. It is desired that only local information is required at each cell so that the array can operate as a distributed system. The assumption is that each cell only has knowledge of its position relative to the object's center of mass as well as the position and orientation of the object relative to the desired origin. This information could be obtained through a centralized vision system broadcasting the object's position and orientation or through local measurements via distributed sensing techniques. This centralized/decentralized approach allows the same information to be passed to each cell from a centralized source and does not overburden a central processor with calculating and disseminating cell-specific information when the array becomes large.

To maintain rolling contact at each wheel, the controller is designed based on kinematic considerations. A proportional controller is applied at each motor in the following form

$$u_{x_k} = -G_x x_{cm} + G_\theta \theta y_{x_k} \tag{11}$$

$$u_{y_l} = -G_y y_{cm} + G_\theta \theta x_{y_l} \tag{12}$$

where G is a gain whose subscript defines the direction.

One appealing property of this controller is that it guarantees rolling contact in some neighborhood of the origin in the six-dimensional phase space. Rolling contact is ensured when the applied force at a given cell is less than the product of the coefficient of static friction and the normal force, or

$$\begin{bmatrix} \alpha\dot{x}_{cm} - \alpha\dot{\theta} y_{x_1} - \beta u_{x_1} \\ \vdots \\ \alpha\dot{x}_{cm} - \alpha\dot{\theta} y_{x_m} - \beta u_{x_m} \\ \alpha\dot{y}_{cm} + \alpha\dot{\theta} x_{y_1} - \beta u_{y_1} \\ \vdots \\ \alpha\dot{y}_{cm} + \alpha\dot{\theta} x_{y_n} - \beta u_{y_n} \end{bmatrix} \leq \mu_s \mathbf{N} \tag{13}$$

where μ_s is the static coefficient of friction and $\mathbf{N}$ is the vector of normal forces. When the control law of (11) and (12) is substituted for the input voltages in (13), each term on the left hand side of the inequality is linear in the states. Thus, as the object approaches the origin, the left hand side approaches the zero vector.

The distribution of voltages that the control law produces is also optimal with respect to the square of the 2-norm of the voltages when the generation of forces and torques are viewed separately. The solution to the optimization problem

$$\begin{aligned} \min \quad & \|u\|_2 \\ s.t.\ \beta \textstyle\sum_{k=1}^{m} u_x &= F_x \end{aligned} \tag{14}$$

is given by

$$\mathbf{u}_x = \begin{bmatrix} 1 \\ \vdots \\ 1 \end{bmatrix} \left([1 \cdots 1] \begin{bmatrix} 1 \\ \vdots \\ 1 \end{bmatrix} \right)^{-1} \frac{F_x}{\beta} = \begin{bmatrix} 1 \\ \vdots \\ 1 \end{bmatrix} \frac{F_x}{\beta m}. \tag{15}$$

Similarly, the optimal solutions for a specified y-force are

$$\mathbf{u}_y = \begin{bmatrix} 1 \\ \vdots \\ 1 \end{bmatrix} \frac{F_y}{\beta n} \tag{16}$$

and the optimal solution for a specified torque is

$$\mathbf{u} = \begin{bmatrix} y_{x1} \\ \vdots \\ y_{xm} \\ -x_{y1} \\ \vdots \\ -x_{yn} \end{bmatrix} \frac{\mathcal{T}}{\beta \left(\sum_{l=1}^{n} x_{y_l}^2 + \sum_{k=1}^{m} y_{x_k}^2 \right)}. \tag{17}$$

Likewise, in the control law, a pure x-direction error creates a uniform field of voltage inputs in the x-direction, a pure y-direction error creates a uniform field in the y-direction, and a pure orientation error generates a field scaled by the actuator locations. Note that while the control law preserves the direction of the force or torque, the magnitude depends on the number of cells covered at a given instant. Applying a force in this manner also generates a disturbance torque, and conversely applying a torque generates a force, due to coupling.

4 Circle Criterion Form

The stability proof of the controlled system under full rolling contact uses the multivariable circle criterion. The multivariable circle criterion proves stability of systems in the form of Figure 3, where G is a linear system and ψ is a sector nonlinearity as defined in [11], by guaranteeing the existence of a common Lyapunov function for all nonlinearities in the sector. The dynamics of an object on the array with the controller applied are described by

$$M\ddot{x}_{cm} + \alpha m \dot{x}_{cm} - \alpha\dot{\theta} \sum_{k=1}^{m} y_{x_k} + \beta m G_x x_{cm} - \beta G_\theta \theta \sum_{k=1}^{m} y_{x_k} = 0 \tag{18}$$

$$M\ddot{y}_{cm} + \alpha n \dot{x}_{cm} + \alpha\dot{\theta} \sum_{l=1}^{n} x_{y_l} + \beta n G_y y_{cm} + \beta G_\theta \theta \sum_{l=1}^{n} x_{y_l} = 0 \tag{19}$$

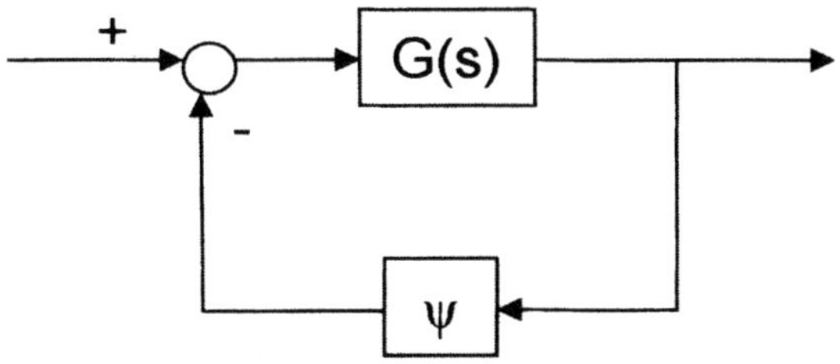

Fig. 3. Circle Criterion Form

$$J\ddot{\theta} + \alpha\dot{y}_{cm}\sum_{l=1}^{n} x_{y_l} + \alpha\dot{\theta}\sum_{l=1}^{n} x_{y_l}^2 - \alpha\dot{x}_{cm}\sum_{k=1}^{m} y_{x_k} + \alpha\dot{\theta}\sum_{k=1}^{m} y_{x_k}^2$$
$$+\beta G_y y_{cm}\sum_{l=1}^{n} x_{y_l} + \beta G_\theta \theta \sum_{l=1}^{n} x_{y_l}^2$$
$$-\beta G_x x_{cm}\sum_{k=1}^{m} y_{x_k} + \beta G_\theta \theta \sum_{k=1}^{m} y_{x_k}^2 = 0. \quad (20)$$

To apply the circle criterion, the dynamics must be divided into linear and nonlinear portions. To do this, it is useful to extract the minimum values m_{min}, n_{min}, and I_{min} of the number of x-acting and y-acting supporting cells and the inertia term $(\sum_{l=1}^{n} x_{y_l}^2 + \sum_{k=1}^{m} y_{x_k}^2)$, respectively, over the entire configuration space for a given object. By defining the operators

$$\mathbf{M}_x = M\frac{d^2}{dt^2} + \alpha m_{min}\frac{d}{dt} + \beta G_x m_{min} \quad (21)$$
$$\mathbf{M}_y = M\frac{d^2}{dt^2} + \alpha n_{min}\frac{d}{dt} + \beta G_y n_{min} \quad (22)$$
$$\mathbf{J} = J\frac{d^2}{dt^2} + \alpha I_{min}\frac{d}{dt} + \beta G_\theta I_{min} \quad (23)$$

the equations of motion can be written in the somewhat more compact form

$$\mathbf{M}_x x(t) + \Delta m(\alpha\frac{d}{dt} + \beta G_x)x(t) - \sum_{k=1}^{m} y_{x_k}(\alpha\frac{d}{dt} + \beta G_x)\theta(t) = 0 \quad (24)$$

$$\mathbf{M}_y y(t) + \Delta n(\alpha\frac{d}{dt} + \beta G_y)y(t) + \sum_{l=1}^{n} x_{y_l}(\alpha\frac{d}{dt} + \beta G_y)\theta(t) = 0 \quad (25)$$

$$\mathbf{J}\theta(t) + \Delta I(\alpha\frac{d}{dt} + \beta G_\theta)\theta(t) - \sum_{k=1}^{m} y_{x_k}(\alpha\frac{d}{dt} + \beta G_x)x(t)$$
$$+\sum_{l=1}^{n} x_{y_l}(\alpha\frac{d}{dt} + \beta G_y)y(t) = 0 \quad (26)$$

where Δ terms represent variations from the minimum values of those variables.

A change of variables is defined by

$$\chi(t) = (\alpha \frac{d}{dt} + \beta G_x) x(t). \tag{27}$$

The expression for $x(t)$ in terms of χ is then

$$x(t) = \frac{1}{\alpha} \int_0^t e^{-\frac{\beta G_x}{\alpha}(t-\tau)} \chi(\tau) d\tau = \mathbf{F}_x \chi(t) \tag{28}$$

where $\mathbf{F}_x$ is a convolution operator. Likewise

$$\gamma(t) = (\alpha \frac{d}{dt} + \beta G_x) y(t)$$
$$\phi(t) = (\alpha \frac{d}{dt} + \beta G_\theta) \theta(t)$$

and

$$y(t) = \mathbf{F}_y \gamma(t)$$
$$\theta(t) = \mathbf{F}_\theta \phi(t)$$

where $\mathbf{F}_y$ and $\mathbf{F}_\theta$ are the appropriate convolution operators. The equations of motion then become

$$\mathbf{M}_x \mathbf{F}_x \chi(t) = -\Delta m \chi(t) + \sum_{k=1}^{m} y_{x_k} \phi(t) \tag{29}$$

$$\mathbf{M}_y \mathbf{F}_y \gamma(t) = -\Delta n \gamma(t) + \sum_{l=1}^{n} x_{y_l} \phi(t) \tag{30}$$

$$\mathbf{J} \mathbf{F}_\theta \phi(t) = -\Delta I \phi(t) + \sum_{k=1}^{m} y_{x_k} \chi(t) - \sum_{l=1}^{n} x_{y_l} \gamma(t). \tag{31}$$

Figure 4 shows the equations in block diagram form, where

$$\psi = \begin{bmatrix} \Delta m & 0 & -\sum_{k=1}^{m} y_{x_k} \\ 0 & \Delta n & \sum_{y=1}^{n} x_{y_l} \\ -\sum_{k=1}^{m} y_{x_k} & \sum_{y=1}^{n} x_{y_l} & \Delta I \end{bmatrix}. \tag{32}$$

The transfer function of the system in the upper block is

$$G(s) = \begin{bmatrix} G_{11}(s) & 0 & 0 \\ 0 & G_{22}(s) & 0 \\ 0 & 0 & G_{33}(s) \end{bmatrix}$$

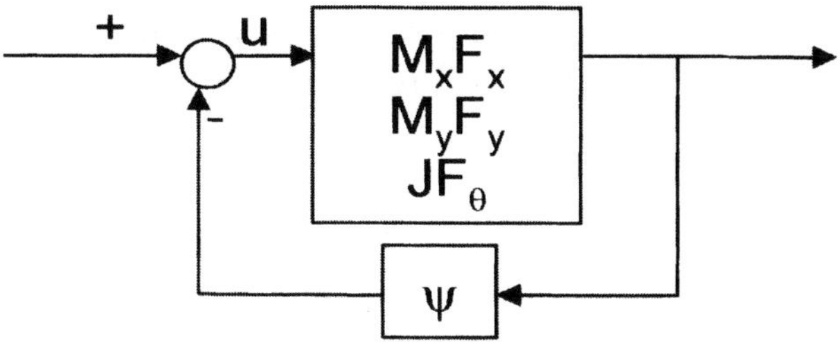

Fig. 4. Block Diagram Form

where the elements are derived from

$$\begin{aligned} u(s) &= \mathscr{L}\{\mathbf{M}_x\mathbf{F}_x\chi(t)\} \\ &= (Ms^2 + \alpha m_{min}s + \beta G_x)\mathscr{L}\{\mathbf{F}_x\chi(t)\} \\ &= \frac{Ms^2 + \alpha m_{min}s + \beta G_x}{\alpha s + \beta G_x}\chi(s), \end{aligned}$$

implying

$$G_{11} = \frac{\alpha s + \beta G_x m_{min}}{Ms^2 + \alpha m_{min}s + \beta G_x m_{min}} \tag{33}$$

$$G_{22} = \frac{\alpha s + \beta G_y n_{min}}{Ms^2 + \alpha n_{min}s + \beta G_y n_{min}} \tag{34}$$

$$G_{33} = \frac{\alpha s + \beta G_\theta I_{min}}{Js^2 + \alpha I_{min}s + \beta G_\theta I_{min}}. \tag{35}$$

The system is now in the analysis form used by the multivariable circle criterion. Note that all coupling terms are now contained in the nonlinearity, which makes analysis of the G matrix much easier.

5 Stability Proof

The multivariable circle criterion states that the system of Figure 3 with sector nonlinearity $\psi \in [0, \infty]$ is globally uniformly asymptotically stable if $G(s)$ is strictly positive real. The matrix ψ will lie in the sector $[0, \infty]$ if it is positive semidefinite. A proper rational transfer function matrix is strictly positive real if and only if [11]

1. $G(s)$ is Hurwitz
2. $G(j\omega) + G^T(j\omega)$ is positive definite $\forall \omega \in R$

3. Either $G(\infty)+G^T(\infty)$ is positive definite or it is positive semidefinite and $\lim_{\omega\to\infty}\omega^2 M^T\left(G(j\omega)+G^T(j\omega)\right)M$ is positive definite for any $p\times(p-q)$ full rank matrix M such that $M^T\left(G(j\omega)+G^T(j\omega)\right)M=0$, where $q=$ rank $\left(G(\infty)+G^T(\infty)\right)$

In the special case that $G(\infty)+G^T(\infty)=0$, the third condition corresponds to a requirement that $\lim_{\omega\to\infty}\omega^2\left(G(j\omega)+G^T(j\omega)\right)$ is positive definite. The proof will proceed by first establishing the definiteness of ψ and then confirming that $G(s)$ is strictly positive real.

As a simple example, consider the one dimensional actuator array described by the equations

$$G(s)=\frac{\alpha s+\beta G_x m_{min}}{Ms^2+\alpha m_{min}s+\beta G_x m_{min}} \tag{36}$$

and

$$\psi=\Delta m. \tag{37}$$

In this SISO case, the circle criterion has a simple graphical interpretation. By definition, ψ is always nonnegative and therefore belongs to the sector $[0,\infty]$. The requirement that G be strictly positive real corresponds to the requirement that the Nyquist plot of G not enter the left half plane as shown in Figure 5. A sufficient condition for the stability of the system under the prescribed controller is

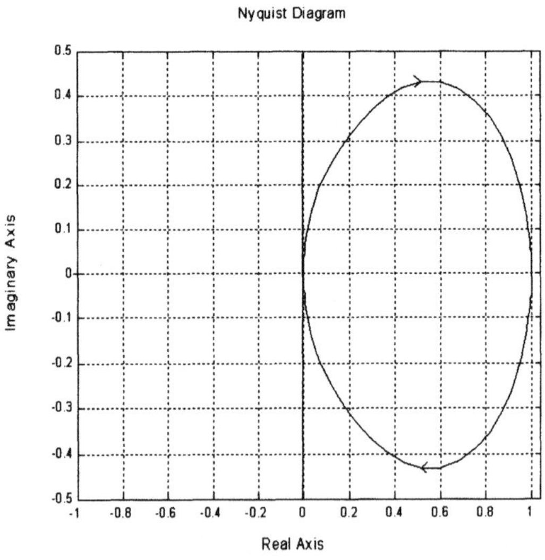

Fig. 5. Circle Criterion for One Dimensional System

$$G_x < \frac{\alpha^2 m_{min}}{\beta M} \tag{38}$$

regardless of how large Δm is. While the multivariable circle criterion does not retain the graphical interpretation, it works on the same principle.

Definiteness of ψ

The definiteness of ψ can be determined by examining its eigenvalues. The eigenvalues of ψ are the roots of its characteristic equation

$$\begin{aligned}
&\lambda^3 - (\Delta n + \Delta m + \Delta I)\,\lambda^2 + [\Delta m \Delta n + \Delta m \Delta n \\
&+\Delta m \Delta I + \Delta n \Delta I - \left(\sum_{k=1}^{m} y_{x_k}\right)^2 - \left(\sum_{l=1}^{n} x_{y_l}\right)^2]\lambda \\
&+\Delta m \left(\sum_{l=1}^{n} x_{y_l}\right)^2 + \Delta n \left(\sum_{k=1}^{m} y_{x_k}\right)^2 - \Delta m \Delta n \Delta I = 0.
\end{aligned} \tag{39}$$

The condition for all eigenvalues to be positive, or ψ to be positive semidefinite, is then that

$$\Delta I - \frac{\left(\sum_{l=1}^{n} x_{y_l}\right)^2}{\Delta n} - \frac{\left(\sum_{k=1}^{m} y_{x_k}\right)^2}{\Delta m} \geq 0. \tag{40}$$

Unfortunately, this will not be true for arbitrary orientations of the object, because ΔI can be zero with all other terms being nonzero, leaving a negative result.

We overcome this difficulty through the loop transformation of Figure 6 [11]. Positive semidefiniteness is guaranteed by choosing

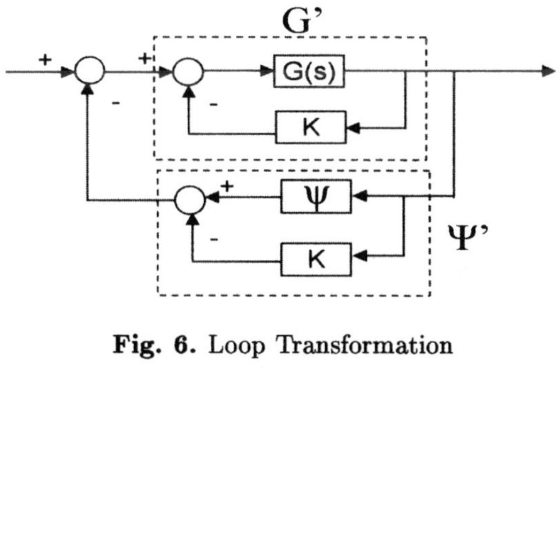

Fig. 6. Loop Transformation

$$\psi' = \begin{bmatrix} \Delta m^\star & 0 & -\sum_{k=1}^{m} y_{x_k} \\ 0 & \Delta n^\star & \sum_{l=1}^{n} x_{y_l} \\ -\sum_{k=1}^{m} y_{x_k} & \sum_{l=1}^{n} x_{y_l} & \Delta I^\star \end{bmatrix}, \tag{41}$$

where $\Delta m^\star = \Delta n^\star = \Delta m + \frac{\max \sum_{k=1}^{m} y_{x_k}}{d}$ and $\Delta I^\star = \Delta I + 2d \max \sum_{k=1}^{m} y_{x_k}$. Subsequently,

$$G'_{11} = \frac{\alpha s + \beta G_x m^\star}{Ms^2 + \alpha m^\star s + \beta G_x m^\star} \tag{42}$$

$$G'_{22} = \frac{\alpha s + \beta G_y n^\star}{Ms^2 + \alpha n^\star s + \beta G_y n^\star} \tag{43}$$

$$G'_{33} = \frac{\alpha s + \beta G_\theta I^\star}{Ms^2 + \alpha I^\star s + \beta G_x I^\star}, \tag{44}$$

where d is the distance between cells, $m^\star = n^\star = \left(m_{min} - \frac{\max \sum_{k=1}^{m} y_{x_k}}{d}\right)$, $I^\star = (I_{min} - 2d \max \sum_{k=1}^{m} y_{x_k})$, and the largest possible diagonal term has been extracted. Note that due to the symmetry of the array, $\max \sum_{k=1}^{m} y_{x_k} = \max \sum_{l=1}^{n} x_{y_l}$ over all possible object configurations.

Strict Positive Realness of G

It is easy to see that if $\left(m_{min} - \frac{\max \sum_{k=1}^{m} y_{x_k}}{d}\right)$ and $(I_{min} - 2d \max \sum_{k=1}^{m} y_{x_k})$ are greater than zero, G' is Hurwitz ($m_{min} = n_{min}$ over all configurations). Because G' is diagonal and all nonzero elements have the same structure, it is sufficient to examine the (1,1) entry to determine the rest of the stability conditions. The second condition for strict positive realness will hold if the real part of each of the diagonal elements is greater than zero. For the first element, the real part is given by

$$Re(G_{11}(j\omega)) = \frac{(\alpha^2 m^\star - M\beta G_x)\,\omega^2 + \beta^2 G_x^2 m^\star}{M\omega^4 + (\alpha^2 m^{\star 2} - 2M\beta m^\star G_x)\,\omega^2 + \beta G_x^2 m^{\star 2}}. \tag{45}$$

From this it is easy to determine a bound on G_x for stability of

$$G_x < \frac{\alpha^2 \left(m_{min} - \frac{\max \sum_{k=1}^{m} y_{x_k}}{d}\right)}{\beta M}. \tag{46}$$

An identical bound is found for G_y, while the bound for rotational gain is given by

$$G_\theta < \frac{\alpha^2 (I_{min} - 2d \max \sum_{k=1}^{m} y_{x_k})}{\beta J}. \tag{47}$$

The third condition for positive realness can be easily verified given the bounds above.

The bounds on the proportional gains can be increased by adding derivative control. A fundamental limitation, however, is encountered by the requirement that $\left(m_{min} - \frac{\max \sum_{k=1}^{m} y_{x_k}}{d}\right)$ and $(I_{min} - 2d \max \sum_{k=1}^{m} y_{x_k})$ be positive. This requirement roughly corresponds to a statement that the linear system pulls the system towards the origin more strongly than the coupling pulls the system away from the origin at all times, and is obviously conservative in nature. An object with its center of mass located far from its centroid will not satisfy this requirement. However, if the center of mass and centroid of an object are collocated, this statement will be true for large enough objects, because $\max \sum_{k=1}^{m} y_{x_k}$ will scale with the length of such an object while m_{min} and $\left(\sum_{l=1}^{n} x_{y_l}^2 + \sum_{k=1}^{m} y_{x_k}^2\right)_{min}$ will scale with the area.

5.1 Simulation

The object's motion on the array under various conditions was simulated in Simulink. The simulated system under the control law of Equations 11 and 12 is always stable, even for very high gains and systems that do not meet the Hurwitz requirement. Figure 7 shows such a simulation in which both the Hurwitz requirement and the gain requirement are violated. Furthermore,

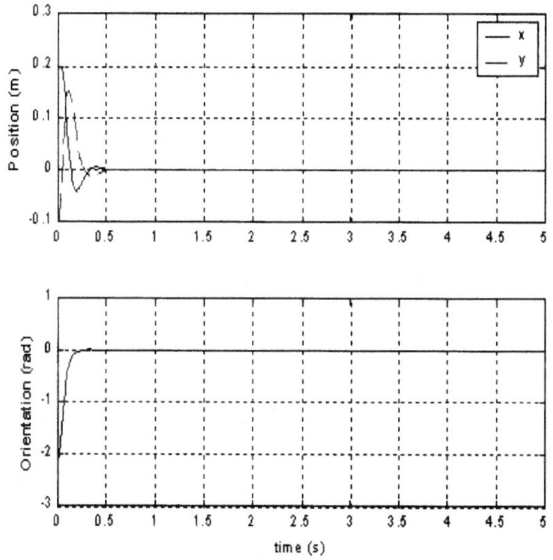

Fig. 7. Full Rolling Contact Simulation

simulations that included stick-slip effects showed the control law to be stable even under changing contact conditions and full-slip contact. Figure 8 shows a simulation of the object's motion under full slip contact and Figure 9 under switching between rolling and slipping contact. Note that Figure 9 is merely a combination of Figures 7 and 8. The object begins in full slipping contact because it is far away from the origin and the control input is proportional to the distance. As the object approaches the origin and slows down, it switches into full rolling contact as predicted by Equation 13. Our efforts to prove this through the multivariable circle criterion have not met with success, however, so we are pursuing other strategies.

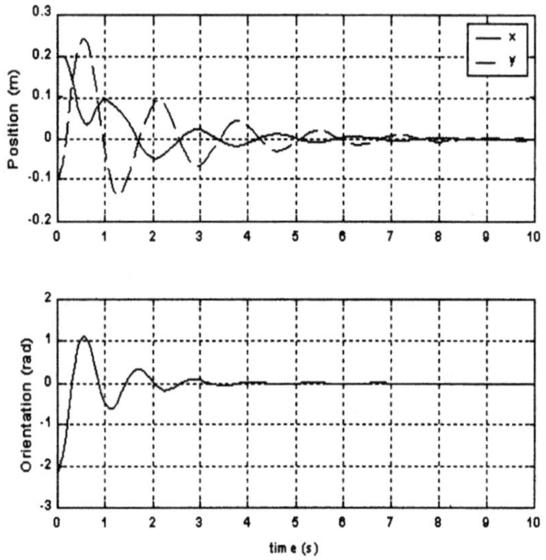

Fig. 8. Full Slip Simulation

6 Conclusions and Future Work

This paper has presented a controller designed to regulate objects on the MDMS under rolling contact. The control design is based primarily on kinematic considerations and the controller maintains rolling contact near the origin and has an optimal voltage distribution with respect to the 2-norm of voltage squared. Conditions under which the control law is stable were developed through the multivariable circle criterion. Finally, simulations demonstrated that the control law is stable outside of the bounds given by the circle criterion and demonstrated stability under stick-slip and full slipping contact.

While the simulation results are suggestive, a proof of the controller's stability under differing contact states is needed. Our attempts to prove stability using the multivariable circle criterion have thus far failed, and we are looking at other techniques for generating common Lyapunov functions. We would also like to extend the gain minimums found here, but the addition of derivative control gives flexibility in choosing gains as is. A further goal

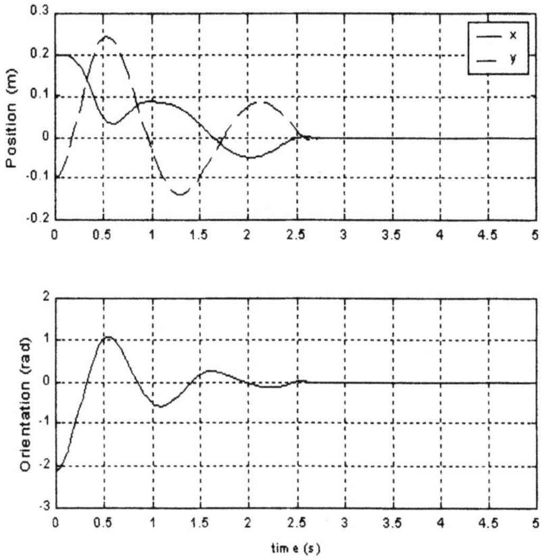

Fig. 9. Stick-Slip Simulation

is to examine the properties of this controller used in conjunction with the feedforward technique presented in [10] to create a stable trajectory following controller. Finally, we are in the process of building a new prototype system to test the controller's effectiveness.

Acknowledgements
The authors gratefully acknowledge NSF support IIS-9872255 for this work. The authors would also like to acknowledge Dr. Jonathan Luntz for creating Figure 1.

References

1. Luntz J.E., Rollins E, and Messner W (1995) The Virtual Vehicle-A Materials Handling System Using Highly Distributed Coordination Control. Proceedings: IEEE Conference on Control Applications 742-7
2. Luntz J.E.,Messner W, and Choset H (1998) Velocity Field Design for the Modular Distributed Manipulator System. In: Agarwal P.K., Kavraki L.E., and Mason M.T.(eds) Robotics: The Algorithmic Perspective. A.K. Peters, Ltd.
3. Akella S, Huang W, Lynch K,and Mason M (1996) Sensorless Parts Feeding with a One Joint Robot. Proceedings: Workshop on the Algorithmic Fundamentals of Robotics
4. Erdmann M (1995) An Exploration of Nonprehensile Two-Palm Manipulation: Planning and Execution. Proceedings: Seventh International Symposium on Robotics Research

5. Goldberg K (1993) Orienting Polygonal Parts Without Sensors. Algorithmica: Special Issue on Computational Robotics. 10(2/3/4)201-25
6. Bohringer K, Donald B, Mihailovich R, and MacDonald N (1994) Sensorless Manipulation Using Massively Parallel Microfabricated Actuator Arrays. Proceedings: IEEE International Conference on Robotics and Automation 826-33
7. Bohringer K, Donald B, Kavraki L, and Lamiraux F (2000) A Distributed, Universal Device for Planar Parts Feeding: Unique Part Orientation in Programmable Force Fields. In: Bohringer K.F. and Choset H (eds) Distributed Manipulation. Kluwer Academic Publishers 1-28
8. Luntz J.E., Messner W, and Choset H (2000) Discreteness Issues in Actuator Arrays. In: Bohringer K.F. and Choset H (eds) Distributed Manipulation. Kluwer Academic Publishers 103-126
9. Murphey T.D. and Burdick J.W. (2001) On the Stability and Design of Distributed Manipulation Control Systems. Proceedings: IEEE International Conference on Robotics and Automation
10. Bedillion M and Messner W (2001) Rolling Contact Dynamics and Optimal Feedforward Control for the Modular Distributed Manipulator System. Proceedings: ASME International Mechanical Engineering Conference and Exhibition
11. Khalil H (2001) Nonlinear Systems. Prentice Hall

Computing Equilibria on Superpositions of Logarithmic-Radial Potential Fields

Hyungpil Moon and Jonathan Luntz

The University of Michigan, Ann Arbor MI 48109, USA

Abstract. Many distributed manipulation systems are capable of generating planar force fields which act over the entire surface of an object to manipulate it to a stable equilibrium within the field. Passive air flow and other physical phenomena, naturally generate force fields through the linear superposition of logarithmically varying radial potential fields. The main advantage of these fields is that they are realizable through very simple actuation. However, they do not lend themselves to analytical prediction of net forces or equilibria. This paper presents an efficient means of numerically computing the net force and moment exerted by such fields on objects composed of multiple simple shapes, as well as efficient means of finding equilibrium points on these fields.

1 Introduction

Distributed manipulation systems induce motions on objects through the application of many external forces. The general task is to impart desired motion to an object, by using either a large number of independently controlled actuators acting at many points on the object, or by exploiting the dynamics of a continuous medium in contact with the object. Many forms of distributed manipulation exist, including vibrating plates [15], actively controlled arrays of air jets [17,20], and planar micro [4,18], mini [19], and macro-mechanical actuator arrays [20,13]. By nature, such systems involve redundant actuation, and provide tremendous manipulation power. Many of the ideas in distributed manipulation stemmed from work in sensorless manipulation using standard robotic manipulators [16,5,8]. Böhringer, Donald, et. al. [4] applied sensorless manipulation strategies to an array of microelectromechanical actuators. Since then, researchers have examined many forms of distributed manipulation, in a wide range of scales.

Much of the prior work [3,2,6] in distributed manipulation dealt with the parts feeding problem: manipulating a single object to a single static position and orientation using a static manipulation "field". While this task represents only a small part of the capabilities of a distributed manipulation system, it is the focus of initial study since it

J.-D. Boissonnat et al. (Eds.): Algorithmic Foundations of Robotics V, STAR 7, pp. 469–485, 2004.

provides a basis for more advanced tasks such as manipulating many objects along arbitrary paths. These analyses typically searched for stable equilibria in the configuration space of the manipulated object. A continuous field approximation of micro-actuator arrays allows the applied forces to be integrated over the object "lifting" the distributed force field to a net force/moment field acting on the geometric center of the object, where conservative distributed force fields provide conservative lifted force fields [2]. Kavraki [10] supplied further analysis of sensorless microactuated systems using potential functions to prove that an elliptic potential field orients objects to symmetry. Along with Böhringer, Donald, and Lamiraux, Kavraki continued this analysis and proved that a "universal feeder orienter" field [3] brings any non-symmetric object to a single predictable orientation.

While these methodologies borrow the sensorless approach from robotic minimalism, they are not minimalistic in a practical sense; to implement these strategies requires a mechanically complex array of actuators, each of which must be individually controllable. We are looking at methods for fields generated by a class of physical phenomena such as planar air flow and electrostatic fields. Specifically this paper describes a method of generating distributed manipulation fields in which passive air flow fields (assumed to be potential flow fields), generated at only a small number of discrete points, apply forces over the entire surface of floating objects. Modeling of these flow fields relies on potential flow theory, where the air velocity is equal to the negative gradient of some potential function. Under this model, flow sinks (suction holes) are points in the plane which generate a rotationally invariant potential function which varies logarithmically with distance from the point. Potential functions (and therefore flow fields) superimpose linearly. This differs from prior work using air to manipulate objects [17,20] in that prior work required independently actuated air jets at each point in the field rather than relying on the fluid dynamics to form an actuation field from only small number of flow sources.

Our approach builds fields from small numbers of flow sinks to provide predictable position and orientation equilibrium for a wide range of part shapes. A properly established field will manipulate the object to a predictable equilibrium position and orientation. In contrast to other work in manipulation using potential fields [6] this approach produces multiple equilibria, such that local perturbation methods are not useful for predicting equilibria, and numerical computation is required. This paper presents methods for reducing computation and narrowing the search for equilibrium.

Section 2 describes a basic of logarithmic potential flow and section 3 shows a belief description of a hardware-testbed. Section 4 reviews the theory of lifting of pointwise potential fields to objects of finite extent and enumerates a few properties of logarithmic potential fields in this context. Section 5 provides efficient means of computing the net force and moment on an object by both reducing the force integration computation and superposing multiple sinks and objects. Section 6 provides some basis for reducing the search for equilibria and proposes an algorithm for efficiently searching for equilibrium.

2 Potential Flow Fields for Manipulation

When a relatively inviscid fluid such as air flows in two dimensions, its flow patterns are predictable using potential flow theory [7] where the two dimensional velocity vector flow field $\boldsymbol{V}$ is equal to the negative gradient of some two dimensional scalar potential function Φ. This theory describes complex fluid flow fields by superimposing the potential functions of a few very simple flow fields.

One standard flow field building block is the flow sink (or equivalently a negative flow source) shown in Figure 1(a). A sink is a point to which fluid flows (and disappears). The potential field and flow field associated with a flow sink are

$$\Phi = \frac{\Lambda}{2\pi} \ln(r) \qquad \text{and} \qquad \boldsymbol{V} = -\boldsymbol{\nabla}\Phi = -\frac{\Lambda}{2\pi}\frac{1}{r}\hat{e}_r, \tag{1}$$

where r is the distance from the sink and Λ is the sink's strength. The equipotential lines of this field form concentric circles and the fluid flows directly toward the sink with increasing velocity.

These flow fields can perform manipulation. Imagine an air-hockey table which is covered by a plate on top. An object is placed between the hockey table and the plate. The flow rate on the bottom of an object is just low only enough to create an air bearing. Suction hoses are placed at the top plate and they generate a manipulating flow field on top of the object. As a first approximation, assume that the traction force (per unit area) is proportional to the velocity of the air flow.

A single sink as shown in Figure 1(a) is of limited use for manipulation since it will pull an object to its center, but cannot orient it due to the field's axi-symmetry. Multiple sinks, however, can combine to form useful manipulation fields. For example, two sinks of equal strength set apart by some distance (Figure 1(b)) will produce a squeeze-like action along the line of the two sinks and tends to orient objects [12].

3 Hardware Testbed

An "air palm" (Figure 2) accomplishes this by generating the manipulation flow field on the top surface of the object while a standard air table generates an air bearing on the bottom. This system acts in three steps to accomplish parts feeding with orientation: 1) The air palm, placed approximately over an object, attracts it to its center and orients it without sensors using a pre-designed air field using flow sinks. 2) After a set time, the object is assumed to be oriented, suction increases, lifting the object and holding it firmly against the palm. 3) The palm carries the object away with known orientation.

Currently, we have built and tested an air manipulation system capable of step 1. A standard toy air-hockey table provides the air bearing, and a plexiglass plate on standoffs with tapped holes is the air palm. Hand valves individually activate the flow sinks which are powered by a shop-vac, which can generate air flow with volume flow rate $\dot{V} = 1.9L/s$ thru each hole. The Reynolds number we get out of this system is $\mathrm{Re} = 1.2 \times 10^4$ which is a reasonable value to apply potential flow theory [7]. The system manipulates small pieces of felt-covered plexiglass (for additional surface drag). While damping in the actual system is quite low, rectangular objects come to equilibrium after approximately 20 seconds, centered on and in line with any pair of activated flow sinks. Switching on a different pair of sinks moves the object from one pose to another.

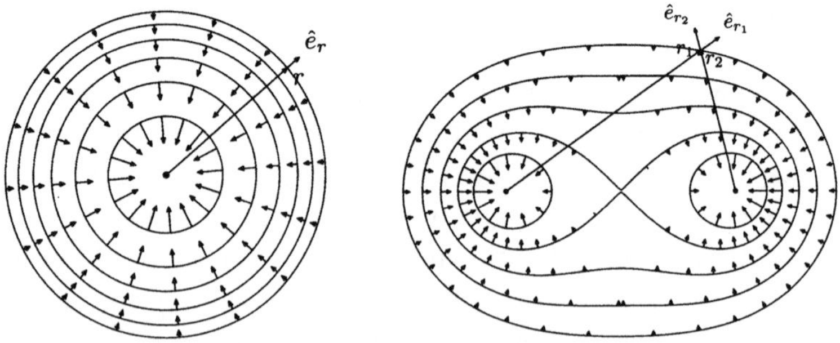

Fig. 1. Logarithmic-radial potential fields. **(a)** A flow sink: equipotential lines form concentric circles; flow moves centrally with increasing velocity. **(b)**Two flow sinks form a useful manipulation field.

4 Properties of Lifted Potential and Lifted Force

The Lifted Potential Function (LPF), introduced by Böhringer, expresses the characteristics of the force and moment a pointwise-defined potential function exerts when integrated over an object of finite extent [11]. For completeness, we restate the following definition.

Definition 1 (Lifted Potential Function). Let u be a potential function and let $q = (x, y, \theta)^T$ be the configuration of an object. The function $U(q) = \int_{D(q)} u(\xi, \eta) d\xi d\eta$ over the configuration space is called the lifted potential field induced by u.

The LPF provides the net force and moment on the object [11].

$$F_x(q) = -\int_{D(q)} \frac{\partial u}{\partial \xi} d\xi d\eta = -\frac{\partial U}{\partial x}(q)$$

$$F_y(q) = -\int_{D(q)} \frac{\partial u}{\partial \eta} d\xi d\eta = -\frac{\partial U}{\partial y}(q)$$

$$M(q) = -\frac{\partial U}{\partial \theta}(q)$$

One property of the LPF is that it is always differentiable (i.e. the net force and moment are continuous over the configuration space), even if the forces in the pointwise field are discontinuous.

Prior analyses examined a unit radial potential field $u = \|r\|$ where u is the potential and r is the distance from the origin of the field [11]. Relying on the monotonicity of the force field guarantees uniqueness of the position (but not orientation) equilibrium on a unit radial field. However, the logarithmic potential field $u = \log(\|r\|)$ does not produce

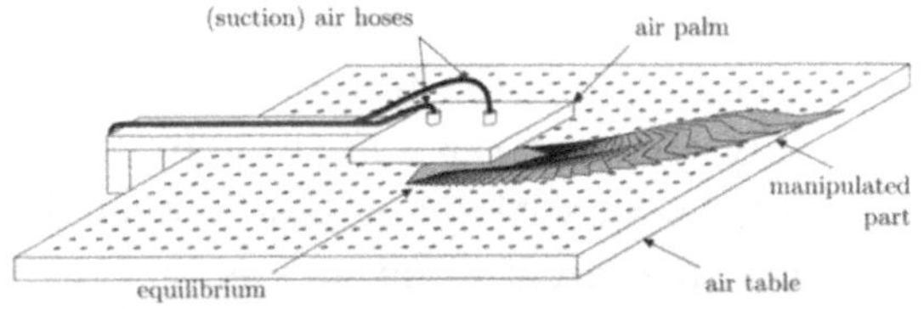

Fig. 2. An air manipulator palm hovers above a standard air table to position, orient, lift, and carry away the object.

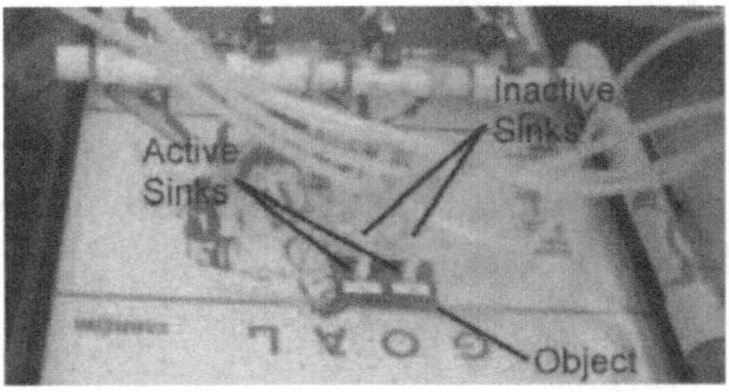

Fig. 3. The experimental air manipulator. Valves activate any combination of four sinks arranged in a square. The rectangular object is shown in the equilibrium pose obtained by activating the indicated sinks.

a monotonic force function (force decreases with distance) so we cannot guarantee unique equilibrium. We will, however limit the locations of equilibria by relying on the fact that the logarithmic potential function is harmonic (a solution of Laplace's equation: $\nabla^2 u = 0$.)

We now establish some smoothness-related properties of the lifted force function for an object of general shape on a logarithmic potential field which we will make use of later in the paper. Note that the logarithmic function has a singularity at the origin of the field. We therefore must examine three cases: the singularity outside of the object, the singularity in the interior of the object, and the singularity on the boundary. We do this in the general case of n superimposed logarithmic fields at different locations.

Proposition 1. *For a given orientation(θ^*), we have the following property depending on the number of sinks which lie within the object.*

$$\frac{\partial^2 U(x,y,\theta^*)}{\partial x^2} + \frac{\partial^2 U(x,y,\theta^*)}{\partial y^2} = 2n\pi$$

where $n \geq 0$ is the number of sinks within the object.

Proof.

This proof relies on the line integral method, based on Green's theorem, which we present in the next section. If there is no singularity inside of the object, it is obvious that the lifted potential function satisfies Laplace equation because the pointwise logarithmic potential is a harmonic function and integration over the area of the object is simply the sum of finite harmonic functions.

Consider a compact object, O_D which contains one sink in its interior. We can always divide the object into a square region (O_{B_ϵ}) centered on the sink and completely contained in the object, and the remainder of the object (O_{D/B_ϵ}). For O_{D/B_ϵ}, the integral of the gradient remains zero, so we need only consider O_{B_ϵ}. Each term of the integral of the second derivatives of the logarithmic potential over the region B_ϵ can be represented by the line integral method as:

$$\frac{\partial^2 U}{\partial x^2} = -\frac{\partial F_x}{\partial x} = \int_S \frac{\partial u}{\partial x} dy, \quad \frac{\partial^2 U}{\partial y^2} = -\frac{\partial F_y}{\partial y} = -\int_S \frac{\partial u}{\partial y} dx$$

where all the equations are evaluated at a given orientation and S is the boundary path of the square piece O_{B_ϵ} with counter-clock-wise(CCW) orientation. We can analytically compute the integral.

$$-\frac{\partial F_x}{\partial x} = \int_S \frac{x}{x^2+y^2} dy = \int_{-\epsilon}^{\epsilon} \frac{2\epsilon}{\epsilon^2+y^2} dy = \pi$$

Similarly, we get $-\frac{\partial F_y}{\partial y} = \pi$. If there exist n sinks inside of the object, we can repeat the same procedure n time and get the result, $2n\pi$. ♠. From this, we can see that the lifted force field is once differentiable everywhere except when the object boundary crosses a sink. We do know, however, that the net force is still continuous [11] and finite.

To demonstrate these properties, consider a rectangular object with fixed orientation in a single sink flow field. Figure 4 shows the lifted force as we vary distance from the single sink. As the object comes closer to the sink, the magnitude of the radial force increases and reaches a maximum when the boundary of the object crosses the sink. Once the sink is inside the object, the magnitude of the lifted force decreases until it reaches equilibrium. This non-monotonicity can generate multiple equilibrium. When the boundary of the object crosses the sink, the magnitude of the radial force is continuous and finite even though the point-wise force is infinite [14].

5 Computation of Force and Moment

It is impossible, in general, to analytically compute the net force and moment an arbitrary arrangement of sinks applies to an arbitrary object, so we must rely on numerical approaches to generate the lifted field. Numerical integration, however, is a time consuming process. We combine two approaches to make this more efficient: 1) integrate the potential function over the 1-D perimeter of the object rather than the force function over the 2-D region. 2) do integration just once for simple objects on a single sink and use some intermediate transformation to produce the force and moment for compound objects on multiple sinks from the one-time integration data.

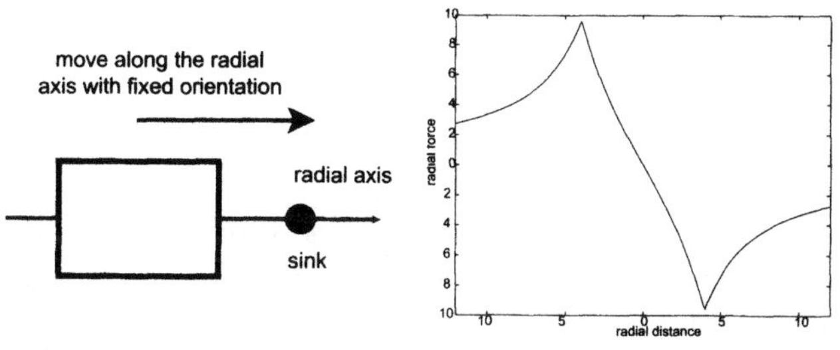

Fig. 4. Radial force : square object moving along the radial axis with fixed orientation

5.1 Force and Moment Obtained by Green's Theorem

Many forms of Green's theorem exist in vector calculus all of which generally relate a line integral with a surface integral. The following lemma is a special form of Green's theorem [9].

Lemma 1. *Let D be a region and S be its boundary. Assume that the orientation of the boundary is positive(CCW). Suppose u is C^1 in the entire domain of D. Then,*

$$\int_S u dx = -\int_D \frac{\partial u}{\partial y} dx dy, \quad \int_S u dy = \int_D \frac{\partial u}{\partial x} dx dy \tag{2}$$

Here we apply Green's theorem for computing net force and moment from the pointwise potential u.

Proposition 2. *Consider a potential function $u(x, y)$ and an object $D(q)$ of which boundary is represented by S. Then the net force vector can be obtained by the following line integral.*

$$(F_x, F_y) = \left(-\int_S u(x,y)dy, \int_S u(x,y)dx\right) = \int_S u(x,y)\hat{n}ds \tag{3}$$

where $\hat{n}$ is the unit normal vector to the boundary. Also, the moment function about an arbitrarily chosen reference point on the object can be obtained as:

$$M = \int_S u(x - x_r) \cdot tds$$

where x_r is an arbitrarily selected on the object and t is the unit tangential vector to the object boundary.

Rather than integrating the net moment, however, we can exploit the fact that for any radially directed force field each point in the field applies no moment about the origin. Therefore, there can be no net moment about the origin, and the moment about the object's reference point is simply : $M = -F_\phi r$.

Green's theorem applies to analytic functions. The fact that the logarithmic potential has a singularity is not a problem since we know the lifted potential is smooth and we can still apply Green's theorem in the case where the singularity lies within the object. One can demonstrate this by removing a small circular region centered on the sink from which we know the net force (and the net line integral) evaluates to zero. If the singularity is on the boundary, we can modify the line integral by taking an ϵ detour around the sink. For detail see [14].

We can also explain the equivalence between the line integral and the surface integral by virtual work analysis. Consider an infinitesimal virtual movement δx of an object. Consider a small segment of the boundary, Δs with slope $\tan(\theta)$ which moves in the positive direction. The incremental energy due to the segment is as following.

$$dU_x = -u \sin\theta \Delta s \delta x, \quad dU_y = u \cos\theta \Delta s \delta y$$

which, in the limit, produces Equation 3 when combined vectorially.

Using this line integral, for a single object near a single sink, we can produce a set of net force functions of two variables, r and θ: the distance between the object's reference point and the origin, and the orientation of the object's body-fixed frame relative to the line connecting the reference and the origin (see Fig. 3, r_i, θ_i in [14]). The object's third degree of freedom: polar angle ϕ is unnecessary due to the rotational invariance of the field. The result is the net radial F_r and tangential F_ϕ force field.

Performing a numerical surface integral requires $O(n^2)$ calculations for a given r and θ (where n is the linear density of numerical grid-points). Building the set of 2-D force functions requires $O(n^4)$ calculations. Using the line integral, however, reduces calculations to $O(n^3)$. Moreover, the line integral allows for analytical integration along straight line segments. Instead of a complete numerical method, we can approximate any object with straight line segments (a 1st order finite-element approach), saving further computation (particularly for polygonal objects). In fact, analytic solutions exist for polygons and circle. However, these analytic solutions are not suitable for locating stable equilibrium points, but only to calculate force and moment.

5.2 An Algorithm for Obtaining Force and Moment

So far, we discussed how we can calculate lifted force and moment at a certain position and orientation using line integral. Here we propose an algorithm to compute lifted force and moment at a range of position and full orientation for an arbitrary arrangement of sinks. Detailed explanation for the steps of joining objects and superposing sinks is given in authors' previous work [12].

Algorithm for obtaining force and moment.

1. Calculate F_r and F_ϕ for each object for a single sink. Note that the radial and tangential forces are calculated and the grid points are generated in polar coordinates. If the base shape of different object is not changed, this step needs not to be repeated.
2. Join objects together to form a compound object.
3. Compute M for the combined object from F_ϕ.
4. Transform F_r and F_ϕ to F_x and F_y on a rectlinear grid.
5. Superpose sinks by shifting operation. Only this step is repeated for rearranging of the sinks.

As Figure 5 shows, there are multiple orderings of these steps. While the relative efficiency depends on the complexity of the object and sinks, computing moment after combining objects generally saves computation (we don't need to combine moment functions). Also, it is preferable to join objects before superposing sinks since objects around a single sink join in 2-D coordinates, while multiple sinks require 3-D coordinates.

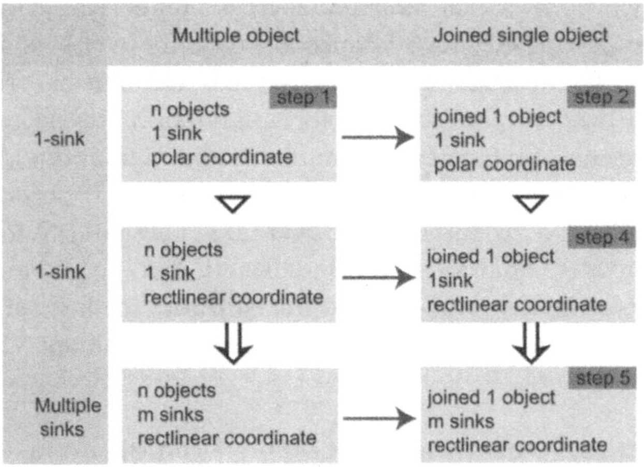

Fig. 5. Joining operation by coordinate transform and interpolation. Multiple simple objects form a compound object, and multiple instances of a single sink form a multi-sink field. → : joining objects, ▽ : coordinate transform, ⇓ : superposing sinks by shifting

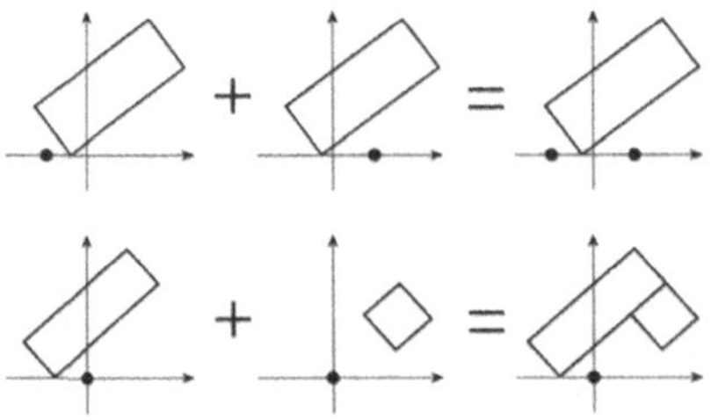

Fig. 6. Two basic operations : Combining fields. Transformed single-sink fields combine to form a multi-sink field (top, ⇓ in Fig. 5). Multiple objects transform and combine to form a single conglomerate object (bottom, → in Fig. 5).

6 Locating Equilibria

In this section, we discuss an algorithm to find equilibrium point(s) from the calculated force and moment matrices which are obtained in the previous sections. A "brute force" approach is to search the 3-D matrices for $F = 0$ and $M = 0$. Our approach reduces the region of searching by considering existence conditions of the equilibrium. In addition, we present an incremental method of searching for equilibrium which applies in some circumstances.

Proposition 3. *For combinations of radially directed fields, there exists at least one equilibrium within the set of all configurations where $O_h \cap S_h \neq \emptyset$. Here O_h means the convex hull of an object and S_h means the convex hull of the set of field origins. Equivalently, that there are no equilibrium points outside of the field-origin-convex-hull.*
Proof.
Given any pose q of the object where $O_h(q) \cap S_h = \emptyset$, there exists a straight line l separating the two convex sets. Define the set of all non intersecting poses, $Q = \{q | O_h(q) \cap S_h = \emptyset\}$. Thus, δQ is the set of all poses where O_h just contacts S_h, forming a two dimensional manifold equivalent to the configuration space "obstacle" generated by the two convex hulls.

At each pose $q \in \delta Q$, the separating line l is tangent to the boundary of the convex-hull of the sinks. At a given orientation θ, by the nature of the construction of δQ, l is also tangent to δQ restricted to θ. Since all sinks lie on the opposite side of l from the object, the net force applied to the object must have a component towards the interior of δQ restricted to θ. Thus, by the definitions and lemmas presented in [11], this is an inward-pointing field and therefore there must exist at least one equilibrium in the interior of δQ (equivalently, in the set of poses where $O_h \cap S_h \neq \emptyset$). Note the results extend [11] to regions with non-smooth boundaries by showing that the force is inward pointing for all possible separating lines at non-smooth points. ♠

Proposition 3 restricts the searching region though it doesn't save enormously. Logarithmic-radial fields allow a stronger statement.

Proposition 4. *For a logarithmic radial field, at least one sink lies inside the object at stable equilibrium pose. Equivalently no stable equilibrium can exist when any object doesn't cover any sink.*
Proof.

Here we show that the two eigenvalues of the Hessian of the lifted potential have different signs, indicating at best a saddle point(unstable equilibrium) when the object does not cover any sink. The characteristic equation of the Hessian is given by $s^2 - trace(H)s + det(H) = 0$. Because the lifted potential is harmonic when no sink exists inside of the object,$U_{xx} + U_{yy} = 0$, which is $trace(H(U))$ where $H(\cdot)$ represents the Hessian matrix. It is easy to see $\det H(U) = U_{xx}U_{yy} - U_{xy}^2 = -U_{xx}^2 - U_{xy}^2 < 0$ if $U_{xy} = U_{yx} \neq 0$. Therefore the eigenvalues have opposite sign. If $U_{xy} = U_{yx} = 0$ the Hessian is not positive definite and such a location can not be a stable equilibrium. ♠

This proposition constrains the region where the equilibrium points may exist, reducing the search. Toward incremental construction, we make the following definitions.

Definition 2. A fixed-orientation equilibrium is the position of the object reference point at a given orientation where $F = 0$ but not necessarily $M = 0$.

Definition 3. The force equilibrium curve set is the set of all fixed-orientation equilibria over all orientations. These curves exist in $\Re^2$.

The following properties are useful to the searching algorithm.

Property 1. At any particular orientation, the set of fixed-orientation equilibria is a set of measure zero for a finite set of sinks *i.e.* finite set of points rather than continuous regions.

Property 2. The force equilibrium curve set for a compact object exists as a finite number of continuous one dimensional curves. Orientation changes as the curves are traversed (although not necessarily monotonically). Each connected component is a closed curve *i.e.* fixed-orientation equilibria can not simply appear or disappear as the object rotates through (some range of) orientations (although the direction of rotation can switch along the curve).

Property 3. Property 2 holds with the additional requirement that orientation spans 2π.

Property 4. Property 3 holds with the additional requirement that orientation changes monotonically along the curves, *i.e.* each curve is a function of orientation.

Property 1 is intuitively understandable from Proposition 1 and we assume it holds for finite numbers of sinks. Properties 2 thru 4 are based on continuity of the lifted force field F over the configuration space and the implicit function theorem [1]. Note that Property 3 implies Property 4 but vice versa. An incremental searching algorithm requires Property 3. Properties 3 and 4 may not hold in all situations.

Based on the propositions and properties we introduce an incremental searching algorithm.

Algorithm 1 for searching for total equilibrium, we can incrementally construct the force equilibrium curve set when the sinks are placed relatively far enough each other as discussed later in detail. Then we can search this curve set for moment equilibrium.

1. For an arbitrary starting orientation, use a brute-force numerical search to find all fixed-orientation equilibria for only those configurations where the object covers at least one sink (based on Proposition 4). This is only a 2-D search. Incremental searching starts at these points.
2. Increment the orientation and search for fixed-orientation equilibria only in the neighborhood of each previous fixed-orientation equilibrium (based on Property 4).
3. Repeat second step until all the curves are searched.
4. Search moment equilibrium along the force equilibrium curves with increasing orientation to find points where $M = 0$ and $\frac{\partial M}{\partial \theta} < 0$.

Example 1. Figure 7 shows the force equilibrium curves and total equilibrium point for a rectangular object. The force equilibrium curves consist of three small "circles", one near each sink where the object overlaps the sink and produces a stable equilibrium, and two larger circles in the center, where the object does not overlap a sink, producing unstable equilibria. Note that all these curves lie completely within the convex hull of the sinks. Total equilibria are obtained by examining the moment along each force equilibrium curve.

The previous algorithm assumed that the force equilibrium curve is a true function of orientation, *i.e.* Property 4. This is only supported by the implicit function theorem when no singularities cross the object boundary during the incremental construction process. Cases do exist where singularity crossing occurs, for which we present the following modified algorithm which works whenever Property 3 holds.

Algorithm 2 modifies steps 2 and 3 from Algorithm 1.

2. Increment the orientation and search for fixed-orientation equilibria only in the neighborhood of each previous fixed-orientation

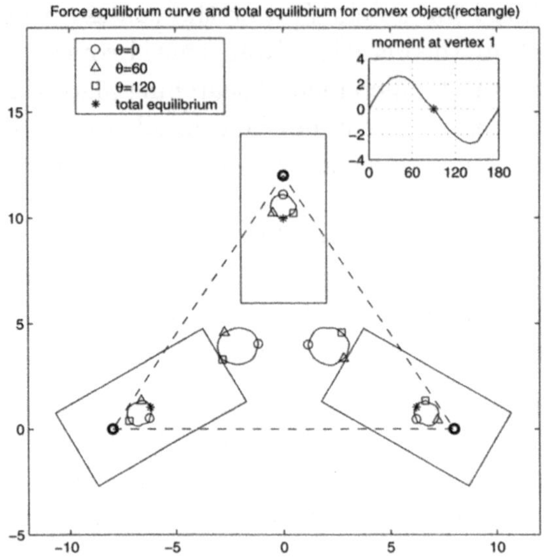

Fig. 7. Example : Convex object(Rectangle : 8 × 4) with three sinks. We use the centroid as a reference. Bold open circles represent sinks connected by their convex hull (dotted lines). Solid lines indicate force equilibrium curves with a few orientations (including total equilibria) marked. The object is draw in each of three stable total equilibrium poses. The moment near vertex one is plotted as a function of orientation.

equilibrium (based on Property 3). Multiple equilibrium points in the neighborhood indicate change in direction of θ along the curve.

3. Repeat second step until all curves are searched, incrementing either forwards or backwards as indicated by the θ direction along the curve.

Example 2. Figure 8 shows the force equilibrium curves and total equilibrium points for an L-shape object. Total stable equilibrium exists on each of three curves near each sink. The "bite" taken out of each of these curves represents the region where the singularity leaves the object and the direction of θ reverses along the curve. Thus these are unstable fixed-orientation equilibria.

Unfortunately cases exist where even Property 3 doesn't hold and we cannot find all the force equilibrium curves from an arbitrary starting orientation. Figure 9 shows the force equilibrium curve and the moment at each force equilibrium curve for such a case. Notice that when the object is close to the sink at $(-6,0)$ (at the left corner), the force equilibrium curve appears only in the orientation range of $80^\circ \sim 170^\circ$. In this case, if we start searching at zero orientation based on the proposed algorithm, we cannot construct the force equilibrium curve near the sink at the left corner. We need to search for force equilibrium points in all possible regions at every orientation.

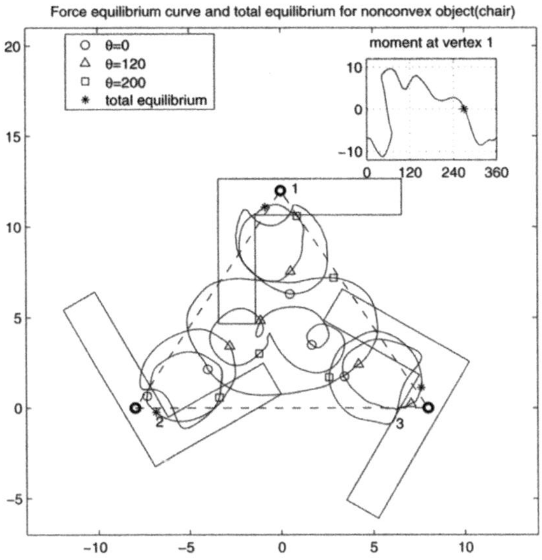

Fig. 8. Example : Non-convex object which is also a composite of two 2 × 8 square objects with three sinks. The combined position equilibrium point is the reference. Each L-shape object at each vertex represents a stable equilibrium pose. Two curves in the middle of the sink convex-hull are unstable according to Proposition 4.

As shown before, the force equilibrium curves fall into three cases, 1) they exist through 2π orientation and orientation changes monotonically as the curves are traversed 2) they exist in full range of 2π orientation but orientation changes back and forth as the curves are traversed 3) they exist only in a part of range of orientation.

Cases 2 and 3 occur when the object gets pulled off the nearest sink by the other sinks. This generates singularity crossings for which the implicit function theorem cannot guarantee that the force equilibrium curve is a function of orientation. To avoid this, sinks can be placed far enough apart so that the infinimum force by the nearest sink is greater than the combined supremum force by the others. Once this condition is satisfied, the implicit function theorem guarantees the incremental construction algorithm.

One way of doing this is as following. Consider an object is placed such that the closest sink is on the boundary and relatively far from other sinks. The minimum pulling force assumes the mass of the object is concentrated at its farthest point from the closest sink. Also,the maximum force by the other sinks assumes the mass is concentrated at the closest point to each other sink. As a conservative bound, if the minimum pulling force by the closest sink is greater than the maximum force of the others, the object is pulled closer the closest sink and trapped. Thus, the force equilibrium curve set exists in whole 2π range,

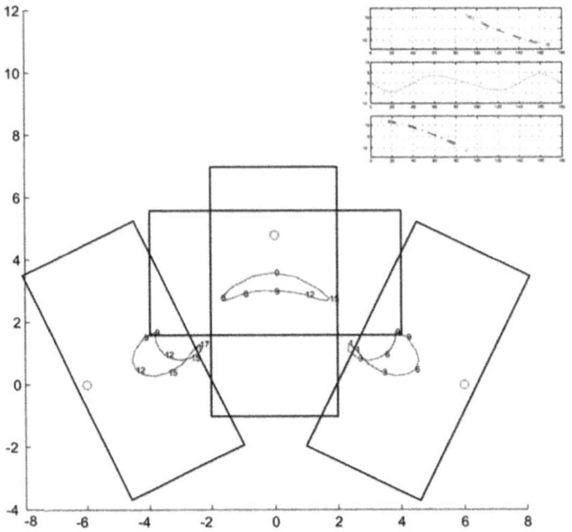

Fig. 9. Example : Convex object(Rectangle : 8 × 4) with three sinks close each other. The graphs on the upper right corner are moment plots as a function of orientation at each vertices. Top one is for the object is at the left corner, middle on is for the middle and the bottom one is for the right corner. The numbers on the force equilibrium curve represent the index of orientation angle.

guaranteeing that Property 4 holds and that Algorithm 1 works. We have no way yet to guarantee that Property 3 holds without Property 4 (where Algorithm 2 works).

7 Conclusions

This paper lays out methodology for dealing with objects being manipulated by linear combinations of radial potential fields, including specific methodology for logarithmic potentials which are the type typically produced by physical phenomena such as air flow and electromagnetic fields. While these sorts of fields have less manipulation power than fields examined by other researchers, they can be produced by physically simple means rather than requiring highly dense arrays of actuators. The main drawback is the necessity to rely on numerical computation to find equilibria on these fields, so we have presented efficiency improvements for computing the forces and the equilibria of objects on these fields.

Providing a design tool to generate useful flow field based on the result presented in this paper is the subject of future work.

Acknowledgement. This work has been supported in part by the NSF under CAREER award RHA-0093312.

References

1. E.K. Blum. *Numerical Analysis and Computation : theory and practice.* Addison-Wesley, 1972.
2. K.F. Böhringer, B.R. Donald, and N.C. MacDonald. Upper and lower bounds for programmable vector fields with applications to mems and vibratory parts feeders. In *WAFR*, 1996.
3. K.F. Böhringer, B.R. Donald, and *et.al.* Part orientation with one or two stable equilibria using programmable vector fields. In *Workshop on Distributed Manipulation at ICRA*, 1999.
4. K.F. Böhringer and *et.al.* Donald. A theory of manipulation and control for microfabricated actuator arrays. In *ICRA*, 1994.
5. M. Erdmann. An exploration of nonprehensile two-palm manipulation: Planning and execution. In *Proceedings. Seventh International Symposium on Robotics Research*, 1995.
6. F.Lamiraux and L.Kavraki. Positioning and Orienting Symmetric and Non-symmetric Parts Using Radial and Constant Force Fields. In *WAFR*, pages 37–50, March 2000.
7. F.M.White. *Fluid Mechanics.* McGrawHill, 3 edition, 1994.
8. K.Y. Goldberg. Orienting polygonal parts without sensors. *Algorithmica: Special Issue on Computational Robotics*, 10:201–225, August 1993.
9. A.J.Tromba J.E.Marsden. *Vector Calculus.* W. H. Freeman, 2 edition, 1980.
10. L. Kavraki. Part orientation with programmable vector fields: Two stable equilibria for most parts. In *ICRA*, 1997.
11. N. C. MacDonald K.F.Böhringer, B.R.Donald. Programmable Vector Fields for Distributed Manipulation, with Applications to MEMS Actuator Arrays and Vibratory Parts Feeders. *International Journal of Robotics Research*, 18(2):168–200, February 1999.
12. J. Luntz and H. Moon. Distributed Manipulation with Passive Air Flow. In *IROS*, 2001.
13. J.E. Luntz, W. Messner, and H. Choset. Closed-loop operation of actuator arrays. In *ICRA*, 2000.
14. H. Moon and J. Luntz. Distributed Manipulation by Superposition of Logarithmic-Radial Potential Field. In *ICRA*, 2002.
15. D. Reznik, E. Moshkoich, and J. Canny. Building a universal planar manipulator. In *Workshop on Distributed Manipulation at ICRA*, 1999.
16. K.Lynch M.Mason S. Akella, W.Huang. Sensorless parts feeding with a one joint robot. In *WAFR*, 1996.
17. H.Fujita S. Konoshi. A Conveyance System Using Air Flow Based on the Concept of Distributed Micro Motion Systems. *Journal of Microelectromechanical Systems*, 3(2):54–58, 1994.
18. J.W. Suh, R.B. Darling, and *et.al.* Cmos integrated organic ciliary actuator array as a general-purpose micromanipulation tool for small objects. In *Workshop on Distributed Manipulation at ICRA*, 1999.
19. S. Tadokoro and T. Takamori. Distributed actuation devices using soft gel actuators. In *Workshop on Distributed Manipulation at ICRA*, 1999.
20. M. Yim and A. Berlin. Contact and non-contact mechanisms for distributed manipulation. In *Workshop on Distributed Manipulation at ICRA*, 1999.

Feedback Control for Distributed Manipulation

Todd D. Murphey and Joel W. Burdick

Department of Mechanical Engineering
California Institute of Technology
Pasadena, CA 91125, USA

Abstract. We consider the feedback stabilization of a part being manipulated by a distributed array of actuators. Previously, the authors showed that open loop methods based on programmable vector fields can lead to unstable dynamics, particularly when the effects of rolling and sliding against actuators play a significant role in the dynamics. We show here how to incorporate these factors into the design of feedback laws in order to stabilize the manipulated object to an equilibrium. Experiments on a test-bed illustrate the results.

1 Introduction

Distributed manipulators typically consist of a large number of similar or identical actuators combined together with a control strategy to create net movement of an object or objects. The goal of most distributed manipulation systems is to allow precise planar positioning of objects from all possible starting configurations. Actuation methods range from air jets and wheels on the macroscale, to microelectromechanical systems (MEMS) and flexible cilia at the microscale. This paper is concerned with a broad class of distributed manipulation problems which involve rolling and sliding contacts between the moving object and actuator surfaces. In such cases friction forces and intermittent contact play an important role in the overall system dynamics. We will use the power dissipation method (PDM), described in Section 1.1, to produce equations of motion which are both first order and implicitly include contact forces of constraint. We will focus on autonomous controllers that stabilize an object to a precise configuration in $SE(2)$ on the array.

Methods to design distributed manipulation control systems have been proposed in several works, including [2,3,7]. These are based on the notion of *programmable vector fields (PVF).* In this methodology, one makes the possibly unrealistic assumption that the array's control capability can be idealized as a continuous distribution of forces across the array surface. In this abstraction, the manipulated object moves under the influence of these forces. The control design problem reduces to the selection of a continuous force field distribution that will transport the object to a prescribed position. To implement the control strategy on the real array, one must adapt the continuous vector field control to the real (and discrete) actuator array.

J.-D. Boissonnat et al. (Eds.): Algorithmic Foundations of Robotics V, STAR 7, pp. 487–503, 2004.

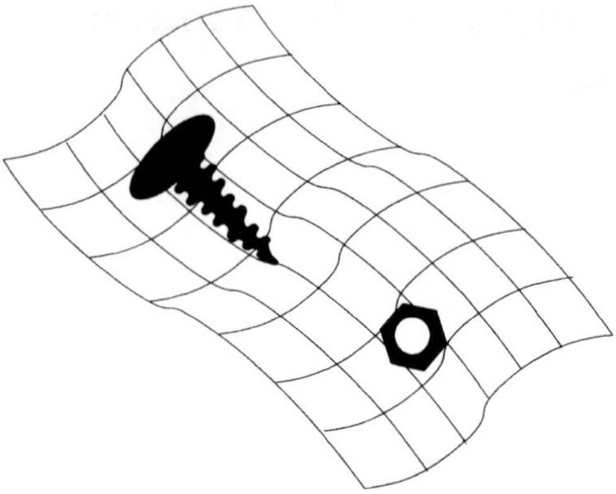

Fig. 1. Parts on a distributed manipulator

The PVF approach is experimentally known to work in some MEMS-fabricated actuator arrays, where the array elements are "small" and "close" together relative to the size of the object being manipulated [8]. It is additionally well suited to distributed air jets, because the aerodynamics effectively "smooth out" the resulting forces on the object. However, in cases where only a small number of actuators are in contact with the object being manipulated (i.e., the continuous actuation approximation is poor) or the coefficient of friction μ is very high, the continuous approximation has been shown experimentally not to work as well. In these cases, the continuous approximation does not adequately incorporate the physics of the actual array and the object/array interface.

This paper makes two main contributions: local stabilization for both fully-actuated and for under-actuated distributed manipulation systems. Section 3.3 shows that if one incorporates the contact states, one can design a local control law based on control Lyapunov methods. Section 4 considers the case of "full actuation"- when all the actuators can be oriented and driven. It gives an extremely simple, scale-able algorithm that is provably globally exponentially stabilizing. Section 5 gives experimental results on the distributed manipulation test-bed we have constructed.

1.1 Background

To analyze control system performance, we seek models for the governing equations of distributed actuation systems that faithfully capture the system's essential physics, and that are tractable and amenable to control and motion planning analysis. In pursuit of this goal, we use a "power dissipation model" (PDM) approach to model the governing dynamics of a discrete actuator/object system. This method produces unique models almost always that are relatively easy to obtain, and to which one can apply control system analysis methods. Since the method is a quasi-static modeling method, it produces first-order governing equations, instead of second order equations that are associated with Lagrange's equations. The primary disadvantage is that the method only applies to quasi-static systems. In [11] we argue that this assumption is quite good for our problems of interest, and for distributed manipulation arrays in general. Here we present an overview of the modeling methodology, and refer the reader to [11] for more details.

Let q denote the configuration of the array/object system, consisting of the object's planar location and the variables that describe the state of each actuator array element. Let us assume that the motion of the actuator array's variables are known. We treat the object and the array element contact as a rigid body contact system (though approximate compliance effects can be easily incorporated into this paradigm). We assume point contacts between the object and the array elements (complex contacts such as line contacts are modeled as a set of point contacts in this approach), and that the object's contact with the manipulating surface is governed by the Coulomb friction law at each contact point.

If $\omega_i(q)$ is the one form describing the kinematic constraint between the part and the ith actuator, then $\omega_i(q)\dot{q}$ describes the relative motion of the contact between the part and the actuator. If $\omega_i(q)\dot{q} = 0$, the contact is not slipping, while if $\omega_i(q)\dot{q} \neq 0$, then $\omega_i(q)\dot{q}$ describes the slipping velocity. In general, the moving part will be in contact with the actuator array at many points. From kinematic considerations, one or more of the contact points must be in a slipping state, thereby dissipating energy. The *power dissipation function* measures the part's total energy dissipation due to contact slippage.

Definition 1. The *Dissipation* or *Friction Functional* for an n-contact state is defined to be

$$\mathcal{D} = \sum_{i=1}^{n} \mu_i N_i \mid \omega(q)\dot{q} \mid \tag{1}$$

with μ_i and N_i being the Coulomb friction coefficient and normal force at the ith contact, which are assumed known.

Since there will generally not exist a motion where all of the contacts can be simultaneously slipless, we are lead to the following concept for finding the governing motions.

Power Dissipation Principle: With $\dot{q}$ small, an object's motion at any given instant is the one that minimizes $\mathcal{D}$.

This method is adapted from the work of [1] on wheeled vehicles. The intuition behind this is based on the fact that both μ_i and N_i typically depend on the configuration, $q \in Q$. Therefore, as the set of supports, coefficient of friction, and center of mass location all change, a different set of contacts are slipping. For each one of these choices of slipping or not slipping the model is different. Moreover, the transitions between these models can be modeled as discontinuous jumps from one model to the next. In the case of distributed manipulation, this switching corresponds to the switching among different contact states (i.e., different sets of slipping contacts) due to variations in contact geometry and surface friction properties. Other examples can be found in [16]. Now we proceed to our formal description of such systems.

Definition 2. A control system Σ is said to be a *multiple model driftless affine system (MMDA)* if it can be expressed in the form

$$\Sigma : \quad \dot{q} = g_1(q)u_1 + g_2(q)u_2 + \cdots + g_m(q)u_m \tag{2}$$

where for any q and t, $g_i \in \{\gamma_{\alpha_i} | \alpha_i \in I_i\}$, with I_i an index set, γ_{α_i} analytic in (q,t) for all α_i, and the controls $u_i \in \mathbb{R}$ piecewise constant and bounded for all i. Moreover, letting σ_i denote the "switching signals" associated with g_i (which will be referred to as "MMDA maps"),

$$\begin{aligned} \sigma_i : Q \times \mathbb{R} &\longrightarrow \mathbb{N} \\ (q,t) &\longrightarrow \alpha_i \end{aligned}$$

then the σ_i are measurable in (q,t).

An MMDA is a driftless affine nonlinear control system where each control vector field may "switch" back and forth between different elements of a finite set. In our case, this switching corresponds to the switching between different contact states between the object and the array surface elements (i.e., different sets of slipping contacts) due to variations in contact geometry, surface friction properties, and normal loading. In [10] it was shown that the PDM generically leads to MMDA systems as in Definition 2.

2 Modeling the Equilibrium Point of a Distributed Manipulation System

Here the PDM technique is applied to a specific example. Consider the actuator array shown in Fig. 2. It has only four wheels whose rims are oriented towards the origin. Actuation is generated by the contact between the rim and the body. The wheel is assumed to make point contact with the body at all times and it is assumed that the wheel's speed u_i is directly controlled. We will use the convention that the ith wheel has a positive input u_i when the top of the wheel is turning towards the the right half plane. This system is under-actuated in the sense that the wheels are in a fixed orientation with respect to the origin. For this model it was shown in [11] that if one uses the mechanical model obtained by the PDM, one gets rotational instability in part placement when the passive programmable vector field approach is applied to this system.

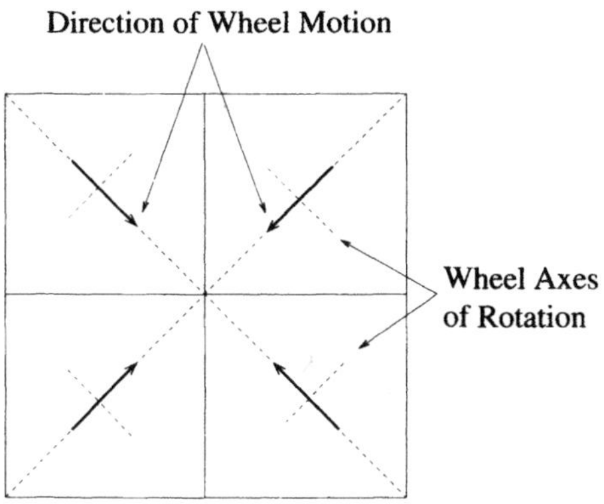

Fig. 2. Four node array centered at the origin

The difference between the stability prediction of the continuous approximation associated with programmable vector fields and the more exact model rests largely on the fact that the continuous approximation does not take any contact mechanics into account, nor does it account for the fact that most real arrays consist of a finite number of discrete actuators. Of course, as previously remarked, the PDM can only be well justified for quasistatic systems where the objects move slowly enough that the contact reaction forces dominate the moments of inertia. This will be true in the case of distributed manipulation, where the local behavior of object motion around an equilibrium point is studied.

Generally, there is no reason to believe that friction at the contact point will be uniform in all directions of the contact plane. Rather, we allow a smooth distribution of coefficient of friction. In particular, we allow the coefficient of friction μ_R in the direction of motion of an actuator to be different from the coefficient of friction μ_S in the direction orthogonal to the actuator motion. (see [5,11] for a discussion of such friction models).

2.1 Equations of Motion

Let us now apply the power dissipation method to the example in Fig. 2. Assume that the desired equilibrium point lies at the intersection point of the lines underlying the wheel rims. With four wheel actuators, there is a potential total of 8 kinematic constraints on the object's motion. However, since the object moves in the plane, at most three of these constraints can be satisfied at any instant–the other constraints are violated via contact slipping. This gives us $\binom{8}{3} = 56$ possible contact states. Using the PDM as the modeling methodology, each has its own first order governing equations. Additionally assume that μ_S and μ_R are uniform across $\mathbb{R}^2$ and that $|\mu_S - \mu_R|$

is sufficiently small that the center of mass completely determines which constraints are satisfied. That is, assume that the constraints are determined by first computing which actuator is the closest to the center of mass, which actuator is the second closest to the center of mass, and using the constraints from these two actuators to determine the first order governing equations. Given this assumption, only 8 of the 56 contact states ever satisfy the minimum of the power dissipation function on a set of full measure, therefore leaving us with a total of 8 possible contact states. Let

$$g_i = \begin{bmatrix} R(\theta_i) & \begin{bmatrix} x_i \\ y_i \end{bmatrix} \\ 0 & 1 \end{bmatrix}$$

be the homogeneous representation of the element of $SE(2)$ going from the origin to the ith actuator node location and orientation, and $R(\cdot)$ is an element of $SO(2)$. Then the constraints associated with each actuator are $\Omega_i(q)\dot{q} = 0$, where

$$\Omega_i(q) = \begin{bmatrix} \left[Ad^T_{g_i^{-1}} \begin{bmatrix} 1 \\ 0 \\ 0 \end{bmatrix} \right]^T & u_i \\ \left[Ad^T_{g_i^{-1}} \begin{bmatrix} 0 \\ 1 \\ 0 \end{bmatrix} \right]^T & 0 \end{bmatrix} \tag{3}$$

$Ad(\cdot)$ is the adjoint transformation which transforms velocities from one coordinate frame to another (see [15] for details). Note that Ω_i are 2×4 matrices where the top row represents the rolling constraint and the bottom row is the side ways constraint.

To apply the PDM, first note that the minimum only occurs when three of the constraints are satisfied, and that moreover, the constraints satisfied are precisely those which would otherwise dissipate the most energy if they were violated. The contact states that dissipate the least amount of energy are those associated with the potential constraints having the largest three $\alpha_i = N_i\mu_i$. Thus, the constraints whose violation will potentially dissipate the most energy are the ones that are satisfied. Based on these observations, if the center of mass determines the normal forces (based on assumptions about surface uniformity, etc.), and if $\mu_S(x,y)$ and $\mu_R(x,y)$ are uniform, then the object's motion satisfies whichever constraints are closest to its center of mass. That is, the particular quadrant in which the center of mass lies determines the first two actively satisfied constraints. The third actively satisfied constraint is one of the two constraints coming from the second closest actuator and is determined by the friction model, i.e., $\mu_S > \mu_R$ or $\mu_S < \mu_R$. Let i be the index of the closest actuator and j be the index of the next closest actuator. If $\mu_S < \mu_R$, and if the two dominating actuators are indexed by i and j, then the first order governing equations derived from the PDM model are

$$\begin{bmatrix} \dot{x} \\ \dot{y} \\ \dot{\theta} \end{bmatrix} = \begin{bmatrix} \frac{u_i[s_j((x_i-x_j)c_i+y_is_i)+c_ic_jy_j]-u_jy_i}{(x_j-x_i)s_j+(y_i-y_j)c_j} \\ \frac{u_jx_i-u_i[c_ic_jx_i+s_i(x_js_j+(y_i-y_j)c_j)]}{(x_j-x_i)s_j+(y_i-y_j)c_j} \\ \frac{u_j-u_i\cos(\theta_i-\theta_j)}{(x_i-x_j)s_j+(y_j-y_i)c_j} \end{bmatrix} \tag{4}$$

where $c_i = \cos(\theta_i)$, $s_i = \sin(\theta_i)$, etc. It should be noted that here the index notation should be thought of as mapping (i, j) pairs to equations of motion in some neighborhood (not necessarily small) around the ith and j^{th} actuator. The transition between the equations of motion determined by actuators i and j to equations of motion determined by actuators k and l will in general be determined by the location of the center of mass. We do not consider the case of $\mu_S > \mu_R$ for reasons discussed in [11].

3 Local Stability

We will see that if one uses feedback, generalized versions of the example in Section 2 are indeed stabilizable to the origin even when the actuators cannot be re-oriented.

3.1 Stability Theorems for Differential Inclusions

First we note that the system in (4) is a differential inclusion of the type found in [4]. I.e. due to the switching of the control vector fields in (2), the governing equations take the form $\dot{x} \in F(t,x)$, where $F(t,x)$ is a (set-valued) multi-function. The following theorem is taken directly from [4,12] and is the generalization of time varying Lyapunov theory to differential inclusions. We remind the reader that the upper and lower derivatives for a function $V(t,x) \in C^1$ are defined by:

$$\dot{V}^* = \sup_{y \in F(t,x)} (V_t + \nabla V y) \qquad \dot{V}_* = \inf_{y \in F(t,x)} (V_t + \nabla V y) \tag{5}$$

Theorem 1. *Let, in a closed domain $D(t_0 \leq t < \infty, |x| \leq \epsilon_0)$, the differential inclusion $\dot{x} \in F(t,x)$ satisfy the basic conditions of existence and $0 \in F(t,0)$; in this domain, let there exist functions $V(t,x) \in C^1, V_0(x) \in C$ for which $V(t,0) = 0, V(t,x) \geq V_0(x) > 0, (0 < |x| < \epsilon_0)$ Then:*
1) If $\dot{V}^* \leq 0$ in D, the solution $x(t) = 0$ of the inclusion $\dot{x} \in F(t,x)$ is stable.
2) If, moreover, there exist functions $V_0(x), V_1(x) \in C$, $W(x) \in C$ and

$$0 < V_0(x) \leq V(t,x) \leq V_1(x), \quad \dot{V}^* \leq -W(x) < 0, \quad V_1(0) = 0$$

then the solution $x(t) = 0$ is asymptotically stable.
3) Moreover, if there exist $k_1, k_2, k_3, c > 0$ such that

$$V_0(x) \geq k_1 ||x||^c, \quad V_1(x) \leq k_2 ||x||^c, \quad W(x) \geq k_3 ||x||^c$$

then the solution $x(t) = 0$ is exponentially stable.

3.2 Stabilizing 2-Dimensional Arrays

In this generalization assume that all the actuators are a finite distance apart and make point contact with the object being manipulated. This is the same theorem as that in [11], but it is constructively proved here, thereby providing control laws we can implement on the system described in Section 5.

Theorem 2. *Given a PVF $\Psi(x,y) : \mathbb{R}^2 \to \mathbb{R}^4$ (where $\mathbb{R}^4$ is the tangent bundle of $\mathbb{R}^2$) which is invariant with respect to the orientation θ of the object, and a discrete planar array geometry indexed by $i, j \in I$ in some neighborhood of the origin, the first order governing equations given by the PDM are not asymptotically stable. Moreover, if $\mu_S < \mu_R$, then such a system is exponentially stabilizable through the use of feedback.*

Proof. To see that the system is not asymptotically stable in θ, it is sufficient to note that $Ad_{g^{-1}}$ in (3) depends only on x and y, therefore leaving the θ dynamics with no dependence on θ. This symmetry implies that the system equations are invariant with respect to initial condition in θ; therefore θ is not asymptotically stable. To see that the system is stabilizable, we explicitly show that the requirements of Theorem 1 are satisfied. First choose the Lyapunov function to be $V(q) = \frac{1}{2}||q||$. Moreover, choose $V_0 = V_1 = W = V$. It must be shown that $\dot{V}^* \leq -W$, where $\dot{V}^*$ was defined in (5). First we must compute $\dot{V}^*$.

Denote $\sin(\cdot)$ by $s(\cdot)$ and $\cos(\cdot)$ by $c(\cdot)$. Recall the governing equations from (4).

$$\dot{q} = \begin{bmatrix} \frac{-(u_j\, y_i)+u_i\, (s(\theta_j)\, (c(\theta_i)\, (x_i-x_j)+s(\theta_i)\, y_i)+c(\theta_i)\, c(\theta_j)\, y_j)}{s(\theta_j)\, (-x_i+x_j)+c(\theta_j)\, (y_i-y_j)} \\ \frac{u_j\, x_i-u_i\, (c(\theta_i)\, c(\theta_j)\, x_i+s(\theta_i)\, (s(\theta_j)\, x_j+c(\theta_j)\, (y_i-y_j)))}{s(\theta_j)\, (-x_i+x_j)+c(\theta_j)\, (y_i-y_j)} \\ \frac{-(c(\theta_i-\theta_j)\, u_i)+u_j}{s(\theta_j)\, (x_i-x_j)+c(\theta_j)\, (-y_i+y_j)} \end{bmatrix}$$

Note that in a region where only one set of governing equations apply

$$\begin{aligned} \dot{V} &= x\dot{x} + y\dot{y} + \theta\dot{\theta} \\ &= \frac{u_j\, (\theta-y\, x_i+x\, y_i)}{s(\theta_j)\, (x_i-x_j)+c(\theta_j)\, (-y_i+y_j)} + \frac{u_i\, (c(\theta_j)\, (y\, s(\theta_i)\, (y_i-y_j)-c(\theta_i)\, (\theta-y\, x_i+x\, y_j)))}{s(\theta_j)\, (x_i-x_j)+c(\theta_j)\, (-y_i+y_j)} \\ &\quad + \frac{u_i\, (s(\theta_j)\, (x\, c(\theta_i)\, (-x_i+x_j)-s(\theta_i)\, (\theta-y\, x_j+x\, y_i)))}{s(\theta_j)\, (x_i-x_j)+c(\theta_j)\, (-y_i+y_j)} \end{aligned}$$

$$\text{or } \dot{V} = u_i f_i(x) + u_j f_j(x). \tag{6}$$

This is split into two coefficients of u_i and u_j, which in turn implies that if one has full knowledge of the state, then one can always choose the inputs so as to make $\dot{V}^* = \dot{V} = 0$. One can in every region that is not a boundary solve the equation $\dot{V} = \frac{\partial V}{\partial q}\dot{q} = -k||q||^2$ for u_i and u_j. Then

$$
\begin{aligned}
u_i &= \frac{u_i^{num}}{u_i^{den}} \quad \text{with} \qquad (7)\\
u_i^{num} &= u_j\,(\theta - y\,x_i + x\,y_i)\\
&\quad + k\left(\theta^2 + x^2 + y^2\right)\left(s(\theta_j)\,(x_i - x_j) + c(\theta_j)\,(-y_i + y_j)\right)\\
u_i^{den} &= s(\theta_j)\,(xc(\theta_i)\,(x_i - x_j) + s(\theta_i)\,(\theta - yx_j + xy_i))\\
&\quad + c(\theta_j)\,(ys(\theta_i)\,(-y_i + y_j) + c(\theta_i)\,(\theta - yx_i + xy_j))
\end{aligned}
$$

where u_j can be chosen arbitrarily. We have to check on the boundary between two different contact states. On the boundary $\dot{V}^* = max\{\frac{\partial V}{\partial q}\dot{q}_i, \frac{\partial V}{\partial q}\dot{q}_j\}$ where

$$
\frac{\partial V}{\partial q}\dot{q}_i = u_{i_1} f_{i_1} + u_{i_2} f_{i_2} \text{ and } \frac{\partial V}{\partial q}\dot{q}_j = u_{j_1} f_{j_1} + u_{j_2} f_{j_2} \qquad (8)
$$

subject to the constraint that $u_{j_1} = u_{i_2}$. Substituting and comparing,

$$
\dot{V}^* = sup\{u_{i_1} f_{i_1} + u_{i_2} f_{i_2}, u_{i_2} f_{j_1} + u_{j_2} f_{j_2}\}
$$

thus implying that despite the constraint both can be made arbitrarily negative, thus ensuring that $\dot{V}^* \leq -V$. Suppose that the system state lies on a boundary between a state where u_j and u_i are the two dominating inputs and a state where u_j and u_k are the dominating actuators. Then one computes the control law on the boundary and gets

$$
\begin{aligned}
u_j &= \frac{u_j^{num}}{u_j^{den}} \quad \text{with} \qquad (9)\\
u_j^{num} &= (-1 + \delta)\,u_k\,(\theta - yx_i + xy_i)\\
&\quad + k\left(\theta^2 + x^2 + y^2\right)\left(s(\theta_j)\,(-x_i + x_j) + c(\theta_j)\,(y_i - y_j)\right)\\
&\quad + \delta u_i\,(s(\theta_j)\,(xc(\theta_i)\,(x_i - x_j) + s(\theta_i)\,(\theta - yx_j + xy_i))\\
&\quad + c(\theta_j)\,(ys(\theta_i)\,(-y_i + y_j) + c(\theta_i)\,(\theta - yx_i + xy_j)))\\
u_j^{den} &= \delta\,(\theta - yx_i + xy_i) + (-1 + \delta)\,s(\theta_j)\,(xc(\theta_i)\,(x_i - x_j)\\
&\quad + s(\theta_i)\,(\theta - yx_j + xy_i)) + (-1 + \delta)\,c(\theta_j)\,(ys(\theta_i)\,(-y_i + y_j)\\
&\quad + c(\theta_i)\,(\theta - yx_i + xy_j))
\end{aligned}
$$

where u_i and u_k are set and $\delta \in [0, 1]$ parameterizes the differential inclusion. The important point here is that one *can* choose a set of inputs that will satisfy Theorem 1. Therefore, the origin is exponentially stable. □

A possible difficulty with this result is that the stabilizing control switches discretely as the contact state switches, which suggests the need for a way of estimating the contact state. Nevertheless, the control laws computed based on the ideas in the preceding section perform quite well, suggesting that a more general proof is possible. The $\mu_S < \mu_R$ assumption only has an obvious physical interpretation for wheel-like contacts, and we do not yet have an interpretation for generic contacts.

Remark 1 (Weakly Stabilizable vs. Strongly Stabilizable). Notice that the control law computed in (9) depends on the parameter δ. This implies that on the boundaries between contact states the system is only ***weakly stabilizable.*** That is, for every selection of the differential inclusion, there exists a control that exponentially stabilizes the system to the origin. However, one can often estimate δ based on the geometry of the contact state boundaries. A static estimation of $\delta = 0.5$ works well for the example in Section 2 in the experiment. Moreover, these boundaries only constitute a set of measure 0 in the state space, and as we will see in the example done in the next section, one can often choose inputs that will make these boundaries ***strongly stabilizable***, but we cannot yet show that it is true in the generic case. To treat cases where the system cannot be made strongly stabilizable and a static estimation of δ will not work, work by Hespanha [6] has been extended in [9].

3.3 Calculations

Here we apply the method presented in the previous section to a situation near the boundary $x = y$ for $x > 0$ and then apply it near the boundary $y = 0$ for $x > 0$. Restricting attention to these regions provides all the salient features of the theory while keeping the problem tractable. Rescaling all the u_i in (4) by $\frac{1}{\sqrt{2}}$, the governing equations in this boundary region $y = 0, x > 0$ are

$$\dot{q} = \begin{bmatrix} -u_1 - u_4 \\ -u_1 + u_4 \\ co\{u_1, u_4\} \end{bmatrix} \tag{10}$$

Do a control Lyapunov design using $V = \frac{1}{2}||(x, y, \theta)||^2$. This means that $\dot{V} = \frac{\partial V}{\partial q}\dot{q} = x(-u_1 + u_4) + y(-u_1 - u_4) + \theta(co\{-u_1, u_4\})$. Rewrite $co\{-u_1, u_4\}$ as $\delta(-u_1) + (1 - \delta)u_4$ thereby parameterizing all selections of $co\{-u_1, u_4\}$ by δ. Solving the equation $\dot{V} = \frac{\partial V}{\partial q}\dot{q} = -k||q||^2$ (with $k > 0$), we see that a choice of

$$\begin{aligned} u_1 &= \tfrac{-k\theta((1-\delta)\theta + x - y) + k||(x,y,\theta)||^2}{\delta\theta + x + y} \\ u_4 &= -k\theta \end{aligned} \tag{11}$$

makes $\dot{V} = -k||q||^2$. Therefore the system is exponentially stabilized. This example should illustrate some of the weaknesses of the theory described in Section 3.2. Since this is only a weakly stable solution, the estimation of δ is necessary. In numerical simulations one knows that $\delta = \frac{1}{2}$, and the simulation works well. However, in a real system δ would have to be estimated on-line. In [9] we extend work in [6] to the case of distributed manipulation so that the feedback law will not be sensitive to variations in δ, and therefore not as sensitive to effects of chattering, etcetera.

Consider the case when $x = y, x > 0$. In this case the governing equations in this boundary region are

$$\dot{q} = \begin{bmatrix} co\{-u_1 - u_4, -u_1 + u_2\} \\ co\{-u_1 + u_4, -u_1 - u_2\} \\ co\{u_2, -u_4\} \end{bmatrix}. \qquad (12)$$

This situation may seem more hopeless than the one in (10). However, as we will see in a moment, this is not the case. In fact, this differential inclusion is strongly stabilizable. Again do a control Lyapunov design using $V = \frac{1}{2}||(x, y, \theta)||^2$. The choice of

$$\begin{aligned} u_1 &= \frac{-(((\delta-1)u_2+\delta u_4)(\theta+x-y))+k(\theta^2+x^2+y^2)}{x+y} \\ u_2 &= -k\theta \\ u_4 &= k\theta \end{aligned} \qquad (13)$$

makes $\dot{V} = -k||q||^2$. Moreover, this choice of $u_2 = -u_4 = -k\theta$ makes (13) reduce to

$$\begin{aligned} u_1 &= \frac{-(-k\theta(\theta+x-y))+k(\theta^2+x^2+y^2)}{x+y} \\ u_2 &= -k\theta \\ u_4 &= k\theta \end{aligned} \qquad (14)$$

which has no dependency on δ. Therefore this differential inclusion is strongly stabilizable. We will use this control strategy in our implementation in Section 5.

4 Aside: Fully-Actuated Distributed Manipulation

We should comment here that in some special cases, the feedback control problem for distributed manipulation is very straightforward. Consider the case where the distributed manipulation system is fully actuated, in the sense that every actuator can exert a force of an arbitrary magnitude in an arbitrary direction while all satisfying the kinematic constraints associated with contact between the actuators and the part. In this case, perhaps not surprisingly, one can effectively reduce such a system from an overactuated system to a system with three inputs and three outputs by enforcing kinematic compatibility constraints between the actuators. (This functions in precisely the same manner as the differential on an automobile, which ensures that both front wheels satisfy the same kinematic constraints to reduce wear on tires due to slipping.) By doing this, one gains direct control over the velocity of the part. This method additionally provides desirable robustness properties with respect to actuator uncertainties (see [14] for details). In fact, the robustness analysis is the only place where we must use properties of the PDM, due to the fact that the system is ideally not slipping at all.

Note that the full actuation assumption is implicit in the programmable vector field approach, so this method should have broad application to other types of devices such as air jets. Others have considered feedback in the case of fully actuation, particularly in [7]. However, their approach differs strongly from the one found here. First, they use components of the open loop theory (using radial fields and rotation fields) to stabilize the system. This involves switching and/or superimposing these fields, leading to significant analytical challenges. The approach here instead takes advantage of the structure of rigid body mechanics and uses techniques from [15] to create control laws.

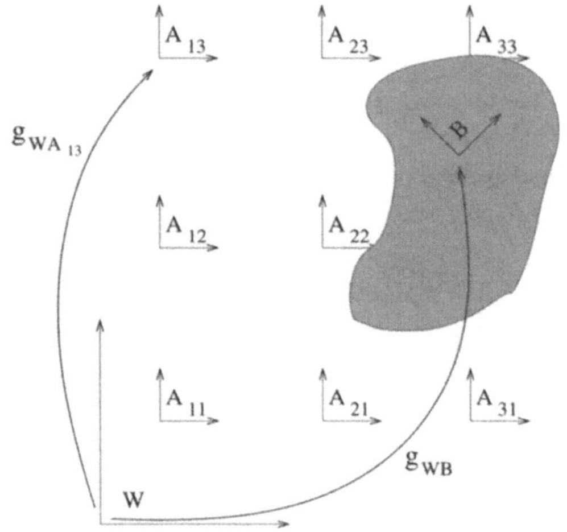

Fig. 3. Rigid body velocities

Consider Fig. 3. In this figure we see an abstraction of the 9 cell experimental system seen in Section 5. There is the world frame (denoted by W), the body frame (denoted by B) attached to the moving object, and an actuator frame (denoted by A_{ij} for an actuator located at (x_i, y_j)) which has fixed orientation with respect to the world frame. The rigid body transformation from the world frame to the body frame is denoted by g_{WB} and the rigid body transformation form the world frame to the actuator frame is denoted by $g_{WA_{ij}}$. Notice that for any body frame motion, there is an equivalent motion in the actuator frame. In particular, we know that the relative velocity $V_{body} = (\dot{x}_{body}, \dot{y}_{body}, \dot{\theta}_{body})$ of a point above actuator A_{ij} on the body is $Ad_{g_{WA_{ij}}} V_{body}$ where in $SE(2)$ the Adjoint operator Ad_g defined earlier in Section 2. This implies we can take the following approach as a control strategy. Suppose we are given a Lyapunov function on the $SE(2)$. As in the previous section, take a "control Lyapunov" approach to the problem by defining the target dynamics to be $\dot{q} = -\frac{\partial V(q)}{\partial q}$ where V is a Lyapunov function yet to be defined. This system is trivially exponentially stable. The velocity $\dot{q}$ is mapped to the actuators in order to get a feedback law. Choose the Lyapunov function to be $V(x,y,\theta) = k_1x^2 + k_2y^2 + k_3\theta^2$ for $k_i > 0$. The Adjoint operator mapping velocities from W to the A_{ij} for actuator frames oriented the same as the world frame is:

$$Ad_{g_{WA_{ij}}} = \begin{bmatrix} Id & \begin{pmatrix} y_j \\ -x_i \end{pmatrix} \\ 0 & 1 \end{bmatrix}$$

Transforming the velocity into the actuator frame yields $Ad_{g_{WA_{ij}}} \cdot (-\frac{\partial V(q)}{\partial q})$. Substituting in for $\frac{\partial V(q)}{\partial q}$, we compute the actuator velocities to be

$$\begin{bmatrix} k_3y_i(\theta - \theta_d) - k_1(x - x_d) \\ -k_3x_i(\theta - \theta_d) - k_2(y - y_d) \\ -k_3\theta \end{bmatrix}$$

where x_d, y_d, θ_d are the desired values and x, y, θ are the state feedback values. To transform this into wheel velocities and wheel orientations for the particular example found here, calculate the magnitude and direction of the (x, y) velocity. This gives for each actuator:

$$\theta_{ij} = \tan^{-1}\left(\frac{-k_3x_i(\theta - \theta_d) - k_2(y - y_d)}{k_3y_i(\theta - \theta_d) - k_1(x - x_d)}\right) \tag{15}$$

and

$$v_{ij} = \sqrt{(-k_3x_i(\theta - \theta_d) - k_2(y - y_d))^2 + (k_3y_i(\theta - \theta_d) - k_1(x - x_d))^2} \tag{16}$$

where θ_{ij} is the orientation of the (i, j) actuator and v_{ij} is the wheel velocity of that actuator. So, given all the actuator locations, one computes (15) and (16) for each actuator, and the feedback law is complete. This control law has several advantages. It is easy to implement and scales nicely with the number of actuators. Moreover, this control lends itself very well to control problems beyond point stabilization. In particular, trajectory tracking can be accomplished with little to no modification. Experimental results in Section 5 illustrate that this method works extremely well. In fact, it seems to be a natural choice of feedback law in the case of full state feedback and full actuation partially because it is empirically very robust and easy to implement.

5 Experiments

An experimental apparatus has been developed at Caltech for testing some of the theories both in this paper and elsewhere. More details of the experimental apparatus can be found in [13]. A photograph can be seen in Fig. 4. This

Fig. 4. Experimental setup

experiment has 9 steer-able wheels which can be controlled from a centralized computer. The system uses visual feedback through a black and white monochrome camera. For the manipulated object a piece of plexiglass is used. The plot of each experiment[1] includes the x, y, and θ trajectories as functions of time and a plot of the (x, y) trajectory in the plane. The following experiments all have relatively little output error because visual feedback is being used. However, there is approximately a 0.5 cm error in the translational directions and an error of 0.05 rad in the estimated orientation. As the theory presented here only considers point stabilization, these are the only experiments included. However, we should point out that the experimental setup is appropriate for doing trajectory tracking and other more challenging problems.

5.1 Under-Actuation

Figure 5 shows an experiment using only four cells to emulate the example in Section 2. In this experiment, the feedback controller from Section 3.3 is used. Since the (x, y) origin is precisely where it is difficult to stabilize an object's orientation, an initial condition with a very high initial θ error value and very small (x, y) error values was given. That is, the goal was to re-orient the object while basically maintaining its position. Notice that the controller design first destabilizes the (x, y) origin because θ is large, and thereby avoids the indeterminacy that occurs at the origin. This experiment used an initial state of approximately $(x_0, y_0, \theta_0) = (-0.1 \text{ m}, 0.1 \text{ m}, 1.5 \text{ rad})$, and the final position was approximately $(x_f, y_f, \theta_f) = (0.01 \text{ m}, 0.01 \text{ m}, 0.05 \text{ rad})$ (the

[1] Movies of these and other experiments can be found at *http://www.cds.caltech.edu/~murphey/experiment/*.

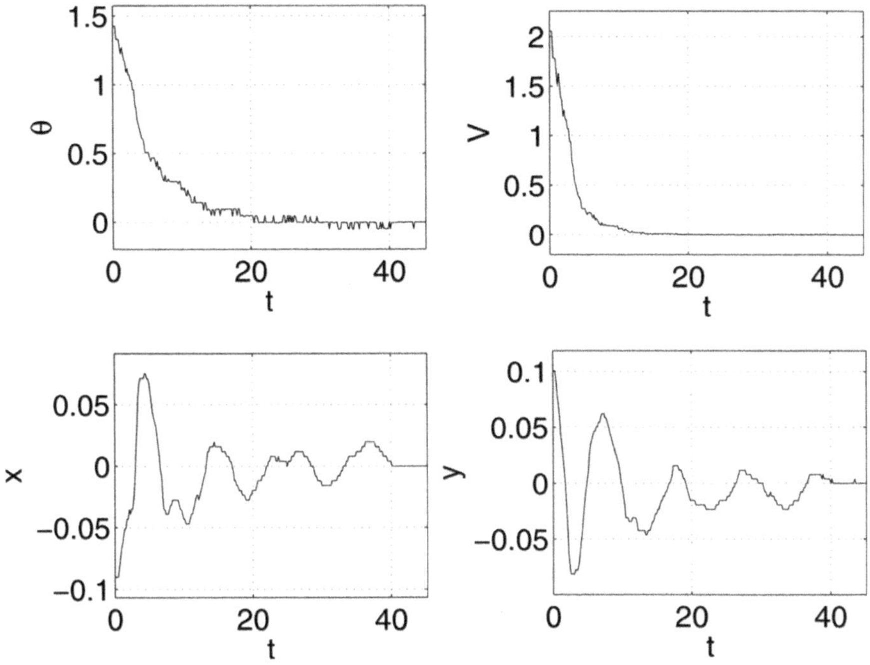

Fig. 5. Under-actuated feedback control

angle cannot currently be resolved any further because of camera pixelization error). We also plot the value of the Lyapunov function versus time to show that, although the individual x, y, and θ trajectories are rather jerky, the Lyapunov function itself decreases quite smoothly, indicating that the control law is performing as expected.

5.2 Full-Actuation

Figure 6 shows the experimental results when this system is controlled using the feedback law described in the Section 4. Notice that the translational stability of the origin is maintained, while the rotational dynamics are stabilized to $\theta = 0$ due to the feedback law. This experiment used an initial state of approximately $(x_0, y_0, \theta_0) = (0.25 \text{ m}, -0.4 \text{ m}, 1.9 \text{ rad})$, and the final position was approximately $(x_f, y_f, \theta_f) = (0.01 \text{ m}, 0.01 \text{ m}, 0.05 \text{ rad})$. This experiment clearly indicates that when a distributed array is fully actuated, the feedback law in (15) and (16) will work extremely well. Note also that the Lyapunov function decreases very smoothly with time in similar fashion to that in Figure 5.

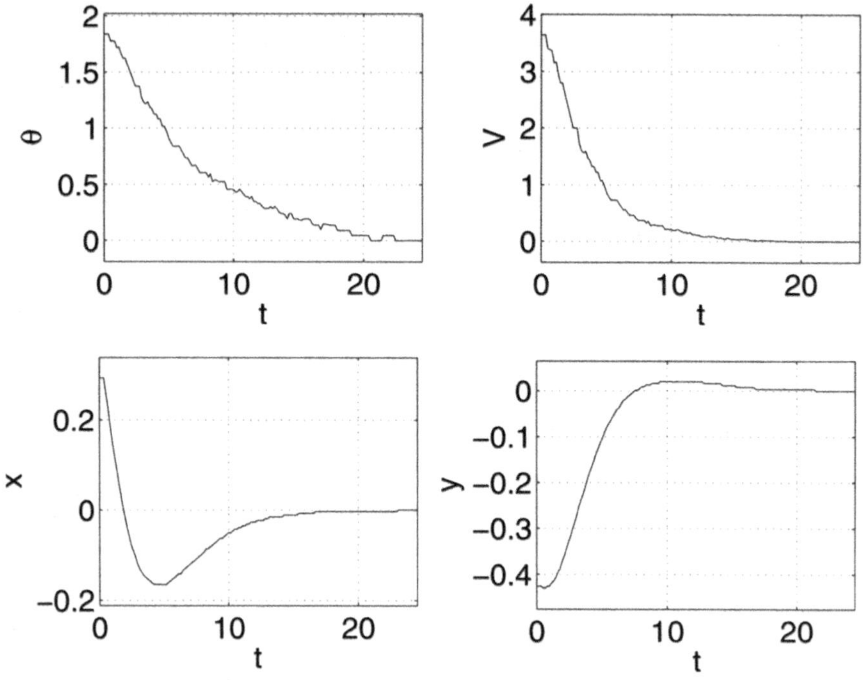

Fig. 6. Fully-actuated feedback control

6 Summary

The work presented in this paper provides a fundamentally different approach to the control of distributed manipulation. Work done by others, particularly the use of programmable vector fields, has traditionally assumed that all the actuators in a distributed array operate at sufficiently high velocities that they were all slipping all the time. This causes significant stress on both the object being manipulated and the actuators themselves. The methods presented here assume that if the actuators are moving at sufficiently slow velocities some, but not necessarily all, actuators are always in contact with the moving body. This requires less energy for a given motion and moreover induces smaller forces on the object and actuators. Moreover, the methods presented here succeed in designing feedback controllers that stabilize an object to a point on a distributed array. It is possible to prove performance bounds on how long it takes an object to reach the goal configuration by showing that exponential convergence can be guaranteed. Most importantly, this was done without explicitly modeling friction, which subsequently leads to relatively simple control laws, improving the likelihood that these control laws can be scaled up to high numbers of actuators.

References

1. J.C. Alexander and J.H. Maddocks. On the kinematics of wheeled vehicles. *The International Journal of Robotics Research*, 8(5):15–27, October 1989.
2. K.F. Böhringer, B.R. Donald, L.E. Kavraki, and F. Lamiraux. A distributed, universal device for planar parts feeding: unique part orientation in programmable force fields. In *Distributed Manipulation*, pages 1–28. Kluwer, 2000.
3. K.F. Böhringer, R.G. Donald, N.C. MacDonald, G.T.A. Kovacs, and J.W. Suh. Computational methods for design and control of mems micromanipulator arrays. *IEEE Computer Science and Engineering*, pages 17–29, 1997.
4. A.F. Filippov. *Differential Equations with Discontinuous Right-Hand Sides.* Kluwer, 1988.
5. S. Goyal, A. Ruina, and J. Papadopoulos. Planar sliding with dry friction. *WEAR*, 143:307–352, 1991.
6. J.P. Hespanha and A.S. Morse. Stability of switched systems with average dwell-time. Technical report, EE-Systems, University of Southern California, 1999.
7. J. E. Luntz, W. Messner, and H. Choset. Distributed manipulation using discrete actuator arrays. *International Journal of Robotics Research*, 20(7), July 2001.
8. H. Fujita M. Ataka, A. Omodaka. A biomimetic micro motion system. In *Transducers - Digest International Conference on Solid State Sensors and Actuators*, pages 38–41, 1993. Pacifico, Yokohama, Japan.
9. T. D. Murphey. *Control of Multiple Model Systems.* PhD thesis, California Institute of Technology, May 2002.
10. T. D. Murphey and J. W. Burdick. Issues in controllability and motion planning for overconstrained wheeled vehicles. In *Proc. Int. Conf. Math. Theory of Networks and Systems (MTNS)*, Perpignan, France, 2000.
11. T. D. Murphey and J. W. Burdick. On the stability and design of distributed systems. In *Proc. IEEE Int. Conf. on Robotics and Automation*, Seoul, Korea, 2001.
12. T. D. Murphey and J. W. Burdick. Global exponential stabilizability for distributed manipulation. In *Proc. IEEE Int. Conf. on Robotics and Automation*, Washington D.C., 2002.
13. T. D. Murphey and J. W. Burdick. Experiments in nonsmooth control of distributed manipulation. In *Submitted to IEEE Int. Conf. on Robotics and Automation*, 2003.
14. T. D. Murphey and J. W. Burdick. Smooth feedback control algorithms for fully actuated distributed manipulators. In *Submitted to IEEE Int. Conf. on Robotics and Automation*, 2003.
15. R.M. Murray, Z. Li, and S.S. Sastry. *A Mathematical Introduction to Robotic Manipulation.* CRC Press, 1994.
16. F. Pfeiffer and C. Glocker. *Multibody Dynamics with Unilateral Controls.* Wiley, Chichester, 1996.

Fast Penetration Depth Estimation Using Rasterization Hardware and Hierarchical Refinement

Young J. Kim, Ming C. Lin, and Dinesh Manocha

University of North Carolina, Chapel Hill, NC, USA
{youngkim,lin,dm}@cs.unc.edu
http://gamma.cs.unc.edu/~PD

Abstract. We present a novel and fast algorithm to estimate penetration depth (PD) between two polyhedral models. Given two overlapping polyhedra, it computes the minimal translational distance to separate them using a combination of discretized computations and hierarchical refinement. The algorithm computes pairwise Minkowski sums of decomposed convex pieces, performs closest point query using rasterization hardware, and refines the estimated PD by incremental walking. It uses bounding volume hierarchies, model simplification, and culling algorithms to further accelerate the computation and refines the estimated PD in a hierarchical manner. We highlight its performance on complex models.

1 Introduction

The problem of computing a distance measure between geometric objects arises in robotics, dynamic simulation, computer gaming, virtual environments, etc. It includes computation of minimal Euclidean or separation distance between disjoint geometric objects as well as a measure of penetration or intersection between two overlapping objects. The separation distance computation problem has been well-studied in the literature and a number of efficient and practical algorithms are known for polyhedral models. On the other hand, there is relatively less work on penetration depth computation between two intersecting objects

Given two inter-penetrating rigid polyhedral models, the penetration measure between them can be defined using different formulations. One of the widely used measures for quantifying the amount of intersection is *penetration depth*, commonly defined as the minimum translational distance required to separate two intersecting rigid models [5,6,8]. Penetration depth (PD) is often used in motion planning [15], contact resolution for dynamic simulation [24,26,30], force computation in haptic rendering [12,25], tolerance verification for virtual prototyping [29], etc.

The PD between two overlapping objects can be formulated based on their *Minkowski sum*. Given two polyhedral models, say P and Q, the PD corresponds to the minimum distance from the origin of the Minkowski sum, $P \oplus (-Q)$, to the surface of this sum. However, the computational complexity of computing the Minkowski sum can be $O(n^6)$, where n is the number of

J.-D. Boissonnat et al. (Eds.): Algorithmic Foundations of Robotics V, STAR 7, pp. 505–521, 2004.

features [8]. In addition to its high computational complexity, the resulting algorithms are also susceptible to accuracy and robustness problems. Hence, no practical algorithms are currently known for accurately computing the PD between general polyhedral models.

Main Results: We present a novel approach to estimate the PD between general polyhedral models using a combination of discretized computations and hierarchical representations. Given the global nature of the PD problem, we systematically decompose the boundary of each polyhedron into convex pieces, compute the pairwise Minkowski sums of the resulting convex polytopes, and use graphics rasterization hardware to perform the closest point query up to a given discretized resolution. The results obtained are refined using a local walking algorithm. To further speed up this computation and improve the estimate, we present a hierarchical refinement technique that takes advantage of geometry culling, model simplification, and local refinement with greedy walking. The overall approach combines discretized closest point queries with geometry culling and refinement at each level of the hierarchy. We also analyze its performance as a function of discretization error.

The resulting algorithm has been implemented and tested on different benchmarks. Depending on the combinatorial complexity of polyhedra and their relative configuration, its performance varies from a fraction of a second to a few seconds on a 1.6 GHz PC with an nVidia GeForce 3 graphics card.

2 Previous Work

In this section, we briefly review previous work related to proximity queries, penetration depth computation, and the use of discretized computations on a geometry SIMD machine for geometric applications.

2.1 Collision and Distance Queries

The problems of collision detection and distance computations are well studied in computational geometry, robotics, and simulated environments. Check out [22] for a survey.

2.2 Penetration Depth Computation

A few efficient algorithms to compute the penetration depth (PD) between convex polytopes have been proposed. The simplest exact algorithm is based on computing their Minkowski sum [13,16] followed by computing the closest point to its boundary from the origin. But its worst case complexity is $O(mn)$, where m and n are the number of features in each polytope. Dobkin et al. computed the directional PD using Dobkin and Kirkpatrick polyhedral hierarchy [8]. For any direction d, it finds the directional PD in $O(\log n \log m)$ time. A randomized algorithm to compute the PD is given in [1]. Its running time is bounded by $O(m^{\frac{3}{4}+\epsilon}n^{\frac{3}{4}+\epsilon} + m^{1+\epsilon} + n^{1+\epsilon})$ for any positive constant ϵ.

Given the worst-case $O(mn)$ complexity of PD computation between convex polytopes, a number of approximation approaches have been proposed for interactive applications. All of them either compute a subset of the boundary or a simpler approximation of the Minkowski sum and compute an upper or lower bound to the PD [4,5,18,27]. Some of them also take advantage of frame-to-frame coherence and perform incremental computations.

Other approximation approaches for general polygonal models are based on discretized distance fields. These include algorithms based on fast marching level-sets for 3D models [10] and others based on graphics rasterization hardware and multi-pass rendering for 2D objects [14].

2.3 Hardware-Assisted Algorithms for Geometric Applications

Interpolation-based polygon rasterization hardware is increasingly being used for geometric applications. A recent survey on different applications is given in [31]. The rasterization hardware performs discretized computations and the resolution is typically determined by the number of pixels in the frame-buffer or depth-buffer. The main benefit comes from speed and dealing with relatively few degenerate configurations.

3 Background and Overview

In this section, we give a brief overview of the PD computation problem and our approach to solve it.

3.1 Penetration Depth and Minkowski Sums

Let P and Q be two intersecting polyhedra. The PD of P and Q, $PD(P,Q)$, is the minimum translational distance that one of the polyhedra must undergo to render them disjoint. Formally, $PD(P,Q)$ is defined as:

$$min\{\parallel \boldsymbol{d} \parallel \mid interior(P+\boldsymbol{d}) \cap Q = \emptyset\} \qquad (1)$$

The Minkowski sum, $P \oplus Q$, is defined as a set of pairwise sums of vectors from P and Q. In other words, $P \oplus Q = \{\boldsymbol{p}+\boldsymbol{q} \mid \boldsymbol{p} \in P, \boldsymbol{q} \in Q\}$. Similarly, $P \oplus (-Q)$ can be defined as $P \oplus (-Q) = \{\boldsymbol{p}-\boldsymbol{q} \mid \boldsymbol{p} \in P, \boldsymbol{q} \in Q\}$.

A general framework to compute the PD is based on Minkowski sums. Without loss of generality, let us assume that two polyhedra P and Q are defined with respect to the global origin $\boldsymbol{O}$. If two polyhedra P and Q intersect, then the origin $\boldsymbol{O}$ is inside $P \oplus -Q$. The $PD(P,Q)$ is defined as a minimum distance from $\boldsymbol{O}$ to the surface of $P \oplus -Q$ [5].

It is relatively easier to compute Minkowski sums of convex polytopes as compared to general polyhedral models. One possible approach for computing Minkowski sums for general polyhedra is based on *decomposition*. It uses the following property of Minkowski computation. If $P = P_1 \cup P_2$, then $P \oplus Q = (P_1 \oplus Q) \cup (P_2 \oplus Q)$. The resulting algorithm combines this property with convex decomposition for general polyhedral models:

1. Compute a convex decomposition for each polyhedron

2. Compute the pairwise convex Minkowski sums between all possible pairs of convex pieces in each polyhedron
3. Compute the union of pairwise Minkowski sums.

After the second step, there can be $O(n^2)$ pairwise Minkowski sums and their union can have $O(n^6)$ complexity [2].

This approach provides an algorithmic framework to compute the Minkowski sum. However, its practical utility is unclear. Besides the combinatorial complexity, it is a major challenge to have a robust implementation of an algorithm for union computations in 3D.

3.2 Our Approach

Our algorithm to estimate the PD is based on the decomposition approach described in Section 3.1. In order to overcome its combinatorial and computational complexity, we use a *surface-based* convex decomposition of the boundary and perform discretized computations and local walking to estimate the PD. We do not explicitly compute the boundary of the union or any approximation to it. Rather, we perform the *closest point query* using hardware-assisted massive ray shooting that estimates the closest point from the origin to the boundary of the union of pairwise Minkowski sums. The resulting maximum depth fragment at each pixel computes an approximation to the PD, up to the pixel resolution used for this computation. Given this PD estimate, we further refine it using an incremental algorithm that performs a local walk on the Minkowski sum. Each step of our approach is relatively simple to implement. However, its worst case complexity can be as high as $O(n^4)$ because of the number of pairwise Minkowski sums and the computational complexity of the closest point query.

We improve the performance of the algorithm using a number of acceleration techniques. These include hierarchical representation based on convex bounding volumes, use of model simplification algorithms, and geometry culling approaches applied to both Minkowski sum computation and hardware assisted ray-shooting. These are explained in detail in Section 5.

3.3 Notation

We use bold-faced letters to distinguish a vector from a scalar value (e.g. the origin, $\boldsymbol{O}$). In Table 1, we enumerate the notations that we use throughout the paper.

4 Penetration Depth Computation

In this section, we present our algorithm for estimating global PD.

4.1 Object Decomposition

We decompose the boundary of each polyhedron P into a collection of convex patches c_i. These c_i's are mutually disjoint, and the union of all the c_i's covers

Table 1. Notation Table

Notation	Meaning
∂P	The boundary of P
C_i^P	A decomposed convex piece of P
$C_i^{P,l}$	A decomposed convex piece of P at level l
M_{ij}	Minkowski sum between C_i and C_j
d_{est}^k	kth refinement of the PD estimation

the entire boundary of P, ∂P. Computing an optimal decomposition is NP-hard and different heuristics to compute such decompositions are well-known [7,9].

Furthermore, we compute a convex hull of each surface patch, c_i, and denote the resulting polytope by C_i. The union of these C_i's completely covers the boundary of the original polyhedron P. Notice that our decomposition strategy is merely a partition of ∂P, not of P. This surface decomposition is sufficient for PD computation, because we are only concerned with the surface of Minkowski sums between polyhedra.

4.2 Pairwise Minkowski Sum Computation

Our PD computation algorithm is based on the decomposition approach described in Section 3.1. The first step involves computing the pairwise Minkowski sums between all possible pairs of convex polytopes, C_i^P and C_j^Q, belonging to P and Q, respectively. Let us denote the resulting Minkowski sum as M_{ij}. Various algorithms are known for computing Minkowski sums of convex polytopes. Using the *topological sweep*, we can compute M_{ij} in $O(n \log n + k)$ time, where n is the number of features in C_i^P and C_j^Q and k is the number of features in M_{ij} [13].

4.3 Closest Point Query Using Rasterization Hardware

Given all the pairwise Minkowski sums, M_{ij}, let

$$M = \bigcup_{ij} M_{ij}. \tag{2}$$

Our goal is to compute the closest point on the boundary of M, i.e. ∂M, from the origin. We perform discretized computations to estimate the closest point on ∂M. It involves using polygon rasterization hardware to perform this query up to a given pixel resolution. The main idea is to rasterize ∂M from the origin without computing a surface representation of ∂M explicitly. After that we compute the closest point, distance and direction.

Rasterization Hardware The discretized computations are performed on two dimensional pixel data in parallel. The underlying model corresponds to an abstract *geometric SIMD machine* which switches between different states [21]. Typical instruction sets allowed in the geometric SIMD model include simple arithmetic, logical and relational operations on the pixel data, also known as the *frame buffer*. Current graphics rasterization hardware is a close realization of this model. From the viewpoint of our PD algorithm, the rasterization hardware is a multi-processor system that performs massive ray-shooting operations onto geometric primitives in parallel. We take advantage of these massive ray-shooting functionality of the rasterization hardware to perform the closest point query.

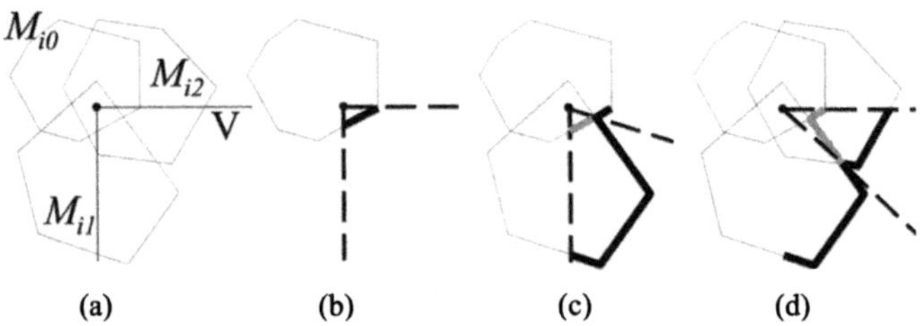

Fig. 1. Rendering the Boundary of the Union From Inside. *In (a), V is the current view-frustum. In (b),* M_{i0} *is rendered, and a new* ∂M *is constructed (thick line). In (c), when* M_{i1} *is rendered, it opens up a new window (dotted line), and the update region (thick gray line) on the current* ∂M *is established. Thus a new* ∂M *(thick line) is constructed. In (d), we perform the same procedure for* M_{i2}.

Rasterizing the Boundary of the Union Our algorithm for rasterizing ∂M from a point inside is essentially a massive ray-shooting procedure from the origin to ∂M by incrementally expanding the front of ∂M. The algorithm can require m^2 passes, where m is the number of convex polytopes, M_{ij}.

The algorithm maintains the current boundary of M, ∂M^k, where k is the current iteration, and incrementally expands it with M_{ij} that intersects ∂M^k. We attempt to add M_{ij} by drawing the front faces of M_{ij}. The front faces that "pierce" the current ∂M^k open up a window through which the origin can see ∂M. After that we draw the backfaces of M_{ij} into the opened window using the maximum depth test.

Computing the Closest Point For a given view, we can compute the closest point on the boundary by simply finding the pixel with the minimum distance value. The algorithm transforms the pixel depth values into distance values based on their (x, y) coordinate positions on the viewing plane. Each pixel depth value is divided by $\cos\theta$, where θ is the angle between the vector to the (x, y) position on the viewing plane and the center viewing direction. The minimum distance and direction to the closest point are derived from the pixel position containing the minimum transformed depth value. In order to examine views in all directions, we construct six views on the faces of a cube around the origin and repeat the operation. For more information about the closest point query, we refer the readers to see our companion paper [17].

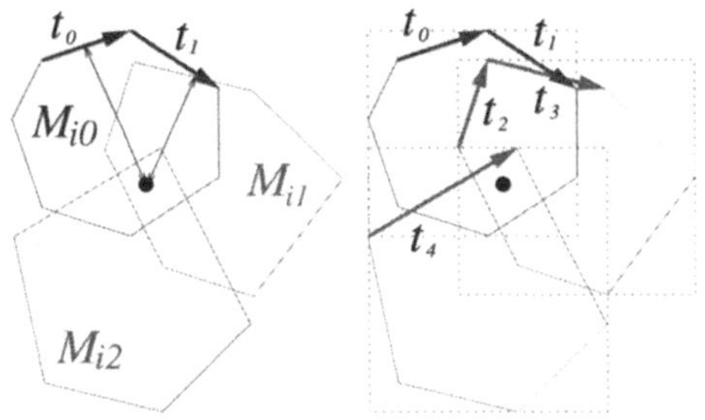

(a) (b)

Fig. 2. Local Refinement by Walking. *We refine the PD by iteratively minimizing the distance between the origin and a line (a triangle in 3D) on the Minkowski sum* M_{i0}. *Thus, in (a), there is a transition from* t_0 *to* t_1, *since the distance from the origin to* t_1 *is smaller than that to* t_0. *In (b), the feature* t_3 *can reduce the PD even further, the transition of the Minkowski sum from* M_{i0} *to* M_{i1} *is followed.*

4.4 Local Refinement

The accuracy of the closest point query and PD estimate is limited by the pixel resolution of the rasterization hardware. We further refine it and improve the PD estimate by performing local *walks* on the boundary of the Minkowski sum of P and Q.

Initially the refinement algorithm starts with identifying the features of P and Q that contribute to the current PD estimate. Each triangle in M_{ij} is generated by only three possible sets of feature combinations from P and Q. These include vertex/face (VF), face/vertex (FV) and edge/edge (EE) combinations [13], and we use that relationship to compute the actual PD features from each polyhedron that correspond to the current PD estimate. At any time, the algorithm also maintains a notion of *current-Minkowski-sum*, which contains the current PD features.

Once the PD features and the Minkowski sum (M_{ij}) which contains them have been identified, the algorithm refines the current PD estimate by locally walking on the surface of M_{ij}, the current-Minkowski-sum. This walk proceeds by iteratively minimizing the distance from the origin to the surface of M_{ij}. We repeat this process until the algorithm reaches a local or global minimum.

As shown in Fig. 2 the algorithm needs to avoid features that are inside the volume of other Minkowski sums. Although it walks towards the interior of the volume, it sets the current-Minkowski-sum accordingly. Therefore, each time the algorithm is walking, it keeps track of which M_{ij}'s might intersect with the current PD features. We accomplish this by keeping track of a subset of Minkowski sums that can potentially intersect with the current PD features and the current-Minkowski-sum.

Let us denote the current-Minkowski-sum as M_{ij}, and also denote the subset of Minkowski sums that potentially intersect with M_{ij} as M_{ij_0}, M_{ij_1}, ...,

M_{ij_l}. Here, we conservatively determine M_{ij_l}'s by intersection checks based on an axis-aligned bounding box (AABB) of the Minkowski sum. Moreover, for each M_{ij_l}, we also keep track of a closest triangle t_{k_l} to the current PD feature t_k in M_{ij}. The overall refinement algorithm proceeds as:

1. Let the triangle t_k in M_{ij} correspond to the PD features computed based on the closest point query. Compute the set of Minkowski sums $M_{ij_0}, M_{ij_1}, ..., M_{ij_l}$ that intersect M_{ij} based on checking their AABBs for overlap. Also compute $t_{k_0}, t_{k_l}, ..., t_{k_l}$, which is a set of triangles respectively on M_{ij_l}'s that is closest to t_k on M_{ij}.
2. Identify the triangles incident to t_k on M_{ij}.
3. Find a neighboring triangle, say t_{k+1}, that results in maximum decrease in the PD estimate and does not intersect with t_{k_l}'s. Change the current PD features from t_k to t_{k+1}. Also update t_{k_l} on each M_{ij_l} to the closest feature to t_{k+1}.
4. If step 4 fails, check whether there exists t_{k_l} in M_{ij_l} such that it intersects with the triangles incident to t_k or t_k itself but reduces the PD. If it exists, repeat the walk from step 1 by setting t_{k_l} as t_k and M_{ij_l} as M_{ij}.
5. Repeat the steps 2-4 until there is no more improvement in the PD.

Eventually the algorithm computes a local minimum on the boundary of the Minkowski sum, M.

5 Acceleration Techniques

The global PD computation algorithm described in Section 4 computes an upper bound on the amount of PD between two polyhedral models. However, its running time can vary based on the underlying models as well as their relative configuration. In this section, we present a number of acceleration techniques to improve its performance. These include hierarchical culling, model simplification, and geometry culling for closest point query.

5.1 Geometry Culling

A significant fraction of the time of the PD estimation algorithm is spent in pairwise Minkowski sum computation. The algorithm presented in Section 4.2 considers all pairs of convex polytopes, C_i^P and C_j^Q, and computes their Minkowski sum, M_{ij}. If we are given an upper bound on the PD, d_{est}, we can eliminate some pairs of convex polytopes without computing their Minkowski sum. This is based on the following lemma:

Lemma 1. *Let d_{ij} be the separation or Euclidean distance between C_i^P and C_j^Q. If $d_{ij} > d_{est}$, then the closest point from the origin to ∂M lies on $\partial(M - M_{ij})$.*

Based on the Lemma 1, we can cull away all pairs of convex polytopes, C_i^P and C_j^Q, whose separation distances are more than d_{est}. Computing separation distance between convex polytopes is relatively cheap as compared to

Minkowski sum computation and a number of efficient algorithms are known [5,23]. The efficiency of this culling approach depends on the quality of the estimate, d_{est}. Furthermore, checking all possible pairs for separation distance can take $O(n^2)$ time. We improve their performance using a bounding volume hierarchy to perform hierarchical culling.

5.2 Bounding Volume Hierarchy

We compute a bounding volume (BV) hierarchy for each polyhedron using a convex polytope as the underlying BV. Each convex polytope obtained using the decomposition algorithm explained in Section 4.1 becomes a leaf node in the hierarchy. We recursively compute the internal nodes in a bottom-up manner, by merging the children nodes and computing the convex hull of the union of their vertices. Let us define the nodes of polyhedron P at level l as $C_i^{P,l}$. The resulting hierarchy is a hierarchy of convex hulls.

This hierarchy is used in our runtime algorithm to speed up the intersection and separation distance queries for the culling algorithm. Furthermore, each level of the hierarchy provides an approximation of the model, which is used by the PD estimation algorithm.

5.3 Hierarchical Culling

We use the BV hierarchy to speed up the performance of the object-space culling algorithm. The goal is to start with an initial estimate to the PD and refine it at every level of the tree. We denote the estimate computed using level k of each BV tree as d_{est}^k.

We initially start with the root nodes of each hierarchy, $C_0^{P,0}$ and $C_0^{Q,0}$, which correspond to the convex hulls of P and Q, respectively. We compute the PD between those convex polytopes [5,4,18] and use that as the estimated PD at level 0. The algorithm proceeds in a hierarchical manner through the levels in each tree:

1. Consider all the pairwise nodes at level k in each tree, $C_i^{P,k}$ and $C_j^{Q,k}$. For each (i, j) pair, compute the separation distance between them. If the nodes overlap, the separation distance is zero.
2. Discard all the node pairs whose separation distances are more than d_{est}^k. Compute the Minkowski sum for the rest of the pairs.
3. Perform the closest point query on the Minkowski sum pairs and compute the new PD estimate, d_{est}^{k+1} using rasterization hardware.
4. Refine the estimate, d_{est}^{k+1} using the object space walking algorithm presented in Section 4.4.

During each iteration, we go down a level in each tree. If we reach the maximum level in one of the trees, we do not traverse down in that tree any further. The algorithm computes an upper bound on the PD in an iterative manner and refines the bound with every traversal as: $d_{est}^0 \geq d_{est}^1 \geq \ldots \geq d_{est}^h$, where h is the maximum height. Finally, the algorithm returns d_{est}^l as the estimated PD between P and Q.

5.4 Model Simplification

Some internal nodes of the hierarchy may have a high number of vertices and that affects the complexity of pairwise Minkowski sum computation. We pre-compute a single convex simplification for each internal node in the BV tree. The simplifications at each level of the BV tree provide a low polygon count approximation to the original models. We compute a simplification for each internal node in the following manner:

1. Simplify the node using any simplification error metric.
2. Compute the convex hull of each simplified node.
3. Scale the resulting convex polytope to enclose the internal node or the underlying geometry as tightly as possible.

We use the simplified BVs to improve the performance of the computations in step 2 (pairwise Minkowski sum computation) and step 3 (closest point query) of the hierarchical culling algorithm presented in Section 5.3. The simplified BVs can increase the estimated PD value, d_{est}, as compared to the original nodes computed by the BV hierarchy computation algorithm. As a result, the number of pairwise Minkowski sums that can be culled at intermediate levels of the hierarchy based on d_{est} may be reduced. However, the running time of the algorithm is significantly reduced. Also, it does not change the accuracy of the final result, as the algorithm does not simplify the leaf nodes in the BV tree.

5.5 Culling for Closest Point Query

The algorithm also spends a considerable fraction of its time in performing the closest point query using the rasterization hardware (as described in Section 4.3). Here we present a number of techniques to improve its performance.

First of all, we compute a subset of the pairs, M_{ij}'s, that contain the origin and render them only once in the algorithm described in Section 4.3. All the pairwise Minkowski sums in this subset have a zero *hop*. We identify this subset, say l out of total of m pairs of M_{ij}'s, by checking whether the corresponding convex polytopes, C_i^P and C_j^Q, overlap [5,9,23]. Once we have computed these l M_{ij}'s, we first render them using the maximum depth test and then the remaining $(m-l)$ pairwise Minkowski sums, M_{ij}'s, $(m-l)$ times using the incremental algorithm.

Secondly, when we repeat the closest point query six times, once for each face of the cube, we apply a culling technique similar to the one discussed in Section 5.1. At each view, the algorithm maintains the current minimum depth value, d_{est}, and then as it proceeds to the next view, it culls away the M_{ij}'s whose distance from the origin is more than d_{est}, as shown in Lemma 1. Finally, for each query, when we render the M_{ij}'s, we perform view-frustum culling by checking whether the axis aligned bounding box of each M_{ij} lies in the current view. This view frustum culling significantly reduces the number of primitives rendered during each iteration of the algorithm.

6 Analysis of PD Algorithm

In this section, we analyze the performance of our PD algorithm and discuss its accuracy.

6.1 Performance Analysis

The basic PD algorithm presented in Section 4 has the following computational complexity at run-time:

- Each non-convex object can have $O(n)$ convex pieces using the convex surface decomposition presented in Section 4.1. Thus, each convex piece has $O(C)$ complexity on the average. In practice, C is a small number of less than 10.
- The pairwise Minkowski sum computation has an input of n^2 combinations of convex pieces from two non-convex objects, and each M_{ij} computation requires $O(C \log C)$ time. Therefore, the overall pairwise Minkowski sum computation requires $O(n^2 C \log C)$ running time.
- The closest point query requires m^2 iterations, where $m = n^2$ is the number of M_{ij}'s. Each iteration requires rasterizing $O(C^2)$ triangles in the worst case, and we assume that the triangle rasterization takes constant time T. The transformation for the perspective correction requires $O(RW^2)$ time, where W is the pixel resolution and R is a cost of a single *read-back* from the frame buffer. Therefore, the total computational complexity of the closest point query is $O(TC^2n^4 + RW^2)$.
- Each refinement walk step requires $O(C^2n^2)$ time in the worst case, since it needs to keep track of all the potential intersectors of the current-Minkowski-sum. In practice, each step requires a small number of constant iterations as opposed to the worst case complexity.
- In summary, the object space computation requires $O(n^2 C \log C)$ time, and the image space computation (i.e. closet point query) requires $O(TC^2n^4 + RW^2)$ time.

The performance of the basic algorithm is improved by different techniques highlighted in Section 5. However, the performance of the resulting algorithm using hierarchical refinement depends heavily on the extent of object-space culling, which is directly related to the amount of inter-penetration between the objects. As a result, when the penetration between two polyhedra is relatively shallow, the algorithm is able to cull away a very high percentage of Minkowski pairs (as shown in Table 2 in Section 7.2) and quite fast in practice. However, it is very hard to analyze the culling performance quantitatively, since the performance depends on various parameters of objects such as its complexity, aspect ratios, the amount of interpenetration between the objects, and their relative configuration.

6.2 Error Analysis

Our algorithm always computes an upper estimate to the PD. In other words, the algorithm may be conservative and the computed answer may be more than the global minimum defined in Equation 1. The rasterization errors and precision of discretized computations governs the tightness of the resulting answer. The main sources of these errors are as follows:

1. The discretization of ray directions to lie on a pixel grid for each view.
2. The fixed precision of the Z-buffer.

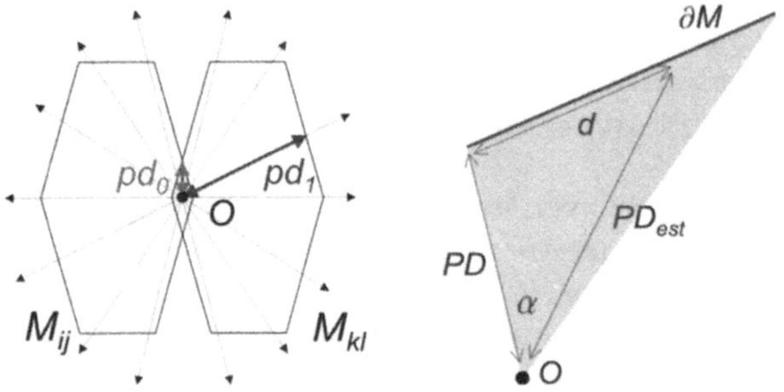

(a) (b)

Fig. 3. (a) Ray Shooting Error. *The true PD value is* pd_0*, which is the minimum distance from* O *to* $\partial(M_{ij} \cup M_{kl})$*. However, due to the discretized ray shooting, the reported PD value* pd_1 *can be arbitrarily larger than* pd_0*.* (b) Error Bound. *PD is optimal, and PD_{est} is closest to PD computed by our algorithm. d is the upper bound of the length of an edge in ∂M. Then,* $PD_{est} \leq PDcos\alpha + \sqrt{d^2 + PD^2 sin^2\alpha}$.

Increasing the resolution of the grid decreases the possibility of the worst-case angular error. Moreover, constructing tighter bounds on the minimum and maximum distances in each view (near and far plane distances) decreases the Z-buffer precision error. However, as illustrated in Fig. 3-(a), the worst case error can be arbitrarily large regardless of the resolution of the grid.

In practice, we can assume that the massive ray-shooting assisted by rasterization hardware is dense enough that every face in M is hit by at least one ray. Furthermore, since we explicitly compute M_{ij}, we know the upper bound d of the length of an edge in ∂M. In this case, as shown in Fig. 3-(b), the upper bound on the estimate, PD_{est} is:

$$PD_{est} \leq PD\ cos\alpha + \sqrt{d^2 + (PD)^2\ sin^2\alpha}$$

by using the cosine law, where α is the smallest angle between rays and PD is the optimal penetration depth between the underlying polyhedra.

We also observed that the PD value is rapidly converging as the pixel resolution increases. The convergence rate also depends on the relative configuration between the objects. Fig. 4 shows a typical convergence behavior of our PD algorithm. The figure shows the convergence rates of two intersecting tori in different configurations, touching and interlocked. More specific data about these models and their relative configuration are given in Fig. 5

and Table 2 in Section 7.2. In practice, we are able achieve a reasonably converging PD value when the pixel resolution is set to 256 × 256.

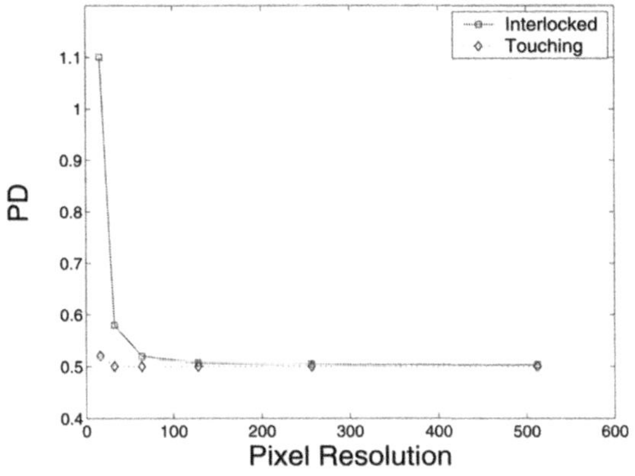

Fig. 4. PD Convergence Rate With Respect To the Pixel Resolution. Two different configurations of two intersected tori, touching and interlocked, are given.

7 Implementation and Results

In this section, we describe the implementation of our PD computation algorithm and demonstrate its performance on different benchmarks and applications. We also refer the readers to see [19,20] where we have used our PD computation algorithm for dynamic simulation of rigid bodies and tolerance verification for rapid prototyping of complex structures.

7.1 Implementation Issues

We use the SWIFT++ implementation of the *Voronoi marching* technique [9] to efficiently perform the separation distance query. It performs distance queries between non-convex polyhedra by using a hierarchy of convex hulls. We use the public domain QHULL package [3] for convex hull computation in 3D. QHULL is particularly efficient for dealing with a relatively small number of points, which is the case in our algorithm. We use the QSlim implementation [11] of the quadric error metric simplification algorithm to ensure that the intermediate nodes of the bounding volume trees do not have more than 50 vertices.

We implement the closest point query operation using OpenGL graphics library. Also, we typically set the screen space resolution to 128 × 128 at the intermediate step of the hierarchical refinement, then at the finest level of the refinement, we set the resolution to 256 × 256. For our benchmarking models, these different resolution schemes provide us with results of a reasonable accuracy, and they also balance the computation time between the object space and the image space.

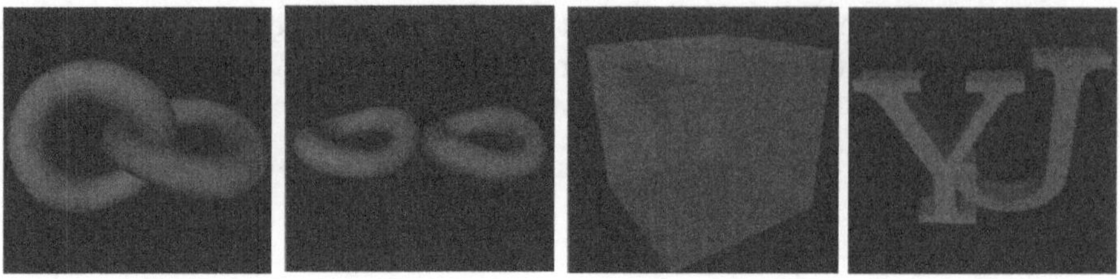

Fig. 5. PD Benchmark Models. *From left to right: interlocked tori, touching tori, interlocked grates, and letters.*

7.2 Benchmark Results

We benchmark our PD algorithm with four models: interlocked tori, touching tori, interlocked "grates" and a pair of alphabet models, with their relative configuration shown in Fig. 5. We used the tori models because it is relatively difficult to compute a good convex decomposition for them. The interlocked "grates" model was chosen because the combinatorial complexity of its exact Minkowski sum is $O(m^3n^3)$ [28]. In our benchmarks, m and n are 1134 and 444, respectively. Therefore, it is a very challenging scenario for any PD computation algorithm. Earlier approaches based on localized computations or convex volumetric decomposition are unable to compute the PD efficiently and accurately on these benchmarks.

We measure the timings on a PC equipped with an Intel Pentium IV 1.6 GHz processor, 512 MB main memory and GeForce 3 graphics card. The complexity of the models varies from a few hundred faces to a few thousand faces. The number of leaf nodes, computed using the convex surface decomposition algorithm, vary from 67 pieces to 409 pieces. The running times vary based on the model complexity and the relative configuration of two polyhedra. It can vary from a fraction of a second, for the touching tori and a pair of alphabet models, to a few seconds for models that have deep penetration (e.g. interlocked tori and interlocked "grates"). Most of the time is spent in pairwise Minkowski sum computations and closest point queries using the graphics hardware. The local refinement based on the walking algorithm is quite fast and takes only a few milliseconds. Detailed timings for some levels of the hierarchy are given in Table 2. The acceleration techniques and hierarchical refinement result in several orders of magnitude improvement in the overall running time. Furthermore, the algorithm is able to compute accurate PD estimates in these cases.

7.3 Performance Speedup by Acceleration Techniques

In Table 3, we have compared the performance of our accelerated PD algorithm presented in Section 5 with the basic algorithm presented in Section 4. As the table illustrates, the basic algorithm suffers from $O(n^4)$ computational costs, and our accelerated algorithm outperforms it by several orders of magnitude. The result is even more dramatic in a very complex scenario such as the interlocking grates model.

Table 2. Benchmark Results. *We show the performance of our PD algorithm for various models. We also break down the performance to the object space culling rate, the pairwise Minkowski computation time and the closest point query time on some of the levels of the hierarchy.*

Level	Cull Ratio	Min. Sum	HW Query	d_{est}
3	31.2 %	0.219 sec	0.220 sec	0.99
5	96.7 %	0.165 sec	0.146 sec	0.53
7	98.3 %	1.014 sec	1.992 sec	0.50

(a) Interlocked Tori (2000 faces, 67 convex pieces each)

Level	Cull Ratio	Min. Sum	HW Query	d_{est}
3	98.4 %	0.135 sec	0.014 sec	0.29
7	99.9 %	0.105 sec	0.032 sec	0.29

(b) Touching Tori (2000 faces, 67 convex pieces each)

Level	Cull Ratio	Min. Sum	HW Query	d_{est}
3	0 %	0.66 sec	0.29 sec	6.41
7	96.9 %	0.43 sec	0.39 sec	0.63
9	99.9 %	0.03 sec	0.07 sec	0.63

(c) Grates (444 & 1134 faces, 169 & 409 pcs)

Level	Cull Ratio	Min. Sum	HW Query	d_{est}
2	50.0 %	0.055 sec	0.021 sec	0.06
4	56.2 %	0.099 sec	0.062 sec	0.03
6	97.6 %	0.080 sec	0.161 sec	0.01

(d) Alphabets (144 & 152 faces, 42 & 43 pcs)

Table 3. Performance Speedup by Acceleration Techniques

Type	Without Accel.	With Accel.
Interlocked Tori	4 hr	3.7 sec
Touching Tori	4 hr	0.3 sec
Grates	177 hr	1.9 sec
Alphabets	7 min	0.4 sec

8 Summary and Future Work

We present a fast, global algorithm to estimate penetration depth between polyhedra using both image-space acceleration techniques and object-space culling and refinement algorithms. The resulting algorithm has been tested on difficult benchmarks.

There are several areas for future work. The performance of our algorithm can be further improved by exploring more optimizations. These include faster implementations of the closest point query using new features of the high-end graphics cards, as well as better hierarchical decompositions. Currently our algorithm only computes the minimum translational distance to separate two overlapping objects. It would be useful to extend it to handle rotational penetration depth.

Acknowledgments.
This research was supported in part by ARO Contract DAAG55-98-1-0322, DOE ASCII Grant, NSF NSG-9876914, NSF DMI-9900157, NSF IIS-982167, NSF ACR-0118743, ONR Contracts N00014-01-1-0067 and N00014-01-1-0496, and Intel. We will like to thank Kenneth Hoff, Miguel A. Otaduy, and Dan Halperin for useful discussions and applications of this algorithm to dynamic simulation and virtual prototyping.

References

1. P. Agarwal, L. J. Guibas, S. Har-Peled, A. Rabinovitch, and M. Sharir. Penetration depth of two convex polytopes in 3D. *Nordic J. Computing*, 7:227–240, 2000.
2. Boris Aronov, Micha Sharir, and Boaz Tagansky. The union of convex polyhedra in three dimensions. *SIAM J. Comput.*, 26:1670–1688, 1997.
3. B. Barber, D. Dobkin, and H. Huhdanpaa. The quickhull algorithm for convex hull. Technical Report GCG53, The Geometry Center, MN, 1993.
4. G. Bergen. Proximity queries and penetration depth computation on 3D game objects. *Game Developers Conference*, 2001.
5. S. Cameron. Enhancing GJK: Computing minimum and penetration distance between convex polyhedra. *Proceedings of International Conference on Robotics and Automation*, pages 3112–3117, 1997.
6. S. Cameron and R. K. Culley. Determining the minimum translational distance between two convex polyhedra. *Proceedings of International Conference on Robotics and Automation*, pages 591–596, 1986.
7. Bernard Chazelle, D. Dobkin, N. Shouraboura, and A. Tal. Strategies for polyhedral surface decomposition: An experimental study. *Comput. Geom. Theory Appl.*, 7:327–342, 1997.
8. D. Dobkin, J. Hershberger, D. Kirkpatrick, and Subhash Suri. Computing the intersection-depth of polyhedra. *Algorithmica*, 9:518–533, 1993.
9. S. Ehmann and M. C. Lin. Accurate and fast proximity queries between polyhedra using convex surface decomposition. *Computer Graphics Forum (Proc. of Eurographics'2001)*, 20(3), 2001.
10. S. Fisher and M. C. Lin. Deformed distance fields for simulation of non-penetrating flexible bodies. *Proc. of EG Workshop on Computer Animation and Simulation*, 2001.
11. M. Garland and P. Heckbert. Surface simplification using quadric error bounds. *Proc. of ACM SIGGRAPH*, pages 209–216, 1997.
12. A. Gregory, A. Mascarenhas, S. Ehmann, M. C. Lin, and D. Manocha. 6-DOF haptic display of polygonal models. *Proc. of IEEE Visualization Conference*, 2000.
13. L. Guibas and R. Seidel. Computing convolutions by reciprocal search. *Discrete Comput. Geom*, 2:175–193, 1987.

14. K. Hoff, A. Zaferakis, M. Lin, and D. Manocha. Fast and simple geometric proximity queries using graphics hardware. *Proc. of ACM Symposium on Interactive 3D Graphics*, 2001.
15. D. Hsu, L. Kavraki, J. Latombe, R. Motwani, and S. Sorkin. On finding narrow passages with probabilistic roadmap planners. *Proc. of 3rd Workshop on Algorithmic Foundations of Robotics*, 1998.
16. A. Kaul and J. Rossignac. Solid-interpolating deformations: construction and animation of PIPS. *Computer and Graphics*, 16:107–116, 1992.
17. Y. Kim, K. Hoff, M. Lin, and D. Manocha. Closest point query among the union of convex polytopes using rasterization hardware. *Journal of Graphics Tools*, 2003. to appear.
18. Y. Kim, M. Lin, and D. Manocha. DEEP: Dual-space Expansion for Estimating Penetration depth between convex polytopes. In *IEEE Conference on Robotics and Automation*, 2002.
19. Y. Kim, M. Otaduy, M. Lin, and D. Manocha. Fast penetration depth computation for physically-based animation. In *ACM Symposium on Computer Animation*, 2002.
20. Y. Kim, M. Otaduy, M. Lin, and D. Manocha. Fast penetration depth computation using rasterization hardware and hierarchical refinement. Technical report, UNC-Chapel Hill TR02-014, 2002.
21. S. Krishnan, N. Mustafa, and S. Venkatasubramanian. Hardware-assisted computation of depth contours. In *ACM-SIAM Symposium on Discrete Algorithms*, 2002.
22. M. Lin and S. Gottschalk. Collision detection between geometric models: A survey. In *Proc. of IMA Conference on Mathematics of Surfaces*, 1998.
23. M.C. Lin and John F. Canny. Efficient algorithms for incremental distance computation. In *IEEE Conference on Robotics and Automation*, pages 1008–1014, 1991.
24. Michael McKenna and David Zeltzer. Dynamic simulation of autonomous legged locomotion. In Forest Baskett, editor, *Computer Graphics (SIGGRAPH '90 Proceedings)*, volume 24, pages 29–38, August 1990.
25. W. McNeely, K. Puterbaugh, and J. Troy. Six degree-of-freedom haptic rendering using voxel sampling. *Proc. of ACM SIGGRAPH*, pages 401–408, 1999.
26. B. Mirtich. Timewarp rigid body simulation. *Proc. of ACM SIGGRAPH*, 2000.
27. C. J. Ong and E.G. Gilbert. Growth distances: New measures for object separation and penetration. *IEEE Transactions on Robotics and Automation*, 12(6), 1996.
28. G. D. Ramkumar. *Tracings and Their Convolution: Theory and Applications.* PhD thesis, Standford, March 1998.
29. A.A.G. Requicha. Mathematical definition of tolerance specifications. *ASME Manufacturing Review*, 6(4):269–274, 1993.
30. D. E. Stewart and J. C. Trinkle. An implicit time-stepping scheme for rigid body dynamics with inelastic collisions and coulomb friction. *International Journal of Numerical Methods in Engineering*, 39:2673–2691, 1996.
31. T. Theoharis, G. Papaiannou, and E. Karabassi. The magic of the Z-buffer: A survey. *Proc. of 9th International Conference on Computer Graphics, Visualization and Computer Vision, WSCG*, 2001.

Generalizing Demonstrated Manipulation Tasks

Nancy S. Pollard[1,3] and Jessica K. Hodgins[2,3]

[1] Brown University, Providence RI 02912, USA
[2] Carnegie Mellon University, Pittsburgh PA 15213, USA
[3] ATR Human Information Science Laboratories, Kyoto 619-0288, Japan

Abstract. Captured human motion data can provide a rich source of examples of successful manipulation strategies. General techniques for adapting these examples for use in robotics are not yet available, however, in part because the problem to be solved by the robot will rarely be the same as that in the human demonstration. This paper considers the problem of adapting a human demonstration of a quasistatic manipulation task to new objects and friction conditions (Figure 1). We argue that a manipulation plan is similar to a demonstration if it involves the identical number of contacts and if the applied contact wrenches follow similar trajectories. Based on this notion of similarity, we present an algorithm that uses the human demonstration to constrain the solution space to a set of manipulation plans similar to the demonstration. Our algorithm provides guarantees on maximum task forces and flexibility in contact placement. Results for the task of tumbling large, heavy objects show that manipulation plans similar to a demonstration can be synthesized for a variety of object sizes, shapes, and coefficients of friction. Experimental results with a humanoid robot show that the approach produces natural-looking motion in addition to effective manipulation plans.

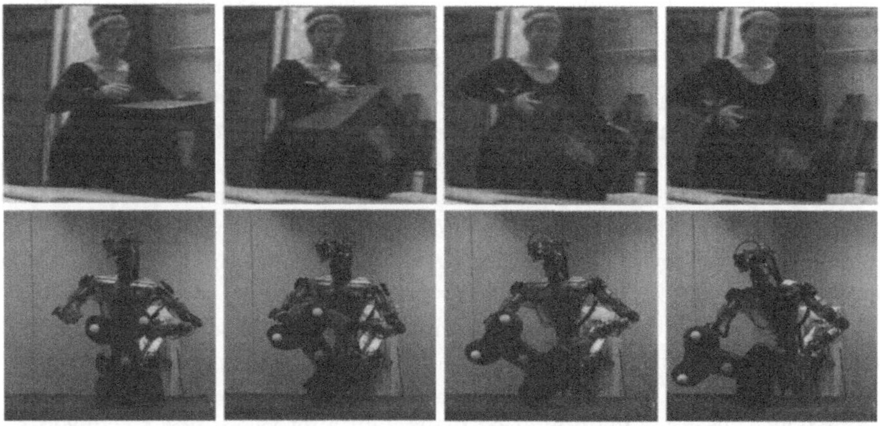

Fig. 1. (Top) Human demonstration of a tumbling task. (Bottom) The demonstration has been adapted to a new object geometry and to the robot kinematics.

J.-D. Boissonnat et al. (Eds.): Algorithmic Foundations of Robotics V, STAR 7, pp. 523–539, 2004.

Fig. 2. Demonstration is a straightforward way to indicate how a task should be performed. Even for a task as simple as tumbling an object from one face to another, a variety of strategies can be observed. These subjects and the subject in Figure 1 have placed their hands very differently while performing this task. All objects are being tumbled from the right to the left.

1 Introduction

Developing robot manipulation skills from human demonstrations is appealing because of the variety of strategies that can be observed in human motion and the ease of conveying a strategy through demonstration (Figure 2). To make human demonstrations a practical way of describing new robot skills, however, we would like to have a system that can take as input a single example and generate behavior similar to that example, gracefully managing differences between the task that was demonstrated and task that the robot must actually perform. Figure 1 shows the type of situation we would like to handle; a significant difference exists between the geometry of the object manipulated by the human demonstrator and the geometry of the object manipulated by the robot.

For quasistatic manipulation tasks, both contact positions and applied forces are important. For this type of task we propose that similarity between a manipulation plan and an example should be based on contact wrenches, i.e., the forces and torques applied to the object at the contact points during the task. In this paper, a demonstration is used to constrain the solution space so that a planner considers only solutions that are similar to that demonstration. Figure 3 shows some results. Compare the contact wrenches applied to the different objects at any instant in time.

The main idea is to use an example to define the number of contacts required for the task and to constrain the roles of those contacts. These constraints can be powerful, and we show how they can be used to define equivalence classes of manipulation plans. Our approach is first to convert the manipulation-planning problem into a set of grasp-planning problems whose solution is a sufficient condition for a manipulation plan to exist. An example-based algorithm is presented to solve this set of grasp-planning problems and

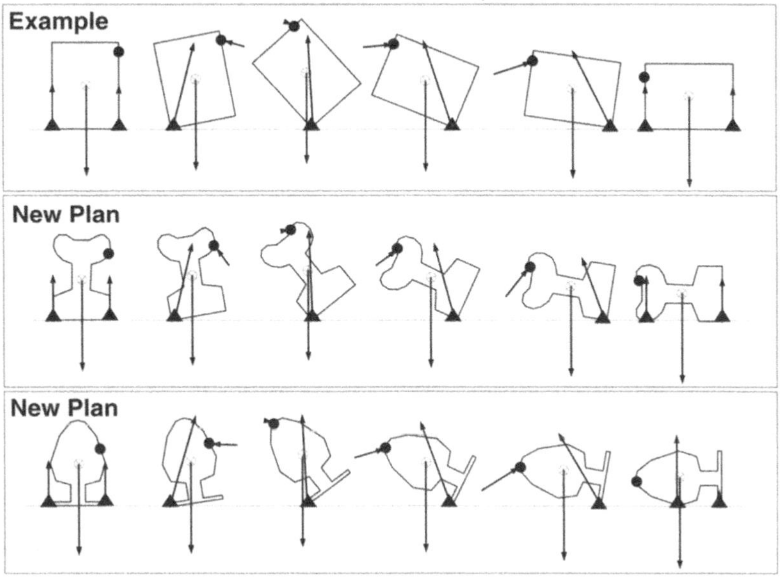

Fig. 3. (Top) The human demonstration is abstracted as an object trajectory and a sequence of contact points. A plausible contact force trajectory is estimated from this information. Snapshots are at intervals of 0.33s. (Middle, Bottom) Our paper presents an algorithm for adapting an example to new objects with known geometries. One goal is that contact wrenches in the new manipulation plans should be similar to those in the example.

select contact points. Given these contact points, a manipulation plan is produced by computing target forces to be applied at the selected contact points over time. Figure 4 shows a block diagram of this approach. Features of this approach include the following:

- **Computational efficiency.** Each step in the process is polynomial in all parameters, including the number of contacts.
- **Task flexibility.** Significant variation in object geometry and friction conditions can be supported.
- **Flexibility in contact placement.** Given the geometry of an object to be manipulated, independent regions of contact are identified so that placing one contact somewhere within each region is sufficient to guarantee that a manipulation plan exists. Examples of such regions are shown in Figure 8.
- **Preservation of force-related features of the example.** Contact wrenches are constrained to a region that contains the example (Figure 3).
- **Quality guarantees.** The regions in Figure 8, for example, were produced with the requirement that the maximum sum of contact normal

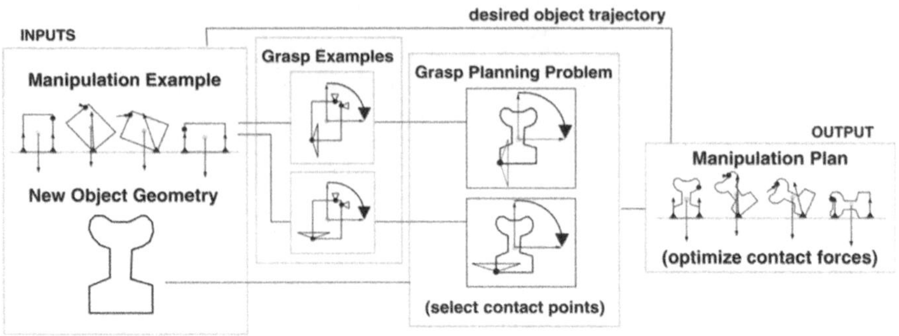

Fig. 4. Block diagram of the planner described in Section 3. The planner takes as input a manipulation demonstration and new object geometry. Contact selection is treated as an example-based grasp-planning problem. A trajectory for the new object is computed from the demonstration. Force computation for the new manipulation plan is accomplished by time sampling and optimization.

forces for manipulating the object should never be greater than twice that required for the example.

- For a humanoid robot, much of the **character of the original human motion** can also be preserved to create a natural-looking appearance (Figure 1).

This approach is suitable for manipulation tasks where the motion is not highly dynamic and where contacts with the object can be abstracted as non-sliding point contacts with friction. The active set of contacts may vary over the course of the motion.

2 Background

This paper draws from research in manipulation, manipulation planning, and grasp planning. Tasks such as tumbling [21] and pivoting [1] have been demonstrated for robot hands, and a number of researchers have explored manipulation planning for tasks similar to tumbling (e.g. [24] [7] [28] [13] [12] [29] [27]). Bicchi [4] provides an overview of grasp-planning research.

Of particular interest here is classification of contact regions. Trinkle and Paul [24] classify contact regions based on object response to squeezing and use these to plan motion to lift an object into a grasp, and Lynch [12] identifies regions where a contact point can be placed so that an object will be toppled by a moving fence. In the work closest to ours, Nguyen [14], Ponce and his colleagues [19] [18] [20] and Chen and Burdick [5] describe algorithms for computing independent contact regions for two-to-four-fingered grasps so that as long as one contact can be placed within each region, a force-closure grasp can be found. van der Stappen et al. [25] and Liu [11] describe how all force closure grasps can be computed for 2D grasps. Independent contact regions could be extracted from the results of their algorithms. Our paper builds on many of these ideas to show how contact regions can be constructed, not from scratch as in previous work, but from an example, so that the contact regions represent a subclass of grasps similar to that example. The key algorithm is an adaptation of Pollard [16] from grasps to manipulation strategies and from frictionless contacts to contacts with friction.

Our work also draws on research in learning from demonstration. Previous work has used one or a combination of two basic approaches. The first is to extract a policy or control strategy directly from the demonstration [2] [10]. The second is to extract enough information about the goal and/or task that a planner or learning algorithm can compute such a policy [9] [8] [26] [22]. The two approaches can be combined. In this case, the planner somehow makes use of the detailed information observed in the human demonstration. For example, in learning a pendulum swingup task, Atkeson and Schaal [3] begin by tracking the hand trajectory of the human demonstrator. They note that merely tracking observed motion was not effective and present an iterative learning algorithm to successfully duplicate pendulum motion from the demonstration.

Our paper relies on a combined approach. In our case, the object trajectory and a set of contact points are extracted from the human demonstration. The example cannot be applied directly to a new object, because identical contact points may not be available; a new set of contact points must be found that are appropriate for the new object. Our approach differs from previous work in that the demonstration is not modified in a local way as in [3], but instead it is used to constrain the space of solutions considered by a global planner. This tactic is practical for tasks that are not highly dynamic. Our approach gives us many of the advantages of a global planner while keeping the results tied to an example and leading to a computationally efficient algorithm.

3 Generalizing a Manipulation Example

An example contains a trajectory for the manipulated object and a trajectory for each contact (Figure 3, top). Our problem is to adapt the example to a new object with known geometry, which requires identifying contact points and solving for contact force targets on the new object. The main idea pursued here is to convert the manipulation problem into a grasp-planning problem such that a solution to the latter is a sufficient condition for a solution to the former to exist. Section 3.1 describes how a manipulation problem is converted to a grasp-planning problem; Sections 3.2 and 3.3 describe how this problem is solved. The frictionless case is presented first, followed by an extension to contacts with friction. Section 3.4 details how these results are converted to a manipulation plan.

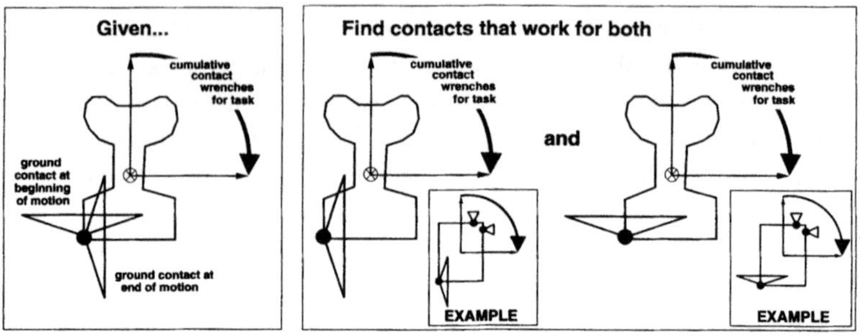

Fig. 5. Grasp-planning problem. Given new object geometry, task wrenches, and a pivot point, shown on the left, find contact points on the new object such that both grasps on the right can achieve all task wrenches. The small insets show the example grasps used as a reference.

3.1 From Manipulation Problem to Grasp-Planning Problem

We wish to convert a manipulation problem where the task is a function of time to a grasp-planning problem where time is not relevant. A first step is to model the task. For the quasistatic problem of tumbling the object 90 degrees to the left, task wrenches are pure forces through the object's center of mass, spanning a 90 degree range as shown in Figure 5.

The next step is to solve for a grasp that can resist these task wrenches. For tumbling, the ground contact at the pivot point can be used as one contact; however, the range of forces available at the ground contact changes with time. If we consider this problem to be two-dimensional, then the ground contact is an edge-vertex contact. Friction cone constraints at this contact are a linear combination of the two extremes at the start and end of the motion. An example of friction cones available at extremes of a tumbling task is shown in Figure 5, left. A sufficient condition for a manipulation plan to exist is to find a set of contact points which, when combined with either extreme ground friction cone, will produce a grasp capable of resisting all task wrenches (Figure 5, right). If the same set of contact points works for each extreme separately, it will also work for intermediate configurations of the ground contact.

To solve the grasp-planning problem posed in Figure 5, we need to find a set of contact points that, for example, produces a force closure grasp with each of the two ground friction cones. If this were the only goal, the demonstration would be unnecessary. The goal that we have, however, is different; we wish contact wrench profiles in the new manipulation plan to be similar to those estimated from the demonstration.

Example grasps extracted from the demonstration can be used to constrain the search for grasps that achieve this goal. The example grasps are created by collecting all contact points observed in the manipulation example and creating grasps from these contact points plus the two extreme ground friction cones. Example grasps for the demonstration explored in this paper are shown as insets in Figure 5. Each of these grasps is capable of achieving all required task wrenches. If this were not true, the task would have to be segmented into subtasks, each covering a smaller range of object motion.

3.2 Placement of Frictionless Contacts

The grasp-planning problem illustrated in Figure 5 must be solved to select a discrete set of contact points. Solving this problem requires several steps. First, independent contact regions are computed for each extreme ground friction cone. Second, these regions are intersected to create independent contact regions that will work for either grasp. Third, discrete contact points are chosen within the regions. After regions have been identified, intersecting them is trivial, and contact points can be selected as desired by the user (e.g., in the centers of the largest connected region or to achieve a kinematic goal such as similarity to the original human motion). The difficult part of the problem is to identify a set of independent contact regions. This section describes how regions are computed for a given ground contact friction cone extreme.

When the grasp is frictionless, the algorithm from Pollard [16] can be used to adapt a grasp to new object geometries by designing wrench space regions around the contact wrenches of the example grasp and then projecting these regions onto a new object. This approach is reviewed below, and Section 3.3 describes how it can be extended to contacts with friction.

Equivalence Classes of Contact Configurations. From a given example grasp, we wish to define equivalence classes of contact configurations on a new object. The information extracted from the example grasp is the particular role ascribed to each contact within that grasp. The idea is that if contacts on a new object can be found to play the same roles, the new grasp is in a way "the same as" the example. This idea can be quantified with the following construction.

Let $\mathbf{g}_i = [\hat{f}_i \quad \gamma(r_i \times \hat{f}_i)]^T$ be the set of contact wrenches for each contact i of the example grasp, formed from unit normal contact forces $\hat{f}_i$ and resulting torque about the object's center of mass, $(r_i \times \hat{f}_i)$, where r_i is the vector from object's center of mass to the contact point. Torque is scaled by some suitable factor γ, for example the distance from the object center of mass to the pivot point.

Form the convex hull of $\mathbf{g}_i$ and the zero wrench. Represent it as a collection of hyperplanes, each expressed as an outward pointing normal $\mathbf{n}_j$ and distance from the wrench space origin d_j:

$$\{\ [\mathbf{n}_1, d_1], [\mathbf{n}_2, d_2], \ldots, [\mathbf{n}_H, d_H]\ \} = CH\{\ \mathbf{0}, \mathbf{g}_1, \mathbf{g}_2, \ldots, \mathbf{g}_N\} \qquad (1)$$

There are N contact points in the example grasp and H facets in the convex hull. An example set of contact points and the convex hull of those points in

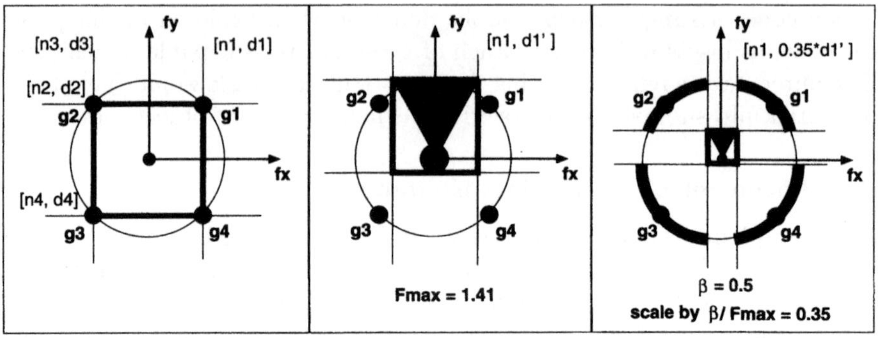

Fig. 6. An example isolates roles of the individual contacts as regions in wrench space. This example corresponds to a frictionless grasp of a circular disc with four evenly spaced contacts. (Left) Contact wrenches $\mathbf{g}_i$ and the convex hull formed from those wrenches. (Center) Move hyperplanes so that they just contain the task wrench space, which is the black shaded area in the figure, oriented toward forces in the y direction. (Right) Scale hyperplanes by β/F_{max}, and highlight unit force wrench space regions that satisfy Equation 5. The bottom two contacts are associated with larger regions due to the anisotropy of the task.

a two-dimensional wrench space (a frictionless grasp of a 2D circular disc) is shown in Figure 6, left.

Index set ρ_i identifies hyperplanes associated with contact i:

$$\rho_i = \{j \ : \ (\mathbf{g}_i \cdot \mathbf{n}_j) = d_j\} \tag{2}$$

For example, in Figure 6 left, $\rho_1 = \{1, 2\}$, because $\mathbf{g}_1$ is contained in hyperplanes 1 and 2.

Adjust each hyperplane along its normal until it just contains all wrenches required for the given task. For a task that can be represented as a collection of points $\mathbf{s}_k$ in wrench space:

$$d_j^{'} = \max\left(0,\ \max_k \left(\mathbf{s}_k \cdot \mathbf{n}_j\right)\right) \tag{3}$$

where $(\mathbf{s}_k \cdot \mathbf{n}_j)$ is the inner product of wrench space vectors $\mathbf{s}_k$ and $\mathbf{n}_j$. An example of this construction is shown in Figure 6 center. The task is the black shaded area, dominated by a range of forces in the positive y direction.

The maximum sum of contact normal forces required for this task, F_{max}, is then:

$$F_{max} = \max_{j=1}^{H} \frac{d_j^{'}}{d_j} \tag{4}$$

The value of F_{max} is 1.41 in the example of Figure 6, because the topmost hyperplane must be scaled by this factor to contain the entire set of task wrenches within its interior.

Any new grasp, defined as the set of contact wrenches $\mathbf{g}_i^{'}$, is in equivalence class β if it satisfies the following collection of inequalities for each contact i:

$$\left(\mathbf{g}_i^{'} \cdot \mathbf{n}_j\right) \geq \frac{\beta}{F_{max}} d_j^{'} \qquad \forall j \in \rho_i \tag{5}$$

This equation states that wrench $\mathbf{g}_i^{'}$ is in the exterior or exactly on all of the hyperplanes that contained point $\mathbf{g}_i$ in the convex hull formed from the example grasp, after these hyperplanes have been adjusted to fit the task and scaled by β/F_{max}. Contact wrench $\mathbf{g}_i^{'}$ is meant to correspond directly to original contact wrench $\mathbf{g}_i$. In other words, the two contacts are intended to play the same roles in their respective grasps. Parameter β allows limits to be placed on the sum of contact normal forces. Setting β to 0.5, for example, guarantees that in the new grasp the sum of contact normal forces will never be greater than twice that required in the example if the task wrenches are the same as those in the example [15]. Figure 6 right shows an example where β is 0.5.

Valid contact regions can now be "painted" onto a new object by sampling the object surface, computing the contact wrench associated with each surface point and testing whether Equation 5 holds. A different region will be computed for each contact of the original grasp. These regions are independent. As long as the new grasp has one contact in each region, we can guarantee that the grasp can exert the task wrenches with a sum of contact normal forces no greater than $1/\beta$ times those of the example grasp. Practically this guarantee is useful for keeping robot joint torques within reasonable limits.

Figure 7 shows an example of a 2D, four-contact, frictionless grasp with a 3D wrench space. The task is to slowly rotate the object 90 degrees to the left, and wrenches required to support the object during this task are pure forces through the object's center of mass spanning a 90 degree range. After the grasp is processed as described in Equations 1 through 4, the set of valid wrenches for each contact, which is defined by Equation 5, can be computed and displayed. Contact force magnitude is arbitrary, and so it is restricted to one to allow contact wrenches to be plotted in a 2D space of force angle vs. torque in the center plots. Wrenches corresponding to frictionless point contact on the edges of any new object can be plotted in this same space. They appear as vertical lines, because force angle does not change as the contact point moves along the edge. Valid contact regions on the object occur when these vertical lines intersect the contact regions. The figures on the right show contact regions for the example object and an object with more complex geometry.

3.3 Placement of Contacts with Friction

The approach just described is not adequate for the example shown in Figure 1, or for the large majority of manipulation tasks because friction is

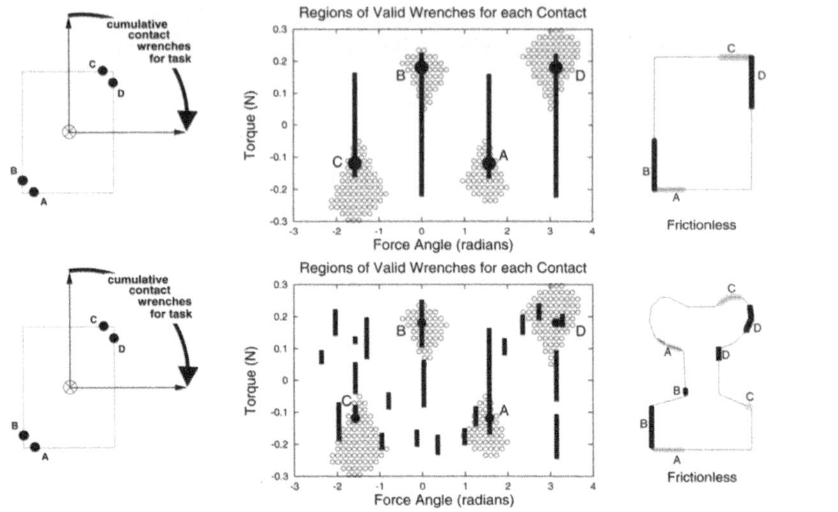

Fig. 7. Results of generalizing a four contact grasp in the frictionless case. (Left) The example grasp. (Middle) Valid wrenches for each contact, along with the vertical lines corresponding to frictionless point contact on the edges of the new object. (Right) Independent contact regions on the new object. Parameter β is 0.5 for these examples. The box is $0.32m \times 0.44m$, γ is 1, the second object is approximately the same size as the box.

important to the success of the strategy. With friction, what is required is that for each extreme of the original friction cone, *some* wrench within the new friction cone can be constructed to play the same role in the grasp. This problem can be solved using linear optimization as follows.

A friction pyramid approximation represents contact wrench $\mathbf{g}_i$ as a linear combination of L extremes, where $\mathbf{g}_i$ and its extremes $\mathbf{g}_{i,l}$ all have unit magnitude force components in the direction of the contact normal.

$$\mathbf{g}_i = \sum_{l=1}^{L} \alpha_l \, \mathbf{g}_{i,l}, \quad \sum_{l=1}^{L} \alpha_l = 1, \quad \alpha_l \geq 0, \quad l = 1, \ldots, L \tag{6}$$

Form the convex hull of all extreme contact wrenches $\mathbf{g}_{i,l}$ and the zero wrench, and express this hull as a set of hyperplanes $[\mathbf{n}_j, d_j]$ as in the frictionless case:

$$\{ \, [\mathbf{n}_1, d_1], [\mathbf{n}_2, d_2], \ldots, [\mathbf{n}_H, d_H] \, \} = CH\{ \, \mathbf{0}, \mathbf{g}_{1,1}, \ldots, \mathbf{g}_{N,L} \} \tag{7}$$

Let index set $\rho_{i,l}$ be defined as follows:

$$\rho_{i,l} = \{ j \; : \; (\mathbf{g}_{i,l} \cdot \mathbf{n}_j) = d_j \} \tag{8}$$

Note that ρ has two indices: i indexes the original contact, and l indexes one of the samples of the friction pyramid at that contact.

Find d_j' and F_{max} as in Equations 3 and 4. Any new grasp, defined as a set of contact wrenches $\mathbf{g}_i'$, with friction pyramid samples $\mathbf{g}_{i,m}'$ is in equivalence class β if the following set of expressions can be solved for all combinations of contacts $i = 1, \ldots, N$ and friction cone extremes $l = 1, \ldots, L$:

$$\left[\left(\sum_{m=1}^{L} \alpha_m \, \mathbf{g}'_{i,m}\right) \cdot \mathbf{n}_j\right] \geq \frac{\beta}{F_{max}} d'_j \qquad \forall j \in \rho_{i,l} \tag{9}$$

$$\sum_{m=1}^{L} \alpha_m = 1 \tag{10}$$

$$\alpha_m \geq 0, \quad m = 1, \ldots, L \tag{11}$$

The first expression states that for contact i in the example grasp, for each friction cone extreme l, proposed contact i on the new object must be able to generate a wrench that plays the same role in the grasp. The second expression maintains unit magnitude for the force component of $\mathbf{g}'_i$ in the normal direction, and the third expression keeps all $\mathbf{g}'_i$ within their friction cones.

For example, a proposed new grasp can be tested for membership in equivalence class β by solving this linear system using the simplex method. To do this, replace β with unknown parameter $\beta_{i,l}$ and maximize $\beta_{i,l}$ for each combination of indices i and l. The proposed grasp is in equivalence class β if:

$$\min_{i=1}^{N}\left(\min_{l=1}^{L} \beta_{i,l}\right) \geq \beta \tag{12}$$

As in the frictionless case, valid contact regions can be "painted" on the new object by sampling the object surface, attempting to solve the system in Equations 9 through 11 for a single contact i, for all friction cone extremes l, and checking that $\sum_l \beta_{i,l} \geq \beta$. All of the examples in Section 4 use this algorithm.

3.4 From Grasp-Planning Results to Manipulation Results

Finally, the manipulation strategy must be extracted from the grasp representation of the problem. Timing is obtained for the new motion by reusing the joint angle trajectories captured from the human actor at their original speed. This decision preserves the dynamics of the robot motion at the possible expense of the dynamics of the manipulated object.

The orientation trajectory of the new object, $\theta'(t)$, can be derived from the orientation trajectory of the example object, $\theta(t)$. Our goal was to duplicate orientation of the example as much as possible, while ensuring that the critical point where the object's center of mass is just above the pivot occurs at the same time in each motion. This critical point is important because it is the

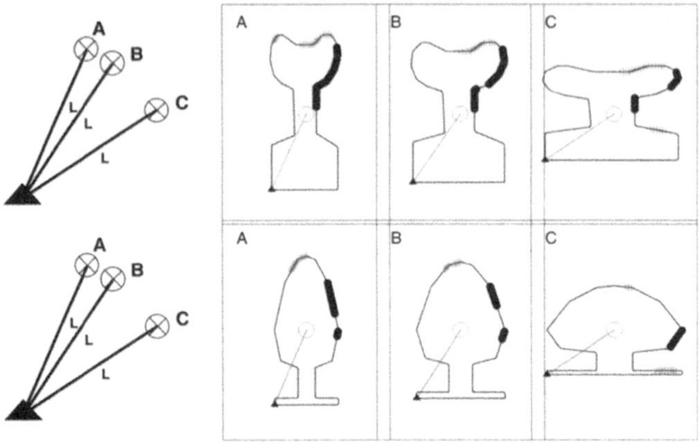

Fig. 8. (Left) Valid aspect ratios for two new objects are represented by a range of angles from pivot point to center of mass. (Right) The pivot point is marked with a solid triangle, and valid independent contact regions for the two contacts are shown. The hand/object coefficient of friction was set to 0.5, and object/ground coefficient of friction was set to 6. (The table was covered with foam, producing a high coefficient of friction.) Quality parameter β was set to 0.5. Note that these results are independent of scale. Length parameter L in the left hand diagram can be set to any value to reflect the actual size of the object to be manipulated.

point at which applied torque about the pivot point changes direction. The following expression is used:

$$\theta'(t) = \left[\psi + \frac{t}{T}(\phi - \psi)\right]\left(\theta(t) - \frac{\pi}{2}\right) + \frac{\pi}{2} \tag{13}$$

where angle θ is measured as the angle from horizontal of the vector from the pivot point to the object's center of mass so that the critical point occurs at $\theta = \pi/2$. Parameters ψ and ϕ are the required scale factors at the beginning and end of the motion, determined from object geometry, and T is the time required for the entire motion.

Forces are estimated by sampling time, computing object orientation from Equation 13, and optimizing contact forces for that orientation and the given contact positions. To optimize contact forces, the sum of normal forces applied at the hand/object contacts is minimized, ignoring the magnitude of the ground reaction force. A conservative estimate of the hand/object coefficient of friction is used, as solutions optimized in this way tend to fall on the extremes of the friction cone.

4 Results

Figures 8 and 9 show contact regions computed for new object geometries by generalizing from the given example. The sizes of the independent contact regions vary with geometry, aspect ratio (Figure 8), coefficient of friction

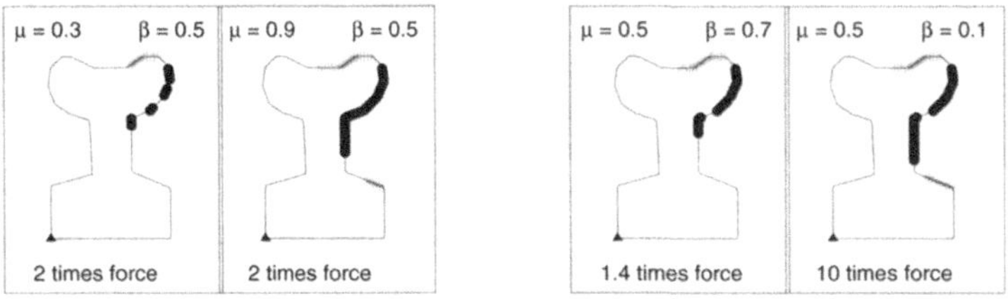

Fig. 9. (Left) Contact regions grow larger as the hand/object coefficient of friction (μ) increases. (Right) Contact regions grow larger as the acceptable amount of contact force ($1/\beta$ times the example) increases.

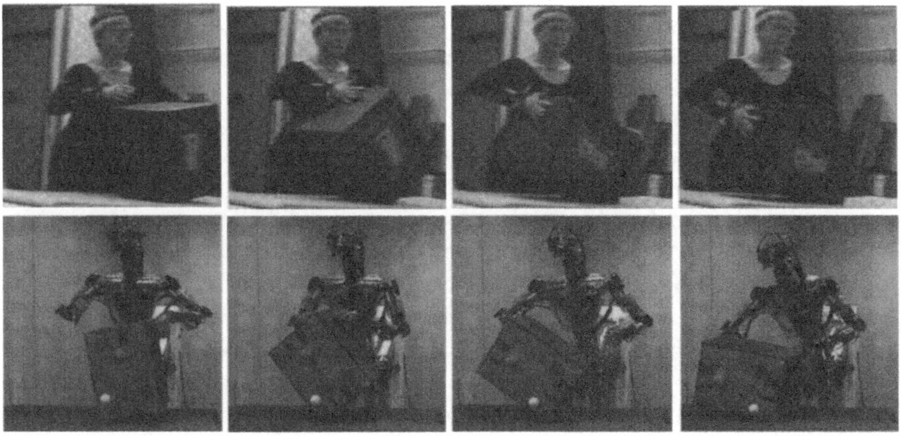

Fig. 10. Comparison of human and humanoid motion. Frames are spaced every 0.33s (top) and 0.67s (bottom). Object weight is between 4kg and 5.4kg.

(Figure 9 left), and maximum allowed contact force (Figure 9 right). Note especially in Figure 8 that the example can be adapted to any scale. The tumbling demonstration of Figure 1 could be used to develop a plan to manipulate a much smaller object with the fingertips of a hand.

Contact forces are computed through time sampling and optimization, as outlined in Section 3.4. Figure 3 shows results for two objects. Despite dramatic changes in object geometry, contact wrench trajectories are similar.

Figures 1 and 10 show comparisons of the original human motion to motion of the Sarcos humanoid robot at ATR [6] tumbling two objects: one similar to that manipulated by the human subject and one less similar. Videos can be found at `http://www.cs.brown.edu/people/nsp/papers/WAFR02.html`. Implementation details can be found in the Appendix. The tumbling task was executed approximately a dozen times for the box without failures when the initial box placement was within a few centimeters of the expected position. This task was executed at half the speed of the human performance to reduce dynamic effects. Tumbling for the object in Figure 1 was slowed to quarter speed to obtain reliable performance.

5 Discussion

This paper describes an approach to adapting a single example of a quasistatic manipulation task to various object geometries and friction conditions. Significant flexibility was obtained both in geometry of the object and in choice of contact locations. Experimental results with a humanoid robot show that the resulting strategies can be executed reliably. They also retain some of the natural feel of the original human motion.

One natural question is "why bother with the example motion?" After the problem has been abstracted to the level of contact positions and forces, it lends itself easily to such alternative approaches as global optimization of contact configurations over time. Our informal observation is that globally optimal solutions frequently do not match strategies actually used by human subjects, perhaps because we have not adequately captured the optimization function, if any, in use by the human subject. We believe by combining planning and example-based approaches, some good features of both can be captured. Using analytic generalization gives us valid contact regions, which provide flexibility to avoid joint limits or other potential problems such as areas where the object is slippery or fragile. Because contact regions are constructed from an example, contact placement is often very similar to that of the example, which may be advantageous when the robot and the human demonstrator have similar kinematics and workspaces. By design of the algorithm, contact wrenches are constrained to be within some region containing those of the example, and so we have a much stronger argument for similarity of the results to the example than do planners which consider task goal only.

This approach will work for any object where Equations 9 through 11 can be solved for all contacts. These equations provide a mapping between a contact index and a set of acceptable object features, although this mapping is not in a practical form at present. Expressing these equations as sets of acceptable object features would be useful for evaluating the generality of a given strategy, for collecting/evaluating a motion database, and for selecting between strategies when faced with a new manipulation problem. We are working toward a practical mapping between wrench space regions and object features.

Computation times for the examples in this paper are on the order of milliseconds. However, sampling the object surface to find good contact regions will be more expensive for truly 3D objects. We are working on a projection algorithm along the lines of Ponce et al. [20] to address this problem.

More complex manipulation tasks will almost certainly require more structure and extensions to the techniques presented here. Manipulation tasks with many contact transitions (e.g., remove a hand and place it somewhere else) may need to be segmented into subtasks. Manipulation tasks with sliding contacts, which are common in human manipulation of objects, cannot be addressed with this approach as it stands. Complex contacts (e.g., contacts in concavities and area contacts) seem to be similar to contact with friction but have not yet been explored.

We are excited about the results to-date and anticipate that the basic approach will prove general: that quasistatic manipulation can be treated as a single task domain, a library of example motions collected within this domain, and these examples adapted to new situations as needed, allowing us to take better advantage of the interesting variety of strategies exhibited in human behavior.

Acknowledgments

We would like to thank Mike Erdmann and Matt Mason for helpful discussions on this problem, Darrin Bentivegna, Gordon Cheng, Marcia Riley, and Stefan Schaal for assistance with the robot experiments, and Chris Atkeson for assistance in both areas. The robot and support for this work were provided by ATR Human Information Science Laboratories. Additional support was provided for Pollard by NSF grant CCR-0093072 and for Hodgins by NSF grant IIS-0196089.

Appendix: Implementation Details

The human demonstration was captured using a commercially available optical motion capture system from Vicon [23]. We used a standard marker set with 35 14mm markers for whole body motion, supplemented with two additional markers on each hand and three markers on the box. The additional markers made it possible to extract hand/object contact points during the motion.

The Sarcos robot at ATR (DB) [6] was used to demonstrate the example and new manipulation plans. To obtain joint trajectories for the robot, the raw marker data from the motion capture session was mapped to a human skeleton, scaled to the robot degrees of freedom and joint limits as described in [17], and adjusted using inverse kinematics to achieve and maintain contact points at desired locations on the object.

The control system used for these examples is shown in Figure 11. The input trajectory was specified as a set of kinematic parameters and a set of parameters relating to contact between the robot and the object. Specifically, inputs are robot joint angles and derivatives ($\theta_{DES}, \dot{\theta}_{DES}, \ddot{\theta}_{DES}$), box orientation ($\theta_{BOX,DES}, \dot{\theta}_{BOX,DES}$), contact forces ($f_{ff}$), and the ratio of contact force magnitudes to torque about the box pivot point (f_{ff}/τ_{BOX}). Two low-gain PD servos are used for control, one that uses visual feedback to maintain box state and one that uses position sensing at the joints to track the joint angle trajectory. The servo associated with box state controls the magnitude of feedforward contact forces to implement a torsional spring and damper about the box pivot point. Out-of-plane orientation errors are ignored. Correction

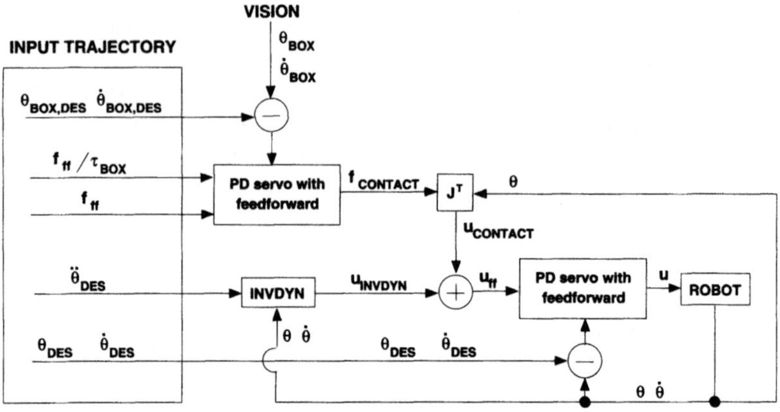

Fig. 11. Controller for the tumbling task. All inputs on the left may have to be adjusted to adapt the input trajectory to a new object and/or new friction conditions. This paper focuses on adjusting contact positions and contact force trajectories f_{ff}.

for such errors would make it possible to manipulate "thinner" objects (i.e., objects with smaller dimension along the axis of rotation).

References

1. Y. Aiyama, M. Inaba, and H. Inoue. Pivoting: A new method of graspless manipulation of object by robot fingers. In *Proc. IEEE/RSJ Intl. Conference on Intelligent Robots and Systems*, 1993.
2. H. Asada and Y. Asari. The direct teaching of tool manipulation skills via the impedance identification of human motion. In *Proc. IEEE Intl. Conference on Robotics and Automation*, 1988.
3. C. G. Atkeson and S. Schaal. Robot learning from demonstration. In *International Conference on Machine Learning*, 1997.
4. A. Bicchi. Hands for dexterous manipulation and robust grasping: A difficult road toward simplicity. *IEEE Transactions on Robotics and Automation*, 16(6):652–662, 2000.
5. I-M. Chen and J. W. Burdick. Finding antipodal point grasps on irregularly shaped objects. *IEEE Transactions on Robotics and Automation*, 9(4):507–512, 1993.
6. DB. http://www.his.atr.co.jp/cyh/.
7. M. Erdmann. An exploration of nonprehensile two-palm manipulation. *International Journal of Robotics Research*, 17(5), 1998.
8. S. B. Kang and K. Ikeuchi. Toward automatic robot instruction from perception–Temporal segmentation of tasks from human hand motion. *IEEE Transactions on Robotics and Automation*, 11(5):670–681, 1995.
9. Y. Kuniyoshi, M. Inaba, and H. Inoue. Learning by watching: Extracting reusable task knowledge from visual observation of human performance. *IEEE Transactions on Robotics and Automation*, 10(6):799–822, 1994.

10. S. Liu and H. Asada. Transferring manipulative skills to robots: Representation and acquisition of tool manipulative skills using a process dynamics model. *Journal of Dynamic Systems, Measurement, and Control*, 114:220–228, June 1992.
11. Y.-H. Liu. Computing n-finger form-closure grasps on polygonal objects. *International Journal of Robotics Research*, 19(2):149–158, February 2000.
12. K. M. Lynch. Toppling manipulation. In *Proc. IEEE Intl. Conference on Robotics and Automation*, 1999.
13. K. M. Lynch and M. T. Mason. Dynamic nonprehensile manipulation: Controllability, planning, and experiments. *International Journal of Robotics Research*, 18(1):64–92, January 1999.
14. V. Nguyen. Constructing force-closure grasps. *International Journal of Robotics Research*, 7(3):3–16, June 1988.
15. N. S. Pollard. Parallel methods for synthesizing whole-hand grasps from generalized prototypes. Technical Report AI-TR-1464, MIT AI Lab., January 1994.
16. N. S. Pollard. Synthesizing grasps from generalized prototypes. In *Proc. IEEE Intl. Conference on Robotics and Automation*, April 1996.
17. N. S. Pollard, J. K. Hodgins, M. J. Riley, and C. G. Atkeson. Adapting human motion for the control of a humanoid robot. In *Proc. IEEE Intl. Conference on Robotics and Automation*, May 2002.
18. J. Ponce and B. Faverjon. On computing three-finger force-closure grasps of polygonal objects. *IEEE Transactions on Robotics and Automation*, 11(6):868–881, 1995.
19. J. Ponce, D. Stam, and B. Faverjon. On computing force-closure grasps of curved two-dimensional objects. *International Journal of Robotics Research*, 12(3):263–273, June 1993.
20. J. Ponce, S. Sullivan, A. Sudsang, J.-D. Boissonnat, and J.-P. Merlet. On computing four-finger equilibrium and force-closure grasps of polyhedral objects. *International Journal of Robotics Research*, 16(1):11–35, February 1997.
21. N. Sawasaki, M. Inaba, and H. Inoue. Tumbling objects using a multi-fingered robot. In *Proc. 20th ISIR*, 1989.
22. S. Schaal, C. G. Atkeson, and S. Vijayakumar. Scalable techniques from nonparametric statistics for real-time robot learning. *Applied Intelligence*, (in press).
23. Vicon Motion Systems. http://www.vicon.com/.
24. J. C. Trinkle and R. P. Paul. Planning for dexterous manipulation with sliding contacts. *International Journal of Robotics Research*, 9(3):24–48, June 1990.
25. A. F. van der Stappen, C. Wentink, and M. H. Overmars. Computing immobilizing grasps of polygonal parts. *International Journal of Robotics Research*, 19(5):467–479, May 2000.
26. R. M. Voyles, J. D. Morrow, and P. K. Khosla. Towards gesture-based programming: Shape from motion primordial learning of sensorimotor primitives. *Robotics and Autonomous Systems*, 22:361–375, 1997.
27. M. T. Zhang and K. Goldberg. Gripper point contacts for part alignment. *IEEE Transactions on Robotics and Automation*, in press.
28. R. Zhang and K. Gupta. Automatic orienting of polyhedra through step devices. In *Proc. IEEE Intl. Conference on Robotics and Automation*, 1998.
29. T. Zhang, K. Goldberg, G. Smith, R.-P. Berretty, and M. Overmars. Pin design for part feeding. *Robotica*, 19(6):695–702, 2001.

Constraint-Based Motion Planning Using Voronoi Diagrams

Maxim Garber and Ming C. Lin

University of North Carolina, Chapel Hill, NC, USA
{garber,lin}@cs.unc.edu
http://gamma.cs.unc.edu/cplan

Abstract. We present a novel algorithm for planning the motion of rigid and articulated robots in complex, dynamic, 3D environments. Our approach is to reformulate the motion planning problem as a simulation of a constrained dynamical system, and guide this system using generalized Voronoi diagrams (GVDs). In our framework, each rigid robot is subject to virtual forces induced by geometric and mechanical constraints. These may include constraints to have a robot follow an estimated path computed using a GVD, constraints to link rigid objects together to represent an articulated robot, or constraints to enforce a spatial relationship between multiple collaborative robots. The resulting algorithm uses all constraint forces to move the robot along an estimated path through the environment, while avoiding collisions with obstacles and enforcing joint and positional constraints. Our algorithm works well in dynamic environments with moving obstacles and is applicable to planning scenarios where multiple robots must move simultaneously to achieve a collision free path.

1 Introduction

Motion planning is one of most fundamental problems in algorithmic robotics. The classic motion planning problem, also referred to as the Piano Mover's problem, can be stated as the following: Given a robot $\mathcal{R}$ and a workspace $\mathcal{W}$, find a path, P, from an initial configuration $\mathcal{I}$ to a goal configuration $\mathcal{G}$, such that $\mathcal{R}$ never collides with any obstacle $\mathcal{O}_i \in \mathcal{W}$. The path P, if such a path exists, is a continuous sequence of positions and orientations (i.e. configurations) of $\mathcal{R}$. Planning tasks are typically characterized by geometric goal regions, a variety of mechanical and geometric constraints, and often a partially known environment with uncertainties and moving obstacles.

Main Results: In this paper, we present a novel motion planning algorithm for rigid and articulated robots. Our algorithmic framework is based on constrained dynamics [27] used for physically-based modeling. We reformulate the motion planning problem as a dynamical system simulation, where constraints are enforced by virtual forces imposed on the system. Each robot is treated as a rigid body, or a collection of rigid bodies, and is moved subject to all types of constraint forces. These may include constraints that are inherent to the planning scenario, such as constraints that enforce the joint angle limits and link connectivity of articulated robots or constraints that enforce

J.-D. Boissonnat et al. (Eds.): Algorithmic Foundations of Robotics V, STAR 7, pp. 541–558, 2004.

some required spatial relationships between multiple collaborative robots. In addition to these required constraints, we introduce constraints designed to guide the robot through the environment to the desired goal configuration. Using global geometric analysis from the generalized Voronoi diagram [12] of the workspace, we define constraints that move the robot to avoid both static and moving obstacles, and also follow an estimated path to the goal. The solution to the motion planning problem is the collection of configurations of the dynamical system that satisfy all geometric and mechanical constraints.

Although our current implementation is only designed to handle certain types of planning scenarios, our proposed framework is rather general. The framework can be extended to incorporate many different constraint types and also many different constraint solving methods, such as optimization, local iteration, symbolic algebraic solvers, etc. Our planning planning framework has the following characteristics:

- **Portable:** It can plan paths for both rigid and articulated robots of any topology, including both revolute and prismatic joints, and of any number of degrees of freedom. It can also be extended to handle robots that deform over time governed by the law of physics.
- **Dynamic:** It can plan collision free motion for an object in the presence of moving, or deforming, obstacles, as well as other collaborating robots, whose motion is not known a priori.
- **General:** It allows the user to specify a wide range of relationships between robots as well as between robots and other objects in the scene. Examples of such relationships include: a robot avoiding other moving robots, multiple robots following a leader, and multiple robots collaborating to manipulate an object.

To demonstrate our framework we have implemented a planner for rigid and articulated robots in dynamic environment that runs at interactive rates for many moderately complex scenes. We demonstrate the effectiveness of our planner for virtual assembly and electronic prototyping. We show that it works well in a changing scenes with moving obstacles and multiple collaborating robots, in applications to assembly line planning, automated car painting, and maintainability studies.

Organization: The rest of the paper is organized as follows. In section 2, we briefly survey related work. Section 3 presents an overview of our planning framework. We describe the detailed formulation of various constraints in section 4. Section 5 explains how planning scenarios are represented in our framework. We demonstrate our planner working on several virtual prototyping problems in section 6. In section 7, we offer a conclusion and suggest some future research directions.

2 Related Work

Motion planning has been extensively studied in robotics, computational geometry and computer-aided manufacturing for more than three decades [19]. Most of the earlier work has focused on the Piano Mover's problem.

2.1 Global vs. Local Planning Methods

There have been two class of approaches to motion planning: global and local methods. Global planning methods, including the first Roadmap Algorithm [6], PRM [14,15] and other geometric or "criticality-based" methods [10,19], are guaranteed to find a path, if one exists, although they may take a long time computing it. Many of the global methods, with the exception to variants of PRM, have been applied with limited success for mostly lower-dimensional planning queries in static environments, due to high computational costs.

On the other hand, local methods such as artificial potential field methods [16], are usually fast, but are not guaranteed to find a path, even if one exists. Algorithms based on artificial potential field methods are frequently used in industrial applications [7,21]. In a dynamic environment, where the motion of obstacles in the scene is not know a priori, it is difficult for global methods to compute the complete solution path in real time to avoid collision with the moving obstacles. Local methods have known limitations as well. For example, potential field methods are known for their entrapment problems at local minima of the potential function.

Our method is an incremental construction of a roadmap, whose the curves locally satisfy all constraints imposed on the robot while remaining maximally clear of nearby obstacles. At the same time, to help avoid the local minima problem, the framework takes global geometric analysis of the environment, obtained from the Generalized Voronoi Diagram, into consideration while performing the local planning.

2.2 Voronoi Diagrams in Motion Planning

Generalized Voronoi diagrams (GVDs) have long been used as a basis for motion planning algorithms [5,8–10,13,22,25,26]. The GVD represents the connectivity of a space but has a dimension lower by one, and (in three dimensions) it is composed of surfaces of maximal clearance. Unfortunately, no good and practical algorithms are known for computing the Voronoi diagrams of large environments. In the worst case the complexity of Voronoi diagrams is $O(n^2)$, where n is the number of polygons in the environment. Moreover, the Voronoi diagram of a polygonal environment is composed of quadric surfaces and degree four curves that meet at junctions whose algebraic degree is eight. It is hard to accurately compute an arrangement of these curves and surfaces using fixed precision arithmetic.

Our approach takes advantage of a method proposed in [12] which uses graphics hardware to quickly compute a discretized, error-bounded approximation to the GVD of the workspace, to provide useful information for motion planning.

2.3 Constraint Solving

Geometric constraint solving has been extensively studied in many different fields, such as CAD/CAM, molecular modeling, and theorem proving [3,4,17]. There are two basic strategies: *instance solvers* and *generic solvers*. Some of the common approaches include numerical algebraic techniques, graph-based algorithms, logical interference and term rewriting, symbolic algebraic solvers, and propagation methods [3]. Any one of these techniques could be used in our planning framework.

In our current implementation, we borrow a combination of ideas from some of these techniques, and specializes them for motion planning. The goal of our implementation is to achieve real-time performance for dynamic scenes with moving obstacles and multiple, rigid or articulated, robots. Since, the constraint solving problem for motion planning can be NP-hard for arbitrarily high degree-of-freedom manipulators, we consider the intended solution and infer certain metric and topological properties of the planning problem as a dynamical system, and deduce a few heuristics that succeed with high probability under the assumptions of compatible constraints and temporal coherence. The details of our constraint solving approach will be given in Sec. 4.

3 Framework Overview

Our approach is an opportunistic planning framework powerful enough to handle dynamic environments, and general enough to incorporate many different types of constraints and global analysis to govern an object's motion. It also allows the incorporation of complex relationships between collaborating entities. In this framework even articulated robots are represented using constraints, linking collections of rigid objects. In addition, this framework allows natural extension to planning of flexible robots and incorporation of dynamics, non-holonomic, and other types of constraints.

3.1 Simulation Framework

The basic essence of our framework is to describe each rigid object in the planning scene as a dynamical system, which is characterized by its state variables (i.e. position, orientation, linear and angular velocity). In this framework, a robot can be a rigid body, or a collection of rigid bodies, subject to the influence of various forces in the workspace, and restricted by various motion constraints. This transforms a motion planning problem into a problem of defining suitable constraints, and then simulating the rigid body dynamics of the scene with each constraint acting as a virtual force on the objects. That is, if q is the configuration of the robot at some time t, then each constraint can be represented as a function of q, $C(q)$. The virtual force induced by each constraint is simply $f_c = \frac{-\partial E(C(q))}{\partial q}$ where the energy function, $E(C(q))$, is defined as $E(C(q)) = \frac{k_s}{2} C(q) \cdot C(q)$ and k_s is a generalized stiffness constant.

We'll use the following notation for the rest of the paper:

- Let $\mathcal{R} = \{\mathcal{R}_1, \ldots, \mathcal{R}_m\}$ be a set of m rigid objects.
- For each $\mathcal{R}_i$ at time t, let a state vector $s_i^t = (pos_i^t, rot_i^t, lin_i^t, ang_i^t)$ represent the objects position, rotation, linear and angular velocity.
- Let S^t be the system state vector, obtained by concatenating the state vectors s_i^t for all i.
- Let $\mathcal{C} = \{\mathcal{C}_1, \mathcal{C}_2, \ldots, \mathcal{C}_n\}$ be a set of n constraints.
- For each constraint $\mathcal{C}_j$, let $F_j(S^t)$ be the force induced by constraint j given the object system state S^t.

The simulation steps from time t to time $t+h$ and updates the state of each object subject to the forces induced by the constraints, using the following steps:

BEGIN LOOP

Update Obstacles: For each dynamic obstacle O_i, update its state in S^t.
Compute Constraint Forces: Summing up all virtual forces, $F_c(S^t) = \sum_{j=1}^{n} F_j(S^t)$.
Update System State: Compute S^{t+h} from S^t subject to $F_c(S^t)$ [27].
Update Object States: For each object $\mathcal{R}_i$, update s_i^{t+h} from S^{t+h}.
Increment Time: $t = t + h$

END LOOP

In this framework the solution to the motion planning problem, for a particular object R_i emerges as the sequence of states, $\{s_i^t, s_i^{t+h}, \ldots, s_i^{t+k*h}\}$, such that the object is in its initial configuration at time t, and achieves the goal configuration at time $t + k * h$. The simulation must run for as many time steps as necessary for all objects for which a planned path is desired, to reach their goal configurations.

3.2 Dynamic Scenes

Since our intent is to handles scenes in which obstacle motion is not known a priori, our planning framework must incorporate an explicit dependence on time. As a consequence, a path computed for a particular time interval may not be valid for the different time interval due to different configurations and trajectories of the obstacles. In the above simulation loop we ensure this interdependence by synchronizing the time step of the planning simulation with the time step that governs the motion of obstacles in the scene. It is important to note that the total number of time steps needed to plan a path through a dynamic scene can vary dramatically with the obstacle motion. This occurs, for example, when the robot must wait for a slow moving obstacle to move out of the way before proceeding to the goal.

3.3 Constraints

To achieve the desired results without robustness problems, we further classify the constraints into soft and hard constraints and treat them differently.

Hard Constraints are those that absolutely must be satisfied at every time step of the simulation. Common examples of hard constraints include:

- objects must stay within the bounds of the scene
- objects must never intersect or penetrate each other %vspace*-0.25em
- links of an articulated robot must remain attached together
- the joint angles in an articulated object must remain within some predefined limit

Soft Constraints serve as guides to *encourage* or influence the objects in the scene to behave in certain ways. They are simulated using penalty forces. Some common examples of soft constraints include:

- an object should move to follow the nearby medial axis of the workspace
- an object should move towards a goal configuration
- an object should move to avoid the nearest obstacles
- an object should move towards goals set by a higher level planner

More details on constraint formulation and constraint solving techniques will be presented in Sec. 4.

3.4 Use of Voronoi Diagrams

In our framework we define constraints that guide the robot towards the planning goal. To obtain these crucial constraints we use information about the environment obtained by Voronoi diagrams. This is done in two ways:

Estimated Paths are used in a manner similar to the basic philosophy taken by [2,8–10,26], to provide global information to the path planner. To obtain this information we use the generalized Voronoi diagram (GVD) of the environment. The GVD computation is accelerated by graphics hardware and can be recomputed as the environment changes. This global information is incorporated into our framework as a constraint that guides the robot along the path computed from the GVD of the environment. The estimated path based on the GVD is not a true collision-free path, but by using other constraints our planner will correct the estimated path, avoiding collisions as the simulation proceeds. This shares the similar theme with OBPRM [1].

Obstacle Avoidance is accomplished using a by-product of the GVD computation. This byproduct is a discretized distance field of the workspace. This distance field can be recomputed in real time around the regions of interests, using hierarchical bounding boxes and parallel computing on today's graphics hardware [12]. It is used to perform distance computation, as well as quick rejection tests, which can be followed by exact collision detection if needed. Using this distance field, we can also define soft constraints to maintain a minimum clearance between a robot and its environment. The penalty forces that result from this constraint will tend to align the robot to the medial axis of the environment.

4 Constraint Formulation

4.1 Solving The Constraints

We use two main constraint solving techniques in the current implementation of our planning framework: penalty forces and iterative relaxation. Penalty forces are used to represent soft constraints, while iterative relaxation is used to enforce hard constraints. Other constraint solving techniques can also be appropriately incorporated. Since we want the object states to satisfy all hard constraints at the end of each time step, we perform relaxation on the hard constraints after applying penalty forces based on soft constraints.

Applying Penalty Forces: We use penalty forces to apply soft constraints to the rigid objects in a planning scene. As we will describe in Sec. 4.3, each type of soft constraint, $\mathcal{C}_i$, has a method $Get\_Penalty\_Force(\mathcal{C}_i, \mathcal{R}_j, S^t, t)$ which returns the force and torque generated by the constraint $\mathcal{C}_i$, on object $\mathcal{R}_j$, at time t, with the scene in the state S^t. The total force on an object is the sum of all penalty forces acting on that object. To update the object state for each object, we integrate this total force using the Midpoint Method [27]. We use this method because some of the soft constraint penalty forces are computationally intensive to evaluate, and the midpoint method provides stable integration with only two force evaluations per time step.

Iterative Relaxation: Once objects in the scene are updated as a result of the penalty forces due to the soft constraints, their state may violate one or more of the hard constraints. For each hard constraint, $\mathcal{C}_h$, we define the residual, $Res(\mathcal{C}_h, S^t)$, to be a real number which represents the degree to which $\mathcal{C}_h$ is violated when the objects are in state S^t. For each type of hard constraint $\mathcal{C}_h$, we require an instance solver $Relax(\mathcal{C}_h, S^t)$, to be described in Sec. 4.2, which returns a new state, S, in which $Res(\mathcal{C}_h, S) = 0$. To efficiently ensure that these constraints are satisfied, we use the well known Nonlinear Gauss-Siedel method [23]. The iterative relaxation method, described in Algorithm. 1, relaxes each constraint in sequence repeatedly, until the objects converge to a state for which the sum of the residuals, over all constraints, is zero.

Relax_Constraints

Input The state S^t, at time t, of all rigid objects, points and vectors in the simulation, the set $\mathcal{C}$ of all hard constraints.
Output New state vector S which satisfies all hard constraints.

Let $S \leftarrow S^t$.
Repeat:{
  for each hard constraint $\mathcal{C}_i$: {
    $S \leftarrow Relax(C_i, S)$.
  }
}**until** $\sum_{i=0}^{n} |Res(C_i, S)| = 0$
return The state S.

ALGORITHM 1: Relax Hard Constraints

Convergence of the Relaxation Method: Our framework allows the use of many different constraint solving techniques. The reasons that we choose, in our current implementation, to use the iterative relaxation method introduced in Sec. 4.1, instead of other techniques (e.g. Lagrangian formalism) to satisfy the hard constraints, is the method's simplicity and because it allows our implementation to achieve interactive performance in most practical scenarios This rapid convergence can be attributed to two factors: temporal coherence and compatible constraints.

Our system is able to take advantage of temporal coherence because, as a physical simulation, it uses small time steps during which the objects in the scene move very little. Moreover, if we assume that all hard constraints are satisfied at the beginning of a time step, the fact that the object motion due to soft constraints is small ensures that the iterative method starts in an initial configuration that is near a valid configuration, if one exists. This all but ensures that the iterative method converges to the solution. It also allows convergence to take place in a relatively small number of iterations, providing stable interactive performance in many practical scenarios. Of course, it is possible for the method to never converge when, for example, there are two incompatible constraints, such that satisfying one necessarily violates the other. This situation does not typically occur in practice because the hard constraints are defined so that they are initially compatible with each other.

4.2 Hard Constraints

To ensure that the simulation enforces the high level hard constraints, such as those mentioned in Sec. 3.3, we use three atomic hard constraints in our current implementation:

- Non-Penetration Constraints
- Point Distance Constraints
- Point Planar Angle Constraints

For each type of hard constraint we will now describe the instance solver $Relax(\mathcal{C}_h, S^t)$. This solver is used in the iterative algorithm of Sec. 4.1 to enforce all of the constraints in the scene. For the constraints that act on rigid objects the solver updates the state, including positions and orientations, of the objects; while for point (distance or angle) constraints, the solver modifies only the positions of the points.

Non-Penetration Constraints: Assuming that all rigid objects in the scene represent closed volumes, we consider a non-penetration constraint between two objects to be satisfied as long as their volumes are disjoint. In most cases, this constraint can be weakened to only require that the objects' boundaries do not penetrate each other. Due to the importance of this constraint for correct motion planning, we use an in-house proximity (collision) query package, PQP [11,18], to detect when two objects penetrate. The residual for a non-penetration constraint is then just 0 if the object are disjoint and 1 if the objects are not. The problem of separating objects that penetrate, in a physical simulation, is one that has been addressed in many ways [27]. The approach that we use to implement the *Relax* solver for non-penetration constraints is the impulse-based rigid body dynamic simulation [27]. The advantages of this method is that objects are guaranteed to be disjoint at the end of every time step, and that objects rebound from collisions in the most natural possible way. In most planning scenarios, non-penetration constraints should be applied between all objects in the scene, ensuring that the objects behave as if they are solid, although it is possible to have objects that are selectively solid, or completely permeable, if desired.

Point Distance Constraints: These constraints enforce a fixed separation between pairs of points. Thus at a time t, given:

- $p_1 \in \mathcal{R}^3$, and its world transformation T_1^t
- $p_2 \in \mathcal{R}^3$, and its world transformation T_2^t
- $d \in \mathcal{R}$, the constraint distance.

the constraint is satisfied when $dist(T_1^t(p_1), T_2^t(p_2)) = d$, where $dist()$ is the Euclidean distance. Given this formulation of a point distance constraint we define the residual of the distance constraint as:

$$Res(C_{dist}, S) = \Delta_d = dist(T_1^t(p_1), T_2^t(p_2)) - d, \tag{1}$$

where Δ_d represents the linear distance that the two points must travel to reach the required separation. To solve the distance constraint we simply move each point a straight line distance of $\Delta_d/2$ towards each other.

Point Planar Angle Constraints These constraints enforce the angle between two points, about a specified axis of rotation. To define these constraints we require, at time t:

- $p_1 \in \mathcal{R}^3$, and its world transformation T_1^t
- $p_2 \in \mathcal{R}^3$, and its world transformation T_2^t
- $o \in \mathcal{R}^3$ the origin of the joint, and its world transformation T_o^t
- $\overrightarrow{axis} \in \mathcal{R}^3$, axis of rotation for the joint, and its world transformation T_a^t
- $\theta_{min}, \theta_{max} \in \mathcal{R}$, the angle limits.

We define the angle θ as the planar angle between the vectors $T_1^t(p_1) - T_o^t(o)$ and $T_2^t(p_2) - T_o^t(o)$ in the plane normal to $T_a^t(\overrightarrow{axis})$. The constraint is satisfied as long as θ is in the interval $[\theta_{min}, \theta_{max}]$. Given this formulation of a point planar angle constraint between points the residual of the angle constraint is then defined as:

$$Res(C_{ang}, S) = \begin{cases} \theta - \theta_{max} & \text{if } \theta > \theta_{max} \\ \theta - \theta_{min} & \text{if } \theta < \theta_{min} \\ 0 & \text{otherwise} \end{cases} \tag{2}$$

To satisfy the angel constrain, if $\Delta_\theta = Res(C_{ang}, t) \neq 0$, the point $T_1^t(p_1)$ is rotated an angle $\Delta_\theta/2$, and $T_2^t(p_2)$ rotates an angle $-\Delta_\theta/2$ about the rotation axis, $T_a^t(\overrightarrow{axis})$.

Relating Constraints: Unlike the non-penetration constraints, the point distance and point planar angle constraints are independent of the object geometry; they instead enforce relationships between points and vectors defined in the scene. These points and vectors can be fixed in world coordinates or expressed relative to the coordinate frames of objects in the scene. In Sec. 5.1 we will address how these point constraints can be combined to achieve the high level hard constraints described in Sec. 3.3.

To allow these constraints between point and vectors to influence the behavior of objects in the scene we use a rigid structure, such as a tetrahedron, which we define using point distance constraints, to represent the coordinate frame of each rigid object. We chose four linearly independent points in the object's coordinate system, and set distance constraints between them to enforce their initial separations. When the object is transformed the rigid structure is also transformed and its distance constraints remain trivially satisfied. At the same time, as long as the constraints that define the rigid structure are satisfied we can uniquely determine the object's transformation from the world locations of the four points of the rigid structure.

To constrain the relative motion of objects in the scene we define constraints between the points of their rigid structures. Thus the order in which the hard constraints are relaxed in Alg. 1 is not arbitrary. At each frame of the simulation the constraints between all of these points are enforced first, possibly changing their locations. Then, from these new point locations, we update the world state of the associated rigid object to a state that respects the constraints. Only after the point constraints have been solved and the associated rigid body states' have been updated do we relax any constraints between the rigid objects themselves.

4.3 Soft Constraints

Soft constraints, in our framework, are constraints that generate penalty forces to *guide* the motion of objects without imposing strict motion restrictions. The types of soft constraints in our current system include:

- Surface repulsion
- Goal attraction
- High-level path following

For each soft constraint, we will describe the function that is used to generate a force along the gradient vector of the constraint.

Surface Repulsion: These constraints use local proximity information, computed as part of the GVD calculation, to allow objects to be repelled from the surfaces of nearby obstacles. To define a surface repulsion constraint for a rigid body, $\mathcal{R}_1$, relative to a second rigid object $\mathcal{R}_2$, we specify distance threshold, δ, and a force coefficient $k \in [0, 1]$. In our method of assigning an estimated path for a robot, $\mathcal{R}_1$, using Voronoi information, the high level planner is able to provide an associated minimum distance tolerance along

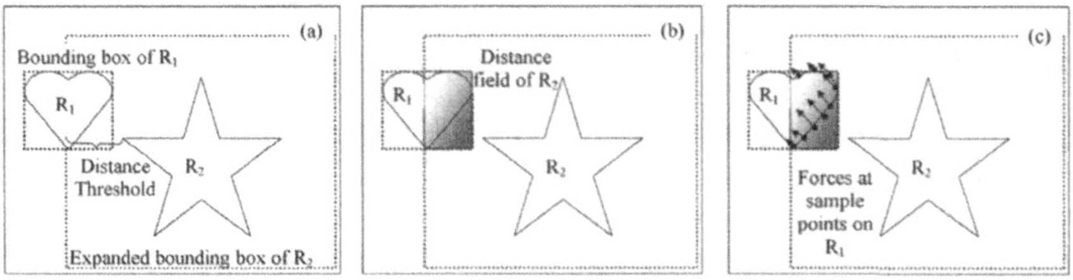

Fig. 1. (a) R_1 has a repulsion constraint repelling it from R_2. (b) A portion of the distance field of R_2. (c) Forces act on R_1 in the gradient direction of the distance field.

the path, which can be used to initialize δ. The priority, k, specifies the relative importance of repulsion constraints, so that if $\mathcal{R}_1$ is trapped between two obstacles, it will give more priority to evading one than the other. Soft constraint priorities are set equal to 1, unless the user specifies another value.

To apply the constraint, we first perform some computations using axis-aligned bounding boxes as approximations for the objects involved. For object $\mathcal{R}_2$ we take the axis-aligned bounding box and expand it by the distance threshold, δ, as shown in Fig. 1 (a). We intersect this expanded bounding box with the bounding box of $\mathcal{R}_1$ to perform a quick rejection test to determine if the two objects are further than the distance threshold δ apart. If this is the case, then we can terminate the computation. If the bounding box test fails, we compute the intersection, call it I, of the two bounding boxes. We then use a hardware accelerated distance field computation [12], a by-product of the GVD computation, to generate the distance values for the surface features of object $\mathcal{R}_2$ in the region I as shown in Fig. 1 (b). The fact that this computation is performed using graphics hardware enables the distance field of the object to be generated in real time without any precomputation or assumptions about the geometry. We intersect this distance field with sample points on the surface of object $\mathcal{R}_1$. For each sample point that lies in I, we check the distance from that point to the nearest point on the surface of $\mathcal{R}_2$ by referencing the distance field. A force is generated, at each sample point, that is in the direction of the gradient of the distance field, proportional to the distance between that sample point and the surface of $\mathcal{R}_2$, as seen in Fig. 1 (c). This force should be zero for sample points beyond the distance threshold, and increase to infinity as the distance between the surfaces decreases. In our implementation the force at each sample point, p_i, is given by:

$$force(p_i, \mathcal{R}_2) = \begin{cases} \frac{\delta^4}{dist(p_i,\mathcal{R}_2)^4} - 1 & \text{if } dist(p_i, \mathcal{R}_2) < \delta \\ 0 & \text{otherwise} \end{cases} \tag{3}$$

The force induced by this constraint on $\mathcal{R}_1$ is the sum of all forces on all l sample points on $\mathcal{R}_1$. This is a force that moves $\mathcal{R}_1$ away from $\mathcal{R}_2$, enforcing the surface repulsion constraint.

Goal Attraction: If the robot has not reached its goal configuration, this constraint generates a penalty force that attracts the robot towards its goal.

This constraint could be applied to one component of an articulated robot, such as the end effector, or to multiple components of the robot.

High Level Path Following Using GVDs: One well-known problem with using the penalty forces to achieve planning goals is that the robot can be caught in a local minima and fail to reach the goal. To address this issue we integrate global geometric analysis, generated by a high-level task planner [20], into our planning framework. There are many well known techniques for obtaining an estimated path for a robot based on the static obstacles in the scene, such as a medial axis based planner [10,12], a Probabilistic Roadmap Planner [14], a binary space partitioning of the workspace [21], or simply by taking input from a user [7]. Any of these can be integrated into our planning framework.

In our current implementation, we use a high level path planner based on the GVD [10]. As described in Sec. 3.4 this planner can be utilized throughout the planning simulation to provide updated estimated paths for the robot. Although this is not a complete solution, it does ensure that the robot is not easily trapped in a local minimum of the constraint force fields. The object's orientation along the estimated path is left arbitrary, so that it can be determined by other soft constraints.

5 Scene Representation

In this section we will briefly outline how to render the given geometry and other sensory input into hard and soft constraints in the constraint-based planning framework. Assume that the geometry representing the robots and obstacles is given, as well as prescribed motion, simulated or scripted, for the obstacles over time. Our system then defines constraints that will restrict the motion of the robots to meet the design specifications, and also guide the robots to complete the planning tasks.

5.1 Hard Constraints and Articulated Robot Joints

We will now illustrate how the connectivity and angle limit constraints of an articulated object can be represented with a combination of point constraints as described in Sec. 4.2. We rely on the ability to use constraints defined between points of a rigid structure of point distance constraints, that represents the coordinate frame of each object, to enforce relationships between the objects themselves, as described in Sec. 4.2.

Ball Joint Example: Suppose that we have two rigid objects, $\mathcal{R}_1$ and $\mathcal{R}_2$, and wish to constrain them to only two relative rotational degrees of freedom, i.e. a ball joint, about some world point o between them. We would then define $p_1 = o$ in the coordinate frame of $\mathcal{R}_1$, and $p_2 = o$ in the coordinate frame of $\mathcal{R}_2$. Then we would link p_1 and p_2 to the rigid structures of $\mathcal{R}_1$ and $\mathcal{R}_2$ respectively, using three linearly independent distance constraints each. These constraints ensure that p_1 is rigidly attached to the structure representing $\mathcal{R}_1$ and p_2 is rigidly attached to the structure representing $\mathcal{R}_2$. We then define a distance constraint, with constraint distance of 0, between p_1 and p_2, so that their world positions are constrained to coincide. This ensures that when all constraints are satisfied the two objects, $\mathcal{R}_1$ and $\mathcal{R}_2$, are rigidly attached to a common rotation point at o, whose position is now expressed in the coordinate frames of the two objects, as shown in Fig. 2.

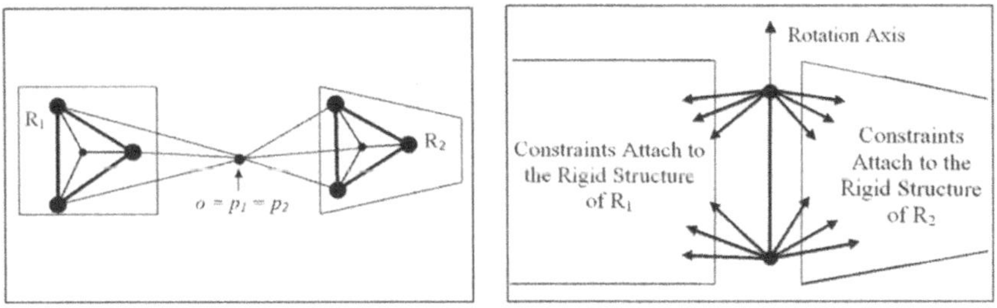

Fig. 2. LEFT: A ball joint, built from distance constraints linking the rigid structures of the two objects. RIGHT: A revolute joint, built from distance constraints linking the rigid structures of the two objects.

Revolute Joint Example: We can extend our formulation of a ball joint to obtain a revolute joint between two rigid objects from the specification of the joint location, axis of rotation, and angle limits. Given such a specification, it is possible to automatically generate the required point distance and planar angle constraints to limit the robot's degrees of freedom to only allow rotation about the joint axis, within the angle limits. To define a revolute joint between two rigid objects, $\mathcal{R}_1$ and $\mathcal{R}_2$, we use two ball joints. As long as the two ball joints have distinct centers of rotation that lie on the intended rotation axis, the constraints will limit the rigid objects to only one relative rotational degree of freedom about the axis. For additional stability we define a redundant distance constraint between the two rotation centers to maintain their separation along the rotation axis.

To limit the angle of the revolute joint, we use a point planar angle constraint, defined to limit the angle about the rotation axis between one point attached to the rigid structure of R_1, and one point attached to the structure of $\mathcal{R}_2$.

There has been previous work, in the field of robot control, on specifying joints using simple constraints [24]. The advantage of our approach is that, using the rigid structure formulation, we reduce the problem of solving constraints between rigid objects to a much simpler problem of solving constraints between points. As we have shown in Sec. 4.1, this allows for a fast and stable constraint solving method to be used.

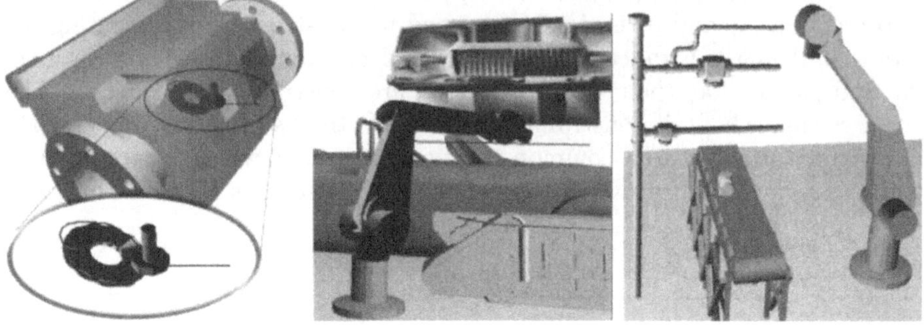

Fig. 3. Left to right: (a) Maintainability Study (Scene 1); (b) Automated Car Painting (Scene 2); (c) Assembly Line Planning (Scene 3).

5.2 Soft Constraints

While the specifications for all hard constraints are provided as part of the specific planning scenario, the soft constraints are defined by our framework to achieve the planning task. In our method, we automatically define soft constraints to both propel the robots toward their goals, and ensure that they do not collide with obstacles. It should be clear from our discussion of soft constraints in Sec. 4.3 how these constraints can be applied to achieve the desired behavior. To move the robots in the scene towards their planning goals we use one of two soft constraints: goal attraction and path following. To accomplish obstacles avoidance, we use surface repulsion constraints defined between the robots and all of the objects in the scene.

6 Implementation and Performance

6.1 Implementation

Our system was implemented in an object-oriented framework using C++. We use the Proximity Query Package [11,18] for collision detection, to enforce non-penetration constraints, and HAVOC3D [12] to generate distance fields for surface repulsion constraints.

6.2 System Demonstration

We have tested our motion planning system on following applications:

Scene 1: Maintainability Study

In assembly maintainability studies, motion planning is used to find whether it is possible to remove a particular part from an assembly, and if so, to find one possible removal path [7]. In our example, shown in Fig. 3(a), a bolt and a washer must avoid each other in the confines of tight compartment inside a pump assembly. The goal, to remove the bolt from the assembly, requires both objects to maneuver around each other without colliding.

Scene 2: Automated Car Painting

In this example seen in Fig. 3(b), an articulated robot arm, with 6 degrees of freedom, is used to trace a path along the body of a car for painting. The robot is composed of rigid components that are held together by constraints. For all of the components of the robot, the planner must compute paths that satisfy the joint constraints, do not collide with the obstacles or the car, and lead the end effector along the prescribed path.

Scene 3: Assembly Line Planning

In this example, shown in Fig. 3(c), the robot arm from scene 2 must access a part moving past it on a conveyer belt. The factory floor contains a piping structure that is moving over the conveyer belt in the opposite direction to the part's movement. The moving obstruction causes the robot to reactively modify its path to avoid collision.

The timings for these scenarios are presented in Table 1. The timings were taken on a PC with a 933MHz Pentium III processor, 256MB RAM and an nVidia GeForce3 graphics card. The motion sequences captured in MPEG are available at: *http://gamma.cs.unc.edu/cplan.*

Table 1. Benchmark timings in seconds on three example scenes. *Poly:* The number of polygons in each scene. *Cons:* The total number of active constraints in each scene. *Per Step:* The average time for the planner to compute one time step of the simulation. *Total:* The total time taken to complete and execute the planning task.

Scene	Poly	Cons	**Per Step**	**Total**
(1) Maintainability	20470	4	0.093 sec	67 sec
(2) Auto Painting	25738	43	0.038 sec	18 sec
(3) Assembly Line	16962	43	0.0085 sec	16 sec

6.3 Discussion

The planning tasks in the example scenes execute, on average, between 10 and 120 time steps per second. The primary bottleneck in our current implementation is the distance field computation used to determine estimated paths and the penalty forces for the surface repulsion constraints. We use a one-level bounding box culling to limit the application of this computation to areas near potential surface collisions. We also use simplified geometry for computing the distance field wherever appropriate to speed up the proximity queries. This approach works well, unless the scene, as in the case of Scene 1, has highly non-convex complex geometry that is poorly approximated by the bounding boxes. In such cases, hierarchical bounding box culling could be used to further limit the application of the distance field computation to increase runtime performance. We are currently working on this optimization, as well as accelerating the 3D distance field computation. With the new optimized implementation, we expect at least an order of magnitude performance improvement.

The constraint solver we have developed for the current system uses an iterative relaxation method that is specialized to provide interactive performance when planning the motion of rigid and articulated robots in dynamic scenes. It works well for overconstrained and consistent systems, such as those produced by our method of modeling robot joints using constraints, and in our planning framework where the dynamic simulation typically advances in small time steps allowing it to take advantage of temporal coherence to achieve performance and stability. However, it is possible for our framework to incorporate other efficient constraint solvers based on the non-trivial extension of [3,4,17], as we extend this work to plan the motion of flexible bodies and to also include different types of geometric and dynamics constraints in our system.

7 Conclusion and Future Work

We reformulate the motion planning problem into a physical simulation where constraints on the robot's motion and the GVD of the workspace guide the robot from its starting to its goal configurations. We have demonstrated its effectiveness and real-time performance on several challenging planning scenarios. The flexibility of our framework offers the possibility of natural extension, including: (a) additional constraints between robots, including constraints for maintaining line of sight contact or other complex geometric relationships, as well as dynamic and non-holonomic constraints; (b) extension to flexible geometry that deforms over time due to contact; and (c) incorporation of direction human interaction and control.

Acknowledgments: This research was supported in part by by ARO DAAG55-98-1-0322, NSF DMI-9900157, NSF IIS-9821067, NSF ACR-00118743, ONR N00014-01-1-0067 and Intel.

References

1. N. Amato, O. Bayazit, L. Dale, C. Jones, and D. Vallejo. Obprm: An obstacle-based prm for 3d workspaces. *Proceedings of WAFR98*, pages 197–204, 1998.
2. O. B. Bayazit, J.M. Lien, and N. M. Amato. Probabilistic roadmap motion planning for deformable objects. *Proceedings of International Conference on Robotics and Automation*, pages 2126–2133, 2002.
3. W. Bouma, X. Chen, I. Fudos, C. Hoffmann, and P. Vermeer. *An Electronic Primer on Geometric Constraint Solving.* http://www.cs.purdue.edu/homes/cmh/electrobook/intro.html, 1990.
4. B. Bruderlin and D. Roller (eds). *Geometric Constraint Solving and Applications.* Spring Verlag, 1998.
5. J. F. Canny and B. Donald. Simplified voronoi diagrams. *Discrete and Computational Geometry*, 3:219–236, 1988.

6. J.F. Canny. *The Complexity of Robot Motion Planning.* ACM Doctoral Dissertation Award. MIT Press, 1988.
7. H. Chang and T. Li. Assembly maintainability study with motion planning. In *Proceedings of International Conference on Robotics and Automation*, 1995.
8. H. Choset and J. Burdick. Sensor based planning, part ii: Incremental construction of the generalized voronoi graph. *IEEE Conference on Robotics and Automation*, 1995.
9. H. Choset and J. Burdick. Sensor based planning: The hierarchical generalized voronoi graph. *Workshop on Algorithmic Foundations of Robotics*, 1996.
10. M. Foskey, M. Garber, M. Lin, and D. Manocha. A voronoi-based hybrid planner. *Proc. of IEEE/RSJ Int. Conf. on Intelligent Robots and Systems*, 2001.
11. S. Gottschalk, M. Lin, and D. Manocha. OBB-Tree: A hierarchical structure for rapid interference detection. In *Proc. of ACM Siggraph'96*, pages 171–180, 1996.
12. K. Hoff, T. Culver, J. Keyser, M. Lin, and D. Manocha. Fast computation of generalized voronoi diagrams using graphics hardware. *Proceedings of ACM SIGGRAPH 1999*, pages 277–286, 1999.
13. K. Hoff, T. Culver, J. Keyser, M. Lin, and D. Manocha. Interactive motion planning using hardware accelerated computation of generalized voronoi diagrams. *IEEE Conference on Robotics and Automation*, pages pp. 2931–2937, 2000.
14. L. Kavraki and J. C. Latombe. Randomized preprocessing of configuration space for fast path planning. *IEEE Conference on Robotics and Automation*, pages 2138–2145, 1994.
15. L. Kavraki, P. Svestka, J. C. Latombe, and M. Overmars. Probabilistic roadmaps for path planning in high-dimensional configuration spaces. *IEEE Trans. Robot. Automat.*, pages 12(4):566–580, 1996.
16. O. Khatib. Real-time obstable avoidance for manipulators and mobile robots. *IJRR*, 5(1):90–98, 1986.
17. G. Kramer. *Solving Geometric Constraint Systems: A case study in kinematics.* MIT Press, 1992.
18. E. Larsen, S. Gottschalk, M. Lin, and D. Manocha. Distance queries with rectangular swept sphere volumes. *Proc. of IEEE Int. Conference on Robotics and Automation*, 2000.
19. J.C. Latombe. *Robot Motion Planning.* Kluwer Academic Publishers, 1991.
20. T. Lozano-Perez and R. Wilson. Assembly sequencing for arbitrary motions. *Proc. IEEE International Conference on Robotics and Automation*, 1993.
21. K. Ahrentsen N. Jacobsen, R. Larsen, and L. Overgaard. Automatic robotwelding in complex shipstructures. *J. Applied Artificial Intelligence*, 1997.
22. C. Ó'Dúnlaing, Micha Sharir, and C. K. Yap. Retraction: A new approach to motion-planning. In *Proc. 15th Annu. ACM Sympos. Theory Comput.*, pages 207–220, 1983.
23. J. M. Ortega and W. C. Rheinboldt. *Iterative Solution of Nonlinear Equations in Several Variables.* Academic Press, 1970.
24. L Overgaard, H. Petersen, and J. Perram. A general algorithm for dynamic control of multilink robots. *Int. J. Robotics Research*, 14(3), 1995.

25. C. Pisula, K. Hoff, M. Lin, and D. Manocha. Randomized path planning for a rigid body based on hardware accelerated voronoi sampling. In *Proc. of 4th International Workshop on Algorithmic Foundations of Robotics*, 2000.
26. Steven A. Wilmarth, Nancy M. Amato, and Peter F. Stiller. Maprm: A probabilistic roadmap planner with sampling on the medial axis of the free space. *IEEE Conference on Robotics and Automation*, 1999.
27. A. Witkin and D. Baraff. *Physically Based Modeling: Principles and Practice*. ACM Press, 1997. Course Notes of ACM SIGGRAPH.

Trajectory Planning for Kinematically Controllable Underactuated Mechanical Systems

Prasun Choudhury and Kevin M. Lynch

Northwestern University, Evanston, IL 60208, USA

Abstract. We develop trajectory planners for a class of second-order underactuated mechanical systems called kinematically controllable systems. For kinematically controllable systems, the problem of planning fast collision-free trajectories can be decoupled into the computationally simpler problems of path planning for a kinematic system followed by time-optimal time scaling. This paper describes efficient path planners using randomized algorithms and dynamic programming to solve the path planning problem for the kinematic system. The resulting kinematic paths are time scaled to produce fast trajectories.

1 Introduction

Underactuated mechanical systems are a subset of mechanical systems that have fewer inputs than degrees of freedom. Example systems include underwater vehicles, spacecraft, three joint robot arms with passive joints, etc. In this paper we focus on the trajectory design problem for underactuated mechanical systems.

The problem of trajectory planning for underactuated systems can be approached in different ways. The trajectory planning problem can be cast as a nonlinear optimization problem, where the control history parameters are coefficients of certain basis functions (e.g., splines, polynomials, truncated Fourier series) and the state trajectory is obtained by integrating the dynamic equations of motion. Such an approach cannot be applied blindly, however; the quality of the solution depends intimately on an initial control guess and the local optima in the design space. The number of local optima, and the size of their basins of attraction, are generally extremely difficult to characterize in advance. Further, constraint and objective function evaluation may involve expensive simulations, resulting in slow convergence times. Gradient-based nonlinear optimization methods are inherently local, most appropriate when an approximate solution is available. On the other hand, dynamic programming methods may provide globally optimal solutions, but they are computationally expensive and not likely amenable to real-time implementations. Heuristic randomized searches [13,9] give up on optimality or completeness in an effort to reduce average computation time.

The key to computationally efficient planning of fast trajectories is to make use of the *structure* of the equations of motion of the underactuated

J.-D. Boissonnat et al. (Eds.): Algorithmic Foundations of Robotics V, STAR 7, pp. 559–575, 2004.

system. As one example, *differentially flat* (or *dynamic feedback linearizable*) systems are a well known class of systems which simplify this problem [7,8,16]. For such systems, it is possible to identify a set of *flat outputs* y equal in dimension to the number of controls such that the state of the system, and the controls, can be expressed as functions of y and its time derivatives. State-to-state trajectory planning for the system reduces to fitting a curve in the output space satisfying initial and terminal constraints on y and its derivatives. The state trajectory and control history are derived directly from this curve.

There are a few limitations to the differential flatness approach to motion planning, however. First, there is no algorithmic procedure to obtain the flat outputs of a system, nor are there general criteria to establish whether flat outputs exist. Second, differential flatness is a non-generic property. In other words, generic perturbations to a flat model of a system render the system "non-flat." Third, incorporating obstacle or control constraints greatly complicates the trajectory generation problem [7].

Recently, Bullo and Lynch have discovered a new class of mechanical systems called *kinematically controllable* systems [5]. Like differentially flat systems, the structure of these systems simplifies the trajectory planning problem. In this case, we can use a *kinematic reduction* of the mechanical system to plan feasible trajectories (accounting for underactuation constraints) directly in the system's n-dimensional configuration space (C-space), rather than in the $2n$-dimensional state space. This is significant for computationally efficient trajectory planning, as complete and resolution-complete algorithms for collision-free motion planning are known to have running times exponential in the dimension of the search space [6,19,20]. In some cases, path planning for kinematically controllable systems can be reduced to closed-form inverse kinematics.

The kinematic reduction allows kinematic constraints (such as obstacles or joint limits) to be dealt with efficiently, perhaps in real-time, as the search for a feasible motion occurs on the n-dimensional C-space instead of the $2n$-dimensional state space. Because a kinematic reduction encodes "preferred" motions for the system, a path found by the motion planner can be turned into the time-optimal trajectory along the path using a time-scaling algorithm incorporating control constraints [3,18]. Thus the basic approach deals with obstacle and control constraints efficiently and provides sub-time-optimal solutions.

In a previous paper [15], a planner has been described for a kinematically controllable three-joint robot arm with the third joint passive. The planner uses a grid-based *dynamic programming* approach, which is difficult (in both running time and storage requirements) to extend beyond a few degrees-of-freedom. In this paper we tailor *Rapidly Exploring Random Trees* (RRT) [13] to kinematically controllable systems and perform a number of experimental studies showing the computational improvements that can be achieved over "generic" RRT methods which do not take advantage of the special structure. Numerical results are presented using the enhanced RRT and modified dynamic programming algorithms and the resulting trajectories have been compared to trajectories found using the original RRT planner.

1.1 Brief Overview of Decoupled Trajectory Planning

An underactuated second order system can be represented as

$$M(q)\ddot{q} + C(q,\dot{q})\dot{q} + G(q) = T(q)\begin{pmatrix} u \\ 0 \end{pmatrix}, \tag{1}$$

where $q \in \Re^n$ is the configuration, $u \in \Re^m$ is the control, $M(q)$ is an $n \times n$ symmetric positive definite inertia matrix, $C(q,\dot{q})\dot{q} \in \Re^n$ is a vector of centrifugal and coriolis forces, and $G(q) \in \Re^n$ is the vector of gravitational forces. $T(q)$ is an $n \times n$ nonsingular matrix whose first m columns describe the action of the actuators. For a manipulator, where the actuators are collocated with the joints, $T(q)$ is the identity matrix. Premultiplying both sides of the equation by $T^{-1}(q)$, we get

$$M'(q)\ddot{q} + C'(q,\dot{q})\dot{q} + G'(q) = \begin{pmatrix} u \\ 0 \end{pmatrix}, \tag{2}$$

where $M' = T^{-1}M, C' = T^{-1}C, G' = T^{-1}G$.

We assume $m < n$ — the system has fewer actuators than degrees-of-freedom. This means that the last $n - m$ rows in equation (2) represent state-dependent constraints on the feasible accelerations of the system. We could write these last $n - m$ rows in the form

$$M_0'(q)\ddot{q} + C_0'(q,\dot{q})\dot{q} + G_0'(q) = 0. \tag{3}$$

These constraints are what make trajectory planning for generic underactuated systems difficult. In general, trajectory planning for an underactuated system must occur in the $2n$-dimensional state space, since the motion constraints described in (3) can only be expressed as constraints on tangent vectors on this space. In contrast, a common way to do sub-optimal trajectory planning for a fully actuated mechanical system is to *decouple* the problem by first planning a path on the configuration space, handling any kinematic constraints such as obstacles and joint limits, and then time scaling the path (subject to actuator limits) to get a trajectory.

Recently Bullo and Lynch [5] have discovered that this decoupled procedure can often be applied to underactuated systems, despite the fact that the constraints described in (3) are second-order. Consider a path on the configuration space $q(s) : \Re \rightarrow \Re^n$ parameterized by $s \in [0,1]$ and a time scaling $s(t)$ which assigns a point on the path for each $t \in [0,T]$. $s(t)$ is twice-differentiable and $\dot{s}(t) > 0$ for all $t \in (0,T)$. Then the trajectory of the system can be written $q(s(t))$. Plugging $q = q(s), \dot{q} = \frac{dq}{ds}\dot{s}, \ddot{q} = \frac{d^2q}{ds^2}\dot{s}^2 + \frac{dq}{ds}\ddot{s}$ into (3), we can write the constraints in the form

$$a(s)\ddot{s} + b(s)\dot{s}^2 + c(s) = 0, \tag{4}$$

where $a(s), b(s), c(s) \in \Re^{n-m}$ are vectors describing the inertial, centrifugal and Coriolis, and gravity terms, respectively, of the system in the constrained directions when restricted to the path. We call the path $q(s)$ a *kinematic motion* if the constraints described in (4) are satisfied for any time scaling, i.e., arbitrary $\dot{s}, \ddot{s}$. In other words, such a path can be followed at any speed without violating the constraints (3).

A velocity vector field $V(q)$ is a *decoupling vector field* if all paths $q(s)$ satisfying

$$\frac{dq(s)}{ds} = V(q(s))$$

are kinematic motions. A *kinematic reduction* of the original dynamic system (1) is any set of decoupling vector fields. Any controlled trajectory of the kinematic reduction can be followed by a controlled trajectory of the original dynamic system. The original dynamic system (1) is *locally kinematically controllable* if there exist p decoupling vector fields such that the kinematic system

$$\dot{q} = \sum_{i=1}^{p} V_i(q)w_i \tag{5}$$

$$(w_1, \ldots, w_p) \in \{(\pm 1, 0, \ldots, 0), (0, \pm 1, 0, \ldots, 0), \ldots, (0, \ldots, 0, \pm 1)\}$$

is locally controllable [5]. Further, these decoupling vector fields can be followed at high speeds, allowing fast motions. The only constraint is that the system must come to rest when switching between the decoupling vector fields, since instantaneous changes in velocity are not possible with finite forces.

For locally kinematically controllable systems, there exists a feasible path between any two zero velocity states in the same open connected component of free C-space, and one can apply any collision-free path planner for the driftless kinematic system (5); see for example [2,12,13]. Because each segment of the resulting path follows one of the decoupling vector fields, the speed along the segment is limited only by actuator limits, not by the underactuation constraints. Switches between decoupling vector fields must occur at zero velocity, so it is appropriate for the path planner to minimize the number of switches to obtain fast trajectories.

1.2 Description of the Robotic Systems

In this paper, we have tested our motion planning algorithms for two kinematically controllable systems:

- A planar rigid body with a force and torque (Figure 1a)
- A 3D rigid body with three thrusters (Figure 1b)

Planar Body with a Force and Torque: Consider a planar body endowed with a control torque and a body-fixed control force applied through the center of mass. The configuration is $(x, y, \theta) \in \Re^2 \times S^1$, and the kinetic energy is $\frac{1}{2}m(\dot{x}^2 + \dot{y}^2) + \frac{1}{2}I\dot{\theta}^2$. The equations of motion for the planar rigid body are given by

$$m\ddot{x} = u_1 \cos\theta, \quad m\ddot{y} = u_1 \sin\theta, \quad I\ddot{\theta} = u_2.$$

Here u_1 is a body-fixed force through the center of mass, u_2 is the torque acting about the center of mass of the rigid body (Figure 1a), m is the mass and I is the moment of inertia of the rigid body. The two decoupling vector fields associated with the 2D rigid body are (a) translational motion along the direction of applied force u_1 (V_1), and (b) rotational motion about the center of mass of the rigid body (V_2). Since the robot has three degrees of freedom, at most three distinct motions (two switches) are required to reach any goal configuration [5] in an obstacle-free environment. For our simulations, we have used different shapes for the planar body robot (see Figure 3).

(a) 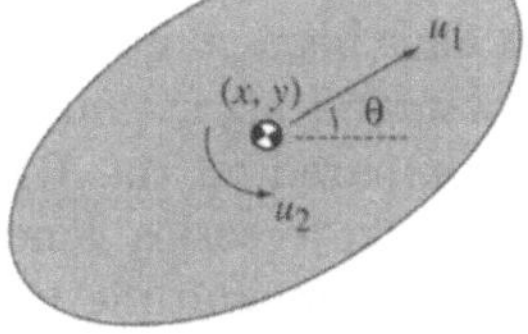

(b) 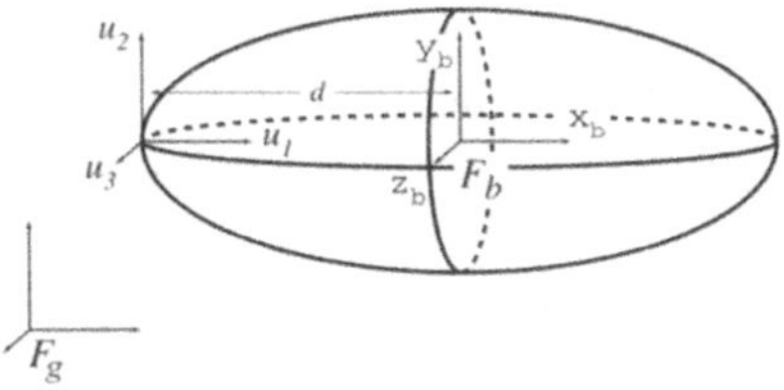

Fig. 1. a) A Planar 2D Rigid Body b) A 3D Spacecraft with Three Thrusters

3D Rigid Body with Three Thrusters: Here a 3D rigid body similar to a spacecraft has been considered with 3 axis-aligned thrusters. The thrusters are located away from the origin at a point along the x_b axis of the 3D rigid body. A frame F_b is attached to the center of mass of the body and aligned with the principal axes of inertia (see Figure 1b). $\omega \in \Re^3$ and $v \in \Re^3$ are the angular and linear velocity of the spacecraft expressed in inertial frame F_g. The equations of motion for the spacecraft are given by

$$m\dot{v} = F, \quad I\dot{\omega} + \omega \times I\omega = N,$$

where m is the mass of the spacecraft and I is the time-varying inertia matrix expressed in a frame aligned with F_g at the center of mass of the spacecraft. The three external control forces u_1, u_2, u_3 act along lines parallel to the x_b, y_b, z_b axes, respectively, at a point $(-d, 0, 0)$, $d > 0$ in F_b. F and N are the input forces and torques due to u_1, u_2 and u_3 expressed in the inertial frame F_g. The decoupling vector fields for the 3D spacecraft are (a) translation along the x_b axis (V_1), (b) rotation about the center of percussion in the x_b-y_b plane (V_2) and (c) rotation about the center of percussion in the x_b-z_b plane (V_3) (see [5] for details). The 3D spacecraft, requires at most six motion primitives (five switches) to reach any goal configuration for obstacle-free environments [5]. For our numerical simulations, we have used different shapes for the spacecraft (see Figure 4).

1.3 Problem Statement, Contribution, and Organization

The problem can be stated as follows: Given a kinematically controllable underactuated system and a geometric description of the environment, efficiently find a fast collision-free trajectory for the robot.

In this paper, a framework is developed for efficient trajectory planning for kinematically controllable systems. Trajectory planning is reduced to path planning for driftless kinematic systems and time-optimal scaling of the resultant paths to obtain the final trajectory. We use enhanced versions of RRT and dynamic programming based planners for path planning for driftless kinematic systems. The final trajectories have been generated from the above paths in C-space using the optimal time-scaling algorithm described in [3,18]. To judge the efficiency of our path planners, the algorithms have been rigorously tested on various types of underactuated mechanical systems and execution-time and run-time performances have been compared with the standard RRT planner. The new trajectory planning algorithm is computationally efficient (20 to 30 times speedup) compared to the RRT based kinodynamic planner. The execution time of the trajectories generated by the algorithm described in this paper is an order of magnitude faster than the trajectories generated by RRT based kinodynamic planner.

Section 2 describes the RRT and dynamic programming based planning algorithms used in the initial phase of path planning for driftless systems. Section 3 briefly covers the time scaling algorithm for various robotic systems. Experimental results related to the above algorithms as applied to different underactuated systems are described in Section 4. Section 5 compares the execution time of the final robot trajectories for RRT-based kinodynamic planners and the trajectory planning algorithm described in the current paper. Finally a summary of the results of the paper and potential areas of future research are discussed in Section 6.

2 Path Planning Algorithms

As mentioned in Section 1.1, motion planning for second-order dynamic systems can be reduced to planning for a kinematic system with the help of decoupling vector fields. This essentially reduces the dimension of the search space on which planning is performed by a factor of two. In general, the dimension of the search space for kinodynamic planning is $2n$ and the dimension of the search for kinematic planners is n. Now we will describe in detail the RRT and dynamic programming based motion planning algorithms which we have used for this purpose.

2.1 Rapidly Exploring Random Trees

The RRT algorithm (pseudo-code shown in Table 1 and explained in Figure 2a) has been presented in detail in [13]. Here M is a $2n$ dimensional manifold representing state space (M is of dimension n for C-space). The initial state of the robot, $\mathbf{x}_{init}$, is chosen as the root of the tree T. For each iteration, a random state $\mathbf{x}_{rand} \in M$ is sampled. The NEAREST-NEIGHBOR function selects the node $\mathbf{x}_{near}$ in the tree T that is closest to $\mathbf{x}_{rand}$. All the possible inputs U are applied to the selected node $\mathbf{x}_{near}$. If a newly generated state $\mathbf{x}_{new}$ is collision-free and the distance $\rho(\mathbf{x}_{new}, \mathbf{x}_{rand})$ is the least for all of the collision free states that have been created by applying every input in U to $\mathbf{x}_{near}$, $\mathbf{x}_{new}$ will be added to the tree T. The CONTROL function returns the input $\mathbf{u}_{best}$ corresponding to the newly added node $\mathbf{x}_{new}$. The algorithm terminates and returns a successful trajectory when a newly generated collision-free node is within an ε neighborhood of the goal state $\mathbf{x}_{goal}$, i.e., $\rho(\mathbf{x}_{new}, \mathbf{x}_{goal}) \leq \varepsilon$. The metric function ρ plays a critical role on the performance of RRT planner. The metric used in the RRT algorithm is a weighted Euclidean metric based on linear and angular configuration (angle of rotation for a planar rigid body and quaternion for a 3D rigid body) of the rigid body.

The algorithm can be extended to two trees (RRT-Dual) instead of one as described above. In that case the two trees will be rooted at the initial state $\mathbf{x}_{init}$ and final state $\mathbf{x}_{goal}$. Both the trees will grow in the above fashion and the algorithm will terminate and return a successful trajectory when the distance between nearest nodes in the two trees is below a threshold ε.

Table 1. Pseudo-Code for the RRT Planner

BUILD-RRT($\mathbf{x}_{init}$)
  T.init($\mathbf{x}_{init}$);
  for k $\leftarrow$ 1 **to** K **do**
    $\mathbf{x}_{rand} \leftarrow$ RANDOM-STATE();
    $\mathbf{x}_{near} \leftarrow$ NEAREST-NEIGHBOR($\mathbf{x}_{rand}, T$);
    $\mathbf{u}_{best}$, $\mathbf{x}_{new}$, success $\leftarrow$ CONTROL($\mathbf{x}_{near}$, $\mathbf{x}_{rand}$, T);
    if success
      Add vertex $\mathbf{x}_{new}$ to tree T;
      Add edge ($\mathbf{x}_{near}$,$\mathbf{x}_{new}$, $\mathbf{u}_{best}$) to tree T
      if ($\rho(\mathbf{x}_{new}$,$\mathbf{x}_{goal}) \leq \varepsilon$
        Return the path from start to goal

In Table 1, CONTROL is the function in the motion planner which selects the best possible input $\mathbf{u}_{best}$ among the set of all possible inputs U. In our revised algorithm (RRT-DualKC), we change the heuristic for selecting the best control input. Typically in the RRT algorithm, control inputs are integrated for a fixed time-step Δt throughout entire planning phase. We modify the control criteria so that one can use adaptive time steps during the planning phase. For our planner, we terminate the integration along a flow field if (a) we hit an obstacle or (b) the flow field returns to a neighborhood of any of its previously reached states or (c) the distance from the new state to the goal (for RRT with a single tree) or the distance from the new state to the nearest node in the other tree (for RRT-Dual) increases after an integration step. The primary motivation for using the above heuristics is to reduce the number of switches, i.e., to reduce the number of changes required between different flow fields in the final trajectory. The computation time is expected to be lower since the dimension of the search space is twice compared to that of kinematic planners.

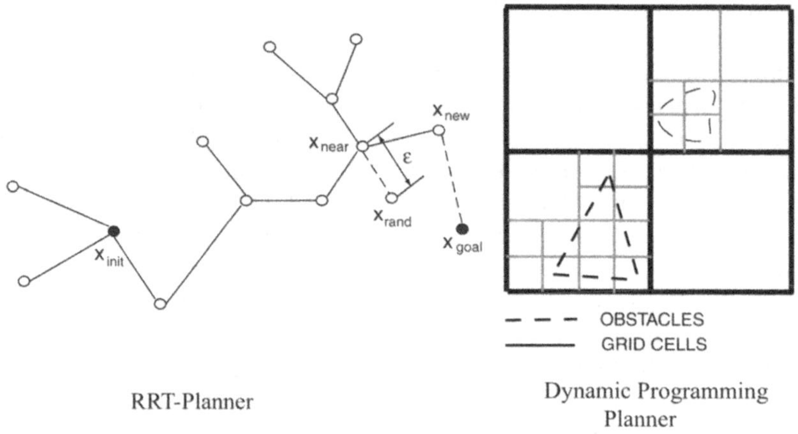

Fig. 2. Rapidly Exploring Random Tree

2.2 Dynamic Programming Based Planning

In this section, we describe in detail a variation of the dynamic programming based motion planner (Multiresolution-DP) which was originally developed by Barraquand and Latombe [2]. The planner is very similar in nature to the A^* search planner using a variation of Dijkstra's algorithm [14]. The overall idea is to discretize the C-space into a grid structure and represent the C-space as a directed graph. The edges between two nodes of the directed graph will represent the cost incurred to reach one state from another. The optimal path in the graph from the start to the goal state will correspond to the least cost path. Dynamic programming based planners return optimal paths based on the cost function for a particular resolution of the grid.

For our planner, a tree T is used to keep track of all the visited grid cells along with the number of switches it took to reach the grid cell from the initial state (here the COST function is the number of switches between decoupling vector fields). A priority queue Q contains all the inserted grid cells based on the cost metric. In our planning algorithm, we use a multi-resolution strategy where a grid cell can be subdivided into finer cells for better accuracy and collision checking. Similar multi-resolution approaches have also been used by Bohlin [4] for lazy PRM's (Probabilistic Roadmaps) and by Reif and Wang [17] for kinodynamic planning. In our algorithm, each grid cell has been represented by CELL data structure which consists of i,j,..... indices (each index corresponds to a dimension of the C-space), pointers to cells of finer resolution (if any), the COST to reach the grid cell from the initial state, the distance of the current grid cell to the goal state, a boolean flag that indicates whether the current grid cell has been visited and a boolean flag that indicates whether the grid cell is in collision or not. The pseudo-code of the dynamic programming based planner is described briefly in Table 2.

Table 2. Pseudo-Code for the Multiresolution Dynamic Programming Based Path Planner

```
DYNAMIC PROGRAMMING PLANNER (x_init)
T ← NULL, Q ← NULL;
GRID-CELLS ← DISCRETIZE-C-Space
Mark each CELL as COLLISION or FREE;
Mark each CELL as UNVISITED;
T.init(x_init); Q.init(x_init);
while numNodes ≤ MAXNODES and Q ≠ NULL, do
    x ← Q.top
    for all inputs
        x_new, ← CONTROL(x);
         if x_new ≠ COLLISION
            x_new ← VISITED and Update COST;
            Add x_new to tree T
            Add x_new to priority queue Q
            if (ρ(x_new,GOAL) ≤ ε
                Return the path from start to goal
       if x_new = COLLISION
            SUBDIVIDE CELL;
            Update COLLISION and COST;
```

The C-space is initially discretized into a relatively coarse resolution grid (dark lines in Figure 2b). All the coarse resolution grid cells are marked UNVISITED and COLLISION or FREE (based on whether a gird cell is in collision with the obstacles). The initial state $\mathbf{x}_{init}$ is added to the tree T and the priority queue Q. While traversing the tree T, we apply all possible inputs to the current grid cell (this action is carried out by the function CONTROL in Table 2). We mark all the new collision-free grid cells that are reached as a result of applying different inputs. Heuristics described in Section 2.1 are used to perform integration with adaptive step-size during the planning phase. After each CONTROL step, we insert the new VISITED grid cells to the priority queue Q based on its COST. We continue the iteration until we reach a user-prescribed maximum number of nodes or we find a path to reach the goal. During each iteration, the least cost grid cell is popped out of the priority queue Q and the CONTROL function is applied to that cell. Grid cells in collision are SUBDIVIDED into finer cells (light lines in Figure 2b) for better accuracy and efficient collision detection. The planner terminates and returns a successful trajectory when a newly generated collision-free node $\mathbf{x}_{new}$ is within an ε neighborhood of $\mathbf{x}_{goal}$.

The storage complexity for the dynamic programming algorithm is exponential in the dimension of the search space and hence the algorithm might not be practical for high-dimensional C-space or fine resolution grids. As an example, even a relatively coarse resolution of 20 divisions along each dimension of 6D C-space results in 64 million cells.

3 Time Scaling

Geometric paths can be time scaled to yield a time-optimal trajectory along the path using algorithms described in [3,18]. For the two examples studied in this paper, the time-scaling problem is trivial. Each motion segment (between decoupling vector field switches) is time-scaled individually. For the 2D rigid body, the time-optimal scalings use bang-bang control profiles. Similarly, for the 3D rigid body, time-optimal translations use a bang-bang profile. For rotations, the trajectory is slightly more complicated. In general, the thruster providing non-zero torque follows a bang-off-bang profile. As the angular velocity increases from zero, the thrust through the center of mass (u_1) increases from zero to maintain the stationarity of the center of percussion. When this thruster is saturated, the torque thruster (u_2 or u_3) turns off. For short rotations, the thruster through the center of mass never saturates, and the torque thruster profile is simply bang-bang.

4 Experimental Results

The enhanced RRT and dynamic programming based planners have been tested on different kinematically controllable systems with environments of varying geometric complexity. We compared the running time and execution time of the planners with the original RRT planner. The motion planning algorithms have been developed on the framework provided within the *Motion*

Strategy Library (*MSL*) [1]. All the algorithms have been implemented on a 400MHz Intel Pentium III PC with 256 MB of memory (we have also implemented the algorithms on a faster 2GHz Pentium IV PC with 512 MB of memory and the nature of the results are similar).

All the motion planning algorithms which have been used for our numerical experiments are enumerated below for clarity.

- RRT-Dual: The original RRT algorithm with two trees rooted at the initial and final states [13]. It performs kinodynamic planning directly on the state space.
- RRT-DualKin: The original RRT algorithm with two trees rooted at the initial and final configurations, used for kinematic planning on C-space.
- RRT-DualKC: Enhanced RRT-DualKin planner (used for kinematic planning) which allows adaptive step size integration (described in Section 2.1)
- Multiresolution-DP: Extension of the dynamic programming based planner [2] (used for kinematic planning) which uses cells of variable size (described in Section 2.2).

The results reported in Section 4.1 compare the efficiency of RRT-DualKC and Multiresolution-DP planner with the RRT-DualKin algorithm for first-order or kinematic systems. The output of these planning algorithms are kinematic paths in C-space. The results described in Section 5 correspond to the performance of the RRT-Dual kinodynamic planner and the trajectory planner consisting of planning in C-space using the RRT-DualKC planner and time scaling the kinematic paths.

4.1 Results of Numerical Simulations

Each planner was run 30 times for a specific robot in a particular environment. The computation times of different planners are described in Table 3. The reported times are average computation times.

The paths obtained using different types of planners for the 2D rigid body for environments with different levels of geometric complexity are shown in Figure 3. For the 2D rigid body in a maze like environment with complicated geometry (not shown in figure), the RRT-DualKin planner was unable to find feasible paths for most cases after generating 50,000 nodes. In general, the number of generated nodes (and hence the computation time) is less for the RRT-DualKC algorithm. On the other hand, the dynamic programming planner which is based on grid search technique is generally slower. The number of switches between the decoupling vector fields are lower for RRT-DualKC and Multiresolution-DP planners as compared to the original RRT-DualKin planner which does not try to minimize switches. As seen from Figure 3, the Multiresolution-DP planner tends to produce paths with the fewest switches between decoupling vector fields, especially for environments cluttered with obstacles. This is also an expected result as dynamic programming planners output optimal paths for the given grid resolution. From the

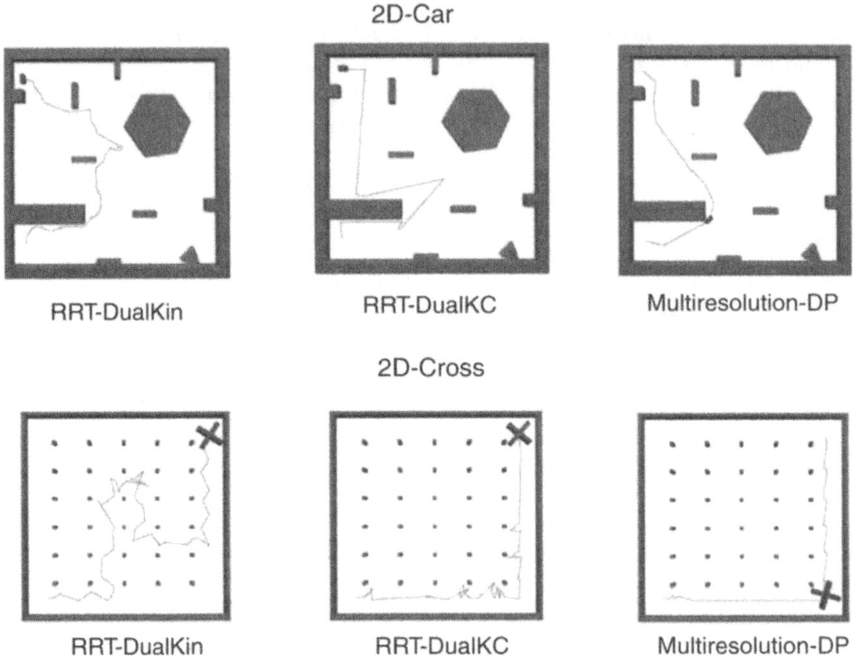

Fig. 3. Motion plans for 2D Rigid Body with various planning algorithms. The number of switches is least for Multiresolution-DP planner.

numerical experiments, we see that the planner took around eight switches for environments with fewer obstacles and the number of switches increased with the complexity of the geometric environment.

Figure 4 shows the output paths for the 3D rigid body in two different environments. The computation times of different planners are described in Table 3. It was not possible to use dynamic programming based planners for 3D spatial environments as the memory required to store the CELL data structure for 6D C-space exceeded the memory available in the computer. Similar to the case for planning for 2D rigid body, RRT-DualKC planner turned out to be more computationally efficient and generated paths with fewer switches between decoupling vector fields. Path shortening techniques based on random sampling algorithms [12] or nonlinear optimization methods [10] can be used as a post-processing step to further improve the quality of the planner.

Table 3. Running Times for Different Planning Algorithms for 2D Rigid Body and 3D Spacecraft

Robot and Environment Description	*Type of Planner RRT-DualKin*			*Type of Planner RRT-DualKC*			*Type of Planner Multiresolution DP*		
	No. of Nodes	*No. of Switches*	*Running Time*	*No. of Nodes*	*No. of Switches*	*Running Time*	*No. of Nodes*	*No. of Switches*	*Running Time*
2D Rigid Body with a force and a torque									
2D-Car	1062	77	4.81	152	8	1.96	4144	9	13.4
2D-Cross	10577	152	951.94	1611	31	26.99	12893	15	1041.52
2D-Maze	—	—	—	1863	22	75.02	4031	11	280.02
3D Spacecraft with 3 Thrusters									
3D-Torus	3027	21	16.21	368	9	2.41	N/A	N/A	N/A
3D-Cage	14450	119	492.56	5061	41	84.54	N/A	N/A	N/A
3D Hole	22315	127	1467.89	4168	47	62.82	N/A	N/A	N/A
3D Passage	17111	271	989.31	2766	91	31.83	N/A	N/A	N/A

5 Trajectory Planning and Comparison with Kinodynamic Planning

In this section, we evaluate the performance of the RRT-Dual kinodynamic planner and the trajectory planner comprised of planning in C-space using the RRT-DualKC planner and time scaling the paths to obtain the final trajectories of the robot. Table 4 compares the computational efficiency of the two planning algorithms. In addition, details are provided for the execution time of the robot for the two planning algorithms. While implementing the time-scaling algorithm for the kinematically controllable systems, certain thrust limits have been chosen for the systems. For the 2D rigid body the magnitude of the maximum force/torque that can be applied by the linear and rotational thruster has been restricted to 100N and 150N-m respectively. The mass and the inertia of the 2D rigid body are 1kg and 1.5kg-m^2. For the 3D spacecraft, the magnitude of maximum force for the linear thruster is 300N and the magnitude of maximum torque that can be applied by the rotational thrusters is 450N-m. The mass of the 3D spacecraft is 1kg, the moment of inertia of the spacecraft is 1.5kg-m^2, and the thrusters are located at a distance of 1.5m away from the origin along x_b axis. We have considered the maximum magnitude of the force/torque as the input for RRT-Dual kinodynamic planner (one actuation input at a time).

For each experiment, the average computation time over 30 trials is much higher for the RRT-Dual kinodynamic planner as compared to the decoupled trajectory planner. For environments with a high degree of geometric

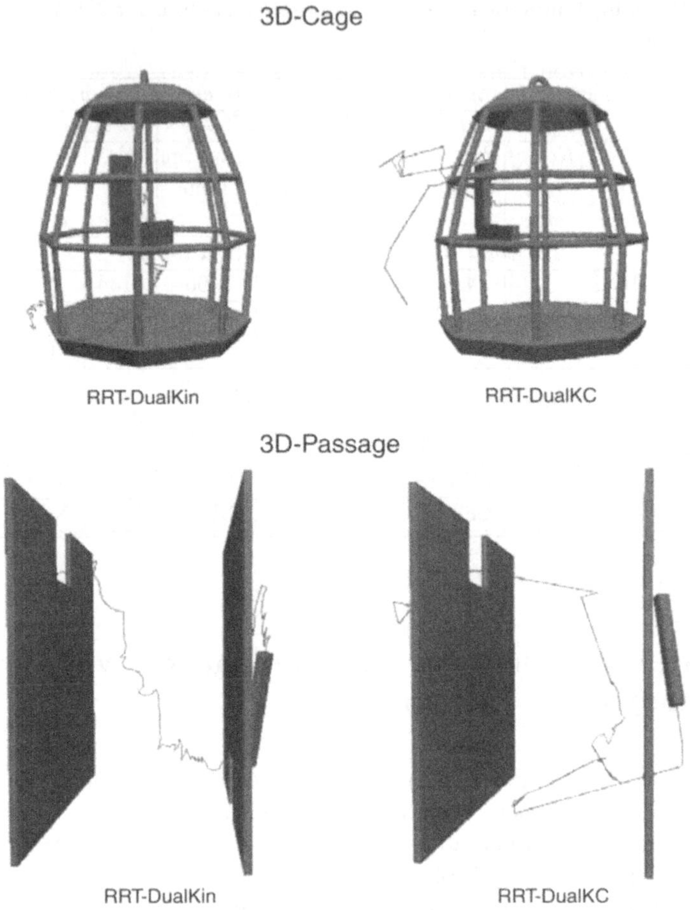

Fig. 4. Motion plans for 3D Spacecraft with three thrusters using various planning algorithms for two complex geometric environments. RRT-DualKC produces paths with lesser number of switches than RRT-DualKin planner.

complexity (maze like environments for 2D rigid body or environments with two narrow passages for 3D rigid body), the RRT-Dual kinodynamic planner was unable to find successful trajectories after generating 50,000 nodes. For the decoupled trajectory planner, computation cost is governed by the time required to plan kinematic paths in C-space (computation time required to time-scale the kinematic paths is negligible).

Execution times i.e., the time the robot will take to execute the trajectory, are much lower for the decoupled trajectory planner than for RRT-Dual kinodynamic planner. Overall, the execution time for the decoupled trajectory planner is roughly 10 times faster and requires much less computation time (20-30 times speedup in computational efficiency) than RRT-Dual kinodynamic planner.

The computation time reported in this paper for kinodynamic planning is substantially higher than the computation time reported in [13] and the algorithm in [13] seem to work for environments with narrow passages. This might be attributed to the fact that the kinodynamic planning algorithm in MSL is not the most optimized implementation of the RRT-Dual algorithm.

Table 4. Execution Time for RRT-based Kinodynamic Planner and Decoupled Trajectory Planner

Robot and Environment Description	*No. of Triangles*		*RRT-Dual Kinodynamic Planner [13]*		*Trajectory Planner: Time Scaling + RRTDualKC*		
	Robot	*Obstacles*	*Running Time(s)*	*Exec. Time(s)*	*Running Time(s)*	*No. of Switches*	*Exec. Time(s)*
Planar Body, 2D Grid	4	228	789.6	31.3	27.9	31	3.1
Planar Body, 2D Maze	2	296	N/A	N/A	77.01	22	2.2
3D Spacecraft, Cage	132	6192	2253.1	27.6	85.31	41	2.69
3D Spacecraft Narrow Slot	66	240	1123.3	38.4	63.11	47	2.91
3D Spacecraft, Narrow Passage	36	384	N/A	N/A	32.43	91	6.97

6 Conclusion

In this paper, we have presented an algorithm for fast trajectory planning for kinematically controllable systems. Trajectory planning has been decoupled into simpler problems of planning kinematic paths in C-space followed by time-optimal time scaling based on the actuator constraints. The trajectory planning algorithm has been compared with original RRT-based kinodynamic planner [13]. Generally, the enhanced RRT based planner turned out to be computationally efficient and generated much faster trajectories.

In future, we would like to study the performance of our trajectory planner for other kinematically controllable systems such as a 3-R robot manipulator with a passive joint [15,5]. In addition, we will explore motion planners based on exact inverse kinematics for point-to-point motions for kinematically controllable systems [5]. These motion planners will use probabilistic roadmaps [11] with a local planner implementing point-to-point motions from inverse kinematics. Along with that, we want to pursue our research in areas related to practical multi-resolution planners for motion planning of kinodynamic systems [17,19,20]. Furthermore, we will focus on smoothing the paths generated by the RRT planner [12,10] to alleviate jerky motions and reduce the time required by the robot to execute motion along collision-free trajectories.

Acknowledgment.
We thank Steve LaValle and Peng Cheng for many stimulating discussions on RRT and motion planning and for their help with the Motion Strategy Library. This work was funded in part by NSF grants IIS-9875469 and IIS-9811571.

References

1. http://msl.cs.uiuc.edu/msl/, "Motion Strategy Library," Dept. of Computer Science, Univ. of Illinois, Urbana-Champaign.
2. J. Barraquand and J. C-Latombe, "Nonholonomic multibody mobile robots: Controllability and motion planning in presence of obstacles," *Algorithmica*, vol. 10, no. 2-3-4, pp. 121-155, 1993.
3. J. E. Bobrow, S. Dubowsky and J. S. Gibson, "Time-optimal control of robotic manipulators along specified paths," *International Journal of Robotics Research*, vol. 4, no. 3, pp. 3-17, 1985.
4. R. Bohlin, "Path planning in practice; lazy evaluation on a multiresolution grid," in *Proc. IEEE Conference on Intelligent Robots and Systems*, October 2001.
5. F. Bullo and K. M. Lynch, "Kinematic controllability for decoupled trajectory planning in underactuated mechanical systems," *IEEE Transactions on Robotics and Automation*, vol. 17, no. 4, pp. 402-412, 2001.
6. B. Donald, P. Xavier, J. Canny and J. Reif, "Kinodynamic motion planning," *Journal of Association for Computing Machinery (ACM)*, vol. 40, no. 5, pp. 1048-1066, 1993.
7. N. Faiz, S. Agrawal, and R. M. Murray, "Differentially flat systems with inequality constraints: An approach to real-time feasible trajectory generation," *AIAA Journal of Guidance, Control and Dynamics*, vol. 24, no. 2, pp. 219-227, 2001.
8. M. Fliess, J. Levine, P. Martin and P. Rouchon, "Flatness and defect of non-linear systems: Introduction, theory and examples," *International Journal of Control*, vol. 61, no. 6, pp. 1327-1361, 1995.
9. D. Hsu, R. Kindel, J.-C. Latombe and S. Rock, "Randomized motion planning with moving obstacles," *International Journal of Robotics Research*, vol. 21, no. 3, pp. 233-255, 2002.
10. T. Karatas and F. Bullo, "Randomized searches and nonlinear programming in trajectory planning," in *Proc. IEEE Conference on Decision and Control*, pp. 5032-5037, 2001.
11. L. E. Kavraki, P. Svestka, J.-C. Latombe, and M. Overmars, "Probabilistic Roadmaps for Path Planning in High Dimensional Configuration Spaces," *IEEE Transactions on Robotics and Automation*, vol. 12, no. 4, pp. 566580, 1996.
12. J.-P. Laumond, P. Jacobs, M. Taix and R. M. Murray, "A motion planner for nonholonomic mobile robots," *IEEE Transactions of Robotics and Automation*, vol. 10, no. 5, pp. 121-155, 1993.

13. S. M. LaValle and J. Kuffner, "Randomized kinodynamic planning," *International Journal of Robotics Research*, vol. 20, no. 5, pp. 378-400, 2001.
14. K. M. Lynch and M. Mason, "Stable pushing: mechanics, controllability and planning," *International Journal of Robotics Research*, vol. 15, no. 6, pp. 533-556, 1996.
15. K. M. Lynch, N. Shiroma, H. Arai and K. Tanie, "Collision free trajectory planning for a 3-dof robot with a passive joint," *International Journal of Robotics Research*, vol. 19, no. 12, pp. 1171-1184, Dec. 2000.
16. P. Martin and P. Rouchon, "Any controllable (driftless) system with 3 inputs and 5 states is flat," *System and Control Letters*, vol. 25, no. 3, pp. 167-173, 1995.
17. J. Reif and H. Wang, "Non-uniform discretization approximations for kinodynamic motion planning and its applications," in *Workshop on the Algorithmic Foundation of Robotics (WAFR,96)*, (J. P. Laumond and M. Overmars eds.), pp. 97-112, 1996.
18. K. G. Shin and N. D. McKay, "Minimum time-control of robotic manipulators with geometric path constraints," *IEEE Transactions on Automatic Control*, vol. 30. no. 6, pp. 531-541, 1985.
19. P. Xavier and B. Donald, "Provably good approximation algorithms for optimal kinodynamic planning: robots with decoupled dynamics bounds," *Algorithmica*, vol. 14, no. 6, pp. 443-479, 1995.
20. P. Xavier and B. Donald, "Provably good approximation algorithms for optimal kinodynamic planning for cartesian robots and open chain manipulators," *Algorithmica*, vol. 14, no. 6, pp. 480-530, 1995.

Springer Tracts in Advanced Robotics

Edited by B. Siciliano, O. Khatib, and F. Groen
Published Titles:

Vol. 1: Caccavale, F.; Villani, L. (Eds.)
Fault Diagnosis and Fault Tolerance for Mechatronic Systems: Recent Advances
191 pp. 2002 [3-540-44159-X]

Vol. 2: Antonelli, G.
Underwater Robots: Motion and Force Control of Vehicle-Manipulator Systems
209 pp. 2003 [3-540-00054-2]

Vol. 3: Natale, C.
Interaction Control of Robot Manipulators: Six-degrees-of-freedom Tasks
120 pp. 2003 [3-540-00159-X]

Vol. 4: Bicchi, A.; Christensen, H.I.; Prattichizzo, D. (Eds.)
Control Problems in Robotics
296 pp. 2003 [3-540-00251-0]

Vol. 5: Siciliano, B.; Dario, P. (Eds.)
Experimental Robotics VIII
685 pp. 2003 [3-540-00305-3]

Vol. 6: Jarvis, R.A.; Zelinsky, A. (Eds.)
Robotics Research – The Tenth International Symposium
580 pp. 2003 [3-540-00550-1]

Printing: Saladruck Berlin
Binding: Stürtz AG, Würzburg

GPSR Compliance

The European Union's (EU) General Product Safety Regulation (GPSR) is a set of rules that requires consumer products to be safe and our obligations to ensure this.

If you have any concerns about our products, you can contact us on ProductSafety@springernature.com

In case Publisher is established outside the EU, the EU authorized representative is:

Springer Nature Customer Service Center GmbH
Europaplatz 3
69115 Heidelberg, Germany

Batch number: 10755310

Printed by Printforce, the Netherlands